Indira's

Objective Agronomy

(Multiple Choice Questions for Competitive Examinations of Agronomy)

3rd Revised Edition

DR. R.L. ARYA
(M.Sc. & Ph.D. in Agronomy, IARI, NEW DELHI)
Ex Senior Scientist
Central Tobacco Research Institute,
Rajahmundry (A.P.)

Published by

Competition Tutor
(Imprint of Scientific Publishers, INDIA)
5 A, New Pali Road, P.O. Box 91
Jodhpur 342 001 (INDIA)
E-mail: info@scientificpub.com
Website : www.scientificpub.com

First Edition: 2014

Second Edition: 2016

ISBN: 978-93-89832-24-2
eISBN: 978-93-89832-25-9

Printed in India

PREFACE

More than 70 per cent of Indian population is dependent on agriculture. Agriculture is a most important source for economic assistance and all round progress of our country. India, with only 2.3 per cent land area, 4 per cent fresh water ad 1 per cent forest cover has to support about 17 per cent world's human population (over 1.18 billion) and 15 per cent livestock (479 million). Science based agriculture adopted for the past four decades in the country has brought appreciable change in the food production scenario of the country. Over the years, a six fold increase in food grain production has transformed food scarcity of early fifties into food sufficiency with ever highest production figure of 257.44 million tones (2011-12). There is still need to increase the total food grain production 300 mt. for ever increasing human population in the country by the year of 2020. The main issues of agricultural growth and development in India which needs scientific redressal are: food self-sufficiency versus food self-reliance; low productivity and declining rate of growth of factor productivity; low and unstable, yields especially in rainfed areas and increasing ecological and environmental insecurity leading to resource wars.

The competitive examinations for Agriculture/Agronomy have become a routine procedure of recruitment and admissions for higher positions and education. Now a days a large number of questions appear in the examination. These questions call for quick answering for success within a specified period of time. A sincere effort has been made by the authors to present them in most easy, short and understandable language for the benefit of scientists, students, farmers and those who are interested in the field of agronomy.

"Indira's Objective Agronomy" 3rd Revised Ed. for competitive exams in agronomy discipline contain 15 chapters covering all related discipline. The chapters included such as: General Agriculture, Historical Perspective of Agronomy, Agricultural Climatology, Tillage Management, Seed and Sowing, Crops and Varieties, Cropping Systems, Agronomic Practices, Nutrient Management, Weed Management, Irrigation Management, Dryland Management, Agroforestry, Plant Protection Measures, Farm Management, Appendix have been given due importance and whole syllabus was covered. Each chapters contains multiple choice questions and total about 8000 objective questions with multiple choice have been framed and arranged sequentially for the easy understanding of the students. Recent information and development in the field of agronomy have been incorporated in the text. Thus this book is based on the syllabus of student of agronomy stream, it may be useful not only to students but also teachers, researchers and development officers for reference and easy answering of many complicated questions.

The chapters are chosen in view to cover the course contents of competitive examinations like IAS, IFS, ARS, PCS and Banking services of agricultural subjects particular in agronomy. The entire book is prepared in most simple, clear and talking language so that the contents could be easily followed by the readers.

Finally, I have belief and hope that this book will fulfill the requirement of our student of agronomy stream for appearing in different competitive examinations.

Being first edition, there could be some shortcomings for which valuable and informative suggestions will be gratefully appreciated, so that these shortcoming may be incorporated in forthcoming edition for the benefit of readers.

Authors shall be grateful to the various books, periodicals, magazines, newsletter, journals etc., rendered help for bringing out of this books. The publisher shall also be grateful for bringing to their notice of any suggestions for further improvement.

DR. R.L. ARYA

CONTENTS

1. **General Agriculture** 1

    Answer Sheet 38

2. **Historical Perspective Of Agronomy** 40

    Answer Sheet 63

3. **Agricultural Climatology** 64

    Answer Sheet 97

4. **Tillage Management** 99

    Answer Sheet 121

5. **Seed and Sowing** 123

    Answer Sheet 133

6. **Crops and Varieties** 135

    Answer Sheet 148

7. **Cropping Systems** 150

    Answer Sheet 166

8. **Agronomic Practices** 167

    Answer Sheet 205

9. **Nutrient Management** 208

    Answer Sheet 293

10. **Weed Management** 298

    Answer Sheet 326

11. **Irrigation Management** 328

    Answer Sheet 367

12.	**Dryland Management**	370
	Answer Sheet	378
13.	**Agroforestry**	379
	Answer Sheet	398
14.	**Plant Protection Measures**	399
	Answer Sheet	437
15.	**Farm Management**	440
	Answer Sheet	465
	Appendix	467

1 GENERAL AGRICULTURE

1. The practice of raising plants and animals is:
(A) Agriculture (B) Horticulture
(C) Agronomy (D) Silviculture

2. Agriculture originated about:
(A) 2300-3000 years back
(B) 20000-30000 years back
(C) 7000-13000 years back
(D) 90000-100000 years back

3. Kung tribe is
(A) Primitive agriculturist
(B) Hunter-gatherer
(C) Advanced agriculturist
(D) Completely dependent on horticulture

4. In various parts of the world, agriculture
(A) Came from Asia Minor
(B) Has spread from Yellow River Valley
(C) Originated independently
(D) Has developed to various degree

5. Fire helped in the cultivation of:
(A) Fruit plants
(B) Fibre Plants
(C) Drug yielding plants
(D) Grain yielding plants

6. Early human civilization was based on:
(A) Development of fibre crops
(B) Fire
(C) Agriculture
(D) Fire and Agriculture

7. A nonagricultural society living today is:
(A) Kung tribe (B) Urban Society
(C) Hintza tribe (D) Kuntizha tribe

8. Early man preferred plants:
(A) Grain yielding
(B) Fruit yielding
(C) Multipurpose
(D) Fibre yielding

9. The famous cultivated plant which developed in China is:
(A) Tea (B) Coffee
(C) Cotton (D) Orange

10. Production centre of Hevea Rubber is:
(A) Malayasia (B) Peruvian Andes
(C) Brazil (D) South West Asia

11. The centre of origin of Wheat is:
(A) South West Asia
(B) South East Asia
(C) Asia minor and Afganistan
(D) Mexico

12. Which is not a new world crop:
(A) Tapioca (B) Peanut
(C) Cashewnut (D) Poppy

13. Which one belong to old world:
(A) Squash (B) Pumpkin
(C) Water Melon (D) Guava

14. Select a crop which belongs to new world:
(A) Jackfruit (B) Papaya
(C) Spinach (D) Pear

15. Which crop belong to new world:
(A) Red Pepper (B) Black pepper
(C) Onion (D) Garlic

16. Which is the crop which originated in Peru and Brazil but is now mainly grown in India is:
(A) Groundnut (B) Potato
(C) Maize (D) None of these

17. The crop which originated in Peru but is most extensively grown in East Europe is:
(A) Coffee (B) Potato
(C) Rubber (D) Maize

18. Name the crop which had its birthplace in Tropical America but the centre of production is now Midwest USA:
(A) Pineapple (B) Oil palm
(C) Cocoa (D) Maize

19. Christopher Columbus landed in island of Bahmas (Carribean sea) in:
(A) 1692 (B) 1592
(C) 1492 (D) 1792

20. Tropical forests are being lost at the annual rate of:
(A) 11 million ha/yr
(B) 110 million ha/yr
(C) 1 million ha/yr
(D) 1100 million ha/yr

21. Post Vavilov megadiversity centre is:
(A) Australia (B) Siberia
(C) Andamana (D) Tibet

22. The number of centres of origin of crop plants recognized by Vavilove is:
(A) 8 (B) 11
(C) 7 (D) 21

23. Guarana is:
(A) Fodder plant (B) Beverage plant
(C) Oil plant (D) Food plant

24. Potato famine of Ireland occurred in:
(A) 1645 (B) 1745
(C) 1845 (D) 1854

25. Potato famine of Ireland was caused by the attack of fungus
(A) *Phytophthora palmivora*
(B) *Plasmopora viticola*
(C) *Puccinia graminis*
(D) *Phytophthora infestans*

26. Natural resistance to late blight of potato is present in:
(A) *Solanum tuberosum*
(B) *Solanum acaule*
(C) *S. demissum*
(D) *S stoloniferum*

27. Gene bank are part of:
(A) Ex situ conservation
(B) In situ conservation
(C) Both ex situ and in situ conservation
(D) Tribal diet

28. *In situ* conservation are carried out through:
(A) Natural parks
(B) Wildlife santuaries
(C) Biosphere reserves
(D) All the above

29. *Ex situ* conservation is carried out through:
(A) Funneling the threatened species into trade
(B) Offsite collection
(C) Gene banks
(D) All the above

30. Gene banks comprise:
(A) Seed banks, orchards, tissue culture and cryopreservation
(B) Seed banks, tissue culture and cryopreservation
(C) Seed banks and cryopreservation
(D) Tissue culture and cryopreservation

31. Orthodox seeds can be preserved for long periods at a temperature of:
(A) -196°C (B) -10 to -20°C
(C) 0-4°C (D) 10 to 15°C

32. Dwarfing gene of wheat is:
(A) Pal -1 (B) Atomita – 1
(C) Norin – 10 (D) Pelita -2

33. Jojoba yields wax for cosmetics from its:
(A) Stem (B) Leaves
(C) Seeds (D) Fruit pulp

34. IR 36 was developed through breeding of:
(A) Six rice varieties and *Oryza nivara*
(B) 13 rice varieties and *Oryza nivara*
(C) *Oryza sativa* and *O. indica*
(D) *Oryza indica* and *O. nivara*

35. Which one is beverage plant:
(A) *Ilex paraguriensis*
(B) *Caltha edulis*
(C) *Poullinia cupana*
(D) All of these

36. The three top most major crops grown in the world in order of total production are:
(A) Wheat>Rice > Maize
(B) Rice>Wheat>Maize
(C) Wheat>Maize>Rice
(D) Rice>Maize>Wheat

37. Potential food plant is:
(A) *Winged bean* (*Psophocarpus tetragonolobus*)
(B) Jajoba (*Simmondsia chinensis*)
(C) Guayule (*Parthenium argentatum*)
(D) *Leucaena*

38. A potential source of rubber is:
(A) Guayule (*Parthenium argentatum*)
(B) *Leucaena leucocephala*
(C) *Cucurbita foetidissima*
(D) *Citrullus colocynthis*

39. Cola, the common beverage, is obtained from:
(A) Seeds (B) Flowers
(C) Leaves (D) Stems

40. Maximum genetic diversity of crop plants occurs where agriculture is:
(A) Run on commercial lines
(B) Advanced
(C) Rainfed (D) Primitive

41. Which is more suitable for germplasm collection:
(A) Centre of production
(B) Homelands
(C) Tropical areas
(D) Temperate areas

42. Which one will be scarched for obtaining maximum genetic diversity:
(A) Recent varieties
(B) Old varieties
(C) Wild relatives
(D) All the above

43. Paraguay Tea is got from:
(A) Leaves of *Ilex*
(B) Seeds of *Ilex*
(C) Leaves of *Paullinia*
(D) Paste of crushed kernels of *Paullinia*

44. Arabian Tea is:
(A) Leaves and buds of *Paullinia*
(B) Seed of Ilex
(C) Leaves and buds of *Catha edulis*
(D) Seeds of Catha

45. Cryopreservation is carried out at:
(A) -10 to -20^{0}C (B) -50 to -60^{0}C
(C) -196^{0}C (D) -100^{0}C

46. Co-varieties of sugarcane obtained red rot resistance from:
(A) *Saccharum munja*
(B) *S. spontaneum*
(C) *S. arundinacearum*
(D) *S. edula*

47. A calcitrant seed will be killed in:
(A) Low temperature (B) Dry air
(C) Further dehydration
(D) All the above

48. A gift from new world to old world is:
(A) Barley (B) Rice
(C) Wheat (D) Maize

49. Which plant was introduced from old to new world:
(A) Potato (B) Rice
(C) Corn (D) Sweet potato

50. Dwarf wheat varieties brought into India from Mexico were:
(A) Sonora-64 and Sonalika
(B) Sharbati Sonora and Pusa Lerma
(C) Sonalika and Sharbati Sonora
(D) Sonora-64 and Lerma Rojo-64

51. Vavilov centres of origin of crop plants are located in:
(A) Tropical plains
(B) Mountains of tropical areas
(C) Mountains of temperate areas
(D) Both (B) and (C)

52. An under utilized oil crop is:
(A) *Brassica campestris*
(B) *Cocos nucifera*
(C) *Helianthus annus*
(D) *Azadiracta indica*

53. Liquid wax is got from:
(A) Jajoba (B) Winged Bean
(C) Guayule (D) Leucaena

54. Wheat varieties collected by Vavilov were:
(A) 26000 (B) 2600
(C) 4600 (D) 260

55. Weight of one cotton bale is equal to:
(A) 160 kg (B) 170 kg
(C) 180 kg (D) 190 kg

56. Germplasm is contained in:
(A) Somatic cells
(B) Growing points
(C) Reproductive cells
(D) Both (A) and (C)

57. Which one will be best for germplasm collection:
(A) Britain (B) France
(C) Portugal (D) Peru

58. Weight of one jute bale is equal to:
(A) 160 kg (B) 170 kg
(C) 180 kg (D) 190 kg

59. In a gene bank, genetic material is stored as:
(A) Genes (B) Seeds
(C) Vegetative material/tissue culture
(D) Both (B) and (C)

60. Genetic erosion is due to:
(A) Deforestation
(B) Shifting cultivation
(C) Adoption of genetically uniform varieties
(D) All the above

61. Exchange of germplasm is carried mostly through shoot tips because it is:
(A) Cheap (B) Small and handy
(C) Virus free (D) Genetically stable

62. Tomato and potato belong to:
(A) North America (B) South America
(C) Asia minor (D) Ethiopia

63. A plant introduced from new world to old world is:
(A) Rice (B) Wheat
(C) Sugarcane (D) Potato

64. National Botanical Research Institute is located at:
(A) Dehradun (B) Delhi
(C) Lucknow (D) Mumbai

65. In India, the largest herbarium is located at:
(A) Dehradun (B) Delhi
(C) Lucknow (D) Kolkata

66. Which of the following country is the centre of origin of rice:
(A) India (B) China
(C) Indo-Malayan region
(D) Russia

67. Central Sugarcane breeding Research Institute is situated at:
(A) Coimbatore (B) Lucknow
(C) Delhi (D) Bhopal

68. A new world spice that has become an essential part of Indian cuisine is:
(A) Red pepper (B) Black Pepper
(C) Cardamom (D) Ginger

69. A social forest tree is:
(A) Teak (B) Sandal wood
(C) Subabul (D) Rosewood

70. A plant species threatened with extinction can be preserved in:
(A) Herbarium
(B) National park
(C) Botanical Garden
(D) Gene Bank

71. A living fossil is:
(A) Gnetum (B) Ginkgo
(C) Riccia (D) Pinus

72. Jojoba is:
(A) Crotolaria (B) Leucaena
(C) Simmondsia (D) Parthenium

73. A place where endangered genetic materials of plants are kept is:
(A) Gene library (B) Gene bank
(C) Gene pool (D) Herbarium

74. Wild varieties of plants must be conserved to:
(A) Feeding wild animals
(B) Maintain ecosystem
(C) Future evolution
(D) Incorporate useful traits in future crop varieties

75. The energy required daily by hunter-gatherer humans was:
(A) 2000-4000 kcal
(B) 6000-10000 kcal
(C) 10000-70000 kcal
(D) 10000-12000 kcal

76. Per capita daily energy consumption in nineteenth century industrialized society was:
(A) 6000 k cal (B) 10000 kcal
(C) 30000 kcal (D) 70000 kcal'

77. Per capita daily energy consumption of USA is:
(A) 10000 kcal (B) 30000 kcal
(C) 70000 k ca (D) 250000 kcal

78. Which is the most important crop in the world:
(A) Maize (B) Sorghum
(C) Barley (D) Wheat

79. The great popularity of wheat is on accounts of:
(A) Well balanced carbohydrates & proteins
(B) Inability for complete mechanized production
(C) Short storage period of grain and flour storage
(D) Limited to narrow range of climate

80. According to Vavilov, the common bread wheat originated in :
(A) The Caucasus-Turkey-Iraq area
(B) Temperate zone of America
(C) Naturally fertile regions of France
(D) Cool areas of China

81. *Acacia senegal* and *Ahlagi camelorum* are the xerophytes belonging to:
(A) Psammophytes
(B) Psychrophytes
(C) Lithophytes (D) Halophytes

82. *Ceina parviflora* is a xerophyte that has the succulent:
(A) Stems (B) Leaves
(C) Roots (D) Fruits

83. "The dominant species with maximum canopy coverage of two or more species" represents a community character known as:
(A) Society (B) Location
(C) Consociation (D) Association

84. In ------- the buds (life forms) are invariably buried in the soil or the substratum as the bulbs and rhizomes:
(A) Chamaephytes
(B) Crypto/geophytes
(C) Hemicryptophytes
(D) Therophytes

85. Collective farming system has originated from:
(A) Bulgaria (B) Poland
(C) Soviet Union (D) France

86. Commercial plantation farming comprises an:
(A) Import-oriented system
(B) Export-oriented system
(C) Indigenous use-oriented system
(D) None of these

87. Agricultural growth is very soon likely to be hampered by scarcity of:
(A) Natural resources
(B) Man-made resources
(C) International cooperation
(D) Local coordination

88. A set of shared, similar characteristics (as applied to logically deduced and symbolically expressed models and observations) may be called as:
(A) Negative isomorphism
(B) Positive isomorphism
(C) Neutral isomorphism
(D) None of these

89. Rubber has probably the -------- range of tolerance for temperature:
(A) Narrow west (B) Widest
(C) Medium (D) None of these

90. The requirement and tolerance of a species may be considered as:
(A) Supplementary
(B) Complementary
(C) Antagonistic
(D) None of these

91. The cactus from North America and the jack rabbit from Europe were introduced into:
(A) Holland (B) Poland
(C) Austria (D) Australia

92. When competition effect is small for both the plant species so that the interspecific effects are ------ the intraspecific effects, both species may be able to live together:
(A) More than (B) Equal to
(C) Less than (D) None of these

93. The self-crowding effect represent the ----------- competition:
(A) Interspecific
(B) Intraspecific
(C) Intergeneric
(D) Intrageneric

94. The interplant competition drastically affects the ------ of the plant:
(A) Size and Shape
(B) Number and Size
(C) Number and Shape
(D) None of these

95. Banana (*Musa sapientum*) plant is an herbaceous -------- which may grow to a height of about 30 feet:
(A) Annual (B) Biennial
(C) Perennial (D) None of these

96. --------- has been the world's leading producer of cane sugar:
(A) India (B) Indonesia
(C) Hawaii (D) Cuba

97. Cotton has been grown in India as a crop of fibre for perhaps ------ years and was known in Peru as early as 1000 BC:
(A) 3000 (B) 4000
(C) 5000 (D) 6000

98. When one set of species react with the environment to facilitate the entry of a new set of species, such a model is known as:
(A) Inhibition model
(B) Tolerance model
(C) Facilitation model
(D) Elucidation model

99. Photoperiodically rice is a type of plant:
(A) Long day
(B) Short day
(C) Day neutral
(D) Both (B) and (C)

100. High yielding dwarf varieties of wheat were developed by:
(A) Dr. B.P. Pal
(B) Dr. Norman E. Borlaug
(C) Dr. E.W. Burton
(D) Dr. M.S. Swaminathan

101. The maximum area under wheat is in:
(A) USSR (east while)
(B) India
(C) USA
(D) China

102. The production of wheat is the highest in:
(A) USSR (eastwhile)
(B) India
(C) China
(D) USA

103. The *Triticum aestivum* is a wheat:
(A) Diploid (B) Tetraploid
(C) Hexaploid (D) Monoploid

104. Maize protein is deficient in:
(A) Betadine
(B) Tryptophane and Lysine
(C) Lysine (D) Tryptophane

105. The country ranking first in respect of area under maize (corn) is:
(A) India (B) China
(C) Brazil (D) USA

106. The largest maize producing country in the world is:
(A) USA (B) Mexico
(C) India (D) Canada

107. The dent corn is also known as:
(A) *Zea mays indentata*
(B) *Zea mays indurate*
(C) *Zea mays tunicate*
(D) *Zea mays everta*

108. All India Co-ordinated Maize Improvement Project was started from:
(A) 1957 (B) 1955
(C) 1965 (D) 1967

109. The place of origin of urdbean is:
(A) India (B) Burma
(C) Bangladesh (D) Ceylon

110. The scientific name of early maturing bold and black seeded urdbean varieties is:
(A) *Vigna mungo* var *viridis*
(B) *Vigna mungo* var *niger*
(C) *Vigna mungo* var *macropode*
(D) *Vigna mungo* var *micropode*

111. Soybean besides good quality protein also contains about-------- per cent of oil:
(A) 10 (B) 20
(C) 30 (D) 40

112. The protein content of soybean is about:
(A) 40% (B) 20%
(C) 30% (D) 10%

113. The grain legume which is also grown as an oilseed is:
(A) Chickpea (B) Urdbean
(C) Lentil (D) Soybean

114. The -----------is considered to be a progenitor of soybean:
(A) *Glycine max* (B) *Glycine sinensis*
(C) *Glycine ussuriensis*
(D) *Glycine usitattisimum*

115. Among the oilseed crops, groundnuts ranks ------ in India:

(A) First (B) Second
(C) Third (D) Fourth

116. Groundnut seed normally contains per cent protein and oil:

(A) 25 and 45 (B) 22 and 43
(C) 45 and 26 (D) 43 and 22

117. Quadricarpellatum capsule of sesamum contains how many carpels:

(A) Two (B) Three
(C) Four (D) Six

118. Sesamum belongs to the family of:

(A) Leguminoceae (B) Papilionaceae
(C) Pedaliaceae (D) Chinipideaceae

119. In south India sesamum is grown in:

(A) Rabi season only
(B) Kharif season only
(C) Rabi and kharif season
(D) Summer season only

120. Castor belongs to family:

(A) Compositeae (B) Euphorbiaceae
(C) Legiminoceae (D) Gramineae

121. The largest producer of rapeseed and mustard in the world is:

(A) China (B) Japan
(C) India (D) Canada

122. Rapeseed and mustard belongs to the family:

(A) Leguminoceaea (B) Compositeae
(C) Euphorbiaceae (D) Crucifereae

123. The 2n chromosome number in Rugosa is:

(A) 20 (B) 24
(C) 32 (D) 36

124. The fruit of rapeseed and mustard is known as:

(A) Pod (B) Silique
(C) Grain (D) Caryopsis

125. The gynophore of groundnut is commonly referred to as the:

(A) Perinth (B) Peg
(C) Ovate (D) Monadelph

126. The protein content in linseed cake has been estimated at:

(A) 12% (B) 24%
(C) 36% (D) 48%

127. The scientific name of erect type of ground nut is:

(A) *Arachis hypogaea* subspecies *fastigiata*
(B) *Arachis hypogeae* subspecies *procumbens*
(C) *Arachis hypogeae* subspecies *oleifera*
(D) *Arachis hypogeae* subspecies *gigantean*

128. Linseed can be used as a crop:

(A) Oilseed
(B) Fibre
(C) Oilseed and fibre
(D) None of these

129. Linseed crop belongs to the family:

(A) Compositeae
(B) Euphorbiaceae
(C) Leguminoceae
(D) Linaceae

130. The fruit of linseed is known as:

(A) Pod (B) Canum
(C) Seedball (D) Dutum

131. Sunflower as an oilseed crop was introduced into India in:

(A) 1965 (B) 1969
(C) 1972 (D) 1985

132. Sunflower is a :

(A) Self pollinated crop
(B) Cross-pollinated crop
(C) Self and cross pollinated crop
(D) Parthenocarpous

133. Cotton belongs to the family of:

(A) Malvaceae (B) Compositeae
(C) Cricifereae (D) Lineaceae

134. The number of known species in genus Gossypium have been estimated at:

(A) 10 (B) 15
(C) 20 (D) 25

135. Jute belongs to the family of:

(A) Malvaceae (B) Compositeae
(C) Lineaceae (D) Tiliaceae

136. Tobacco has been classified as a crop:

(A) Cash (B) Catch
(C) Inter (D) Relay

137. Tobacco belongs to the family:

(A) Compositeae (B) Solanaceae
(C) Crucifereae (D) Leguminaceae

138. *Nicotiana rustica* has been derived as an amphidiploid from a cross between:
(A) *Nicotiana tabacum* x *Nicotiana paniculata*
(B) *Nicotiana paniculata* x *Nicotiana sylvestris*
(C) *Nicotiana undulata* x *Nicotiana sylvestris*
(D) *Nicotiana undulata* x *Nicotiana paniculata*

139. Nicotine content of the leaves of *Nicotiana rustica* ranges between:
(A) 2.0 – 5.5 (B) 3.0-4.5
(C) 3.5-8.0 (D) 5.0 – 10.5

140. Nicotine content of the leaves of *Nicotiana tabacum* ranges between:
(A) 2.0 – 5.5 (B) 3.0-4.5
(C) 3.5-8.0 (D) 5.0 – 10.5

141. Out of the following -------- is commonly cultivated potato:
(A) *Solanum demissum*
(B) *Solanum stenotonum*
(C) *Solanum tuberosum*
(D) *Slanum andigenum*

142. The family of sugarbeet is a:
(A) Chenopodiaceae
(B) Crucifereae
(C) Euphorbiaceae
(D) Compositeae

143. A tuber represents:
(A) Enlarged underground root
(B) Enlarged underground stem
(C) Enlarged underground nodule
(D) None of these

144. Potato belongs to the family:
(A) Solanaceae (B) Compositeae
(C) Legminaceae (D) Euphorbiaceae

145. Department of Revenue, agriculture and Commerce was set up in India on:
(A) April 17, 1871
(B) April 27, 1871
(C) April 27, 1872
(D) April 17, 1872

146. Agriculture colleges were opened in India first of all in year:
(A) 1905 (B) 1900
(C) 1907 (D) 1911

147. The Department of Revenue, Agriculture and Commerce was set up during the reign of:
(A) Lord Curzon (B) Lord Mayo
(C) Lord Wellington (D) None of these

148. Department of agriculture in provinces were set up in:
(A) 1871 (B) 1881
(C) 1891 (D) 1911

149. First Inspector General of Agriculture was appointed in year:
(A) 1901 (B) 1902
(C) 1900 (D) 1907

150. Imperial Agricultural Research Institute was set up in Pusa, Bihar in year:
(A) 1905 (B) 1901
(C) 1871 (D) 1872

151. IARI was shifted to New Delhi in year:
(A) 1925 (B) 1926
(C) 1936 (D) 1947

152. Imperial Council of Agricultural Research was set up in a year of:
(A) 1901 (B) 1926
(C) 1929 (D) 1936

153. Who was the first president of ICAR:
(A) Dr. B.P. Pal
(B) Khan Bahadur Sir Mod. Habibullah
(C) Dr. M.S. Swaminathan
(D) Dr. Panjab Singh

154. Indian Council of Agricultural Research came into existence in:
(A) 1929 (B) 1936
(C) 1946 (D) 1947

155. Total net sown area of India during 2009-10 is:
(A) 120 mha (B) 142 mha
(C) 180 mha (D) 95 mha

156. Total cropped area of India during 2009-10 is?
(A) 120 mha (B) 142 mha
(C) 190 mha (D) 95 mha

157. The headquarter of National Bureau of Soil Survey and Land Use Planning is situated at:
(A) Nagpur (B) Pune
(C) New Delhi (D) Hyderabad

158. The headquarter of National Bureau of Plant Genetic Resources is situated at:
(A) Nagpur (B) Pune
(C) New Delhi (D) Hyderabad

159. Match the name of ICAR institute and their headquarters:

Institutes	Headquarter	
1. IGFRI	i.	Jhansi
2. CAZRI	ii.	Cuttack
3. CRRI	iii.	Jodhpur
4. IISR	iv.	Lucknow

Which of the match is correct?
(A) 1. i 2. ii 3. iii 4. iv
(B) 1. i 2. ii 3. iii 4. iv
(C) 1. i 2. iii 3. ii 4. iv
(D) 1. i 2. iv 3. iii 4. ii

160. IIHR stands for:
(A) Indian Institute of Horticultural Research
(B) Indian Inland Horticulture Research
(C) Institute of Inland Horticulture Research
(D) None of these

161. Which one of the following is matched correctly
(A) CIAE – Bhopal
(B) CSWCTI – Jodhpur
(C) CTCRI – Dehradun
(D) NAARM – New Delhi

162. CRIDA is situated at:
(A) Hyderabad (B) New Delhi
(C) Bangalore (D) Lucknow

163. The ICAR institutes in 'A' and its head-quarters in 'B' are given below:

A	B
1. CTRI	i. Trivendrum
2. CTCRI	ii. Rajahmundry
3. IASRI	iii. Nagpur
4. CICR	iv. New Delhi

Which of the match is correct?
(A) 1. i 2. ii 3 iii 4. iv
(B) 1. i 2. iv 3. iii 4. ii
(C) 1. ii 2. i 3. iv 4. iii
(D) 1. iii 2. iv 3. i 4. ii

164. CIMMYT is working mainly for:
(A) Wheat (B) Maize and Wheat
(C) Rice (D) Cotton

165. The name of International Research Institutes and its head quarters are given below:

Institutes	Head quarters
1. ICRISAT	Hyderabad
2. IRRI	Philippines
3. ICARDA	Syria
4. IBPGR	Rome

Which of the above pairs is matched correctly?
(A) 1 (B) 3
(C) 4 (D) All of the above

166. Zein is known as:
(A) Rice protein (B) Wheat protein
(C) Maize protein
(D) Millet protein

167. Seed ball is a fruit of:
(A) Safflower (B) Sunflower
(C) Linseed (D) Sugarbeet

168. For which type of fertilizer, India is fully depenent on imports:
(A) N- fertilizer (B) P-fertilizer
(C) K-fertilizer (D) None of these

169. Pulses are rich source of:
(A) Fat (B) Carbohydrate
(C) Protein (D) Vitamin

170. Who introduce tobacco cultivation in India?
(A) ITC
(B) Portugeese merchants
(C) Chinese (D) Persians

171. Central Tobacco Research Institute is situated in:
(A) UP (B) MP
(C) AP (D) Karnataka

172. Removal of terminal bud in tobacco is called:
(A) Topping (B) Desuckering
(C) Nipping (D) Curing

173. The dehulled rice grain is known as:
(A) White rice (B) Brown rice
(C) Grey rice (D) None of these

174. Agricultural Research Institute in relation to Jute is situated in:
(A) Karnataka (B) Andhra Pradesh
(C) Tamilnadu (D) WB

175. The process by which the fibre in the bark get loosened and separated from the woody stalk is called as:
(A) Retting (B) Extraction
(C) Both (A) & (B) (D) None of these

176. Wheat protein is called as:
(A) Zein (B) Gluten
(C) Dhurin (D) Ricin

177. Which of the following states has the highest area of groundnut?
(A) Maharastra (B) Karnataka
(C) Punjab (D) Gujarat

178. The rice variety containing 'Dee-geo-woo-gen' gene is:
(A) Indrasan (B) Barmoti
(C) Tilak (D) IR 8

179. The origin place of soybean is:
(A) India (B) China
(C) America (D) Brazil

180. Soybean is a:
(A) Day – neutral plant
(B) Long day plant
(C) Short day plant
(D) None of these

181. Origin of berseem crop is:
(A) USSR (B) Egypt
(C) India (D) China

182. *Phalaris minor* is also known as:
(A) Barmuda grass (B) Canery grass
(C) Kans (D) Water Hyacinth

183. How much sugar can be prepared from 1000 quintals of sugar cane?
(A) 125 q (B) 100 q
(C) 80 q (D) 50 q

184. The crop in which maximum pesticides are used?
(A) Tomato (B) Potato
(C) Sugarcane (D) Cotton

185. Central Rice Research Institute is situated at:
(A) Cuttuck (B) Karnal
(C) Mumbai (D) Manila

186. One acre is equivalent to:
(A) 1.8 ha (B) 0.652 ha
(C) 0.404 ha (D) 0.038 ha

187. Potato is a:
(A) Modified stem (B) Modified root
(C) Modified leaf (D) Modified seed

188. Yellow revolution indicates the production of:
(A) Oilseeds (B) Fruits production
(C) Sugar crops (D) Pulses production

189. First Director General of ICAR was:
(A) Dr. M.S. Swaminathan
(B) Dr. N.S. Randhawa
(C) Dr. G.S. Randhawa
(D) Dr. B.P. Pal

190. Total food grain production in India in 08-09 is:
(A) 207 mt (B) 275 mt
(C) 300 mt (D) 234.47 mt

191. White revolution indicates the production of:
(A) Eggs production
(B) Sugar production
(C) Fruits production
(D) Pulses production

192. Green revolution indicates the production of:
(A) Foodgrain production
(B) Sugar production
(C) Fruits production
(D) Pulses production

193. Blue revolution indicates the production of:
(A) Fish production
(B) Sugar production
(C) Fruits production
(D) Pulses production

194. First State Agricultural University of India is:
(A) Gujarat Agricultural University
(B) Punjab Agricultural University
(C) G.B. Pant University of Agricl. & Tech.
(D) Hisar Agricultural University

195. Inflorescence of sugarcane is known as:
(A) Arrow (B) Tillers
(C) Silk (D) Tassel

196. The highest productivity of wheat (2011-12) is in:
(A) Haryana (B) Punjab
(C) UP (D) Bihar

197. Predominant source of net domestic product in India over the years is:
(A) Industry (B) Agriculture
(C) Service sector (D) All of these

198. The principle male sterile line introduced for hybrid pearlmillet production in India was:
(A) Tift-23 A (B) BJ-104
(C) MHB -118 (D) CO-2

199. Essential oils are used in:
(A) Cosmetics and Toiletries
(B) Food beverage
(C) Paint industry
(D) All of these

200. First Face Pack was developed in:
(A) France (B) Rome
(C) Greece (D) None of these

201. Oil content in fresh herbage in menthol mint is:
(A) 0.1-0.5% (B) 0.5-0.8%
(C) 1.0-1.5% (D) 1.5 – 2.0%

202. Major constituent of *Occimum cannum* is:
(A) Methyl chavical (B) Thymal
(C) Linalool (D) Engenol

203. Origin of cow pea crop is:
(A) India (B) Pakistan
(C) China (D) Japan

204. Sunflower crop belongs to the family of:
(A) Crucifereae
(B) Compositeae
(C) Chenopodiaceae
(D) Liliaceae

205. What is the optimum pulse requirement per capita per day?
(A) 20 g (B) 40 g
(C) 80 g (D) 100 g

206. Origin of French bean is:
(A) America (B) India
(C) China (D) Pakistan

207. Protein content in rajmash is:
(A) 20.9 % (B) 22.9 %
(C) 30.9 % (D) 32.9 %

208. Carbohydrate content in rajmash is:
(A) 50.6 % (B) 60.6 %
(C) 70.6 % (D) 80.6 %

209. Protein content in pea is:
(A) 20 % (B) 22 %
(C) 25 % (D) 32 %

210. Carbohydrate content in potato is:
(A) 60.3 % (B) 70.3 %
(C) 81.3 % (D) 91.3 %

211. India occupies position in vegetable production in world:
(A) First (B) Second
(C) Third (D) Fourth

212. India occupies position in production of onion in world:
(A) First (B) Second
(C) Third (D) Fourth

213. Total cattle population in India during 2006-07 is:
(A) 199 million (B) 196 million
(C) 201 million (D) 210 million

214. Total buffalo population in India during 2006-07 is:
(A) 80 million (B) 90 million
(C) 95 million (D) 105 million

215. Total goat population in India during 2006-07 is:
(A) 100 million (B) 115 million
(C) 140 million (D) 135 million

216. Total sheep population in India during 2006-07 is:
(A) 35 million (B) 51 million
(C) 58 million (D) 71 million

217. Total camel population in India during 2006-07 is:
(A) 0.52 million (B) 0.75 million
(C) 0.76 million (D) 0.63 million

218. Total milk production in India during 2008-09 is:
(A) 77.7 million tonnes
(B) 80.6 million tonnes
(C) 85.8 million tonnes
(D) 112.2 million tones

219. The contribution of livestock to total agricultural output is:
(A) 20 % (B) 25 %
(C) 30 % (D) 35 %

220. The contribution of livestock sector to total GDP is:
(A) 1 % (B) 3 %
(C) 6 % (D) 8 %

221. India's share in total milk production of world is
(A) 10 % (B) 13 %
(C) 15 % (D) 21 %

222. The average productivity of Indian cow is:
(A) 810 kg/lactation
(B) 987 kg/lactation
(C) 1200 kg/lactation
(D) 1500 kg/lactation

223. Dairy sector provides employment to the total work force of:
(A) 5 % (B) 10 %
(C) 15 % (D) 20 %

224. India's position in total cattle population is:
(A) First (B) Second
(C) Third (D) Fourth

225. Central Sheep and Wool Research Institute is situated at:
(A) Avikanagar (Raj)
(B) Bangalore (KN)
(C) Bikaner (Raj)
(D) Pusa (New Delhi)

226. Which of the following state has highest sheep production in India:
(A) Rajasthan
(B) Madhya Pradesh
(C) Gujarat
(D) Uttar Pradesh

227. India's share in goat population in the world is:
(A) 15% (B) 20%
(C) 25% (D) 30%

228. Which of the following state is leading in goat and sheep population?
(A) Madhya Pradesh
(B) Rajasthan
(C) J & K
(D) Gujarat

229. The meat contribution by goat to total meat production in India is:
(A) 20 % (B) 30 %
(C) 37 % (D) 48 %

230. The highest percentage of total goat production is found in the region:
(A) Western and semi-arid region
(B) South humid tropical region
(C) Eastern region
(D) North temperate region

231. The contribution of goat in total milk production of India is only:
(A) 1.7% (B) 3.2%
(C) 5.7% (D) 6.1%

232. The mean rate of goat slaughtering is:
(A) 25% (B) 30%
(C) 36% (D) 41%

233. Which of the state is leading in poultry production?
(A) UP (B) AP
(C) Karnataka (D) Maharastra

234. The total egg production during 2008-09 in India is:
(A) 20.2 million (B) 35.1 million
(C) 55.6 million (D) 40.3 million

235. The highest poultry production in India is:
(A) UP (B) AP
(C) MP (D) WB

236. Which of the country has the highest population of poultry birds?
(A) USA (B) UK
(C) China (D) India

237. Per capita availability of animal protein in Indian diet is:
(A) 10 g/day (B) 15 g/day
(C) 18 g/day (D) 20 g/day

238. The protein per cent in egg yolk is:
(A) 5% (B) 11%
(C) 16% (D) 18%

239. The highest duck producing state is:
(A) UP (B) WB
(C) Bihar (D) Assam

240. Turkey production is highest in the world is:
(A) USA (B) India
(C) Brazil (D) China

241. Per capita availability of milk in India is:
(A) 200 g/day (B) 226 g/day
(C) 230 g/day (D) 240 g/day

242. India's rank in world egg production is:
(A) Ist (B) IInd
(C) IIIrd (D) Vth

243. Total wool production in India during 2008-09 is:
(A) 40 million kg (B) 50 million kg
(C) 60 million kg (D) 42.7 million kg

244. Total fish production in India during 2008-09 is:
(A) 76.37 million tones
(B) 70.12 million tones
(C) 75.63 million tones
(D) 80.06 million tones

245. Highest inland fish production during 2008-09 is
(A) 40.25 million tones
(B) 46.60 million tones
(C) 42.68 million tones
(D) 45.69 million tones

246. Highest marine fish production during 2008-09 is:
(A) 21.36 million tones
(B) 25.36 million tones
(C) 29.77 million tones
(D) 32.69 million tones

247. The number of milk recording centres at national level are:
(A) 30 (B) 50
(C) 94 (D) 115

248. The number of Central Cattle Breeding Centres in India are:
(A) 3 (B) 5
(C) 7 (D) 9

249. Central Cattle Breeding Centre is situated at:
(A) Suratgarh (Rajasthan)
(B) Andeshnagar (UP)
(C) Alamadi (TN)
(D) All of these

250. Central Frozen Semen Production and Training Institute is located at:
(A) Hisar (B) Suratgarh
(C) Hessarghatta (D) Bangalore

251. Central Sheep Breeding Farm is located at:
(A) Hisar (B) Suratgarh
(C) Hesarghatta (D) Bangalore

252. How many pig breeding farms have been established in India?
(A) 101 (B) 158
(C) 260 (D) 275

253. Central Fodder Seed Production Farm is located at:
(A) Jhansi (B) Hessarghatta
(C) Hyderbad (D) Hisar

254. India's share in world buffalo population is:
(A) 20 % (B) 40 %
(C) 57 % (D) 75 %

255. India's share in world cattle population is:
(A) 10% (B) 12%
(C) 15% (D) 18%

256. Central Institute of Fisheries, Nautical and Engineering Training is located at:
(A) Kochi
(B) Mumbai
(C) Vishakhapattanam
(D) Kolkata

257. Central Institute of Coastal Engineering for Fisheries is located at:
(A) Kochi (B) Mumbai
(C) Bangalore (D) Kolkata

258. The number of major fishing harbours in India is:
(A) 2 (B) 4
(C) 6 (D) 8

259. Total honey production in India is:
(A) 7500 tonnes
(B) 10200 tonnes
(C) 13000 tonnes
(D) 15000 tonnes

260. Which country has highest milk production is:
(A) India (B) Denmark
(C) Switzerland (D) USA

261. India's share in world bovine population is:
(A) 17% (B) 19%
(C) 21% (D) 23%

262. Buffalo constitute the total bovine population of:
(A) 35% (B) 40%
(C) 45% (D) 47%

263. What is the contribution of dairy sector to the employment:
(A) 5% of total work force
(B) 8% of total work force
(C) 9% of total work force
(D) 12% of total work force

264. Milk production in India during 2010-11 is:
(A) 72 m tones (B) 121.8 m tonnes
(C) 77 m tones (D) 100 m tones

265. Largest milk colony of Asia is:
(A) Aere milk colony
(B) Allahabad dairy farm
(C) Rajasthan dairy farm
(D) Kanpur diary farm

266. First Dairy farm in India was opened in year of:
(A) 1870 (B) 1877
(C) 1889 (D) 1890

267. First Dairy farm in India was opened at:
(A) Anand (B) Allahabad
(C) Karnal (D) New Delhi

268. Contribution of dairy sector to the total agricultural output is nearly:
(A) 21% (B) 25%
(C) 29% (D) 31%

269. Contribution of dairy sector to the GDP is around:
(A) 4% (B) 6%
(C) 8% (D) 10%

270. Before 1951, India's milk production was:
(A) Less than 20 mt
(B) 20-30 mt
(C) 30-40 mt (D) More than 40 mt

271. Wind blown materials, comprised primarily of silt with some fine sand and clay, are called:
(A) Eclial (B) Loess
(C) Marine (D) None of these

272. Indian Dairy Corporation was setup mainly for:
(A) Success of Operation of Flood Program
(B) Implementation of Operation Flood Program
(C) Trade exchange related to Operation Flood Program
(D) All of these

273. The plan investment in Animal Husbandry and Dairying is about:
(A) 4% (B) 5%
(C) 6% (D) 8%

274. India's share in Asian bovine population is:
(A) 27% (B) 38%
(C) 51% (D) 60%

275. According to ICAR, minimum requirement of milk/day/head is:
(A) 200 g (B) 280 g
(C) 290 g (D) 300 g

276. National Agricultural Insurance Scheme launched in India in year:
(A) 1972 (B) 1985
(C) 1999 (D) 2000

277. Comprehensive Crop Insurance Scheme was launched in year:
(A) 1975 (B) 1985
(C) 1995 (D) 2001

278. Central Institute of Agricultural Engineering is situated at:
(A) Kanpur (B) Bhopal
(C) Ludhiana (D) Nagpur

279. The intermediaries came into Indian Agriculture in which year:
(A) 1793 (B) 1783
(C) 1773 (D) 1763

280. The first bill for abolition of intermediaries or zamidari introduced in:
(A) 1946, UP (B) 1948, Bihar
(C) 1971, TN (D) 1967, Gujarat

281. Central Farm Machinery Training and Testing Institute is situated at:
(A) Budni, MP
(B) Ludhiana, Punjab
(C) Kanpur, UP
(D) Jabalpur, MP

282. Zamidary system was introduced in year:
(A) 1792 (B) 1793
(C) 1794 (D) 1795

283. Rayatwari system was introduced by:
(A) Thomas Munro and Captain Read
(B) Lord Powel
(C) Lord Curzon
(D) Lord Cornwalish

284. Land Reform Implementation Committee was set up in 1963 under the chairmanship is:
(A) G.L. Nanda (B) Prof. Dantwola
(C) Arvind Ghosh (D) M.K. Gandhi

285. The minimum wages Act (Labour) was enacted during:
(A) 1932 (B) 1936
(C) 1948 (D) 1952'

286. The Employees Provident Fund and Family Pension Act was enforced in:
(A) 1952 (B) 1962
(C) 1972 (D) 1982

287. The payment of Bonus Act was implemented in:
(A) 1948 (B) 1958
(C) 1965 (D) 1968

288. The payment of Gratuity Act was implemented in:
(A) 1952 (B) 1962
(C) 1972 (D) 1982

289. The Industrial Disputes Act was enacted in:
(A) 1937 (B) 1947
(C) 1957 (D) 1967

290. The Trade Union Act was enacted in:
(A) 1926 (B) 1936
(C) 1946 (D) 1956

291. The Bonded Labour System (Abolition) ordinance came into enforce in:
(A) 1971 (B) 1973
(C) 1975 (D) 1978

292. The Livestock Importation Act was implemented in:
(A) 1862 (B) 1872
(C) 1898 (D) 1903

293. The Agriculture Produce (grading) and Marketing Act was passed in:
(A) 1927 (B) 1937
(C) 1947 (D) 1957

294. The Cattle Trespass Act was implemented in:
(A) 1781 (B) 1881
(C) 1981 (D) 1991

295. The prevention of Cruelty to animals Act, 1950 was implemented in:
(A) 1857 (B) 1891
(C) 1907 (D) 1950

296. Dangerous Machines (Regulation) Act was passed in:
(A) 1971 (B) 1981
(C) 1983 (D) 1993

297. The Land Improvement Act was initiated in:
(A) 1869 (B) 1870
(C) 1871 (D) 1872

298. The Land Improvement Loan Act was started in:
(A) 1871 (B) 1876
(C) 1883 (D) 1891

299. The Agriculturists Loans Act was implemented in:
(A) 1881 (B) 1882
(C) 1883 (D) 1884

300. The National Cooperative Development Corporation Act was started in:
(A) 1947 (B) 1959
(C) 1962 (D) 1971

301. The Agricultural Produce (Development and Ware-housing) Corporation Act was implemented in:
(A) 1947 (B) 1956
(C) 1962 (D) 1971

302. The National Co-operative Development Corporation was set up in:
(A) 1952 (B) 1962
(C) 1963 (D) 1964

303. The Agricultural Refinance and Development Corporation Act came into existence in:
(A) 1961 (B) 1962
(C) 1963 (D) 1964

304. The Warehouse Corporation Act was implemented in:
(A) 1960 (B) 1961
(C) 1962 (D) 1963

305. The Regional Rural Banks Act was enforced in India in:
(A) 1971 (B) 1974
(C) 1976 (D) 1981

306. The Multi-Unit Cooperative Societies Act was implemented in:
(A) 1940 (B) 1941
(C) 1942 (D) 1943

307. The All India Rural Credit Survey Committee was constituted by RBI in:
(A) 1931 (B) 1941
(C) 1951 (D) 1961

308. National Development Council Resolution was passed in
(A) 1949 (B) 1959
(C) 1969 (D) 1979

309. The New Cooperative Policy Resolution was adopted in:
(A) 1975 (B) 1976
(C) 1977 (D) 1978

310. National Cooperative Union of India was set up in:
(A) 1951 (B) 1953
(C) 1957 (D) 1961

311. Agricultural Finance Corporation was set up in:
(A) April 10, 1965
(B) April 10, 1966
(C) April 10, 1968
(D) April 10, 1969

312. Agricultural Refinance and Development Corporation (ARDC) was setup in:
(A) 1961 (B) 1962
(C) 1963 (D) 1964

313. Tribal Area Development Project was launched in:
(A) 1971-72 (B) 1973-74
(C) 1976-77 (D) 1978-79

314. State Trading Corporation Ltd was constituted in:
(A) 1956 (B) 1957
(C) 1958 (D) 1959

315. Indian Standard Institute was established in:
(A) 1945 (B) 1946
(C) 1947 (D) 1948

316. Marketing and Designing centre is situated at:
(A) Pune (B) Faridabad
(C) Nagpur (D) New Delhi

317. Central Agmark Laboratory is situated at:
(A) New Delhi
(B) Nagpur
(C) Bangalore
(D) Kanpur

318. Indian Standard Institute is situated at:
(A) Kolkata (B) New Delhi
(C) Madras (D) Hyderabad

319. Agricultural Produce (Grading and Marketing) Act was established in:
(A) 1932 (B) 1937
(C) 1939 (D) 1940

320. Central Marketing department was opened in:
(A) 1932 (B) 1035
(C) 1937 (D) 1941

321. The per capita availability of land has been decreased from 0.5 ha in 1950-51 to 0.14 ha in 2000 and will further reduced by 2020:
(A) 0.08 ha (B) 1.08 ha
(C) 1.18 ha (D) 1.28 ha

322. First Agricultural Marketing Advisor was appointed on:
(A) January 1, 1935
(B) January 26, 1935
(C) October 2, 1935
(D) August 15, 1935

323. Agricultural Price Commission was constitutes in:
(A) 1955 (B) 1965
(C) 1975 (D) 1985

324. Essential Commodities Act was passed on:
(A) 1954 (B) 1955
(C) 1956 (D) 1957

325. Cold Storage Order came into existence in:
(A) 1962 (B) 1963
(C) 1964 (D) 1965

326. Meat Food Products order was passed on:
(A) 1971 (B) 1972
(C) 1973 (D) 1974

327. Export (Quality Control and Inspection) Act was passed on:
(A) 1965 (B) 1968
(C) 1969 (D) 1971

328. Head office of Directorate of Marketing and Inspection is situated at:
(A) New Delhi (B) Faridabad
(C) Gurgaon (D) Ghaziabad

329. Food Corporation of India (FCI) was came into existence on:
(A) 1965 (B) 1966
(C) 1967 (D) 1968

330. A expert committee to review the Cooperative Marketng was set up in 1964 under the chairmanship of:
(A) Dr. M.L. Dantewala
(B) R.N. Mirdha
(C) Raj Gopalacharya
(D) R.L. Mehta

331. The Agricultural Finance sub committee was constituted under the chairmanship of:
(A) V.L. Mehta
(B) D.R. Gadgil
(C) Sardar Ballabh Bhai Patel
(D) J.L. Nehru

332. Annual loss due to pests, diseases and weeds annually is estimated nearly:
(A) >50000 crores
(B) <50000 crores
(C) > 1.0 crores (D) < 1.0 crores

333. India is using biopesticides presently:
(A) 1% of total pesticide
(B) 5% to total pesticide
(C) 10% of total pesticide
(D) 15% of total pesticide

334. Average biopesticide use globally is:
(A) 5% of total pesticide
(B) 10% of total pesticide
(C) 12% of total pesticide
(D) 15% of total pesticide

335. Total pesticide consumption in India during 2012 is :
(A) 35.65 million tonnes
(B) 50.58 million tonnes
(C) 39.26 million tonnes
(D) 42.56 million tonnes

336. Destructive Insects and Pest Act was launched in India in:
(A) 1907 (B) 1914
(C) 1921 (D) 1946

337. Directorate of Plant Protection Quarantine and Storage was started in:
(A) 1907 (B) 1914
(C) 1921 (D) 1946

338. Which of the airport does not have plant quarantine station?
(A) Lucknow (B) Kolkata
(C) New Delhi (D) Madras

339. Headquarter of Central Silk Board is:
(A) Lucknow (B) Bangalore
(C) Hyderabad (D) Behrampur

340. Central Sericulture Research Institute and Technology is situated at:
(A) New Delhi (B) Kolkata
(C) Behrampur (D) Madras

341. Headquarter of Tobacco Board is:
(A) Rajahmundry (B) Guntur
(C) Vijaywada (D)Vishakhapatannam

342. What is origin place of cauliflower:
(A) Canada (B) Italy
(C) India (D) China

343. Antisenescent polyhormone is represented by:
(A) Auxins (B) Cytokinins
(C) Gibberellin (D) Ethylene

344. What is origin place of bottle guard:
(A) Canada (B) USA
(C) India (D) China

345. What is origin place of pumpkin:
(A) Canada (B) America
(C) India (D) China

346. What is origin place of carrot:
(A) Canada (B) USA
(C) Asia (D) China

347. What is origin place of Sweet potato:
(A) North America (B) South America
(C) Australia (D) China

348. What is origin place of chilli:
(A) Brazil (B) Chile
(C) Gutemala (D) China

349. APEDA refers to:
(A) Agricultural and Processed Food Products Export Development Authority
(B) Agricultural Production and Export Development Authority
(C) Agricultural Production and Export Development Agency
(D) Asian Policy and Export Development Association

350. India is leading in production of:
(A) Turmeric (B) Black pepper
(C) Ginger (D) All of these

351. Total production of vegetable in India during 2010 is:
(A) 93.92 mt (B) 100 mt
(C) 102.11 mt (D) 120.71 mt

352. What is the contribution of India in world vegetable production:
(A) 10 % (B) 15 %
(C) 20 % (D) 23 %

353. Which of the following vegetable is produced maximum in India is:
(A) Cauliflower (B) Cabbage
(C) Tomato (D) Brinjal

354. Which of the following spices is most important in India is:
(A) Chilli (B) Turmeric
(C) Garlic (D) All of these

355. India's share in chilli production is:
(A) 27 % (B) 33.7 %
(C) 45 % (D) 51.2 %

356. Minimum requirement of fruit per head per day for balanced diet is:
(A) 60 gm (B) 80 gm
(C) 85 gm (D) 120 gm

357. CFTRI is situated at:
(A) Mysore (B) Kolkata
(C) New Delhi (D) Bangalore

358. First fruit product research laboratory was opened in India at:
(A) Kodour (B) Pune
(C) Lucknow (D) Bhopal

359. The origin place of banana is:
(A) South-East Asia
(B) South-West Asia
(C) South Africa (D) South America

360. The origin place of loquat is:
(A) China (B) India
(C) South Africa (D) South America

361. The origin place of jackfruit is:
(A) China (B) Mexico
(C) India (D) South America

362. The origin place of grape is:
(A) Asia (B) China
(C) South Africa (D) South America

363. The origin place of peach is:
(A) Europe (B) China
(C) South Africa (D) South America

364. The origin place of pear is:
(A) Iran (B) Asia
(C) Africa (D) America

365. The origin place of gladiolus is:
(A) China (B) Egypt
(C) South Africa (D) South America

366. India's place in fruit production in world is:
(A) First (B) Second
(C) Third (D) Fourth

367. The fruit production in India is:
(A) 26 mt (B) 40 mt
(C) 75 mt (D) 80 mt

368. The area under fruit production in India is:
(A) 1.5 m ha (B) 2.5 m ha
(C) 3.8 m ha (D) 8.5 m ha

369. What is the total production of horticultural crops in India during 2011 is:
(A) 152.5 mt (B) 227.0 mt
(C) 171.0 mt (D) 185.13 mt

370. What is the share of India in total world fruit production?
(A) 5% (B) 12.5%
(C) 15% (D) 20%

371. Which of the production in included in food grain production?
(A) Pulse crops only
(B) Coarse cereal crops only
(C) Cereal (Rice and wheat) crops only
(D) All cereals (Rice and wheat), pulses and coarse cereals

372. The area under herbal garden in India is:
(A) 200 ha (B) 226 ha
(C) 310 ha (D) 510 ha

373. The fruits are arranged in ascending order in respect of its area. Which of the following sequence is correct:
(A) Mango>Citrus>Banana>Apple
(B) Mango>Apple>Banana>Citrus
(C) Guava>Mango>Banana>Citrus
(D) Apple>Mango>Citrus>Litchi

374. What is the share of India in total mango production of world?
(A) 23% (B) 42%
(C) 54% (D) 60%

375. Presently the total production of cut flowers in India during 2008-09 is:
(A) 550 million (B) 780 million
(C) 1952 million (D) 910 million

376. Which of the following fruit has its highest productivity in India?
(A) Grape (B) Mango
(C) Banana (D) Guava

377. Ethopia is the origin home of:
(A) Coffee (B) Maize
(C) Rice (D) Cabbage

378. Cabbage originated in:
(A) Mexico and Central America
(B) Mediterranean
(C) Peruvian Andes
(D) South west Asia

379. Maize evolved in:
(A) Brazil
(B) Mexico and Central America
(C) USA
(D) Asia minor and Afganistan

380. The centre of origin of Almond and Apple is:
(A) South West Asia
(B) South East Asia
(C) China
(D) Asia minor and Afganistan

381. The origin home of rice is:
(A) South West Asia
(B) South East Asia
(C) Mediterranean
(D) China

382. Peruvian Andes are believed to be centre of origin of:
(A) Tomato (B) Potato
(C) Tomato and Potato
(D) Tomato, Potato and Chillies

383. The origin of sunflower is believed to be in:
(A) Mexico and Central America
(B) Brazil
(C) USA
(D) Peruvian Andes

384. South East Asia is thought to be centre for the origin of:
(A) Rice (B) Rice & Sugarcane
(C) Rice, Sugarcane and Mango
(D) Rice, Sugarcane, Mango & Banana

385. A multipurpose plant of India is
(A) Walnut (B) Almond
(C) Apple (D) Coconut

386. A grain producing plant where spathe covering helped in domestication
(A) Wheat (B) Maize
(C) Rice (D) Barley

387. National Agriculture Policy aims the growth rate of:
(A) >2.5%pa (B) >3.0%pa
(C) >3.5%pa (D) >4.0%pa

388. The total food grain production in India during 2007-08 was:
(A) 202 million tones
(B) 220 million tones
(C) 225 million tones
(D) 230 million tones

389. Agriculture became state subject in year:
(A) 1910 (B) 1919
(C) 1947 (D) 1952

390. Indian Agricultural Services came to function in the year:
(A) 1919 (B) 1921
(C) 1924 (D) 1947

391. Contribution of flag leaf in photosynthates is about:
(A) 40% (B) 48%
(C) 35% (D) 52%

392. What was the name of ICAR in initial years of its establishment?
(A) Indian Council of Agricultural Research
(B) Imperial Council of Agricultural Research
(C) Imperial Centre of agricultural Research
(D) Indian Centre of Agricultural Research

393. The establishment if ICAR is based on the recommendation of which commission?
(A) Royal Commission on Agriculture
(B) Famine Commission
(C) Grow More Food Commission

(D) Balwant Rai Mehta Commission

394. ICAR was renamed Indian Council of Agricultural Research in which year?

(A) 1929 (B) 1946
(C) 1949 (D) 1952

395. Who was the first Director General of ICAR and Secretary to Ministry of Agriculture?

(A) Dr. M.S. Swaminathan
(B) Dr. B.P. Pal
(C) Dr. O.P. Gautam
(D) Dr. R.S. Paroda

396. When did ARS service start?

(A) 1947 (B) 1952
(C) 1973 (D) 1982

397. How many Deputy Director Generals are working with ICAR?

(A) 05 (B) 06
(C) 07 (D) 08

398. Who works as President of ICAR?

(A) The Prime Minister of India
(B) The Union Agriculture Minister
(C) The State Minister of Agriculture
(D) Director General, ICAR

399. What is the main mandate of ICAR?

(A) To plan aid, undertake, promote and coordinate agriculture research in the country
(B) To plan, aid undertake, promote and coordinate agricultural research in the country
(C) To plan, aid undertake, promote and coordinate agricultural educa-tion and research in the country
(D) To plan, aid, undertake, promote and coordinate agricultural research, education and extension in the country

400. The growth of ICAR started with the establishment of which of the following institutes, being the first one?

(A) Imperial Bacteriological Laboratory, Pune
(B) Agricultural Research Institute, Pusa, Bihar
(C) Indian Agricultural Research Institute, New Delhi
(D) Agriculture College, Kanpur

401. How many National Institutes are working in the ICAR system?

(A) 04 (B) 05
(C) 06 (D) 07

402. Which of the following was the first institute/university to be establishment only land Grant College System in 1958?

(A) Indian Agricultural Research Institute, New Delhi
(B) G.B. Pant University of Agriculture & Technology, Pantnagar
(C) College of Agriculture, Coimbatore
(D) College of Agriculture, Kanpur

403. NAARM (National Academy of Agricultural Research Management), Hyderabad was established to train scientists in which year?

(A) 1949 (B) 1973
(C) 1976 (D) 1986

404. AICRP stands for?

(A) All India Crop Research Programme
(B) All India Coordinated Research Project
(C) All India Cropping System Research Project
(D) None of these

405. MANAGE, Hyderabad is an institution working in the area of:

(A) Agricultural research management
(B) Agricultural extension mangement
(C) Agricultural Education Management
(D) Agricultural research, education and extension management

406. What for ASRB stands?

(A) Agricultural Services Recruitment Board
(B) Agricultural Scientist Recruitm- ent Board
(C) Agricultural Science Research Bureau
(D) Agricultural Scientist Research Bureau

407. How many Trainers Training Centres (TTCs) are in India?

(A) 05 (B) 08
(C) 13 (D) 21

408. IVLP (Institution Village Linkage Programme) centres are functioning at how many places?

(A) 70 (B) 79
(C) 83 (D) 89

409. How many Zonal Directors have been function as KVKs also?
(A) 50 (B) 53
(C) 57 (D) 59

410. How many central institutes of ICAR are working in agricultural science?
(A) 10 (B) 36
(C) 73 (D) 97

411. How many Bureaus have been established in India?
(A) 01 (B) 02
(C) 06 (D) 11

412. Origin of *Corchorus capsularis* (white jute) is believed to be in the:
(A) Africa
(B) Indo-Myanmar region
(C) China (D) USA

413. The sub-species of *Oryza sativa* species is grown:
(A) Japonica in Japan
(B) Jawanica in Indonesia
(C) Indica in India
(D) All of these

414. National Research Centre for Women in Agriculture is established at which place?
(A) Calcutta
(B) Bhubaneshwar
(C) Jorhat
(D) New Delhi

415. National Centre of Agricultural Economics & Policy Research is established at which place?
(A) New Delhi (B) Hyderabad
(C) Bangalore (D) Madras

416. How many state agricultural universities are functional in India?
(A) 25 (B) 30
(C) 29 (D) 54

417. How many central agricultural universities are established in India?
(A) 01 (B) 02
(C) 03 (D) 04

418. How many ICAR institutes have been upgraded to the status of Deemed universities?
(A) 02 (B) 04
(C) 05 (D) 12

419. Sea island cotton belongs to:
(A) *Gossypium arboreum*
(B) *Gossypium herbaceum*
(C) *Gossypium hirsutum*
(D) Gossypium barbadense

420. ICRISAT stands for:
(A) International Crops Research Institute for the Semi Arid Tropics
(B) International Centre for Research in Sugarcane & Tobacco
(C) Indian Crop Research Institute for Semi Arid Tropics
(D) International Cotton Research for Institute for Semi Arid Tropics

421. Agriculture is the subject of which of the following?
(A) State Govt.
(B) Central Govt.
(C) State & Central Govt.
(D) None of these

422. ICAR/SAUs conduct what kind of extension?
(A) Public extension
(B) Frontline extension
(C) Private extension (D) None of these

423. What is the main mandate of ICAR/SAUs in the area of agricultural extension?
(A) Technology transfer
(B) Technology assessment on refinement
(C) Technology assessment and refinement
(D) None of these

424. Extension Education Institutes (EEIs) impart training in the area of:
(A) Subject matter
(B) Training and Communication methodology
(C) Subject matter, training and communication methodology
(D) None of these

425. The most serious economic problem of India are:
(A) Poverty and Unemployment
(B) Unemployment and stagnation
(C) Underdevelopment and Poverty
(D) Corruption and Unemployment

426. The percentage of people below poverty line in India is:
(A) 25.1 (B) 27.8
(C) 20.3 (D) 21.45

427. By eradicating poverty, who wished to "wipe out every tear from every eye":
(A) Jawahar Lal Nehru
(B) Mahatma Gandhi
(C) Rabindra Nath Tagore
(D) Vinoba Bhave

428. The poverty lines in rural area on the basis of an average daily intake:
(A) 2100 calories (B) 2300 calories
(C) 2400 calories (D) 3400 calories

429. The 'Garibhi hatao' slogan was coined by:
(A) Jawahar Lal Nehru
(B) Mahatma Gandhi
(C) Rabindra Nath Tagore
(D) Indira Gandhi

430. The proportion of Indian population living in the villages:
(A) One-third (B) Half
(C) Two-third (D) Three-fourth

431. Who said 'If the villages perish, India will perish:
(A) Jawahar Lal Nehru
(B) Mahatma Gandhi
(C) Rabindra Nath Tagore
(D) Indira Gandhi

432. Rural development can be defined as a strategy designated to improve:
(A) Economic and social life of rural poor
(B) Condition of all section of people
(C) Economic condition of rural society
(D) Physical infrastructure

433. The basic guide, philosopher and friend in rural development:
(A) BDO (B) VDO
(C) Pradhan (D) Pramukh

434. Rural development programmes are related to:
(A) Self employment
(B) Employment generation
(C) Poverty alleviation
(D) All of the above

435. The body responsible for rural development at base level:
(A) Panchayat (B) Block
(C) District (D) State

436. The basic operational unit for rural development in India is:
(A) Village (B) District
(C) Tehsil (D) Block

437. Who coined the term Green revolution:
(A) Williamn Gadd
(B) M.S. Swaminathan
(C) Norman E. Borlaug
(D) C. Subramanian

438. Agricultural Scientists Recruitment Board (ASRB) was set up based on the recommendations of the committee headed by:
(A) Mohan Sinha Mehta
(B) M S Swaminathan
(C) Gajendra Gadkar
(D) B P Sinha

439. Agricultural Scientists Recruitment Board (ASRB) was set up in the year:
(A) 1973 (B) 1975
(C) 1977 (D) 1979

440. Agriculture and livestock was published by:
(A) Ministry of Agriculture & Cooperation
(B) Ministry of Rural Development
(C) Indian Council of Agricultural Research
(D) Ministry of Animal Husbandry

441. The concept of agro-climatic zones were introduced during:
(A) NATP (B) NARP
(C) NREP (D) NAIP

442. The concept of agro-eco system analysis was developed during:
(A) Late 60s (B) Late 70s
(C) Late 80s (D) Late 90s

443. The concept of agro-eco system analysis was developed in:
(A) USA (B) Philippines
(C) Thailand (D) Israel

444. The concept of agro-eco system was basically originated at University of Chiang Mai by:
(A) Tonnis Fermetty (B) Adam Groves
(C) Levis Carol (D) Gordon Conway

445. The first farm magazine in India is:
(A) Kheti (B) Phool Phul
(C) Yojana (D) Krishi Chayanika

446. The general micro-unit of an agro-ecosystem is:
(A) District (B) Village
(C) Farmer (D) Household

447. The headquarter of Food and Agricultural Organization are at:
(A) New York (B) New Delhi
(C) Geneva (D) Rome

448. The ideal types of social norms are:
(A) Traditional (B) Modern
(C) Both (A) & (B) (D) None of these

449. The II most important hygiene factor according to Heirzberg is:
(A) Policy & administration
(B) Achievement (C) Salary
(D) Working conditions

450. The impact of green revolution was largely noted in:
(A) Rice only (B) Wheat only
(C) Rice and Wheat
(D) Rice, Wheat & Pulses

451. The National Steering Committee (NSC) is chaired by:
(A) Director General, ICAR
(B) Union Minister of Agriculture
(C) Minister of State for Agriculture
(D) Principle Secretary (Agriculture)

452. The president of ICAR is:
(A) Prime Minister
(B) Agriculture Minister
(C) Director General
(D) Finance Minister

453. Scientific name of Flint corn is:
(A) *Zea mays indurata*
(B) *Zea mays tunicata*
(C) *Zea mays saccharata*
(D) *Zea mays ceratina*

454. Scientific name of Pod corn is:
(A) *Zea mays tunicate*
(B) *Zea mays indurate*
(C) *Zea mays saccharata*
(D) *Zea mays ceratina*

455. Scientific name of Sweet corn is:
(A) *Zea mays indurata*
(B) *Zea mays tunicate*
(C) *Zea mays saccharata*
(D) *Zea mays ceratina*

456. Scientific name of Waxy corn is:
(A) *Zea mays indurate*
(B) *Zea mays tunicata*
(C) *Zea mays saccharata*
(D) *Zea mays ceratina*

457. Scientific name of Popcorn is:
(A) *Zea mays indurata*
(B) *Zea mays tunicata*
(C) *Zea mays everta*
(D) *Zea mays ceratina*

458. Scientific name of Soft or flour corn is:
(A) *Zea mays indurata*
(B) *Zea mays amylacea*
(C) *Zea mays saccharata*
(D) *Zea mays ceratina*

459. In America which crop is known as yellow gold:
(A) Maize (B) Rice
(C) Soybean (D) Wheat

460. Scientific name of green varieties of black gram is:
(A) *Vigna mungo var viridis*
(B) *Vigna mungi var niger*
(C) *Vigna mungo var macropode*
(D) *Vigna mungi var micropode*

461. Scientific name of bunch type of ground nut is:
(A) *Arachis hypogea sp fastigiata*
(B) *Arachis hypodea sp procumbens*
(C) *Arachis hypogea sp viridis*
(D) *Arachis hypogea sp niger*

462. Scientific name of spreading type of ground nut is:
(A) *Arachis hypogea sp fastigua*
(B) *Arachis hypodea sp procumbens*
(C) *Arachis hypogea sp viridis*
(D) *Arachis hypogea sp niger*

463. Scientific name of desi cotton is:
(A) *Gossypium herbacium*
(B) *Gossypium arboreum*
(C) Both (A) and (B)
(D) None of these

464. Scientific name of American cotton is:
(A) *Gossypium barbedensii*
(B) *Gossypium hirsutum*
(C) Both (A) and (B)
(D) None of these

465. Diploid species of cotton are:
(A) Desi (B) American
(C) Both (A) & (B) (D) None of these

466. Tetraploid species of cotton are:
(A) Desi (B) American
(C) Both (A) & (B) (D) None of these

467. Poor man's timber is:
(A) Mango wood (B) Bamboo
(C) Acacia wood (D) Eucalyptus wood

468. Tendu/Kendu leaves used as wrappers for bidis are got from:
(A) *Diospyros ebenum*
(B) *Dipterocarpus indicus*
(C) *Dipterocarpus tuberculatus*
(D) *Diospyros melonxylon*

469. Conservation is:
(A) Protection of natural resources
(B) Management of natural resources
(C) Proper use of natural resources
(D) All of these

470. MAB is:
(A) Man and Botany
(B) Man and Biosphere
(C) Man and Biotic community
(D) Man, Antibiotics and Bacteria

471. Silent valley has been preserved as it has:
(A) Recreational value
(B) Valuable timber plants
(C) Natural forest
(D) Rare plant and animals

472. Silent valley having rare plants and animals is located in:
(A) Kerala
(B) Karnataka
(C) Jammu and Kashimir
(D) Andhra Pradesh

473. Which one is connected with conservation of forests?
(A) Gir (B) Silent valley
(C) Kaziranga (D) Bharatpur

474. New approach to conservation is establishment of:
(A) Reserve forests
(B) Santuries
(C) Biosphere reserves
(D) National parks

475. Social forestry is:
(A) Management of forests by cooperative societies
(B) Growing and management of useful plants on common lands
(C) Growing one type of trees
(D) Growing different types of plants together

476. Decrease in species diversity in tropical countries is mainly due to:
(A) Urbanisation
(B) Pollution
(C) Deforestation
(D) Soil erosion

477. Niger is a:
(A) Pulse crop (B) Fibre crop
(C) Cereal crop (D) Oilseed crop

478. Tobacco is extensively grown in:
(A) Andhra Pradesh and Gujarat
(B) Jummu and Kashmir
(C) M.P. and Maharastra
(D) North-Eastern states

479. Cotton is cultivated on commercial lines in:
(A) U.P. and Bihar
(B) A.P., Karnataka and Maharastra
(C) Gujarat and Rajasthan
(D) None of these

480. Sunhemp is largely grown for fibre purpose in:
(A) Eastern U.P. (B) Eastern Bihar
(C) Eastern M.P. (D) Rajasthan

481. Emergence of flag leaf indicates emergence of:
(A) Panicle (B) Inflorescence
(C) Ear (D) Tillers

482. In wheat crop, crown roots emerge:
(A) 2 weeks after sowing
(B) 3 weeks after sowing
(C) 4 weeks after sowing
(D) 5 weeks after sowing

483. In view of area and production of opium the first place is occupied by:
(A) M.P. (B) U.P.
(C) Bihar (D) Rajasthan

484. Food grain production in India during 2010-11 increased in comparison to 2009-10 to the tune of :

(A) 13 million tones
(B) 14 million tones
(C) 15 million tones
(D) 26 million tones

485. Four major mustard producing states in India are:

(A) Rajasthan, U.P., M.P. and Haryana
(B) Rajasthan, M.P., U.P. and Hrayana
(C) Rajasthan, Haryana, U.P. and M.P.
(D) Rajasthan, Punjab U.P. and Haryana]

486. Adulteration of mustard oil with *Argemone mexicana* seed oil caused in the recent past:

(A) Tuberculosis (B) Diarrhoea
(C) Dropsy (D) None of these

487. Wheat is a:

(A) Monoecious plant
(B) Diocious plant
(C) Hermaphrodite plant
(D) None of these

488. First of all coloured cotton developed in India in the state of:

(A) Madhya Pradesh (B) Maharashtra
(C) Gujarat (D) Andhra Pradesh

489. Botanical name of kodon is:

(A) *Echinochloa frumentacea* L.
(B) *Eleusine coracana* Gaertn
(C) *Paspalum scrobiculatum*
(D) None of these

490. First hybrid variety of sorghum was evolved in the year:

(A) 1946 (B) 1956
(C) 1964 (D) 1974

491. First hybrid variety of bajra was evolved in the year:

(A) 1946 (B) 1956
(C) 1965 (D) 1974

492. According to Manchurian classification soybean has been divided into:

(A) Two groups (B) Three groups
(C) Four groups (D) Only one group

493. Substances responsible for bread making quality of wheat is:

(A) Gluten (B) Globulin
(C) Glycine (D) Lycine

494. Jute is divided into:

(A) Two groups (B) Thtee groups
(C) One group (D) None of these

495. The rice varieties grown in India belongs to:

(A) Indica (B) Japonica
(C) Javanica (D) None of these

496. In alkaline soil which can be successfully grown?

(A) Potato (B) Paddy
(C) Banana (D) Cauliflower

497. Pegging of groundnut refers to:

(A) Flower initiation
(B) Initiation of secondary branches
(C) Entering of fertilized ovary in soil
(D) Grain development

498. The edible part of cabbage is:

(A) Root (B) Flower
(C) Leaf (D) Fruit

499. Krishak Bharti periodical is published from:

(A) New Delhi (B) Lucknow
(C) Pantnagar (D) Hisar

500. Directorate of Knowledge Management in Agriculture is situated at:

(A) Pusa, New Delhi
(B) Pusa, Samasti pur
(C) Hyderabad, AP (D) Hisar, Haryana

501. Fakhurruddin Ali Ahmad award relate to:

(A) Best research work on Tribal people upliftment
(B) Team work in agricultural development
(C) Best technical book in hindi
(D) None of these

502. Hari Om Trust Award related to:

(A) Best research work on Tribal people upliftment
(B) Team work in agriculture development
(C) Best technical book in hindi
(D) None of these

503. Dr. Rajendra Prasad award relate to:

(A) Best research work on Tribal people upliftment
(B) Team work in agriculture development
(C) Best technical book in hindi
(D) None of these

504. Evergreen revolution is related to:
(A) Potato protection
(B) Food grain production
(C) Milk production
(D) Increasing production in all together

505. Rainbow revolution is related to:
(A) Potato protection
(B) Food grain production
(C) Milk production
(D) Increasing production in all together

506. Green revolution is related to:
(A) Potato protection
(B) Food grain production
(C) Milk production
(D) Increasing production in all together

507. Brown /Black revolution is related to:
(A) Potato protection
(B) Food grain production
(C) Milk production
(D) Non conventional energy

508. Round revolution is related to:
(A) Potato protection
(B) Food grain production
(C) Milk production
(D) Increasing production in all together

509. White revolution is related to:
(A) Potato protection
(B) Food grain production
(C) Milk production
(D) Increasing production in all together

510. Blue revolution is related to:
(A) Potato protection
(B) Food grain production
(C) Milk production
(D) Fish production

511. Grey revolution is related to:
(A) Fertilizer protection
(B) Food grain production
(C) Milk production
(D) Increasing production in all together

512. Red revolution is related to:
(A) Potato protection
(B) Meat/tomato production
(C) Milk production
(D) Increasing production in all together

513. Golden revolution is related to:
(A) Potato protection
(B) Food grain production
(C) Horticultural crop
(D) Increasing production in all together

514. Pink revolution is related to:
(A) Potato protection
(B) Food grain production
(C) Milk production
(D) Pawn production

515. Silver revolution is related to:
(A) Chicken/eggs production
(B) Food grain production
(C) Milk production
(D) Increasing production in all together

516. Yellow revolution is related to:
(A) Potato protection
(B) Oilseed production
(C) Milk production
(D) Increasing production in all together

517. The agricultural Price Commission was set up to evolve balanced and integrated price structure in the year:
(A) 1962 (B) 1963
(C) 1964 (D) 1965

518. In Uttar Pradesh Krishi Utpadan Mandi Adhiniyam was passed in:
(A) 1962 (B) 1963
(C) 1964 (D) 1965

519. International financing agency to support National Agricultural Technology Project in India is:
(A) FAO (B) World Bank
(C) Asian Bank (D) UNDP

520. Which year is declared as International year of rice:
(A) 2004 (B) 2005
(C) 2006 (D) 2007

521. Which year is declared as International year of parthenium/neem patented :
(A) 2004 (B) 2005
(C) 2006 (D) 2007

522. Which year is declared as International year of Desert and desertification:
(A) 2004 (B) 2005
(C) 2006 (D) 2007

523. Which year is declared as International year of water:

(A) 2004 (B) 2005
(C) 2006 (D) 2007

524. Which year is declared as International year of potato:

(A) 2008 (B) 2005
(C) 2006 (D) 2007

525. Which year is declared as International year of Astronomy:

(A) 2008 (B) 2009
(C) 2010 (D) 2011

526. Which year is declared as International year of bio-diversity:

(A) 2008 (B) 2009
(C) 2010 (D) 2011

527. Which year is declared as International year of forest:

(A) 2008 (B) 2009
(C) 2010 (D) 2011

528. Who says "Everything else can wait, but not agriculture"?

(A) Shri. J.L. Nehru
(B) Dr. A.P.J. Abul Kalam
(C) Dr. H. Rembold
(D) Mahatma Gandhi

529. Who says "Agriculture is the culture of the times as/or all culture"?

(A) Shri. J.L. Nehru
(B) Dr. A.P.J. Abul Kalam
(C) Dr. H. Rembold
(D) Mahatma Gandhi

530. Who says "We have to have the second green revolution for ever increasing population of the country"?

(A) Shri. J.L. Nehru
(B) Dr. A.P.J. Abul Kalam
(C) Dr. H. Rembold
(D) Mahatma Gandhi

531. Who says "India is sitting on gold mines will all its Neem trees"?

(A) Shri. J.L. Nehru
(B) Dr. A.P.J. Abul Kalam
(C) Dr. H. Rembold (D) Mahatma Gandhi

532. Total food grain production in India during 2018-19 was:

(A) 235.00 mt (B) 250.55 mt
(C) 284.80 mt (D) 267.00 mt

533. Total pulses production in India during 2018-19 was:

(A) 23.4 mt (B) 18.2 mt
(C) 19.2 mt (D) 20.2 mt

534. Total Oilseeds production in India during 2018-19 was:

(A) 23.5 mt (B) 32.3 mt
(C) 35.1 mt (D) 35.5 mt

535. Total wheat production in India during 2018-19 was:

(A) 102.2 mt (B) 95.54 mt
(C) 95.98 mt (D) 96.31 mt

536. Total rice production in India during 2018-19 was:

(A) 85.00 mt (B) 90.54 mt
(C) 95.98 mt (D) 116.4 mt

537. Food grain production in India during 2011-12 increased in comparison to 2010-11 was:

(A) 12.95 mt (B) 13.95 mt
(C) 14.95 mt (D) 15.95 mt

538. Total coarse cereal production in India during 2018-19 was:

(A) 37.24 mt (B) 43.0 mt
(C) 39.24 mt (D) 40.24 mt

539. Total cotton production in India during 2018-19 was:

(A) 23.5 mt (B) 25.0 mt
(C) 28.7 mt (D) 35.5 mt

540. Wheat production in India during 2018-19 increased in comparison to 2017-18 was:

(A) 4.03 mt (B) 5.03 mt
(C) 6.03 mt (D) 2.50 mt

541. Rice production in India during 2018-19 increased in comparison to 2017-18 was:

(A) 3.50 mt (B) 9.34 mt
(C) 10.34 mt (D) 11.34 mt

542. Area under sugarcane in India during 2018-19 was:

(A) 315.05 mha (B) 400.00 mha
(C) 428.04 mha (D) 430.28 mha

543. Area under total pulses in India during 2011-12 was:

(A) 23.28 mha (B) 26.40 mha
(C) 24.78 mha (D) 25.85 mha

544. Area under total oilseeds in India during 2011- 12 was:
(A) 27.56 mha (B) 25.96 mha
(C) 27.22 mha (D) 24.44 mha

545. Area under wheat in India during 2011- 12 was:
(A) 29.90 mha (B) 29.07 mha
(C) 28.46 mha (D) 27.75 mha

546. Area under rice in India during 2011- 12 was:
(A) 35.85 mha (B) 40.50 mha
(C) 43.97 mha (D) 45.85 mha

547. The concept of Q/I relationship was developed by:
(A) Mitscherlich (B) Beckett
(C) V.L. Chopra (D) Schofield

548. The varieties which belong to species *Oryza glaberrima* are found in:
(A) Myanmar (B) Asia
(C) America (D) Africa

549. Tick the correct agro-climatic zone of UP which has highest average population density per km^2:
(A) Bundelkhand zone
(B) Central plains zone
(C) Hills zone
(D) North-eastern plains zone

550. National Food Security Mission started in which five year plan:
(A) 9th (B) 10th
(C) 11th (D) 12th

551. Bharat Ratna Dr. C. Subramanian Award is related to:
(A) Best agriculture teacher
(B) Best technical book
(C) Tribal people upliftment
(D) Best research in agriculture

552. In India, during 1960 milk production was 2 crore metric tones, it was rose in 2010-11 to:
(A) 10.2 crore metric tones
(B) 11.2 crore metric tones
(C) 12.2 crore metric tones
(D) 13.2 crore metric tones

553. Father of White Revolution in India is:
(A) Dr. M.S. Swaminathan
(B) Dr. Vergeese Kurien
(C) Dr. R.S. Paroda (D) Dr. B.P. Pal

554. National Dairy Development Board was established in:
(A) 1965 (B) 1966
(C) 1967 (D) 1968

555. Operation of Flood was launched in:
(A) 1969 (B) 1970
(C) 1971 (D) 1972

556. Gujarat Cooperative Milk Marketing Federation (GCMMF) was established in:
(A) 1971 (B) 1972
(C) 1973 (D) 1074

557. India Rank in milk production during 2010-11 was:
(A) First (B) Second
(C) Third (D) Fourth

558. Marketing of AMUL is to be done by:
(A) AMUL it self
(B) GCMMF
(C) Government of India
(D) Gujarat State

559. Who is known as milk man of India and Amul man is?
(A) Dr. M.S. Swaminathan
(B) Dr. Vergeese Kurien
(C) Dr. R.S. Paroda
(D) Dr. B.P. Pal

560. Growth rate of agricultural development in 9th five year plan was:
(A) 2.5% (B) 2.4%
(C) 3.3% (D) 4.0%

561. Growth rate of agricultural development in10th five year plan was:
(A) 2.5% (B) 2.4%
(C) 3.3% (D) 4.0%

562. Growth rate of agricultural development in 11th five year plan was:
(A) 2.5% (B) 2.4%
(C) 3.3% (D) 4.0%

563. Growth rate of agricultural development in 12th five year plan was:
(A) 2.5% (B) 2.4%
(C) 3.3% (D) 4.0%

564. The wheat production during 1965-66 was 12.57 mt but now rose during 2011-12 to:
(A) 90 mt (B) 92 mt
(C) 94 mt (D) 96 mt

565. The increase in production of pulses by 2020 AD to be:
(A) 29.0 mt (B) 30.3 mt
(C) 31.0 mt (D) 32.0 mt

566. The increase in production of pulses by 2030 AD to be:
(A) 29.0 mt (B) 30.3 mt
(C) 31.0 mt (D) 32.0 mt

567. Agriculture word derived from:
(A) Latin (B) French
(C) Greek (D) English

568. Agriculture denotes:
(A) Agri + cultura
(B) Ager + cultura
(C) Agri + culture
(D) Agris + cultura

569. Agriculture is the branch of:
(A) Social science (B) Basic science
(C) Applied science (D) None of these

570. Central Upland Rice Research Station located at:
(A) Cuttak (B) Phillipines
(C) New Delhi (D) Masodha

571. National Cooperative Development Corporation established in:
(A) 1962 (B) 1963
(C) 1964 (D) 1965

572. Who is the chairman of Niti Ayog:
(A) President
(B) Prime Minister
(C) Agriculture Minister
(D) D.G. of ICAR

573. Name of agriculture scientist who won Nobel prize:
(A) N.E. Borlaug (B) M.S. Swaminathan
(C) B.P. Pal (D) P.J. Kurien

574. Which state has the largest area and production of sugarcane in country:
(A) Tamilnadu (B) Karnataka
(C) Uttar Pradesh (D) Rajasthan

575. Which of the following food grains is most coarse:
(A) *Panicum miliaceum*
(B) *Echinochloa frumentacea*
(C) *Setaria italica*
(D) *Paspalum scrobiculatum*

576. Remote sensing helps in studying:
(A) Cropped area
(B) Soil characters
(C) Underground water

577. (D) All of theseIARI got status of deemed university on:
(A) 1958 (B) 1959
(C) 1960 (D) 1961

578. IARI established during:
(A) Lord Curzon (B) Lord Macnaro
(C) Lord Carbvalis (D) None of these

579. Who was first secretary of ICAR:
(A) Mohd. Habibulla
(B) S.A. Haidery
(C) B.P.Pal
(D) M.S. Swaminathan

580. Who was first chairman of ICAR:
(A) Mohd. Habibulla
(B) S.A. Haidery
(C) B.P.Pal
(D) M.S. Swaminathan

581. When ICAR declared an autonomous body?
(A) 1964 (B) 1965
(C) 1966 (D) 1967

582. First Director General of ICAR was:
(A) Mohd. Habibulla
(B) S.A. Haidery
(C) B.P.Pal
(D) M.S. Swaminathan

583. Who was first Indian director of ICAR:
(A) B. Vishwanath
(B) S. A. Haidery
(C) B. P. Pal
(D) M. S. Swaminathan

584. 'A' value concept was given by:
(A) Sorenson (B) Beckett
(C) Schofield (D) Fried & Dean

585. First irrigation commission was established on:
(A) 1898 (B) 1899
(C) 1900 (D) 1901

586. Central Agriculture Commission was established on:
(A) 1969 (B) 1970
(C) 1971 (D) 1972

587. Indian National Sugarcane Committee was established on:
(A) 1944 (B) 1945
(C) 1946 (D) 1947

588. First five year plan was started on:
(A) 1947-1951 (B) 1948-1952
(C) 1951-1956 (D) 1950-1955

589. NABARD established on:
(A) 1981 (B) 1982
(C) 1983 (D) 1984

590. LEISA is related to:
(A) Organic farming
(B) Inorganic farming
(C) Natural farming
(D) None of these

591. ICAR foundation day celebrated on:
(A) 16 July (B) 16 August
(C) 16 September (D) 16 October

592. National Agriculture Insurance was started on:
(A) 1998-1999 (B) 1999-2000
(C) 2000-2001 (D) 2001-2002

593. Now-a-days wheat cultivation started in:
(A) West Bengal (B) Andhra Pradesh
(C) Maharastra
(D) North-eastern states

594. Maximum water used for irrigation in which river:
(A) Yamuna (B) Ganga
(C) Narmada (D) Ghaghara

595. Hand Book of Agriculture published from:
(A) New Delhi (B) Lucknow
(C) Bhopal (D) Jaipur

596. Rothamsted Agricultural Research Station started by:
(A) Larves and Gilbert
(B) Phiphs
(C) Bourlag (D) Swaminathan

597. Maximum consumption of edible oils in India:
(A) Mustard (B) Coconut
(C) Groundnut (D) Sesame

598. Maximum agriculture culturable area in which country:
(A) America (B) Australia
(C) Canada (D) India

599. Queen of food grain is:
(A) Wheat (B) Rice
(C) Maize (D) Potato

600. Which crop is declared as a gold in America:
(A) Wheat (B) Rice
(C) Sorghum (D) Maize

601. Weight of a cotton bale is:
(A) 150 kg (B) 160 kg
(C) 170 kg (D) 180 kg

602. Queen of spices is:
(A) Ilaichi (B) Turmeric
(C) Chilli (D) Coriander

603. King of fruit is:
(A) Banana (B) Papaya
(C) Guava (D) Mango

604. Pseudo cereal is:
(A) Duram wheat (B) Buck wheat
(C) Macroni wheat (D) None of these

605. Rice grown in Indonesia belongs to:
(A) *O. glaberrima*
(B) *O. sativa indica*
(C) *O. sativa japonica*
(D) *O. sativa javanica*

606. Maximum potato production in:
(A) Punjab
(B) Haryana
(C) Uttar Pradesh
(D) Madhya Pradesh

607. Maximum rice production in:
(A) Tamilnadu (B) West Bengal
(C) Uttar Pradesh (D) Orissa

608. Maximum sugarcane production in:
(A) Punjab
(B) Haryana
(C) Uttar Pradesh
(D) Madhya Pradesh

609. Which state is most affected by water erosion:
(A) West Bengal (B) Uttar Pradesh
(C) Rajasthan (D) Madhya Pradesh

610. Which state is most affected by wind erosion:
(A) West Bengal (B) Uttar Pradesh
(C) Rajasthan (D) Madhya Pradesh

611. Which state has maximum area under alkali soils in India:
(A) West Bengal (B) Uttar Pradesh
(C) Gujarat (D) Bihar

612. Which state has maximum area under acidic soils in India:
(A) West Bengal (B) Uttar Pradesh
(C) Gujarat (D) Bihar

613. Maximum fertilizer consumption per hectare is:
(A) Punjab (B) Pondicherry
(C) Haryana (D) Uttar Pradesh

614. Maximum per capita availability of land (Approximately 57 ha) is:
(A) Canada (B) Australia
(C) America (D) England

615. Maximum area under non reporting land is:
(A) Arunachal Pradesh
(B) Meghalaya
(C) Tripura
(D) J & K

616. Area under problematic land is:
(A) 187 mha (B) 189 mha
(C) 195 mha (D) 198 mha

617. Per capita availability of pulses during 2011-12 is:
(A) 27 g/day (B) 28 g/day
(C) 29 g/day (D) 30 g/day

618. Maximum cigarette tobacco production is:
(A) Andhra Pradesh (B) Gujarat
(C) Bihar (D) Uttar Pradesh

619. Maximum Bidi tobacco production is:
(A) Andhra Pradesh (B) Gujarat
(C) Bihar (D) Uttar Pradesh

620. Maximum chewing tobacco production is:
(A) Andhra Pradesh
(B) Gujarat
(C) Bihar (D) Uttar Pradesh

621. Highest productivity of wheat is:
(A) Uttar Pradesh (B) Punjab
(D) Haryana (D) Rajasthan

622. Highest total production of wheat in India is:
(A) Uttar Pradesh (B) Madhya Pradesh
(C) Rajasthan (D) Gujarat

623. Highest total production of pulses in India is:
(A) Uttar Pradesh (B) Madhya Pradesh
(C) Rajasthan (D) Gujarat

624. Highest total production of mustard in India is:
(A) Uttar Pradesh (B) Madhya Pradesh
(C) Rajasthan (D) Gujarat

625. The cohesiveness and tenderness of cooked rice depends largely on:
(A) The percentage of amylose
(B) The percentage of amylopectin
(C) The proportion of amylose and protein
(D) The proportional amylose and amylopectin

626. In functional allelopathy:
(A) Toxic substances are released as such from the plant
(B) A precursor is released which is converted into active substances by some microorganisms
(C) There is no question of release of any toxic substance
(D) Release of nitrogen from nodule of legume take place

627. Non traditional areas of rice cultivation include:
(A) Punjab and Haryana
(B) Uttar Pradesh and Bihar
(C) West Bengal and Orissa
(D) Tamil nadu and Andhra Pradesh

628. All the grain legume have a high photo-respiration because of:
(A) C_3 mechanism
(B) More vegetative growth
(C) Indeterminate
(D) C_4 mechanism

629. Highest total production of rubber in India is:
(A) West Bengal (B) Madhya Pradesh
(C) Bihar (D) Kerala

630. Highest total production of onion in India is:
(A) Maharastra
(B) Madhya Pradesh
(C) Bihar
(D) Gujarat

631. Maximum production of sugarcane in the world is:
(A) Brazil (B) China
(C) India (D) France

632. Maximum production of cotton in the world is:
(A) Brazil (B) China
(C) India (D) France

633. Maximum production of tobacco in the world is:
(A) Brazil (B) China
(C) India (D) France

634. Maximum production of tea in the world is:
(A) Brazil (B) China
(C) India (D) France

635. Maximum production of coffee in the world is:
(A) Brazil (B) China
(C) India (D) France

636. Maximum production of wheat in the world is:
(A) Brazil (B) China
(C) India (D) France

637. Maximum production of rice in the world is:
(A) Brazil (B) China
(C) India (D) France

638. Maximum production of groundnut in the world is:
(A) Brazil (B) China
(C) India (D) France

639. Maximum production of mustard in the world is:
(A) Brazil (B) China
(C) India (D) France

640. Maximum production of onion in the world is:
(A) Brazil (B) China
(C) India (D) France

641. Maximum production of fruits in the world is:
(A) Brazil (B) China
(C) India (D) France

642. Maximum production of potato in the world is:
(A) Brazil (B) China
(C) India (D) France

643. Maximum production of fishes in the world is:
(A) Brazil (B) China
(C) India (D) France

644. Maximum area under macroni wheat in India is:
(A) Punjab (B) West Bengal
(C) Punjab and West Bengal
(D) Uttar Pradesh

645. Bowl of rice is known as:
(A) West Bengal (B) Kerala
(C) Tamilnadu (D) Karnataka

646. Bowl of spices is known as:
(A) West Bengal (B) Kerala
(C) Tamilnadu (D) Karnataka

647. City of gardens is known as:
(A) Bangalore (B) J & K
(C) New Delhi (D) Hyderabad

648. Minimum area under forest is:
(A) Rajasthan (B) Madhya Pradesh
(C) West Bengal (D) Uttar Pradesh

649. Maximum area under forest is:
(A) Rajasthan (B) Madhya Pradesh
(C) West Bengal (D) Uttar Pradesh

650. Presently, which of the following chemical fertilizer is not produced in India in being imported?
(A) Triple Surper phosphate
(B) Potassium sulphate
(C) Ammonium sulphate
(D) All of these

651. For providing inputs like quality seeds, fertilizers and pesticides, like agency present at the village Panchayat Samiti level is:
(A) Nationalised Banks
(B) Cooperative Society
(C) Insurance companies

652. (D) NABARDPea cultivation is not recommended in:
(A) Normal soils
(B) Acidic soils
(C) Alluvial soils
(D) Alkali soils

653. Raring of sheep was started on:
(A) 8700 AD (B) 7700 AD
(C) 7500 AD (D) 6000 AD

654. Raring of goat was started on:
(A) 8700 AD (B) 7700 AD
(C) 7500 AD (D) 6000 AD

655. Cultivation of wheat and barley was started on:
(A) 8700 AD (B) 7700 AD
(C) 7500 AD (D) 6000 AD

656. Raring of cow and pig was started on:
(A) 8700 AD (B) 7700 AD
(C) 7500 AD (D) 6000 AD

657. Raring of silkworm was started on:
(A) 1700 AD (B) 2700 AD
(C) 3500 AD (D) 4000 AD

658. Cultivation of maize was started on:
(A) 4400 AD (B) 2900 AD
(C) 2300 AD (D) 2200 AD

659. Knowledge gain for ploughing and irrigation was on:
(A) 4400 AD (B) 2900 AD
(C) 2300 AD (D) 2200 AD

660. Cultivation of arhar, pine apple, mustard and cotton was started on:
(A) 4400 AD (B) 2900 AD
(C) 2300 AD (D) 2200 AD

661. Raring of poultry, buffalo and elephant was started on:
(A) 4400 AD (B) 2900 AD
(C) 2300 AD (D) 2200 AD

662. Cultivation of ragi and mandua was started on:
(A) 1800 AD (B) 1725 AD
(C) 1500 AD (D) 1400 AD

663. Cultivation of sorghum was started on:
(A) 1800 AD (B) 1725 AD
(C) 1500 AD (D) 1400 AD

664. Cultivation of sugarcane was started on:
(A) 1800 AD (B) 1725 AD
(C) 1500 AD (D) 1400 AD

665. Cultivation of potato was started on:
(A) 3500 AD (B) 4400 AD
(C) 4500 AD (D) 7500 AD

666. Imprtance of animals and raring of animals was known as era of:
(A) Pastoral age (B) Neolithic age
(C) Bronze age (D) None of these

667. Which type of cultivation practices mainly responsible for agricultural development in India:
(A) Precision farming
(B) Shifting cultivation
(C) Mutistoryed cultivation
(D) Organic farming

668. Cotton is commercially grown in:
(A) Andhra Pradesh
(B) Karnataka
(C) Maharastra
(D) All of these

669. Sorghum is known as camel due to:
(A) Important crop of millet
(B) Drought tolerant
(C) Important feed of camel
(D) Important crop of dry areas

670. Cultivation of duram wheat mainly in:
(A) Central and southern India
(B) North India
(C) North-eastern India
(D) Dry region of India

671. Spot satellite services was started by:
(A) UK (B) USA
(C) India (D) China

672. Which country is origin of tea but totally stopped for its cultivation:
(A) China (B) India
(C) Srilanka (D) Bhuan

673. Area and production of posta maximum in which state:
(A) Bihar
(B) West Bengal
(C) Uttar Pradesh
(D) Haryana

674. Panchgavya is an organic product has the potential to play the role of promoting:
(A) Growth
(B) Productivity
(C) Growth and productivity
(D) None of these

675. Panchgavya consist of:
(A) Cow dung and cow urine
(B) Cow milk and cow curd
(C) Cow ghee
(D) All of these

676. Sorghum is considered as the king of millets also known as:
(A) Camel crop
(B) King of cereals
(C) Queen of cereals
(D) Important crop of kharif

677. Total area under fodder production is:
(A) 7.3 mha (B) 8.3 mha
(C) 9.3 mha (D) 10.3 mha

678. Requirement of concentrate as feed for dairy in India is:
(A) 570 mt (B) 923 mt
(C) 1025 mt (D) 1135 mt

679. Requirement of green fodder as feed for dairy in India is:
(A) 570 mt (B) 923 mt
(C) 1025 mt (D) 1135 mt

680. Requirement of dry fodder as feed for dairy in India is:
(A) 570 mt (B) 923 mt
(C) 1025 mt (D) 1135 mt

681. Availability of concentrate as feed for dairy in India is:
(A) 45 mt (B) 390 mt
(C) 443 mt (D) 501 mt

682. Availability of green fodder as feed for dairy in India is:
(A) 45 mt (B) 390 mt
(C) 443 mt (D) 501 mt

683. Availability of dry fodder as feed for dairy in India is:
(A) 45 mt (B) 390 mt
(C) 443 mt (D) 501 mt

684. IRRI is the abbreviation for:
(A) International Raw Rice Institute
(B) Institute of Rice Research in India
(C) Indian Rice Research Institute
(D) International Rice Research Institute

685. The typical odour of garlic is due to:
(A) Sinigrin (B) Ficin
(C) Allicin (D) Alliin

686. Which crop is called as king of cereal crops?
(A) Rice (B) Wheat
(C) Oat (D) Barley

687. The term 'Tropical oils' refers to:
(A) Palm oil and coconut oil
(B) Olive oil and almond oil
(C) Corn oil and soyabean oil
(D) Sunflower oil and safflower oil

688. 'Touch me not' plant exhibits a reflex action that is due to:
(A) Special knee joint
(B) Sudden temperature control
(C) Exceptional nervous conduction
(D) Powerful water pressure alteration

689. "Tension wood" is the name given to that wood which reacts to:
(A) Drought
(B) Low oxygen
(C) Drooping
(D) Disease

690. Button lac is manufactured from:
(A) Molten lac
(B) Crushed lac
(C) Shel lac
(D) Seed lac

691. The cropped area of India under double crop is:
(A) 35 mha (B) 45 mha
(C) 55 mha (D) 65 mha

692. Before independence ICAR was known as:
(A) Independence Council of Agricultural Research
(B) Imperial Council of Agricultural Research
(C) Initial Council of Agricultural Research
(D) Indian Council of Agricultural Research

693. After independence ICAR is known as:
(A) Independence Council of Agricultural Research
(B) Imperial Council of Agricultural Research
(C) Initial Council of Agricultural Research
(D) Indian Council of Agricultural Research

694. NAARM was originally started as:
(A) Central Staff College for Agriculture
(B) Central Recuitment for Agriculture
(C) Research Centre
(D) Extension centre

695. DR. Norman E. Borlaug was rewarded by:
(A) Padma Vibhushan
(B) Padam Shree
(C) Bharat Ratna
(D) None of these

696. AICRP means:
(A) All India Coordinated Research Project
(B) All India Cooperative Research Project
(C) All India Corporation Research Programme
(D) All India Coordinated Research Priority

697. It unertakes research and coordinates activities in germplasm collection, introducti on and exchange of seeds and plant materials; and characterization, documentaion, maintenance & conservation of genetic resources for utilization in crop management:
(A) National Bureau of Plant Genetic Resources, New Delhi
(B) National Plant Genetic Resources Bureau, New Delhi
(C) National Bureau of Crop Genetic Resources, New Delhi
(D) National Bureau Genetic Resources, New Delhi

698. NARP means:
(A) National Agronomy Research Project
(B) National Agricultural Reserch Programme
(C) National Agricultural Regional Programme
(D) National Agricultural Research Programme

699. Which of the following is deemed to be University?
(A) Central Institute of Fisheries Education, Mumbai
(B) Indian Veterinary Research Institute, Izatnagar, Bareilly
(C) National Dairy Research Institute, Karnal
(D) All of these

700. Which of the following is deemed to be University?
(A) Central Institute of Fisheries Education, Mumbai
(B) Allahabad Agricultural Institute, Naini, Allahabad
(C) Indian Agricultural Research Institute, New Delhi
(D) All of these

701. Which of the following is central University with agricultural faculty?
(A) Banaras Hindu University, Varanasi
(B) Aligarh Muslim University, Aligarh
(C) Vishwa Bhati, Shantiniketan
(D) All of these

702. Which of the following is central university with agricultural faculty?
(A) Nagaland University, Medziphema
(B) Osmania University, Hyderabad
(C) Jawahar Lal Nehru University, New Delhi
(D) None of these

703. CGIAR means:
(A) The Consultative Group on International Agricultural Research
(B) The Cooperative Group on International Agricultural Research
(C) The Consultative Guidance on International Agricultural Research
(D) The Consultative Group on Indian Agricultural Research

704. CGIAR is a part of the system:
(A) Global Agriculural Research System
(B) National Agricultural Research System
(C) Global Agricultural Extension System
(D) Global Agricultural Education System

705. Major activities of CGIAR includes:
(A) Food for people
(B) Environment for People
(C) Policies for people
(D) All of these

706. How many centres together maintain major role in maintaining genebanks under CGIAR?
(A) 10 (B) 11
(C) 15 (D) 20

707. How many CGIAR supported international agricultural research centres are working world wide?
(A) 10 (B) 11
(C) 15 (D) 20

708. Which one of the following states has the largest area under Kharif Maize?
(A) Uttar Pradesh
(B) Rajasthan
(C) Bihar
(D) Madhya Pradesh

709. CIMMYT is:
(A) International Centre for Improvmenet of Maize, Wheat and Barley
(B) Centre International de Majormiento de Maizy Trigo
(C) Both (A) & (B)
(D) None of these

710. CIAT is:
(A) International Centre for Tropical Agriculture
(B) Centre of International level for Tropical Agriculture
(C) Centre of International Agricultural Technology
(D) Interdisciplinary Centre for Tropical Agriculture

711. CIFOR is:
(A) Centre for International Forestry Output Research
(B) Centre for International Forestry Research
(C) Centre for Indian Forestry Research
(D) Centre for Initiatives of Forestry output Research

712. CIP is:
(A) International Potato Centre
(B) Centro Internationally De La Papa
(C) Centre for Potato Improvement
(D) Both (A) and (B)

713. IFPRI is:
(A) Indian Food Policy Research Institute
(B) International Food Policy Research Institute
(C) International Forestry Policy Research Institute
(D) International Financial Policy Refinement Institute

714. IITA is:
(A) Indian Institute of Tropical Agriculture
(B) International Investment for Tropical Agriculture
(C) International Institute for Training of Agro chemical
(D) International Institute of Tropical Agricultrue

715. Headquarter of CAB International is located:
(A) UK (B) USA
(C) Australia (D) India

716. WFC is:
(A) World Fish Centre
(B) Wild Fish Centre
(C) World-class Fish Centre
(D) World Finance Centre

717. ILRI is:
(A) Initiatives of Livestock Research Institute
(B) International Lion Research Institute
(C) International Livestock Research Institute
(D) Indian Livestock Research Institute

718. IWMI is:
(A) Innovative Water Management Institute
(B) International Water Management Institute
(C) Indian Water Management Institute
(D) International Wildlife Management Institute

719. Headquarter of the World Bank is located:
(A) London (B) Toranto
(C) Beijing (D) Washington DC

720. Headwaurter of FAO is:
(A) London (B) Toranto
(C) Italy (D) Washington DC

721. FAO is:
(A) Food and Agriculture Organization
(B) Forest and Agriculture Organization
(C) Food and Aquaculture Organization
(D) Fertilizer & Agriculture Organization

722. Which states are the highest and lowest producers of potato crop?
(A) UP and HP
(B) Gujarat & Assam
(C) UP and Jharkhad
(D) Bihar & Kashmir

723. Which is not prepared from potato?
(A) Paper
(B) Farina and Alcohol
(C) Wine
(D) Acetic Acid

724. Which of the following award is associated with Agriculture?
(A) Bourley Award
(B) Shanti Swaroop Bhatnagar Award
(C) Arjun Award
(D) Vyasa Samman

725. Which in the most widely used green manuring crop in Assam, West Bengal, Bihar, Haryana, Punjab, Tamilnadu and Uttar Pradesh:
(A) Sunnhemp (B) Dhaincha
(C) Senji (D) Egyptian Clover

726. The Depog method of raising rice nursery was introduced in India from:
(A) Myanmar (B) Japan
(C) China (D) Philippines

727. The queen of cereal is
(A) Rice (B) Mustard
(C) Maize (D) Pearlmillet

728. Brown manuring is generally recommended in which crop?
(A) Wheat (B) Mustard
(C) Rice (D) Potato

729. One Angstrom is a unit of length equal to:
(A) 10^{-2} cm (B) 10^{-4} cm
(C) 10^{-8} cm (D) 10^{-10} cm

730. The term "evergreen revolution" has been given by
(A) Dr. R.S. Paroda
(B) Dr. M.S. Swaminathan
(C) Dr. A.S. Faroda
(D) Dr. V.L. Chopra

731. 'Dwarf Wheat' (Mexican) is introduced in
(A) Dr. N.E. Barloug
(B) Dr. M.S. Swaminathan
(C) Dr. B.P. Pal
(D) Dr. C. Subramanian

732. Idea of Super rice was given by :
(A) V.L. Chopra (B) G.S. Kushi
(C) Yoshida (D) G.H. Shull

733. The practice of cross ploughing the young crop of rice to reduce weeds and vegetative growth of crop is called :
(A) Puddling (B) Beushening
(C) Khedua (D) Taungya

734. Legume crop which responds to N application as much as cereal crops :
(A) Fieldpea (B) Guar
(C) French bean (D) Cowpea

735. Maize grain is mainly used for :
(A) Poultry feed
(B) Cattle feed
(C) Food to human consumption
(D) Starch making

736. Under droughts, the sorghum plants synthesize Dhurin in :
(A) Roots
(B) Shoots
(C) Leaves
(D) All the plant parts

737. Dropsy disease in human beings is caused by which weed?
(A) Pluchea lanceolate
(B) Argemone mexicana
(C) Digera arvensis
(D) Convolvulus arvensis

738. Most of the wheat varieties are :
(A) Quantitative long dry plants
(B) Qualitative long dry plants
(C) Day neutral
(D) Short day plants

739. The chemical responsible for lathyrism in mammals in :
(A) BOAA (B) HCN
(C) NAA (D) None of these

ANSWER SHEET
1. General Agriculture

1. (A)	2. (C)	3. (B)	4. (C)	5. (D)	6. (D)	7. (A)	8. (B)	9. (A)	10. (A)
11. (A)	12. (D)	13. (C)	14. (B)	15. (A)	16. (A)	17. (B)	18. (D)	19. (C)	20. (A)
21. (A)	22. (B)	23. (B)	24. (C)	25. (D)	26. (C)	27. (A)	28. (D)	29. (D)	30. (A)
31. (B)	32. (C)	33. (C)	34. (B)	35. (D)	36. (C)	37. (A)	38. (A)	39. (A)	40. (D)
41. (B)	42. (D)	43. (A)	44. (A)	45. (C)	46. (B)	47. (D)	48. (D)	49. (B)	50. (D)
51. (B)	52. (D)	53. (A)	54. (A)	55. (B)	56. (C)	57. (D)	58. (C)	59. (D)	60. (D)
61. (C)	62. (B)	63. (D)	64. (C)	65. (D)	66. (A)	67. (A)	68. (A)	69. (C)	70. (C)
71. (B)	72. (C)	73. (B)	74. (D)	75. (A)	76. (D)	77. (D)	78. (D)	79. (A)	80. (A)
81. (A)	82. (C)	83. (D)	84. (B)	85. (C)	86. (B)	87. (A)	88. (B)	89. (A)	90. (B)
91. (D)	92. (C)	93. (B)	94. (A)	95. (C)	96. (D)	97. (B)	98. (C)	99. (B)	100. (B)
101. (A)	102. (A)	103. (C)	104. (B)	105. (D)	106. (A)	107. (A)	108. (A)	109. (A)	110. (B)
111. (B)	112. (A)	113. (D)	114. (C)	115. (A)	116. (A)	117. (C)	118. (C)	119. (C)	120. (B)
121. (D)	122. (D)	123. (D)	124. (B)	125. (B)	126. (C)	127. (A)	128. (C)	129. (D)	130. (C)
131. (B)	132. (B)	133. (A)	134. (C)	135. (D)	136. (A)	137. (B)	138. (D)	139. (C)	140. (A)
141. (C)	142. (A)	143. (B)	144. (A)	145. (B)	146. (B)	147. (B)	148. (A)	149. (A)	150. (A)
151. (C)	152. (C)	153. (B)	154. (C)	155. (B)	156. (C)	157. (A)	158. (C)	159. (C)	160. (A)
161. (A)	162. (A)	163. (C)	164. (B)	165. (D)	166. (C)	167. (C)	168. (C)	169. (C)	170. (B)
171. (C)	172. (A)	173. (B)	174. (D)	175. (A)	176. (B)	177. (D)	178. (D)	179. (B)	180. (C)
181. (B)	182. (B)	183. (D)	184. (D)	185. (A)	186. (C)	187. (A)	188. (A)	189. (D)	190. (D)
191. (A)	192. (A)	193. (A)	194. (C)	195. (A)	196. (B)	197. (B)	198. (A)	199. (D)	200. (B)
201. (B)	202. (C)	203. (A)	204. (A)	205. (C)	206. (A)	207. (B)	208. (B)	209. (B)	210. (B)
211. (B)	212. (B)	213. (A)	214. (D)	215. (C)	216. (D)	217. (A)	218. (D)	219. (C)	220. (C)
221. (B)	222. (B)	223. (A)	224. (A)	225. (A)	226. (A)	227. (B)	228. (B)	229. (C)	230. (A)
231. (B)	232. (D)	233. (B)	234. (C)	235. (B)	236. (A)	237. (A)	238. (C)	239. (B)	240. (A)
241. (B)	242. (C)	243. (D)	244. (A)	245. (B)	246. (C)	247. (C)	248. (C)	249. (D)	250. (C)
251. (A)	252. (B)	253. (B)	254. (C)	255. (C)	256. (A)	257. (C)	258. (C)	259. (B)	260. (A)
261. (B)	262. (B)	263. (A)	264. (B)	265. (A)	266. (C)	267. (B)	268. (B)	269. (B)	270. (A)
271. (B)	272. (D)	273. (B)	274. (C)	275. (B)	276. (C)	277. (B)	278. (B)	279. (A)	280. (A)
281. (A)	282. (B)	283. (A)	284. (A)	285. (C)	286. (C)	287. (C)	288. (C)	289. (B)	290. (A)
291. (C)	292. (C)	293. (B)	294. (A)	295. (D)	296. (C)	297. (C)	298. (C)	299. (D)	300. (C)
301. (B)	302. (C)	303. (C)	304. (C)	305. (C)	306. (C)	307. (C)	308. (B)	309. (C)	310. (C)
311. (C)	312. (C)	313. (A)	314. (A)	315. (C)	316. (B)	317. (B)	318. (B)	319. (B)	320. (B)
321. (A)	322. (A)	323. (B)	324. (B)	325. (C)	326. (C)	327. (B)	328. (B)	329. (A)	330. (A)
331. (B)	332. (A)	333. (A)	334. (C)	335. (B)	336. (B)	337. (D)	338. (A)	339. (B)	340. (C)
341. (B)	342. (B)	343. (C)	344. (C)	345. (B)	346. (C)	347. (B)	348. (A)	349. (A)	350. (D)
351. (B)	352. (B)	353. (A)	354. (A)	355. (B)	356. (C)	357. (A)	358. (A)	359. (A)	360. (A)
361. (C)	362. (A)	363. (B)	364. (A)	365. (C)	366. (B)	367. (C)	368. (C)	369. (B)	370. (B)

371. (D) 372. (B) 373. (A) 374. (C) 375. (C) 376. (A) 377. (A) 378. (B) 379. (B) 380. (D)

381. (B) 382. (D) 383. (C) 384. (D) 385. (D) 386. (B) 387. (D) 388. (D) 389. (B) 390. (C)

391. (D) 392. (B) 393. (A) 394. (B) 395. (A) 396. (C) 397. (D) 398. (B) 399. (C) 400. (A)

401. (B) 402. (A) 403. (C) 404. (B) 405. (B) 406. (B) 407. (B) 408. (A) 409. (B) 410. (D)

411. (C) 412. (B) 413. (D) 414. (B) 415. (A) 416. (D) 417. (A) 418. (B) 419. (D) 420. (A)

421. (A) 422. (B) 423. (C) 424. (B) 425. (A) 426. (B) 427. (B) 428. (C) 429. (D) 430. (D)

431. (B) 432. (A) 433. (B) 434. (D) 435. (A) 436. (D) 437. (A) 438. (C) 439. (A) 440. (C)

441. (B) 442. (B) 443. (C) 444. (D) 445. (A) 446. (B) 447. (D) 448. (C) 449. (C) 450. (C)

451. (A) 452. (B) 453. (A) 454. (A) 455. (C) 456. (D) 457. (C) 458. (B) 459. (A) 460. (A)

461. (A) 462. (B) 463. (C) 464. (C) 465. (A) 466. (B) 467. (B) 468. (D) 469. (D) 470. (B)

471. (D) 472. (A) 473. (B) 474. (C) 475. (B) 476. (C) 477. (D) 478. (A) 479. (B) 480. (A)

481. (C) 482. (B) 483. (C) 484. (D) 485. (A) 486. (C) 487. (A) 488. (B) 489. (C) 490. (C)

491. (C) 492. (B) 493. (A) 494. (A) 495. (A) 496. (B) 497. (C) 498. (C) 499. (C) 500. (A)

501. (A) 502. (B) 503. (C) 504. (D) 505. (D) 506. (B) 507. (D) 508. (A) 509. (C) 510. (D)

511. (A) 512. (B) 513. (C) 514. (D) 515. (A) 516. (B) 517. (D) 518. (C) 519. (B) 520. (A)

521. (B) 522. (C) 523. (D) 524. (A) 525. (B) 526. (C) 527. (D) 528. (A) 529. (A) 530. (B)

531. (C) 532. (C) 533. (A) 534. (B) 535. (A) 536. (D) 537. (A) 538. (B) 539. (C) 540. (D)

541. (A) 542. (B) 543. (C) 544. (D) 545. (A) 546. (C) 547. (B) 548. (D) 549. (B) 550. (C)

551. (A) 552. (C) 553. (B) 554. (A) 555. (B) 556. (C) 557. (A) 558. (B) 559. (B) 560. (A)

561. (B) 562. (C) 563. (D) 564. (C) 565. (B) 566. (D) 567. (A) 568. (B) 569. (C) 570. (D)

571. (B) 572. (B) 573. (A) 574. (C) 575. (D) 576. (D) 577. (B) 578. (A) 579. (B) 580. (A)

581. (C) 582. (C) 583. (A) 584. (D) 585. (D) 586. (B) 587. (A) 588. (C) 589. (B) 590. (A)

591. (A) 592. (B) 593. (D) 594. (B) 595. (A) 596. (A) 597. (C) 598. (B) 599. (C) 600. (D)

601. (C) 602. (A) 603. (D) 604. (B) 605. (D) 606. (C) 607. (B) 608. (C) 609. (A) 610. (C)

611. (C) 612. (A) 613. (B) 614. (B) 615. (D) 616. (A) 617. (C) 618. (A) 619. (B) 620. (C)

621. (B) 622. (A) 623. (B) 624. (C) 625. (D) 626. (B) 627. (A) 628. (A) 629. (D) 630. (A)

631. (A) 632. (B) 633. (B) 634. (B) 635. (A) 636. (B) 637. (B) 638. (B) 639. (B) 640. (B)

641. (B) 642. (B) 643. (B) 644. (C) 645. (A) 646. (B) 647. (A) 648. (A) 649. (B) 650. (B)

651. (D) 652. (D) 653. (A) 654. (B) 655. (C) 656. (D) 657. (B) 658. (A) 659. (B) 660. (C)

661. (D) 662. (A) 663. (B) 664. (C) 665. (A) 666. (A) 667. (B) 668. (D) 669. (B) 670. (A)

671. (B) 672. (C) 673. (A) 674. (C) 675. (D) 676. (A) 677. (B) 678. (B) 679. (C) 680. (A)

681. (A) 682. (B) 683. (C) 684. (D) 685. (C) 686. (B) 687. (A) 688. (D) 689. (C) 690. (A)

691. (A) 692. (B) 693. (D) 694. (A) 695. (A) 696. (A) 697. (A) 698. (B) 699. (D) 700. (D)

701. (D) 702. (A) 703. (A) 704. (A) 705. (D) 706. (B) 707. (C) 708. (A) 709. (C) 710. (A)

711. (B) 712. (D) 713. (B) 714. (D) 715. (A) 716. (A) 717. (C) 718. (B) 719. (A) 720. (C)

721. (A) 722. (C) 723. (A) 724. (A) 725. (A) 726. (D) 727. (C) 728. (C) 729. (C) 730. (B)

731. (A) 732. (B) 733. (B) 734. (C) 735. (A) 736. (A) 737. (B) 738. (B) 739. (A)

2 HISTORICAL PERSPECTIVE OF AGRONOMY

1. Agronomy is originated from:
   (A) Greek words (B) Latin words
   (C) German words (D) English words
2. Agronomy is made of two words namely:
   (A) Agro + Nomous
   (B) Agros + Nomous
   (C) Agros + Nomos
   (D) None of these
3. Literary meaning of agronomy is:
   (A) Nutrients management
   (B) Soil management
   (C) Crop management
   (D) All the above
4. The branch of agriculture which deals with the principles and practices of crop production and field management is called:
   (A) Agronomy (B) Agroforestry
   (C) Agriculture (D) Sericulture
5. First agronomic research was conducted in:
   (A) 1832 (B) 1834
   (C) 1932 (D) 1934
6. First agronomic research was conducted at:
   (A) Assells (B) Brussels
   (C) USA (D) India
7. The separate branch of agronomy was originated from agriculture in:
   (A) 1900 (B) 1901
   (C) 1902 (D) 1903
8. Agronomy is a:
   (A) Applied science
   (B) Basic science
   (C) Informative science
   (D) All of these
9. The expansion of area for increasing total agricultural production is called:
   (A) Horizontal growth
   (B) Vertical growth
   (C) Linear growth (D) Cubical growth
10. The increase in production of per unit time and area from intensive cropping system is called:
    (A) Horizontal growth
    (B) Vertical growth
    (C) Linear growth (D) Cubical growth
11. The food requirement of ever increasing population can be met out by:
    (A) Horizontal growth
    (B) Vertical growth
    (C) Linear growth
    (D) Cubical growth
12. Department of Agricultural Research and Education (DARE) was started in:
    (A) 1971 (B) 1972
    (C) 1973 (D) 1974
13. The East India company brought 12 American Cotton producer to impart training on cotton production practices in:
    (A) 1839 (B) 1840
    (C) 1841 (D) 1842
14. Sustainable agriculture can be achieved by:
    (A) More resources
    (B) More subsidiers
    (C) More technological support
    (D) Better prices of the production
15. Cultivation of hybrid rice is most popular in:
    (A) Japan (B) Korea
    (C) China (D) India

16. Green revolution in wheat is basically attributed to:
(A) Greater use of seeds of HYV
(B) More use of NPK fertilizer
(C) Greater availability of irrigation
(D) All of these

17. The farming is a more precise farm management practices taking into account the variability in soil and micro climatic conditions that occurs within the field:
(A) Precision farming
(B) Multistoried farming
(C) Conservation agriculture
(D) All of these

18. The cultivation practices is based on GIS, GPS and RSS is called:
(A) Precision farming
(B) Multistoried farming
(C) Conservation agriculture
(D) All of these

19. To study the relation of agricultural crops and environment is called:
(A) Agro-ecology
(B) Urban agriculture
(C) Per urbon agriculture
(D) Conservation agriculture

20. The practice of cultivating, processing and distributing food in or around (peri-urbon), village, town or city is called:
(A) Urban agriculture
(B) Peri urbon agriculture
(C) Both (A) and (B)
(D) Conservation agriculture

21. Urbon agriculture contributes to:
(A) Food security
(B) Food safety
(C) Both of these
(D) None of these

22. The ability of a farm to produce food in definitely without causing severe or irregular damage to eco-system health is called:
(A) Sustainable agriculture
(B) Conservation agriculture
(C) Organic agriculture
(D) Urban agriculture

23. The production system that sustain the health of soils, ecosystems and peoples is called:
(A) Sustainable agriculture
(B) Conservation agriculture
(C) Organic agriculture
(D) Urban agriculture

24. The concept for resource-saving agricultural crop productive that strives to achieve acceptable profits together with high and sustained production levels while concurrently conserving the environment is referred as:
(A) Sustainable agriculture
(B) Conservation agriculture
(C) Organic agriculture
(D) Urban agriculture

25. Sustainable agriculture integrates main goals are:
(A) Environmental stewardship
(B) Farm profitability
(C) Prosperous farming communities
(D) All of these

26. The modern farming that refers to the industrialized production of livestock, poultry, fish and crops is called:
(A) Industrial agriculture
(B) Precision agriculture
(C) Bio-dynamic agriculture
(D) Community supported agriculture

27. New technology that allows farmers to look at their fields more site specifically than before and apply inputs in a manner more specific than a blanket application is called:
(A) Industrial agriculture
(B) Precision agriculture
(C) Bio-dynamic agriculture
(D) Community supported agriculture

28. A method of organic farming that has its basis in a spiritual word-view, treats farms as unified and individual organisms, emphasizing balancing the holistic development and interrelationship of the soil, plants, animal as a closed, self nourishing system is called:
(A) Industrial agriculture
(B) Precision agriculture
(C) Bio-dynamic agriculture
(D) Community supported agriculture

29. The farmer's grows food for a group of share-holders who pledge to buy a portion of the farm's crop that season is called:
(A) Industrial agriculture
(B) Precision agriculture
(C) Bio-dynamic agriculture
(D) Community supported agriculture

30. Cutting and burning of forests or wood-lands to create field for agriculture or pasture for livestock, or for a variety of other purposes is called:
(A) Industrial agriculture
(B) Slash and Burn agriculture
(C) Bio-dynamic agriculture
(D) Community supported agriculture

31. Principles of conservation agriculture is/ are:
(A) Minimal disturbances of the soil
(B) Permanent soil cover with organic materials
(C) Precision land leveling
(D) All of these

32. The harmful constituents in tobacco are:
(A) Nicotine and Tar
(B) Carbon mono oxide and Nitric oxides
(C) Nitrosamines and benzopyrenes
(D) All of these

33. The harmful constituents in tobacco can be minimized by application of:
(A) Potassic fertilizers
(B) Phosphatic fertilizers
(C) Nitrogenous fertilizers
(D) None of these

34. The quality neutral filler tobacco is known as:
(A) Karnataka light soil
(B) Northern light soil
(C) Southern light soil
(D) Black soil

35. By which process the harvested tobacco leaf is made ready for the market?
(A) Bulking
(B) Curing
(C) Loading in the barn
(D) None of these

36. When the leaves are harvested in moist conditions and loaded in the barn, there is probable infection of bacteria and fungi on either side of veins is called:
(A) Burn rot (B) Barn scaled
(C) Cooked leaf (D) Runback leaf

37. If the barn temperature is raised to more than 130^0 F before leaf drying, leaf becomes dark brown due to translocation of cell sap from the midrib to lamina portion is called:
(A) Burn rot (B) Barn scaled
(C) Cooked leaf (D) Runback leaf

38. Either due to increase in temperature beyond 160^0F during midrib drying or due to the loading of leaf near to the flue pipes, leaf tips and margins will be burnt is called:
(A) Burn rot (B) Barn scaled
(C) Cooked leaf (D) Runback leaf

39. During midrib drying, if temperature drops below required limits, the cell sap translocated from the midrib to the lamina change to in brown colour on either side of midrib, which later turns to black is called:
(A) Burn rot (B) Barn scaled
(C) Cooked leaf (D) Runback leaf

40. Nicotine sulphate and solanesol can be obtained from:
(A) Tobacco (B) Tea
(C) Coffee (D) None of these

41. Nicotine based pesticides are:
(A) Non persistent
(B) Non systematic
(C) Contact insecticide
(D) All of these

42. Tobacco oil can be safely used as:
(A) For edible
(B) For paint industry
(C) Both of these
(D) None of these

43. An excellent production of tobacco for phyto chemicals are:
(A) Nicotine (B) Solanesol
(C) Crude protein (D) All of these

44. The recovery of oil from chewing tobacco is:
(A) 30% (B) 35 %
(C) 37% (D) 40 %

45. The processes like steeping, retting and extraction are the important processes in:
(A) Tobacco (B) Jute
(C) Sugarcane (D) Rice

46. The process of dipping of jute bundles in water is called:
(A) Steeping (B) Retting
(C) Extraction (D) None of these

47. Under and over retting of jute results into:
(A) Poor quality fibre
(B) Good quality fibre
(C) Croppy fibre
(D) Dazed week fibre

48. Under retting of jute results into:
(A) Good quality fibre
(B) Poor quality fibre
(C) Croppy fibre
(D) Dazed week fibre

49. Over retting of jute results into:
(A) Good quality fibre
(B) Poor quality fibre
(C) Croppy fibre
(D) Dazed week fibre

50. *Triticum aestivum* is a variety of wheat is known as:
(A) Bread wheat (B) Hard wheat
(C) Emmer wheat (D) None of these

51. *Triticum durum* is a variety of wheat is known as:
(A) Bread wheat (B) Hard wheat
(C) Emmer wheat (D) None of these

52. *Triticum dicocum* is a variety of wheat is known as:
(A) Bread wheat
(B) Hard wheat
(C) Emmer wheat
(D) None of these

53. *Oryza japonica* is mainly grown in:
(A) Japan (B) Indonesia
(C) Western Africa (D) None of these

54. *Oryza javanica* is mainly grown in:
(A) Japan (B) Indonesia
(C) Western Africa (D) None of these

55. *Oryza glaberrima* is mainly grown in:
(A) Japan (B) Indonesia
(C) Western Africa (D) None of these

56. Top grade of bidi tobacco is called:
(A) Bukha (B) Geran
(C) Galia (D) Rago

57. Top grade of jati tobacco is called:
(A) Special grade (B) No. 1 grade
(C) Khunda (D) Fired leaves

58. The wild species of jute is:
(A) *Corchorus antichorous*
(B) *Corchorus acutangulus*
(C) Both of these
(D) None of these

59. The most common and largely grown species of jute is:
(A) *Corchorus antichorous*
(B) *Corchorus acutangulus*
(C) *Corchorus capsularis*
(D) *Corchorus olitorius*

60. Which jute species are called Tosha pat, Sona pat and Meena pat is:
(A) *Corchorus antichorous*
(B) *Corchorus acutangulus*
(C) *Corchorus capsularis*
(D) *Corchorus olitorius*

61. Soil amendment used to reclaim usar land is:
(A) Calcium sulphate
(B) Sodium sulphate
(C) Potassium sulphate
(D) Calcium carbonate

62. Which jute species are called Kakya Bombay, deshi jute, safed jute, Sada pat, Virajganj and dhaka jute is:
(A) *Corchorus antichorous*
(B) *Corchorus acutangulus*
(C) *Corchorus capsularis*
(D) *Corchorus olitorius*

63. Which microorganism is present in large quantities in soil:
(A) Algae
(B) Bacteria
(C) Fungi
(D) Protozoa

64. Which one is a insect trap crop is:
(A) Arhar
(B) Paddy
(C) Bhindi
(D) Potato

65. The term "Evergreen revolution" has been given by:
(A) Dr A.S. Faroda
(B) Dr. M.S. Swaminathan
(C) Dr. V.L. Chopra
(D) Dr. R.S. Paroda

66. Ratio of grain to stalk in mungbean is:
(A) 1 : 2 (B) 1 : 3
(C) 1 : 10 (D) 1 : 6-8

67. Which one of the following is correct:
(A) P_2O_5 x 2.29 = P
(B) P_2O_5 x 1.20 = P
(C) P_2O_5 x 0.44 = P
(D) P_2O_5 x 0.83 = P

68. Which one of the following is correct:
(A) K x 2.29 = K_2O
(B) K x 1.20 = K_2O
(C) K x 0.44 = K_2O
(D) K x 0.83 = K_2O

69. Lucerne (*Medicago sativa*) is a native of:
(A) Persia
(B) India
(C) Taiwan
(D) Afganisthan

70. Which one is correct:
(A) N x 0.286 = NH_4
(B) N x 1.286 = NH_4
(C) N x 4.43 = NH_4
(D) N x 3.00 = NH_4

71. Which one is correct:
(A) N x 4.43 = NO_2
(B) N x 0.43 = NO_2
(C) N x 1.43 = NO_2
(D) N x 3.00 = NO_2

72. Which one is correct:
(A) OC x 0.724 = OM
(B) OC x 3.00 = OM
(C) OC x 1.724 = OM
(D) OC x 4.43 = OM

73. Which one is correct:
(A) Ca x 0.40 = CaO
(B) Ca x 2.40 = CaO
(C) Ca x 3.40 = CaO
(D) Ca x 1.40 = CaO

74. Which one is correct:
(A) S X 3.00 = SO_4
(B) S X 4.00 = SO_4
(C) S X 5.00 = SO_4
(D) S X 6.00 = SO_4

75. The process of desuckering is mainly essential in:
(A) Potato (B) Sugarcane
(C) Tobacco (D) Maize

76. Delay in harvesting of jute is:
(A) Increase in fibre quality
(B) Decrease in fibre quality
(C) Uncharged
(D) Unknown

77. *Eleucine corocana* is a botanical name of:
(A) Finger millet (Ragi)
(B) Proso millet (Cheena)
(C) Foxtail millet (Kakun)
(D) Kodon millet (Kodon)

78. *Panicum miliceum* is a botanical name of:
(A) Finger millet (Ragi)
(B) Proso millet (Cheena)
(C) Foxtail millet (Kakun)
(D) Kodon millet (Kodon)

79. *Panicum italicum* is a botanical name of:
(A) Finger millet (Ragi)
(B) Proso millet (Cheena)
(C) Foxtail millet (Kakun)
(D) Kodon millet (Kodon)

80. *Paspalum scrobiculatum* is a botanical name of:
(A) Finger millet (Ragi)
(B) Proso millet (Cheena)
(C) Foxtail millet (Kakun)
(D) Kodon millet (Kodon)

81. Echinoclova crusgalli var Frumentaceae is a botanical name of:
(A) Finger millet (Ragi)
(B) Proso millet (Cheena)
(C) Barnyad millet (Sawan)
(D) Kodon millet (Kodon)

82. American society of Agronomy is established in year:
(A) 1831 (B) 1873
(C) 1908 (D) 1941

83. Indian Society of Agronomy came into existence in years:
(A) 1908 (B) 1941
(C) 1955 (D) 1971

84. First agronomic research was started by:
(A) J.B. Bossinggault (B) Leibig
(C) Borlaug (D) None of these

85. Agronomy journal is published from:
(A) India (B) China
(C) USA (D) UK

86. The land management system in which agriculture and forest crops are grown in combined form on same unit of the land is called:
(A) Conservation farming system
(B) Ley farming system
(C) Alley cropping system
(D) Agroforestry system

87. The land on which production of crop is economical and practical both is known as:
(A) Productive land
(B) Economical land
(C) Conserved land
(D) Arable land

88. For which type of fertilizer India depends on imports?
(A) Nitrogenous fertilizers
(B) Phosphatic fertilizers
(C) Potassic fertilizers
(D) All of these

89. Active factor of production is:
(A) Land (B) Labour
(C) Capital (D) All of these

90. Total cropped area includes:
(A) Net area sown + area sown more than once
(B) Net area sown + fallow land
(C) Net area sown + fallow land + area under forest
(D) None of these

91. The total dry matter produced by a crop is known as:
(A) Dry matter yield
(B) Biological yield
(C) Crop yield
(D) Economic yield

92. The maximum dry weight in plants is formed by:
(A) Carbon & Oxygen
(B) Carbon & Hydrogen
(C) Nitrogen & Carbon
(D) Hydrogen & Oxygen

93. Energy metabolism in plants is performed by the elements:
(A) Potassium (B) Magnesium
(C) Phosphorus (D) Iron

94. One kilogram of nitrogen is equivalent to the following:
(A) 2.17 kg urea (B) 3.22 kg urea
(C) 4.22 kg urea (D) 1.76 kg urea

95. Hay is prepared from crops rich in:
(A) Sugar (B) Protein
(C) Fat (D) Crude fibre

96. Green revolution has been most successful in:
(A) Wheat and Potato
(B) Wheat and Rice
(C) Wheat amd Maize
(D) Rice and Barley

97. Berseem was introduced into India from which country in 1904:
(A) South Africa
(B) Armania
(C) Egypt
(D) USA

98. Azotobactor is useful in fixing of nitrogen from air in:
(A) Non leguminous crops like wheat & rice
(B) Leguminous
(C) Soybean
(D) Rajmah

99. Pulse crop which does not fix atmospheric nitrogen:
(A) Arhar (B) Pea
(C) Soybean (D) Rajmash

100. Azolla (a nitrogen fixing organism) is a symbiotic association with:
(A) Fern and BGA
(B) Bryophyta & BGA
(C) Fungi and BGA
(D) Nostoc and BGA

101. Phosphorus is taken by plants in the form of:
(A) Phosphorus chloride
(B) Phosphoric acid
(C) Phosphorus sulphate
(D) Phosphorus carbonates

102. When organic matter is added to the soil
(A) Bulk density is increased
(B) Bulk density is decreased
(C) Bulk density does not changed
(D) None of these

103. Rock phosphate is used in:
(A) Neutral soil (B) Acidic soil
(C) Alkaline soil (D) Saline soil

104. The maximum area under hybrid rice in the world, is found in which country:
(A) Indonesia (B) China
(C) India (D) Pakistan

105. The hilly soils of Uttar Pradesh are formed by:
(A) Granite and quartzite rocks
(B) Schist and biotite rocks
(C) Sand stone
(D) Lime stone

106. The process of topping is practiced in tobacco to remove:
(A) Additional branches
(B) Top leaves
(C) Inflorescence
(D) None of these

107. Pegging is useful phenomenon in:
(A) Topioca (B) Groundnut
(C) Sweet potato (D) Sugarcane

108. Arrowing is known as:
(A) Tillering in sugarcane
(B) Emergence of inflorescence in sugarcane
(C) Arrow like shape of sugarcane leaf
(D) None of these

109. Decomposition of organic matter in submerged soil is carried out by:
(A) Bacteria (B) Actinomycetes
(C) Fungi (D) Earthworm

110. Under high pH soil a successfully grown crop is:
(A) Pea (B) Barley
(C) Sorghum (D) Bajra

111. Applying fertilizer either in bunds or near the plants is called:
(A) Hill placement
(B) Band placement
(C) Plough sole placement
(D) None of these

112. The well decomposed, more or less stable part of the organic matter of the soil is called:
(A) Humus
(B) Organic substances
(C) Both of these
(D) None of these

113. A compacted layer at the bottom of the zone of ploughing is called:
(A) Plough sole
(B) Furrow
(C) Both of these
(D) None of these

114. A method of fertilizer application in which fertilizer is placed in a continuous band in the bottom of the furrow is called:
(A) Hill placement
(B) Band placement
(C) Plough sole placement
(D) None of these

115. Government of India started subsidy on fertilizers with effect from:
(A) Nov. 1, 1966
(B) Nov. 1, 1969
(C) Nov. 1, 1977
(D) Oct. 2, 1988

116. Motihari tobacco mainly grown in:
(A) West Bengal
(B) Tamilnadu
(C) Andhra Pradesh
(D) Gujarat

117. Jati tobacco mainly grown in:
(A) West Bengal
(B) Tamilnadu
(C) Andhra Pradesh
(D) Gujarat

118. Bidi tobacco mainly grown in:
(A) West Benga
(B) Tamilnadu
(C) Andhra Pradesh
(D) Gujarat

119. Lanka tobacco mainly grown in:
(A) West Bengal
(B) Tamilnadu
(C) Andhra Pradesh
(D) Gujarat

120. Cigar wrapper tobacco mainly grown in:
(A) West Bengal
(B) Tamilnadu
(C) Andhra Pradesh
(D) Gujarat

121. Cigar filler tobacco mainly grown in:
(A) West Bengal
(B) Tamilnadu
(C) Andhra Pradesh
(D) Gujarat

122. Cigerette tobacco mainly grown in:
(A) West Bengal
(B) Tamilnadu
(C) Andhra Pradesh
(D) Gujarat

123. Burley tobacco mainly grown in:
(A) West Bengal
(B) Tamilnadu
(C) Andhra Pradesh
(D) Gujarat

124. HDBRG tobacco mainly grown in:
(A) West Bengal
(B) Tamilnadu
(C) Andhra Pradesh
(D) Gujarat

125. Natu tobacco mainly grown in:
(A) West Bengal
(B) Tamilnadu
(C) Andhra Pradesh
(D) Gujarat

126. Pikka tobacco mainly grown in:
(A) West Bengal
(B) Orissa
(C) Andhra Pradesh
(D) Gujarat

127. Tick the correct agro-climatic zone of UP which has highest average population density per km^2?
(A) Bundelkhand zone
(B) Central plain zone
(C) Hill zone
(D) North-eastern plains zone

128. Which country is the place of origin of potato:
(A) South America
(B) South America
(C) South Africa
(D) South China

129. The crop grown in Uttar Pradesh during all the three seasons is:
(A) Urd (B) Sorghum
(C) Moong (D) Maize

130. The ginning percentage in cotton can be worked out by the formula:

(A) $\frac{\text{Weight of cotton seed}}{\text{Weight of lint}} \times 100$

(B) $\frac{\textbf{Weight of lint}}{\textbf{Weight of cotton seed}} \times \mathbf{100}$

(C) $\frac{\text{Weight of lint}}{\text{Weight of seed cotton}} \times 100$

(D) $\frac{\text{Weight of seed cotton}}{\text{Weight of lint}} \times 100$

131. The lint index of cotton can be worked out by the formula:

(A) $\frac{\textbf{Weight of 100 seed} \times \textbf{Ginning percentage}}{\textbf{100 - Ginning percentage}} \times \mathbf{100}$

(B) $\frac{\text{Weight of lint}}{\text{Weight of seed cotton}} \times 100$

(C) $\frac{\text{Weight of lint} \times \text{Ginning percentage}}{\text{Weight of seed cotton}} \times 100$

(D) $\frac{\text{Weight of seed cotton} \times \text{Ginning percentage}}{100 - \text{Ginning percentage}} \times 100$

132. Punjab and Haryana have high percentage of net sown area under farming because of:
(A) Fertile soil
(B) Perennial rivers and canals
(C) Modern farming method
(D) All of these

133. In which of the following states is the black soil found?
(A) Uttarakhand (B) Punjab
(C) Rajasthan (D) Madhya Pradesh

184. Hybrid rice for commercial production was first evolved in:
(A) India
(B) China
(C) Indonesia
(D) USA

135. Napier grass (*Pennisetum purpureum*) was introduced to India in 1912 from:
(A) Zimbabwe
(B) South Africa
(C) Iraq
(D) Tanzania

136. Greening of potato results in:
(A) Increase in nutritional quality
(B) Decrease in nutritional quality
(C) Increase in disease resistance
(D) Decrease in diseases resistance

137. Ratio of grain to stalk in pigeonpea is:
(A) 1 : 2 (B) 1 : 3
(C) 1 : 6-8 (D) 1 : 10

138. Ratio of grain to stalk in urdbean is:
(A) 1 : 1.2 (B) 1 : 3
(C) 1 : 6-8 (D) 1 : 10

139. Ratio of grain to stalk in chickpea is:
(A) 1 : 2 (B) 1 : 5
(C) 1 : 6-8 (D) 1 : 10

140. Ratio of grain to stalk in wheat is:
(A) 1 :1.25 (B) 1 : 3
(C) 1 : 6-8 (D) 1 : 10

141. Ratio of grain to stalk in jowar is:
(A) 1 : 2 (B) 1 : 3
(C) 1 : 6-8 (D) 1 : 10

142. Ratio of grain to stalk in bajra is:
(A) 1 : 2 (B) 1 : 3
(C) 1 : 8 (D) 1 : 10

143. Coarse grained crops are grown in:
(A) High rainfed areas
(B) Waterlogged areas
(C) Lowland areas
(D) Upland areas

144. Groundnut pegs when developed in the soil from:
(A) Juber
(B) Fruits
(C) Stems
(D) Roots

145. Fire curing is followed in:
(A) Bidi tobacco
(B) Hookah tobacco
(C) Cigeratte tobacco
(D) Chewing type tobacco

146. Seed plant technique is followed in:
(A) Wheat
(B) Bajra
(C) Potato
(D) Paddy

147. The group of organisms which converts ammonia into nitrites and nitrite to nitrate are respectively:
(A) Nitrosomonas, nitrobactor
(B) Pseudomonas, Azatobactor
(C) Streptomyces, Bacillus anthracis
(D) Micromonospora, nocardis

148. Pulse crop which does not fix nitrogen from atmosphere easily is:
(A) Arhar
(B) Pea
(C) Soybean
(D) Frenchbean

149. Following are the millet's grown in India:
(A) Jowar, bajra, rice, maize
(B) Bajra, wheat, jowar, rice
(C) Jowar, bajra, ragi, maize
(D) Jowar, bajra, ragi, kodon

150. The leading producing states of jowar, bajra and ragi respectively are:
(A) Maharastra, Rajasthan and Karnataka
(B) Karnataka, Rajasthan and Maharastra
(C) Karnataka, Rajasthan and Maharastra
(D) Karnataka, Rajasthan and Maharastra

151. The millet is gown both as fodder and food crops:
(A) Wheat (B) Rice
(C) Maize (D) Bajra

152. India is a major consumer of pulses because:
(A) Pulses are rich in dietary fibre
(B) Pulses are the major source of proteins
(C) Pulses are tasty to eat
(D) Pulses are inexpensive

153. Tur, moong and urd are mainly grown in:
(A) Rabi (B) Kharif
(C) Zaid (D) Rabi and Zaid

154. Moong and urd can be grown in:
(A) Rabi (B) Kharif
(C) Zaid (D) Kharif and Zaid

155. Following crops require manual labour:
(A) Tea, Coffee, rice and maize
(B) Tea, Coffee, sugarcane and cotton
(C) Wheat, jowar, bajra and cotton
(D) Tea, coffee, maize and ragi

156. Cultivation of coffee was introduced on which of the following hills?
(A) Nilgiri hills
(B) Palni hills
(C) Baba Budan hills
(D) Khasi hills

157. Indian coffee is known in the world for its:
(A) Flavour (B) Colour
(C) Aroma (D) Strong taste

158. Frequant showers, well distributed over the years are required by this crop:
(A) Cotton (B) Coffee
(C) Tea (D) Sugarcane

159. Frequent showers are essential for tea as it ensures:
(A) Growth of tender leaves
(B) Growth of tea bush
(C) Development of flavour
(D) Growth of tea trees

160. Which of the following oilseeds are grown both as rabi and kharif crops?
(A) Mustard and groundnut
(B) Sunflower and groundnut
(C) Caster and sesame
(D) Soyabean and cotton seeds

161. Why should the pulses be grown in rotation with other crops:
(A) They increase the taste of pulses
(B) Being leguminous crop, pulses help in restoring soil fertility
(C) They require less water for irrigation if grown in rotation with ccrops
(D) None of these

162. Millets are important for us to grow for the following seasons:
(A) Coarse grains
(B) Rich in dietary fibre
(C) Rich in nutrients
(D) All of these

163. Which one of the following groups of food grains is an appropriate indicator of the first three major food grains grown in India?
(A) Wheat, jowar and rice
(B) Rice, wheat and jowar
(C) Rice, wheat and bajra
(D) Wheat, bajra and rice

164. In which of the following states, three crops of paddy in Aus, Aman and Boro are grown in India:
(A) Assam, West Bengal and Orissa
(B) Punjab, Haryana and Uttar Pradesh
(C) Bihar, Assam and Tamilnadu
(D) Punjab, Assam and Orissa

165. These crops require frost free climatic conditions for their growth:
(A) Sugarcane and coffee
(B) Rice and pulses
(C) Wheat and rice
(D) Cotton and tea

166. Following states are the leading producers of rubber, coffee, cotton and jute, respectively:
(A) Tamilnadu, Kerala, Gujarat and Assam
(B) Kerala, Karnataka, Maharastra and West Bengal
(C) Karnatáka, Kerala, Gujarat and Orissa
(D) Kerala, Karnataka, Gujarat and Assam

167. The four major fibre crops grown in India are:
(A) Rope, Coffee, Silk and Wool
(B) Wool, Silk, Tea and Rice
(C) Cotton, Jute, Hemp and Dhaincha
(D) Cotton, Silk, Jute and Rope

168. Which of the following crops require more than 200 cm of annual rainfall:
(A) Tea (B) Rice
(C) Coffee (D) Rubber

169. Sugarcane, cotton and wheat require approximately this much amount of annual rainfall:
(A) 50 to 75 cm (B) 75 to 125 cm
(C) 75 to 110 cm (D) 100 to 125 cm

170. What is sericulture?
(A) It is the art of making saries
(B) It is the culture of wearing saries
(C) It is the rearing of silkworms for silk fibre
(D) It is rearing of sheep

171. Which of the following reforms in agriculture do not pertain to technological reforms?
(A) Use of HYV's for higher production
(B) Introduction of Kissan Credit Card and Personal Accident Insurance Scheme
(C) Application of fertilizer and manures for raising crop production
(D) Mechanization in agriculture and pest control measures

172. Development in agriculture is decelerating not due to one of the following reasons:
(A) Reduction in public investment in agriculture
(B) Reduction in import duties on agriculture products
(C) Rise in subsidy on fertilizers, power and irrigation
(D) Rise in the cost of production

173. Which one of the following has not contributed to the development and mechanization of agriculture?
(A) Establishment of agriculture universities and research farms
(B) Modern technologies inputs
(C) Development of agriculture infrastructures
(D) Reduced import duties and decrease in subsidies

174. Bloodless revolution pertains to which one of the following?
(A) Land donations
(B) Green revolution
(C) Agriculture reforms
(D) White revolution

175. Jute cann't be grown in any soil other than this type of soil:
(A) Bangar of alluvial soils
(B) Black soils of Deccan
(C) Soils of flood plains
(D) Sandy soils of Rajasthan

176. Profit making is the sole aim of this type of farming:
(A) Subsistence farming
(B) Intensive subsistence farming
(C) Shifting cultivation
(D) Commercial farming

177. The major consumer of rubber in India is:
(A) Latex form
(B) Auto tyres and tubes
(C) Foot wear
(D) Cycle tyres

178. The success of rabi crops depends on:
(A) Heavy cyclonic rainfall
(B) Intense irrigation
(C) Light winter rainfall
(D) All of these

179. Which one of the following is not the characteristics of commercial farming?
(A)Crop is grown for self-consumption
(B) Modern inputs of farming are used
(C) Monoculture is practiced
(D) Crops are grown with the purpose of selling in market

180. Hoe, das and digging sticks are used in which of the following farming types?
(A) Subsistence farming
(B) Intensive subsistence farming
(C) Shifting cultivation
(D) Commercial farming

181. Farm productivity is low in shifting cultivation because of which one of the following reasons?
(A) Size of farm is small
(B) Farmers uses fertilizers irrationally
(C) He does not use modern inputs in farming
(D) Mechanisation in agriculture

182. Land holdings are uneconomical in intensive farming because:
(A) Fragmentation of land holdings
(B) Ceiling on landholdings
(C) Zamidari system
(D) Use of inputs in farming

183. Which one of the following is a rabi crop?
(A) Rice (B) Millets
(C) Chickpea (D) Coffee

184. Which one of the following describes a system of agriculture where a single crop is grown on a large area?
(A) Shifting agriculture
(B) Plantation agriculture
(C) Horticulture
(D) Intensive agriculture

185. Which one of the following is a leguminous crop?
(A) Niger (B) Millet
(C) Jowar (D) Groundnut

186. Which one of the following is a oilseeds crops?
(A) Niger (B) Cowpea
(C) Jute (D) Sunhemp

187. Which one of the following is a sugar crop?
(A) Sugarbeet (B) Sorghum
(C) Clusterbean (D) Sweet potato

188. Which one of the following is a narcotic crop?
(A) Tea (B) Coffee
(C) Poppy (D) All of these

189. Which one of the following prices is announce by the government in support of a crop?
(A) Maximum Support Price
(B) Minimum Support Price
(C) Moderate Support Price
(D) Influential Support Price

190. Which of the following about pulses is not true?
(A) Help fixing nitrogen in soil
(B) Protein rich
(C) Restore soil fertility
(D) A major import item

191. Punjab-Haryana are rice surplus states because:
(A) Rainfall is more than 100 cm
(B) It is not the stable diet of people
(C) They are the leading producers
(D) They only grow rice through the year

192. Punjab is the leading producer of:
(A) Wheat (B) Rice
(C) Sorghum (D) Bajra

193. West Bengal produces the maximum amount of:
(A) Jute (B) Cotton
(C) Sunhemp (D) Linseed

194. Limited crops are grown in:
(A) Kharif season (B) Rabi season
(C) Zaid season (D) None of these

195. Major rice producing state in India is:
(A) Uttar Pradesh (B) West Bengal
(C) Gujarat (D) Punjab

196. Major wheat producing state in India is:
(A) Uttar Pradesh (B) West Bengal
(C) Gujarat (D) Punjab

197. Major cotton producing state in India is:
(A) Uttar Pradesh (B) West Bengal
(C) Gujarat (D) Punjab

198. Major jute producing state in India is:
(A) Uttar Pradesh (B) West Bengal
(C) Gujarat (D) Punjab

199. Major sugarcane producing state in India is:
(A) Uttar Pradesh (B) West Bengal
(C) Gujarat (D) Punjab

200. The state is leading in coffee production:
(A) Karnataka (B) Assam
(C) Tamilnadu (D) Kerala

201. The state is leading in tea production:
(A) Karnataka (B) Assam
(C) Tamilnadu (D) Kerala

202. Largest groundnut producing state in India:
(A) Karnataka (B) Assam
(C) Gujarat (D) Kerala

203. Largest maize producing state in India:
(A) Karnataka (B) Assam
(C) Gujarat (D) Kerala

204. What are the most important beverage crops in India:
(A) Tea (B) Coffee
(C) Both (A) & (B) (D) None of these

205. In Punjab, rice is grown as an irrigated crop, but in West Bengal it is a:
(A Irrigated (B) Rainfed
(C) Both (A) and (B) (D) None of these

206. Jhumming is a form of shifting cultivation practiced in:
(A) North-Eastern states
(B) South India
(C) Western India
(D) North India

207. Ganga-Satluj plains are suitable for growing wheat while Ganga delta and eastern coastal plains are suitable for:

(A) Cotton (B) Sugarcane
(C) Rice (D) Bajra

208. Cotton, Jute and Sugarcane are:

(A) Commercial crop
(B) Vegetable crops
(C) Oilseed crops (D) Fibre crop

209. 'Slash and burn' agriculture is also termed as in North-East India is:

(A) Jhumming (B) Kuruwa
(C) Dahiya (D) Bringa

210. 'Slash and burn' agriculture is also termed as in Jharkhand is:

(A) Jhumming (B) Kuruwa
(C) Dahiya (D) Bringa

211. 'Slash and burn' agriculture is also termed as in Madhya Pradesh is:

(A) Jhumming (B) Kuruwa
(C) Dahiya (D) Bringa

212. 'Slash and burn' agriculture is also termed as in Orissa is:

(A) Jhumming (B) Kuruwa
(C) Dahiya (D) Bringa

213. Empirical formula of Crop Equivalent Yield is:

(A) $\sum_{i=1}^{n} (Y_i - e\,i)$

(B) $La + Lb = \frac{Ya}{Sa} + \frac{Yb}{Sb}$

(C) $\frac{\text{Mixture yield of `a'}}{\text{Pure stand yield of `a' - Mixture yield of `a'}}$

(D) $\frac{\text{Mixture yield of `a'}}{\text{Pure stand yield of `a' - Mixture yield of `b'}}$

214. Empirical formula of Land Equivalent Ratio is:

(A) $\sum_{i=1}^{n} (Y_i - e\,i)$

(B) $La + Lb = \frac{Ya}{Sa} + \frac{Yb}{Sb}$

(C) $\frac{\text{Mixture yield of `a'}}{\text{Pure stand yield of `a' - Mixture yield of `a'}}$

(D) $\frac{\text{Mixture yield of `a'}}{\text{Pure stand yield of `a' - Mixture yield of `b'}}$

215. Empirical formula of Relative Crowding Coefficient, if mixture is 50:50 is:

(A) $\sum_{i=1}^{n} (Y_i - e\,i)$

(B) $La + Lb = \frac{Ya}{Sa} + \frac{Yb}{Sb}$

(C) $\frac{\text{Mixture yield of `a'}}{\text{Pure stand yield of `a' - Mixture yield of `a'}}$

(D) $\frac{\text{Mixture yield of `a'}}{\text{Pure stand yield of `a' - Mixture yield of `b'}}$

216. Empirical formula of Relative Crowding Coefficient for a mixing differing from 50:50 is:

(A) $\frac{Yab \times Z\,ba}{(Yaa - Yab) \times Zab}$

(B) $La + Lb = \frac{Ya}{Sa} + \frac{Yb}{Sb}$

(C) $\frac{\text{Mixture yield of `a'}}{\text{Pure stand yield of `a' - Mixture yield of `a'}}$

(D) $\frac{\text{Mixture yield of `a'}}{\text{Pure stand yield of `a' - Mixture yield of `b'}}$

217. Empirical formula of Aggressivity is:

(A) $\sum_{i=1}^{n} (Y_i - e\,i)$

(B) $La + Lb = \frac{Ya}{Sa} + \frac{Yb}{Sb}$

(C) $\frac{\text{Mixture yield of `a'}}{\text{Pure stand yield of `a' - Mixture yield of `a'}}$

(D) $\frac{\text{Mixture yield of `a'}}{\text{Pure crop yield of `a'}} - \frac{\text{Mixture yield of `b'}}{\text{Pure crop yield of `b'}}$

218. Empirical formula of Aggressivity is:

(A) $\frac{Yab}{Yaa \times Zab} - \frac{Yba}{Ybb \times Zba}$

(B) $La + Lb = \frac{Ya}{Sa} + \frac{Yb}{Sb}$

(C) $\frac{\text{Mixture yield of `a'}}{\text{Pure stand yield of `a' - Mixture yield of `a'}}$

(D) $\frac{\text{Mixture yield of `a'}}{\text{Pure stand yield of `a' - Mixture yield of `b'}}$

219. Empirical formula of Competition Index is:

(A) $\frac{\textbf{(Yab – Yaa × (Yba – Ybb)}}{\textbf{(Yaa – Ybb)}}$

(B) $\frac{(Yab}{(Yaa - Zab)} \div \frac{Yba}{(Ybb - Zba)}$

(C) $\frac{\text{Relative crowding coefficient of a specific species}}{\text{Total relative crowding coefficient of all crops in the mixture}} \times 100$

(D) Value of combined intercrop yield × $\frac{\text{LER-1}}{\text{LER}}$

220. Empirical formula of Competition Ratio is:

(A) $\frac{(Yab - Yaa \times (Yba - Ybb)}{(Yaa - Ybb)}$

(B) $\frac{\textbf{Yab}}{\textbf{(Yaa - Zab)}} \div \frac{\textbf{Yba}}{\textbf{(Ybb - Zba)}}$

(C) $\frac{\text{Relative crowding coefficient of a specific species}}{\text{Total relative crowding coefficient of all crops in the mixture}} \times 100$

(D) Value of combined intercrop yield × $\frac{\text{LER-1}}{\text{LER}}$

221. Empirical formula of Competition Ratio is:

(A) $\frac{(Yab - Yaa \times (Yba - Ybb)}{(Yaa - Ybb)}$

(B) $\frac{(Yab}{(Yaa - Zab)} \div \frac{Yba}{(Ybb - Zba)}$

(C) $\frac{\text{Relative crowding coefficient of a specific species}}{\text{Total relative crowding coefficient of all crops in the mixture}} \times 100$

(D) $\left(\frac{Yab}{Yaa} \div \frac{Yba}{Zba} \times \frac{Zba}{Zab}\right) = \left[\frac{LERa}{LERb}\right] \times \frac{Zba}{Zab}$

222. Empirical formula of Competition Coefficient is:

(A) $\frac{(Yab - Yaa) \times (Yba - Ybb)}{(Yaa - Ybb)}$

(B) $\frac{(Yab}{(Yaa - Zab)} \div \frac{Yba}{(Ybb - Zba)}$

(C) $\frac{\text{Relative crowding coefficient of a specific species}}{\text{Total relative crowding coefficient of all crops in the mixture}} \times 100$

(D) Value of combined intercrop yield × $\frac{\text{LER-1}}{\text{LER}}$

223. Empirical formula of Monetary Advatage is:

(A) $\frac{(Yab - Yaa) \times (Yba - Ybb)}{(Yaa - Ybb)}$

(B) $\frac{(Yab}{(Yaa - Zab)} \div \frac{Yba}{(Ybb - Zba)}$

(C) $\frac{\text{Relative crowding coefficient of a specific species}}{\text{Total relative crowding coefficient of all crops in the mixture}} \times 100$

(D) Value of combined intercrop yield × $\frac{\text{LER-1}}{\text{LER}}$

224. In a intercropping system, the chickpea and mustard yield was obtained 1800 and 800 kg/ha, where sale price was Rs. 12 and 15/kg. Findout chickpea equivalent yield.

(A) 2800 (B) 2900
(C) 3000 (D) 3100

225. The pure crop yield of chickpea and mustard is 2000 and 1800 kg/ha, respectively. When these crops grown in intercropping the chickpea and mustard yield was obtained 1500 and 1200 kg/ha respectively. Findout the LER:

(A) 0.42 (B) 1.42
(C) 2.42 (D) 3.42

226. If Competition Index is one for each component species it means that:

(A) No effect in mixing
(B) Less advantage in mixing
(C) Advantage in mixing
(D) Unknown

227. Which agronomic assessment is used for calculation of relative crowding coefficient of all crops in the mixture, thereby maximum production can be obtained without leaving any harmful effect.

(A) Competition Index
(B) Competition coefficient
(C) Competition ratio
(D) Monetary advantage

228. If a given species Competition Index is less than one (<1) it means that:

(A) No effect in mixing
(B) Less advantage in mixing
(C) Advantage in mixing
(D) Unknown

229. If the competition index is more than one (>1) it means that:

(A) No effect in mixing
(B) Less advantage in mixing
(C) Advantage in mixing
(D) Unknown

230. Relative Crowding Index was proposed by:

(A) Hall (1974)
(B) De Wil (1960)
(C) Blackman (1919)
(D) Turners (1986)

231. Relative Crowding Index was examined by:

(A) Hall (1974) (B) De Wil (1960)
(C) Blackman (1919) (D) Turners (1986)

232. Relative Growth Rate was proposed by:

(A) Hall (1974)
(B) De Wil (1960)
(C) Blackman (1919)
(D) Turners (1986)

233. Relative Water Content was proposed by:

(A) Hall (1974) (B) De Wil (1960)
(C) Blackman (1919) (D) Turners (1986)

234. Net Assimilation Rate is expressed as:

(A) $g/m^2/day$
(B) kg/kg nutrient applied
(C) day or week
(D) cm^2/g

235. Agronomic Efficiency is expressed as:

(A) $g/m^2/day$
(B) kg/kg nutrient applied
(C) day or week
(D) cm^2/g

236. Leaf Area Duration is expressed as:

(A) $g/m^2/day$
(B) kg/kg nutrient applied
(C) day or week
(D) cm^2/g

237. Specific Leaf Area is expressed as:

(A) $g/m^2/day$
(B) kg/kg nutrient applied
(C) day or week
(D) cm^2/g

238. An index of the amount of growing material per unit dry weight of plant per unit time is called:

(A) Relative Growth Rate
(B) Specific Leaf Area
(C) Net assimilation Rate
(D) Lead Area Duration

239. The ratio of area of the leaf to the weight of the leaf is called:

(A) Relative Growth Rate
(B) Specific Leaf Area
(C) Net assimilation Rate
(D) Lead Area Duration

240. Relative Yield Total was proposed by:

(A) Hall (1974)
(B) De Wil (1960)
(C) de Wit and Van Deb Burgh
(D) Turners (1986)

241. Empirical formula of Multiple Cropping Index is:

(A) $\dfrac{\sum_{i=1}^{n} ai}{A} \times 100$

(B) $\dfrac{\sum_{i=1}^{n} ai\ ti}{AoT + \sum_{J=1}^{m}}$

(C) $\dfrac{\text{Area under Kharif + Rabi + Zaid}}{\text{Area under actual cultivation}} \times 100$

(D) $\dfrac{\text{Yield obtained in fertilized plot - Yield obtained in unfertilized plot}}{\text{Quantity of nutrient applied in plot}}$

242. Empirical formula of Multiple Cropping Index is:

(A) $\frac{\text{Total cropped area in a year}}{\text{Net cultivated area}} \times 100$

(B) $\frac{\textbf{Total number of crops} \times \textbf{Total area under crops}}{\textbf{Total land available for cropping}} \times \mathbf{100}$

(C) $\frac{\text{Area under Kharif + Rabi + Zaid}}{\text{Area under actual cultivation}} \times 100$

(D) $\frac{\text{Yield obtained in fertilized plot - Yield obtained in unfertilized plot}}{\text{Quantity of nutrient applied in plot}}$

243. Empirical formula of Cropping Intensity Index is:

(A) $\frac{\sum_{i=1}^{n} ai}{A} \times 100$

(B) $\frac{\sum_{i=1}^{n} ai\ ti}{AoT + \sum_{j=1}^{m}}$

(C) $\frac{\text{Area under Kharif + Rabi + Zaid}}{\text{Area under actual cultivation}} \times 100$

(D) $\frac{\text{Yield obtained in fertilized plot - Yield obtained in unfertilized plot}}{\text{Quantity of nutrient applied in plot}}$

244. Empirical formula of Cropping Intensity is:

(A) $\frac{\sum_{i=1}^{n} ai}{A} \times 100$

(B) $\frac{\sum_{i=1}^{n} ai\ ti}{AoT + \sum_{j=1}^{m}}$

(C) $\frac{\textbf{Area under Kharif + Rabi + Zaid}}{\textbf{Area under actual cultivation}} \times \mathbf{100}$

(D) $\frac{\text{Yield obtained in fertilized plot - Yield obtained in unfertilized plot}}{\text{Quantity of nutrient applied in plot}}$

245. Empirical formula of Cropping Intensity is:

(A) $\frac{\textbf{Total cropped area in a year}}{\textbf{Net cultivated area}} \times \mathbf{100}$

(B) $\frac{\text{Total number of crops} \times \text{Total area under crops}}{\text{Total land available for cropping}} \times 100$

(C) $\frac{\text{Area under Kharif + Rabi + Zaid}}{\text{Area under actual cultivation}} \times 100$

(D) $\frac{\text{Yield obtained in fertilized plot - Yield obtained in unfertilized plot}}{\text{Quantity of nutrient applied in plot}}$

246. Empirical formula of Agronomic Efficiency is:

(A) $\frac{\sum_{i=1}^{n} ai}{A} \times 100$

(B) $\frac{\sum_{i=1}^{n} ai\ ti}{AoT + \sum_{j=1}^{m}}$

(C) $\frac{\text{Area under Kharif + Rabi + Zaid}}{\text{Area under actual cultivation}} \times 100$

(D) $\frac{\textbf{Yield obtained in fertilized plot - Yield obtained in unfertilized plot}}{\textbf{Quantity of nutrient applied in plot}}$

247. Empirical formula of Crop Growth Rate:

(A) $\frac{\mathbf{W_2 - W_1}}{\mathbf{t_2 - t_1}}$

(B) $\frac{\text{Log } W_2 - \text{Log } W_1}{t_2 - t_1}$

(C) $\frac{\text{Total leaf area}}{\text{Ground area}}$

(D) LAI x M

248. Empirical formula of Relative Growth Rate:

(A) $\frac{W_2 - W_1}{t_2 - t_1}$

(B) $\frac{\mathbf{Log\ W_2 - Log\ W_1}}{\mathbf{t_2 - t_1}}$

(C) $\frac{\text{Total leaf area}}{\text{Ground area}}$

(D) LAI x M

249. Empirical formula of Leaf Area Index:

(A) $\frac{W_2 - W_1}{t_2 - t_1}$

(B) $\frac{\text{Log } W_2 - \text{Log } W_1}{t_2 - t_1}$

(C) $\frac{\textbf{Total leaf area}}{\textbf{Ground area}}$

(D) LAI x M

250. Empirical formula of Leaf Area Duration:

(A) $\frac{W_2 - W_1}{t_2 - t_1}$

(B) $\frac{\text{Log } W_2 - \text{Log } W_1}{t_2 - t_1}$

(C) $\frac{\text{Total leaf area}}{\text{Ground area}}$

(D) LAI x M

251. Empirical formula of Leaf Area Ratio:

(A) $\frac{L_2 - L_1}{\text{Log } L_2 - \text{Log } L_1} \times \frac{\text{Log } W_2 - \text{Log } W_1}{W_2 - W_1}$

(B) $\frac{\text{Area of the lead (cm}^2)}{\text{Weight of the leaf (g)}}$

(C) $\frac{(W_2 - W_1)(\text{Log } L_2 - \text{Log } L_1)}{(T_2 - T_1)(L_2 - L_1)}$

(D) $\frac{FW - DW}{TW - DW} \times 100$

252. Empirical formula of Specific Leaf Area:

(A) $\frac{L_2 - L_1}{\text{Log } L_2 - \text{Log } L_1} \times \frac{\text{Log } W_2 - \text{Log } W_1}{W_2 - W_1}$

(B) $\frac{\textbf{Area of the lead (cm}^2)}{\textbf{Weight of the leaf (g)}}$

(C) $\frac{(W_2 - W_1)(\text{Log } L_2 - \text{Log } L_1)}{(T_2 - T_1)(L_2 - L_1)}$

(D) $\frac{FW - DW}{TW - DW} \times 100$

253. Empirical formula of Net Assimilation Rate:

(A) $\frac{L_2 - L_1}{\text{Log } L_2 - \text{Log } L_1} \times \frac{\text{Log } W_2 - \text{Log } W_1}{W_2 - W_1}$

(B) $\frac{\text{Area of the lead (cm}^2)}{\text{Weight of the leaf (g)}}$

(C) $\frac{\mathbf{(W_2 - W_1)(Log\ L_2 - Log\ L_1)}}{\mathbf{(T_2 - T_1)(L_2 - L_1)}}$

(D) $\frac{FW - DW}{TW - DW} \times 100$

254. Empirical formula of Relative Water Content:

(A) $\frac{L_2 - L_1}{\text{Log } L_2 - \text{Log } L_1} \times \frac{\text{Log } W_2 - \text{Log } W_1}{W_2 - W_1}$

(B) $\frac{\text{Area of the lead (cm}^2)}{\text{Weight of the leaf (g)}}$

(C) $\frac{(W_2 - W_1)(\text{Log } L_2 - \text{Log } L_1)}{(T_2 - T_1)(L_2 - L_1)}$

(D) $\frac{\mathbf{FW - DW}}{\mathbf{TW - DW}} \times \mathbf{100}$

255. The information technology and mechotronics have provided a strong sophisticate technology base to support precision farming such as:

(A) Global Positioning System (GPS)

(B) Geographical Information System (GIS)

(C) Simulation Modeling for Decision Support System (DSS)

(D) All of these

256. Which system provides the accurate positional information, which will be more useful in location the spatial variability with accuracy:

(A) GPS (B) GIS

(C) DSS (D) All of these

257. The inherent accuracy of the GPS is about 20 m, which is based on a 95% probability that the position given will be within:

(A) 10 m (B) 20 m

(C) 30 m (D) 40 m

258. Development of precision GIS/GPS auto navigation systems increased the efficiency of the field operations in:
(A) Precision Farming
(B) Conservation agriculture
(C) Multistory farming
(D) Cooperative farming

259. In precision farming a software system, designed to provide the tools to manipulate and display spatial data:
(A) Global Positioning System (GPS)
(B) Geographical Information System (GIS)
(C) Simulation Modeling for Decision Support System (DSS)
(D) All of these

260. In precision farming a deterministic model is needed that reflects different soil conditions within fields rather than a breed regional based generalized capacity models:
(A) Global Positioning System (GPS)
(B) Geographical Information System (GIS)
(C) Simulation Modeling
(D) Decision Support System (DSS)

261. A precision farming decision support system could link:
(A) Site specific attribute data models
(B) GIS technology
(C) Both (A) and (B) (D) None of these

262. DSS integrate information from many sources such as:
(A) Synthesize management plans
(B) Evaluates them
(C) Both (A) and (B) (D) None of these

263. Growing trees along with agricultural crops is known as:
(A) Silviculture (B) Horticulture
(C) Social forestry (D) Agroforestry

264. The type of forests in which Teak plants grow is:
(A) Alpine (B) Deciduous
(C) Evergreen (D) Scrub jungle

265. The process of cultivating wild species to suit human needs is called:
(A) Hybridization (B) Acclimatization
(C) Selection (D) Domestication

266. "Block wood" is obtained from:
(A) Acacia (B) Pongamia
(C) Dalbergia (D) Albizia

267. Type of nursery used in paddy transplantor is:
(A) Protray (B) Mat
(C) Conventional (D) Raised bed

268. Arrowing is known as:
(A) Tillering in sugarcane
(B) Emergence of inflorescence in sugarcane
(C) Arrow like shape of sugarcane leaf
(D) None of these

269. How many factors are identified for influencing plant growth till now?
(A) 50 (B) 52
(C) 55 (D) 60

270. At the vegetative growth stage flowering is stopped in food grain crops, known as:
(A) Sigmoid growth curve
(B) Determinate growth
(C) Indeterminate growth
(D) Grand growth period

271. Which one of the following can be assessed by using the following equation?
A=Economic production/Biomass production
(A) Panicle emergence rate
(B) Rate of flowering
(C) Harvest Index
(D) Leaf production rate

272. The crop with yarns of sunhemp is prepared in:
(A) 12-15 weeks (B) 15-17 weeks
(C) 10-12 weeks (D) 8-10 weeks

273. Which is not an inorganic matter in th following?
(A) Magnesium (B) Iron
(C) Fat (D) Iodine

274. Which is the highest digestible protein non-leguminous crop among the following?
(A) Napier (B) Maize silage
(C) Maize (D) Jowar

275. Which is not matched in relation to sugarcane?
(A) Seed treatment - Agalol 3%
(B) For seed. - 6-7 month old crop
(C) Sowing - upper portion is more used
(D) After flowering - used for sowing

276. Which is not basic principle of Agronomy?
(A) To select appropriate materials for seed and sowing
(B) Management of soil and climate
(C) Appropriate intercropping activities management for crop
(D) Livestock management for agriculture

277. At which stage of crop, there is no competition for light, moisture and nutrients?
(A) Node formation stage
(B) Seedling stage
(C) Grain formation stage
(D) Before maturity stage

278. Due to which reason is the Napier grass mixed with cluster been or cowpea for breeding?
(A) Due to high crude fibre
(B) Due to oxalic acid
(C) Due to more carbohydrates
(D) Due to more HCN content

279. More than how much percentage of loss of urea should not be during silage fermentation?
(A) 20-22% (B) 15-18%
(C) 10-12% (D) 8-10%

280. The plant population for 9000 m^2 rice plot, if transplanting to be done at a spacing of 20 cm x 10 cm and two seedlings used at one hill. How much number of seedlings are required?
(A) 800000 (B) 900000
(C) 1000000 (D) 1100000

281. Wheat variety HD2329 sown at a distance of 20 cm x 5 cm, seed germination per cent is 80, purity percentage is 70 and test weight of sed is 50 g. Calculate the seedrate of wheat is kg/ha:
(A) 89.28 (B) 80.28
(C) 79.28 (D) 75.28

282. Upland rice grown under late sown conditions having following details.
(i) Spacing = 20 cm x 4 cm
(ii) Germination percengate of seed = 85
(iii) Purity percentage of seed = 75
(iv) Test weight of seed = 22 g
(v) Number of damaged seeds by birds which need to be replaced by another seed = 10 seeds/m^2 area

Calculate the seed rate in kg/ha of rice:
(A) 30.46 (B) 35.46
(C) 40.46 (D) 45.46

Statement for linked Answer Questions 283 and 284

Groundnut grown under Kharif season at a spacing of 50 cm x 20 cm. The germination percentage of seed is 90, purity percentage of seed is 85, test weight of seed is 700 g ad shelling per cent is 63.

283. The seedrate of groundnut be calculated in terms of weight of kernels in kg/ha:
(A) 83.8 (B) 84.2
(C) 84.8 (D) 85.1

284. The seedrate of groundnut to be calculated in terms of unshelled pods in kg/ha:
(A) 130 (B) 133
(C) 140 (D) 143

285. Calculate the weight of potato tubers required to be taken out in q/ha from the cold storage for sowing of one hectare area. The spacing is 60 cm row to row and 20 cm plant to plant distance apart. The average weight of tuber is 25 g and rottage loss of tubers fromt he cold storage to the field in terms of percent weight of tubers used for sowing in 5%:
(A) 20.00 (B) 20.80
(C) 21.84 (D) 22.84

Statement for linked Answer Questions 286-290

Cultivation of autumn planted sugarcane to be done with following details
(i) Average number of nodes on the cane = 45
(ii) Types of setts to be used for sowing = 3 budded
(iii) Average number of damaged buds on each cane = 4
(iv) Average weight of cane = 950 g
(v) Average length of each budded sett = 30 cm
(vi) Spacing between two rows = 90 cm
(vii) System of planting in horizontal position: end to end and bud to bud

286. Total number of setts needed/ha in end to end system of planting:
(A) 37037 (B) 40037
(C) 45037 (D) 47037

287. Total number of canes need for one hectare area by end to end system of planting:

(A) 2010 (B) 2710
(C) 3010 (D) 3710

288. Weight of canes needed for planting of one hectare area in q/ha by end to end system of planting:

(A) 15.75 (B) 20.75
(C) 25.75 (D) 30.75

289. In case of bud to bud system of planting usually each sett is overlapped to the extent of 2/3 of its length. Therefore, the total number of setts needed:

(A) 55556 (B) 50556
(C) 48556 (D) 45556

290. Weight of canes needed for planting of one hectare area in q/ha by bud to bud system of planting:

(A) 28.63 (B) 38.63
(C) 32.63 (D) 29.63

Statement for linked Answer Question 291 & 292.

A farmer of western Uttar Pradesh grown wheat crop under irrigated conditions. Where spacing is 22 cm x 5 cm, average number of effective tillers/plant is 5, average number of grains/spike is 30, test weight is 44 g and grain : straw ratio is 1: 1.3:

291. Calculate the yield of wheat grain in q/ha:

(A) 50 (B) 60
(C) 70 (D) 80

292. Calculate the yield of wheat straw in q/ha:

(A) 70 (B) 75
(C) 78 (D) 80

293. A progressive farmer of Kanput grown hybrid maize crop with the following details as given below:

(i) Spacing = 60 cm x 30 cm
(ii) Average number of cobs/plant = 2.2
(iii) Average number of grain rows/cob = 12
(iv) Test weight = 220 g
(v) Average number of grains/row = 29

Calculate the expected yield of maize in q/ha

(A) 93.57 (B) 92.57
(C) 91.57 (D) 90.57

294. Calculate the expected yield of fine Basmati rice in q/ha with the average number of panicles/m^2 is 300, average number of filled grains/panicle is 130 and test weight is 20 g:

(A) 70 (B) 74
(C) 78 (D) 82

Statement for linked Answer Question 295 to 298.
Good crop of hybrid rice having following details:

(i) Spacing = 20 cm x 10 cm
(ii) Average number of effective tillers/ hill = 7
(iii) Average number of grain / panicle = 150
(iv) Average number of unfilled grains/ panicle = 14
(v) Test weight = 24 g
(v) Yield of unknown rice = 105 q/ha

295. Calculate the yield of rough hybrid rice in q/ha:

(A) 125 (B) 126
(C) 127 (D) 128

296. Calculate hulling percentage of hybrid rice:

(A) 81.33 (B) 82.33
(C) 83.33 (D) 84.33

297. Calculate fertility ratio of hybrid rice:

(A) 10.30 (B) 10.50
(C) 10.60 (D) 10.71

298. Calculate percentage of unfilled grains:

(A) 9.33 (B) 9.00
(C) 8.50 (D) 8.33

299. Calculate the potato tuber yield (q/ha) with spacing 60 cm x 20 cm, average number of tuber/plant is 16 and average weight of tuber is 42 g.

(A) 500 (B) 520
(C) 540 (D) 560

Statement for linked Answer Question 300 to 302.

The following particulars of groundnut crop grown in Gujarat.

(i) Spacing = 60 cm x 20 cm
(ii) Average number of matured pods/plant = 30
(iii) Average number of kernels/Plant = 3.6
(iv) Test weight of Kernels = 550 g
(v) Shelling percentage = 62
(vi) Percentage of soil in Kernels = 45

300. Yield of groundnut Kernels in q/ha

(A) 49.5 (B) 50.0
(C) 50.5 (D) 51.0

301. Yield of unshelled (pods) pods in q/ha:

(A) 78.84 (B) 79.84
(C) 80.34 (D) 81.84

302. Yield of groundnut in q/ha:

(A) 20.27 (B) 21.27
(C) 22.27 (D) 23.27

303. What would be the yield of Jute seed in q/ha based on spacing in 20 cm x 10 cm, test weight of seed is 7.5 g and average number of seeds/plant is 200?

(A) 6.0 (B) 6.5
(C) 7.0 (D) 7.5

304. Jute crop grown at a spacing of 25 cm x 15 cm, however, average weight of mature plant (at harvesting stage) is 80 g and extractable fibre percentage in 8.25. Find out the fibre yield of Jute in q/ha:

(A) 19.6 (B) 17.4
(C) 17.2 (D) 17.0

Statement for linked Answer Question 305 to 308
A good crop of American cotton grown in Vidarbh region having following informations:

(i) Spacing = 80 cm x 60 cm
(ii) Average number of sympodial branches/plant = 5.5
(iii) Average number of bolls/branche = 4.3
(iv) Number of locules/boll = 4.6
(v) Average number of seeds/locule = 9.0
(vi) Seed to lint ratio = 5.3
(vii) Test weight of cotton seed = 140 g

305. Calculate the yield of cotton seed in q/ha:

(A) 28.56 (B) 29.01
(C) 29.56 (D) 29.96

306. Calculate the lint in q/ha:

(A) 17.00 (B) 17.14
(C) 17.54 (D) 18.04

307. Calculate the yield of seed cotton in q/ha:

(A) 43.70 (B) 44.70
(C) 45.70 (D) 46.70

308. Calculate the ginning percentage:

(A) 34.51 (B) 35.51
(C) 36.51 (D) 37.51

Statement for linked Answer Question 309 to 312
Sugarcane var CO1158 crop grown in Western Uttar Pradesh having following details:

(i) Average number of milliable canes per clumps = 13
(ii) Average weight of cane = 810 g
(iii) Spacing of clump = 90 cm x 60 cm
(iv) Percent extractable juice in canes = 76
(v) Percent sugar in juice = 19
(vi) Percent Jaggary in juice = 20

309. Calculate yield of sugarcane in tonnes/ha:

(A) 234 (B) 240 (C) 248 (D) 260

310. Calculate total amount of expected juice in cane (t/ha):

(A) 175.85 (B) 177.84
(C) 180.84 (D) 190.84

311. Calculate total amount of expected Jaggary (t/ha):

(A) 32.59 (B) 33.57
(C) 34.57 (D) 35.57

312. Calculate total amount of expected cane sugar(t/ha):

(A) 31.79 (B) 32.76
(C) 33.79 (D) 35.79

313. Calculate the partial factor productivity of nitrogen (kg rice grain/ kg of applied N) if the rate of N application to the hybrid rice was 170 kg N/ha and the productivity of rice was 105.5 q/ha:

(A) 62.06 (B) 63.06 (C) 64.06 (D) 65.06

314. Calculate the agronomic efficiency of applied nitrogen (kg yield increase/ kg N applied) if the yield of maize in N treated plot was 72.5 q/ha, and the control treatment with no N maize grain yield was 35.8 q/ha and the amount of N fertilizer applied was 150 kg N/ha:

(A) 23.5 (B) 24.5 (C) 25.5 (D) 26.5

315. Calculate the crop recovery efficiency of N applied for rice crop, if the total plant N uptake in above ground biomass (grain + straw) at maturity (kg/ha) in N treated plot was 200 kg/ha and the total N uptake in above ground biomass (grain + straw) at maturity (kg/ha) in a plot that did not receive N was 125 kg/ha. The amount of fertilizer N applied was 150 kg/ha:

(A) 40% (B) 45% (C) 50% (D) 55%

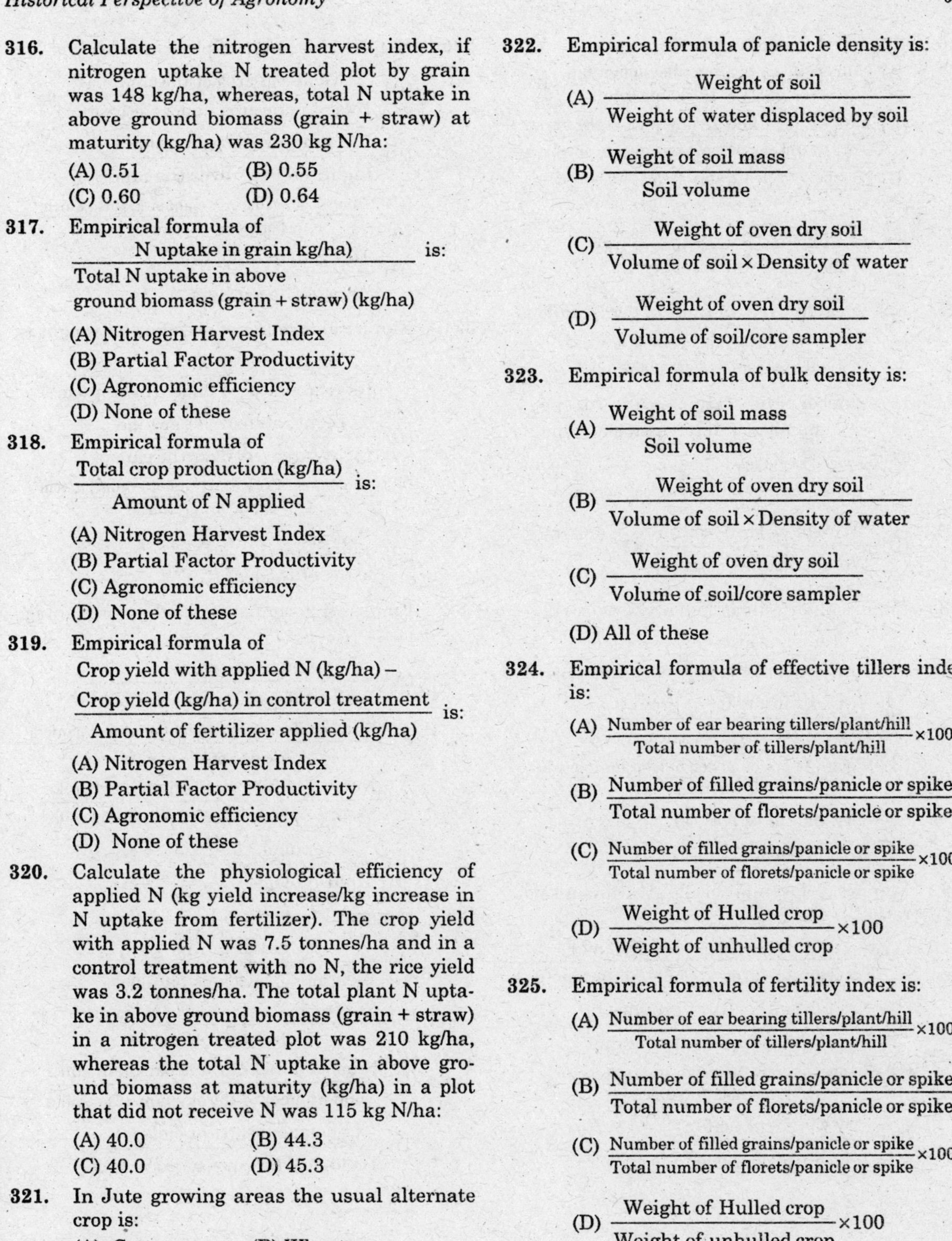

316. Calculate the nitrogen harvest index, if nitrogen uptake N treated plot by grain was 148 kg/ha, whereas, total N uptake in above ground biomass (grain + straw) at maturity (kg/ha) was 230 kg N/ha:

(A) 0.51 (B) 0.55
(C) 0.60 (D) 0.64

317. Empirical formula of $\frac{\text{N uptake in grain kg/ha)}}{\text{Total N uptake in above ground biomass (grain + straw) (kg/ha)}}$ is:

(A) Nitrogen Harvest Index
(B) Partial Factor Productivity
(C) Agronomic efficiency
(D) None of these

318. Empirical formula of $\frac{\text{Total crop production (kg/ha)}}{\text{Amount of N applied}}$ is:

(A) Nitrogen Harvest Index
(B) Partial Factor Productivity
(C) Agronomic efficiency
(D) None of these

319. Empirical formula of $\frac{\text{Crop yield with applied N (kg/ha)} - \text{Crop yield (kg/ha) in control treatment}}{\text{Amount of fertilizer applied (kg/ha)}}$ is:

(A) Nitrogen Harvest Index
(B) Partial Factor Productivity
(C) Agronomic efficiency
(D) None of these

320. Calculate the physiological efficiency of applied N (kg yield increase/kg increase in N uptake from fertilizer). The crop yield with applied N was 7.5 tonnes/ha and in a control treatment with no N, the rice yield was 3.2 tonnes/ha. The total plant N uptake in above ground biomass (grain + straw) in a nitrogen treated plot was 210 kg/ha, whereas the total N uptake in above ground biomass at maturity (kg/ha) in a plot that did not receive N was 115 kg N/ha:

(A) 40.0 (B) 44.3
(C) 40.0 (D) 45.3

321. In Jute growing areas the usual alternate crop is:

(A) Cotton (B) Wheat
(C) Rice (D) Sugarcane

322. Empirical formula of panicle density is:

(A) $\frac{\text{Weight of soil}}{\text{Weight of water displaced by soil}}$

(B) $\frac{\text{Weight of soil mass}}{\text{Soil volume}}$

(C) $\frac{\text{Weight of oven dry soil}}{\text{Volume of soil} \times \text{Density of water}}$

(D) $\frac{\text{Weight of oven dry soil}}{\text{Volume of soil/core sampler}}$

323. Empirical formula of bulk density is:

(A) $\frac{\text{Weight of soil mass}}{\text{Soil volume}}$

(B) $\frac{\text{Weight of oven dry soil}}{\text{Volume of soil} \times \text{Density of water}}$

(C) $\frac{\text{Weight of oven dry soil}}{\text{Volume of soil/core sampler}}$

(D) All of these

324. Empirical formula of effective tillers index is:

(A) $\frac{\text{Number of ear bearing tillers/plant/hill}}{\text{Total number of tillers/plant/hill}} \times 100$

(B) $\frac{\text{Number of filled grains/panicle or spike}}{\text{Total number of florets/panicle or spike}}$

(C) $\frac{\text{Number of filled grains/panicle or spike}}{\text{Total number of florets/panicle or spike}} \times 100$

(D) $\frac{\text{Weight of Hulled crop}}{\text{Weight of unhulled crop}} \times 100$

325. Empirical formula of fertility index is:

(A) $\frac{\text{Number of ear bearing tillers/plant/hill}}{\text{Total number of tillers/plant/hill}} \times 100$

(B) $\frac{\text{Number of filled grains/panicle or spike}}{\text{Total number of florets/panicle or spike}}$

(C) $\frac{\text{Number of filled grains/panicle or spike}}{\text{Total number of florets/panicle or spike}} \times 100$

(D) $\frac{\text{Weight of Hulled crop}}{\text{Weight of unhulled crop}} \times 100$

326. Empirical formula of fertility ratio is:

(A) $\frac{\text{Number of ear bearing tillers/plant/hill}}{\text{Total number of tillers/plant/hill}} \times 100$

(B) $\frac{\text{Number of filled grains/panicle or spike}}{\text{Total number of florets/panicle or spike}}$

(C) $\frac{\text{Number of filled grains/panicle or spike}}{\text{Total number of florets/panicle or spike}} \times 100$

(D) $\frac{\text{Weight of Hulled crop}}{\text{Weight of unhulled crop}} \times 100$

327. Empirical formula of hulling percentage is:

(A) $\frac{\text{Number of filled grains/panicle or spike}}{\text{Total number of tillers/plant/hill}} \times 100$

(B) $\frac{\text{Number of filled grains/panicle or spike}}{\text{Total number of florets/panicle or spike}}$

(C) $\frac{\text{Number of filled grains/panicle or spike}}{\text{Total number of florets/panicle or spike}} \times 100$

(D) $\frac{\text{Weight of Hulled crop}}{\text{Weight of unhulled crop}} \times 100$

328. Empirical formula of leaf area index is:

(A) $\frac{\text{Total leef area of the crop (cm}^2\text{)}}{\text{Total ground area under the crop (cm}^2\text{)}}$

(B) $\frac{\text{Light intensity at ground}}{\text{Light intensity over the canopy}}$

(C) $\frac{\text{Number of grains or weight of grains/panicle}}{\text{Length of the panicle}}$

(D) $\frac{\text{Dry weight of leaf (g)}}{\text{Area of the leaf (cm}^2\text{)}}$

329. Empirical formula of light transmission ratio:

(A) $\frac{\text{Total leef area of the crop (cm}^2\text{)}}{\text{Total ground area under the crop (cm}^2\text{)}}$

(B) $\frac{\text{Light intensity at ground}}{\text{Light intensity over the canopy}}$

(C) $\frac{\text{Number of grains or weight of grains/panicle}}{\text{Length of the panicle}}$

(D) $\frac{\text{Dry weight of leaf (g)}}{\text{Area of the leaf (cm}^2\text{)}}$

330. Empirical formula of panicle density is:

(A) $\frac{\text{Total leef area of the crop (cm}^2\text{)}}{\text{Total ground area under the crop (cm}^2\text{)}}$

(B) $\frac{\text{Light intensity at ground}}{\text{Light intensity over the canopy}}$

(C) $\frac{\text{Number of grains or weight of grains/panicle}}{\text{Length of the panicle}}$

(D) $\frac{\text{Dry weight of leaf (g)}}{\text{Area of the leaf (cm}^2\text{)}}$

331. Empirical formula of specific leaf weight is:

(A) $\frac{\text{Total leef area of the crop (cm}^2\text{)}}{\text{Total ground area under the crop (cm}^2\text{)}}$

(B) $\frac{\text{Light intensity at ground}}{\text{Light intensity over the canopy}}$

(C) $\frac{\text{Number of grains or weight of grains/panicle}}{\text{Length of the panicle}}$

(D) $\frac{\text{Dry weight of leaf (g)}}{\text{Area of the leaf (cm}^2\text{)}}$

332. Empirical formule of shelling percentage is:

(A) $\frac{\text{Weight of grains}}{\text{Weight of pods/cobs/panicle}}$

(B) $\frac{\text{Number of grains or weight of grains/panicle}}{\text{Length of the panicle}}$

(C) $\frac{\text{Number of filled grains/panicle of spike}}{\text{Yield number of florets/panicle or spike}}$

(D) $\frac{\text{Economic yield}}{\text{Total Biological yield}}$

333. Empirical formula of harvest index is:

(A) $\frac{\text{Weight of grains}}{\text{Weight of pods/cobs/panicle}}$

(B) $\frac{\text{Number of grains or weight of grains/panicle}}{\text{Length of the panicle}}$

(C) $\frac{\text{Number of filled grains/panicle of spike}}{\text{Yield number of florets/panicle or spike}}$

(D) $\frac{\text{Economic (grain yield)}}{\text{Biological (grain + straw) yield}}$

ANSWER SHEET
2. Historical Perspective of Agronomy

1. (A) 2. (C) 3. (D) 4. (A) 5. (B) 6. (A) 7. (A) 8. (A) 9. (A) 10. (B)
11. (B) 12. (C) 13. (A) 14. (C) 15. (C) 16. (D) 17. (A) 18. (A) 19. (A) 20. (C)
21. (C) 22. (A) 23. (C) 24. (B) 25. (D) 26. (A) 27. (B) 28. (C) 29. (D) 30. (B)
31. (D) 32. (D) 33. (A) 34. (A) 35. (B) 36. (A) 37. (B) 38. (C) 39. (D) 40. (A)
41. (D) 42. (C) 43. (D) 44. (C) 45. (B) 46. (A) 47. (A) 48. (C) 49. (D) 50. (A)
51. (B) 52. (C) 53. (A) 54. (B) 55. (C) 56. (A) 57. (A) 58. (C) 59. (C) 60. (D)
61. (A) 62. (C) 63. (B) 64. (C) 65. (B) 66. (A) 67. (C) 68. (B) 69. (A) 70. (B)
71. (A) 72. (C) 73. (D) 74. (A) 75. (C) 76. (B) 77. (A) 78. (B) 79. (C) 80. (D)
81. (C) 82. (C) 83. (C) 84. (A) 85. (C) 86. (D) 87. (A) 88. (B) 89. (D) 90. (A)
91. (B) 92. (C) 93. (C) 94. (A) 95. (D) 96. (B) 97. (C) 98. (A) 99. (D) 100. (A)
101. (B) 102. (B) 103. (B) 104. (B) 105. (B) 106. (B) 107. (B) 108. (B) 109. (A) 110. (B)
111. (A) 112. (A) 113. (A) 114. (C) 115. (C) 116. (A) 117. (A) 118. (D) 119. (C) 120. (A)
121. (B) 122. (C) 123. (C) 124. (C) 125. (C) 126. (B) 127. (B) 128. (A) 129. (D) 130. (B)
131. (A) 132. (D) 133. (D) 134. (D) 135. (B) 136. (B) 137. (D) 138. (A) 139. (B) 140. (A)
141. (D) 142. (C) 143. (D) 144. (B) 145. (C) 146. (D) 147. (A) 148. (D) 149. (D) 150. (A)
151. (D) 152. (B) 153. (B) 154. (D) 155. (B) 156. (C) 157. (A) 158. (C) 159. (A) 160. (B)
161. (B) 162. (D) 163. (B) 164. (A) 165. (D) 166. (B) 167. (C) 168. (D) 169. (A) 170. (C)
171. (B) 172. (C) 173. (D) 174. (A) 175. (C) 176. (D) 177. (B) 178. (C) 179. (A) 180. (C)
181. (C) 182. (A) 183. (C) 184. (B) 185. (D) 186. (A) 187. (A) 188. (D) 189. (B) 190. (D)
191. (B) 192. (A) 193. (A) 194. (C) 195. (B) 196. (A) 197. (C) 198. (B) 199. (A) 200. (A)
201. (B) 202. (C) 203. (A) 204. (C) 205. (B) 206. (A) 207. (C) 208. (A) 209. (A) 210. (B)
211. (C) 212. (D) 213. (A) 214. (B) 215. (C) 216. (A) 217. (D) 218. (A) 219. (A) 220. (B)
221. (D) 222. (C) 223. (D) 224. (A) 225. (B) 226. (A) 227. (B) 228. (B) 229. (C) 230. (B)
231. (A) 232. (C) 233. (D) 234. (A) 235. (B) 236. (C) 237. (D) 238. (A) 239. (B) 240. (C)
241. (A) 242. (B) 243. (B) 244. (C) 245. (A) 246. (D) 247. (A) 248. (B) 249. (C) 250. (D)
251. (A) 252. (B) 253. (C) 254. (D) 255. (D) 256. (A) 257. (B) 258. (A) 259. (B) 260. (C)
261. (C) 262. (C) 263. (D) 264. (B) 265. (D) 266. (A) 267. (B) 268. (B) 269. (B) 270. (B)
271. (C) 272. (A) 273. (C) 274. (D) 275. (D) 276. (D) 277. (D) 278. (A) 279. (D) 280. (D)
281. (A) 282. (B) 283. (A) 284. (B) 285. (C) 286. (A) 287. (B) 288. (C) 289. (A) 290. (B)
291. (B) 292. (C) 293. (A) 294. (C) 295. (B) 296. (C) 297. (D) 298. (A) 299. (D) 300. (A)
301. (B) 302. (C) 303. (D) 304. (A) 305. (A) 306. (B) 307. (C) 308. (D) 309. (A) 310. (B)
311. (D) 312. (C) 313. (A) 314. (B) 315. (C) 316. (D) 317. (A) 318. (B) 319. (C) 320. (D)
321. (C) 322. (A) 323. (B) 324. (A) 325. (B) 326. (C) 327. (D) 328. (A) 329. (B) 330. (C)
331. (D) 332. (A) 333. (D)

3 AGRICULTURAL CLIMATOLOGY

1. The physical state of the atmosphere at a given place and time is referred to as:
(A) Weather (B) Climate
(C) Both of these (D) None of these

2. The first meteorological department of India was started at Shimla in:
(A) 1863 (B) 1864
(C) 1865 (D) 1866

3. Climate word originated from:
(A) Greek (B) Latin
(C) English (D) Parsian

4. Climate word derived from:
(A) Klima+ Logos
(B) Klima+Logo
(C) Clima + Logo
(D) Clima+Logos

5. The basic meaning of klima is:
(A) Angle incidence of sun light
(B) Changes in temperature
(C) Incidence of humidity
(D) All of these

6. The science that seeks to describe and explain the nature of climate is called:
(A) Climatology (B) Meteorology
(C) Both of these (D) None of these

7. The study of physical status of climate is called:
(A) Physical climatology
(B) Statistical climatology
(C) Descriptive climatology
(D) Applied climatology

8. To study the physical atmospheric factors viz., temperature, relative humidity, clouds, wind velocity and direction is called:
(A) Physical climatology
(B) Statistical climatology
(C) Descriptive climatology
(D) Applied climatology

9. To study the different factors of climate such as temperature and rainfall, are divided into state level is called:
(A) Physical climatology
(B) Statistical climatology
(C) Descriptive climatology
(D) Applied climatology

10. To study the relationship with other branch of climatology is called:
(A) Physical climatology
(B) Statistical climatology
(C) Descriptive climatology
(D) Applied climatology

11. The weather information collected from observatory are depicted in tables is called:
(A) Weather chart (B) Weather map
(C) Both of these (D) None of these

12. The weather information are depicted through map is called:
(A) Weather chart (B) Weather map
(C) Both of these (D) None of these

13. World Meteorological Organisation (WMO) was established at:
(A) Geneva (Switzerland)
(B) Spain
(C) USA (D) Canada

14. World Meteorological Organisation (WMO) was established in:
(A) 1950 (B) 1951
(C) 1952 (D) 1953

15. International Meteorological Organisation was established in:
(A) 1877 (B) 1878
(C) 1879 (D) 1880

16. The science of climatological condition which have direct relation or relevance to agriculture is called:
(A) Agro-meteorology (B) Agro-climatology
(C) Meteorology (D) Climatology

17. The composite or generalization of the variety of day to day weather conditions is called:
(A) Climate (B) Weather
(C) Season (D) All of these

18. How many years of weather is optimum for climate:
(A) 30 years (B) 31 years
(C) 32 years (D) 33 years

19. The observations recorded from open area of 2 meter height of soil surface is called:
(A) Macro climate (B) Micro climate
(C) Topo climate (D) Local climate

20. When weather data collected form sloppy land is called:
(A) Macro climate (B) Micro climate
(C) Topo climate (D) Local climate

21. When weather data collected form standing crop is called:
(A) Macro climate (B) Micro climate
(C) Topo climate (D) Local climate

22. The data recorded in macro climate is called:
(A) Synoptic Meteorological observation
(B) Asynoptic Meteorological observation
(C) Both of these
(D) None of these

23. Upto 15 km above the earth is called:
(A) Troposphere (B) Mesosphere
(C) Stratosphere (D) None of these

24. 50 km above the earth is called:
(A) Troposphere (B) Mesosphere
(C) Stratosphere (D) None of these

25. In atmosphere mainly which gas is determine the atmospheric activities is:
(A) O_2 (B) Ozone
(C) Both of these (D) None of these

26. In atmosphere gasses found in:
(A) Mixed status
(B) Alone
(B) Without mixed status
(D) None of these

27. Four gasses- nitrogen, oxygen, argon and CO_2 accounts:
(A) 97% (B) 98%
(C) 99 % (D) 100 %

28. Nitrogen alone constituents nearly:
(A) One-third by volume
(B) Two-third by volume
(C) Three-fourth by volume
(D) Four-fifth by volume

29. Oxygen alone constituents nearly:
(A) One-third by volume
(B) Two-third by volume
(C) One-fifth by volume
(D) Four-fifth by volume

30. The blanket of gases, suspended liquids and solids the entirely envelops the earth is called:
(A) Atmosphere (B) Climate
(C) Clouds (D) Gasses

31. The atmosphere over one square foot of earth's surface at the Equator will contain approximately ------ pounds of water vapour:
(A) 20 (B) 30
(C) 40 (D) 50

32. It has been found that peas will remain in good harvest condition twice as long at --------^{0}F than at 80 ^{0}F:
(A) 32 (B) 42
(C) 52 (D) 62

33. The changes that accompany hardening of plants include:
(A) Increased sugar concentration
(B) Decreased soluble proteins
(C) Decreased osmotic pressure
(D) Decreased bound water

34. Cocoa, a tropical plant has been found to produce the maximum growth in approximately ------ of full sunlight:
(A) 5% (B) 10%
(C) 15% (D) 25%

35. When white light is passed through a prism, it is dispersed into different wavelengths of various colours and the orange colour appears at:
(A) 435-490 (B) 574-595
(C) 400-435 (D) 595-626

36. It has been demonstrated that the ------ light present in the white light prevents maximal growth rate:
(A) Blue (B) Red
(C) Green (D) Violet

37. Wavelengths longer than ----- mμ are not visible to the eye, and are called infrared:

(A) 450 (B) 550
(C) 650 (D) 750

38. It has been observed that in a number of outbreeding populations of rye grass, *Lolium perenne* could be brought about at low temperature (0-3^{0}C) or --------:

(A) Long day
(B) Short day
(C) Intermediate day
(D) None of these

39. In --------, long days promote bulb formation whereas, short days promote flowering and seed production:

(A) Sweet potato (B) Artichoke
(C) Cassava (D) Onion

40. The ------- variety Maryland Mommoth in the latitude of Maryland (USA) produces only leaves and needs shorter days for seed production:

(A) Tomato (B) Carrots
(C) Radish (D) Tobacco

41. Beyond the lower atmosphere, extending out into space, is the stratosphere, in which temperature remains relatively constant from:

(A) -15 to -25 (B) -45 to -75
(C) -80 to -95 (D) -5 to -10

42. One of the characteristics of the stratosphere is the occurrence of a well-marked layer having an increased concentration of -------- at about 20 miles (32 km) above sea level:

(A) Carbon dioxide
(B) Carbon monoxide
(C) Ozone (D) Hydrogen

43. In the ozone layer, the ozone does not exceed ------- ppm at 20 miles (32 km) above sea level (stratosphere):

(A) 5 (B) 10
(C) 15 (D) 20

44. The gaseous mixing in the atmosphere is, in general, similar at sea level and at heights upto ----km.

(A) 20 (B) 40
(C) 60 (D) 80

45. Ozonated hexane led to distruption of ----- in plants:

(A) Membrane permeability
(B) Protein synthesis
(C) Lipid synthesis
(D) Stem elongation

46. Animals and nongreen plants continuously use oxygen and liberate ------ into the air:

(A) Sulfur dioxide (B) Carbon monoxide
(C) Carbon dioxide (D) Nitrous oxide

47. The rate of photosynthesis under the conditions of high light intensity may be controlled by:

(A) Light (B) Carbon dioxide
(C) Water (D) None of these

48. North of the equator, surface winds are known as:

(A) Northeast trade winds
(B) Southeast winds
(C) Westerlies (D) Easterlies

49. In the humid tropics and subtropics the process of weathering (soil formation) is known as:

(A) Podzolization (B) Gleization
(C) Latosolization (D) None of these

50. The range of coffee production is limited approximately by the ------- ^{0}F of isotherms for the coldest month:

(A) 25 (B) 35
(C) 45 (D) 55

51. The diversity pertaining to a plant composition in several plant communities in one range of environment is known as:

(A) Alpha index of diversity
(B) Beta index of diversity
(C) Gamma index of diversity
(D) None of these

52. Among the types of natural succession, the one which begins in water is known as:

(A) Xerarch (xerosere)
(B) Halosere
(C) Hydrarch (hydrosere)
(D) Psammosere

53. In the earth's atmosphere at ----- cal/cm^2/annum of solar energy is received:

(A) 15 x 10^6 (B) 15 x 10^8
(C) 15 x 10^{10} (D) 15 x 10^{12}

54. Which of the following crops utilizes even upto 10-12 per cent of the total solar energy:

(A) Sorghum (B) Rice
(C) Maize (D) Sugarcane

55. Fixation of each gramme of carbon from carbon dioxide to organic compounds during photosynthesis traps -------- of solar energy:

(A) 124 k cal (B) 134 k cal
(C) 144 k cal (D) 154 k cal

56. C_3 plants show a light saturation upto 40000 to 50000 lux, while C_4 plants upto full sun in tropics upto:

(A) 60000-70000 (B) 70000-80000
(C) 90000-100000 (D) 115000-120000

57. ------ pollutants are very harmful for human health & therefore such polluted water should essentially be purified before their supply for drinking purposes:

(A) Biological (B) Physical
(C) Mechanical (D) Metallic

58. Ionising -------- are very important and harmful pollutants:

(A) Reflection (B) Deflection
(C) Radiation (D) Absorptions

59. The sound beyond --------dB are regarded as pollutant as it harms hearing ability:

(A) 60 (B) 80
(C) 100 (D) 120

60. If all the moisture present in the air at any time to converted into precipitation, it would be equal to 2.5 cm or about -------of water supply by precipitation (average residence time):

(A) 5 days (B) 10 days
(C) 15 days (D) 20 days

61. About 2.5 cm rainfall sufficient to produce over 10 tonnes of stream flow water on a watershed area of about -------cm^2:

(A) 3000 (B) 4000
(C) 5000 (D) 6000

62. Fresh water bodies like rivers and streams are known as:

(A) Palustrine
(B) Lotic or riverine
(C) Lentic or lacustrice
(D) None of these

63. On the basis of temperature variation in the vertical column of water, or thermal stratification, lake ecosystem have been divided into the following zones wherein ----- comes in direct contact of sun and wind:

(A) Thermocline
(B) Hypolimnion
(C) Epilimnion
(D) None of these

64. Mercury level of ----- μg/g of fresh weight of fish is considered unsafe for human consumption:

(A) 0.5-1.0 (B) 1.6-2.0
(C) 2.1-2.5 (D) 2.6-3.0

65. Fish absorbs and may accumulate zinc in muscles amounting approximately upto how many time more than that in sea water:

(A) 3700 (B) 4700
(C) 5700 (D) 6700

66. Permissible limit of iron in drinking water is:

(A) 1.0 ppm (B) 1.5 ppm
(C) 2.0 ppm (D) 2.5 ppm

67. Potassium in natural waters never assume toxic proportions and remain less than ------ mg/litre:

(A) 10 (B) 15
(C) 20 (D) 25

68. When with chloride ion in drinking water, the dominant cation is potassium or magnesium, the salty taste is not felt even upto ---------- mg/litre:

(A) 500 (B) 750
(C) 1000 (D) 1250

69. The characteristic pungent odour taste of onion is due to :

(A) Magnesium (B) Calcium
(C) Potassium (D) Sulphur

70. Many factories waters contain how much mg/litre of nitrate nitrogen:

(A) 20 (B) 30
(C) 40 (D) 50

71. Chloride added to drinking water supply when encountered with ------ gets neutralized and chloroamines are formed:

(A) Nitrous oxide (B) Carbon dioxide
(C) Mathane (D) Ammonia

72. The nitrate content say in the range of 30 to 40 mg/litre lead to poor ionation of blood haemoglobin and blood turns:
(A) Blackish (B) Whitish
(C) Bluish (D) Yellowish

73. In a study at USA, the earthworm inhabiting a foliarly fed DDT-tree, contained upto 80 ppm of DDT even after how months:
(A) 5 (B) 7.5 (C) 10 (D) 10.5

74. The coliform bacteria count in drinking water is regarded safe if it is (per litre):
(A) 5 or less (B) 10 or less
(C) 15 or less (D) 20 or less

75. The PSI value range ------ has been classified as hazardous (air) causing premature on set of certain diseases and aggravation of symptoms:
(A) 200-299 (B) 300-399
(C) 100-199 (D) 50-100

76. Methane gas once released in the atmosphere stays for how many years:
(A) 3-4 (B) 4-5
(C) 5-7 (D) 8-10

77. The earth during the last one century has warmed up by---------- ^{0}C as estimated:
(A) 0.3-0.7 (B) 0.8-1.0
(C) 1.2-1.3 (D) 1.4-1.5

78. The earth temperature during the past 20000 years has risen by --- ^{0}C:
(A) 2-3 (B) 4-5
(C) 6-7 (D) 8-9

79. Utraviolet radiation are of similar wavelengths than of visible light and those ranging from 320-400 nm are called:
(A) UV-C (B) UV-B
(C) UV-A (D) UV-D

80. CFCs methane and nitrogen oxide gases have the capacity to react with ozone to convert into:
(A) Nitrogen (B) Hydrogen
(C) Oxygen (D) None of these

81. The estimates reveal that the maximum NO_2 emi-ssion in European countries is in ----- in the range of 2T g/annum (Tetagramme = 10^{12} grammes):
(A) France (B) Germany
(C) England (D) None of these

82. The threshold concentration upto which no visible injury symptoms appear on plants due to NO_2 is 1 mg/m^2 on:
(A) Daily average (24 hrs)
(B) Weekly average
(C) Monthly average
(D) Annual average

83. The suggested threshold level of fluoride pollution in atmosphere is how much quantity (ppm):
(A) 10-40 (B) 50-80
(C) 90-100 (D) 110-115

84. Uranium contribute the radioactivity of about – pci/litre of ocean water:
(A) 2 (B) 3
(C) 4 (D) 5

85. The Department of Environment in independence India was formed in November:
(A) 1960 (B) 1970
(C) 1980 (D) 1990

86. Serious efforts on Environment protection in India began with the establishment of NCEPC by the Department of Science and Technology in February:
(A) 1952 (B) 1962
(C) 1972 (D) 1982

87. Environment Protection Act 1986 came into force in India on --------- 1968:
(A) October 19
(B) November 19
(C) September 19
(D) August 19

88. In December, 1989 standards were laid for the noise limits in day and night time in various area and silence zones upto -------- around educational institutions, hospitals and courts in India:
(A) 50 m (B) 100 m
(C) 150 m (D) 1200 m

89. The optimum temperature range for ideal germination of wheat varied from:
(A) 25-30 ^{0}C (B) 20-25 ^{0}C
(C) 15-20 ^{0}C (D) 10-15 ^{0}C

90. The optimum rainfall range for growing maize is:
(A) 360-500 mm (B) 560-700 mm
(C) 300-400 mm (D) 460-600 mm

91. The arhar (pigeonpea) is commonly grows well in:
(A) Tropical and subtropical climate
(B) Temperate climate
(C) Humid and temperate climate
(D) Sub-humid and temperate climate

92. For urdbean, the suitable climate is:
(A) Hot and humid (B) Cool and humid
(C) Cool and dry (D) Hot and dry

93. High temperature above 41^{0}C at flowering of castor results in:
(A) Rotting of pollens and poor fertilization
(B) Blasting of flowers and poor seed set
(C) Delayed pollination and poor fertilization
(D) Reduced pollination and rotting of seed

94. The crop of ----- is more liable to suffer from frost and cold:
(A) Toria (B) Pahari rai
(C) Sarson (D) Banarasi rai

95. Linseed is a which season crop:
(A) Warm (B) Cool
(C) Hot (D) Dry

96. Safflower is a which season crop:
(A) Dry (B) Summer
(C) Arid (D) Cool

97. The sunflower crop requires cool climate from:
(A) Germination to seedling stage
(B) Seedling to flowering stage
(C) Flowering to maturity stage
(D) Seedling to maturity stage

98. If temperature is high during maturation of sunflower, the linoleic acid:
(A) Remain unaffected
(B) Increases with slow rate
(C) Decreases
(D) Increases with rapid rate

99. Tobacco requires temperature for uniformly rapid germination:
(A) 17-22^{0}C (B) 19-25^{0}C
(C) 27-32^{0}C (D) 29-42^{0}C

100. The soil temperature for satisfactory growth of potato tuber should be:
(A) 7-9^{0}C (B) 11-13 ^{0}C
(C) 17-19 ^{0}C (D) 20-23 ^{0}C

101. Optimum temperature for germination of pigeonpea crop is:
(A) 15-20^{0}C (B) 20-25^{0}C
(C) 30-35^{0}C (D) 40-45^{0}C

102. Temperature required for good sugarcane germination is:
(A) 20-25^0 C (B) 25-32^0 C
(C) 35-41^0 C (D) 38-45^0 C

103. Optimum temperature for vegetative growth of gram is:
(A) 10-15^0 C (B) 15-20^0 C
(C) 20-25^0 C (D) 25-30^0 C

104. Troposphere of the atmosphere extending upto a height from ground to:
(A) 8-18 km (B) 50 km
(C) 40 km (D) >50 km

105. Fog, dew, mist, rain etc. occur in the layer of:
(A) Thermosphere (B) Mesosphere
(C) Stratosphere (D) Troposphere

106. Troposphere is the densest part of atmosphere and contains about:
(A) 55% of atmospheric mass
(B) 75% of atmospheric mass
(C) 85% of atmospheric mass
(D) 95% of atmospheric mass

107. The layer separated the stratosphere and mesosphere is called:
(A) Tropopause (B) Stratopause
(C) Mesopause (D) None of these

108. Inoosphere is present in:
(A) Troposphere (B) Stratosphere
(C) Mesosphere (D) Thermosphere

109. Heat can flow between substance or within substance by:
(A) Conduction (B) Convection
(C) Radiation (D) All of these

110. The reason(s) for the decline in temperature with increase altitude is:
(A) Air is heated by the earth
(B) Water vapour content of air decreases with increase in altitude
(C) Expansion of air at higher elevation
(D) All of these

111. Vertical temperature decrease is called lapse rate and the normal lapse rate is:
(A) 4.5^0 C/km (B) 6.5^0 C/km
(C) 8.5^0 C/km (D) None of these

112. Heat transfer into the soil is mainly by:
(A) Conduction (B) Convection
(C) Radiation (D) None of these

113. The rate of heat flux into soil depends on:
(A) Temperature gradient
(B) Thermal conductivity
(C) Both (A) & (B) (D) None of these

114. Thermal conductivity of soil decrease with:
(A) Increase porosity
(B) Increase organic matter
(C) Decrease soil moisture
(D) All of these

115. What will be the effect of tillage operation on thermal conductivity of soil?
(A) Remain unchanged
(B) Decrease (C) Increase
(D) First increase then decrease

116. Minimum temperature for the germination of cotton seed is:
(A) 10^{0}C (B) 16^{0}C
(C) 20^{0}C (D) 26^{0}C

117. Optimum temperature for retting of jute is:
(A) 20^{0}C (B) 24^{0}C
(C) 30^{0}C (D) 34^{0}C

118. The optimum temperature for germination of soybean is:
(A) 10^{0}C (B) 22^{0}C
(C) 17^{0}C (D) 35^{0}C

119. South-west monsoon contributes how much per cent of rainfall?
(A) 50 % (B) 75 %
(C) 80 % (D) 95 %

120. Average rainfall in India is around:
(A) 2120 mm (B) 1120 mm
(C) 850 mm (D) 500 mm

121. Annual rainfall required for sugarcane crop is:
(A) 50-100 mm (B) 50-100 cm
(C) 100-125 cm (D) >125 cm

122. Annual rainfall required for cotton crop is:
(A) 25-50 cm (B) 50-75 cm
(C) 75-115 cm (D) >115 cm

123. Annual rainfall required for gram crop is:
(A) 40-50.cmm (B) 60-90 cm
(C) 100-110 cm (D) >110 cm

124. Annual rainfall required for sugarbeet crop is:
(A) 50 cm (B) 75 cm
(C) 100 cm (D) 125 cm

125. Which of the following disorder in mango is caused by low temperature?
(A) Black heart (B) Crotch injury
(C) Collor injury (D) All of these

126. How much rainfall is required for the cultivation of banana?
(A) 100-120 cm (B) 125-150 cm
(C) 170-200 cm (D) 300-350 cm

127. Rainfall intensity is generally higher on:
(A) Northern hemisphere
(B) Southern hemisphere
(C) Near the equator
(D) None of these

128. Environment can be improved by:
(A) Wind erosion (B) Water erosion
(C) Conservation (D) None of these

129. Smog is produced due to:
(A) Air and metal dust
(B) Water and nitrogen oxide
(C) Smoke and fog (D) None of these

130. The photochemical smog is formed by:
(A) Nitrogen oxides
(B) Hydrocarbons
(C) Nitrogen oxides and hydrocarbons
(D) Solar radiations upon nitrogen oxides and hydrocarbons

131. O_3 promotes:
(A) Oxidation of biochemicals
(B) Destruction of chlorenchyma
(C) Both (A) and (B)
(D) Growth of plants

132. Fertilizers, pesticides and insecticides may cause:
(A) Air pollution
(B) Water pollution
(C) Both (A) and (B)
(D) None of these

133. Photochemical smog brings about:
(A) Defoliation
(B) Bleaching of foliage
(C) Floral bud shedding
(D) Discolouration and curling of petals

134. Water soacked appearance of leaves due to sulphur dioxide pollution is caused by:
(A) Destruction of chlorophyll
(B) Damage to membrane
(C) Both (A) and (B)
(D) Corrosive action over metals

135. Addition of phosphates and nitrate fertilizers into water leads to:
(A) Increased growth decompositors
(B) Reduced algal growth
(C) Increased algal growth
(D) Nutrient enrichment/ eutrophication

136. Major reason for soil erosion in India is:
(A) Tilling of land
(B) Deforestation
(C) Overgrazing and deforestation
(D) Over grazing

137. Domestic water is:
(A) Effluent
(B) Biodegradable
(C) Nonbiodegradable
(D) Air pollutant

138. Inadequate drainage in a soil will lead to:
(A) Floods (B) Soil erosion
(C) Aridity (D) Salination

139. Jhuming cultivation generally practiced in India in:
(A) Hilly areas (B) Madhya Pradesh
(C) North-East states
(D) Malabar area

140. Ecological balance of a mining area is restricted through:
(A) Conversion into agricultural area
(B) Revegetation of area
(C) Prevention of soil erosion
(D) Protection against grazing

141. Ultra-violet light causes:
(A) Formation of pyrimidines
(B) Sticky metaphases
(C) Photodynamic action
(D) Destruction of hydrogen bonds between complementary DNA strands

142. Sulphur dioxide affects:
(A) Cell wall (B) Plasmodesmata
(C) All membrane system
(D) Nucleus

143. Ultra-violet radiation are lethal because of inactivation of:
(A) Proteins, nucleic acids and pigments
(B) Minerals, air and water
(C) Carbohydrates, fats and vitamins
(D) Water, carbon dioxide and oxygen

144. SO_2 and NO_2 produce pollution by increasing:
(A) Alkalinity (B) Acidity
(C) Neutrality (D) Buffer action

145. Spraying of DDT produces pollution of:
(A) Air
(B) Air and water
(C) Air and soil
(D) Air, soil & water

146. Most abundant water pollutant is:
(A) Detergents
(B) Pesticides
(C) Industrial wastes
(D) Ammonia

147. Ozone depletion in stratosphere shall result in
(A) Forest fires
(B) Increased incidence of skin cancer
(C) Global warming
(D) None of these

148. Temperature requirement for proper grain filling of wheat is:
(A) 23-25^{0}C (B) 20-23^{0}C
(C) 15-20^{0}C (D) 16-20^{0}C

149. The best suited temperature for growth of sugarcane lies between:
(A) 15-20^{0}C (B) 20-25^{0}C
(C) 26-32^{0}C (D) 38-45^{0}C

150. The most suitable temperature for maize growth is:
(A) 21^{0}C (B) 26^{0}C
(C) 29^{0}C (D) 32^{0}C

151. Which of the following is the ideal temperature range for tillering stage in wheat:
(A) 16-20^{0}C (B) 28-40^{0}C
(C) 20-23^{0}C (D) 23-25^{0}C

152. Which causes water pollution?
(A) Automobile exhausts
(B) Smoke/flyash
(C) 2,4-D and pesticides
(D) Aeroplanes

153. Drawback of DDT as pesticide is:
(A) It becomes ineffective after some time
(B) It is less effective than others
(C) It is not easily/rapidly degraded in nature
(D) Its high cost

154. Which one is found in higher concentration in the atmospheric?
(A) Ozone
(B) Nitrous oxides
(C) Carbon dioxide, carbon monoxide and sulphur dioxide
(D) All of these

155. Carbon dioxide is a major pollutant in:
(A) Air (B) Soil
(C) Water (D) Noise

156. Major source of methane in India is:
(A) Rice fields (B) Sugarcane fields
(C) Wheat fields (D) Fruit orchards

157. Phosphate pollution is caused by:
(A) Weathering of phosphate rocks only
(B) Agricultural fertilizers
(C) Phosphate rocks and sewage
(D) Sewage and agricultural fertilizers

158. Which pigment protects plants from UV damage?
(A) Chlorophyll (B) Xanthophyll
(C) Carotenoids (D) Phycocyanin

159. Indescriminate use of fertilizers causes:
(A) Air pollution (B) Water pollution
(C) Land pollution (D) All of these

160. SO_2 pollution causes:
(A) Plasmolysis
(B) Destruction of cell wall
(C) Destruction of Golgi bodies
(D) Destruction of chlorophyll

161. Depletion of ozone layer is due to:
(A) Oxides of nitrogen
(B) Oxides of carbon
(C) Oxides of sulphur
(D) None of these

162. Chipko movement originated in the district:
(A) Allahabad (B) Chamoli
(C) Hardwar (D) Dehradun

163. Chipko movement originated in Gopeshwar in:
(A) 1953 (B) 1963
(C) 1973 (D) 1083

164. A movement similar to Chipko which is spread in south is:
(A) Uliso (B) Appiko
(C) Belesu (D) Balasu

165. Chipko movement originated in:
(A) Silent valley of Kerala
(B) Kangra valley of Himanchal Pradesh
(C) Tehri Garwal of Uttarakhand
(D) Panchmari of MP

166. Extinction of numerous wild animals is due to:
(A) Hunting (B) Deforestation
(C) Overcrowding (D) Scarcity of food

167. Chipko movement is connected with:
(A) Conservation of natural resources
(B) Plant/forest conservation
(C) Plant Breeding
(D) Project tiger

168. Forest control drought through:
(A) Lot of water plants
(B) Increasing rainfall
(C) Retention of water and prevention of erosion
(D) Functioning as water shed

169. Overgrazing causes:
(A) Negative pollution
(B) Positive pollution
(C) Reduction in crop yields
(D) Soil erosion

170. Deforestation brings about:
(A) Increased sunlight
(B) Increased grazing area
(C) Weed control
(D) Soil erosion

171. Deforestation will decrease:
(A) Soil erosion (B) Land slides
(C) Soil fertility (D) Rainfall

172. Deforestation is the major causal agent of:
(A) Depletion of natural resources
(B) Environmental pollution
(C) Desertification of habitat
(D) Genetic erosion

173. Deforestation does not lead to:
(A) Quick nutrient cycling
(B) Soil erosion
(C) Alteration of local weather conditions
(D) Destruction of natural habitat of wild animals

174. Forest destruction results in:
(A) Loss of wild life
(B) Floods and drought
(C) Soil erosion
(D) All of these

175. Afforestation should be with:
(A) Exotic species (B) Indigenous species
(C) Bamboos (D) Eucalyptus

176. If we remove half of the forest cover of earth, the crises that will occur:
(A) Many species would become extinct
(B) Population, pollution and ecological imbalance will rise
(C) Energy crises will commence
(D) The remaining forest will correct the imbalance

177. Decreased rainfall in northern areas is due to:
(A) Change in sun
(B) Global phenomenon
(C) Deforestation in catchment areas
(D) Deforestation in evaporating areas near equator

178. Increased incidence of floods in the plains of northern India are due to:
(A) Increased deforestation in catchment areas
(B) Increase in incidence of rainfall
(C) Silting of dams
(D) More area under cultivation

179. Soil erosion can be prevented by:
(A) Restricted human activity
(B) Good plant cover
(C) Checking movement of animals
(D) Wind screen alone

180. Adherence to two-children norms shall help in:
(A) Improving food web
(B) Checking pollution
(C) Stabilising ecosystem
(D) Fertility of soil

181. Retting temperature for jute is:
(A) 14^0C (B) 20^0C
(C) 28^0C (D) 34^0C

182. Fog is formed by:
(A) Smoke
(B) Condensation of water vapours
(C) Condensation of water vapours over dust particles
(D) Both (A) and (C)

183. Succession on sand dune is:
(A) Lithosere (B) Psychrosere
(C) Psammosere (D) Halosere

184. The amount of biogenic nutrient present in the abiotic environment is called:
(A) Standing state (B) Standing crop
(C) Detritus (D) Reserve nutrient

185. Phosphorus has:
(A) Imperfect nutrient cycle
(B) Perfect nutrient cycle
(C) Gasseous cycle
(D) Mixed cycle

186. Carbosphere comprises:
(A) Atmosphere (B) Lithosphere
(C) Hydrosphere (D) All of these

187. Adiabatic expansion resulting in cooling is called:
(A) Coriolis effect
(B) Joule-Kelvin effect
(C) Le Chatelier Principle
(D) Principle of Parsimony

188. Cyclic/frontal rainfall generally occurs:
(A) All over the world
(B) Tropical areas
(C) Temperate areas
(D) Boundary of warm tropical air and cold polar air

189. Which area has four seasons:
(A) Tropics (B) Monsoonic
(C) Temperate (D) Arctic

190. The land of wind mills is:
(A) UK
(B) USA
(C) Netherlands
(D) HImanchal Pradesh

191. The shining and thunder in clouds during rainy season:
(A) Start simulataneosly
(B) Occur thunder first and then shining starts
(C) Occur shining first and then thunder starts
(D) Occur mostly no thunder and no Shining

192. Daily weather forecast is obtained in the country from:
(A) Pune (B) New Delhi
(C) Mumbai (D) Kolkata

193. The layer of atmosphere having uniform temperature is:
(A) Stratosphere (B) Exosphere
(C) Thermosphere (D) Troposphere

194. Layer of ozone is present in:
(A) Stratosphere (B) Thermosphere
(C) Troposphere (D) None of these

195. The atmosphere stratum nearest to earth is:
(A) Thermosphere (B) Stratosphere
(C) Troposphere (D) None of these

196. Part of earth covered by water is about:
(A) 73% (B) 50%
(C) 92% (D) 87%

197. Floods can be checked by:
(A) Making dams (B) Deforestation
(C) Reforestation (D) None of these

198. Percentage of fresh water in the hydrosphere is:
(A) 3% (B) 30%
(C) 13% (D) 33%

199. Percentage of fresh water found in lakes, ponds, swamp, streams and rivers is:
(A) 22.4% (B) 77.2%
(C) 7.7% (D) 0.36%

200. Of the total fresh water, ground water constitutes:
(A) 90% (B) 10%
(C) 22.4% (D) 44.2%

201. Of the total water evaporated from oceans, precipitation returns:
(A) 100% (B) 90%
(C) 75% (D) 50%

202. Number of agro climatic zones in Uttar Pradesh are:
(A) 6 (B) 7
(C) 8 (D) 9

203. Planning Commission agro-climatic region the whole country has been divided into:
(A) 15 (B) 20
(C) 25 (D) 30

204. The concept of agro-climatic zones were introduced during:
(A) NATP (B) NARP
(C) NREP (D) NAIP

205. The number of agro-climatic zones under NARP:
(A) 127 (B) 146
(C) 152 (D) 164

206. The number of agro-ecosystems identified under NATP:
(A) 6 (B) 9
(C) 14 (D) 21

207. Movement of air occurs in the atmosphere upto height of:
(A) 100km (B) 50 km
(C) 25 km (D) 10 km

208. Peak concentration of ozone in the stratosphere (ozonosphere) lies at a height of:
(A) 10 km (B) 25 km
(C) 40 km (D) 50 km

209. Planning Commission advice that the agro-climatic zones in which plan:
(A) Fifth Five Year Plan
(B) Sixth Five Year Plan
(C) Seventh Five Year Plan
(D) Eight Five Year Plan

210. The estimates postulate of carbon dioxide content in atmosphere has increased from 280 parts per million to ---- during last three decades:
(A) 300 ppm (B) 350 ppm
(C) 400 ppm (D) None of these

211. Nitrate levels in drinking water above ------ mg per litre are considered a human health hazard:
(A) 5 (B) 10
(C) 15 (D) 20

212. The discovery of oxygen was done by which scientist who proved the keystone of other discoveries that went far toward unlocking the mystery of plant life:
(A) John Mayow (B) John Woodward
(C) Benjamin Franklin
(D) Priestly

213. There was not only one "Principle of vegetation" as reflected by agricultural writings during 17th and 18th centuries, was observed by:
(A) William Michael (B) Guy Baird
(C) Francis Home (D) Jone Cummings

214. More than which per cent of earth's surface is covered by water in the form of seas, lakes and rivers:
(A) 40 (B) 50
(C) 60 (D) 70

215. Depending upon the temperature factor, the upper warmer region of lacks is called:
(A) Epilimnion (B) Hypolimnion
(C) Thermocline (D) None of these

216. Secondary consequences of increasing elevation, both agronomically & economically significant are:
(A) Decreased temperature
(B) Decreased precipitation
(C) Decreased wind velocity
(D) Increased soil fertility

217. "Diversion of water from mountain streams by diversion dams to canals of different length and breadth & ditches of varying size, and finally directed to individual fields" is an important practice in:
(A) China (B) Egypt
(C) Philippines (D) Japan

218. The success or failure of the cropping season is determined by the intensity of:
(A) Climatic factors (B) Political factors
(C) Social factors (D) None of these

219. The ideal conditions of temperature for crop production are between:
(A) 12-18 ºC (B) 18-24ºC
(C) 24-30 ºC (D) 30-36ºC

220. The direct effect of strong winds on plants is of which nature:
(A) Chemical (B) Biological
(C) Mechanical (D) None of these

221. Altitude, because of it's ---- effects, induces a zonation of vegetation similar to that found through a wide range of latitude:
(A) Humidity (B) Radiation
(C) Temperature (D) None of these

222. Rice, maize and sugarcane are replaced by temperate zone species at elevations of possibly:
(A) 1000-3000' (B) 4000-6000'
(C) 7000-9000' (D) None of these

223. At 0ºC water is in equilibrium with a water vapour pressure of ----- mm of mercury:
(A) 2.58 (B) 3.58
(C) 4.58 (D) 5.58

224. The powerful oxidizing agents and cause a threat to life and property when concentrations become excessive is:
(A) Ozone (B) Nitrogen
(C) Oxygen (D) Carbon di oxide

225. The lower portion of the atmosphere, extending upto about 8 km at the poles and 18 km at the equator is called:
(A) Troposphere (B) Stratosphere
(C) Ionosphere (D) Exosphere

226. The upper portion of the atmosphere, extending upto 50-60 km is called:
(A) Troposphere (B) Stratosphere
(C) Ionosphere (D) Exosphere

227. The upper portion of the stratosphere, extending upto 60-500 km is called:
(A) Troposphere (B) Stratosphere
(C) Ionosphere (D) Exosphere

228. The upper portion of the atmosphere, extending upto 500-1000 km is called:
(A) Troposphere (B) Stratosphere
(C) Ionosphere (D) Exosphere

229. Ozone gas mainly found in which region:
(A) Troposphere (B) Stratosphere
(C) Ionosphere (D) Exosphere

230. High air pressure and its affected area is called:
(A) Anticyclone (B) Cyclone
(C) Depression (D) Rainfall

231. Low air pressure and its affected area is called:
(A) Anticyclone (B) Fog
(C) Cyclone & Depression
(D) Rainfall

232. The unit of atmospheric pressure is:
(A) Millibar (B) Bar
(C) Pascal (D) Gram

233. What per cent of isolation is absorbed in the atmosphere by gasses, clouds and suspended solids?
(A) 19% (B) 17%
(C) 15% (D) 13%

234. Oxygen and ozone at high levels absorb most of which radiation:
(A) Ultra violet radiation
(B) Gama radiation
(C) Blue radiation (D) All of these

235. The ratio of the amount of radiation reflected to the amount received on a surface is termed as:
(A) Albedo (B) Radiation
(C) Reflection (D) All of these

236. The ability to admit most of the insolation yet retard losses by reradiation from the earth's surface commonly is known as:
(A) Greenhouse effect
(B) Atmosphere effect
(C) Both of these (D) None of these

237. The highest per cent of albedo is found:
(A) Desert (B) Tropical forest
(C) Deciduous forest
(D) Coniferous forest

238. The instrument which measure sunshine and radiation is:
(A) Total radiation pyranometer
(B) Net radiometer
(C) Foster switch sunshine duration sensor
(D) All of these

239. The ratio of the mass of water vapour actually in the air to a unit of mass of air, including the water vapour is called:
(A) Specific humidity
(B) Relative humidity
(C) Both of these (D) None of these

240. The ratio of the amount of water vapour actually in a volume occupied by air to the amount the space could contain at saturation is called:
(A) Relative humidity
(B) Specific humidity
(C) Both of these (D) None of these

241. Early morning hours when temperatures are low the relative humidity will be:
(A) More (B) Less
(C) Not change (D) Unknown

242. The visible aggregates of water droplets, ice particles, or a mixture of both along-with varying amounts of dust particles is called:
(A) Fog (B) Frost
(C) Clouds (D) All of these

243. The clouds are a cirriform layer or patch of small white flakes or tiny globules arranged in distinct group or lines is called:
(A) Cirrostratus (B) Cirrocumulus
(C) Altocumulus (D) Altostratus

244. The clouds is a thin white veil of cirrus is called:
(A) Cirrocumulus (B) Cirrostratus
(C) Altocumulus (D) Altostratus

245. The middle clouds predominate at height ranging from 2 to 8 km above ground levels is called:
(A) Altocumulus (B) Altostratus
(C) Both of these (D) None of these

246. The banner clouds above peaks are of which type clouds is called:
(A) Cirrocumulus (B) Cirrostratus
(C) Altocumulus (D) Altostratus

247. When the cloud layer lowers to become some what thicker and darker and falling rains or snow obscures its base it is called:
(A) Cirrocumulus (B) Cirrostratus
(C) Altocumulus (D) Nimbostratus

248. The clouds are broken into fragments by the wind it is called:
(A) Cirrocumulus (B) Cirrostratus
(C) Fractostratus (D) Altostratus

249. The clouds are dense, dome shaped and have flat base is called:
(A) Cumulus (B) Cirrostratus
(C) Altocumulus (D) Altostratus

250. The heavy showers of rain, snow or hail, often accompanied by lighting and thunder occurs by:
(A) Cirrocumulus (B) Cumulonimbus
(C) Altocumulus (D) Altostratus

251. The optimum temperature for retting of sunhemp is:
(A) 14-21⁰C (B) 21-27⁰C
(C) 30-34⁰C (D) 35-38⁰C

252. Less than 30 per cent rains occurs by:
(A) South-west monsoon
(B) North-East monsoon
(C) Both of these (D) None of these

253. North-East monsoon also called:
(A) Post monsoon period
(B) Post rainy season
(C) Both of these (D) None of these

254. A counter clock wise rotation of wind in northern hemisphere is:
(A) Cyclone (B) Anticyclone
(C) Rainfall (D) Fog

255. A counter clock wise rotation of wind in southern hemisphere is:
(A) Cyclone (B) Anticyclone
(C) Rainfall (D) Fog

256. Wind speed measured by:
(A) Windvane (B) Anemometer
(C) Both of these (D) None of these

257. The barometric pressure is highest at its centre and decreased out ward is indication of:
(A) Cyclone (B) Anticyclone
(C) Rainfall (D) Fog

258. The common pressure that initiates thunder storms development are:
(A) Heating and convection in moist air over warm land surfaces
(B) Passage of cold, moist air over warm water
(C) Radiational cooling at upper levels and Advection of cold air aloft
(D) All of these

259. The height of a thunderstorms is related to:
(A) Latitude (B) Season
(C) Both of these (D) None of these

260. Thunderstorms acused by surface heating over land are most common in summer in the:
(A) Afternoon (B) Early morning
(C) Both of these (D) None of these

261. Most violent storms of the lower troposphere is:
(A) Tornado (B) Watersprout
(C) Both of these (D) None of these

262. The moisture index is more in:
(A) Perhumid (B) Humid
(C) Semiarid (D) Arid

263. The moisture index is less in:
(A) Perhumid (B) Humid
(C) Semiarid (D) Arid

264. Human health and comfort suggest another possible approach to defining:
(A) Climatic types (B) Relative humidity
(C) Temperature (D) None of these

265. The summation of all atmospheric events an infinite time span does not changes is called:
(A) Climate (B) Weather
(C) Both of these (D) None of these

266. The periods of atmospheric changes on the order of millions of years is:
(A) Paleoclimates (B) Taiga climate
(C) Polar climate (D) Ocean climate

267. The mean monthly temperatures are below 0⁰C (snow, ice or barren rocks covers such areas) and vegetation is entirely lacking is called:
(A) Paleoclimates (B) Taiga climate
(C) Polar climate (D) Ocean climate

268. The key element in the water budget in which is also the link between moisture and energy exchange is:
(A) Evaporation (B) Transpiration
(C) Evapo-transpiration
(D) None of these

269. Evapotraspiration can be measured by:
(A) Lysimeter
(B) Evapotranspirometer
(C) Both of these (D) None of these

270. The frost occurs on clear nights with a temperature inversion and usually results in formation of ice crystals on bold objects is called:
(A) Advection frost
(B) Radiation frost
(C) Black frost
(D) None of these

271. Advection frost is also called:
(A) Radiation frost (B) Advection frost
(C) Black frost (D) None of these

272. The frost which results when the temperature at the surface in an air mass is below freezing is called:
(A) Radiation frost (B) Advection frost
(C) Black frost (D) None of these

273. The deficit that results when soil moisture is in-sufficient to meet the demands of potential evapotranspiration is:
(A) Drought (B) Rainfed
(C) Dryland (D) None of these

274. Drought may be classified as:
(A) Permanent drought may be associated with arid climate
(B) Seasonal drought, which occurs in climate with distinct amount periods of dry weather
(C) Drought due to variability of precipitation
(D) All of these

275. Combating the drought effects are:
(A) Growing of less water requiring crops
(B) Decreasing the water need of crops
(C) Increasing the water supply
(D) All of these

276. When the annual rainfall is less than the normal over an area for a prolonged period is called:
(A) Meteorological drought
(B) Hydrological drought
(C) Agricultural drought
(D) None of these

277. When a prolonged drought results in drying of the available water sources like streams, tanks, reservoir, well etc. is called:
(A) Meteorological drought
(B) Hydrological drought
(C) Agricultural drought
(D) None of these

278. Shortage of supply of water to meet the demand for crop maturity is called:
(A) Meteorological drought
(B) Hydrological drought
(C) Agricultural drought
(D) None of these

279. The ideal temperature for tuberization in potato is:
(A) 14^0C (B) 18^0C
(C) 21^0C (D) 30^0C

280. During snowfall, the trees:
(A) Do not respire
(B) Do not photosynthesis
(C) Show minimum life function
(D) Show maximum transpiration

281. The pollutants which are already present in nature, but are released in substantial amounts by man are known as:
(A) Quantitative pollutants
(B) Qualitative pollutants
(C) Degradable pollutants
(D) Primary pollutants

282. Stevenson screen is related to:
(A) Bacteriology
(B) Biotechnology
(C) Agrometeorology
(D) Microbiology

283. Atmosphere is essential for :
(A) Winds (B) Cloud-formation
(C) Weather phenomena
(D) All of these

284. Anemometer measures :
(A) Wind direction (B) Relative humidity
(C) Net radiation (D) Wind velocity

285. Potato tuber growth is stopped at what temperature?
(A) 40-42°C (B) 30-32°C
(C) 35-37°C (D) 38-40°C

286. The optimum cardinal tempeature point for germination of rice seed is :
(A) 20-25°C (B) 18-22°C
(C) 30-32°C (D) 37-39°C

287. In metropolitan cities the automobiles cause air pollution:
(A) 80% (B) 60%
(C) 50% (D) 100%

288. Huge quantity of sewage is dumped in a river. The BOD will:
(A) Increase
(B) Decrease
(C) Slightly increase
(D) Remain unchanged

289. Jet aeroplane releases pollutants in air is called:
(A) Smog
(B) Photo-chemical oxidants
(C) Aerosols
(D) Algae and bacteria

290. Fertilizer, pesticides and insecticides may cause:
(A) Air pollution (B) Water pollution
(C) Both (A) & (B) (D) None of these

291. Pollutants directly emanating from human activities are:
(A) Quantitative pollutants
(B) Qualitative pollutants
(C) Air pollutants
(D) Water pollutants

292. Atmospheric pollutants formed by human activities constitute of the total:
(A) 95% (B) 75%
(C) 25% (D) 0.05%

293. Most atmospheric pollutants do not rise above:
(A) 600 m (B) 1000 m
(C) 2000 m (D) 200 m

294. Aerosol is a system of colloidal particles having a size of:
(A) 5 μm
(B) 1 – 2 μm
(C) Less than 1 μm
(D) More than 10 μm

295. A secondary pollutants is:
(A) O_3 (B) CO
(C) CH_4 (D) Pb

296. A colourless, nonirritant, highly toxic gas that impairs respiration is:
(A) SO_2 (B) NO_2
(C) CO_2 (D) CO

297. Lead pollution in Indian cities is mainly due to:
(A) Lead pencils (B) Batteries
(C) Automobiles (D) Electronic gadgets

298. Pollutants likely to deplete ozone layer is:
(A) CO
(B) Nitrogen oxides
(C) Chloroflurocarbons
(D) Both (B) and (C)

299. Acid rain is caused by:
(A) Excess production of coal gas
(B) Excess release of CO_2 due to increasing combustion and respiration
(C) Excess release of SO_2 and NO_2 from burning fossil fuels
(D) Excess production of gaseous hydrocarbons

300. SO_2 pollution is indicated by destruction of:
(A) Climbers (B) Lichens
(C) Mosses (D) Grasses

301. Freom is:
(A) Ferrous sulphide pollutant
(B) Chloroflurocarbon
(C) Iron containing polycyclic hydrocarbons
(D) Metal fluoride

302. Maximum atmospheric NO_2 pollution is found in:
(A) Baroda (B) Chennai
(C) Nagpur (D) Mumbai

303. O_3 promotes:
(A) Oxidation of biochemicals
(B) Destruction of chlorenchyma
(C) Both (A) and (B)
(D) Growth of plants

304. Water soaked appearance of leaves due to sulphur dioxide pollution is caused by:
(A) Destruction of chlorophyll
(B) Damage to membranes
(C) Both (A) and (B)
(D) Corrrosive action over metals

305. In addition to SO_2, metals and textiles are damaged by:
(A) Nitrogen oxides (B) Fluorides
(C) Carbon monoxide (D) Aldehydes

306. Photochemical reaction of nitrogen oxides with unsaturation hydrocarbons produces:
(A) Methyl isocyanate
(B) Phosgene
(C) Benzpyrene
(D) Peroxy-acyl nitrates

307. Level of atmospheric CO_2 has increased from:
(A) 290 to 340 ppp (B) 250 to 300 ppm
(C) 300 to 350 ppm (D) 330 to 350 ppm

308. CO_2 content of the atmosphere is likely to double by the year:
(A) 2001 (B) 2010
(C) 2020 (D) 2050

309. Gravity setting tanks/porous filters are used to remove particles larger than:
(A) 2 μm (B) 50 μm
(C) 20 μm (D) 10 μm

310. Electronic precipitators are employed to remove particulate pollutants of the size of:
(A) 5-20 μm (B) 20-50 μm
(C) 50 μm & larger (D) 0.5-5.0 μm

311. B.O.D. is related to:
(A) Detergents
(B) Putrescibility
(C) Inorganic pollutants
(D) Organic pollutants

312. A persistent insecticide is:
(A) Organophosphates
(B) Carbamates
(C) Thiocarbamates
(D) Chlorinated hydrocarbons

313. An organoclorine insecticides is:
(A) DDT/aldrin (B) Malathion
(C) Tetraethyl pyrophosphate
(D) Thiocarbate

314. Terrestrial life would become impossible with:
(A) Change in orbit of earth
(B) Change on atmosphere permitting all solar radiations to reach lithosphere
(C) Disappearance of moon
(D) Decrease in mean annual temperature by 10^0C

315. Disease aggravated by pollution is:
(A) Scurvy (B) Rheumation
(C) Haemophilia (D) Bronchitis

316. 'Ozone –hole' means:
(A) Large sized hole in the ozone layer
(B) Thinning of the ozone layer
(C) Small holes scattered in the ozone layer
(D) Thickening of ozone in the ozone layer

317. Ozone layer is getting depleted because of:
(A) Excessive use of automobiles
(B) Excessive formation of industrial units
(C) Excessive use of man made compounds containing both fluorine and chlorine
(D) Excessive deforestation

318. Viruses, which cause hepatitis, are transmitted through:
(A) Air (B) Water
(C) Food (D) Personal contact

319. Vectors can be defined as:
(A) Animals carry the infecting agents from risk person to another healthy person
(B) Microorganisms which cause many diseases
(C) Infected person
(D) Diseased plants

320. Which of the following is a recently originated problem of environment?
(A) Ozone layer depletion
(B) Green house effect
(C) Global warming
(D) All of these

321. When we breathe in air, nitrogen also goes inside along with oxygen. What is the fate of this nitrogen?
(A) It moves alongwith oxygen into the cells
(B) It comes out with the CO_2 during exhalation
(C) It is absorbed only by the nasal cells
(D) Nitrogen concentration already move in the cells so it is not at all absorbed

322. One of the following processes is not a step involved in the water cycle operating in nature:
(A) Evaporation (B) Temperature
(C) Precipitation (D) Photosynthesis

323. The term "water – pollution" can be defined in several ways, which of the following statements does not give the correct definitions?
(A) The addition of undesirable substances to water bodies
(B) The removal of desirable substances from water bodies
(C) A change in pressure of the water bodies
(D) A change in temperature of the water bodies

324. Which step is not involved in the carbon-cycle?
(A) Photosynthesis (B) Transpiration
(C) Respiration (D) Burning of fossil fuels

325. South-west monsoon contributes how much per cent of rainfall?
(A) 50 % (B) 75 %
(C) 80 % (D) 95 %

326. A source of nonbiological oxygen is:
(A) Photodissociation
(B) Photolysis
(C) Formation of fats from carbohydrates
(D) Both (B) and (C)

327. A process that occurs without loss or gain of heat is:
(A) Coriolis effect (B) Eustatic effect
(C) Adiabatic (D) All of these

328. Which area has four seasons:
(A) Tropics (B) Monsoonic
(C) Temperate (D) Arctic

329. Praying Mantis is green coloured predator insect that shows mimicry of the type:
(A) Concealing-protective
(B) Warning-protective
(C) Alluring-aggressive
(D) Concealing-aggressive

330. Male Atlas Moth has huge antennae that can detect pheromone released by female moth from a distance of:
(A) 1 km (B) 2 km
(C) 5 km (D) 8 km

331. The biological amplication of DDT in the various trophic levels is known as:
(A) Green house effect
(B) Biomagnification
(C) Entraphication
(D) Pollution

332. Taj may be destroyed by:
(A) Flood in Yamuna
(B) Temperature mediated spoilage of marble
(C) Air pollutants from Mathura refinery
(D) All of these

333. Smog occurs in places of:
(A) Excess SO_2
(B) Low temperature
(C) Excess NH_3 (D) High temperature

334. Major pollution causing agents is:
(A) Hydrocarbons (B) Animals
(C) Human (D) Birds

335. Pollution is:
(A) Removal of top soil
(B) Release of toxic/undesirable materials in environment
(C) Conservation of energy
(D) All of these

336. Effect of pollution is most marked on:
(A) Natural balanced of nature
(B) Natural geochemical cycles
(C) Natural flora of place
(D) All of these

337. Chief sources of water and soil pollution is:
(A) Mining
(B) Thermal power plant
(C) Agroindustry (D) All of these

338. Ultimate environment hazards to mankind is due to:
(A) Nuclear pollution
(B) Water pollution
(C) Air pollution
(D) Noise pollution

339. Air pollution is not caused by:
(A) Pollen grains
(B) Hydroelectric power
(C) Industries (D) Automobiles

340. Pollution is controlled by:
(A) Banning atomic blasts
(B) Use of electrically operated automobiles
(C) Sewage treatment
(D) All of these

341. Carbon monoxide is harmful to human being as it is:
(A) Carcinogenic
(B) Antagonistic to CO_2
(C) With higher affinity for haemoglobin as compared to oxygen
(D) Destructive to O_3

342. Fish die in water bodies polluted by sewage due to:
(A) Pathogens
(B) Clogging of gills by silt
(C) Reduction in oxygen
(D) Foul smell

343. Lichens indicate pollution by:
(A) O_3 (B) SO_2
(C) NO_2 (D) CO

344. Drained sewage has B.O.D.:
(A) More than that of water
(B) Less than that of water
(C) Equal to that of water
(D) None of these

345. Which one is the most important causative pollutants of soil?
(A) Glass junk (B) Iron junk
(C) Detergents (D) Plastics

346. BOD is:
(A) Biological oxygen deficit
(B) Biosphere oxygen demand
(C) Biological oxygen demand
(D) None of these

347. Pollution can bring about change in:
(A) Biogeochemical cycling
(B) Abiotic environment
(C) Biotic environment
(D) None of these

348. Which one is the major reason for pollution in big cities?
(A) Fossil fuel (B) Acid rain
(C) Heat dispersion (D) None of these

349. World environment day is:
(A) 28th Feb. (B) 5th June
(C) 7th Aug. (D) 10th April

350. Biodegradable pollutants is:
(A) Plastic (B) Asbestos
(C) Sewage (D) Mercury

351. Water pollution is due to:
(A) Sulphur dioxide
(B) Carbon dioxide
(C) Oxygen
(D) Industrial discharges

352. The nitrogen molecules present in air can be converted into nitrates and nitrites by:
(A) A biological process of nitrogen fixing bacteria present in soil
(B) A biological process of carbon fixing factor present in soil
(C) Any of the industries manufacturing nitrogenous compounds
(D) The plants used as cereal crops in field

353. Phosphate pollution is caused by:
(A) Weathering of phosphate rocks only
(B) Agricultural fertilizers
(C) Phosphate rocks and sewage
(D) Sewage and agricultural fertilizers

354. Chief air pollutant which is likely to deplete ozone layer is:
(A) Sulphur dioxide
(B) Carbon dioxide
(C) Carbon monoxide
(D) Nitrogen oxides and fluorocarbons

355. Minamata disease is due to pollution of:
(A) Organic waste into drinking water
(B) Oil spill in water
(C) Industrial waste mercury into fishing water
(D) Arsenic into the atmosphere

356. Cyclone collector is used for minimising:
(A) Radioactive pollution
(B) Air pollution
(C) Noise pollution
(D) Water pollution

357. Which abiotic components of environment contributes in various ways to fulfillment of human needs?
(A) Air
(B) Water
(C) Soil
(D) All of these

358. Environment consists of:
(A) Sum total of biotic factors
(B) Sum total of abiotic factors
(C) Both (A) and (B)
(D) Both (A) and (B) minus humans

359. Which constitutes environmental biotechnology?
(A) Waste treatment
(B) Waste disposal
(C) Technologies generating less waste
(D) Both (A) and (B)

360. Which constitutes 'end of pipe' technology?
(A) Waste treatment
(B) Waste disposal
(C) Technologies generating less waste
(D) Both (A) and (B)

361. Which constitutes 'front of pipe' technology?
(A) Waster treatment
(B) Waste disposal
(C) Technologies generating less waste
(D) Both (A) and (B)

362. Recalcitrants xenobiotic compounds are highly stable due to the presence of which substitution groups?
(A) Halogens and carbamyl groups
(B) Nitro- and sulphonate groups
(C) Amino- and methoxy groups
(D) All of these

363. Bioaccumulation/biomagnifications may occurs due to:
(A) Xenobiotic compounds are continuously taken up from the environment & accumulated in the lipid deposits of the body
(B) Consumption of these organisms by other organisms leads to further accumulation of xenobiotic compounds in the bodies of latter
(C) The xenobiotic compounds are recalcitrant and lipid soluble
(D) All of these

364. Which diseases spread through contaminated water/food?
(A) Amoebiosis and typhoid
(B) Cholera and diarrhea
(C) Hepatitis (except B)
(D) All of these

365. Which is not an example of physical pollutants?
(A) Dust and mechanical stress
(B) Sound pollution and electrical storms
(C) Heavy metals
(D) Radiation and fires

366. Presence of coliform bacteria in water is indicative of contamination by mammalian excreta and suggestive of the presence of which pathogenic bacteria?
(A) *Vibrio cholera*
(B) *Salmonella*
(C) *Shigella*
(D) All of these

367. Which pollutants pose immediate threat to humans and is included in the black list?
(A) Oils and hydro carbons from crude oils
(B) Heavy metals, metaloides and compounds containing them
(C) Organophosphates, cyanides and asbestos
(D) All of these

368. Various approaches to waste treatment may be grouped into which categories?
(A) Anaerobic digestion
(B) Biofilters and aerobic digestion
(C) Landfill and burning
(D) All of these

369. Landfil sites can be useful in which ways?
(A) Reclamation of derelict sites to develop landscaped gardens, etc
(B) As a source of biogas
(C) As a source of organic fertilizers
(D) Both (A) and (B)

370. Which microorganisms is the commonly found inaerobic digestion systems?
(A) Fungi (B) Bacteria
(C) Protozoa (D) Algae

371. Bioremeddiation of inorganic contaminants is primarily based on:
(A) Animal species
(B) Microorganisms
(C) Plant species
(D) All of these

372. The water (leachate) from landfills may contain :
(A) COD value of 1000-2000
(B) Total solids content upto 6000-7000 mg/l
(C) Sodium upto 800 mg/l
(D) All of these

373. Which statements is correct about landfill management?
(A) The risk due to methane can be removed by burning or tapping
(B) The risks of fires, offensive odours and increased vector populations may be circumvented by covering the waste with soil on daily basis
(C) The risk of damage from, landfill leachate may be avoided by lining the pit with an impermeable material
(D) All of these

374. The various parameters used to determine the level of pollution in water bodies, usually, determine:

(A) Amounts of organic matter present in the water

(B) Amounts of important inorganic nutrients

(C) Numbers of bacteria per mililitre of waste water

(D) Both (A) and (B)

375. Treatment of the water proceeds through five stages; select the correct order of different stages of water treatment:

(A) Sludge treatment → preliminary treatment → primary treatment → secondary treatment → tertiary treatment

(B) Preliminary treatment → primary treatment → secondary treatment→ tertiary treatment → Sludge treatment

(C) Preliminary treatment → Sludge treatment → primary treatment secondary treatment → tertiary treatment

(D) Preliminary treatment → secondary treatment→ tertiary treatment → primary treatment → Sludge treatment

376. Which waste water treatments is an anaerobic process?

(A) Primary treatment

(B) Secondary treatment

(C) Tertiary treatment

(D) Sludge treatment

377. Preliminary treatment for waste water is necessary to prevent?

(A) Clogging of pipelines

(B) Damage to pump

(C) Interference with subsequent process

(D) All of these

378. Which is correct about aerobic digestion of waste water?

(A) Bacteria are the most common organisms

(B) Bacterial population may be more than 10^9 cells/ml

(C) Bacteria are responsible for the removal of about 80-90% of the BOD

(D) All of these

379. Practical application of microbes for xenobiotic degradation is facilitated by:

(A) Supply of sufficient nutrients or cometabolites

(B) Maintenance of the xenobiotic compounds at nontoxic levels

(C) Provision of microbial population or inoculums

(D) All of these

380. The quality of treated effluent from the digester depends mainly on :

(A) Sludge loading rate

(B) Organic loading rate

(C) Residence time

(D) All of these

381. NO_3^- is removed from water by the action of which bacteria?

(A) *Pseudomonas* (B) *Micrococcus*

(C) Alcaligenes (D) All of these

382. Sewage contains which pathogenic organisms?

(A) Bacteria causing typhoid, cholera, diarrhea

(B) Protozoa causing amoebic dysentery, giardiasis

(C) Viruses causing poliomyletis, hepatitis, gastroenteritis

(D) All of these

383. Protozoa are represented in waste waters by:

(A) Amoebal forms (B) Ciliates

(C) Flagellates (D) All of these

384. The most common type of protozoa found in waste water:

(A) Amoebal forms (B) Ciliates

(C) Flagellates (D) All of these

385. Which form of protozoa is most diverse group?

(A) Amoebal forms (B) Ciliates

(C) Flagellates (D) All of these

386. Efficiency of anaerobic digestion can be enhanced by:

(A) By using solid support to retain biomas

(B) By recycling the active biomass

(C) By limiting the biomass loss to a rate compatible with high population density in the digester

(D) Any of these

387. Bioremediation of organic contaminants is primarily based on:

(A) Microorganisms naturally present at the site

(B) Microbial inoculants developed in the laboratory and introduced at the site

(C) Suitable plants planted at the site

(D) All of these

388. Oil spills cause severe damage to the ecosystems and pose which threats?

(A) Ground water pollution due to percolation

(B) Air pollution due to exoporation

(C) Fire

(D) All of these

389. Which is not correct?

(A) The US Environment Protection Agency and the Exxon Campany used microorganisms to clean up Alaskan beaches contaminated by the Valdez oil spill

(B) In India, a consortium of bacterial species has been developed to combat oilspills and oily sludge; the inoculants is aptly called 'oilzapper'

(C) Inoculation with Oilzapper reduced oil-sludge contamination in soil to merely 0.5% in 360 days from the initial 13.41%

(D) Oilzapper has not been effective in field trials

390. The anaerobic digestion process involves a wide variety of organisms which digest the organic molecules, such as lipids, proteins, carbohydrates etc., into mainly :

(A) NO_3^- and PO_4 (B) CH_4 and CO_2

(C) SO_4^{--} and CO (D) All of these

391. Which heterotrophic bacteria are responsible for oxidation of organic molecules?

(A) Pseudomonas and Escherichia

(B) Streptococcus abd Salmonella

(C) Shigella and Aerobacter

(D) All of these

392. For phytoremediation of flyash, seedings of which of the following trees are established at abandoned flyash ponds/dams?

(A) Eucalyptus (B) Melia

(C) Poplar (D) All of these

393. Which is not correct about halocarbons?

(A) They contain different numbers of halogen atoms in the place of He atoms

(B) They are used as solvents, in condenser units of cooling systems, etc

(C) The C_1-C_2 haloalkanes escape into the atmosphere where they destroys the protective ozone layer

(D) Pesticides leach into water bodies, where many of them are subjected to biomagnifications

394. Which is correct about alkylbenzyl sulphonates?

(A) These are surface-active detergents

(B) Sulphonate group present at one end resists microbial degradation

(C) At present, alkylbenzyl sulphonates are biodegraded by ß–oxidation from their alkyl ends

(D) All of these

395. Phytoremediation may occur due to:

(A) Accumulation of the organic waste in plant tissues

(B) Translocation of the organic waste to leaf and volatilization from there

(C) The organic waste or in the rhizosphere itself

(D) Any of these

396. Continued exposure of microorganisms to xenobiotic compounds can often lead to the evolution of metabolic processes needed to wholly or partly degrade the xenobiotic. These capabilities may arise due to:

(A) Mutation

(B) Transfer of plasmid –borne genes

(C) Transformation

(D) Both (A) and (B)

397. Continued exposure of microorganisms to xenobiotic compounds can often lead to the evolution of metabolic processes needed to wholly or partly degrade the xenobiotic. These can be expected to modify the active site of an enzyme so that it has an increased affinity for the xenobiotic compounds?

(A) Mutation

(B) Transfer of plasmid –borne genes

(C) Transformation

(D) Both (A) and (B)

398. Alicyclic hydrocarbons are present naturally in:
(A) Wax from plants
(B) Microbial lipids
(C) Crude oils
(D) All of these

399. Which is not related to green-house gas?
(A) CH_2 (B) CO_2
(C) O_2 (D) CFC

400. Fluoride pollution mainly affects:
(A) Heart (B) Kidney
(C) Brain (D) Teeth

401. Biochemical oxygen demand measures:
(A) Industrial pollution
(B) Air pollution
(C) Dissolved O_2 needed by microbes to decompose organic waste
(D) Oxygen need to support microorganism growth

402. Smog is the product of:
(A) Smoke (B) Fog
(C) Soot (D) Both (A) and (B)

403. Green house effect is related to:
(A) Global warming
(B) Global afforestation
(C) Deforestation
(D) Both (A) & (C)

404. Increasing skin cancer and rate of mutation are the result of:
(A) Acid rain (B) Ozone depletion
(C) SO_2 pollution (D) CO_2 pollution

405. Which diseases is caused by eating fish inhabitating mercury contaminated water?
(A) Minamata disease
(B) Osteosclerosis
(C) Hashimoto disease
(D) Brain fever

406. Depletion of ozone layer is caused by:
(A) CO_2 (B) SO_2
(C) CFC (D) Both (B) and (C)

407. The atmosphere of metro cities is polluted by:
(A) Pesticides
(B) Automobile exhausts
(C) House-hold wastes
(D) Radioactive fallout

408. Which associates with the blood haemoglobin more rapidaly than oxygen?
(A) CO (B) CO_2
(C) SO_4 (D) NO

409. Which decibles of sound becomes hazardous moise pollution?
(A) >20 (B) >80
(C) >120 (D) >140

410. Which plants can be used as indicators of SO_2 pollution of air?
(A) Liverworts
(B) Hornworts
(C) Epiphytic lichens
(D) Fern

411. Attack of Asthama in certain persons may be due to:
(A) Inhalation of some airborne pollution
(B) Exposure to cold temperature
(C) Exposure to certain pollen grains
(D) All of these

412. Which gases contributes the maximum to the 'green – house effect'?
(A) CO_2 (B) CFC
(C) CH_4 (D) N_2O

413. Which poses the greatest risk?
(A) Nuclear plants
(B) Thermal power plants
(C) Biogas plants
(D) Hydroelectric plants

414. Which is an example of nonrenewable resources?
(A) Petroleum (B) Biological agents
(C) Natural gas (D) Both (A) and (C)

415. Maximum deposition of DDT will occur in:
(A) Phytoplankton (B) Fish
(C) Crab (D) Sea gull

416. Which green-house gases is being generated by agricultural fields?
(A) Sulphur dioxide
(B) SO_3
(C) Nitrous oxide
(D) Ammonia

417. Ozone day is observed on:
(A) September 16 (B) June 5
(C) May 1 (D) December 25

418. Which is an example of a renewable resource?

(A) Plants
(B) Animals
(C) Microorganisms
(D) All of these

419. Bioremediation of organic wastes is generally based on:

(A) Bacteria (B) Animal species
(C) Plant species (D) Both B and C

420. Ozone depletion in the stratosphere will cause:

(A) Gobal warming
(B) Increased sea level
(C) Increased incidence of skin cancer
(D) All of these

421. Which is most ecofriendly?

(A) Bacteria
(B) Animal species
(C) Plant species
(D) Both B and C

422. Ultra-violet light causes:

(A) Formation of pyrimidines
(B) Sticky metaphases
(C) Photodynamics action
(D) Destruction of hydrogen bonds between complementary DNA strands

423. Sulphur dioxide affects:

(A) Cell wall
(B) Plasmodesmata
(C) All membrane systems
(D) Nucleus

424. Fluoride pollution produces necrosis or chlorosis of:

(A) Leaf margin and tip
(B) Stem
(C) Petiole but not lamina
(D) Midrib

25. Air pollutants photochemical oxidants include:

(A) Nitrous oxide, nitric acid and nitric oxide
(B) Oxygen, chlorine and nitric acid
(C) Ozone, peroxyacetyl nitrate and aldehydes
(D) Carbon monoxide and sulphur dioxide

426. Ultraviolet radiations are lethal because of inactivation of:

(A) Proteins, nucleic acids and pigments
(B) Carbohydrates, fats and vitamins
(C) Minerals, air and water
(D) Water, carbon dioxide ad oxygen

427. Nonionising radiations with specific biological effects are:

(A) UV radiations (B) Beta-ray
(C) Gamma rays (D) X-rays

428. Carbon monoxide is pollutant as it:

(A) Inactibates nerves
(B) Inhibits glycolysis
(C) Combines with oxygen
(D) Combines with haemoglobin

429. Air pollutants that produce photochemical oxidants include:

(A) CO_2, CO and SO_2
(B) Nitrous oxide, nitric oxide and nitric acid
(C) Oxygen, clorine and nitric acid
(D) Ozone, chlorine and sulphur dioxide

430. Most abundant water pollution is:

(A) Detergents
(B) Pesticides
(C) Industrial waste
(D) Ammonia

431. Atmospheric pollutants is:

(A) CO_2 (B) CO
(C) O_2 (D) N_2

432. Burning of fossil fuels is the main source of pollution:

(A) Nitrogen oxide (B) Nitric oxide
(C) Nitrous oxide (D) Sulphur dioxide

433. Modes of controlling pollution in large cities include:

(A) Cleanliness and less use of insecticides
(B) Proper disposal of organic wastes, sewage and industrial effluents
(C) Broader roads and shifting of factories out of the residential areas
(D) All of these

434. Domestic waste constitutes:

(A) Nonbiodegradable pollution
(B) Biodegradable pollution
(C) Effluents
(D) Air pollution

435. SO_2 pollution is indicated by:
(A) Deschampsia (grasses)
(B) Sphagnum (mosses)
(C) Usnea (lichens)
(D) Cucurbita (climbers)

436. Increased asthmatic attacks in certain seasons are related to:
(A) Inhalation of seasonal pollen
(B) Eating of seasonal vegetables
(C) Low temperature
(D) Wet and dry environment

437. Which one of the following dissolves more rapidly in blood haemoglobin than oxygen?
(A) Ozone
(B) Nitrous oxide
(C) Sulphur dioxide
(D) Carbon monoxide

438. Which of the following forms a toxic substance in blood by combining with haemoglobin?
(A) Carbon dioxide
(B) Carbon monoxide
(C) Oxygen
(D) Methane

439. Ultraviolet radiation from sunlight causes a reaction that produces:
(A) Fluorides
(B) Carbon monoxide
(C) Sulphur dioxide
(D) Ozone

440. Noise pollution is measured in:
(A) Hertz (B) Fathoms
(C) Nanometres (D) Decibels

441. Most hazardous metal pollutant of automobile exhausts is:
(A) Mercury (B) Cadmium
(C) Lead (D) Copper

442. Pollution is rising due to:
(A) Rains
(B) Research institutes
(C) Population explosion
(D) Automobiles and industries

443. Pollution can be checked through:
(A) Reducing discharge of waste
(B) Increasing industrial development
(C) Throwing wastes in open spaces
(D) All of these

444. Continuous sewage flow into a stream would lead to:
(A) Eutrophication
(B) Algal bloom
(C) Depletion of oxygen
(D) Increase in temperature

445. Sewage infection of water will be demonstrated by cysts of:
(A) Entamoeba
(B) Escherichia coli
(C) Pseudomonas
(D) Leishmania

446. Measurement of rate of oxygen utilization by a unit volume of water over a period of time is to measure:
(A) Fermentation
(B) Biogas generation
(C) Biosynthetic pathway
(D) Biological oxygen demand

447. Sudden mass death of fish is more likely to occur in:
(A) Eutrophic lake
(B) Mesotrophic lake
(C) Oligotrophic lake
(D) Oxalotrophic lake

448. Which pollution is causes jaundice?
(A) Water (B) Air
(C) Land (D) Thermal

449. Which pollution causes leaf curling?
(A) CO (B) H_2S
(C) O_3 (D) SO_2

450. Water pollution causes:
(A) Increased photosynthesis
(B) Increased deoxygenation and turbidity
(C) Decreased turbidity
(D) Increased oxygenation

451. Naural sink of stratospheric ozone is:
(A) Sulphuric dioxide
(B) Sulphur fluxes of oceans
(C) Freons
(D) HCFCs

452. Ozone hole is maximum over:
(A) Europe
(B) Antarctica
(C) India
(D) Africa

453. In coming years, skin related disorders will become more common due to:
(A) Excessive use of detergents
(B) Water pollution
(C) Air pollution
(D) Depletion of ozone layer

454. Common indicator organism of water pollution is:
(A) *Escherichia coli*
(B) *Cholera vibrio*
(C) *Salmonells typhi*
(D) *Entamoeba histolytica*

455. Component of smog is:
(A) O_3
(B) PAN
(C) PPN and PBN
(D) Both (A) & (B)

456. Indiscriminate use of fertilizers causes:
(A) Air pollution (B) Water pollution
(C) Land pollution (D) All of these

457. Most harmful type of environmental pollutants are:
(A) Human organic wastes
(B) Natural nutrients in excess
(C) Waste animal feed
(D) Nonbiodegradable chemicals

458. Which is not dangerous for life?
(A) Nuclear blast (B) Deforestation
(C) Ozone layer (D) Biopollutants

459. Spraying DDT on crops shall produce pollution of:
(A) Air and water
(B) Air and soil
(C) Crops and air
(D) Soil and water

460. Sewage water is purified by:
(A) Microorganisms
(B) Light
(C) Fishes
(D) Aquatic plants

461. Sewage treatment in which a portion of decomposer bacteria present in waste is recycled into the beginning of the process:
(A) Cyclic treatment
(B) Activated sludge treatment
(C) Primary treatment
(D) Tertiary treatment

462. Ozone hole refers to:
(A) Hole in ozone layers
(B) Reduction in thickness of ozone layer in stratosphere
(C) Reduction of thickness of ozone in troposphere
(D) Increase concentration of ozone

463. Photochemical smog is related to pollution of:
(A) Air (B) Water
(C) Soil (D) Nostoc

464. Among pollutants which has the most lasting effect:
(A) Carbon monoxide
(B) Pesticides (C) Sulphur dioxide
(D) Smoke from chimneys

465. Pollution indicators plants:
(A) Are resistant to pollution
(B) Can purify the atmosphere
(C) Are very sensitive to pollutants
(D) Can indicate onset of rains

466. Depletion of ozone layer is due to:
(A) Oxides of nitrogen
(B) Oxides of carbon
(C) Oxides of sulphur
(D) None of these

467. Use of lichens in case of pollution is:
(A) Treatment of pollutant water
(B) Bioindicators of pollution
(C) Promote pollution
(D) Lichens have no relation with pollution

468. Peeling of ozone umbrella is due to:
(A) CFCs (B) PAN
(C) CO_2 (D) Coal burning

469. Anxiety and stress are caused by:
(A) Air pollution
(B) Noise pollution
(C) Water pollution
(D) Nuclear pollution

470. Ozone layer of stratosphere requires protection from indiscriminate use of:
(A) Fungicides, insecticides, bactericides and medicines
(B) Aerosols and high flying jets
(C) Atomic explosions and industrial wastes
(D) Baloons and turbo-prop aeroplanes

471. Environmental pollution affects:
(A) Biotic components
(B) Plants only
(C) Man only
(D) Biotic and abiotic components of environment

472. Water pollution is due to:
(A) Agricultural discharges
(B) Sewage and other wastes
(C) Industrial effluents
(D) All of these

473. Carbon monoxide kills because it destroys:
(A) Haemoglobin (B) Phytochrome
(C) Cytochrome (D) Both (A) and (B)

474. Water is often treated with chlorine to:
(A) Increase oxygen content
(B) Kill germs (C) Remove hardness
(D) Remove suspended particles

475. Water blooms are formed by:
(A) Lemna (B) Hydrilla
(C) Water Hyacinth
(D) Planktonic algae

476. DDT concentration recorded in human body has been:
(A) 0.3 – 0.7 ppm (B) 1.3 – 3.1 ppm
(C) 13 – 31 ppm (D) 30 - 70 ppm

477. Mottling of teeth is due to presence of an element in drinking water:
(A) Mercury (B) Fluorine
(C) Boron (D) Chlorine

478. Which one of the following is not pollutant?
(A) CO_2 (B) CO
(C) SO_2 (D) NO_2

479. As it passes into food chain, the concentration of DDT:
(A) Remains same (B) Decreases
(C) Increases (D) Unpredictable

480. NEERI is:
(A) National Environmental Engineering Research Institute
(B) National Ecological and Environmental Research Institute
(C) National Ethological and Ecological Research Institute
(D) National Eugenics and Ecological Research Institute

481. A person has impaired nervous system and signs of madness due to continued intake of metal contaminated water. The metal is:
(A) Mercury (B) Calcium
(C) Manganese (D) Lead

482. Wildsterben is an effect of:
(A) Ozone
(B) Nitrogen oxide and sulphur oxide
(C) Acid rain
(D) All of these

483. Waldsterben is:
(A) Decline in forests
(B) Development of fog
(C) Harm to sculptures, status & buildings
(D) All of these

484. Ozone is:
(A) Pale blue
(B) Highly reactive
(C) With penetrating odour
(D) All of these

485. An ozone depleting chemical which is used in fire-extinguishers is:
(A) Nitrogen oxide
(B) Methyl bromide
(C) Halon
(D) Chlorofluorocarbon

486. Smoked foods are not favoured now because of the possible toxicity of:
(A) Benzpyrene
(B) Nitrogen oxides
(C) Sulphur dioxide
(D) All of these

487. Benzpyrene is an ingredient of:
(A) Automobile exhausts
(B) Wood smoke and tobacco smoke
(C) Coal tar
(D) All of these

488. Smoke or dust dispersed in air is called:
(A) Fog (B) Haze
(C) Mist (D) Smog

489. Mist is fog with horizontal visibility of:
(A) 100 m (B) 200 m
(C) 500 m (D) 1000 m

490. The term green house effect was given by:
(A) Fourier (B) Tyndall
(C) Arrhenius (D) Linnaeus

491. Polychlorinated biphenyls are chemicals useful as:
(A) Fire – resistant
(B) Fertilizers
(C) Pesticides
(D) Germicides

492. Use of polychlorinated biphenyl is restricted because of their:
(A) Extreme toxicity
(B) Persistent toxicity
(C) Biomagnification
(D) Both (B) and (C)

493. Ecofriendly detergents do not contain:
(A) Phosphates and bleaches
(B) Alcohols
(C) Water softners (D) Perborates

494. Sudden rattling of doors and windows is due to:
(A) Earthquake
(B) Sonic boom
(C) Movements of crustal layers of earth
(D) Slipping of glaciers on the nearby mountain

495. Plantation of rows of trees and shrubs along roads, rails, around residential complexes, educational institutes, etc is called:
(A) Green muffler (B) Shelter belt
(C) Wind break (D) Both (B) & (C)

496. Exposure of UV radiations causes:
(A) Damage of eye sight
(B) Sunburn
(C) Skin cancer
(D) All of these

497. Third pollution is:
(A) Controlled tipping
(B) Land or landscape pollution
(C) Noise pollution
(D) UV pollution

498. Sewage is broken down in:
(A) Primary treatment
(B) Secondary treatment
(C) Tertiary treatment
(D) All of these

499. Tipping is a technique of:
(A) Sewage disposal
(B) Waste disposal
(C) Smoke treatment
(D) Effluent

500. Photocatalysis is useful in breakdown of:
(A) Acids and alkalies
(B) Chemicals containing heavy metals
(C) Toxic chemicals
(D) All of these

501. Chlorination of treated waste water required for:
(A) Killing all microbes
(B) Removing suspended materials
(C) Complete breakdown of organic matter
(D) All of these

502. Excess nitrate present in drinking water, vegetables and fruits shall result in:
(A) Methaemoglobinaemia
(B) Mottling of teeth
(C) Minamata disease
(D) Itai-itai

503. Bag filters are used for removal of pollutants from:
(A) Exhaust emissions
(B) Effluent water
(C) Sewage water
(D) Both (B) and (C)

504. Thermal plants have installed:
(A) Cyclone collector
(B) Porous filters
(C) Electrostatic precipitators
(D) Adsorbents

505. Substitute of chlorofluorocarbon is:
(A) Hydrofluorocarbon
(B) Hydrochlorofluorocarbon
(C) Fluorocarbon
(D) Both (A) and (B)

506. Smog that occurs due to low temperature and has reducing environment is:
(A) Photochemical smog
(B) Los Angeles smog
(C) Nonsulphurous smog
(D) London smog

507. Green audit is:
(A) Estimation of crop yield and crop area
(B) Estimation of forested area and pasture land
(C) Checking of impact of an establishment over environment
(D) Effect of environment over vegetation of an area

508. Runoff from fields is:
(A) Point source pollution
(B) Line source pollution
(C) Area source pollution
(D) Diffuse source pollution

509. Stick and leaf insects resembles the surroundings. They undertake mimicry of the type:
(A) Concealing-protective
(B) Warning – protective
(C) Alluring-aggressive
(D) Concealing-aggressive

510. Nitrogen oxides released in the stratosphere cause ozone depletion due to:
(A) Catalysing conversion of ozone oxygen
(B) Binding ozone
(C) Reacting with ozone
(D) All of these

511. A succession that does not follow the natural course due to human activity is:
(A) Subsere (B) Plagiosere
(C) Prisere (D) Asere

512. In foggy and hazy environment, birds are able to locate position of sun and their path by means of:
(A) Infra-red wavelength
(B) Ultra-violet radiations
(C) Magnetic system
(D) Pheromones

513. The amount of biogenic nutrients present in the abiotic environment is called:
(A) Standing state
(B) Standing crop
(C) Detritus
(D) Reserve nutrient

514. Carbon dioxide is being added to the cycling pool by:
(A) Respiration and combustion
(B) Sedimentation of carbonates
(C) Precipitation, skeletons & fossilization
(D) Both (B) and (C)

515. A source of nonbiological carbon dioxide is:
(A) Volcanic eruptions
(B) Hot springs
(C) Weathering
(D) All of these

516. Use of green plants and their associated microorganisms, soil amendments, and agronomic practices to remove, contain or render harmless environmental contaminants is known as:
(A) Phytoremediation
(B) Phytostabilization
(C) Both (A) & (B) (D) None of these

517. Use of plants to sequester contaminants is cell wall lignins of root tissues, or to promote its binding to humus is known as:
(A) Phytoremediation
(B) Phytostabilization
(C) Both (A) & (B) (D) None of these

518. Suitable temperature for tuberization in potato is:
(A) 12-15^{0}C (B) 18-20^{0}C
(C) 25-27^{0}C (D) 30-32^{0}C

519. The atmosphere of the earth is heated by radiations which are mainly:
(A) Radiated by the sun
(B) Re-radiated by land
(C) Re-radiated by water
(D) Re-radiated by land and water

520. If there were no atmosphere around the earth, the temperature of the earth will:
(A) Increase
(B) Go on decreasing
(C) Increase during day and decrease during night
(D) Be unaffected

521. What would happen, if all the oxygen present in the environment is converted to ozone:
(A) We will be protected more
(B) It will become poisonous and kill living forms
(C) Ozone is not stable, hence it will be toxic
(D) It will help harmful sun radiation to reach earth & damage many life forms

522. The two forms of oxygen found in the atmosphere are:
(A) Water & Ozone
(B) Water & Oxygen
(C) Ozone and Oxygen
(D) Water and Carbon dioxide

523. Rainfall pattern depend on:
(A) The underground watertable
(B) The density pattern of human population in an area
(C) The number of water bodies in an area
(D) The prevailing season in an area

524. Among the given options, which one is not correct for the use of large amount of fertilizers and pesticides?
(A) They are eco-friendly
(B) The fields barren after some time
(C) The adversely affect the useful component from the soil
(D) They destroy the soil fertility

525. Which of the following ranges of temperature is ideal for wheat seed germination?
(A) 25-30^{0}C (B) 20-25^{0}C
(C) 15-20^{0}C (D) 10-15^{0}C

526. Which is not the cause of heavy rainfall?
(A) Flood
(B) Destruction of crops
(C) Drought
(D) Soil erosion

527. Desert regions receive the rainfall:
(A) More than 50 cm
(B) Less than 50 cm
(C) 80-100 cm
(D) None of these

528. High velocity wind whirls are called:
(A) Monsoon (B) Cyclones
(C) Earthquake (D) None of these

529. Which is the effect of anticyclone?
(A) Pleasant weather
(B) Heavy rainfall
(C) Hail
(D) Snow fall

530. The formation of upholds by horizontal compression is called:
(A) Anticyclones (B) Synclines
(C) Rift valley (D) None of these

531. For converting gifts of nature into a resource, man requires:
(A) Skill and technology
(B) Machines
(C) Capital and labour
(D) All of these

532. Which of the following is not a community resources?
(A) Fruit orchard (B) A village pond
(C) Burial (D) Play ground

533. Which one of the following is a natural resources:
(A) An apartment (B) A river
(C) A factory (D) A car

534. Rajasthan and Gujarat are the store houses of:
(A) Solar and wind energy
(B) Coal and iron ore
(C) Tidal energy (D) Hydel energy

535. Which type of resource is solar energy?
(A) Exhaustive resource
(B) Soil resource (C) Biotic resource
(D) Non-exhaustive resource

536. Which one of the following type of resource is iron ore?
(A) Renewable (B) Biotic
(C) Flow (D) Non renewable

537. Under which of the following type of resource, the tidal energy can be put?
(A) Exhaustible (B) Human-made
(C) Abiotic (D) Non-exhaustible

538. Which one of the following types of resources is forest resources?
(A) Biotic (B) Non-renewable
(C) Stock (D) Individual

539. Between 1951 and 1980, according to Forest Survey of India, how much of the forest land was brought under farming?
(A) 27200 sq. km
(B) 22200 sq km
(C) 26200 sq km
(D) 28300 sq km

540. When was the maximum damage of forest cover inflincted in India?
(A) After independence
(B) While developing commercial activities
(C) During mughal period
(D) During British period

541. Himalayan Yew is in trouble because of:
(A) Over- exploitation
(B) Spread of diseases
(C) Fuel-wood collection
(D) Grazing

542. When was the Indian Wildlife Protection Act implemented?
(A) 1962 (B) 1972
(C) 1982 (D) 1992

543. Which one of the following is not the aim of Indian Wildlife Protection Act?
(A) Encourage hunting
(B) Establish Wildlife sanctuaries
(C) Initiate projects for saving specific animals
(D) Legal protection to the habitats of animals

544. Which one of the following conservation strategies does not directly involve community participation?
(A) Joint Forest Management
(B) Chipko movement
(C) Beej Bachao Andolan
(D) Demarcation of wildlife sanctuaries

545. Which of these statements is not a valid reason for the depletion of flora and fauna?
(A) Agricultural expansion
(B) Grazing and fuel wood collection
(C) Large scale development projects
(D) Rapid industrialization and urbanization

546. Which one of the following statements about forest cover is wrong?
(A) North-eastern states and parts of Gujarat have a high percentage of un-classed forests
(B) J & K, AP, Kerala, Uttarakhand and WB have a high percentage of resourced forest
(C) Bihar, Punjab, Himanchal Pradesh & Haryana have permanent forests estates
(D) Madhya Pradesh has the largest area under permanent forests

547. Which one of the following facts is wrong, about the role of community in conserving nature?
(A) Villagers in Rajasthan have encouraged the government officials to practice mining to saves their livelihood
(B) Each community has their own set of traditional methods of conserving nature
(C) Certain communities worship some particular plants and animals which they have preserved from time to time
(D) Tribal belief have preserved several virgin forests in pristine form called 'Sacred groves'

548. What measures is not clear from the dynamics of environment destruction and conservation from the following?
(A) Adopt strategies to save nature at any cost
(B) Accept only those economic activities that are people centric, environment-friendly and economically rewarding
(C) Local communities should not be involved in any resource management
(D) None of these

549. Which factors does not cause depletion of biodiversity?
(A) Environment pollution
(B) Expansion on animals
(C) Judicious way of collecting products from forests
(D) Excessive hunting

550. Fog is a colloid of:
(A) Solid in gas (B) Liquid in gas
(C) Gas in liquid (D) Liquid in gas

551. The --------------- zone is absent in ponds:
(A) Sublittoral (B) Limnetic
(C) Profundal (D) Littoral

552. Most stable ecosystem is:
(A) Desert (B) Mountain
(C) Ocean (D) Forest

553. Which part of India has very high rainfall?
(A) Extreme east (B) Extreme west
(C) Extreme north (D) Extreme south

554. Whcih part of India has very low rainfall?
(A) Extreme east (B) Extreme west
(C) Extreme north (D) Extreme south

555. Annual rainfall of nearly one-third of India:
(A) Below 55 cm (B) Below 65 cm
(C) Below 75 cm (D) Below 85 cm

556. India occupancy of arid zone :
(A) 520000 sq. km (B) 6320000 sq. km
(C) 7320000 sq. km (D) 320000 sq. km

557. States of India under occupancy of arid zone:
(A) Rajasthan (B) Gujarat
(C) Haryana (D) All of these

558. How much area of total land of India has serious limitation for crop production?
(A) One-third (B) One-sixth
(C) One-fourth (D) One-fifth

559. It is area under cold desert in sq. km. in India:
(A) 84000 (B) 44000
(C) 884000 (D) 64000

560. Estimated area in million hectares of surface water in India :
(A) 176 (B) 276
(C) 376 (D) 476

561. Estimated area in million hectares of ground water in India:
(A) 57 (B) 67
(C) 77 (D) 87

562. As against an ecological safe level of 30 per cent, how many per cent of the land area is under forest coverage in India?
(A) 5 (B) 10
(C) 15 (D) 20

563. Where and when was first Meterological Organization established?
(A) New York - 1980
(B) Washington - 1978
(C) Geneva - 1978
(D) Rome - 1976

564. Effective cause of atmospheric pressure is not :
(A) Temperature
(B) Altitude from the see level
(C) Rotation of earth
(D) Soil erosion

565. Which is not correct in the following?
(A) X-rays & Gamma rays - 97% of energy
(B) Visible lighting rays - 41% of energy
(C) Infrared rays - 50% of energy
(D) Ultraviolet rays - 10% of energy

566. Match List-I with the List-II and select the correct answer using the codes given below:
List - I - Crop
(a) Paddy (B) Wheat
(C) Maize (D) Groundnut
List - II -Temperature for germination
1. 20°C 2. 22°C 3. 25°C 4. 26°C
Codes

	(a)	(b)	(c)	(d)
(A)	4	2	3	1
(B)	2	1	4	3
(C)	3	4	2	1
(D)	1	3	4	2

567. How much minimum temperature is essential for wheat germination and optimum temperature for crop growth in degree cellsius :
(A) 3.5-5.5°C and 20-22°C
(B) 5.6-7.5°C and 16-18°C
(C) 8.0-10.0°C and 30-35°C
(D) 10-12°C and 25-30°C

568. The acidity of normal rain water is due to:
(A) CO_2 (B) Cl_2
(C) NO_2 (D) SO_2

569. Percent of water on the world's surface representing fresh water is:
(A) 97 (B) 50
(C) 10 (D) 3

570. Concentration of CO_2 in present day atmosphere is:
(A) ~ 220 ppm (B) ~ 280 ppm
(C) ~ 360 ppm (D) ~ 390 ppm

571. In biological systems sulphur is largely bound in:
(A) Lipids (B) Proteins
(C) Nucleic acid
(D) Proteins and nucelic acids

572. The source of thermal pollution are:
(A) Power plant (B) Cooling forests
(C) Industrial effluents
(D) All of the above

573. Which one of the following is a waste recycling method of solid waste management?
(A) Pelletisation (B) Composting
(C) Incineration (D) Sanitary landfil

574. Natural gas is composed primarily of:
(A) Methane (B) n-butane
(C) n-octane
(D) Mixture of n-butane and n-octane

575. The principal components of photochemical smog in urban areas are:
(A) SO_2 and NO_2 (B) SPM and CO
(C) SPM and NO_2
(D) Hydrocarbons and ozone

576. Lightening in the atmosphere produces:
(A) NO (B) CO
(C) CO_2 (D) NH_3

577. The Global Warming Potential (GWP) is the least for which of the following green house gases:
(A) CH_4 (B) CO_2
(C) N_2O (D) SF6

578. Scales of meterology are in the following order starting from the least:
(A) Macro -, Micro -, Meso-
(B) Macro -, Meso -, Micro -
(C) Meso-, Macro -, Micro -
(D) Micro-, Meso-, Macro -

579. The uppermost zone of atmosphere of earth, where shorwave ultra violate radiations are absorbed is:
(A) Troposphere
(B) Stratosphere
(C) Mesosphere
(D) Thermosphere

580. In which part of the atmosphere the momentum, heat flux and moisture content are conserved?
(A) Surface boundary layer
(B) Mesophere
(C) Troposphere
(D) Stratosphere

581. Perma frost represents
(A) Permanently frozen sub surface soil
(B) Frozen leaves of oak trees
(C) Frozen needles of pine trees
(D)Temporarily frozen subsurface soil

582. Soil air contains CO_2 (%):
(A) 0.25 (B) 0.03
(C) 2.50 (D) 3.00

583. The CO_2 content of soil air is:
(A) 0.03% (B) 0.30%
(C) 21% (D) None of these

584. Atmosphere extends above mean sea level to a height of about:
(A) 160 km (B) 1600 km
(C) 1600 miles (D) 1600 m

585. Densest part of atmospheric strata is:
(A) Stratosphere (B) Troposphere
(C) Mesosphere (D) thermosphere

586. Global Warming is attributed to increase in concentration of green hours gases like:
(A) CO_2 (B) CFCs
(C) CH_4 (D) All of these

587. Under seventh approximation soil classification, the number of soil order is:
(A) 8 (B) 11
(C) 14 (D) 15

588. The number of Agro-climate zones of India in:
(A) 7 (B) 10
(C) 15 (D) 17

589. Solar constant is equal to (in cal/cm^2/min):
(A) 0.194 (B) 1.94
(C) 19.40 (D) 19.4

590. The grand period of rainfall in India is:
(A) South-West monsoon
(B) North-east monsoon
(C) Cold weather period
(D) Post monsoon

591. Horse latitude lies in between north and south latitudinal belt of equator:
(A) 10 to 15° (B) 0 to 5°
(C) 20 to 25° (D) 30 to 35°

592. Which of the following gases contributes maximum to global warming?
(A) CFC (B) CO_2
(C) Methane (D) N_2O

593. Dangerous gas for depletion of Ozone layer is:
(A) Methane (B) CFC
(C) Ethane (D) Carbon-dioxide

594. The present level of carbon-dioxide in atmospheric is:
(A) 420-460 ppm (B) 490 ppm
(C) 295-300 ppm (D) 190 ppm

595. Impact of green house effect is:
(A) Sea formation
(B) Ice melting
(C) Global warming
(D) Sea level rise

596. Absolute zero is:
(A) 273°K (B) 273°F
(C) –273°C (D) 273°C

597. Continuous variation in atmospheric pressure is recorded by:
(A) Thermograph (B) Barograph
(C) Thermometer (D) Hydrograph

ANSWER SHEET
3. Agricultural Climatology

1. (A)	2. (B)	3. (A)	4. (A)	5. (A)	6. (A)	7. (A)	8. (B)	9. (C)	10. (D)
11. (A)	12. (B)	13. (A)	14. (B)	15. (B)	16. (A)	17. (A)	18. (B)	19. (A)	20. (C)
21. (B)	22. (A)	23. (A)	24. (C)	25. (C)	26. (A)	27. (C)	28. (D)	29. (C)	30. (A)
31. (D)	32. (D)	33. (A)	34. (D)	35. (D)	36. (C)	37. (D)	38. (B)	39. (D)	40. (D)
41. (B)	42. (C)	43. (B)	44. (D)	45. (A)	46. (C)	47. (B)	48. (A)	49. (C)	50. (D)
51. (B)	52. (C)	53. (B)	54. (D)	55. (C)	56. (C)	57. (D)	58. (C)	59. (B)	60. (B)
61. (B)	62. (B)	63. (C)	64. (A)	65. (D)	66. (A)	67. (A)	68. (C)	69. (D)	70. (B)
71. (D)	72. (C)	73. (C)	74. (C)	75. (B)	76. (B)	77. (B)	78. (C)	79. (B)	80. (C)
81. (C)	82. (B)	83. (D)	84. (B)	85. (B)	86. (C)	87. (C)	88. (B)	89. (B)	90. (D)
91. (A)	92. (A)	93. (B)	94. (A)	95. (B)	96. (D)	97. (A)	98. (C)	99. (B)	100. (C)
101. (C)	102. (B)	103. (B)	104. (A)	105. (D)	106. (B)	107. (B)	108. (D)	109. (D)	110. (D)
111. (B)	112. (A)	113. (B)	114. (D)	115. (B)	116. (B)	117. (D)	118. (B)	119. (B)	120. (B)
121. (D)	122. (C)	123. (B)	124. (C)	125. (D)	126. (C)	127. (C)	128. (C)	129. (C)	130. (D)
131. (C)	132. (C)	133. (B)	134. (C)	135. (D)	136. (C)	137. (B)	138. (D)	139. (C)	140. (B)
141. (D)	142. (C)	143. (A)	144. (B)	145. (D)	146. (C)	147. (B)	148. (A)	149. (C)	150. (D)
151. (A)	152. (C)	153. (C)	154. (D)	155. (B)	156. (A)	157. (D)	158. (C)	159. (B)	160. (D)
161. (A)	162. (B)	163. (C)	164. (B)	165. (C)	166. (B)	167. (B)	168. (D)	169. (A)	170. (D)
171. (D)	172. (C)	173. (A)	174. (D)	175. (B)	176. (B)	177. (C)	178. (A)	179. (B)	180. (C)
181. (D)	182. (C)	183. (C)	184. (B)	185. (A)	186. (D)	187. (B)	188. (D)	189. (C)	190. (C)
191. (B)	192. (A)	193. (C)	194. (A)	195. (C)	196. (A)	197. (C)	198. (A)	199. (D)	200. (C)
201. (B)	202. (B)	203. (A)	204. (B)	205. (A)	206. (A)	207. (D)	208. (B)	209. (C)	210. (B)
211. (B)	212. (D)	213. (C)	214. (D)	215. (A)	216. (A)	217. (C)	218. (A)	219. (B)	220. (C)
221. (C)	222. (B)	223. (C)	224. (A)	225. (A)	226. ((B)	227. (C)	228. (D)	229. (B)	230. (A)
231. (C)	232. (D)	233. (A)	234. (A)	235. (A)	236. (C)	237. (A)	238. (D)	239. (A)	240. (A)
241. (A)	242. (C)	243. (B)	244. (B)	245. (C)	246. (C)	247. (D)	248. (C)	249. (A)	250. (B)
251. (B)	252. (B)	253. (C)	254. (A)	255. (B)	256. (C)	257. (A)	258. (D)	259. (C)	260. (C)
261. (C)	262. (A)	263. (D)	264. (A)	265. (A)	266. (A)	267. (C)	268. (C)	269. (B)	270. (B)
271. (C)	272. (B)	273. (A)	274. (D)	275. (D)	276. (A)	277. (B)	278. (C)	279. (B)	280. (C)
281. (A)	282. (C)	283. (C)	284. (D)	285. (B)	286. (A)	287. (A)	288. (A)	289. (C)	290. (C)
291. (B)	292. (D)	293. (A)	294. (C)	295. (A)	296. (D)	297. (C)	298. (D)	299. (C)	300. (B)
301. (B)	302. (A)	303. (C)	304. (C)	305. (A)	306. (D)	307. (A)	308. (C)	309. (B)	310. (A)
311. (D)	312. (D)	313. (A)	314. (B)	315. (D)	316. (B)	317. (C)	318. (B)	319. (A)	320. (D)
321. (B)	322. (D)	323. (C)	324. (B)	325. (B)	326. (A)	327. (C)	328. (C)	329. (D)	330. (D)

331. (B) 332. (C) 333. (A) 334. (C) 335. (B) 336. (A) 337. (D) 338. (A) 339. (A) 340. (D)
341. (C) 342. (C) 343. (B) 344. (A) 345. (C) 346. (C) 347. (C) 348. (A) 349. (B) 350. (C)
351. (D) 352. (A) 353. (D) 354. (D) 355. (C) 356. (B) 357. (D) 358. (D) 359. (D) 360. (D)
361. (C) 362. (D) 363. (D) 364. (D) 365. (C) 366. (D) 367. (D) 368. (D) 369. (D) 370. (B)
371. (C) 372. (D) 373. (D) 374. (D) 375. (B) 376. (D) 377. (D) 378. (D) 379. (D) 380. (D)
381. (D) 382. (D) 383. (D) 384. (C) 385. (B) 386. (D) 387. (D) 388. (D) 389. (D) 390. (B)
391. (D) 392. (D) 393. (A) 394. (D) 395. (D) 396. (D) 397. (A) 398. (D) 399. (C) 400. (D)
401. (C) 402. (D) 403. (D) 404. (B) 405. (A) 406. (C) 407. (B) 408. (A) 409. (B) 410. (C)
411. (D) 412. (A) 413. (A) 414. (D) 415. (D) 416. (C) 417. (A) 418. (D) 419. (A) 420. (C)
421. (C) 422. (D) 423. (C) 424. (A) 425. (C) 426. (A) 427. (A) 428. (D) 429. (B) 430. (C)
431. (B) 432. (D) 433. (D) 434. (B) 435. (C) 436. (A) 437. (D) 438. (D) 439. (D) 440. (D)
441. (C) 442. (D) 443. (A) 444. (C) 445. (A) 446. (D) 447. (A) 448. (A) 449. (D) 450. (B)
451. (C) 452. (B) 453. (D) 454. (A) 455. (D) 456. (B) 457. (D) 458. (C) 459. (D) 460. (A)
461. (B) 462. (B) 463. (A) 464. (B) 465. (C) 466. (A) 467. (B) 468. (A) 469. (B) 470. (B)
471. (D) 472. (D) 473. (A) 474. (B) 475. (D) 476. (C) 477. (B) 478. (A) 479. (A) 480. (A)
481. (D) 482. (D) 483. (A) 484. (D) 485. (C) 486. (A) 487. (D) 488. (B) 489. (D) 490. (C)
491. (A) 492. (D) 493. (A) 494. (B) 495. (A) 496. (D) 497. (B) 498. (B) 499. (B) 500. (C)
501. (A) 502. (A) 503. (A) 504. (C) 505. (D) 506. (D) 507. (C) 508. (D) 509. (A) 510. (C)
511. (B) 512. (B) 513. (A) 514. (A) 515. (D) 516. (A) 517. (B) 518. (B) 519. (D) 520. (C)
521. (B) 522. (C) 523. (C) 524. (A) 525. (C) 526. (C) 527. (B) 528. (B) 529. (A) 530. (A)
531. (A) 532. (A) 533. (B) 534. (A) 535. (A) 536. (D) 537. (D) 538. (A) 539. (C) 540. (D)
541. (A) 542. (B) 543. (A) 544. (D) 545. (B) 546. (C) 547. (A) 548. (C) 549. (C) 550. (A)
551. (C) 552. (B) 553. (A) 554. (B) 555. (C) 556. (D) 557. (D) 558. (B) 559. (A) 560. (A)
561. (A) 562. (D) 563. (C) 564. (D) 565. (A) 566. (B) 567. (D) 568. (C) 569. (D) 570. (D)
571. (B) 572. (D) 573. (B) 574. (A) 575. (D) 576. (A) 577. (B) 578. (D) 579. (D) 580. (A)
581. (A) 582. (A) 583. (A) 584. (B) 585. (B) 586. (D) 587. (B) 588. (C) 589. (B) 590. (A)
591. (D) 592. (B) 593. (B) 594. (C) 595. (C) 596. (A) 597. (B)

4 TILLAGE MANAGEMENT

1. Mechanical manipulations of the soil to provide necessary soil conditions favourable for the growth of crop is called:
(A) Tillage (B) Hoeing
(C) Sowing (D) All of these

2. The objective of tillage is:
(A) Seed bed preparation
(B) Weed control
(C) Soil and water conservation
(D) All of these

3. Tillage operation to create desired soil configurations is called:
(A) Land forming (B) Land grading
(C) Land leveling (D) Land planning

4. Tillage operation that cuts and moves small layers of soil to provide a smooth and refined surface is called:
(A) Land forming (B) Land grading
(C) Land leveling (D) Land planning

5. Tillage operation employing rotary action to cut, break and mix soil is called:
(A) Ridge terrace (B) Ridge planting
(C) Rotary tillage (D) None of these

6. Tillage operations which constitute the initial major soil working operation is called:
(A) Primary tillage
(B) Secondary tillage
(C) Preparatory tillage
(D) Reduced tillage

7. The tillage operations following primary tillage to create a good seed bed for proper seeding/planting is called:
(A) Primary tillage
(B) Secondary tillage
(C) Preparatory tillage
(D) Reduced tillage

8. Various tillage operations carried out on the farm before the crops are sown to make a fine seedbed is called:
(A) Primary tillage
(B) Secondary tillage
(C) Preparatory tillage
(D) Reduced tillage

9. A tillage system in which the primary tillage operation is performed in conjunction with special planting procedure in order to reduce or eliminate secondary tillage operations is called:
(A) Primary tillage
(B) Secondary tillage
(C) Preparatory tillage
(D) Reduced tillage

10. Reduction in the specific volume of soil by means of mechanical manipulation is called:
(A) Primary tillage
(B) Secondary tillage
(C) Preparatory tillage
(D) Soil compaction

11. Which of the following is not a tillage conservation technique?
(A) Zero tillage
(B) Reduced tillage
(C) Minimum tillage
(D) All of these

12. Soil are tilled to create soil physical conditions which facilitates:
(A) Seed germination
(B) Seedling emergence
(C) Development of rooting
(D) All of these

13. Central Institute of Agricultural Engineering is situated at:
(A) Bhopal (B) Ludhiana
(C) New Delhi (D) Kolkata

14. The constituent of producer gas which occupies the highest percentage by volume and helps in increasing its overall calorific value is:

(A) CO (B) CO_2
(C) H_2 (D) CH_4

15. During field operation, the shank of a tractor drawn rigid tyne sweep type cultivator is mainly subjected to:

(A) Bending (B) Shear
(C) Torsion (D) Bending and torsion

16. Central Institute of Post Harvest Engineering and Technology is situated at:

(A) Bhopal (B) Ludhiana
(C) New Delhi (D) Kolkata

17. The power developed and the exhaust gas temperature of a diesel engine compared to a spark ignition engine of the same size and running at the same speed respectively, are:

(A) Higher and lower
(B) Higher and higher
(C) Lower and higher
(D) Lower and lower

18. Tilth is related to:

(A) Shape of the soil aggregates
(B) Size distribution of the soil aggregates
(C) Arrangement of the soil aggregates
(D) All of these

19. Blind hoeing refers to:

(A) Summer ploughing
(B) Hoeing in standing crop rows
(C) Primary tillage
(D) Hoeing before germination

20. Zero tillage system was first used successfully in 1950 in pasture removation in:

(A) Germanry (B) Japan
(C) United Kingdom (D) USA

21. Zero-till seed drill is used for sowing of:

(A) Wheat in maize-wheat cropping system
(B) Wheat in mungbean –wheat cropping system
(C) Wheat in rice-wheat cropping system
(D) Wheat in Fodder sorghum – wheat cropping system

22. Bund former is used for making:

(A) Furrows (B) Ridges/bunds
(C) Soil smooth (D) All of these

23. The force of attraction binds the molecules of the same kind is:

(A) Adhesion (B) Matric force
(C) Cohesion (D) None of these

24. Raising of crop with the least tillage operations is called:

(A) Zero tillage
(B) Minimum tillage
(C) Medium tillage
(D) Heavy tillage

25. In a semi-modular outlet, the discharge:

(A) Is independent of water levels in the distributary and the water course
(B) Depends upon the water levels of both distributary and water course
(C) Depends upon the water level in the distributary
(D) Depends upon the water level in the water course

26. A trapezoidal grassed waterway is constructed along a longitudinal gradient of 4%. If the cross-sectional area of flow is 1.52 m^2, wetted perimeter is 12.5 m and Manning's n for the waterway is 0.04 $m^{-1/3}$ s, the flow through the waterway in $m^3 s^{-1}$ is:

(A) 1.9 (B) 2.1
(C) 2.3 (D) 2.5

27. How much horse power (h.p.) is in one kW (Kilowatt) ?

(A) 1.34 (B) 1.90
(C)2.00 (D) 3.80

28. A tractor power take-off (PTO) driven stationary peg tooth type wheat thresher operating at a cylinder speed of 540 rpm requires a torque of 250 Nm at PTO. The minimum net engine power required, in kW is:

(A) 12-14 (B) 15-17
(C) 17-19 (D) 22-22

29. Man power is considered equal to how much horse power (h.p.) ?

(A) 0.1 (B) 0.2
(C) 0.3 (D) 0.4

30. The draft of 'Victory plough' is:

(A) 80-100 kg (B) 40-50kg
(C) 70-80 kg (D) 10-20 kg

31. A two wheel drive tractor, weighing 15.84 kN with a wheel base of 2160 mm, has the static weight divided between the front and rear axles in the ratio of 30 : 70 on a horizontal level surface. The hitch point is at a height of 700 mm from the ground and at a horizontal distance of 120 mm to the rear side from the center of the rear axle. Pull acts at an angle of 12° downwards from the horizontal. The maximum pull in kN, when the front wheels would just start rising from the ground is:

(A) 1.48 (B) 14.46
(C) 39.04 (D) 85.54

32. A horizontal axis drag type wind mill with square blades and a horizontal axis lift type wind mill with airfoil section blades having same rotor size are installed at a height of 10 m above the ground. The average wind speed is 25 km h^{-1}. The maximum power coefficient for drag type and lift type wind mills is 0.148 and 0.593, respectively. If the maximum power extracted by drag type wind mill is 5 kW, the corresponding power extracted by lift type wind mill, in kW is:

(A) 8.43 (B) 12.63
(C) 18.03 (D) 20.03

33. The thresher of a wheat combine harvester has an optimal throughput capacity of 2400 kg (crop) per hour. The harvester has a forward velocity of 4.5 km h^{-1}. Sample tests have revealed that the yield of crop in the field is 3000 kg (grain) per ha. Grain to straw ratio is 60 : 40. If the above through-out is to be maintained, the width of cut of the harvester in m, neglecting turning losses, is:

(A) 0.71 (B) 1.07
(C) 1.78 (D) 2.96

34. In a disc clutch, the inside and outside radii of the clutch plate are 50 and 100 mm, respectively. If the axial force exerted on the disc is 4 kN, the maximum pressure experienced by the clutch plate in N mm-2 under uniform wear conditions is:

(A) 0.13 (B) 0.17
(C) 0.25 (D) 0.51

35. The length of 'Gunter chain' is:

(A) 66 ft (B) 100 ft
(C) 20 ft (D) 90 ft

36. It is proposed to construct bench terraces on a 10% hill slope. If the batter slope is ½ H : 1 V, the percentage area that will be lost for cultivation due to bench terracing is:

(A) 4.68 (B) 5.47
(C) 6.25 (D) 6.78

Common Data for Questions 37 and 38:

37. A diesel engine running in dual fuel mode with diesel as pilot fuel and producer gas as primary fuel produces 3.5 kW at rated engine speed and is coupled directly to a generator for producing electricity. The amount of diesel and producer gas consumed per hour is 460 ml and 12.5 m^3, respectively. Assuming calorific value of diesel and producer gas as 35280 and 3.97 MJ m^{-3}, respectively, the brake thermal efficiency of the engine in percentage is

(A) 17.19 (B) 19.13
(C) 22.79 (D) 25.32

38. If generator efficiency is 90%, the maximum electricity produced, in kW is:

(A) 2.85 (B) 3.00
(C) 3.15 (D) 3.50

Statement for Linked Answer Questions 39 and 40:

Soybean is to be planted with a precision planter that meters 54 seeds per revolution of the metering disc powered from a ground wheel of diameter 490 mm. The desired plant population is 44800 per ha with a row to row spacing of 0.75 m. The germination percentage is 84. The planter is to be operated at 2.5 km h^{-1} with a 10% skid of ground wheel.

39. The angular speed of ground wheel in rpm is:

(A) 20.3 (B) 24.6
(C) 28.3 (D) 32.6

40. The angular speed ratio of metering disc to ground wheel for obtaining the desired plant population is:

(A) 0.125:1 (B) 0.150:1
(C) 0.225:1 (D) 0.250:1

Statement for Linked Answer Questions 41 and 42:

A 1 hp motor is used for running a dual cylinder reciprocating compressor of a refrigeration system based on R-134a refrigerant having 185 kJ kg^{-1} cooling capacity. COP of the system is 4.2 and overall efficiency of the compressor is 80%. Specific volume of the refrigerant vapour at suction temperature is 0.15 m^3 kg^{-1}. The compressor with bore diameters of 40 mm each runs at 1440 rpm.

41. The mass flow rate of the refrigerant in kg min^{-1}:
(A) 1.634 (B) 1.090
(C) 0.813 (D) 0.240

42. The compressor stroke length in mm is:
(A) 16.8 (B) 33.7
(C) 50.5 (D) 67.4

43. During the testing of a spike-tooth type thresher for wheat crop, the throughput and the thresher capacity were found to be 750 kg h^{-1} and 445 kg h^{-1}, respectively. The grain-straw ratio (grain : straw) of the crop is 1.5 : 1. Material other than grain (MOG) collected at the main grain outlet is 0.5%. The total grain loss in percentage from all the sources will be:
(A) 0.33 (B) 0.50
(C) 1.11 (D) 1.61

44. The rear furrow wheel in a tractor mounted disc plough is provided to:
(A) Reduce the frictional power loss
(B) Maintain the uniform depth of cut
(C) Reduce the side draft
(D) Improve the penetration of the plough

45. The cumulative discharge of a tractor mounted hydraulic sprayer having 7 nozzles is 2.0 L min^{-1} when the tractor is operated at 4 km h^{-1}. If the nozzle spacing is 0.3 m, the discharge in L ha^{-1} is:
(A) 1000.02 (B) 285.72
(C) 166.67 (D) 142.86

46. A two-wheel drive tractor, while negotiating a terrain, indicates 100% slip of one of the rear wheels. Under such a condition, the use of differential lock causes:
(A) Equal speed and equal power distribution to both the drive wheels
(B) Equal speed and equal torque distribution to both the drive wheels
(C) Equal power and equal torque distribution to both the drive wheels
(D) Equal speed and unequal power distribution to both the drive wheels

47. Which of the following is best for driving machine from low power to slow speed?
(A) Belts & pulley
(B) Spur gear
(C) Toothed wheel & chains
(D) None of the above

48. The horse power of the motor running the compressor of a refrigerator having COP of 4.5 and extracting 200 kJ kg^{-1} of evaporating heat with 1.5 kg min^{-1} refrigerant flow rate is:
(A) 0.5 (B) 1.0
(C) 1.5 (D) 2.0

49. The gridiron pipe drainage system is more economical than the herringbone pipe drainage system because:
(A) It is adopted in the fields which do not require complete drainage
(B) The number of main or sub-main lines is reduced
(C) The number of junctions and the double-drained area are reduced
(D) It has only main or sub-main lines

50. The Rational method is used to estimate:
(A) Runoff volume
(B) Peak runoff rate
(C) Runoff depth
(D) Direct surface runoff

51. A 16 m high wind break is constructed to protect soil from wind erosion due to wind velocity of 18 m s^{-1} at 15 m height. The minimum wind velocity at 15 m height capable of moving the soil fraction is 8m s^{-1}. The angle of deviation of prevailing wind direction from the perpendicular to the wind break is 30^0. The distance of full protection from the wind break in m is:
(A) 60.44 (B) 104.69
(C) 306.00 (D) 530.01

52. Animal driven cane planter is developed by:
(A) ICAR (B) CIAE
(C) IISR (D) IARI

53. A 10 cm wide belt is running at a speed of 900 r.p.m. If the 15 H.P. is available, then how much plies of belt ?

(A) 3 (B) 10
(C) 6 (D) 4

54. How much width belt would be needed if 20 H.P. is to be transferred through a belt of 10 plies, where speed of belt is 1460 r.p.m.?

(A) 4 cm (B) 8cm
(C) 5 cm (D) 6 cm

55. A 10 cm wide belt, having the speed of 1350 meter per minute, would be appropriate to transfer the how much H.P.?

(A) 20 (B) 15
(C) 25 (D) 40

56. A general farmer used deshi plough for the purpose of:

(A) Land ploughing
(B) Hoeing
(C) Making soil powdery
(D) Above all works

57. A two-wheel drive tractor pulls an implement that has a draft force of 11.5 kN. The total motion resistance of the tractor is 2.5 kN. Under these circumstances, the slip of the drive wheels is 20%. If the power loss in transmission is 20%, the percentage of power lost in converting engine power into drawbar power is:

(A) 37.4-47.5 (B) 47.4-47.5
(C) 57.4-57.5 (D) 67.4-67.5

58. A tractor mounted off-set type reciprocating mower is driven by the PTO shaft. The maximum inertia force of 3.2 kN occurs along the pitman at 32^0 crank angle and 27^0 pitmàn angle with the horizontal plane. The knives of the cutterbar are riveted to the slider. If each of the allowable tensile and compressive stresses of the slider material is 50 MPa, the minimum cross-sectional area of the slider in mm^2 is:

(A) 29.05 (B) 33.91
(C) 54.27 (D) 57.02

59. Nozzle spacing on the boom in power operated sprayer depends upon:

(A) Spray angle
(B) Height of the boom from the target
(C) Diameter of spray pattern
(D) All of these

60. To overcome rearward deflection the load given to outer end of the cutter bar is about:

(A) 2.5 cm/m (B) 2.0 cm/m
(C) 4.0 cm/m (D) 4.5 cm/m

61. Speed of the cutter bar in case of combine (grain) is:

(A) 100-250 rpm (B) 500-800 rpm
(C) 800-1000 rpm (D) 400-550 rpm

62. Compression pressure in diesel engine cylinder is:

(A) 6-10 kg/cm^2 (B) 10-20 kg/cm^2
(C) 35-45 kg/cm^2 (D) None of these

63. Compression pressure in diesel engine cylinder is:

(A) 6-10 kg/cm^2 (B) 10-20 kg/cm^2
(C) 35-45 kg/cm^2 (D) 20-35 kg/cm^2

64. Injection pressure of the fuel nozzle in diesel engine is:

(A) 50-100 kg/cm^2
(B) 150-200 kg/cm^2
(C) 100-200 kg/cm^2
(D) None of these

65. Per cent of wheel slip allowed for field operation to steel is in the range of:

(A) 20-25 (B) 10-15
(C) 15-20 (D) 5-10

66. The lower value of the ratio of drawbar height of length of tractor is better:

(A) Variation in weight transfer increased
(B) Variation in weight transfer decreased
(C) Risk increases while ascending a gradient
(D) Variation in lateral stability is reduced

67. Turbo charges are driven by:

(A) Direct connection with engine power
(B) Exhaust gases from engine
(C) Electric current from battery
(D) None of these

68. Inflation pressure in rear wheel of tractor:

(A) 0.8 – 1.2 kg/cm^2
(B) 1.2 – 2.0 kg/cm^2
(C) 2.0-2.5 kg/cm^2
(D) None of these

69. The lateral stability of a four wheel tractor in a turning situation can be increased by:
(A) Increasing the radius of turn
(B) Attaching a front end loader
(C) Increasing the height of the centre of gravity
(D) Increasing the total weight of the tractor

70. The main difference between flywheel and governer is:
(A) Flywheel is heavier than governer
(B) Flywheel is fixed to the crankshaft while governer is not
(C) Flywheel absorbs energy and governer controls engine speed
(D) None of these

71. Gears commonly used on fuel injection pumps are:
(A) Spiral (B) Helical
(C) Rack & pinion (D) None of these

72. In a forced feed lubricantioin system of an IC engine, the oil pump is operated by:
(A) Cam shaft (B) Crank shaft
(C) Fan belt (D) None of these

73. The choke in carburetor is provided to:
(A) Cut off fuel supply
(B) Control air supply
(C) Cut off air fuel mixture
(D) Control fuel mixture

74. BHP of an engine indicates:
(A) Power in cylinder
(B) Power on flywheel
(C) friction power
(D) Power at PTO pulley

75. Which of the following is a transmission dynamometer?
(A) Torsion dynamometer
(B) Froude's hydraulic dynamometer
(C) Belt dynamometer
(D) Prony brake dynamometer

76. The injection pressure if CI engine used on tractor is in the range of:
(A) 0-6 kg/cm^2
(B) 60-120 kg/cm^2
(C) 120-200 kg/cm^2
(D) 300 – 400 kg/cm^2

77. The operation of surface finishing of the engine cylinder is called:
(A) Reboring (B) Scuffing
(C) Honing (D) Scavenging

78. A six cylinder engine can have .... possible number of firing orders:
(A) 4 (B) 3
(C) 2 (D) 8

79. Cold spark plug is used on:
(A) Heavy engines (B) Small engines
(C) IC engines (D) All of these

80. The use of tractor is considered economical when it is utilized for:
(A) 1500 hrs/year
(B) 1200 hrs/year
(C) 1000 hrs/year
(D) 1400 hrs/year

81. Normal consumption of fuel in litres/hour by a 35 hp tractor is:
(A) 3.0 (B) 4.5
(C) 3.6 (D) 6.0

82. In a CI engine, the compressor pressure inside the cylinder normally in the range of:
(A) 35-45 kg/cm^2 (B) 25-35 kg/cm^2
(C) 10-16 kg/cm^2 (D) 45-55 kg/cm^2

83. Thermal efficiency of a tractor is in proportion to specific fuel consumption:
(A) Direct (B) Inverse
(C) Unaffected (D) None of these

84. In an offset harrow, the number of gang is:
(A) 4 (B) 2
(C) 1 (D) 3

85. The flywheel of a hand operated chaff cutter with two cutting knives is rotated at 30 rpm and is connected to a worm gear assembly for driving the feed rollers. The number of teeth of the worm gear is 24 and number of starts (threads) of the worm is 2. If diameter of each of the feed rollers is 15 cm, the chaff length in mm will be:
(A) 9.8 (B) 12.8
(C) 19.6 (D) 39.2

86. The cutter bar of a tractor operated onower makes .... strokes/min.:
(A) 1200-1400 (B) 400-850
(C) 800-1200 (D) 450-600

87. A 4-cylinder, 4-stroke compression ignition engine has piston stroke of 10.5 cm and cylinder bore of 11 cm. At a mean piston speed of 7 m s^{-1}, the developed brake mean effective pressure is 650 kPa. The brake power in kW developed by the engine is:

(A) 39.40 (B) 43.24
(C) 86.48 (D) 172.96

88. A centrifugal pump having an overall efficiency of 75% requires 6 kW power at 1450 rpm to deliver water against a suction head of 5 m and a delivery head of 12 m. If the pump runs at 1650 rpm and frictional head losses are negligible, the total head developed by the pump in metres will be:

(A) 22.01 (B) 25.05
(C) 29.35 (D) 31.72

89. A 100 ha watershed received rainfall at a rate of 5 cm h^{-1} for 2 hours. If the runoff generated by the storm was at the rate of 1 m^3 s^{-1} for 10 hours, the runoff coefficient for the watershed would be:

(A) 3.6×10^{-3} (B) 6.0×10^{-2}
(C) 0.36 (D) 36

90. The two types of cylinders used on power wheat threshers ae:

(A) Hammer mill and fails
(B) Serrated disc and rasp bar
(C) Spike tooth and hammer mill
(D) None of these

91. A watershed, with an area of 360 km^2, has a triangular shaped 4^{-h} unit hydrograph with a base of 50 hours. The peak discharge of direct runoff hydrograph due to 3 cm of rainfall-excess in 4 hours from the watershed in m^3 s^{-1} is:

(A) 13.33 (B) 40.00
(C) 120.00 (D) 160.00

92. A rotary dryer is used to dry 1200 kg h^{-1} of paddy containing 30% moisture (wet basis) to give a product containing 15% moisture (wet basis). Alternately, a portion of the dry product may be recycled and mixed with the fresh feed such that the mixed feed enters the dryer with moisture content of 20% (wet basis). The moisture evaporation rate without recycle and the paddy recycle rate in kg h^{-1} respectively in the dryer are:

(A) 211.76 and 2400
(B) 211.76 and 600
(C) 256.50 and 2400
(D) 256.50 and 600

93. The work index of a material is 6.25. If 80% of the feed and 80% of the product pass through IS Sieve No. 340 (3.25 mm opening) and IS Sieve No. 40 (0.42 mm opening), respectively, the power consumed in kW to crush 5000 kg h^{-1} of sorghum is:

(A) 9.77 (B) 20.49
(C) 26.29 (D) 32.29

94. In a screen separator, the mass flow rates of feed, overflow and underflow are 150, 140 and10 kg h^{-1}, respectively. Mass fraction of material in the feed and overflow are 0.9 and 0.96, respectively. The effectiveness of separation in percentage is:

(A) 32 (B) 42
(C) 52 (D) 62

95. A milk fat globule of 2 μm diameter is rising in whole milk of density 1030 kg m^{-3} and coefficient of viscosity 10^{-3} Poise. If the fat density is 950 kg m^{-3}, the time needed to rise 10 mm for this fat globule in min is:

(A) 0.57 (B) 34.57
(C) 35.57 (D) 95.57

Common Data for Questions 96 and 97:

A 9 × 20 cm fluted roller type seed drill is operated at a forward speed of 3 km h^{-1} in a field of size 120 m × 90 m. The effective ground wheel diameter of the seed drill is 0.5 m and the ratio of ground wheel rpm to the fluted roller rpm is 2. For one complete rotation of each fluted roller, 6.8 g seed is transferred from the seed box to the seed tube. The average time taken for each turn while operating length-wise is 50 s and the total time wasted in refilling the seed box for sowing the entire field is 40 min.

96. The seed rate in kg ha^{-1} will be:

(A) 108.32 (B) 122.55
(C) 136.99 (D) 240.71

97. The actual field capacity of the machine in ha hr^{-1} is:

(A) 0.30 (B) 0.32
(C) 0.36 (D) 0.40

Common Data for Questions 98 and 99:

Specific heat capacity of dry air and water vapour are 1.005 and 1.88 kJ kg^{-1} K^{-1}, respectively. In an energy conserving system, 1 kg s^{-1} air at 30°C with constant absolute humidity of 0.02 kg water (kg dry air)$^{-1}$ is heated up to 65°C. Water at the wet bulb temperature of air is then sprayed into the air so that the final temperature of the air-water vapour mixture is 40°C. Latent heat of vapourization of water at 70°C and 40°C are 2334 and 2407 kJ kg^{-1}, respectively.

98. The thermal energy supplied per second during heating in kW is:
(A) 18.2 (B) 36.5
(C) 101.0 (D) 166.8

99. The absolute humidity of the exhaust air from the spray chamber in kg water (kg dry air)$^{-1}$ is:
(A) 0.027 (B) 0.029
(C) 0.031 (D) 0.033

Statement for Linked Answer Questions 100 and 101:

A two-wheel drive tractor weighing 21 kN has a wheel base of 2.1 m. The CG of the tractor is 0.7 m ahead of rear axle centre. The tractor is pulling a single-axle trailer with gross trailer weight of 50 kN on a leveled concrete road while maintaining the line of pull parallel to the ground surface. The tractor hitch point is 42 cm behind the rear axle centre and 52.5 cm above the ground surface. During operation, 20% of the gross trailer weight is transferred to the tractor hitch point. If the coefficient of rolling resistance for each of the tractor and trailer wheels is taken as 0.04, and their ground reactions are assumed to pass through their respective wheel centres,

100. The dynamic ground reaction against the tractor rear wheels in kN is:
(A) 14.0 (B) 24.0
(C) 24.5 (D) 26.4

101. The gross traction ratio developed by the tractor is:
(A) 0.061 (B) 0.082
(C) 0.108 (D) 0.123

102. The unt draft of an indigenous plough making a 20 cm wide and 10 cm deep V shaped furrow in 70 kg per sq. cm. The draft is:
(A) 40 kg (B) 50 kg
(C) 70 kg (D) 60 kg

103. Thermal efficiency .... with increase in stroke : bore ratio:
(A) Decreases
(B) Increases
(C) Remains the same
(D) None of these

104. High compression petrol engines are used in some tractors and have high performance in:
(A) U.S.A. (B) Germany
(C) Japan (D) Holland

105. The distribution of fertilizers by aircraft is widely practised in:
(A) England (B) France
(C) Germany (D) New Zealand

106. A 2- row potato harvester, working in good conditions with 4 to 5 men on the machine, can do how much hectare per day ?
(A) 0.5-0.6 ha (B) 1.2-1.4 ha
(C) 2.5 ha (D) 1.8-1.9 ha

107. Mole drainage is practised extensively in country:
(A) New Zealand (B) Japan
(C) U.S.A. (D) China

108. Which is not included in the sources of energy, getting for agricultural work ?
(A) Diesel engine (B) Electric motor
(C) Bullocks (D) Cow

109. Which is included in dairy equipments ?
(A) Threshers (B) Lactometer
(C) Cane Planter (D) Winnowers

110. In our country, for the manufacturing of agricultural implements tools, mostly wood is used, because:
(A) Easy available at all the places
(B) Wood is cheaper
(C) Easy to repare tools made-up of wooden, compared to metal tools
(D) Above (A) to (C) are correct

111. Which of the following wood is not used for making handle of spade ?
(A) Sheesham (B) Babool
(C) Neem (D) Mango

112. Soil turning plough makes the furrow of which type (shape) ?
(A) 'V' shape (B) 'L' shape
(C) 'O' shape (D) No definite shape

113. Which of the following is one (single) handed soil turning plough ?
(A) Punjab plough (B) Praja plough
(C) Victory plough (D) U. P. No.1 plough

114. The purpose of tillage is/are:
(A) Soil clods breaking and suppressing in soil
(B) Eradication of weeds
(C) Leveling of soil
(D) Above (A), (B) and (C)

115. The main function of cultivator is:
(A) To turn the soil
(B) To make furrow in soil
(C) To pulverize the soil
(D) All above three functions

116. Bakhar is generally used in:
(A) U. P.
(B) Bundelkhand
(C) Vindhya Pradesh
(D) In all the States/parts of India (in All India)

117. Harrow is drawn by:
(A) Bullocks (B) Tractor
(C) Diesel
(D) Bullocks and Tractor both
(E) It is automatic tool

118. Which of the following is not a secondary tillage implement ?
(A) Cultivators (B) Harrow
(C) Hoe (D) Meston plough

119. The term 'Olpad' in 'Olpad Thresher' is named on the name of:
(A) Scientist (B) Village
(C) Labourer (D) Farmer

120. Which of the following 'hoe is bullock drawn ?
(A) Akola hoe (B) Sharma hoe
(C) Wheel hoe (D) Naini type hoe

121. The land leveling implement is:
(A) Patela (B) Roller
(C) Scrapper (D) All of these

122. *Patela* is used for:
(A) Sowing (B) Levelling
(C) Earthing (D) Weeding

123. Which of the material is not used for making rollers ?
(A) Wood (B) Stone
(C) Iron (D) Steel

124. In which condition, roller is used ?
(A) Where soil is wet
(B) Where clods are present in soil
(C) Where soil is friable
(D) None of these

125. Which one of the following work is not done by scrappers ?
(A) For leveling of soil
(B) For making irrigation channels
(C) For covering the sown-seeds by earth
(D) For making ridges

126. Dibbler is used for:
(A) Ploughing
(B) Seed sowing
(C) Leveling of land
(D) Interculture

127. The minimum expenditure is incurred by implements for sowing seeds among the following:
(A) Deshi plough (B) Cultivator
(C) Dibbler (D) Seed-drill

128. Which of the following tool is used for measuring the draft of agricultural implements?
(A) Dynamometer (B) Hydrometer
(C) Galvanometer (D) Barometer

129. Which of the following does not affect the draft of ploughs ?
(A) Width of furrow
(B) Depth of furrow
(C) Length of furrow
(D) Soil moisture

130. Most commonly transmission system used in India is:
(A) Synchronoesh (B) Planetary gear
(C) Selective gear (D) Hydrostatic

131. Best speed to get minimum brake specific fuel consumption is:
(A) Minimum (B) Maximum
(C) Average (D) None of these

132. maximum number of tractor being manufactured in India are in the hp range of:
(A) 15-20 (B) 25-36
(C) 30-40 (D) 51-65

133. Traction and transmission coefficienct is maximum for:
(A) Two-wheel drive tractor
(B) Four wheel drive tractor
(C) Track type tractor
(D) None of these

134. The working efficiency per day of deshi plough is:
(A) 0.3 ha (B) 0.4 ha
(C) 0.6 ha (D) 0.8 ha

135. Which of the method of ploughing is mostly practised ?
(A) Outside to inside ploughing
(B) Inside to outside ploughing
(C) Ploughing by putting furrow from one side of field
(D) Ploughing by halai making

136. The best method of ploughing through deshi plough is:
(A) Outside to inside ploughing
(B) Inside to outside ploughing
(C) Ploughing by making halai
(D) Ploughing by making furrow from one side of field

137. How much is the working efficiency of dibbler (ha per day) ?
(A) 0.15 (B) 0.25
(C) 0.35 (D) 0.45

138. Tillage includes:
(A) Ploughing of land
(B) Keep the land free from weeds
(C) Make the soil leveled
(D) All of the above operations

139. Function of the seed-drill is:
(A) Making furrow
(B) Dropping seeds
(C) Covering the seeds in furrow
(D) All of these

140. Wrought iron contains carbon (per cent) :
(A) 0.05-1.0% (B) 1-2%
(C) 2-3% (D) 3-4%

141. The mould-board of a tractor drawn soil turning plough is the type of:
(A) General purpose
(B) Stubble
(C) Sod (breaker)
(D) High speed

142. Among the following, ridger is not used in crop:
(A) Maize (B) Gram
(C) Potato (D) Sugarcane

143. Belt mostly used is of which types:
(A) Rubber (B) Kirmich
(C) Leather (D) Cotton thread

144. The major defects of rubber belts are:
(A) Costly (B) Slips on wet
(C) Early rubbed
(D) Increased due to heat

145. How many mm are in one foot length ?
(A) 304.8 (B) 404.8
(C) 204.8 (D) 104.8

146. The grooved pulleys are made-up of:
(A) Steel (B) Cast iron
(C) Wooden
(D) Above (A) and (B) both

147. During a test, sound level was measured as 90 dB in the operator's cabin on a tractor. Taking reference sound pressure as 2 X 10^{-5} N m^{-2}, the measured RMS sound pressure in N m^{-2} is:
(A) 6.32 (B) 6.32 X 10^{-1}
(C) 1.8 X 10^{-3} (D) 6.32 X 10^{-10}

148. During land leveling of agricultural land for irrigation and drainage purposes, the acceptable deviation in elevation from the design value in metre is:
(A) 0.015 (B) 0.025
(C) 0.055 (D) 0.150

149. Reapers are used for:
(A) Crop cutting
(B) Threshing of harvested crop produce (lank)
(C) Seeds sowing
(D) Fodder cutting

150. Threshers (except Olpad threshers) are driven by:
(A) One pair of bullocks
(B) Two pair of bullocks
(C) Diesel
(D) All of the above

151. Winnowing it called:
(A) Cutting a crop
(B) To separate straw etc. from threshed lank
(C) To thresh
(D) Cutting the fodder

152. Chaff-cutter is driven by:
(A) Hand (B) Bullocks
(C) Electric power (D) All of these

153. Sugarcane juice is extracted (%) from canes through bullock-drawn cane-crusher:
(A) 50-55 (B) 70-75
(C) 25-30 (D) 60-65

154. 'Olpad' thresher is used in:
(A) Oil extraction from mustard, toria etc.
(B) Extracting juice from cane
(C) Threshing of wheat, barley, pea etc.
(D) All (A), (B) and (C)

155. 'Seed dresser' is used for:
(A) Mixing/treating seeds with chemicals (pesticides/fungicides etc.)
(B) Sowing seeds at proper distance
(C) Making seeds of high grade
(D) Keeping seeds effective upto longer period

156. 'Try square' is used by:
(A) Blacksmith (B) Carpenter
(C) Potter (D) Farmers

157. Which type of saw is not included ?
(A) Cross cut saw (B) Deshi saw
(C) Tenon saw (D) Teething saw

158. Which type of saw, used for cutting round shape in hole ?
(A) Fret saw (B) Penal saw
(C) Tenon saw (D) Key hole saw

159. 'Draw-Knife' is used for:
(A) To fit galua in grinder
(B) Making round the corners of wood
(C) Smoothing the base of wood
(D) Sharping teeth of saw

160. The tool used for making deep pit in wood is:
(A) Adge (B) Saw
(C) Draw knife (D) Chisel

161. The main work of 'scrapper' is:
(A) For cutting the wood
(B) For scrapping the wood
(C) For smoothing the wood
(D) All of the above

162. 'File' is used for:
(A) Scrapping the wood
(B) Cutting the wood
(C) Forcing the wood
(D) Making equal by scrapping the wood

163. The type of file is generally:
(A) Round (B) Triangular
(C) Flat (D) Above all types

164. Negative slip of tractor is obtained in field operation of:
(A) Ploughing (B) Interculture
(C) Spraying (D) Rotatilling

165. Clutch in the tractors is designed on the basis of:
(A) Constant torque
(B) Road
(C) Constant pressure
(D) Constant power

166. 'Bar cramp' is a tool of:
(A) Wood cutting
(B) Catching tightly wood
(C) Beating tool
(D) Smoothing tool

167. The anti-knock quality of petrol fuels is given by:
(A) Octane number (B) Cetane number
(C) Butane number (D) Carbon dioxide

168. Tractor operation in a lower gear causes:
(A) More pull (B) Reduced pull
(C) Less torque (D) All of these

169. In a tractor, the centre of gravity should be located approximately .... s head of rear axle:
(A) 1/3 of the wheel base
(B) 1/2 of the wheel base
(C) 2/3 of the wheel base
(D) 3/4 of the wheel base

170. The most commonly used pumps in tractor hydraulic system are:
(A) Centrifugal type (B) Gear type
(C) Vane type (D) None of these

171. The maximum overall efficiency of a hydrostatic drive is:
(A) 90% (B) 75%
(C) 80% (D) 70%

172. The roller of cane-crusher is made-up of which steel ?
(A) Cast iron (B) Gun metal
(C) High carbon steel
(D) Wrought iron

173. The tool use for catching (handling) the claw and turning it hither and thither during beating is called:
(A) Sewage block (B) Chisel
(C) Plier (D) Tongs

174. In combines straw racks generally oscillate at ------ strokes/min:
(A) 100-200 (B) 200-300
(C) 300-450 (D) 150-200

175. Agricultural tractors are generally having horse power (h.p.) :
(A) 20-50 (B) 50-70
(C) 10-15 (D) 20-25

176. A three-tine furrow plough is ploughing at a speed of 2 km/hr and the each furrow size is 20 cm wide and 12 cm deep furrow. How much time, it would require to plough 5 hectare land ?
(A) 41 hours 40 minutes
(B) 40 hours 55 minutes
(C) 20 hours 55 minutes
(D) None of these

177. If a gear 'A', having 50 teeth and running at a speed of 200 r.p.m. is driving to another gear 'B' having 125 r.p.m. How many number of teeth in gear' B' ?
(A) 500 (B) 400
(C) 80 (D) 60

178. The PTO HP of a tractor is:
(A) Less than drawbar horsepower
(B) More than engine horsepower
(C) More than drawbar horsepower
(D) Less than friction horsepower

179. The basic function of a governer in an engine is to regulate:
(A) Engine speed (B) Fuel consumption
(C) Power developed
(D) None of these

180. Turning space and turning circle of the tractor test must be preferred at a tractor speed of:
(A) At any speed
(B) Not exceeding 2 km/hr
(C) Exceeding 2 km/hr
(D) None of these

181. Soil erosion by wind brings about the serious damage in soil by changing the:
(A) Soil permeability
(B) Soil structure
(C) Soil texture
(D) Soil plasticity

182. A tractor driving pulley has its 25 cm diameter and revolving at a speed of 960 r.p.m. If on the shaft of a thresher, an attached pulley is revolving at a speed of 1600 r.p.m. what would be the diameter of this pulley ?
(A) 16 cm (B) 15 cm
(C) 12 cm (D) 20 cm

183. A pulley of 21 cm diameter and revolving at 1600 r.p.m. Find out the speed of the belt running over it:
(A) 1120 m/minute (B) 1200 m/minute
(C) 1056 m/minute (D) None of these

184. If the distance between two pullies, having the diameter of 45 cm and 39 cm, is 3.2 metre, then what would be length of required flate belt ?
(A) 6.8 m (B) 7.72 m
(C) 9.4 m (D) 7.2 m

185. Two pair of bullocks are ploughing a field by victory ploughs at a speed of 2 km/hr. If the average width of each furrow is 25 cm. How much time will be required to plough one hectare land? Where after each two hours working, 20 minutes rest is required to provide?
(A) 10 hours
(B) 8 hours
(C) 11 hours 40 minutes
(D) 8 hours 20 minutes

186. Among the following, ridger is not used in crop:

(A) Potato (B) Pea
(C) Sugarcane (D) Maize

187. A 'chain' contains how much metre ?

(A) 20.1168 (B) 21.1168
(C) 15.1168 (D) 18.1168

188. The length of 'Engineer chain' is:

(A) 100 ft (B) 80 ft
(C) 20 ft (D) None of these

189. Thermal efficiency of a diesel engine varies between:

(A) 32-38% (B) 25-31%
(C) 20-25% (D) 38-45%

190. The compression raio of diesel engines are in the order of:

(A) 21 to 30 :1 (B) 9 to 13 : 1
(C) 14 to 20 : 1 (D) None of these

191. Standard PTO speed is:

(A) 536 rev/min. (B) 100 rev/min
(C) 1440 rev/min (D) 2000 rev/min

192. The indamped natural frequency of wheel of tractors lies in the range of:

(A) 5 to 10 Hz (B) 15 to 20 Hz
(C) 10 to 25 Hz (D) 20 to 25 Hz

193. Tractive and transmission co-efficient is:

(A) Ratio of bhp and Ihp
(B) dbph and ihp
(C) Grawbar power to PTO power
(D) None of these

194. Moving the centre of gravity of a tractor towards its front wheel creates the problem of:

(A) Instability (B) Steering
(C) Over turning (D) None of these

195. A tractor operated single acting trailing type disc harrow has 8 discs on each gang. The gang angle of both the gangs is maintained at 35^0. The horizontal component of resultant soil reaction force on each disc is 600 N and it makes an angle of 30^0 with the gang axis. If the speed of operation is 6 km h^{-1}, the required drawbar power in kW to operate the harrow will be:

(A) 14.4-14.6 (B) 15.4-15.6
(C) 16.4-16.6 (D) 17.4-17.6

196. Optical pyrometer is used to measure:

(A) Light intensity
(B) Low temperature
(C) High temperature
(D) Low Light intensity and high temperature

197. In mould board plough, soft centred stell is used for making of..........:

(A) Frog (B) Mould Board
(C) Land Side (D) None of these

198. Though tilth is dynamic in nature, can be measured by:

(A) Aggregate analysis
(B) Chemical analysis
(C) Biological analysis
(D) None of these

199. According to CRIDA, Hyderabad, 15-20 cm deep ploughing is known as:

(A) Shallow (B) Medium
(C) Deep (D) None of these

200. Tilt angle of disc furrow is:

(A) 20^0 (B) 5^0
(C) 45^0 (D) 0^0

201. Loosing of the top soil hastens the development of a dry layer at the surface which reduces further loss of water from soil is known as:

(A) Shallow tillage (sand mulch)
(B) Traditional tillage
(C) Conventional tillage
(D) Minimum tillage

202. The tillage practices is also associated with faster decomposition of soil organic matter, greater erosion risk and degradation of soil structure is known as:

(A) Shallow tillage
(B) Traditional tillage
(C) Conventional tillage
(D) Minimum tillage

203. The tillage envisages minimum disturbance of soil, residues are left in the field and weeds are killed with herbicides is called:

(A) Shallow tillage
(B) Traditional tillage
(C) Conventional tillage
(D) Minimum tillage

204. Tillage operations are considerably reduced to the minimum required for good seed bed preparation is called:
(A) Shallow tillage
(B) Traditional tillage
(C) Conventional tillage
(D) Minimum tillage

205. Extremes of minimum tillage is called:
(A) Zero tillage
(B) Traditional tillage
(C) Conventional tillage
(D) Minimum tillage

206. Sowing without any tillage operation is called:
(A) Shallow tillage
(B) Zero tillage
(C) Conventional tillage
(D) Minimum tillage

207. The tillage practices combined with crop residues, dry weeds and other types of vegetative material spread on the soil surface is called:
(A) Shallow tillage (B) Traditional tillage
(C) Mulch tillage (D) Minimum tillage

208. A tillage system in which only isolated bunds of soil are tilled is called:
(A) Strip tillage
(B) Sub soiling
(C) Sub soil placement
(D) Surface tillage

209. Breaking of compact sub soils without inverting them with a special narrow cultivator, showed or chisel at depth usually from 30-60 cm from the surface is called:
(A) Strip tillage
(B) Sub soiling
(C) Sub soil placement
(D) Surface tillage

210. Placement of fertilizers in the subsoil with the help of heavy power machinery usually followed in humid and sub humid region is known as:
(A) Strip tillage
(B) Sub soiling
(C) Sub soil placement
(D) Surface tillage

211. To till the surface without turning the tilled layers is called:
(A) Strip tillage
(B) Sub soiling
(C) Sub soil placement
(D) Surface tillage

212. To plough or cultivate soil for seeding is called:
(A) Strip tillage (B) Sub soiling
(C) Till (D) Surface tillage

213. The action of a tillage tool in executing a specific form of soil manipulation such as soil cutting, pulverizing or inversion is called:
(A) Tillage action (B) Tillage depth
(C) Tillability (D) Vertical mulching

214. Vertical distance from the initial soil surface to a specific point of penetrations of the tool is called:
(A) Tillage action (B) Tillage depth
(C) Tillability (D) Vertical mulching

215. The degree of ease with which a soil may be manipulated for a specific purpose is called:
(A) Tillage action (B) Tillage depth
(C) Tillability (D) Vertical mulching

216. Machine cuts the soil, pulverizes the soil to powder and spreads uniformly to level the field is:
(A) Rotavator
(B) Ridge and furrow makers
(C) Harrow
(D) None of these

217. A machine is used for making irrigation channel and boundaries of a field is:
(A) Rotavotor
(B) Ridge and furrow makers
(C) Harrow
(D) None of these

218. The speed of rotavotor is:
(A) 100-200 rpm (B) 200-300 rpm
(C) 300-400 rpm (D) 400-500 rpm

219. Harrows are tillage implement of:
(A) Primary tillage
(B) Secondary tillage
(C) Both (A) and (B)
(D) None of these

220. A ridger plough consists of:
(A) V – shaped (B) U – shaped
(C) L – shaped (D) M – shaped

221. An implement used for puddling the paddy field before transplanting is:
(A) Rotavotor (B) Ridge and furrow makers
(C) Harrow (D) Puddler

222. A two bladed flywheel type forage cutter rotates at 60 rpm. At a feed rate of 1-26 meters per minutes, theoretical length of cut of the forage will be:
(A) 21.0 mm (B) 42.0 mm
(C) 5.3 mm (D) 10.5 mm

223. The poisonous exhaust gas from a petrol engine is:
(A) Hydrochloric acid
(B) Carbon monoxide
(C) Carbon dioxide
(D) None of these

224. The oil that can be used in the braking system of a vehicle is:
(A) Of plant origin (B) Engine oil
(C) Mineral oil (D) None of these

225. The main advantage of 'PVC' pipes for drainage is the feasibility of:
(A) Mechanical laying
(B) Physical laying
(C) Chemical laying
(D) Physical, chemical laying both

226. 'ESCORT' tractor is manufactured at:
(A) Faridabad (B) Ghaziabad
(C) Kanpur (D) Chennai

227. The type of pump used in forced water cooling system of a tractor engine is:
(A) Piston (B) Centrifugal
(C) Gear (D) Vane

228. Which one of the following statements is NOT appropriate regarding cone index:
(A) It reflects strength of soil
(B) It is a composite parameter
(C) It is dimensionless
(D) It is measured at a constant penetration rate of 30 mm/s

229. The draft and total power requirement of a rotary cultivator operating in concurrent mode as compared to a spring tyne cultivator of equal cutting width under the same operating conditions, respectively are:
(A) Higher and higher
(B) Lower and lower
(C) Lower and higher
(D) Higher and lower

230. The soil erodibility factor needs to be determined for use in the universal soil loss equation. The length, in m and slope, in % of the experimental plot to be used for this purpose, respectively are:
(A) 19, 12 (B) 21, 11
(C) 22, 9 (D) 23, 8

231. The difference between Fore Bearing and Back Bearing of a traverse line is:
(A) Exactly 90^0 (B) Less than 180^0
(C) Exactly 180^0 (D) Greater than 180^0

232. A pumping device that combines the advantages of both centrifugal and reciprocating pumps is known as:
(A) Air lift pump (B) Hydraulic ram
(C) Jet pump (D) Rotary pump

233. If v is the kinematic viscosity of air – water vapour mixture and is D_{AB} the mass diffusivity of water vapour in air then the ratio V/D_{AB} is known as:
(A) Stanton number
(B) Prandtl number
(C) Schmidt number
(D) Sherwood number

234. Work index in size reduction can be obtained by multiplying Bond's energy constant with:
(A) 10 (B) √10
(C) 3√10 (D) √10

235. Approximate percentage of scores that fall within ±σ (standard deviation) of the mean in a normal distribution is:
(A) 34 (B) 68
(C) 95 (D) 99

236. Central Institute of Research on Jute and Allied Fibre Technology is situated at:
(A) Bhopal (B) Ludhiana
(C) New Delhi (D) Kolkata

237. Central Institute for Cotton technology is situated at:
(A) Bhopal (B) Ludhiana
(C) Mumbai (D) Kolkata

238. Indian Lac Research Institute is situated at:
(A) Ranchi (B) Ludhiana
(C) New Delhi (D) Kolkata

239. According to CRIDA, Hyderabad, less than 15 cm deep ploughing is known as:
(A) Shallow (B) Medium
(C) Deep (D) Very deep

240. According to CRIDA, Hyderabad, more than 15 cm deep ploughing is known as:
(A) Shallow (B) Medium
(C) Deep (D) Very deep

241. Rotavator mainly used for:
(A) Threshing (B) Ploughing
(C) Levelling
(D) Ridge & furrow maker

242. Desi plough make the furrow of which shape?
(A) 'V' shapes (B) 'L' shapes
(C) 'O' shapes (D) No definite shapes

243. 'Mahindra' tractor is manufactured at:
(A) Hoshiarpur (B) Rudrapur & Jaipur
(C) Faridabad (D) Gurgaon

244. 'Massey furgusion' tractor is manufactured at:
(A) Hoshiarpur
(B) Bangalaore
(C) Madurai & Alwar
(D) Gurgaon

245. 'Eicher' tractor is manufactured at:
(A) Hoshiarpur (B) Bangalore
(C) Gurgaon (D) Faridabad & Bhopal

246. 'Ford' tractor is manufactured at:
(A) Hoshiarpur (B) Faridabad
(C) Guragaon (D) New Delhi

247. 'Swaraj' tractor is manufactured at:
(A) Hoshiarpur (B) Mohali
(C) Ludhiana (D) Gurgaon

248. 'Sonalika' tractor is manufactured at:
(A) Hoshiarpur (B) Ludhiana
(C) Faridabad (D) Gurgaon

249. Ferti cum seed drill is used for:
(A) Fertilizer application
(B) Seed sowing
(C) Both (A) and (B)
(D) None of these

250. Lesser land leveler is a machine used for:
(A) Ploughing (B) Leveling
(C) Harrowing (D) Lifting of water

251. Vikas and Vijay is a name of seed drill is manufactures at:
(A) Hoshiar pur (B) Meerut
(C) Chennai (D) Khorai

252. The draft of deshi plough is:
(A) 80-100 kg (B) 70-80 kg
(C) 50-60 kg (D) 30-35 kg

253. The draft of meston plough is:
(A) 80-100 kg (B) 70-80 kg
(C) 50-60 kg (D) 30-35 kg

254. Central Farm Machinery Training and Testing Institute us situated at:
(A) Bhopal, MP (B) Budni, MP
(C) Ludhiana, Pb (D) Jabalpur, MP

255. 'John Deere' tractor is manufactured at:
(A) Nabha, Pb (B) Faridabad
(C) Pune (D) Mohali

256. For a deep ploughing (more than 15 cm) which plough is used?
(A) Victory plough (B) Deshi plough
(C) Chisel plough (D) Rotavator

257. To break the hard pan of sub soil, which plough is used:
(A) Deshi plough (B) Chisel plough
(C) Rotavator (D) Victory plough

258. Chisel plough is required tractor of:
(A) 35 HP (B) 40 HP
(C) 45 HP (D) More than 50 HP

259. The draft of bullock drawn seed cum ferti drill plough is:
(A) 34-36 kg (B) 40-44 kg
(C) 50-54 kg (D) 60-64 kg

260. The draft of 'wheel hand hoe' is:
(A) 9-9.5 kg (B) 10-10.5 kg
(C) 11-11.5 kg (D) 12-12.5 kg

261. How much working efficiency of bullock drawn cum ferti drill (ha per hour):
(A) 0.30-0.40 (B) 0.50-0.80
(C) 0.80-0.90 (D) 1.0-1.2

262. How much working efficiency of wheel hand hoe (ha per hour):
(A) 0.10-0.12 (B) 0.20-0.25
(C) 0.30-0.40 (D) 0.40-0.50

263. The draft of 'Power drawn eight tined paddy transplanter' is:
(A) 250 kg (B) 275 kg
(C) 325 kg (D) 350 kg

264. How much is the working efficiency of power drawn eight tined paddy transplanter (ha/day):
(A) 0.5 (B) 0.6
(C) 0.7 (D) 0.8

265. Rotavator works with the help of:
(A) Hydraulic pressure
(B) PTO shaft
(C) Both (A) and (B)
(D) None of these

266. Rotavator of 1.5 meter size can easily run by the tractor of:
(A) 40 HP (B) 45 HP
(C) 50 HP (D) 55 HP

267. Rotavator of 1.75 meter size can easily run by the tractor of:
(A) 40 HP (B) 45 HP
(C) 50 HP (D) 55 HP

268. Rotavator of 2.0 meter size can easily run by the tractor of:
(A) 40 HP (B) 45 HP
(C) 50 HP (D) 55 HP

269. Rotavator of 2.25 meter size can easily run by the tractor of:
(A) 40 HP (B) 45 HP
(C) 50 HP (D) 55 HP

270. The prime objective of conservation tillage is:
(A) Partial retention of plant residues
(B) Removal of plant residues
(C) Complete retention of plant residues
(D) All of these

271. In no tillage systems, the surface soil layers have:
(A) Lower bulk density
(B) Higher bulk density
(C) Lower water infiltration
(D) Higher total pore space

272. Conservation tillage tends to encourages:
(A) Higher microbial population
(B) Lower number of earthworms
(C) Residues soil fauna
(D) None of these

273. Winnowing of threshed crop can be done with the help of:
(A) Puddler (B) Thresher
(C) Winnower fan (D) Reapers

274. Interculture operations can be done in the standing crop:
(A) Cultivator (B) Hand hoe
(C) Harrow (D) None of these

275. The physical condition of the soil resulting from tillage is known as:
(A) Tilth (B) Till
(C) Hoeing (D) None of these

276. Whom soil is brought to such a condition ideal for crop growth, it is called:
(A) Seed bed (B) Good tilth
(C) Both (A) and (B)
(D) None of these

277. Findout the area sprayed by a full tank of a sprayer, if the volume of the sprayer tank is 225 litres and rate of spray application/ha is 159 litres:
(A) 1.25 ha (B) 1.50 ha
(C) 1.75 ha (D) 2.00 ha

278. A soil clod having oven dry weight of 35 g ws resin/wax cooled and dipped into a measuring cylinder containing initial level of water 300 ml. The height of water level in the measuring cinder increased to 330 ml after dipping of clod. Findout the bulk density (g/cm^3) of clod:
(A) 1.40 (B) 1.50
(C) 1.17 (D) 1.08

279. The oven dry weight of soil per unit of its bulk volume is called:
(A) Bulk density
(B) Particle density
(C) Both (A) and (B)
(D) None of these

280. The mass per unit volume of soil particle is known as:
(A) Bulk density
(B) Particle density
(C) Both (A) and (B)
(D) None of these

281. Bulk density of soil is expressed as:
(A) g/cm^3 (B) g/cm^2
(C) g/cm (D) g/m

282. Particle density of soil is expressed as:
(A) g/cm^2 (B) g/cm^3
(C) g/cm (D) g/m

283. Soil having low bulk density exhibits:
(A) Favourable for good physical conditions
(B) Favourable for poor physical conditions
(C) No effect on physical conditions
(D) Unknown for physical conditions

284. Soil having high bulk density exhibits:
(A) Favourable for good physical conditions
(B) Favourable for poor physical conditions
(C) No effect on physical conditions
(D) Unknown for physical conditions

285. Bulk density of soil is also known as:
(A) Apparent specific density
(B) Particle density
(C) Either (A) or (B)
(D) Neither (A) nor (B)

286. How much is the working efficiency of 'Shallow Persian wheel' is?
(A) 0.43 ha (B) 0.33 ha
(C) 0.23 ha (D) 0.20 ha

287. How much is the working efficiency of 'Deep Persian wheel' is?
(A) 0.15 ha (B) 0.24 ha
(C) 0.30 ha (D) 0.35 ha

288. How much is the working efficiency of 'Single chain pump' is?
(A) 0.15 ha (B) 0.20 ha
(C) 0.27 ha (D) 0.35 ha

289. How much is the working efficiency of 'Double chain pump' is?
(A) 0.15 ha (B) 0.20 ha
(C) 0.30 ha (D) 0.40 ha

290. How much is the working efficiency of 'Washer Persian wheel' is?
(A) 0.23 ha (B) 0.20 ha
(C) 0.15 ha (D) 0.10 ha

291. How much is the working efficiency of 'Egyptian screw' is?
(A) 0.40 ha (B) 0.30 ha
(C) 0.20 ha (D) 0.15 ha

292. How much is the working efficiency of 'Swing basket' is?
(A) 0.08 ha (B) 0.12 ha
(C) 0.18 ha (D) 0.28 ha

293. How much is the working efficiency of 'Charsa or mot' is?
(A) 0.28 ha (B) 0.22 ha
(C) 0.18 ha (D) 0.08 ha

294. How much is the working efficiency of 'Jal chakra' is?
(A) 0.80 ha (B) 0.70 ha
(C) 0.50 ha (D) 0.40 ha

295. How much is the working efficiency of 'Don' is?
(A) 0.10 ha (B) 0.18 ha
(C) 0.28 ha (D) 0.38 ha

296. How much is the working efficiency of 'Mayadas lift (single)' is?
(A) 0.36 ha (B) 0.26 ha
(C) 0.16 ha (D) 0.10 ha

297. How much is the working efficiency of 'Mayadas lift (double)' is?
(A) 0.20 ha (B) 0.25 ha
(C) 0.30 ha (D) 0.40 ha

298. How much is the working efficiency of 'centrifugal pump (12.5 cm)' is?
(A) 1.13 ha (B) 1.0 ha
(C) 0.90 ha (D) 0.80 ha

299. How much is the working efficiency of 'Hand pump' is?
(A) 0.40 ha (B) 0.30 ha
(C) 0.20 ha (D) 0.10 ha

300. 'New Holland' is manufactured at:
(A) Nabha, Pb (B) Pune, MH
(C) Buddi, HP (D) Mohali , HP

301. 'Indo Farm' is manufactured at:
(A) Nabha, Pb (B) Pune, MH
(C) Baddi, HP (D) Mohali , HP

302. 'HMT' tractor is manufactured at:
(A) Pinjore (B) Panchkula
(C) Hyderabad (D) All of these

303. Which seeding and fertilizer application equipment is developed by CCSHAU, Hisar centre?
(A) Ridge seeder
(B) Two-row bullock drawn FESPO plough
(C) Tractor drawn planter
(D) Malviya seed cum fertilizer drill

304. Mechanisation of agriculture is required for:
(A) Better agricultural yield
(B) Multiple cropping
(C) Generation of employment
(D) All of these

305. How much farm power is available per hectare?
(A) 0.5 KW (B) 1.0 KW
(C) 1.5 KW (D) 2.0 KW

306. Level of farm mechanization is affected by:
(A) Size of holding
(B) Labour requirement
(C) Intensification of crop production
(D) All of these

307. Which of the following state is most mechanization in India?
(A) UP (B) Punjab
(C) Haryana (D) Kerala

308. Raingun mainly used for:
(A) Irrigation method
(B) Shoting (C) Ploughing
(D) None of these

309. Animal drawan cane planter is develpd by:
(A) NBRI (B) IARI
(B) CFTRI (D) IISR

310. Length of cutter bas for the tractor operated mower or vertical conveyor reaper is:
(A) 1.5 – 3.0 m (B) 3.0 – 4.5 m
(C) 4.5 – 5.5 m (D) 5.5 – 6.5

311. The ratio of rearward deflection the load given to outer end of the cutter bar is about:
(A) Velocity index
(B) Angular velocity index
(C) Reel speed index
(D) Kinematic index

312. Speed of the cuter bar in case of combine (grain) is:
(A) 100-250 rpm (B) 500-800 rpm
(C) 800-1000 rpm (D) 400-550 rpm

313. Unthresheld seed discharged from the rear of the machine either in straw or in the material from the cleaning shoe is called:
(A) Cylinder loss (B) Walker loss
(C) Shoe loss (D) None of these

314. Free seed carried over the walker or rack in straw and discharged from the rear of machine is called:
(A) Cylinder loss (B) Gathering loss
(C) Walker loss (D) Shoe loss

315. Free seed carried over the straw discharge from the rear of shoe is called:
(A) Cylinder loss (B) Walker loss
(C) Shoe loss (D) Gathering loss

316. Axial flow rasp bar cylinder gives better tthreshing efficiency than cross-flow rasp bar due to:
(A) Axial path increases residence time and number of impact on material
(B) Greater peripheral speed results in higher threshing efficiency
(C) Lesser concave clearance residence time
(D) All of these

317. A combine harvestor with 4 m cutter bar and 4 km/hr speed of operation will harvest ------ ha/hr:
(A) 1.2 (B) 2.5
(C) 1.6 (D) 1.0

318. Plough generally used for breaking impervious layer just below ploughing layers with help of number of tynes is called:
(A) Sub soiler (B) Chisel plough
(C) Disc plough (D) Indebenous plough

319. Rotavator working in reverse direction for compacting the soil and giving good touch to inter culturing is called:
(A) Rotary tiller (B) Power tiller
(C) Power harrow (D) Treader

320. The rotary or oscillating tool which does not turn the soil upsid down is:
(A) Power tiller (B) Rotavator
(C) Power harrow
(D) Treader or rotary tiller

321. V-shaped sweeps are best suited for:
(A) Primary tillage
(B) Stubble mulch tillage
(C) Rotary tillage
(D) Minimum tillage

322. Which stage represent optimum conditions for tillage?
(A) Plastic (B) Hard (cemented)
(C) Friable (crumby) (D) None of these

323. Rotary tiller are wide adopted in irce cultivation due to:
(A) Excessive pulverization
(B) Suitability for puddling paddy field
(C) Reduced draft requirement
(D) Complete inversion of soil

324. What refers to the surface roughness of the tool?
(A) Micro shape (B) Gross shape
(C) Edge shape (D) Macro shape

325. The movement of soil across a tool surface without sticking is called as:
(A) Pulverisation (B) Inversion
(C) Scouring (D) Turning

326. Penetration of disc plough can be improved by:
(A) Increasing disc angle
(B) Decreasing tilt angle
(C) Both (A) and (B)
(D) None of these

327. The power required to operate a country plough at a speed of 3 km/hr with 90 kg pull 300 angle from the horizontal is:
(A) 2.5 hp (B) 0.75 hp
(C) 0.5 hp (D) 0.85 hp

328. The water horse power required to discharge the liquid through a sprayer at 30 lit/min. rate and 30 kg/cm^2 pressure will be:
(A) 3.5 hp (B) 4 hp
(C) 2 hp (D) 2.5 hp

329. Centre of resistance of a mould board plough lies at distance equal to:
(A) 1/4th the size of plough from the share wing
(B) 3/4th the size of plough from the share wing
(C) 2/3rd the size of plough from the share wing
(D) None of these

330. Drillability of fertilizers is .... Proportional to the kinetic angle of repose:
(A) Inversely (B) Directly
(C) Unaffected (D) None of these

331. Large tilt angle is best for:
(A) Acidic soil
(B) Sticky and non-scoring soil
(C) Dry and cemented soil
(D) None of these

332. The plough used for maximum moisture conservation is:
(A) Disc harrow (B) Cultivator
(C) Chisel plough (D) None of these

333. Total grain losses in a combine increase with:
(A) Increase in forward speed
(B) Decrease in stubble height
(C) Increase in concave clearance
(D) All of these

334. Width of cut of disc plough is increased by:
(A) Increasing disc angle
(B) Increasing tilt angle
(C) Addition of more weight
(D) Decreasing disc angle

335. Sub soiler are operated at maximum depth of:
(A) 45-75 cm (B) 20-25 cm
(C) 75-90 cm (D) 40-60 cm

336. For operating rotavator in trashy soils, which lines are better:
(A) L-shaped (B) Hook shaped
(C) Backward curved
(D) T-shaped

337. Soil strength is determined by:
(A) Penetrometer (B) Micrometer
(C) Hydrometer (D) pH meter

338. Power sprayer are operated at a pressure of:
(A) 15-20 kh/m^2 (B) 10-20 kg/m^2
(C) 0-65 kg/m^2 (D) 20-55 kg/m^2

339. In electrostatic spraying, the total volume of liquid to be sprayed over one hectare is:
(A) < 1 litre (B) 20 litres
(C) 80 litres (D) >100litres

340. The power developed by average par of bullocks varies from:
(A) 0-0.5 hp (B) 0.5-2.0 hp
(C) 0.5-1.0 hp (D) 1.5-2.0 hp

341. The size of MB plough is expressed in terms of its:
(A) Width of cut (B) Depth of cut
(C) Length of share
(D) None of these

342. Per hectare application rate of a ULV sprayer is:
(A) upto 5 litres (B) 5 – 100 litres
(C) 400 litres (D) 250-350 litres

343. Horizontal plate seed metering device is used in:
(A) Planter
(B) Seed drill
(C) Seed cum fertilizer drill
(D) None of these

344. The depth of penetration of a disc harrow is increased by:
(A) Tilt angle (B) Disc angle
(C) Gang angle (D) All of these

345. Sub soiler plough is best suited for:
(A) Making ditched
(B) Breaking hard pan
(C) Inter cultivation
(D) None of these

346. In check rowing, the seed spacing is:
(A) Constant (B) Continuous
(C) Same as rows (D) None of these

347. Seed drill have slip equal to:
(A) Zero (B) Positive slip
(C) Negative slip (D) None of these

348. Most mechanized crop in India is presently:
(A) Cotton (B) Wheat
(C) Maize (D) Pea

349. Tilt angle of a disc plough is generally:
(A) 10^0 (B) 25^0
(C) 20^0 (D) 15^0

350. The throat of a reaper cutter bar is:
(A) 5 cm (B) 8 cm
(C) 8.5 cm (D) 15 cm

351. Disc angle in disc plough is:
(A) 42-45 (B) 34-37
(C) 38-40 (D) 25-30

352. Land scraper is an equipment best used for:
(A) Reclamation (B) Cultivation
(C) Soil movement (D) None of these

353. In size of a seed drill is doubled and speed os halved, then coverage will be:
(A) Doubled (B) Halved
(C) Unchanged (D) None of these

354. Type of threshing drum in paddy thresher is:
(A) Peg type
(B) Resp bar type
(C) Hammer mill type
(D) None of these

355. Efficiency of a pair of bullocks is .... the efficiency of a single bullock in field:
(A) Less than (B) Greater than
(C) Twice (D) None of these

356. Traction of wheel can be improved by:
(A) Using high hitch point
(B) Ballasting rear wheels
(C) Using high hitch point as well as ballasting rear wheels
(D) None of these

357. The traction co-efficient is maximum in the field when:
(A) It is ploughed
(B) It is irrigated
(C) It is dry
(D) It is ploughed irrigated

358. The most used and least effieicnt power outlet of a tractor is:
(A) PTO shaft in the front
(B) PTO shaft in the rear
(C) Drawbar in the rear
(D) None of these

359. An indigenous plough is:
(A) A wetland puddler
(B) A primary tillage implement
(C) A secondary tillage implement
(D) None of these

360. Land plan is ideally suited for:
(A) Land leveling (B) Earth moving
(C) Top finishing (D) None of these

361. Vertical suction of a plough influences:
(A) Pulverization (B) Depth of cut
(C) Width of cut (D) Direction of pull

362. The type of starting aid generally used in diesel power tiller is:
(A) Thermostat
(B) Glow plug
(C) Decompression level
(D) None of these

363. Fly wheel type chaff cutter used:
(A) Spiral gear (B) Bevel gear
(C) Worm gear (D) Both (A) and (B)

364. Power operated paddy transplanter is being manufactured in India by:
(A) M/s Escort Ltd
(B) M/s Mitsubishi Ltd
(C) M/s Eicher Tractors Ltd
(D) M/s H.M.T.

365. Hand operated sprayers are operated at .... kg/cm^2:
(A) 2-8 (B) 7-10
(C) 1-7 (D) 10-15

366. For spray application, the quantity to be applied in a field will depend upon:
(A) Forward speed (B) No. of nozzles
(C) Pressure (D) All of these

367. Heavy draft of a disc plough is due to:
(A) Blunt disc (B) Forrows too wide
(C) Loose bearings (D) None of these

368. Puddling is done to:
(A) Reduce percolation of water
(B) Kill weeds
(C) Both A & B
(D) None of these

369. A stubble type mould board gives:
(A) Medium pulverisation
(B) Through pulverisation
(C) Little pulverisation
(D) None of these

370. The size of a seed drill is expressed by:
(A) Area covered per unit time
(B) Length x width of the mechine
(C) The number of furrow openers x distance between two furrow openers
(D) None of these

371. The size of the planter is given by:
(A) Length x height of the mechine
(B) Number of furrow operners x spacing between them
(C) Length x width of the mechine
(D) Area covered per unit time

372. Working life of a tractor drawn cultivator is usually:
(A) 10000 hrs (B) 2500 hrs
(C) 15000 hrs (D) 8500 hrs

373. The type of furrow opener recommended for use in hard and trashy ground and also in wet, dry soil in:
(A) Single disc type
(B) Hoe type
(C) Stub runner type
(D) Curved runner type

374. Most suitable furrow opener for seeding in a trashy and hard seed bed:
(A) Curved type (B) Shovel type
(C) Disc type (D) Hoe type

375. Accurate and inclined placement of seeds to provide rows in two perpendicular direction is known as:
(A) Drilling
(B) Planting
(C) Precision seeding
(D) Check-row planting

376. Raising of crop with the least tillage operations is called:
(A) Zero tillage
(B) Minimum tillage
(C) No tillage
(D) Heavy tillage

377. Wet tillage :
(A) Destroy soil texture
(B) Improve soil texture
(C) Destroy soil structure
(D) Improve soil structure

378. Continuous zero-tillage in the crop field encourages the preponderance of :
(A) Annual weeds
(B) Binnial weeds
(C) Perennial weeds
(D) Parasitic weeds

379. Blind tillage refers to :
(A) Summer ploughing
(B) Hoeing in standing crop rows
(C) Primary tillage
(D) Hoeing before germination

380. Zero tillage system was first weed successfully in 1950 in pasture renovation in :
(A) Germany
(B) Japan
(C) United Kingdom
(D) USA

381. Zero-tillage seed drill is used for sowing of:
(A) Wheat in maize-wheat cropping system
(B) Wheat in maize-wheat cropping system
(C) Wheat in rice-wheat cropping system
(D) None of these

ANSWER SHEET
4. Tillage Management

1. (A)	2. (D)	3. (A)	4. (D)	5. (C)	6. (A)	7. (B)	8. (C)	9. (D)	10. (D)
11. (A)	12. (D)	13. (A)	14. (A)	15. (A)	16. (B)	17. (D)	18. (B)	19. (D)	20. (D)
21. (C)	22. (B)	23. (C)	24. (B)	25. (C)	26. (A)	27. (A)	28. (B)	29. (A)	30. (A)
31. (B)	32. (D)	33. (B)	34. (C)	35. (A)	36. (B)	37. (B)	38. (C)	39. (B)	40. (A)
41. (C)	42. (B)	43. (D)	44. (C)	45. (D)	46. (D)	47. (A)	48. (C)	49. (C)	50. (B)
51. (B)	52. (C)	53. (D)	54. (B)	55. (B)	56. (D)	57. (B)	58. (D)	59. (D)	60. (B)
61. (D)	62. (C)	63. (A)	64. (C)	65. (D)	66. (B)	67. (B)	68. (A)	69. (A)	70. (C)
71. (C)	72. (A)	73. (B)	74. (B)	75. (C)	76. (C)	77. (C)	78. (C)	79. (A)	80. (C)
81. (A)	82. (A)	83. (B)	84. (B)	85. (C)	86. (C)	87. (B)	88. (A)	89. (C)	90. (C)
91. (C)	92. (A)	93. (A)	94. (D)	95. (D)	96. (A)	97. (B)	98. (B)	99. (C)	100. (D)
101. (C)	102. (C)	103. (B)	104. (A)	105. (D)	106. (B)	107. (A)	108. (D)	109. (B)	110. (D)
111. (A)	112. (B)	113. (B)	114. (D)	115. (C)	116. (B)	117. (E)	118. (D)	119. (B)	120. (A)
121. (D)	122. (B)	123. (D)	124. (D)	125. (C)	126. (B)	127. (D)	128. (A)	129. (C)	130. (C)
131. (B)	132. (B)	133. (C)	134. (B)	135. (D)	136. (C)	137. (A)	138. (D)	139. (D)	140. (A)
141. (D)	142. (B)	143. (A)	144. (B)	145. (A)	146. (D)	147. (B)	148. (A)	149. (A)	150. (C)
151. (B)	152. (D)	153. (D)	154. (C)	155. (A)	156. (B)	157. (D)	158. (D)	159. (B)	160. (D)
161. (C)	162. (D)	163. (D)	164. (D)	165. (C)	166. (B)	167. (A)	168. (A)	169. (A)	170. (B)
171. (C)	172. (A)	173. (D)	174. (B)	175. (A)	176. (A)	177. (C)	178. (C)	179. (A)	180. (B)
181. (C)	182. (C)	183. (B)	184. (B)	185. (C)	186. (B)	187. (A)	188. (A)	189. (A)	190. (C)
191. (B)	192. (A)	193. (C)	194. (B)	195. (A)	196. (C)	197. (B)	198. (A)	199. (B)	200. (A)
201. (A)	202. (B)	203. (C)	204. (D)	205. (A)	206. (B)	207. (C)	208. (A)	209. (B)	210. (C)
211. (D)	212. (C)	213. (A)	214. (B)	215. (C)	216. (A)	217. (B)	218. (B)	219. (B)	220. (A)
221. (D)	222. (D)	223. (B)	224. (B)	225. (A)	226. (A)	227. (B)	228. (C)	229. (C)	230. (C)
231. (C)	232. (D)	233. (C)	234. (C)	235. (B)	236. (D)	237. (C)	238. (A)	239. (A)	240. (C)
241. (B)	242. (A)	243. (B)	244. (C)	245. (D)	246. (B)	247. (B)	248. (C)	249. (C)	250. (B)
251. (D)	252. (D)	253. (A)	254. (B)	255. (C)	256. (C)	257. (B)	258. (D)	259. (A)	260. (C)
261. (B)	262. (A)	263. (C)	264. (D)	265. (B)	266. (A)	267. (B)	268. (C)	269. (D)	270. (A)
271. (B)	272. (A)	273. (C)	274. (B)	275. (A)	276. (C)	277. (B)	278. (C)	279. (A)	280. (B)

281. (A) 282. (B) 283. (A) 284. (B) 285. (A) 286. (A) 287. (B) 288. (C) 289. (D) 290. (A)
291. (B) 292. (C) 293. (D) 294. (A) 295. (B) 296. (C) 297. (D) 298. (A) 299. (B) 300. (A)
301. (C) 302. (D) 303. (C) 304. (D) 305. (D) 306. (D) 307. (B) 308. (A) 309. (D) 310. (A)
311. (C) 312. (D) 313. (A) 314. (C) 315. (C) 316. (A) 317. (C) 318. (B) 319. (D) 320. (C)
321. (B) 322. (C) 323. (B) 324. (A) 325. (C) 326. (B) 327. (B) 328. (C) 329. (B) 330. (A)
331. (B 332. (C) 333. (D) 334. (A) 335. (A) 336. (A) 337. (A) 338. (D) 339. (A) 340. (C)
341. (A) 342. (A) 343. (A) 344. (C) 345. (B) 346. (C) 347. (B) 348. (B) 349. (D) 350. (A)
351. (A) 352. (C) 353. (C) 354. (A) 355. (A) 356. (C) 357. (C) 358. (C) 359. (B) 360. (C)
361. (B) 362. (C) 363. (B) 364. (A) 365. (B) 366. (A) 367. (C) 368. (C) 369. (A) 370. (C)
371. (B) 372. (D) 373. (A) 374. (C) 375. (D) 376. (B) 377. (C) 378. (C) 379. (D) 380. (D)
381. (C)

5 SEED AND SOWING

1. Seed is a:
(A) Developed ovule
(B) Fertilised and developed ovule
(C) Developed ovary
(D) Fertilised and developed ovary

2. The seeds of kind/species state by the sender, or found the predominance in the test is known as:
(A) Coleoptile
(B) Pure seed
(C) Germination index
(D) Purity analysis

3. Artificial seed refers to:
(A) Seeds of a plant with desired characteristics synthesized with recombinant DNA technology
(B) Somatic embryos encapsulated in a gel
(C) Seeds of synthetic plants
(D) All of these

4. Terminator seeds produced by seed companies using techniques of genetic engineering are in fact:
(A) Hybrid seeds which germinate in the farmer's field but do not produce seeds in the next generation
(B) Hybrid seeds which give a very high yield
(C) Seeds which germinate and produce incompatible gametes
(D) Seeds that develop into fertile plants which produce non germinable seeds

5. Synthetic seeds are:
(A) Protoplasts (B) Somaclones
(C) Cybris (D) Somatic embryos

6. A transgenic rice that accumulates β-carotene in its seeds; expected to combat vitamin A deficiency areas where rice is the staple food is:
(A) Golden rice (B) Hybrid rice
(C) Boiled rice (D) Desi rice

7. During seed germination, seed coat ruptured due to:
(A) Differentiation of cotyledons
(B) Massive glycolysis in endosperm and cotyledons
(C) Massive imbibitions of water
(D) Sudden increase in cell division

8. Jalna city of Maharastra is known as:
(A) Seed city
(B) Education city
(C) Tourist city
(D) Highly developed city

9. The purest seed, genetically is:
(A) Breeder seed
(B) Nucleus seed
(C) Foundation seed
(D) Certiified seed

10. Breeder seed is produced by:
(A) Plant breeder
(B) Government agencies
(C) Registered growers
(D) N.S.C.

11. Foundation seed is produced from:
(A) Nucleus seed (B) Breeder seed
(C) Quality seed (D) Certified seed

12. Certified seed is produced from:
(A) Nucleus seed
(B) Breeder seed
(C) Foundation seed
(D) Certified seed

13. Breeder seed is produced by:
(A) Nucleus seed (B) Breeder seed
(C) Quality seed (D) Certified seed

14. Quality seed is produced by:
(A) Nucleus seed (B) Breeder seed
(C) Quality seed (D) Certified seed

15. Basic input of agriculture is:
(A) Seed (B) Fertilizer
(C) Soil (D) Irrigation

16. How many primary samples will be taken from 3 containers of 100 kg?
(A) 3 (B) 5
(C) 15 (D) 9

17. Difference between seed and grain is:
(A) Seed processing
(B) Genetic purity
(C) Seed quality (D) Cost

18. Embryo exised test ----- used for surface disinfestation:
(A) NaOCl (B) Lactophenol
(C) $AlK(SO_4)_2$ (D) All of these

19. Fanning mill is:
(A) Air screen cleaner
(B) Hammer mill
(C) Blower (D) Rice mill

20. The optimum seed rate for sunflower in (kg/ha):
(A) 8-10 (B) 10-15
(C) 18-20 (D) 30-35

21. Seed lot certificate colour is
(A) Orange (B) Green
(C) Purple (D) Orange or green

22. National Nematode collection of India is located at
(A) New Delhi (B) Dharwad
(C) Anand (D) Bangalore

23. Leverage ratio is:
(A) Owner's equity/Total assets
(B) Owner's equity/Current liabilities
(C) Current liabilities/Owner's equity
(D) Total debt/Owner's equity

24. Seed sample certificate colour is:
(A) Blue (B) Orange
(C) Purple (D) White

25. Hot water treatment is used to break dormancy in:
(A) Lentil (B) Cherry
(C) Cotton (D) None

26. Tarai Development Corporation (TDC) was established in:
(A) 1965 (B) 1969
(C) 1970 (D) 1971

27. Alkaloid test is used for:
(A) Lucern (B) Lupins
(C) Crysanthemum (D) Cereals

28. Average diameter of bulbet of onion (seed standard) shall not be less than:
(A) 2.5 cm (B) 1 cm
(C) 3 cm (D) 2 cm

29. Pure seed fraction in *Abelmosschus esculentus*:
(A) 98 (B) 95
(C) 99 (D) 96

30. Harvesting index is maximum in:
(A) Carrot (B) Pigeonpea
(C) Rice (D) Wheat

31. ---------- sugar found in germinating seed in large amount:
(A) Maltose (B) Sucrose
(C) Cellulose (D) Lactose

32. Match the following:

List I	List II
A. Accelerating Ageing test	a. Maize
B. Cold test	b. Soybean
C. A_6 Test	c. Virus Y
D. M_1 Test	d. Leaching Test
E. Cool Germination Test	e. Groundnut
F. Mitochondrial Activity	f. Cotton

	A	B	C	D	E	F
(A)	a	b	c	d	e	f
(B)	a	b	c	d	f	e
(C)	b	a	c	d	e	f
(D)	b	a	c	d	f	e

33. In maize, a foundation single cross production field must be isolated by not less than ------ from maize with icernal colour or texture different from that of seed part:
(A) 400 mts (B) 600 mts
(C) 200 mts (D) 250 mts

34. A group of similar plants in structural features and performance is known as:
(A) Variety (B) Clone
(C) Pure line (D) Inbred line

35. Possible reason for seed dormancy is:
(A) Presence of pathogens
(B) Cracking of hulls
(C) Immature embryo
(D) Green distoration

36. A sound seed certification programme requires:
(A) Direct participation of breeder
(B) Use of sophisticated equipment
(C) Classification of seed
(D) Support of law

37. The seed forms which generally used are called:
(A) Breeder Seed (B) Foundation seed
(C) Certified seed (D) Nucleus seed

38. The rudimentary root of the seed or seedling that forms the primary root of the young plant is known as:
(A) Rachis (B) Radicle
(C) Rachilla (D) Raceme

39. Seed drying is very important to maintain its:
(A) Viability and vigour
(B) Protein content
(C) Oil content
(D) Chemical composition

40. The two most important factors influencing the life span of seeds under storage are:
(A) Relative humidity and temperature
(B) Rainfall and Relative humidity
(C) Climate and temperature
(D) Seed size and moisture content

41. Maximum moisture content for safe storage in wheat seed is:
(A) 12 (B) 15
(C) 7 (D) None of these

42. Determination of the physical composition of the seed samples in terms of the percentage (by weight) of seeds of the kind under consideration (pure seed), other crop seeds, weed seeds and inert matter is known as:
(A) Coleoptile
(B) Coleorhiza
(C) Germination index
(D) Purity analysis

43. In India, normally how many generation system seed are produced?
(A) 1 (B) 2
(C) 3 (D) 4

44. Off types are removed by:
(A) +ve selection
(B) -ve selection
(C) Pure line selection
(D) Mass selection

45. Seed certification involves:
(A) 1 phase (B) 3 phase
(C) 5 phase (D) 7 phase

46. Slow drying seed is:
(A) Cereals (B) Rapeseed and mustard
(C) Grass (D) Pulses

47. Seed dormancy are shown by:
1 Nigella 2. Xanthium
3. Tomato 4. Lettuce
(A) Only 1 & 2 (B) Only 1, 2 and 3
(C) Only 1, 2 & 4 (D) Only 2

48. +ve photoblastic seeds are:
(A) Nigella (B) Silene
(C) Nemophills (D) Arraranutum

49. Epigeal germination is found in:
(A) Pea (B) Gram
(C) Cucumber (D) Mango

50. Germination is inhibited by:
(A) Red light (B) Blue light
(C) IR light (D) UV light

51. Tetrazolium test determine the level of activities of enzyme:
(A) ATPase (B) Dehydrogenase
(C) Carboxylase (D) Isomerase

52. Production of seedless grapes require:
(A) Gibberellins (B) NAA
(C) Ethylene (D) IAA

53. Which of the following vigour test method estimates decarboxylation activities?
(A) Leachate test (B) GADA test
(C) ATPase test (D) Exhaustion test

54. ISTA was established in the year:
(A) 1871 (B) 1876
(C) 1921 (D) 1924

55. While sampling from a seed lot of 50 bags, the minimum number of primary sample should be:
(A) 5 (B) 10
(C) 15 (D) 20

56. Certified seed of cotton should have minimum germination of:
(A) 50% (B) 60%
(C) 70% (D) 80%

57. Half gram is working sample is required for purily analysis of:
(A) Tomato (B) Onion
(C) Berseem (D) Tobacco

58. The most suitable range of temperature for the germination of rabi crop seeds is:
(A) 5-10°C (B) 20-30°C
(C) 31-40°C (D) 41-50°C

59. Specific combining ability of the parents should be high in order to develop:
(A) Pure line (B) Synthetics
(C) Composites (D) Hybrids

60. Seed viability is tested by use of:
(A) Bromide (B) Iodide
(C) Chloride (D) None of these

61. For seed samples kept in an incubator for germination test, light is:
(A) Always essential (B) Never essential
(C) Not harmful (D) Harmful

62. Development of embryo without fertilization is:
(A) Apomixis (B) Amghimixis
(C) Parthenocarpy (D) None of these

63. Male sterility is used to develop:
(A) Inbreds (B) Hybrids
(C) Synthetic (D) None

64. Occurrence of more than one embryo in seed is known as:
(A) Polysomy (B) Polyembryony
(C) Apogamy (D) None of these

65. The seed rate of hybrid maize is:
(A) 20-25 kg/ha (B) 15-20 kg/ha
(C) 35-40 kg/ha (D) 10-12 kg/ha

66. The GoT is done for verification of:
(A) Physical purity (B) Genetic purity
(C) Germination % (D) None of these

67. In T_2 test colour of living tissues of seed changed to:
(A) Red (B) Blue
(C) Yellow (D) Green

68. The minimum number of counts to be taken for an area upto 2 ha in wheat design any field inspection are:
(A) 7 (B) 5
(C) 3 (D) 10

69. The standard for germination % (min.) for certified seed of gram is:
(A) 85 (B) 90
(C) 80 (D) 75

Match the following

70.

List – I	List – II
a. Stratification	1. Peas
b. Bleached seed	2. Thermodormancy
c. Hybrex	3. Father of seed technology
d. Nobbe	4. Wheat

Codes:

	A	B	C	D
(A)	1	2	3	4
(B)	2	1	4	3
(C)	2	3	4	1
(D)	4	2	3	1

71.

List – I	List – II
a. Cotton	1. Bolting
b. Reddish	2. Heterostyly
c. Tobacco	3. Delinting
d. Brinjal	4. Steckling

Codes:

	A	B	C	D
(A)	1	4	3	2
(B)	1	2	3	4
(C)	3	4	1	2
(D)	3	2	4	1

72.

List – I	List – II
a. Genetic purity	1. Cross polination
b. Scarification	2. Offtype
c. Dormancy	3. Seed dormancy
d. Roguing	4. Somatic mutation

Codes:

	A	B	C	D
(A)	4	3	2	1
(B)	1	3	4	2
(C)	4	1	3	2
(D)	2	1	3	4

73. List – I | List – II

List – I	List – II
a. Seed research	1. Certification
b. Poor storage	2. Polyembryony
c. Notified variety	3. ISST
d. Citrus	4. Low germination

Codes:

	A	B	C	D
(A)	3	2	1	4
(B)	3	1	2	4
(C)	2	1	4	3
(D)	3	4	1	2

74.

List – I	List – II
a. Orthodex seed	1. Legume
b. Recalcitrant seed	2. Sulphuric acid
c. Hard seed	3. Wheat
d. Scarification	4. Cocca

Codes:

	A	B	C	D
(A)	1	2	3	4
(B)	3	4	1	2
(C)	1	4	2	3
(D)	4	3	2	1

75.

List – I	List – II
a. Germination test	1. Other crop seed
b. Moisture test	2. Formazan
c. Purity test	3. PUG
d. T2 test	4. Pre drying

Codes:

	A	B	C	D
(A)	2	4	3	1
(B)	3	4	1	2
(C)	1	3	2	3
(D)	3	1	2	4

76.

List – I	List – II
a. Vigour test	1. Trier
b. Nobbe	2. Hiltners test
c. Ganga 5	3. Hybrid maize
d. Quick viability	4. T2 test

Codes:

	A	B	C	D
(A)	4	1	2	3
(B)	1	2	3	4
(C)	2	1	3	4
(D)	3	4	2	1

77.

List – I	List – II
a. Seed Act	1. 1968
b. NSC	2. 1961
c. Seed rules	3. 1966
d. World seed day	4. 1963

Codes:

	A	B	C	D
(A)	2	4	3	1
(B)	3	4	1	2
(C)	4	1	3	2
(D)	1	2	3	4

78.

List – I	List – II
a. NSP	1. 1961
b. UPOV	2. 1983
c. Seed control order	3. 1976
d. New Seed Act	4. 2006

Codes:

	A	B	C	D
(A)	2	4	3	1
(B)	1	2	3	4
(C)	4	2	1	3
(D)	3	1	2	4

79.

List – I	List – II
a. Breeder seed	1. Blue
b. Foundation seed	2. Buff
c. Nucleous seed	3. Golden yellow
d. Certified seed	4. White

Codes:

	A	B	C	D
(A)	1	3	2	4
(B)	3	4	2	1
(C)	2	3	4	1
(D)	1	4	2	1

80. There are ------ seed certification agencies in India at present:

(A) 11 (B) 15
(C) 19 (D) 23

81. The recommended isolation distance (meters) for foundation seed production of radish, turnip and cauliflower is:

(A) 800 (B) 1200
(C) 1600 (D) 2000

82. The isolation distance required to produce the certified seed of chickpea, lentil, mungbean and urdbean is ------ meters:

(A) 25 (B) 15
(C) 10 (D) 5

83. The number of plants per hectare to be observed in taking one count in seed production of pearlmillet, sorghum, rice and wheat is ----as recommended:

(A) 500 (B) 1000
(C) 1500 (D) 2000

84. The sampling lot in agricultural seed of large size should not exceed ----- kg except that of maze (where 40000 kg is allowed):

(A) 5000 (B) 10000
(C) 15000 (D) 20000

85. Use of linear measurements of plumule growth as a vigour test for cereals and sugarbeet was first suggested by:

(A) S. Mathews in 1929
(B) J.S. Burris in 1939
(C) Germ in 1949
(D) Hegarty in 1959

86. The seed, the progeny of which comes from foundation seed or registered seed, is handled to maintain genetic purity, approved and certified by certifying agency is called:

(A) Certified seed (B) Foundation seed
(C) Nucleus seed (D) Registered seed

87. The seed of oil palm require as high as --------°C temperature to break the dormancy:

(A) 30-40 (B) 40-50
(C) 50-60 (D) 60-70

88. The seeds of ---- are positively photoblastic:

(A) Tomato (B) Tobacco
(C) Chillies (D) Cucurbits

89. Seed plot technique is followed in:

(A) Wheat (B) Paddy
(C) Potato (D) Bajra

90. A seed vigour test that assesses the potential of a seed lot in prolonged ambient storage by:

(A) Accelerated ageing test
(B) Cold test
(C) Conductivity test
(D) Excised embryo test

91. A seed vigour test that measures the potential of a seed lot subjected to low temperature during emergence by:

(A) Accelerated aging test
(B) Cold test
(C) Conductivity test
(D) Excised embryo test

92. A seed vigour test designated to measure the electric conductivity of leachates from seeds after soaking in water by:

(A) Accelerated aging test
(B) Cold test
(C) Conductivity test
(D) Excised embryo test

93. A quick test for evaluating the growth potential of a root-shoot axis that has been detached from the remainder of the seed by:

(A) Accelerated aging test
(B) Cold test
(C) Conductivity test
(D) Excised embryo test

94. A bio-chemical vigour test which measure the activity of GAD enzyme present in the seed by:

(A) Accelerated aging test
(B) Cold test
(C) Hilter test
(D) Glutamic acid decarboxylase activity test

95. A vigour test designated to assess the potential of a cereal seedlings to emerge from a specified layer of brick gravel by:

(A) Accelerated aging test
(B) Cold test
(C) Hilter test
(D) Glutamic acid decarboxylase activity test

96. Trueness to type or cultivar is called:

(A) Genetic purity
(B) Geotropism
(C) Seed vigour
(D) Seed vigour index

97. Plant growth response to gravity is called:

(A) Genetic purity
(B) Geotropism
(C) Seed vigour
(D) Seed vigour index

98. If an estimate is significantly either poorer or better than a specification by:

(A) One way test
(B) Two way test
(C) Paper piercing test
(D) TX test

99. If an value is significantly either better or poorer than the other by:
(A) One way test
(B) Two way test
(C) Paper piercing test
(D) TX test

100. A seed vigour test which measures the potential of emerging seedlings to pierce the appears of a specified standard by:
(A) One way test
(B) Two way test
(C) Paper piercing test
(D) TX test

101. To determine seed viability by using 2,3,-5, tri phenyl tetrazolium chloride or bromide by:
(A) One way test
(B) Two way test
(C) Paper piercing test
(D) TX test

102. A bio-chemical vigour test which measures the intensity of dehydrogenase group of enzymes present in the seed by:
(A) Tetrazolium vigour test
(B) Two way test
(C) Paper piercing test
(D) TX test

103. Sample is a small portion taken from one point into lot is called:
(A) Primary sample
(B) Composite sample
(C) Sub sample
(D) Submitted sample

104. Sample is formed by combining and mixing all, the primary samples taken from the lot is called:
(A) Primary sample
(B) Composite sample
(C) Sub sample
(D) Submitted sample

105. The portion of a sample obtained by educing the composite sample using one of the prescribed sampling methods is called:
(A) Primary sample
(B) Composite sample
(C) Sub sample
(D) Submitted sample

106. Sample submitted to the seed testing laboratory is called:
(A) Primary sample
(B) Composite sample
(C) Sub sample
(D) Submitted sample

107. The number of containers/bags samples for obtaining primary samples in relation to the size of whole seed lot is called:
(A) Primary sample
(B) Composite sample
(C) Sampling intensity
(D) Submitted sample

108. Sub sample taken from the submitted sample in the laboratory for a given quality test is called:
(A) Primary sample
(B) Composite sample
(C) Working sample
(D) Submitted sample

109. Sum total of those proportion of the seed which determine the level of activity and performance of the seed lot during germination and seedling emergence is called:
(A) Seed vigour
(B) Seedling growth test
(C) Seed vigour index
(D) Viable seed

110. A vigour test designated to assess the rapidly of seedling growth by measuring seedling length or dry weight under specified conditions by:
(A) Seed vigour
(B) Seedling growth test
(C) Seed vigour index
(D) Viable seed

111. A product of germination percentage multiplied by seedling length of a seed lot is called:
(A) Seed vigour
(B) Seedling growth test
(C) Seed vigour index
(D) Viable seed

112. Seed is capable of germinating under proper conditions is called:
(A) Seed vigour
(B) Seedling growth test
(C) Seed vigour index
(D) Viable seed

113. The process of mechanically scarring or rubbing the hard seed coat to make it more permeable to water is known as:
(A) Scarification (B) Stratification
(C) Both of these (D) None of these

114. Exposing seed imbibing water to low temperature conditions for a few days prior to germination for breaking dormancy is known as:
(A) Scarification (B) Stratification
(C) Both of these (D) None of these

115. Summation of number of seedlings emerging on a specific day divided by the number of day on which the germination was counted is called:
(A) Coleoptile
(B) Coleorhiza
(C) Germination index
(D) Purity analysis

116. Viability of the seed is tested with:
(A) Triphenyl chloride
(B) Indole acetic acid
(C) Mercuric chloride
(D) 2,4-D

117. Seed dormancy may due to:
(A) Permeable seed coat
(B) Hard impermeable seed coat
(C) Tin seed coat (D) Lack of reserve food

118. The hormone which can break seed dormancy is:
(A) Coumarin (B) Ferulic acid
(C) ABA (D) GA

119. Mechanical injuring of seed coat to break dormancy is called:
(A) Scarification (B) Stratification
(C) Impaction (D) Compaction

120. Thiourea is used in overcoming seed dormancy by:
(A) Chemical scarification
(B) Counteracting inhibitors
(C) Inducing cell division
(D) Develop osmotic pressure

121. The most important external factor for seed germination is:
(A) Light (B) Soil
(C) Oxygen (D) Water

122. The first process which occurs when the seed is placed in the soil is:
(A) Photosynthesis
(B) Respiration
(C) Imbibition
(D) Solubilisation of food

123. Seeds placed deep in the soil do not germinate because they are:
(A) Unable to get sufficient oxygen
(B) Without sufficient food to bring the seedling to the surface
(C) Under pressure of overlying soil layers
(D) Unable to get light

124. During germination, micropyle of seed takes part in:
(A) Forming weak point for emergence of radical
(B) Absorption of water
(C) Passage of gases
(D) Leaching of inhibitors

125. Which are the external conditions required for seed germination:
(A) Oxygen, carbon dioxide and suitable temperature
(B) Oxygen, light and suitable temperature
(C) Light, moisture and suitable temperature
(D) Oxygen, moisture and suitable temperature

126. A seed which does not require oxygen for germination is:
(A) Pea (B) Rice
(C) *Typha* (D) Both (B) and (C)

127. A seed which is unable to germinate in the presence of light is:
(A) *Viscum* (B) Onion
(C) Bean (D) Maize

128. The seed in which dormancy can be broken by red light is:
(A) Pea (B) Gram
(C) Lattuce (D) Castor

129. Food is stored in albuminous seed in:
(A) Testa (B) Cotyledon
(C) Endosperm (D) Persperm

130. Vivipary is:
(A) Seed germination with subterranean cotyledons
(B) Seed germination with epiterranean cotyledons
(C) Fruit development without pollination
(D) Seed germination inside the fruit while attached to the plant

131. In hypogeal germination, the cotyledons do not turn green because they:
(A) Lack chloroplasts
(B) Contain inhibitors
(C) Remain below soil
(D) Abscise early

132. Dry seeds can endure high temperature and other adverse conditions better than germinating seeds due to:
(A) Hydration makes the enzymes sensitive to temperature
(B) Seeds take part in perennation
(C) Dry seeds possess more reserve food
(D) Dry seeds possess thick seed coats

133. Plant which buries its seeds in the soil in:
(A) Peanut (B) Potato
(C) Sweet potato (D) Cashewnut

134. Oil reserve of groundnut is present in:
(A) Embryo (B) Cotyledons
(C) Endosperm (D) Underground tubers

135. The blackish brown covering around the endosperm containing kernel of coconut is:
(A) Seed coat (B) Epicarp
(C) Mesocarp (D) Endocarp

136. An albuminous seed showing hypogeal germination is:
(A) Castor (B) Bean
(C) Gram (D) Maize

137. Oil is stored in the endosperm of:
(A) Groundnut (B) Soybean
(C) Coconut (D) Cashewnut

138. Germination of seed within fruit is:
(A) Ovipary (B) Vivipary
(C) Hypogeal (D) Epigeal

139. Alerurone layer helps in:
(A) Storage of food in endosperm
(B) Protection of embryo
(C) Utilization of stored food
(D) All of these

140. In the legume seed, food is stored in:
(A) Cotyledons (B) Endosperm
(C) Perisperm (D) Seed coats

141. Epithelial layer of maize scutellum secretes the hormone:
(A) Auxin (B) Cytokinin
(C) Gibberellin (D) Ethylene

142. Shield-shaped cotyledon/scutellum occurs in:
(A) Maize (B) Gram
(C) Pea (D) Cucumber

143. Outgrowth developing alongwith helium of the seed is:
(A) Plumule (B) Radicle
(C) Strophiole (D) Perisperm

144. A gas required for germination of pea seed is:
(A) Nitrogen (B) Oxygen
(C) Hydrogen (D) Water vapours

145. In groundnut seeds are:
(A) Geocarpic (B) Amphicarpic
(C) Photocarpic (D) Epicarpic

146. Which enzyme is active in Maize?
(A) Maltase (B) Zymase
(C) Diastase (D) Urease

147. Part of Castor seed that yields oil is:
(A) Caruncle (B) Endosperm
(C) Nucellus (D) Cotyledon

148. Which one is essential for seed germination in pea?
(A) O_2 (B) H_2
(C) N_2 (D) CO_2

149. Protective covering over radical during seed germination is:
(A) Coleorhiza (B) Coleoptile
(C) Suspensor (D) Epithelium

150. After removal of covering in pea, the seed consists of:
(A) Cotyledons
(B) Embryo
(C) Cotyledons + Endosperm
(D) Cotyledons + Endosperm + Pericarp

151. The character found only in halophytes is:
(A) Vivipary (B) Velamen
(C) Heterophylly (D) Sunken stomata

152. Endosperm is formed in angiosperms due to double fertilization. It is however, absent in certain seeds due to lack of:
(A) Certain enzymes
(B) Growth hormone
(C) Dicotyledonous hormone
(D) Nutrients

153. Coleorhiza is cap-like covering over:
(A) Radicle in dicots
(B) Plumule in dicots
(C) Plumule in monocots
(D) Radicle in monocots

154. Seed dormancy allows the plants to:
(A) Overcome unfavourable climatic conditions
(B) Develop healthy seeds
(C) Reduce viability
(D) Prevent deterioration of seeds

155. Which of the following is associated with dormancy of seeds?
(A) IBA (B) GA
(C) Ethylene (D) Ferulic acid

156. Germination is epigeal in:
(A) *Zea mays* (B) *Helianthus*
(C) *Mangifera* (D) *Pisum*

157. Among the following which compound can induce seed dormancy:
(A) Gibberellins (B) Caffeine
(C) ABA (D) Potassium nitrate

158. Fruit of peanut (*Arachis hypogeal*) is:
(A) Pod (B) Achene
(C) Caryopsis (D) Drupe

159. A method of breaking dormancy and allowing ample absorption of water is:
(A) Stratification (B) Scarification
(C) Vernalisation (D) Devernalisation

160. Which one of the following is not correctly matched:

Crop	Test weight in gm
(A) Pearmillet	5-7
(B) Wheat	20-22
(C) Mungbean	34-36
(D) Sunflower	60-62

161. The reserve food in albuminous seeds is stored in:
(A) Testa (B) Plumule
(C) Cotyledons (D) Endosperm

162. Which one of the following substances releases seeds from their dormant state?
(A) Ethylene (B) Gibbrellin
(C) Cytokinin (D) Auxin

163. The major controlling factor is separation of grain by fluidized bed separator is:
(A) Air velocity and weight of grain
(B) Size and aerodynamic properties of grain
(C) Shape and size of grains
(D) Shape and texture of grain

164. Cytoplasmic hybrids are known as:
(A) Hybrid
(B) Cybrids
(C) Both (A) and (B)
(D) None of these

165. Which phytohormone is used for development of synthetic seeds?
(A) IAA (B) GA
(C) ABA (D) Ethylene

166. Which of the following is involved in seedless watermelon production?
(A) Apomixes (B) Interspecific crosses
(C) Tetraploidy (D) None of these

167. Seed work board is required for:
(A) Viability test of seed
(B) Germination test of seed
(C) Purity test of seed
(D) Blending of seed

168. The seedrate of rice cultivation in broadcasting and drilling method, respectively should be:
(A) 100, 60 kg/ha (B) 60, 100 kg/ha
(C) 45, 100 kg/ha (D) 100, 45 kg/ha

169. How much quantity of true potato seed is required for one hectare planting:
(A) 100 kg (B) 300 kg
(C) 400 kg (D) 500 kg

170. Tetrazoloin test is used to determine:
(A) Seed purity (B) Seed germination
(C) Seed viability (D) All of these

171. The type of germination in mungbean is known as:
(A) Epigeal (B) Hypogeal
(C) Hypoepigeal (D) None of these

172. Optimum seed rate of soybean is (kg/ha):
(A) 40-50 (B) 70-80
(C) 80-90 (D) 100-110

173. The optimum seed rate (kg/ha) for getting desired population of castor is:
(A) 12-15 (B) 18-20
(C) 20-25 (D) 25-30

174. The optimum seedrate for berseem should be (kg/ha):
(A) 10-20 (B) 20-25
(C) 25-30 (D) 30-40

175. How much seed of legume crop should be treated with one packet of Rhizobium culture:
(A) 5 kg (B) 10 kg
(C) 15 kg (D) 20 kg

176. The depth of sowing of bajra is:
(A) 1-2 cm (B) 4-6 cm
(C) 4-5 cm (D) 6-8 cm

177. STP techniques of planting is used in:
(A) Sweet Potato (B) Potato
(C) Sugarcane (D) Jute

178. Optimum seed rate of sugarbeet is:
(A) 8-10 kg/ha (B) 18-20 kg/ha
(C) 28-30 kg/ha (D) 38-40 kg/ha

179. Seed viability can be tested by using the solution of :
(A) Tetrazolium chloride
(B) Hydrochloride
(C) Potatssium chloride
(D) Tetrazolian nitrate

180. The crop stage most affected in alkaline soil:
(A) Germination (B) Maturity
(C) Tillering (D) Flowering

181. Crop lodging has been most widely used for nutrient management in:
(A) Cotton (B) Rice
(C) Sugarcane (D) Alfalfa

182. Nursery area required for one hectare rice transplanting is:
(A) 0.05 ha (B) 0.10 ha
(C) 0.15 ha (D) 0.20 ha

183. Mat type nursery is related to:
(A) Tobacco crop (B) Paddy crop
(C) Onion crop (D) Brinjal

184. The optimum stage of rice seedling for transplanting under irrigated condition during kharif reason is:
(A) Two leaf stage
(B) Four leaf stage
(C) Six leaf stage
(D) Eight leaf stage

185. The planting value (%) of a seed lot having purity 90% and germination 80% will be:
(A) 52 (B) 62
(C) 72 (D) 82

186. Seed rate of transplanted basmati rice is:
(A) 25-30 kg/ha (B) 35-40 kg/ha
(C) 45-60 kg/ha (D) 55-70 kg/ha

187. Optimum seed rate of soybean is (kg/ha) :
(A) 50-60 (B) 70-80
(C) 80-90 (D) 100-110

188. Optimum seed rate of sunflower is (kg/ha):
(A) 8-10 (B) 10-15
(C) 18-20 (D) 30-35

189. Optimum seed rate for berseen should be (kg/ha):
(A) 15-20 (B) 20-25
(C) 25-30 (D) 30-40

190. For ideal plant population, the seed rate of sugarbeet should be (kg/ha):
(A) 4-8 (B) 8-10
(C) 10-15 (D) 25-30

ANSWER SHEET
5. Seed and Sowing

---

1. **(B)**	2. **(B)**	3. **(B)**	4. **(D)**	5. **(D)**	6. **(A)**	7. **(C)**	8. **(A)**	9. **(B)**	10. **(A)**
11. **(B)**	12. **(C)**	13. **(A)**	14. **(D)**	15. **(A)**	16. **(D)**	17. **(B)**	18. **(A)**	19. **(A)**	20. **(A)**
21. **(D)**	22. **(A)**	23. **(D)**	24. **(A)**	25. **(A)**	26. **(B)**	27. **(B)**	28. **(D)**	29. **(C)**	30. **(A)**

31. **(A)**	32. **(D)**	33. **(A)**	34. **(A)**	35. **(C)**	36. **(A)**	37. **(C)**	38. **(B)**	39. **(A)**	40. **(A)**
41. **(A)**	42. **(D)**	43. **(C)**	44. **(B)**	45. **(C)**	46. **(D)**	47. **(A)**	48. **(D)**	49. **(C)**	50. **(C)**
51. **(B)**	52. **(A)**	53. **(C)**	54. **(D)**	55. **(D)**	56. **(B)**	57. **(D)**	58. **(B)**	59. **(D)**	60. **(C)**
61. **(C)**	62. **(A)**	63. **(B)**	64. **(B)**	65. **(A)**	66. **(B)**	67. **(A)**	68. **(B)**	69. **(A)**	70. **(B)**
71. **(C)**	72. **(C)**	73. **(D)**	74. **(B)**	75. **(B)**	76. **(C)**	77. **(B)**	78. **(D)**	79. **(B)**	80. **(C)**
81. **(C)**	82. **(D)**	83. **(B)**	84. **(D)**	85. **(C)**	86. **(D)**	87. **(C)**	88. **(B)**	89. **(C)**	90. **(A)**
91. **(B)**	92. **(C)**	93. **(D)**	94. **(D)**	95. **(C)**	96. **(A)**	97. **(B)**	98. **(A)**	99. **(B)**	100. **(C)**
101. **(D)**	102. **(A)**	103. **(A)**	104. **(B)**	105. **(C)**	106. **(D)**	107. **(C)**	108. **(C)**	109. **(A)**	110. **(B)**
111. **(C)**	112. **(D)**	113. **(A)**	114. **(B)**	115. **(C)**	116. **(A)**	117. **(B)**	118. **(D)**	119. **(A)**	120. **(B)**
121. **(D)**	122. **(C)**	123. **(A)**	124. **(B)**	125. **(D)**	126. **(D)**	127. **(B)**	128. **(C)**	129. **(C)**	130. **(D)**
131. **(C)**	132. **(C)**	133. **(A)**	134. **(B)**	135. **(A)**	136. **(D)**	137. **(C)**	138. **(B)**	139. **(C)**	140. **(A)**
141. **(C)**	142. **(A)**	143. **(C)**	144. **(B)**	145. **(A)**	146. **(C)**	147. **(B)**	148. **(A)**	149. **(A)**	150. **(B)**
151. **(A)**	152. **(B)**	153. **(D)**	154. **(A)**	155. **(D)**	156. **(B)**	157. **(C)**	158. **(A)**	159. **(B)**	160. **(C)**
161. **(D)**	162. **(B)**	163. **(A)**	164. **(B)**	165. **(C)**	166. **(A)**	167. **(D)**	168. **(A)**	169. **(A)**	170. **(C)**
171. **(A)**	172. **(B)**	173. **(B)**	174. **(C)**	175. **(B)**	176. **(A)**	177. **(C)**	178. **(A)**	179. **(A)**	180. **(A)**
181. **(C)**	182. **(A)**	183. **(B)**	184. **(B)**	185. **(C)**	186. **(B)**	187. **(B)**	188. **(A)**	189. **(C)**	190. **(B)**

6 CROPS AND VARIETIES

1. The biennial crop is:
(A) Paddy (B) Sugarbeet
(C) Tobacco (D) Castor

2. Which variety of groundnut is spreading in nature:
(A) T-64 (B) Kaushal
(C) Amber (D) Chitra

3. Avarodhi variety of chickpea is resistant against disease of:
(A) Rust (B) Blight
(C) Wilt (D) Powdary mildew

4. UPAS 120 is a short duration variety of:
(A) Gram (B) Wheat
(C) Arhar (D) Soybean

5. Which paddy variety is specially recommended for usar soils:
(A) Lakara (B) Madhukar
(C) Mahsuri (D) Swarna

6. Which one is the scented variety of rice:
(A) CSR-10 (B) N-12
(C) CSR-13 (D) T-22 A

7. The cash crop is:
(A) Lentil (B) Paddy
(C) Sugarcane (D) Mustard

8. The stimulant crop is:
(A) Maize (B) Potato
(C) Tobacco (D) Til

9. The cereal crop is:
(A) Linseed (B) Wheat
(C) Castor (D) Cowpea

10. The legume crop treated as oilseed crop is:
(A) Chickpea (B) Groundnut
(C) Til (D) Lathyrus

11. Which one is late sown variety of wheat:
(A) UP 2003 (B) HD 2329
(C) K 9107 (D) K 7903

12. Which one is the recent released variety of Arhar:
(A) Bahar (B) Amar
(C) UPAS 120 (D) Laxmi

13. The most suitable variety for late sown gram is:
(A) K 850 (B) KPG 59
(C) KWR 108 (D) Sadabahar

14. The scientific name of field pea is:
(A) *Pisum sativum var. arvense*
(B) *Pisum sativum var hortense*
(C) *Pisum sativum var leguminosarum*
(D) *Pisum sativum var esculentum*

15. The scientific name of vegetable pea is:
(A) *Pisum sativum var. arvense*
(B) *Pisum sativum var hortense*
(C) *Pisum sativum var leguminosarum*
(D) *Pisum sativum var esculentum*

16. The type of germination in field pea is called:
(A) Hypogeal (B) Epigeal
(C) Hypo-epigeal (D) Epi-hypogeal

17. Arhar (pigeonpea) commonly known as:
(A) Red gram (B) Black gram
(C) Bengal gram (D) Horse gram

18. The first most important Indian pulse crop is:
(A) Chickpea (B) Pigeonpea
(C) Cow pea (D) Field pea

19. The second most important Indian pulse crop is:
(A) Chickpea (B) Pigeonpea
(C) Cow pea (D) Field pea

20. The temporary roots in maize root system are:
(A) Seminal roots (B) Aerial roots
(C) Crown roots (D) Coronal roots

21. The pH range for successful cultivation of maize was started from:
(A) 6.5-8.5 (B) 7.5-8.5
(C) 7.5-9.5 (D) 5.5-7.5

22. The type of germination in mungbean is known as:
(A) Epigeal (B) Hypogeal
(C) Hypo-epigeal (D) Epi-hypogeal

23. The pH range suitable for soybean cultivation is:
(A) 5.0-6.0 (B) 7.0-8.5
(C) 5.5-6.5 (D) 6.0-7.5

24. Kalyan sona is a:
(A) Tall wheat
(B) Triple gene dwarf
(C) Single gene dwarf
(D) Double gene dwarf

25. PBW 343 is the variety of:
(A) Rice (B) Barley
(C) Oat (D) Wheat

26. Jawahar is the variety of maize:
(A) Composite (B) Open pollinated
(C) Hybrid (D) Desi

27. Himalayan -123 is the variety of maize:
(A) Composite (B) Open pollinated
(C) Hybrid (D) Desi

28. Recently released hybrid sorghum variety for saline soil is:
(A) CSH 29 (B) CSH 19
(C) CSH 9 (D) CSH 1

29. Which of the following is a cash crop?
(A) Potato (B) Tobacco
(C) Cotton (D) All of these

30. Morden is the variety of:
(A) Sunflower (B) Rapeseed& mustard
(C) Castor (D) Linseed

31. Sunflower crop is grown in all season because of:
(A) Determinate growth habit
(B) Photosensitivity
(C) Photo insensitivity
(D) None of these

32. RMO 40 is a variety of:
(A) Mungbean (B) Sweet pea
(C) Horsebean (D) Mothbean

33. Which variety of the following is related to fibre crop?
(A) Nalanda Sanni (B) Sonali
(C) Mahalaxmi (D) All the above

34. Which of the following is commonly used to green manuring?
(A) Potato (B) Dhaincha
(C) Green gram (D) Both (B) and (C)

35. The root emerging from a sugarcane sett are known as:
(A) Sett roots (B) Shoot roots
(C) Primary roots (D) Secondary roots

36. The recently developed terminator technology has been used in:
(A) Rice (B) Cotton
(C) Tobacco (D) Wheat

37. Sugarcane is a:
(A) Vegetable crop (B) Forge crop
(C) Tuber crop (D) Sugar crop

38. Sugarbeet is a:
(A) Annual plant (B) Biennial plant
(C) Perennial plant
(D) None of these

39. The sucrose found in sugar beet:
(A) Stem (B) Leaves
(C) Roots (D) Inflorescence

40. Sugarbeet is grown during:
(A) Rabi (B) Kharif
(C) Zaid (D) Rabi and kharif

41. Sunflower is sown in:
(A) Kharif season (B) Rabi season
(C) Zaid season (D) All the season

42. During kharif season sunflower takes days to mature:
(A) 80-90 (B) 100-110
(C) 120-130 (D) 105-130

43. The optimum pH range for sunflower is:
(A) 5.5-7.5 (B) 6.5-8.5
(C) 7.5-9.5 (D) 7.0-9.0

44. The best pH range for jute cultivation is:
(A) 5.0-6.5 (B) 6.0-7.5
(C) 7.0-8.5 (D) 8.0-9.5

45. Normally tobacco crop is grown in which season:
(A) Rabi (B) Kharif
(C) Summer (D) Spring

46. In Karnataka tobacco crop is grown in which season:

(A) Rabi (B) Kharif
(C) Summer (D) Spring

47. Autumn rice in W. B. is known as:

(A) AUS (B) Mun
(C) Boro (D) Aman

48. Spring rice in W. B. is known as:

(A) AUS (B) Mun
(C) Boro (D) Aman

49. Kharif rice in W. B. is known as:

(A) AUS (B) Mun
(C) Boro (D) Aman

50. Aromatic grasses belongs to the family of:

(A) Malvaceae (B) Poaceae
(C) Leguminoceae (D) Tiliaceae

51. Which species of sugarcane is referred as noble cane?

(A) *Saccharum harberi*
(B) *Saccharum officinarum*
(C) *Saccharum spontanium*
(D) *Saccharum sinense*

52. Which crop is grown in utera cultivation:

(A) Gram (B) Lentil
(C) Pea (D) Sorghum

53. Which crop is not suitable for green manuring?

(A) Berseem (B) Moong
(C) Sanai (D) Dhainc

54. Varuna is a variety of:

(A) Indian mustard (B) Rape seed
(C) Black sarson (D) Brwon sarson

55. Kiran is a variety of:

(A) *Brassica juncia*
(B) *Brassica napus*
(C) *Brassica carinata*
(D) *Brassica compestris*

56. Horse Gram is also known as:

(A) Chickpea (B) Pigeonpea
(C) Mungbean (D) Sword bean

57. Arkel is a variety of:

(A) Groundnut (B) Arhar
(C) Pea (D) Toria

58. The crop grown in all the three seasons under UP conditions is:

(A) Paddy (B) Wheat
(C) Sunflower (D) Groundnut

59. The crop sown in the month of July is called:

(A) Summer season crop
(B) Kharif season crop
(C) Rabi season crop
(D) Spring season crop

60. Which of the following crop fixes nitrogen?

(A) Alfalfa (B) Red clover
(C) Cowpea (D) All of these

61. Canola is a group of plants belongs to:

(A) Mustarrd (B) Safflower
(C) Niger (D) None of these

62. Pulses fit well in cropping system as they are:

(A) Short duration crops
(B) Disease-resistant crops
(C) Long duration crops
(D) Moisture stress resistant crops

63. In which of the following, composite and synthetic cultivars are used:

(A) Rice (B) Wheat
(C) Maize (D) Cotton

64. Which one is not the variety of American cotton?

(A) Sujata (B) Digvijaya
(C) Laxmi (D) Mahalaxmi

65. Which one is the species of American cotton?

(A) *Gossypium hirsutum*
(B) *Gossypium barbadence*
(C) Both of these (D) None of these

66. *Brassica compestris* is known as:

(A) Sarson (B) Toria
(C) Mustard (D) Taramira

67. The botanical name of mustard is:

(A) *Brassica compestris*
(B) *Brassica napus*
(C) *Brassica juncea* (D) None of these

68. Kranti is the variety of:

(A) Mustard (B) Sarson
(C) Toria (D) Taramira

69. The word Fuzz refers to:
(A) Course fibre (B) Fine fibre
(C) Jute fibre (D) Sanai fibre

70. Major constitute is of cotton fibre:
(A) Fat (B) Lignin
(C) Cellulose (D) Protein

71. Nicotine content in cigarette tobacco is:
(A) 1-2 % (B) 3-4 %
(C) 5-6 % (D) 6-7 %

72. Nicotine content in chewing tobacco is:
(A) 1-2 % (B) 3-4 %
(C) 5-6 % (D) 8-9 %

73. Nicotine content in cigar tobacco is:
(A) 1-2 % (B) 3-4 %
(C) 5-6 % (D) 6-7 %

74. Nicotine content in burley tobacco is:
(A) 1-2 % (B) 3-4 %
(C) 5-6 % (D) 6-7 %

75. Nicotine content in bidi tobacco is:
(A) 1-2 % (B) 3-4 %
(C) 5-6 % (D) 6-7 %

76. Which is an insect trap crop:
(A) Paddy (B) Arhar
(C) Bhindi (D) Potato

77. First dwarf variety of rice developed in India is:
(A) Jaya (B) Prasad
(C) PD-4 (D) None of these

78. Aroma in scented rice is due to the presence of:
(A) 1Acetyle -1pyralin
(B) 2 Acetyle -1 pyralin
(C) 1 Acetyle -2pyralin
(D) 2 Acètyle -2pyralin

79. Recently released Basmati variety of rice is:
(A) Vasumati
(B) Pusa-suganda-2
(C) Pusa-suganda-3
(D) All the above

80. In North India, maize crop is mainly sown in:
(A) Kharif (B) Rabi
(C) Spring (D) Rabi and Kharif

81. Which one is the vegetable type variety of soybean?
(A) Willomel (B) Green soya
(C) Imperical (D) All of these

82. Pungency of mustard oil is caused by presence of:
(A) Amino acid (B) Erucic acid
(C) None of these (D) Sinigrin

83. Which one is a multicut variety of sorghum?
(A) UP Chari-1 (B) Pant Chari -3
(C) MP Chari (D) Pusa Chari -1

84. Which one is the dual purpose variety of sorghum?
(A) CSV-15 (B) Pant Chari -5
(C) CSH -13R (D) All the above

85. Crude protein content in Hybrid Napier grass is:
(A) 5.2 % (B) 10.2 %
(C) 14.2 % (D) 18.2 %

86. Hybrid napier is a cross of:
(A) Sorghum x Bajra
(B) Bajra x Napier
(C) Napier x Sudan grass
(D) Napier x Sorghum

87. Napier grass is multiples through:
(A) Seed (B) Stem
(C) Root (D) Both (B) and (C)

88. *Pennisetum pedicellatum* is a botanical name of:
(A) Napier grass (C) Dinanath grass
(B) Sudan grass (D) Phulva grass

89. Which one is the blast resistant variety of rice:
(A) Pant Dhan 4 (B) Govind
(C) Jaya (D) All of these

90. Which one is not the scented variety of rice?
(A) T-3 (B) Basmati – 1
(C) N- 12 (D) T-23

91. Wilt resistant variety of gram is:
(A) JG – 315 (B) VG – 244
(C) Pusa – 212 (D) Al of these

92. Which one is not the late maturing variety of gram?
(A) Radhey (B) Pant G 114
(C) Avrodhi (D) Pusa 372

93. Which crop is grown in Rabi season?
(A) Okra (B) Pea
(C) Potato (D) Both (B) and (C)

94. Which of the following is not a kabuli gram variety?
(A) L-550 (B) Pusa 267
(C) Pragati (D) All of these

95. Which of the following is not a late maturing kabuli gram variety?
(A) L-550 (B) Pusa 267
(C) Phule G 0517 (D) All of these

96. Which of the following is a gram variety lodging tolerant is?
(A) Pusa 256 (B) Sadbhavna
(C) DCP 92-3 (D) All of these

97. Which of the following is not a pea variety?
(A) Shikha (B) Sapna
(C) Swati (D) Mallika

98. Which of the following is a lentil variety?
(A) Mallika (B) K-75
(C) Priya (D) All of these

99. Which of the following variety of pea is suitable for mountain agriculture?
(A) VL Matar 3 (B) VL Ageti Matar 7
(C) Vivek Matar 8 (D) All of these

100. Which one is not a spreading type variety of groundnut?
(A) Jyoti (B) Jawahar 2
(C) T-32 (D) None of these

101. Kadiri-3 is the variety of:
(A) Groundnut (B) Pea
(C) Lentil (D) Sesamum

102. Pragati is a variety of:
(A) Pigeonpea (B) Gram
(C) Pea (D) Lentil

103. Day neutral crop is:
(A) Rice (B) Wheat
(C) Sunflower (D) All of these

104. Which one is the bunching type variety of ground nut?
(A) T-25 (B) T-28
(C) T-33 (D) T-99

105. Which one is the first dwarf variety of rice in India?
(A) IR-8 (B) Balu
(C) Jaya (D) Ratna

106. Specific leaf weight is equal to:
(A) Leaf weight/leaf area
(B) Leaf weight/plant weight
(C) Plant weight/leaf weight
(D) None of these

107. Which one is the single cut variety of oat?
(A) Kent (B) OS-6
(C) OS 7 (D) None of these

108. What is the economic part of poppy?
(A) Root (B) Stem
(C) Capsule (D) Leaves

109. The recommended variety of pigeonpea for pigeonpea-wheat cropping system is:
(A) T 7 (B) T 17
(C) Bahar (D) UPAS 120

110. Parodovic is the variety of:
(A) Soyabean (B) Sunflower
(C) Sugarcane (D) Sesame

111. Wheat variety suitable for rainfed areas is:
(A) HD 2329 (B) HD 2285
(C) Lok 1 (D) PBW 343

112. ICPL 187 is a variety of:
(A) Chickpea (B) Pigeonpea
(C) Cowpea (D) Lentil

113. Gauri, Sweta and Versha are newly released varieties of:
(A) Rice (B) Hybrid rice
(C) Maize (D) Sorghum

114. Pusa RH 10, HRI – 120, PHV 71 and APHR 1 are newly released hybrid varieties of:
(A) Rice (B) Hybrid rice
(C) Maize (D) Sorghum

115. Pusa hybrid -1, Shaktiman-1 and Shaktiman-2 varieties of maize are rice in:
(A) Protein (B) Fat
(C) Carbohydrate (D) Starch

116. Kanchan (NLS 4) is a prominent variety of tobacco for :
(A) Northern light soil
(B) Southern light soil
(C) Black soil (D) All of these

117. Hema and Gauthami are prominent varieties of tobacco for:
(A) Northern light soil
(B) Southern light soil
(C) Black soil (D) All of these

118. Anand 119 (A-119) is a prominent variety is:
(A) Bidi (B) Flue cured variety
(C) Cigar (D) Chewing

119. Gujarat tobacco 9 (GT 9) is a prominent variety is:
(A) Bidi (B) Flue cured variety
(C) Cigar (D) Chewing

120. Manasi is recently released chewing tobacco variety of:
(A) Motihari (B) Jati
(C) Both of these (D) None of these

121. Torsa is recently released chewing tobacco variety of:
(A) Motihari (B) Jati
(C) Both of these (D) None of these

122. Dharla is a chewing tobacco variety of:
(A) Hemti (Late type)
(B) Bitri (early type)
(C) Both of these (D) None of these

123. Vijay, Amber, Sona and Kisan is composite varieties of:
(A) Wheat (B) Maize
(C) Rice (D) Jowar

124. Salt tolerant barley varieties are:
(A) Amber (B) Jyoti
(C) Azad (D) All of these

125. 6 rows tolerant barley varieties are:
(A) C 138 (B) PL 56
(C) C 164 (D) All of these

126. Which barley varieties are most suited for malt purposes:
(A) Amber (B) Jyoti
(C) Azad (D) All of these

127. Huskless barley varieties are:
(A) Karan - 19 (B) Ratna
(C) Amber (D) All of these

128. Dual purpose and high yielding varieties of Bajra is:
(A) PCB 164 (B) BJ 104
(C) BK 560 (D) HB 68

129. Indaf-1, Indaf-5 and sharda are important varieties of:
(A) Finger millet (Ragi)
(B) Proso millet (Cheena)
(C) Barnyard millet (Sawan)
(D) Kodon millet (Kodon)

130. CO1, MS 4872, MS 4884 and Bhavna are important varieties of:
(A) Finger millet (Ragi)
(B) Proso millet (Cheena)
(C) Barnyard millet (Sawan)
(D) Kodon millet (Kodon)

131. ISC 700, ISC 701, Pratap Kangni 1 and Nischal are important varieties of:
(A) Finger millet (Ragi)
(B) Proso millet (Cheena)
(C) Foxtail millet (Kakun)
(D) Kodon millet (Kodon)

132. UP 319 is a variety of:
(A) Paddy (B) Wheat
(C) Jowar (D) Bajra

133. Niwas 1, Dindori 73, Pali and JK 76 are important varieties of:
(A) Finger millet (Ragi)
(B) Proso millet (Cheena)
(C) Barnyard millet (Sawan)
(D) Kodon millet (Kodon)

134. Anurag, Chandan, IPM 149 and IPM 151 are important varieties of:
(A) Finger millet (Ragi)
(B) Proso millet (Cheena)
(C) Barnyard millet (Sawan)
(D) Kodon millet (Kodon)

135. Bragg, Lee, Clark 63 and Improved pelicone are important varieties of:
(A) Soyabean (B) Wheat
(C) Sunflower (D) Safflower

136. Shilajeet, Ankur, Alankar and Gaurav are important varieties of:
(A) Soyabean (B) Wheat
(C) Sunflower (D) Safflower

137. Sonali, Sabujsona, Shyamli, Reshma, Padma and Baldev are important varieties of which jute species:
(A) *Corchorus antichorous*
(B) *Corchorus acutangulus*
(C) *Corchorus capsularis*
(D) *Corchorus olitorius*

138. Which one is a barley variety :
(A) Sonalika (B) Naveen
(C) Azad (D) Sona

139. Basudev, Naveen, Mahadev, Savitri and Rabti are important varieties of which jute species:

(A) *Corchorus antichorous*
(B) *Corchorus acutangulus*
(C) *Corchorus capsularis*
(D) *Corchorus olitorius*

140. Which one is a oat variety :

(A) Sonalika (B) Naveen
(C) Azad (D) Kent

141. Which of the following variety are related to oilseeds?

(A) CSH 4 (B) Janak
(C) Ganga 5 (D) Varuna

142. Rajkoo is a variety of:

(A) Paddy (B) Wheat
(C) Jowar (D) Bajra

143. Which wheat variety is suitable for sodic soil:

(A) K-68 (B) K 8027
(C) HP 1102 (D) Raj 3077

144. Chickpea variety is suitable for sodic soil is:

(A) Karnal chana (B) DCP 92-3
(C) KWR 108 (D) JG 315

145. Which one of the following crops causes maximum reduction in soil alkalinity:

(A) Paddy (B) Maize
(C) Cowpea (D) Wheat

146. Chandra, TMV 3, Type 28 and C 501 are important varieties of groundnut is:

(A) Spreading type
(B) Bunch type
(C) Semi spreading type
(D) None of these

147. TMV 1, TMV 2, TMV 9, Jyoti and Kuber are important varieties of groundnut is:

(A) Spreading type
(B) Bunch type
(C) Semi spreading type
(D) None of these

148. Sangam, Bhawani and Shivani are important varieties of:

(A) Toria (B) Yellow sarson
(C) Brown sarson (D) Rai or Raya

149. K 88, Pusa Karishma and Pusa Bahar are important varieties of:

(A) Toria (B) Yellow sarson
(C) Brown sarson (D) Rai or Raya

150. Pusa Kalyani, Suphla, BS 70 and BSH 1 are important varieties of:

(A) Toria (B) Yellow sarson
(C) Brown sarson (D) Rai or Raya

151. Sita, Varuna, Rohini, Vardan, Shekhar, Prakash, Pusa Jai Kisan and Vaibhav are important varieties of:

(A) Toria (B) Yellow sarson
(C) Brown sarson (D) Rai or Raya

152. Japani rape, Chineese rape, FS 9002 and FS 908 are important varieties of:

(A) Toria (B) Yellow sarson
(C) Fodder sarson (D) Rai or Raya

153. Pusa aghani and NRCDR 02 are important varieties of rai for:

(A) Normal sowing (B) Early sowing
(C) Late sowing (D) None of these

154. DRSF 108, UPS 2, UPS 5 and Tara are important varieties of:

(A) Sunflower (B) Safflower
(C) Maize (D) Sorghum

155. KBSH 1, Jwala mukhi, Sungene 85, Pro sun 09 are important varieties of sunflower is:

(A) Hybrid (B) Deshi
(C) Composite (D) None of these

156. Chambal, Sweta, Shubhra, Neelam are important varieties of:

(A) Linseed (B) Lentil
(C) Mustard (D) Sunflower

157. Gaurav, Jeevan, Nagarkot and Parvati are important dual purpose varieties of:

(A) Linseed (B) Lentil
(C) Mustard (D) Sunflower

158. Aruna, Bhagya, Soubhagya, Jyoti and Gauch 1 are important dual purpose varieties of:

(A) Linseed (B) Lentil
(C) Mustard (D) Castor

159. Chandana, Prachi, Gauri, Pratap and Madhvi are important varieties of:

(A) Linseed (B) Sesame
(C) Castor (D) Safflower

160. Type 65, Malviya 305, Type 56, Type 9 are important varieties of:
(A) Sesame (B) Safflower
(C) Castor (D) Lentil

161. Phule G 95311, Jawahar Chana-1, KAK-2, Chamtkar, Ujjawala and Dhawla are important varieties of:
(A) Desi chickpea (B) Kabuli chickpea
(C) Pigeonpea (D) Lentil

162. Anvita, RSG 963, Phule G 12 and PBG 3 are important varieties of:
(A) Desi chickpea (B) Kabuli chickpea
(C) Pigeonpea (D) Lentil

163. Newly released of lathyrus which contains less toxic substances is:
(A) Ratan (B) B-1
(C) Pusa 24 (D) Nirmal

164. Ujjawala and Dhawla are newly released varieties of:
(A) Bold seeded desi chickpea
(B) Small seeded desi chickpea
(C) Bold seeded kabuli chickpea
(D) Small seeded kabuli chickpea

165. KPG 59 (Udai) variety of desi chickpea is recommended for:
(A) Early sown conditions
(B) Late sown conditions
(C) Salt affected soils
(D) None of these

166. UPAS 120, Manak, Pragati and Jagrati are important varieties pigeonpea of:
(A) Early maturing
(B) Medium maturing
(C) Late maturing (D) None of these

167. Malviya arhar 3, Maruti, Asha and BSMR 736 are important varieties pigeonpea of:
(A) Early maturing
(B) Late maturing
(C) Medium maturing
(D) None of these

168. Bahar, Narendra arhar-1, Amar, Azad and Malviya chamatkar are important varieties of pigeonpea is:
(A) Early maturing
(B) Medium maturing
(C) Late maturing
(D) None of these

169. Meha, Samrat, HUM 16, COGG 912 and HUM 12 are important varieties of:
(A) Mungbean (B) Urdbean
(C) Pigeonpea (D) Cowpea

170. Uttara, KU 96-3, NDU 99-2, KU 300 and Barkha are important varieties of:
(A) Mungbean (B) Urdbean
(C) Lentil (D) Arhar

171. Priya, Sheri, Noori, Narendra masor-1 and Garima are important varieties of lentil is:
(A) Small seeded (B) Medium seeded
(C) Bold seeded (D) None of these

172. Pant masoor 406, Pant masoor 4, Pant masoor 639, Ranjan and Asha are important varieties of lentil is:
(A) Small seeded (B) Medium seeded
(C) Bold seeded (D) None of these

173. IPFD 1-10, IPFD 99-13, Jayanti, KPMR 522 and KPMR 400 are important varieties of:
(A) Fieldpea (B) Cowpea
(C) Chickpea (D) Pigeonpea

174. Dual purpose varieties of fieldpea is:
(A) Uttara (B) Shikha
(C) Jayanti (D) Alankar

175. Arkil, Early December, Early Bejor and Bonville are important varieties of:
(A) Fieldpea (B) Garden pea
(C) Chickpea (D) Pigeonpea

176. Containder, Mizoram local, Naga local, Meghalaya local and Local purple pod are important varieties of:
(A) French bean (B) Rice bean
(C) Mung bean (D) Urd bean

177. PDR 14, HUR 15, HUR 137, IIPR 96-4 and Utkarsh are important varieties of:
(A) French bean (B) Rice bean
(C) Mung bean (D) Urd bean

178. Pusa barsati, Pusa rituraj, Gomti and FS 68 are important varieties of:
(A) Pigeonpea (B) Cowpea
(C) Chickpea (D) Field pea

179. HFO 114, UPO 94 and Bundel Jai 851 are important varieties of:
(A) Berseem (B) Oat
(C) Lucern (D) None of these

180. Rambler, Sirsa Type 9, Anand 2, NDRI Selection 1 and S 244 are important varieties of:

(A) Cowpea (B) Oat
(C) Berseem (D) Lucern

181. MP chari, Pusa chari and J 69 are important varieties of:

(A) Single cut (B) Two cut
(C) Multi cut (D) None of these

182. Reo, Pusa chari 2, Pusa chari 6 and Haryana chari are important varieties of:

(A) Single cut (B) Two cut
(C) Multi cut (D) None of these

183. African tall is a important varieties of:

(A) Chari (B) Fodder Maize
(C) Bajra (D) None of these

184. Improved chari is a important varieties of:

(A) Chari (B) Maize
(C) Bajra (D) Teosinte

185. FOS 217, Pusa Sona and Durgapura safed are important varieties of:

(A) Cluster bean (B) Rice bean
(C) Mung bean (D) Urd bean

186. Hamil, Gatton, Sigor, Nchise and Kingranch are important varieties of:

(A) Guinea grass (B) Sudan grass
(C) Para grass (D) Napier grass

187. Wheeler, Sweet sudan, Tift sudan and 59-3 are important varieties of:

(A) Guinea grass (B) Sudan grass
(C) Para grass (D) Napier grass

188. Pusa Giant and NB 21 are important varieties of:

(A) Guinea grass (B) Sudan grass
(C) Para grass (D) Napier grass

189. IGFRI 2808, PP 10, PP 15, JP 12, Pusa 3, Pusa 38, PP 47, IGFRI 2-2-2 and IGFRI 32-11 are important varieties of:

(A) Guinea grass (B) Sudan grass
(C) Dinanath grass (D) Napier grass

190. Red colour potato varieties are:

(A) Kufri Sinduri (B) Kufri Lalima
(C) Kufri Kisan (D) All of these

191. Potato varieties are suitable for chips making:

(A) Chipsona – 1 (B) Chipsona -2
(C) Both of these (D) None of these

192. Which variety of sugarcane is known as wonder cane of North India:

(A) CO 1148 (B) COS 767
(C) CO 91010 (D) CO 87044

193. BO 91, CO 740 and CO 453 sugarcane varieties are suitable for:

(A) Sodic soils
(B) Waterlogged conditions
(C) Dryland conditions
(D) Highest sucrose content

194. CO 1007, CO 1157, CO 785, CO 975, BO 91, COS 8118 sugarcane varieties are suitable for:

(A) Sodic soils
(B) Waterlogged conditions
(C) Dryland conditions
(D) Highest sucrose content

195. CO 421, CO 449, COS 510 and COL 8001 sugar-cane varieties are suitable for:

(A) Sodic soils
(B) Waterlogged conditions
(C) Dryland conditions
(D) Highest sucrose content

196. CO 775, COL 7901, COL 7810 and COS 510 sugarcane varieties are suitable for:

(A) Sodic soils
(B) Waterlogged conditions
(C) Dryland conditions
(D) Highest sucrose content

197. BO 17 and BO 32 sugarcane varieties are suitable for:

(A) Khader areas of rivers
(B) Waterlogged conditions
(C) Dryland conditions
(D) Highest sucrose content

198. CO 740, COS 510, COL 8504, COS 8207 and COS 8016 sugarcane varieties are suitable for:

(A) Ratooning
(B) Waterlogged conditions
(C) Dryland conditions
(D) Highest sucrose content

199. CO 94008 (Shyama) sugarcane variety is suitable for:

(A) Sodic soils
(B) Waterlogged conditions
(C) Dryland conditions
(D) Early maturing high yielding variety of Penninsular region

200. Ramonskaya 60, IISR composite 1, LS 6, Pant S 10, Moribo and Magnapoly are important varieties of:
(A) Sugarbeet (B) Sugarcane
(C) Cotton (D) Jute

201. Shyamali, Lohit, Moulisoni, Banila, Vijay and Dharwad – 1 are important varieties of:
(A) Desi cotton (B) American cotton
(C) Misri cotton (D) None of these

202. Bikaneri Narma, Mahalaxmi and Bhagya are important varieties of:
(A) Desi cotton (B) American cotton
(C) Misri cotton (D) BT cotton

203. Sea Island, Andruse, Sujata, Savin and Gija 45 are important varieties of:
(A) Desi cotton (B) American cotton
(C) Misri cotton (D) BT cotton

204. MICH 12, MICH 162 and MICH 185 are important varieties of:
(A) Desi cotton (B) American cotton
(C) Misri cotton (D) BT cotton

205. Hybrid 4, JK hybrid, Godavari, Barlaxmi and Savitri 45 are important varieties of:
(A) Desi cotton (B) American cotton
(C) Hybrid cotton (D) BT cotton

206. K 12, M 18, M 19, M 35, ST 119 and D-9 are important varieties of:
(A) Sunhemp (B) Dhaincha
(C) Jute (D) None of these

207. Gandak bahar, Sona, Prabha, PT 76, Vaishali special and Litchhavi are important varieties in Bihar of:
(A) Chewing tobacco (B) Bidi tobacco
(C) FCV tobacco (D) Cigar tobacco

208. A-145, GT 6, DK and GT 8 are important varieties in Gujarat of:
(A) Chewing tobacco
(B) Bidi tobacco
(C) FCV tobacco
(D) Cigar tobacco

209. Meenakshi, BR 2, I-64, S – 1, PV 2, AD 1, Vairam, Dhangam and Bhagyalakshmi are important varieties in Tamilnadu of:
(A) Chewing tobacco (B) Bidi tobacco
(C) FCV tobacco (D) Cigar tobacco

210. NP 219, C 2, C 304, Patiyali, ST-1 and SK 417 are important varieties in UP of:
(A) Chewing tobacco (B) Bidi tobacco
(C) FCV tobacco (D) Hukka tobacco

211. DD 437, DD 407, DD 411 and DD 412 are important varieties in UP of:
(A) Chewing tobacco (B) Bidi tobacco
(C) FCV tobacco (D) Hukka tobacco

212. Sonar Motihari, DD 402, DD 403 and DD 404 are important varieties in West Bengal of:
(A) Chewing tobacco
(B) Bidi tobacco
(C) FCV tobacco
(D) Hukka tobacco

213. VV 2, Olor 10, Pennbel 69, Rg and Krishna are important varieties in Tamilnadu of:
(A) Chewing tobacco
(B) Cigar tobacco
(C) FCV tobacco
(D) Hukka tobacco

214. Burley 21, KY 416, D1 and D2 are important varieties in Andhra Pradesh of:
(A) Burley tobacco
(B) Bidi tobacco
(C) FCV tobacco (D) Hukka tobacco

215. Rustica tobacco is also called :
(A) Calcutti tobacco
(B) Bidi tobacco
(C) FCV tobacco (D) Hukka tobacco

216. GCT -1, GCT 2 and GCT 3 are important varieties of chewing tobacco mainly grown in Gujarat of:
(A) Rustica type (B) Tabacum type
(C) Both of these (D) None of these

217. Lanka special is an important variety in Andhra Pradesh of:
(A) Lanka tobacco
(B) Natu tobacco
(C) Cigar wrapper tobacco
(D) Hukka tobacco

218. Natu special, Prabhat, Vishwanath and Takaku are important varieties in Andhra Pradesh of:
(A) Lanka tobacco
(B) Natu tobacco
(C) Cigar wrapper tobacco
(D) Hukka tobacco

219. Calcuttia, Farrukhabad local and Kampla are important varieties in Uttar Pradesh of:
(A) Lanka tobacco (B) Natu tobacco
(C) Cigar wrapper tobacco
(D) Hukka tobacco

220. Dixie shade and S 5 are important varieties in West Bengal of:
(A) Lanka tobacco (B) Natu tobacco
(C) Cigar wrapper tobacco
(D) Hukka tobacco

221. Gajapati is a improved variety of:
(A) Lanka tobacco (B) Natu tobacco
(C) Pikka tobacco
(D) Hukka tobacco

222. ST-1 is a newly released variety of:
(A) Lanka tobacco
(B) FCV tobacco
(C) Chewing tobacco
(D) Hukka tobacco

223. Bread making vareties of wheat are:
(A) Lok 1 (B) K 9107
(C) HUW 549 (D) All of these

224. Biscuit making vareties of wheat are:
(A) PBW 175 (B) PBW 498
(C) UP 2525 (D) All of these

225. Paddy varieties are suitable for usar soils are:
(A) Saket 4 (B) Jaya
(C) Sona 349 (D) All of these

226. Drought tolerant rice varieties are:
(A) Nagina 22 (B) Bala
(C) Kaveri (D) All of these

227. Badsah bhog, Pasand bhog and malbhog rice varieties mainly grown in:
(A) Assam (B) UP
(C) MP (D) Tamilnadu

228. Rice varieties are suitable for deep waterlogged areas are:
(A) Jalmagna (B) Sighara
(C) Chakia (D) All of these

229. Bhudev is a newly released rice variety is suitable for:
(A) Flash flood water
(B) Deep water
(C) Medium deep water
(D) Very deep water

230. Which one of the following is not correctly matched?

Crop	Variety
(A) Mustard	Basanti
(B) Groundnut	Chandra
(C) Lentil	Malika
(D) Arhar	Neelam

231. The huskless variety of barley is:
(A) Manjula (B) KK 441
(C) Dolma (D) RS 6

232. Which of the following variety of chickpea is recommended for late sowing?
(A) KPG 59 (B) Sadabahar
(C) Pragati (D) K 850

233. Aruna is a mutant variety of:
(A) Pea (B) Cotton
(C) Castor (D) Soyabean

234. Variety of arhar suitable for double cropping system is:
(A) Bahar (B) T17
(C) Type 7 (D) UPAS 120

235. Vijay, Umber, Sona and Kisan are composite varieties of:
(A) Wheat (B) Rice
(C) Maize (D) Jowar

236. Which one is a barley variety:
(A) Sonalika (B) Naveen
(C) Azad (D) Sona

237. Which variety of the following relates to oilseeds?
(A) CSH4 (B) Janak
(C) Ganga 5 (D) Varuna

238. UP 319 is a variety of:
(A) Paddy (B) Wheat
(C) Jowar (D) Bajra

239. Ujjawla is a recently released variety of:
(A) Desi chana (B) Kabuli chana
(C) Horse gram (D) Bengal gram

240. Jal Lahri is a rice variety suitable for:
(A) Shallow low land ecosystem
(B) Semi-deep ecosystem
(C) Deep water ecosystem
(D) Flash flood ecosystem

241. Jal Priya is a rice variety suitable for:
(A) Shallow low land ecosystem
(B) Semi-deep ecosystem
(C) Deep water ecosystem
(D) Flash flood ecosystem

242. Jalnidhi is a rice variety suitable for:
(A) Shallow low land ecosystem
(B) Semi-deep ecosystem
(C) Deep water ecosystem
(D) Flash flood ecosystem

243. Barani deep is a rice variety suitable for:
(A) Shallow low land ecosystem
(B) Semi-deep ecosystem
(C) Rainfed upland ecosystem
(D) Flash flood ecosystem

244. Barh Avrodhi is a rice variety suitable for:
(A) Shallow low land ecosystem
(B) Semi-deep ecosystem
(C) Deep water ecosystem
(D) Flash flood ecosystem

245. Specific feature of rice variety Narendra Lalmoti is:
(A) Red kernels
(B) Pinkish kernels
(C) Yellow kernel
(D) White kernels

246. African tall is a high yielding variety of:
(A) Maize (B) Fodder maize
(C) Sorghum (D) Fodder sorghum

247. Macchari is a cross of:
(A) Maize (B) Cheri (sorghum)
(C) Both (A) & (B) (D) None of these

248. Rajko is a variety of:
(A) Fodder bajra (B) Chari
(C) Fodder sorghum (D) Fodder maize

249. Most suitable fodder grass for sodic soils is:
(A) Rhodes grass (B) Setaria grass
(C) Para grass (D) Guinea grass

250. Most suitable fodder grass for flash flood conditions is:
(A) Rhodes grass (B) Setaria grass
(C) Para grass (D) Guinea grass

251. Most suitable fodder grass for deep water logging conditions is:
(A) Rhodes grass (B) Setaria grass
(C) Para grass (D) Guinea grass

252. Nandi is a variety of:
(A) Rhodes grass (B) Setaria grass
(C) Para grass (D) Guinea grass

253. Hamil, Reverse date and macuni are important varieties of:
(A) Rhodes grass (B) Setaria grass
(C) Para grass (D) Guinea grass

254. Dolma is a important variety of:
(A) Oat (B) Barley
(C) Berseem (D) Lobia

255. Japanese rai mixed with which crop to increase the total yield:
(A) Oat (B) Lobia
(C) Berseem (D) Maize

256. Fodder sorghum can be mixed with:
(A) Lobia (B) Cowpea
(C) Guar (D) All of these

257. Barano and scabra are important varieties of:
(A) Rhodes grass (B) Setaria grass
(C) Para grass (D) Stylo

258. Kent is a important variety of:
(A) Oat (B) Barley
(C) Berseem (D) Cowpea

259. Salt resistant variety of rice is:
(A) Jaya (B) Padma
(C) IR – 8 (D) All of these

260. Salt resistant variety of wheat is:
(A) Sonalika (B) Kalyansona
(C) Hira (D) All of these

261. Salt resistant variety of rai is:
(A) Varuna (B) RS3
(C) Pilibhit (D) All of these

262. Salt resistant variety of sugarcane is:
(A) Co205 (B) Co286
(C) Co280 (D) All of these

263. Basantic is a high yielding variety of:
(A) Brown sesson (B) Black sesson
(C) Toria (D) Yellow sesson

264. Narendra Swarna Rai is a newly released variety of:
(A) Rai (B) Toria
(C) Yellow sesson (D) None of these

265. Urbasi is a high yielding variety of:
(A) Rai (B) Toria
(C) Taramira (D) Linseed

266. Rohini, Varuna and Kranti are high yielding varieities of:
(A) Toria (B) Til
(C) Sesson (D) Linseed

267. Ashirwad and Verdan are high yielding varieties of serson suitable for:
(A) Early sowing (B) Timely sowing
(C) Late sowing (D) Off season sowing

268. Bhavani is a newly released high yielding variety of toria suitable for:
(A) Ist fortnight of September
(B) IInd fortnight of September
(C) Ist fortnight of October
(D) IInd fortnight of October

269. Which sorghum variety is not multi-cut?
(A) Pusa chari-2 (B) MP chari-2
(C) MP chari (D) UP Chari-1

270. Which variety of Barley is huskless?
(A) Pragati (K508)
(B) Ritambhara (K551)
(C) Geetanjali (K1149)
(D) Karan-3

271. Which does not match in relation to cotton?
(A) C-520 – Diploid
(B) G-27 – Diploid
(C) Vikas – American
(D) Ranivan – American

272. Which is the perennial variety of Elephant grass (Napier grass)?
(A) Pusa Giant (B) NB-21
(C) Pusa Giant Napier
(D) Napier - 2

273. Which is the suitable rabi maize variety?
(A) Sharadmani (B) Azad uttam
(C) Naveen (D) Ganga 5

274. Which of the following is not matched?
(A) Varuna - Irrigated or unirrigated condition
(B) Vardan - Unirrigated condition
(C) Narendra 85 - Saline soil
(D) K-88 - Black grain

275. Which is a variety of Oat?
(A) K-12 (B) Naveen
(C) LD 491 (D) Kent

276. Which is not fodder variety of cowpea?
(A) Russian Giant (B) K-585
(C) Sirsa-10 (D) Azad-1

277. Which is not a variety of vegetable pea?
(A) Arkel (B) Azad pea 3
(C) Pant Matar 2 (D) Aparna

278. Which rice variety is not susceptible to insects and diseases?
(A) Pant Dhan-10 (B) Govind
(C) Narendra Dhan 2
(D) Saket -4

279. Which Paddy variety is not suitable for upland condition?
(A) Narendra Dhan 1
(B) Renu
(C) Saraju 52
(D) Narendra Dhan-118

280. Which variety of groundnut is grown in summer?
(A) Avtar (B) Chitra
(C) Chandra (D) Amber

281. Dhaman, Karad and Anjan are the:
(A) Varieties of sheep in Rajasthan
(B) Varieties of caster seed in Gujarat
(C) Varieties of grass in Rajasthan
(D) Three heros of Gawari dance

282. The Jawahar and Amber are considered as important and good varieties of maize used for:
(A) Fodder (B) Orissa
(C) Quality Protein maize
(D) Popcorn

283. Ratna is a variety of:
(A) Wheat (B) Barley
(C) Maize (D) Rice

284. Prabhat is an early short duration variety of:
(A) Black gram (B) Red gram
(C) Green gram (D) Horse gram

285. Pusa Jai Kisan is a somaclone of?
(A) Indian mustard
(B) Khesari
(C) Citronella java
(D) Basmati rice

286. The crop grown in UP during all the three season is:
(A) Rice (B) Sorghum
(C) Moong (D) Maize

287. The first wheat variety having shor plant height, lodging resistance and higher grain yield was:
(A) Dee-gee-woo-gen (B) Norin-10
(C) Lerma Rojo 64A (D) Sonara-64

288. Which one of the following is TPS variety of potato:
(A) JH 222 (B) HPS 1/113
(C) PJ 376 (D) Jr 5857

289. Which one of the following varieties of barley is nematode resisance:
(A) Jyoti (B) Ratna
(C) Karan -19 (D) Raj kiran

290. Opeque-2 maize composites are richer as compared to normal maize in:
(A) Tryptophan (B) Lysine
(C) Proline (D) Protein

291. Which one of the following pulse crops is used as a pulse, a fodder and a green manure crop:
(A) Moong (B) French bean
(C) Cowpea (D) Lentil

292. Scientific name of sawan grass is:
(A) *Lasiurus sindicus*
(B) *Panicum maximum*
(C) *Cynodon dactylon*
(D) *Cenchrus ciliaris*

293. Sugarbeet (*Beta vulgaris*) originated in Mediterranean region belongs to:
(A) Gramineae
(B) Convolvulaceae
(C) Chenopodiaceae
(D) Malvaceae

294. *Brassica juncea* has been evolved after hybridization between:
(A) *B. nigra* and *B. compestris*
(B) *B. nigra* and *B. oleracea*
(C) *B. compestris* and *B. oleracea*
(D) B. compestris and *B. carinata*

295. Botanical name of ragi is:
(A) *Eleusine coracana*
(B) *Echinochloa frumentacea*
(C) *Panicum miliacium*
(D) None of these

296. Which one of the following pairs is not correctly matched:
(A) Bajra - Tift 23 D2A
(B) Maize - T Cytoplasm
(C) Wheat - Rht1 Rht2
(D) Rice - Norin 10

ANSWER SHEET
6. Crops and Varieties

1. (B)	2. (C)	3. (C)	4. (C)	5. (A)	6. (B)	7. (C)	8. (C)	9. (B)	10. (B)
11. (D)	12. (B)	13. (B)	14. (D)	15. (B)	16. (A)	17. (A)	18. (A)	19. (B)	20. (A)
21. (D)	22. (A)	23. (D)	24. (C)	25. (D)	26. (A)	27. (C)	28. (A)	29. (D)	30. (A)
31. (C)	32. (D)	33. (D)	34. (D)	35. (A)	36. (B)	37. (D)	38. (A)	39. (C)	40. (A)
41. (D)	42. (A)	43. (B)	44. (B)	45. (A)	46. (B)	47. (A)	48. (D)	49. (C)	50. (B)
51. (B)	52. (B)	53. (A)	54. (A)	55. (C)	56. (A)	57. (C)	58. (C)	59. (B)	60. (D)
61. (B)	62. (A)	63. (C)	64. (B)	65. (C)	66. (A)	67. (C)	68. (A)	69. (B)	70. (C)
71. (A)	72. (D)	73. (D)	74. (B)	75. (C)	76. (C)	77. (A)	78. (B)	79. (D)	80. (A)
81. (A)	82. (D)	83. (C)	84. (D)	85. (B)	86. (B)	87. (D)	88. (C)	89. (D)	90. (D)
91. (A)	92. (C)	93. (D)	94. (D)	95. (C)	96. (C)	97. (D)	98. (D)	99. (D)	100. (D)
101. (A)	102. (B)	103. (C)	104. (C)	105. (C)	106. (B)	107. (D)	108. (C)	109. (D)	110. (B)

111. (C)	112. (B)	113. (A)	114. (A)	115. (A)	116. (A)	117. (C)	118. (A)	119. (A)	120. (B)
121. (A)	122. (A)	123. (B)	124. (D)	125. (D)	126. (A)	127. (A)	128. (A)	129. (A)	130. (B)
131. (C)	132. (B)	133. (D)	134. (C)	135. (A)	136. (A)	137. (C)	138. (C)	139. (D)	140. (D)
141. (D)	142. (D)	143. (D)	144. (A)	145. (A)	146. (A)	147. (B)	148. (A)	149. (B)	150. (C)
151. (D)	152. (C)	153. (B)	154. (A)	155. (A)	156. (A)	157. (A)	158. (D)	159. (B)	160. (B)
161. (B)	162. (A)	163. (A)	164. (C)	165. (B)	166. (A)	167. (C)	168. (C)	169. (A)	170. (B)
171. (C)	172. (A)	173. (A)	174. (C)	175. (B)	176. (A)	177. (A)	178. (B)	179. (B)	180. (D)
181. (C)	182. (A)	183. (B)	184. (D)	185. (A)	186. (A)	187. (B)	188. (D)	189. (C)	190. (D)
191. (C)	192. (A)	193. (A)	194. (B)	195. (C)	196. (D)	197. (A)	198. (A)	199. (D)	200. (A)
201. (A)	202. (B)	203. (C)	204. (D)	205. (C)	206. (A)	207. (A)	208. (B)	209. (A)	210. (D)
211. (D)	212. (D)	213. (B)	214. (A)	215. (A)	216. (A)	217. (A)	218. (B)	219. (D)	220. (C)
221. (C)	222. (C)	223. (D)	224. (D)	225. (D)	226. (D)	227. (A)	228. (D)	229. (C)	230. (D)
231. (C)	232. (A)	233. (C)	234. (D)	235. (C)	236. (C)	237. (D)	238. (B)	239. (B)	240. (A)
241. (B)	242. (C)	243. (C)	244. (D)	245. (B)	246. (B)	247. (C)	248. (A)	249. (A)	250. (B)
251. (C)	252. (B)	253. (D)	254. (B)	255. (C)	256. (D)	257. (D)	258. (A)	259. (D)	260. (D)
261. (D)	262. (D)	263. (D)	264. (C)	265. (A)	266. (C)	267. (C)	268. (B)	269. (D)	270. (D)
271. (D)	272. (C)	273. (D)	274. (D)	275. (B)	276. (C)	277. (D)	278. (D)	279. (A)	280. (A)
281. (D)	282. (C)	283. (D)	284. (B)	285. (A)	286. (D)	287. (B)	288. (B)	289. (D)	290 (B)
291. (C)	292. (A)	293. (C)	294. (C)	295. (A)	296 (A)				

7 CROPPING SYSTEMS

1. Criteria for assessing yield advantages are:
(A) Where intercropping must give full yield of a main crop and some yield of a second crop
(B) Where the combined intercrop yield must exceed the higher sole crop yield
(C) Where the combined intercrop yield must exceed a combined sole crop yield
(D) All of these

2. The conversion of crop yield into one form to compare the crops grown in mixed/inter/ sequentially cropped is called:
(A) Crop Equivalent Yield
(B) Land Equivalent Ratio
(C) Relative Crowding Coefficient
(D) Monetary Advantage

3. The relative land area under sole crops that is required to produce the yields achieved in inter-cropping is called:
(A) Crop Equivalent Yield
(B) Land Equivalent Ratio
(C) Relative Crowding Coefficient
(D) Monetary Advantage

4. Each species has its own coefficient (K) which gives a measure of whether the species has produce more or less yield than expected is called:
(A) Crop Equivalent Yield
(B) Land Equivalent Ratio
(C) Relative Crowding Coefficient
(D) Monetary Advantage

5. The common practice of examining intercropping advantages by expressing yields in monetary term is called:
(A) Crop Equivalent Yield
(B) Land Equivalent Ratio
(C) Relative Crowding Coefficient
(D) Monetary Advantage

6. If LER is 1.15, it mean the increase in yield by:
(A) 10 % (B) 15 %
(C) 20 % (D) 25 %

7. If LER is more than 1.0 it mean that the intercropping is:
(A) Profitable (B) Not profitable
(C) No change (D) None of these

8. If LER is less than 1.0 it mean that the intercropping is:
(A) Profitable (B) Not profitable
(C) No change (D) None of these

9. Relative crowding coefficient assumes that mixture treatments from a:
(A) Replacement series
(B) Additive series
(C) Both of these (D) None of these

10. Relative crowding coefficient is less than one, it means it has produced:
(A) Less yield (B) More yield
(C) Same yield (D) Unknown

11. Relative crowding coefficient is more than one, it means it has produced:
(A) Less yield (B) More yield
(C) Same yield (D) Unknown

12. Relative crowding coefficient is equal, it means it has produced:
(A) Less yield (B) More yield
(C) Same yield (D) Unknown

13. It assume that mixture from a replacement series and it gives a simple measure of how much the relative yield increase in species a is greater than that for species b is known as:
(A) Aggressivity
(B) Land Equivalent Ratio
(C) Relative Crowding Coefficient
(D) Monetary Advantage

14. An aggressivity value is zero indicates that the component species are:
(A) Not competitive
(B) Equally competitive
(C) Less competitive
(D) None of these

15. Land Equivalent Ratio (LER) can be used in:
(A) Replacement series
(B) Additive series
(C) Both of these
(D) None of these

16. Aggressivity can be used in:
(A) Replacement series
(B) Additive series
(C) Both of these
(D) None of these

17. The index is the product of the two equivalent factors is called:
(A) Competition index
(B) Monetary advantage
(C) Multiple cropping index
(D) Intensity index

18. The sum of area planted to different crops and harvested in single year divided by the total cultivated area times 100 is called:
(A) Competition index
(B) Monetary advantage
(C) Multiple cropping index
(D) Intensity index

19. The ratio between total cropped area and actual net cultivated area is called:
(A) Cropping intensity (B) Intensity index
(C) Both of these (D) None of these

20. The competition index is more than 1 (>1) there has been an:
(A) Advantage of mixing
(B) No advantage in mixing
(C) No change in mixing
(D) Unknown

21. The competition index is less than 1 (<1) there has been an:
(A) Advantage of mixing
(B) No advantage in mixing
(C) No change in mixing
(D) Unknown

22. The ratio of relative crowding coefficient of a specific and total relative crowding coefficient of all crops in the mixture is called:
(A) Competition coefficient
(B) Competition ratio
(C) Both of these
(D) None of these

23. Competition coefficient is used for calculation of relative crowding coefficient of all crops in the mixture, thereby maximum production can be obtained without leaving any:
(A) Equal (C) Beneficial effect
(C) Unknown (D) Harmful effect

24. The amount of area-time devoted to each crop or group of crops compared to total area time available to the farms for crop production is termed as:
(A) Specific cropping index
(B) Diversity index
(C) Relative yield index
(D) Relative spread index

25. The mean yield for a crop in a region divided by the mean of all India yields of the same crop times hundred is known as:
(A) Specific cropping index
(B) Diversity index
(C) Relative yield index
(D) Relative spread index

26. The multiplicity of farm products which are planted by computing the reciprocal of sum of share of gross revenue received from each individual farm enterprises in a single year is called:
(A) Specific cropping index
(B) Diversity index
(C) Relative yield index
(D) Relative spread index

27. An area of the crop expressed as percentage of the total cultivated area in the zone divided by the area of the crop expressed as percentage of the total cultivated area in the country, times hundred is called:
(A) Specific cropping index
(B) Diversity index
(C) Relative yield index
(D) Relative spread index

28. The relative yields, the value in a total which has been called:
(A) Specific cropping index
(B) Diversity index
(C) Relative yield total
(D) Relative spread index

29. The amount of area time allocated to one crop or a group of crops relative to the area – time actually used in production of all crops is called:

(A) Relative crop intensity index
(B) Crop growth rate
(C) Relative growth rate
(D) Net assimilation rate.

30. The dry weight gained by a unit area of crop in a unit time is called:

(A) Relative crop intensity index
(B) Crop growth rate
(C) Relative growth rate
(D) Net assimilation rate

31. Crop growth rate expressed as:

(A) g/day/m^2 (B) g/g/day
(C) cm^2/g (D) mm/day

32. The increase in dry weight in unit time over unit original weight of the plant is called:

(A) Relative crop intensity index
(B) Crop growth rate
(C) Relative growth rate
(D) Net assimilation rate

33. The increase in dry weight of plant per unit leaf area per unit time is called:

(A) Relative crop intensity index
(B) Crop growth rate
(C) Relative growth rate
(D) Net assimilation rate

34. Relative growth rate expressed as:

(A) g/day/m^2 (B) g/g/day
(C) cm^2/g (D) mm/day

35. Leaf area ratio expressed as:

(A) g/day/m^2 (B) g/g/day
(C) cm^2/g (D) mm/day

36. The ratio of leaf area to covered ground area is called:

(A) Leaf Area Index
(B) Leaf Area Ratio
(C) Leaf Area Duration
(D) None of these

37. The ability of the plant produce and maintain leaf area and is obtained by integrating the leaf area index over crop growth period is called:

(A) Leaf Area Index
(B) Leaf Area Ratio
(C) Leaf Area Duration
(D) None of these

38. The ratio of leaf area to dry weight of plant is called:

(A) Leaf Area Index
(B) Leaf Area Ratio
(C) Leaf Area Duration
(D) None of these

39. Which one of the following is the example of parallel cropping:

(A) Maize + Soybean
(B) Sugarcane+Mustard
(C) Potato + Mustard
(D) Sugarcane + Potato

40. Maize-toria-wheat-moong rotation is termed as a system of:

(A) Mixed cropping
(B) Relay cropping
(C) Multiple cropping
(D) Intercropping

41. Growing plants of different heights in the same field at the same time is known as:

(A) Multiple cropping
(B) Multistoreyed cropping
(C) Relay cropping
(D) Mixed cropping

42. Planting of succeeding crop before harvest of standing crop is known as:

(A) Mixed cropping
(B) Relay copping
(C) Inter cropping
(D) Multiple cropping

43. If on a 20 acre farm, crops are raised on 15 acres in rabi, 15 acres in kharif and 20 acres in zaid in a year, what shall be the cropping intensity:

(A) 250% (B) 400%
(C) 150% (D) 350%

44. In sugarcane + wheat intercropping, sugarcane is planted in the month of:

(A) October (B) February
(C) March (D) April

45. Which one of the following is relay cropping system:

(A) Maize-mustard-paerlmillet+cowpea
(B) Maize-potato-wheat-greengram
(C) Blackgram-wheat-greengram
(D) Pearlmillet-mustard-greengram

46. The intensity of following one year crop rotation Maize-potato-onion is:
(A) 100% (B) 150%
(C) 200% (D) 300%

47. Intercropping refers to:
(A) Growing two or more crops simultaneously and intermingled
(B) Growing two or more crops simultaneously in definite row proportions
(C) Growing more than one crop on same land in one year
(D) None of these

48. Which one of the following is the best example of catch crop:
(A) Rice (B) Mustard
(C) Toria (D) Wheat

49. Practice of raising forest or fruit trees on the society land to reduce atmospheric pollution is known as:
(A) Farm forestry (B) Silvipasture
(C) Agroforestry (D) Social forestry

50. Concept of agro-climatic zoning was developed by:
(A) World bank (B) USA
(C) FAO (D) ICAR

51. Lay farming refers:
(A) Growing of field crops in rotation with grasses and fodder legumes
(B) Growing trees in rotation with grasses and fodder legumes
(C) Growing of pastures in rotation with grasses and fodder legumes
(D) Growing of grasses and fodder legumes in forest areas

52. Alley cropping means:
(A) Growing of pastures in between two widely spaced rows of fast growing trees
(B) Growing of field crops in between two widely spaced rows of fast growing trees/shrubs
(C) Growing of only short duration crops in between two widely spaced rows of fast growing tree
(D) Growing of only fodder crops in between two widely spaced rows of fast growing trees

53. The intensity of groundnut + arhar – sugarcane crop rotation would be:
(A) 100 % (B) 150 %
(C) 200 % (D) 250 %

54. The intensity of paddy – pea –sugarcane-fallow-lobia – wheat is:
(A) 100 % (B) 150 %
(C) 166.6 % (D) 200 %

55. The intensity of jowar+arhar–fallow–wheat is:
(A) 100 % (B) 150 %
(C) 200 % (D) 250 %

56. The intensity of fallow – sugarcane –rotoon is:
(A) 66.6 % (B) 100 %
(C) 125 % (D) 150 %

57. The intensity of maize (cob) –early potato – late potato – sugarcane- ratoon - berseem is:
(A) 100 % (B) 150 %
(C) 200 % (D) 250 %

58. The intensity of jowar+guar– berseem + oat is:
(A) 100 % (B) 150 %
(C) 200 % (D) 250 %

59. The intensity of sunhemp (GM)–sugarcane is:
(A) 100% (B) 25%
(C) 50 % (D) 75%

60. Multiple cropping index proposed by:
(A) Dalrimpal (1971)
(B) Rao and Willey (1980)
(C) Monge and Associates (1978)
(D) Monge (1978)

61. Competition ratio proposed by:
(A) Dalrimpal (1971)
(B) Rao and Willey (1980)
(C) Monge and Associates (1978)
(D) Monge (1978)

62. Cropping intensity index proposed by:
(A) Dalrimpal (1971)
(B) Rao and Willey (1980)
(C) Monge and Associates (1978)
(D) Monge (1978)

63. Specific cropping intensity index proposed by:
(A) Dalrimpal (1971)
(B) Rao and Willey (1980)
(C) Monge and Associates (1978)
(D) Monge (1978)

64. Cultivated land utilization index proposed by:
(A) Choung (1973)
(B) Wong, You and Strout (1978)
(C) Donald (1963)
(D) Mc Gilchrist (1965)

65. Competition index proposed by:
(A) Choung (1973)
(B) Wong, You and Strout (1978)
(C) Donald (1963)
(D) Mc Gilchrist (1965)

66. Diversity index proposed by:
(A) Choung (1973)
(B) Wong, You and Strout (1978)
(C) Donald (1963)
(D) Mc Gilchrist (1965)

67. Aggressivity proposed by:
(A) Choung (1973)
(B) Wong, You and Strout (1978)
(C) Donald (1963)
(D) Mc Gilchrist (1965)

68. Net assimilation rate proposed by:
(A) Rao and Willey (1980)
(B) Dalrimpal (1971)
(C) Evans (1982)
(D) Monge (1978)

69. Leaf area index proposed by:
(A) Choung (1973)
(B) Radford (1967)
(C) Donald (1963)
(D) Mc Gilchrist (1965)

70. Growing of crops one after the other without seasonal fallowing is called:
(A) Continuous cropping
(B) Contingency cropping
(C) Contour cropping
(D) Conservation cropping

71. Growing of crops in aberrant situations like drought and floods is called:
(A) Continuous cropping
(B) Contingency cropping
(C) Contour cropping
(D) Conservation cropping

72. The cultivation of crops along the contours of a slope is called:
(A) Continuous cropping
(B) Contingency cropping
(C) Contour cropping
(D) Conservation cropping

73. A way of farming that aims to maximize the protection against erosion that can be achieved through soil and crop management for sustained farm productivity is called:
(A) Continuous cropping
(B) Contingency cropping
(C) Contour cropping
(D) Conservation cropping

74. As the ability to regenerate by shoot or root suckers, of a forest to established is called:
(A) Coppice farming
(B) Cover crop
(C) Crop competition
(D) Crop canopy

75. The practice of intensive exploitation of land capability by regular harvesting of such trees which produce numerous shoots from the stump after cutting is called:
(A) Coppice farming
(B) Cover crop
(C) Crop competition
(D) Crop canopy

76. Crops which are grown primarily to cover the soil and to reduce the loss of moisture due to leaching and erosion by wind and water is called:
(A) Coppice farming (B) Cover crop
(C) Crop competition (D) Crop canopy

77. Two plants are in competition with each other when the growth of either one or both of them is reduced or their form modified as compared with their growth or form in isolation is called:
(A) Coppice farming
(B) Cover crop
(C) Crop competition
(D) Crop canopy

78. It is the structures of aerial vegetative parts with special reference to the size, orientation, depth, density and arrangement of leaves influencing the penetration and interception of radiant energy is called:
(A) Coppice farming
(B) Cover crop
(C) Crop competition
(D) Crop canopy

79. The concept, approach, method and process of growing more than one crop per year in the same field by increasing cropping intensity is called:
(A) Crop intensification
(B) Crop modeling
(C) Cropping pattern
(D) Cropping scheme

80. How inputs or variable enter, move, interact and are controlled within the system and how the output of the system is generated and affected by other inputs and controlled variables is called:
(A) Crop intensification
(B) Crop modeling
(C) Cropping pattern (D) Cropping scheme

81. Yearly sequence and spatial arrangement of crops & fallow, on a given area is called:
(A) Crop intensification
(B) Crop modeling
(C) Cropping pattern
(D) Cropping scheme

82. The plan according to which crops are raised on individual plots of a farm with an object of getting the maximum returns from each crops without impairing fertility of the soil is called:
(A) Crop intensification
(B) Crop modeling
(C) Cropping scheme
(D) Cropping pattern

83. Pattern of crops taken up for a given piece of land, or other in which the crops are cultivated on piece of land over a fixed period and other interaction with farm resources and other farm enterprises and available technology which determine their make up is called:
(A) Cropping system
(B) Crop production
(C) Sequential cropping
(D) Crop productivity

84. The exploitation of plant morphological and plant physiological responses within a soil and atmospheric environment to produce a high yield per unit area of land is called:
(A) Cropping system
(B) Crop production
(C) Sequential cropping
(D) Crop productivity

85. The growing of different crops on a piece of land in a pre-planned succession is called:
(A) Cropping system
(B) Crop production
(C) Crop productivity
(D) Crop rotation/Sequential cropping

86. Production of plant product of economic importance, expressed in standard units per unit area is called:
(A) Sequential cropping
(B) Crop production
(C) Crop productivity
(D) Cropping system

87. Comprises all components required for the production of a particular crop and the inter-relationship between them and environment is called:
(A) Crop system (B) Crop technology
(C) Diversified cropping
(D) Diversified farm

88. The theoretical and practical approach to crop production in which concepts of applied science and advanced techniques are related to the financial consequences of their use is called:
(A) Crop system
(B) Crop technology
(C) Diversified cropping
(D) Diversified farm

89. A cropping plan in which no single crop contributes 50% or more towards the total crop production or monetary income annually is called:
(A) Crop system
(B) Crop technology
(C) Diversified cropping
(D) Diversified farm

90. A farm on which no single product or source of income equals as much as 50% of the total receipt and on such farm for farmer depends on several sources of income is called:
(A) Crop system
(B) Crop technology
(C) Diversified cropping
(D) Diversified farm

91. A cropping system in which the same crop if grown year after year in the same field is known as:
(A) Mono cropping
(B) Double cropping
(C) Quandruple cropping
(D) Triple cropping

92. Taking two crops in a year in sequence from the same piece of land is called:
(A) Mono cropping
(B) Double cropping
(C) Quandruple cropping
(D) Triple cropping

93. Growing three crops in a year in sequence is called:
(A) Mono cropping
(B) Double cropping
(C) Triple cropping
(D) Quandruple cropping

94. Growing four crops in a year in sequence is called:
(A) Mono cropping
(B) Double cropping
(D) Quandruple cropping
(C) Triple cropping

95. Growing two or more crops consecutively on the same field in the same year is called:
(A) Multiple cropping
(B) Diversified cropping
(C) Relay cropping
(D) Opportunity cropping

96. Temporal conplementarity in inter-cropping results from:
(A) The growth patterns of the component crops differing in time
(B) The growth patterns of the component crops differing in height
(C) Difference in yield
(D) Differences in cost of cultivation

97. The main objective of growing a catch crop is to:
(A) Add more residues to the soil
(B) Prevent cracking of soil
(C) Suppress weeds
(D) Get an additional income without further investment

98. A companion crop which nourishes the main crop by way of nitrogen fixation and/ or adding the organic matter in to the soil is called:
(A) Nurse crop
(B) Truck crop
(C) Catch crop
(D) Guard crop

99. Growing main crop in the centre surrounded by hardy and thorny crop such as safflower around pea, mesta around sugarcane and sorghum around maize is called:
(A) Nurse crop
(B) Truck crop
(C) Catch crop
(D) Guard crop

100. The practice of placing primary emphasis on the use of stored soil moisture while determining whether or not to establish a crop is called:
(A) Multiple cropping
(B) Diversified cropping
(C) Relay cropping
(D) Opportunity cropping

101. The range of environmental factors over which an organism or a physiological process can function is called:
(A) Ecological amplitudes
(B) Ecological equivalence
(C) Ecological niche
(D) Eco-type

102. The situation in which two or more species can replace one another in the same ecological niche because of the similarity in their ecological amplitude is called:
(A) Ecological amplitudes
(B) Ecological equivalence
(C) Ecological niche
(D) Eco-type

103. The position or functional status of an organism within its community and eco-system, resulting in organisms structural adaptation, physiological response and specific behaviour is called:

(A) Ecological amplitudes
(B) Ecological equivalence
(C) Ecological niche
(D) Eco-type

104. Plant type or strain within a species, resulting from exposure to a particular environment is called:

(A) Ecological amplitudes
(B) Ecological equivalence
(C) Ecological niche
(D) Eco-type

105. It is a cropping zone in which relative yield index is 200% and relative spread indices vary from 90-200% is called:

(A) Efficient cropping zone
(B) Relative yield index
(C) Relative spread index
(D) None of these

106. A concept involving the farming of fast growing plants/trees for the purpose of providing biomass that can be used directly as fuel or converted in to other forms of fuel or energy products is called:

(A) Energy farming
(B) Organic farming
(C) Diversified farming
(D) Energy productivity ratio

107. Ratios showing food yield per unit energy inputs and are most often used in evaluating the energetics of agricultural systems is called:

(A) Energy farming
(B) Organic farming
(C) Diversified farming
(D) Energy productivity ratio

108. It is the practice of allowing crop land to lie idle during a growing season to build up the soil moisture and fertility content so that a better crop can be produced the following year is called:

(A) Fallow
(B) Fallow land
(C) Fallow system
(D) None of these

109. Land left unfarmed for one or more growing seasons to kill weeds, make the soil richer is called:

(A) Fallow
(B) Fallow land
(C) Fallow system
(D) None of these

110. The sequence of crop years with fallow years is called:

(A) Fallow
(B) Fallow land
(C) None of these
(D) Fallow system

111. Extensive fallow systems are:

(A) Shifting cultivation system
(B) Bush fallow or grass fallow systems
(C) Fallow system
(D) None of these

112. Intensive fallow systems are:

(A) Shifting cultivation system
(B) Bush fallow or grass fallow systems
(C) Fallow system
(D) None of these

113. The crops are grown in parallel strips laid out across the general slope but which do not fallow the contour is known as:

(A) Field strip cropping
(B) Contour strip cropping
(C) Both of these
(D) None of these

114. Yield of the plant parts of economic interest (economic yield) as percentage of total biological yield is called:

(A) Harvest index
(B) Harvest ratio
(C) Both of these
(D) None of these

115. Maximum use of the land by means of frequent succession of harvested crop is called:

(A) Intensive cropping
(B) Intercropping
(C) Mixed cropping
(D) Mixed inter cropping

116. Growing two or more generally dissimilar crop simultaneously on the same piece of land, base crop necessarily in distinct row arrangement is called:
(A) Intensive cropping
(B) Intercropping
(C) Mixed inter cropping
(D) Mixed cropping

117. Growing of two or more crops simultaneously on the same piece of land, without any definite row arrangement is called:
(A) Intensive cropping
(B) Intercropping
(C) Mixed inter cropping
(D) Mixed cropping

118. Growing two or more crops simultaneously with no distinct row management is called:
(A) Intensive cropping
(B) Intercropping
(C) Mixed inter cropping
(D) Mixed cropping

119. The crops which are grown around the field boundaries in narrow strips is called:
(A) Outer crop
(B) Guard crop
(C) Border crop
(D) All of these

120. Plants tolerant to high acidic soil conditions is called:
(A) Oxylophytes
(B) Acedophytes
(C) Basilophytes
(D) None of these

121. Crop sown broad cast in the standing crop of low land rice before its harvest where the residual moisture is used for the establishment of sown crop is called:
(A) Paira crop
(B) Utera crop
(C) Both of these
(D) None of these

122. A rotation of arable crops requiring annual cultivation and artificial pastures occupying field for two years or longer is called:
(A) Ley farming
(B) Alley cropping
(C) Silvi pastoral system
(D) None of these

123. Yield advantage in an intercropping system occurs due to the development of:
(A) Temporal complementarity
(B) Spatial complementarity
(C) Both (A) and (B)
(D) Competitive relationship

124. Crops which are grown during the rainy season is called:
(A) Kharif crops (B) Rabi crops
(C) Zaid crops (D) Perennial crops

125. Crops which are grown during the winter season is called:
(A) Kharif crops (B) Rabi crops
(C) Zaid crops (D) Perennial crops

126. Crops which are grown during the summer season is called:
(A) Kharif crops (B) Rabi crops
(C) Zaid crops (D) Perennial crops

127. Crops requires seasons two years is called:
(A) Kharif crops (B) Rabi crops
(C) Zaid crops (D) Binnial crops

128. Crops which are grown during the more than two years is called:
(A) Kharif crops (B) Rabi crops
(C) Zaid crops (D) Perennial crops

129. The emergence of new tillers from the root/ crown of a plant specially as in sugarcane, sorghum etc. is called:
(A) Ratoon
(B) Ratoon cropping
(C) Rangeland management
(D) Recession farming

130. The cultivation of an additional crop from the regrowth of stubbles of previous main crop after its harvest, thereby avoiding reseeding/ replanting is called:
(A) Ratoon
(B) Ratoon cropping
(C) Rangeland management
(D) Precession farming

131. Management of a range to produce maximum forage for sustained use without jeopardizing the other resources or uses of land is called:
(A) Ratoon
(B) Ratoon cropping
(C) Rangeland management
(D) Recession farming

132. System in which crops are planted in flooded areas as the rainy season ends and water recedes is called:

(A) Recession farming
(B) Diara land farming
(C) Both of these
(D) None of these

133. The population of a particular species expressed as a percentage of all the species in a habitat is called:

(A) Relative dominance
(B) Relative density
(C) Relative spread index
(D) Relative yield index

134. The actual area occupied by a species expressed as a percentage of total area occupied by all species in a given land area is called:

(A) Relative dominance
(B) Relative density
(C) Relative spread index
(D) Relative yield index

135. LAI refers to:

(A) Lower Available Index
(B) Light Available Intensity
(C) Leaf Area Index
(D) Lux Area Index

136. The rotation intensity of maize-potato-sugarcane-sunflower is:

(A) 100% (B) 150%
(C) 200% (D) 250%

137. The mean of plant relative yield (ratio of the per plant yields in a mixture) to that in monoculture will have a value close to unity is called:

(A) Relative Yield total
(B) Relative crop intensity index
(C) Relative spread index
(D) Relative yield index

138. Land Equivalent ratio (LER) is required in:

(A) Inter-cropping system
(B) Sequential cropping system
(C) Mixed cropping system
(D) Sole cropping system

139. The leaf area per unit leaf weight expressed as cm^2g^{-1} is called:

(A) Relative Yield total
(B) Specific leaf area
(C) Relative spread index
(D) Relative yield index

140. Growing two or more crops simultaneously where one or more crops are planted in rows is called:

(A) Row intercropping
(C) Strip cropping
(B) Row mixed cropping
(D) None of these

141. These crops which are preserved in a succulent condition by partial fermentation in a tight receptacle is called:

(A) Silage crop (B) Sod culture
(C) Soilage crop (D) None of these

142. The practice of cultivating clearings scattered in the reservoir of natural vegetation and of abandoning them as soon as the soil is exhausted; and this includes the practice of shifting homesteads in order to fallow the cultivator's search for new fertile land is called:

(A) Shifting cultivation
(B) Row intercropping
(C) Strip cropping
(D) None of these

143. A system of soil management wherein the plants are grown in permanent grass without tillage is called:

(A) Silage crop (B) Sod culture
(C) Soilage crop (D) None of these

144. Forage cut green and fed to livestock while in the fresh conditions are known as:

(A) Silage crop (B) Sod culture
(C) Soilage crop (D) None of these

145. A crop grown in pure stand at optimum population is called:

(A) Sole crop (B) Intercrop
(C) Strip crop (D) None of these

146. Growing soil covering and soil depleting crops in alternate strips running perpendicular to the slope of the land or to the direction of prevailing winds for the purpose of reducing erosion is called:

(A) Sole cropping
(B) Intercropping
(C) Strip cropping
(D) None of these

147. Growing two or more crops simultaneously in different strips wide enough to permit independent cultivation but narrow enough for the crop to interact agronomically is called:

(A) Sole cropping
(B) Intercropping
(C) Strip intercropping
(D) None of these

148. The crop which is grown under certain problematic conditions when no other crop can be grown as floating rice in flood prone areas is called:

(A) Subsistence crop
(B) Subsistence farming
(C) Substitution farming
(D) None of these

149. Certain fast growing crops which work as support to vine crop is called:

(A) Sugar crop (B) Support crop
(C) Nurse crop (D) Guard crop

150. It is farming enterprise which provides food and commodities just sufficient for the farming family, and there are no surplus to sell is called:

(A) Subsistence farming
(B) Subsistence crop
(C) Substitution farming
(D) None of these

151. Substitution of existing inefficient crop with an identified efficient crop in a zone or region is called:

(A) Subsistence farming
(B) Subsistence crop
(C) Substitution farming
(D) None of these

152. The crops grown for the purpose of producing sugar from them is called:

(A) Sugar crop
(B) Support crop
(C) Nurse crop
(D) Guard crop

153. Crops which yield in tones and grown for different market requiring heavy transport is called:

(A) Truck crops
(B) Guard crops
(C) Nurse crops
(D) None of these

154. A type of agro-forestry system in which trees are planted with food crops which serve as a nursery for the young tree seedlings is called:

(A) Taungya cultivation
(B) Shifting cultivation
(C) Both of these
(D) None of these

155. Growing of two crops in a definite row proportion on the same piece of land is called:

(A) Parallel cropping
(B) Inter-cropping
(C) Mixed cropping
(D) Sole cropping

156. Intercropping is advantageous when LER is:

(A) 1 (B) Less than 1
(C) Zero (D) More than 1

157. The best example of synergistic cropping system is:

(A) Pigeonpea + Green gram
(B) Potato + Wheat
(C) Autumn planted sugarcane + mustard
(D) Autumn planted sugarcane + potato

158. Multi-intercropping includes:

(A) 2 crops in intercropping
(B) Relay cropping
(C) Mixed cropping
(D) More than 2 crops in intercropping

159. The plant geometry of wheat crop is:

(A) Square
(B) Spherical
(C) Extreme rectangular
(D) None of these

160. Number of steps taken in preparation of cropping scheme are:

(A) 2 (B) 4
(C) 6 (D) 8

161. Yearly sequence and spatial arrangement of crops on a given land area is called:

(A) Cropping system
(B) Cropping pattern
(C) Sequential cropping
(D) Crop rotation

162. Crop should be harvested at:
(A) Physiological maturity
(B) Real maturity
(C) Over maturity
(D) None of the above

163. Number of crops taken in relay cropping should be:
(A) 1 (B) 2
(C) 3 (D) 4

164. Evaluation of intercropping is done by:
(A) Rotational intensity
(B) Cropping intensity
(C) Land Equivalent ratio
(D) None of the above

165. The recommended row ratio for inter-cropping of wheat + mustard is:
(A) 3 : 1 (B) 5 : 1
(C) 9 : 1 (D) 7 : 1

166. The soybean crop can be grown with:
(A) Wheat (B) Chickpea
(C) Maize (D) Tobacco

167. Safflower can be grown as an intercrop with:
(A) Wheat (B) Urdbean
(C) Maize (D) Sorghum

168. How many crops should be grown in two years rotation to findout the rotation intensity of 200%:
(A) 4 (B) 3
(C) 2 (D) 1

169. The example of intercropping in replacement series is:
(A) Sugarcane + Potato
(B) Sugarcane +Wheat
(C) Pigeonpea + Urdbean
(D) Potato +Mustard

170. Multiple cropping increases over single cropping:
(A) Soil fertility
(B) Soil productivity
(C) Production of crops
(D) Autotoxins

171. Main pre-requisite for adoption of multiple cropping is:
(A) Longer duration varieties
(B) Soil conservation
(C) Irrigation facility
(D) Weed control chemicals

172. Crop rotation intensity is equal to:
(A) $\frac{\text{No. of crops in a rotation}}{\text{Duration of crop rotation}} \times 100$
(B) $\frac{\text{No. of crops in a rotation}}{\text{Number of crop rotation}} \times 100$
(C) $\frac{\text{Duration of rotation}}{\text{Number of crops in rotation}} \times 100$
(D) None of these

173. Cropping intensity is equal to:
(A) $\frac{\text{No. of crops in a rotation}}{\text{Net available area}} \times 100$
(B) $\frac{\text{No. of crop sown}}{\text{Net available area}} \times 100$
(C) $\frac{\text{Total cropped area}}{\text{Duration of rotation}}$
(D) None of these

174. Relay cropping means:
(A) Growiing of more than one crop in the same land in a year
(B) Growing of three crops on the same land in a year
(C) Growing of four crops on the same land in a year in such a manner that the following crop is sown before the harvest of preceeding crop
(D) None of these

175. Rotation intensity of maize-sugarcane-ratoon is:
(A) 100% (B) 150%
(C) 200% (D) 300%

176. Cropping intensity of Paddy – Wheat –Urd bean rotation is:
(A) 100% (B) 200%
(C) 300% (D) 400%

177. Presently the cropping intensity of India is:
(A) 100% (B) 127%
(C) 137% (D) 150%

178. Which one of the following is the example of parallel intercropping?
(A) Sugarcane + Potato
(B) Sugarcane + Mustard
(C) Sugarcane + Soybean
(D) Potato + Mustard

179. Monoculture is always harmful due to:
(A) Development of harmful soil organisms
(B) Reduction in soil fertility
(C) Accumulation of allelochemicals
(D) All of these

180. In India, a high percentage of cropped area is covered by rice because:
(A) There is adequate rainfall
(B) Rice can be grown in different season
(C) Rice is adoptable to most type of land and agro-climatic zones
(D) Large population of India is rice eater

181. Multiple cropping means:
(A) Growing > 2 crops in a year
(B) Growing > 3 crops in a year
(C) Growing > 1 crops in a year
(D) All of these

182. For getting 10% advantage, the LER should be:
(A) 1.0 (B) 0.90
(C) 1.10 (D) 110

183. Cropping intensity of the rotation chari – Berseem – Rice – Pea is:
(A) 100 % (B) 200 %
(C) 300 % (D) 400 %

184. Cropping intensity of the rotation green manure-wheat is:
(A) 50 % (B) 100 %
(C) 150 % (D) 200 %

185. Rotation Rice – Pea – Sugarcane-Ratoon is of:
(A) 1 year (B) 2 year
(C) 3 year (D) 4 year

186. Mixed cropping refers to:
(A) Growing of two crops together
(B) Growing of two crops together in different lines
(C) Sowing of mixed seeds of two crops
(D) All of these

187. NAR refers to:
(A) Net Assimilation Rate
(B) Night Assessment Ratio
(C) Net Assimilation Ratio
(D) New American Resistant

188. Which of the following cropping system is most prevalent in India:
(A) Rice – Wheat
(B) Rice – Soybean
(C) Maize – Gram
(D) Maize-Sugarcane-Ratoon-Wheat

189. Alley cropping means:
(A) Growing of pastures in between two widely spaced rows of fast growing trees
(B) Growing of field crops in between two widely spaced rows of fast growing trees
(C) Growing of only short duration crops in between two widely spaced rows of fast growing trees
(D) Growing of only fodder crops in between two widely spaced rows of fast growing trees

190. Which one of the example of relay cropping?
(A) Paddy-Wheat
(B) Soybean – Wheat
(C) Maize – Wheat
(D) Maize – Potato – Cucurbit

191. In wheat, mustard may be grown as inter crop in the row ratio of:
(A) 6 : 1 (Six rows of Wheat + One row of Mustard)
(B) 9 : 1 (Nine rows of Wheat + One row of Mustard)
(C) 12 : 2 (Twelve rows of Wheat + Two rows of Mustard)
(D) 18 : 3 (Eighteen rows of Wheat + Three rows of Mustard)

192. Rice – Rice – Rice rotation indicates:
(A) Monoculture (B) Duoculture
(C) Polyculture (D) None of these

193. Sowing of toria in standing crop of maize is:
(A) Mixed cropping
(B) Mixed intercropping
(C) Alley cropping
(D) Relay cropping

194. The highest yield of potato is obtained when cultivated in the rotation:
(A) Maize – Potato – Wheat
(B) Maize – Potato – Sunflower
(C) Dhaincha – Potato – Sunflower
(D) Moong – Potato – Sunflower

195. Which value can be assessed by using equation

$$= \frac{\text{Total number of crops} \times \text{Their respective area}}{\text{Total area available for cropping pattern}} \times 100$$

(A) Diversity index
(B) Simultaneous cropping index
(C) Multiple cropping index
(D) Cropping intensity index

196. The intensity of groundnut + arhar + sugarcane crop rotation would be:
(A) 150% (B) 200%
(C) 250% (D) 300%

197. Which one of the following planting geometry is recommended to optimize system productivity in intercropping system in semi-arid and sub-humid ecosystem:
(A) Normal plating of intercrops between two rows of normal planted main crop
(B) Skip row planting
(C) Three rows of main crop and two rows of intercrops
(D) Four rows of main crop and one row of intercrops

198. Relay cropping means:
(A) Growing of more than one crop on the same land in a year
(B) Growing of four crops on the same land in a year
(C) Growing of four crops on the same land in a year in such a manner that the following crop is sown before the harvest of preceding crop
(D) None of these

199. A short duration crop in between two main seasonal crops is termed as:
(A) Cash crop
(B) Inter crop
(C) Catch crop
(D) Companion crop

200. Which of the following intercropping system has synergetic effect on component crop:
(A) Autumn planted sugarcane + mustard
(B) Autumn planted sugarcane + chickpea
(C) Autumn planted sugarcane + wheat
(D) Autumn planted sugarcane + potato

201. Maize (for green cobs) –Autumn cane – ratoon–wheat will have cropping intensity of:
(A) 133% (B) 233%
(C) 333% (D) 433%

202. Which one of the following can be expressed by using equation

$$= \frac{\text{Harvest Diversity Index} \times 100}{\text{Multiple cropping index}}$$

(A) Diversity index
(B) Multiple cropping index
(C) Simultaneous cropping index
(D) Cropping intensity index

203. A short duration crop in between two main seasonal crops is termed as:
(A) Mixed crop (B) Intercrop
(C) Companian crop (D) Catch crop

204. Which is the correct row ratio for intercropping of mustard with gram?
(A) 6 : 1 (B) 6 : 2
(C) 9 : 1 (D) 9 : 2

205. The prevalent intercropping system in Bundelkhand region is:
(A) Sugarcane + Mustard
(B) Potato + Wheat
(C) Jowar + Arhar
(D) Cotton + Onion

206. In an intercropping system, if increase in the output of one crop help increase in the output of other crop, it is known as:
(A) Competitive (B) Complementary
(C) Supplementary
(A) Mutual inhibitive

207. Sesamum is mainly grown in UP as a:
(A) Paira crop (B) Mixed crop
(C) Inter crop (D) Pure crop

208. Read the following crop rotation and mark the correct answer:
(A) Arhar (Type 7) – Wheat (K 8020)
(B) Arhar (Bahar) - Wheat (K 8020)
(C) Arhar (Type 17) – Wheat (HD 2285)
(D) Arhar (UPAS 120) – Wheat (Golden Halna)

209. Mark the Land Equivalent Ratio is beneficial intercropping:
(A) 0 (Zero) (B) Less than 1
(C) 1 (One) (D) More than 1

210. Read the following statements and mark the recommended row proportion of component crops in intercropping:
(A) Potato + mustard (2:1 row ratio)
(B) Potato + mustard (3:1 row ratio)
(C) Potato + mustard (4:1 row ratio)
(D) Potato + mustard (5:1 row ratio)

211. Intercropping of mustard with potato is recommended in:
(A) Replacement series
(B) Additive series
(C) Replacement cum additive series
(D) None of these

212. The rotation intensity of maize – wheat + gram – moong is:
(A) 200% (B) 250%
(C) 300% (D) 400%

213. Intercropping of mustard with potato is recommended in:
(A) Replacement series
(B) Additive series
(C) Replacement cum additive series
(D) None of these

214. Which one of the following is the example of parallel intercropping:
(A) Maize+soybean
(B) Sugarcane + mustard
(C) Potato + mustard
(D) Sugarcane + potato

215. Maize – toria – wheat - moong rotation is termed as a system of:
(A) Mixed cropping (B) Relay cropping
(C) Intercropping (D) Multiple cropping

216. An example of commpanian cropping is:
(A) Sugarccane + potato
(B) Potato + mustard
(C) Potato + radish
(D) Wheat + mustard

217. Productivity of an intercrop per unit area of ground compared with that expected from sole crop sown in the same proportions is termed as:
(A) Land equivalent ratio
(B) Comparative ratio
(C) Land equivalent coefficient
(D) Crop performance ratio

218. Which one of the following fertilizer schedule is recommended to optimize the productivity in intercropping system:
(A) Full recommended dose of both the main crop and intercrop
(B) Full recommended dose of main crop and half of intercrop
(C) Half recommended dose of main crop and intercrop
(D) Full recommended dose of main crop only

219. Growing of coconut, black pepper and ginger simultaneously in the same field is called:
(A) Relay cropping
(B) Multiple cropping
(C) Mixed cropping
(D) Multistoried cropping

220. Crop rotation helps to:
(A) Lesser use of pesticides
(B) More crop yield
(C) Produce a greater choice of plant products
(D) Eliminate parasites which have selective hosts

221. Which one of the following makes a case for intensive, modern farming?
(A) Cropping pattern
(B) Higher output using organic method
(C) Remunerative price
(D) None of these

222. Which one of the following can be suitable for cropping as a wheat mixed crop?
(A) Cabbage (B) Cotton
(C) Mustard (D) Jowar

223. The cropping intensity of maize-potato-toabcco is:
(A) 200% (B) 300%
(C) 100% (D) 150%

224. These are crops, which provide a good harvest along with enrichment or restoration or amelioration of the soil is called:
(A) Restorative crops
(B) Exhaustive crops
(C) Paira crops
(D) Brake crops

225. These are crops, which on growing leave the field exhausted because of a more aggressive nature is called:

(A) Restorative crops
(B) Exhaustive crops
(C) Paira crops
(D) Brake crops

226. These are crops plants, which are sown a few days or weeks before the harvesting of thes tanding mature crop is called:

(A) Restorative crops
(B) Exhaustive crops
(C) Paira crops
(D) Brake crops

227. These crop plants belong to the grass family and have sod type tillers with matted foliage and roots close to the soil surface is called:

(A) Restorative crops
(B) Exhaustive crops
(C) Pairá crops
(D) Brake crops

228. These crops plants belong to the grass family and have sod type fillers with matted foliage and roots close to the soil surface is called:

(A) Sod or turf crops
(B) Cleaning crops
(C) Ware crops
(D) Soiling crops

229. These are crop plants whose agronomic practices make the field clean from weeds and stubble is called:

(A) Sod or turf crops
(B) Cleaning crops
(C) Ware crops
(D) Soiling crops

230. These crop plants are grown for temporary storing as intect in ware houses for future use or safe in called:

(A) Sod or turf crops
(B) Cleaning crops
(C) Ware crops
(D) Soiling crops

231. These crops plants are grown to harvest which they are still green and fed fresh to live stock in stalls in called:

(A) Sod or turf crops
(B) Cleaning crops
(C) Ware crops
(D) Soiling crops

232. In intercrop situations when subsidiary crops are usually of a shorted duration in a long duration main crop i.e. The main crop gets the company of a short duration crop for a certain period at the early stages of its growth in called:

(A) Companion crops
(B) Parallel crops
(C) Riparian crops
(D) Avenue crops

233. These consist of two or more crops which are grown simultaneously but do not have inter-competitive effects i.e. they are parallel with respect to competition is called:

(A) Companion crops
(B) Parallel crops
(C) Riparian crops
(D) Avenue crops

234. These crop plants are grown along the farm roads and fences is called:

(A) Companion crops
(B) Parallel crops
(C) Riparian crops
(D) Avenue crops

235. These crop plants are grwon along irrigation and drainage channels or water bodies is called:

(A) Companion crops
(B) Parallel crops
(C) Riparian crops
(D) Avenue crops

236. When subcrops are sown to supplement the yield of the main crops, the subcrops are known as :

(A) Augmentation crops
(B) Avenue crops
(C) Companion crops
(D) Riparian crops

237. The formula LER-1/LER x value of combined intercrop represents:

(A) MAI
(B) LER
(C) IER
(D) None of these

238. The ratio between gross and net cropped area is called an index of:
(A) Cropping intensity
(B) Multiple cropping
(C) High intensity cropping
(D) None of these

239. The rotation intensity of maize-wheat + gram – moong is:
(A) 200% (B) 400%
(C) 300% (D) 100%

240. Relationship between water and fertilizer as production factors in crops is:
(A) Additive (B) Synergetic
(C) Antagonistic (D) None of these

241. Which one of the following crop rotation is the best for maintaining soil fertility:
(A) Maize-toria-wheat
(B) Paddy-wheat-cowpea
(C) Paddy-potato-greengram
(D) Soybean-wheat-greengram

ANSWER SHEET
7. Cropping Systems

1. (D) 2. (A) 3. (B) 4. (C) 5. (D) 6. (B) 7. (A) 8. (B) 9. (A) 10. (A)
11. (B) 12. (C) 13. (A) 14. (B) 15. (C) 16. (A) 17. (A) 18. (C) 19. (C) 20. (A)
21. (B) 22. (A) 23. (D) 24. (A) 25. (C) 26. (B) 27. (D) 28. (C) 29. (A) 30. (B)
31. (A) 32. (C) 33. (D) 34. (B) 35. (C) 36. (A) 37. (C) 38. (B) 39. (C) 40. (C)
41. (B) 42. (B) 43. (A) 44. (A) 45. (B) 46. (D) 47. (B) 48. (C) 49. (D) 50. (D)
51. (A) 52. (B) 53. (B) 54. (C) 55. (B) 56. (A) 57. (C) 58. (C) 59. (C) 60. (A)
61. (B) 62. (C) 63. (D) 64. (A) 65. (C) 66. (B) 67. (D) 68. (C) 69. (B) 70. (A)
71. (B) 72. (C) 73. (D) 74. (A) 75. (A) 76. (B) 77. (C) 78. (D) 79. (A) 80. (B)
81. (C) 82. (C) 83. (A) 84. (B) 85. (D) 86. (C) 87. (A) 88. (B) 89. (C) 90. (D)
91. (A) 92. (B) 93. (C) 94. (D) 95. (A) 96. (A) 97. (D) 98. (A) 99. (D) 100. (D)
101. (A) 102. (B) 103. (C) 104. (D) 105. (A) 106. (A) 107. (D) 108. (A) 109. (B) 110. (D)
111. (A) 112. (B) 113. (A) 114. (A) 115. (A) 116. (B) 117. (D) 118. (C) 119. (D) 120. (A)
121. (C) 122. (A) 123. (C) 124. (A) 125. (B) 126. (C) 127. (D) 128. (D) 129. (A) 130. (B)
131. (C) 132. (C) 133. (B) 134. (A) 135. (C) 136. (C) 137. (A) 138. (A) 139. (B) 140. (A)
141. (A) 142. (A) 143. (B) 144. (C) 145. (A) 146. (C) 147. (C) 148. (A) 149. (B) 150. (A)
151. (C) 152. (A) 153. (A) 154. (A) 155. (B) 156. (D) 157. (D) 158. (D) 159. (C) 160. (C)
161. (C) 162. (A) 163. (D) 164. (C) 165. (C) 166. (C) 167. (A) 168. (A) 169. (D) 170. (C)
171. (C) 172. (A) 173. (A) 174. (C) 175. (A) 176. (C) 177. (C) 178. (C) 179. (B) 180. (A)
181. (B) 182. (C) 183. (B) 184. (B) 185. (C) 186. (C) 187. (A) 188. (A) 189. (B) 190. (D)
191. (B) 192. (A) 193. (D) 194. (C) 195. (C) 196. (A) 197. (B) 198. (C) 199. (C) 200. (D)
201. (A) 202. (C) 203. (D) 204. (B) 205. (C) 206. (B) 207. (D) 208. (D) 209. (D) 210. (C)
211. (A) 212. (D) 213. (A) 214. (A) 215. (B) 216. (A) 217. (A) 218. (D) 219. (D) 220. (D)
221. (A) 222. (C) 223. (B) 224. (A) 225. (B) 226. (C) 227. (D) 228. (A) 229. (B) 230. (C)
231. (D) 232. (A) 233. (B) 234. (D) 235. (C) 236. (A) 237. (A) 238. (A) 239. (C) 240. (B)
241. (D)

8 Agronomic Practices

1. The method of sowing of wheat requiring minimum seed rate per unit area is:
(A) Regular method
(B) Border method
(C) Skip method
(D) Dibbling method

2. What is the optimum seedling age of rice for transplanting in usar soils:
(A) 40-45 days (B) 21-25 days
(C) 28-30 days (D) 25-30 days

3. The optimum spacing for paddy transplanting is:
(A) 30 x 15 cm (B) 20 x 10 cm
(C) 15 x 5 cm (D) 25 x 15 cm

4. The N, P, K requirement in high yielding rice is:
(A) 60 : 30 : 30 kg/ha
(B) 80 : 40 : 40 kg/ha
(C) 120 : 60 : 60 kg/ha
(D) 100 : 30 : 30 kg/ha

5. The recommended number of seedlings/hill for rice transplanting is:
(A) 2-3 (B) 2-5
(C) 1-2 (D) only one

6. The recommended seed rate for late sown wheat is:
(A) 100 kg/ha (B) 80 kg/ha
(C) 125 kg/ha (D) 150 kg/ha

7. The recommended dose of N P K fertilizers for late sown wheat is:
(A) 100 : 50 : 50 kg/ha
(B) 120 : 60 : 60 kg/ha
(C) 80 : 40 : 40 kg/ha
(D) 100 : 80 : 40 kg/ha

8. Sowing of water soaked seed of wheat in late sown condition improves the:
(A) Germination (B) Tillers
(C) Growth (D) None of the above

9. The optimum sowing time of normal sown irrigated wheat in central UP is:
(A) 1 Dec. to 25 Dec.
(B) 5 Nov. to 25 Nov.
(C) 25 Dec. to 5 Jan.
(D) None of the above

10. Recommended seed rate of dhaincha for green manuring in sodic soil is:
(A) 20 kg/ha (B) 40 kg/ha
(C) 60 kg/ha (D) 60 kg/ha

11. Urea hydrolysis in the soil produces:
(A) Ammonium carbonate
(B) Uric acid
(C) Ammonium carbamate
(D) Amides

12. Zinc sulphate (Zn SO_4) should not be mixed with:
(A) Urea (B) MOP
(C) DAP (D) FYM

13. What is the form in which zinc is absorbed by the plants:
(A) Zn^+ (B) Zn^{++}
(C) Zn^{+++} (D) All of these

14. For the maximum yield of wheat, the ear bearing tillers should be:
(A) 100-150/m2 (B) 200-250/m2
(C) 300-350/m2 (D) 400-450/m2

15. Maximum yield of mustard is obtained at a plant geometry of:
(A) 45 x 20 cm (B) 60 x 30 cm
(C) 90 x 30 cm (D) 105 x 30 cm

16. Sulphur coated urea contains nitrogen:
(A) 36 % (B) 25 %
(C) 40 % (D) 46 %

17. For ideal plant population, the seed rate of sugarbeet should be (kg/ha):
(A) 4-6 (B) 8-10
(C) 10-15 (D) 20-25

18. Soil brought by Ganga-Yamuna and Indus rivers and their tributaries are deep and are known as:

(A) Bhabar soils (B) Vindhyan soils
(C) Desert soils (D) Alluvial soils

19. The optimum period of sowing for Capsularis types (white jute) is:

(A) March-April (B) April-May
(C) July-August (D) January-February

20. For the transplanting of rice under SRI method, nursery should be:

(A) 5 days old (B) 6-8 days old
(C) 8-10 days old (D) 10-12 days old

21. The seed rate of rice under SRI method is:

(A) 5 kg/ha (B) 6 kg/ha
(C) 8 kg/ha (D) 10 kg/ha

22. The appropriate time for planting mentha in plains is:

(A) June-July
(B) September-October
(C) January-February
(D) April-May

23. The optimum plant population of deshi maize is:

(A) 80000-90000 plants/ha
(B) 90000-100000 plants/ha
(C) 100000-110000 plants/ha
(D) More 120000 plants/ha

24. The optimum plant population of hybrid maize is:

(A) 80000-85000 plants/ha
(B) 90000-100000 plants/ha
(C) 100000-110000 plants/ha
(D) More 120000 plants/ha

25. The optimum plant population of composite maize is:

(A) 85000-90000 plants/ha
(B) 90000-100000 plants/ha
(C) 100000-110000 plants/ha
(D) More 120000 plants/ha

26. Optimum sowing time of rabi maize in north India is:

(A) First week of October to second week of October
(B) Mid of October to End of October
(C) Last week of October to First week of November
(D) First week of November to Second week of November

27. Optimum sowing time of rabi maize in north India is:

(A) First week of October to second week of October
(B) Mid of October to Mid of November
(C) Last week of October to First week of November
(D) First week of November to Second week of November

28. Fertilizer requirement of hybrid rabi maize is:

(A) 120 kg N + 60 kg P_2O_5+40 kg K_2O/ha
(B) 100 kg N + 50 kg P_2O_5 + 50 kg K_2O/ha
(C) 120 kg N + 60 kg P_2O_5 + 50 kg K_2O/ha
(D) 100 kg N + 50 kg P_2O_5 + 40 kg K_2O/ha

29. Fertilizer requirement of composite rabi maize is:

(A) 120 kg N + 60 kg P_2O_5 + 40 kg K_2O/ha
(B) 100 kg N + 50 kg P_2O_5+50 kg K_2O/ha
(C) 120 kg N + 60 kg P_2O_5 + 50 kg K_2O/ha
(D) 100 kg N + 50 kg P_2O_5 + 40 kg K_2O/ha

30. Reasons for higher yields in rabi maize is:

(A) Establishment of better plant stand
(B) Better weed control measures
(C) Poor infestation of insect/pest and diseases
(D) All of these

31. Optimum seed rate of barley under sodic soil is:

(A) 75 kg/ha (B) 100 kg/ha
(C) 125 kg/ha (D) 150 kg/ha

32. Optimum seed rate of barley under normal irrigated is:

(A) 75 kg/ha (B) 100 kg/ha
(C) 125 kg/ha (D) 150 kg/ha

33. Optimum seed rate of barley under normal unirrigated conditions is:

(A) 75 kg/ha (B) 100 kg/ha
(C) 125 kg/ha (D) 150 kg/ha

34. Fertilizer requirement of barley under sodic soil is:

(A) 30-40 kg N + 20 kg P_2O_5+20 kg K_2O/ha
(B) 40 kg N + 20 kg P_2O_5 + 20 kg K_2O/ha
(C) 60 kg N + 40 kg P_2O_5 + 30 kg K_2O/ha
(D) 70-80 kg N + 30-40 kg P_2O_5 + 30-40 kg K_2O/ha

35. Fertilizer requirement of malt barley is:
(A) 30-40 kg N + 20 kg P_2O_5 + 20 kg K_2O/ha
(B) 40 kg N + 20 kg P_2O_5 + 20 kg K_2O/ha
(C) 60 kg N + 40 kg P_2O_5 + 30 kg K_2O/ha
(D) 70-80 kg N + 30-40 kg P_2O_5 + 30-40 kg K_2O/ha

36. Fertilizer requirement of barley under unirrigated conditions is:
(A) 30-40 kg N+20 kg P_2O_5 + 20 kg K_2O/ha
(B) 40 kg N + 20 kg P_2O_5 + 20 kg K_2O/ha
(C) 60 kg N + 40 kg P_2O_5 + 30 kg K_2O/ha
(D) 70-80 kg N + 30-40 kg P_2O_5 + 30-40 kg K_2O/ha

37. Fertilizer requirement of barley under irrigated conditions is:
(A) 30-40 kg N + 20 kg P_2O_5 + 20 kg K_2O/ha
(B) 40 kg N + 20 kg P_2O_5 + 20 kg K_2O/ha
(C) 60 kg N + 40 kg P_2O_5 + 30 kg K_2O/ha
(D) 70-80 kg N + 30-40 kg P_2O_5 + 30-40 kg K_2O/ha

38. The optimum plant population of rabi sorghum is:
(A) 85000-90000 plants/ha
(B) 90000-100000 plants/ha
(C) 100000-110000 plants/ha
(D) 150000 plants/ha

39. Optimum sowing time of rabi sorghum in north India is:
(A) Second week of September to second week of October
(B) Mid of October to End of October
(C) Last week of October to First week of November
(D) First week of November to Second week of November

40. Fertilizer requirement of desi sorghum is:
(A) 30-40 kg N+20 kg P_2O_5 + 20 kg K_2O/ha
(B) 40 kg N + 20 kg P_2O_5 + 20 kg K_2O/ha
(C) 50-60 kg N + 30-40 kg P_2O_5 + 30-35 kg K_2O/ha
(D) 70-80 kg N + 30-40 kg P_2O_5 + 30-40 kg K_2O/ha

41. Fertilizer requirement of hybrid sorghum is:
(A) 30-40 kg N+20 kg P_2O_5 + 20 kg K_2O/ha
(B) 40 kg N + 20 kg P_2O_5 + 20 kg K_2O/ha
(C) 50-60 kg N + 30-40 kg P_2O_5 + 30-35 kg K_2O/ha
(D) 100-120 kg N + 50-60 kg P_2O_5 + 40-50 kg K_2O/ha

42. Optimum seed rate of bajra is:
(A) 4-5 kg/ha
(B) 5-6 kg/ha
(C) 6-7 kg/ha
(D) 7-8 kg/ha

43. Optimum seed rate of bajra in mixed cropping is:
(A) 2-3 kg/ha
(B) 3-4 kg/ha
(C) 4-5 kg/ha (D) 5-6 kg/ha

44. Optimum sowing time of bajra is:
(A) First fort night of July to second fortnight of July
(B) Mid of June to End of June
(C) Last week of June to First week of July
(D) Last week of July to First week of Aug.

45. Optimum sowing time of ragi in north India is:
(A) May to June
(B) June to July
(C) July to August
(D) June to August

46. Optimum sowing time of ragi in south India is:
(A) July to August
(B) August to September
(C) September to October
(D) October to November

47. Optimum sowing time of Cheena is:
(A) First fortnight of July
(B) Last week of June to Second week of July
(C) Second week of July to Third week of July
(D) Third week of July to Last week of July

48. Optimum sowing time of Kakun is:
(A) Second week of June to First week of July
(B) Last week of June to Second week of July
(C) Second week of July to Third week of July
(D) Third week of July to Last week of July

49. Optimum sowing time of Kodon is:
(A) Second week of June to First week of July
(B) Mid June to Mid July
(C) Second week of July to Third week of July
(D) Third week of July to Last week of July

50. Optimum sowing time of Sawan is:
(A) Second week of June to First week of July
(B) Last week of June to Second week of July
(C) Second week of July to Third week of July
(D) Third week of July to Last week of July

51. Optimum seed rate of ragi is:
(A) 8-10 kg/ha (B) 10-12 kg/ha
(C) 12-15 kg/ha (D) 15-20 kg/ha

52. Optimum seed rate of Cheena is:
(A) 8-10 kg/ha (B) 8-12 kg/ha
(C) 12-15 kg/ha (D) 15-20 kg/ha

53. Optimum seed rate of kakun is:
(A) 8-10 kg/ha (B) 10-12 kg/ha
(C) 12-15 kg/ha (D) 15-20 kg/ha

54. Optimum seed rate of kodon is:
(A) 8-10 kg/ha (B) 10-12 kg/ha
(C) 12-15 kg/ha (D) 15-20 kg/ha

55. Optimum seed rate of sawo is:
(A) 8-10 kg/ha (B) 10-12 kg/ha
(C) 12-15 kg/ha (D) 15-20 kg/ha

56. Optimum spacing of ragi is:
(A) 20-30 x 10-12 cm
(B) 25-30 x 8-10 cm
(C) 20-30 x 8-10 cm
(D) 20-30 x 10-15 cm

57. Optimum spacing of cheena is:
(A) 20-30 x 10-12 cm
(B) 25-30 x 8-10 cm
(C) 20-30 x 8-10 cm
(D) 20-30 x 10-15 cm

58. Optimum spacing of kakun is:
(A) 20-30 x 10-12 cm
(B) 25-30 x 8-10 cm
(C) 25-30 x 10-15 cm
(D) 20-30 x 10-15 cm

59. Optimum spacing of kodon is:
(A) 20-30 x 10-12 cm
(B) 25-30 x 8-10 cm
(C) 20-30 x 8-10 cm
(D) 20-30 x 10-15 cm

60. Optimum spacing of sawo is:
(A) 20-30 x 10-12 cm
(B) 25-30 x 8-10 cm
(C) 20-30 x 8-10 cm
(D) 20-30 x 10-15 cm

61. Fertilizer requirement of ragi is:
(A) 50-60 kg N + 30-40 kg P_2O_5 +20-30 kg K_2O/ha
(B) 40-60 kg N + 30-40 kg P_2O_5 + 20-30 kg K_2O/ha
(C) 40-60 kg N + 30 kg P_2O_5 + 20-30 kg K_2O/ha
(D) 40 kg N + 40 kg P_2O_5 + 20 kg K_2O/ha

62. Fertilizer requirement of cheena is:
(A) 50-60 kg N + 30-40 kg P_2O_5 +20-30 kg K_2O/ha
(B) 40-60 kg N + 30-40 kg P_2O_5 + 20-30 kg K_2O/ha
(C) 40-60 kg N + 30 kg P_2O_5 + 20-30 kg K_2O/ha
(D) 40 kg N + 40 kg P_2O_5 + 20 kg K_2O/ha

63. Optimum seed rate of soybean grown in kharif season is:
(A) 50-60 kg/ha
(B) 70-75 kg/ha
(C) 70-100 kg/ha
(D) 100-120 kg/ha

64. Fertilizer requirement of kakun is:
(A) 50-60 kg N + 30-40 kg P_2O_5 +20-30 kg K_2O/ha
(B) 40-60 kg N + 30-40 kg P_2O_5 + 20-30 kg K_2O/ha
(C) 40-60 kg N + 30 kg P_2O_5 + 20-30 kg K_2O/ha
(D) 40 kg N + 40 kg P_2O_5 + 20 kg K_2O/ha

65. Fertilizer requirement of kodon is:
(A) 50-60 kg N + 30-40 kg P_2O_5 +20-30 kg K_2O/ha
(B) 40-60 kg N + 30-40 kg P_2O_5 + 20-30 kg K_2O/ha
(C) 40-60 kg N + 30 kg P_2O_5 + 20-30 kg K_2O/ha
(D) 40 kg N + 40 kg P_2O_5 + 20 kg K_2O/ha

66. Fertilizer requirement of sawan is:
(A) 50-60 kg N + 30-40 kg P_2O_5 +20-30 kg K_2O/ha
(B) 40-60 kg N + 30-40 kg P_2O_5 + 20-30 kg K_2O/ha
(C) 40-60 kg N + 30 kg P_2O_5 + 20-30 kg K_2O/ha
(D) 40-50 kg N + 30-40 kg P_2O_5 + 20-30 kg K_2O/ha

67. Optimum sowing time of soybean is:
(A) Second week of June to First week of July
(B) Third week of June to Second week of July
(C) Second week of July to Third week of July
(D) Third week of July to Last week of July

68. Optimum seed rate of soybean grown in spring season is:
(A) 50-60 kg/ha
(B) 70-75 kg/ha
(C) 70-100 kg/ha
(D) 100-120 kg/ha

69. Fertilizer requirement of soybean without rhizobium culture is:
(A) 100-120 kg N + 30-40 kg P_2O_5 +20-30 kg K_2O/ha
(B) 120-150 kg N + 60-80 kg P_2O_5 + 40-60 kg K_2O/ha
(C) 100-120 kg N + 50-60 kg P_2O_5 + 20-30 kg K_2O/ha
(D) 120-150 kg N + 30-40 kg P_2O_5 + 40-60 kg K_2O/ha

70. Optimum sowing time of groundnut in north India during kharif season is:
(A) May to June
(B) Third week of June to Second week of July
(C) Second week of July to Third week of July
(D) Third week of July to Last week of July

71. Optimum sowing time of groundnut in south India during kharif season is:
(A) May to June
(B) Third week of June to Second week of July
(C) Second week of July to Third week of July
(D) Third week of July to Last week of July

72. Optimum sowing time of groundnut during rabi season is:
(A) November-December
(B) October-November
(C) December-January
(D) January-February

73. Optimum seed rate of spreading type of groundnut is:
(A) 50-60 kg/ha (B) 60-80 kg/ha
(C) 80-100 kg/ha (D) 100-120 kg/ha

74. Optimum seed rate of bunch type of groundnut is:
(A) 50-60 kg/ha (B) 60-80 kg/ha
(C) 80-100 kg/ha (D) 100-120 kg/ha

75. Optimum spacing of spreading type of groundnut is:
(A) 20-30 x 10-12 cm
(B) 30-40 x 10-15 cm
(C) 45-60 x 15-20 cm
(D) 20-30 x 10-15 cm

76. Optimum spacing of bunch type of groundnut is:
(A) 20-30 x 10-12 cm
(B) 30-40 x 10-15 cm
(C) 45-60 x 15-20 cm
(D) 20-30 x 10-15 cm

77. Optimum seed rate of sarson in Tarai region is:
(A) 2-3 kg/ha (B) 5-6 kg/ha
(C) 6-7 kg/ha (D) 7-8 kg/ha

78. Fertilizer requirement of groundnut is:
(A) 15-20 kg N + 30-40 kg P_2O_5 +20-30 kg K_2O/ha
(B) 15-20 kg N + 60-80 kg P_2O_5 + 40-60 kg K_2O/ha
(C) 15-20 kg N + 50-60 kg P_2O_5 + 30-40 kg K_2O/ha
(D) 15-20 kg N + 30-40 kg P_2O_5 + 40-60 kg K_2O/ha

79. Optimum seed rate of toria is:
(A) 2-3 kg/ha (B) 5-6 kg/ha
(C) 6 kg/ha (D) 7-8 kg/ha

80. Optimum seed rate of sarson in mixed/ intercropping is:
(A) 2-3 kg/ha (B) 5-6 kg/ha
(C) 6-7 kg/ha (D) 7-8 kg/ha

81. Optimum seed rate of rai is:
(A) 2-3 kg/ha (B) 5-6 kg/ha
(C) 6-7 kg/ha (D) 7-8 kg/ha

82. Optimum sowing time of toria is:
(A) Last week of September to first week of October
(B) First week of October to Second week of October
(C) First week of October to Third week of October
(D) Last week of October to First week of November

83. Optimum sowing time of rai is:
(A) Last week of September to first week of October
(B) First week of October to Second week of October
(C) First week of October to Third week of October
(D) Last week of October to First week of November

84. Fertilizer requirement of toria is:
(A) 60 kg N + 30 kg P_2O_5 +30 kg K_2O/ha
(B) 50 kg N + 40 kg P_2O_5 + 40kg K_2O/ha
(C) 60 kg N + 40 kg P_2O_5 + 40 kg K_2O/ha
(D) 80 kg N+40-60 kg P_2O_5+40-50 kg K_2O/ha

85. Fertilizer requirement of rai is:
(A) 60 kg N + 30 kg P_2O_5 +30 kg K_2O/ha
(B) 50 kg N + 40 kg P_2O_5 + 40kg K_2O/ha
(C) 60 kg N + 40 kg P_2O_5 + 40 kg K_2O/ha
(D) 80 kg N + 40 kg P_2O_5 + 40 kg K_2O/ha

86. Fertilizer requirement of sarson is:
(A) 60 kg N + 30 kg P_2O_5 +30 kg K_2O/ha
(B) 50 kg N + 40 kg P_2O_5 + 40kg K_2O/ha(C) 60 kg N + 40 kg P_2O_5 + 40 kg K_2O/ha
(D) 60-80 kg N + 40-60 kg P_2O_5 + 40-50 kg K_2O/ha

87. Optimum seed rate of hybrid sunflower is:
(A) 5-6 kg/ha (B) 6-7 kg/ha
(C) 7-8 kg/ha (D) 10-12 kg/ha

88. Optimum seed rate of composite sunflower is:
(A) 5-6 kg/ha
(B) 6-7 kg/ha
(C) 7-8 kg/ha
(D) 10-12 kg/ha

89. Optimum sowing time of sunflower during kharif season is:
(A) Last week of June to First week of July
(B) First week of July to Second week of July
(C) Second week of July to Third week of July
(D) Last week of July to First week of August

90. Optimum sowing time of sunflower during rabi season is:
(A) Last week of September to First week of October
(B) First week of October to Second week of October
(C) First week of October to Third week of October
(D) First week of November to Last week of November

91. Optimum sowing time of sunflower during zaid/ summer season is:
(A) Second week of February to First week of March
(B) First week of March to Second week of March
(C) Second week of March to Third week of March
(D) First week of April to Last week of April

92. Fertilizer requirement of sunflower is:
(A) 60 kg N + 30 kg P_2O_5 +30 kg K_2O/ha
(B) 50 kg N + 40 kg P_2O_5 + 40kg K_2O/ha
(C) 60 kg N + 40 kg P_2O_5 + 40 kg K_2O/ha
(D) 80 kg N + 60 kg P_2O_5 + 40 kg K_2O/ha

93. Optimum seed rate of bold seeded linseed is:
(A) 15-20 kg/ha (B) 20-25 kg/ha
(C) 25-30 kg/ha (D) 40 kg/ha

94. Optimum seed rate of small seeded linseed is:
(A) 15-20 kg/ha (B) 20-25 kg/ha
(C) 25-30 kg/ha (D) 40 kg/ha

95. Optimum seed rate of linseed sown under utera cultivation is:
(A) 15-20 kg/ha (B) 20-25 kg/ha
(C) 25-30 kg/ha (D) 40 kg/ha

96. Optimum sowing time of linseed is:
(A) Second week of September to Last week of September
(B) Last week of September to First week of October
(C) First week of October to First week of November
(D) First week of November to Second week of November

97. Optimum sowing time of linseed under utera cultivation is:
(A) Last week of September
(B) Second week of September to Last week of September
(C) First week of October to Second week of October
(D) First week of November to Last week of November

98. Fertilizer requirement of linseed under irrigated conditionsis:
(A) 30-40 kg N + 20-25 kg P_2O_5 +30 kg K_2O/ha
(B) 50 kg N + 40 kg P_2O_5 + 40kg K_2O/ha
(C) 20 kg N + 20 kg P_2O_5 + 20 kg K_2O/ha
(D) 80 kg N + 60 kg P_2O_5 + 40 kg K_2O/ha

99. Fertilizer requirement of linseed under unirrigated conditions is:
(A) 30-40 kg N + 20-25 kg P_2O_5 +30 kg K_2O/ha
(B) 50 kg N + 40 kg P_2O_5 + 40kg K_2O/ha
(C) 20 kg N + 20 kg P_2O_5 + 20 kg K_2O/ha
(D) 80 kg N + 60 kg P_2O_5 + 40 kg K_2O/ha

100. Optimum seed rate of sesame is
(A) 2-3 kg/ha
(B) 3-5 kg/ha
(C) 5-6 kg/ha
(D) 6-7 kg/ha

101. Fertilizer requirement of linseed under utera conditions is:
(A) 30-40 kg N + 20-25 kg P_2O_5 +30 kg K_2O/ha
(B) 50 kg N + 40 kg P_2O_5 + 40kg K_2O/ha
(C) 20 kg N + 20 kg P_2O_5 + 20 kg K_2O/ha
(D) 80 kg N + 60 kg P_2O_5 + 40 kg K_2O/ha

102. Optimum seed rate of castor is:
(A) 15-20 kg/ha (B) 20-25 kg/ha
(C) 25-30 kg/ha (D) 40 kg/ha

103. Optimum sowing time of castor during kharif season is:
(A) May to June
(B) June to July
(C) Second fortnight of July to First fortnight of August
(D) First fortnight of August to Second fortnight of August

104. Optimum sowing time of castor during rabi season is:
(A) Second fortnight of July to First fortnight of August
(B) First fortnight of August to Second fortnight of August
(C) September to October
(D) October to November

105. Optimum sowing time of castor during zaid season is:
(A) December (B) January
(C) February (D) March

106. Optimum spacing of castor is:
(A) 75 x 75 cm (B) 90 x 90 cm
(C) 90 x 45 cm (D) 75 x 60 cm

107. Fertilizer requirement of castor is:
(A) 30-40 kg N + 20-25 kg P_2O_5 +30 kg K_2O/ha
(B) 50 kg N + 40 kg P_2O_5 + 40kg K_2O/ha
(C) 60 kg N + 40 kg P_2O_5 + 40 kg K_2O/ha
(D) 80 kg N + 60 kg P_2O_5 + 40 kg K_2O/ha

108. Optimum sowing time of sesame during kharif season is:
(A) First week of June to Last week of June
(B) Last week of June to First week of July
(C) Second fortnight of July to First fortnight of August
(D) First fortnight of August to Second fortnight of August

109. Fertilizer requirement of castor under unirrigated conditions is:
(A) 30-40 kg N + 20-35 kg P_2O_5 +20-30 kg K_2O/ha
(B) 50 kg N + 40 kg P_2O_5 + 40kg K_2O/ha
(C) 20 kg N + 20 kg P_2O_5 + 20 kg K_2O/ha
(D) 80 kg N + 60 kg P_2O_5 + 40 kg K_2O/ha

110. Fertilizer requirement of safflower under irrigated conditions is:
(A) 50-60-kg N + 40-50 kg P_2O_5 +30-40 kg K_2O/ha
(B) 50 kg N + 40 kg P_2O_5 + 40kg K_2O/ha
(C) 20 kg N + 20 kg P_2O_5 + 20 kg K_2O/ha
(D) 80 kg N + 60 kg P_2O_5 + 40 kg K_2O/ha

111. Optimum sowing time of sesame during rabi season is:
(A) First fortnight of August to Second fortnight of August
(B) September to October
(C) Mid October to First week of November
(D) Last week of October to First week of November

112. Fertilizer requirement of castor under irrigated conditions is:
(A) 30-40 kg N + 20-25 kg P_2O_5 +30 kg K_2O/ha
(B) 50 kg N + 40 kg P_2O_5 + 40kg K_2O/ha
(C) 50-60 kg N + 40-50 kg P_2O_5 + 30-40 kg K_2O/ha
(D) 80 kg N + 60 kg P_2O_5 + 40 kg K_2O/ha

113. Optimum sowing time of safflower is:
(A) First fortnight of August to Second fortnight of August
(B) September to October
(C) Mid October to First week of November
(D) Last week of October to First week of November

114. Fertilizer requirement of safflower under unirrigated conditions is:
(A) 30-40 kg N + 30-40 kg P_2O_5 +20-30 kg K_2O/ha
(B) 50 kg N + 40 kg P_2O_5 + 40kg K_2O/ha
(C) 20 kg N + 20 kg P_2O_5 + 20 kg K_2O/ha
(D) 80 kg N + 60 kg P_2O_5 + 40 kg K_2O/ha

115. Optimum seed rate of safflower is
(A) 5-6 kg/ha
(B) 6-8 kg/ha
(C) 8-10 kg/ha
(D) 10-12 kg/ha

116. Optimum sowing time of chickpea under unirrigated conditions is:
(A) First fortnight of August to Second fortnight of August
(B) September to October
(C) First fortnight of October
(D) Last week of October to First week of November

117. Optimum sowing time of chickpea under irrigated conditions is:
(A) First fortnight of August to Second fortnight of August
(B) September to October
(C) Mid October to First week of November
(D) Last week of October to First week of November

118. Optimum seed rate of desi chickpea is
(A) 50-60 kg/ha (B) 70-80 kg/ha
(C) 80-100 kg/ha (D) 100-120 kg/ha

119. Optimum seed rate of kabuli chickpea is
(A) 50-60 kg/ha
(B) 70-80 kg/ha
(C) 80-100 kg/ha
(D) 100-120 kg/ha

120. Optimum spacing of chickpea is:
(A) 25-30 x 10-15 cm
(B) 25-30 x 22.5-25.0 cm
(C) 30-45 x 10-15 cm
(D) 30-45 x 22.5-25.0 cm

121. Toxic substances of lathyrus is:
(A) BOAC (B) ODAP
(C) Both of these (D) None of these

122. Continuous consumption of lathyrus caused:
(A) Paralysis (B) Anemia
(C) Dehydration (D) Night blindness

123. Application of 20 kg S/ha in oilseeds increases:
(A) Fat content (B) Protein content
(C) Oil content (D) Starch content

124. Application of 20 kg S/ha in pulses increases:
(A) Fat content (B) Protein content
(C) Oil content (D) Starch content

125. Optimum sowing time of early maturing varieties of pigeonpea is:
(A) Last week of May to First week of June
(B) First week of June to Last week of June
(C) First week of July to Second week July
(D) Third week of July to Last week of July

126. Optimum sowing time of late maturing varieties of pigeonpea is:
(A) Last week of May to First week of June
(B) First week of June to Last week of June
(C) First week of July to Second week of July
(D) Third week of July to Last week of July

127. Optimum seed rate of pigeonpea is
(A) 10-12 kg/ha (B) 12-15 kg/ha
(C) 15-17 kg/ha (D) 18-20 kg/ha

128. Optimum seed rate of kharif sown mungbean is
(A) 10-12 kg/ha (B) 12-15 kg/ha
(C) 15-17 kg/ha (D) 18-20 kg/ha

129. Optimum seed rate of rabi sown mungbean is
(A) 10-12 kg/ha (B) 12-15 kg/ha
(C) 15-17 kg/ha (D) 20-25 kg/ha

130. Optimum sowing time of kharif sown mungbean is:
(A) Last week of May to First week of June
(B) First week of June to Last week of June
(C) Second fortnight of June to First fortnight of July
(D) Third week of July to Last week of July

131. Optimum sowing time of rabi sown mungbean is:
(A) Mid September to Last September
(B) October to November
(C) November to December
(D) Mid December to Mid January

132. Optimum spacing of kharif mungbean is:
(A) 30 x 10 cm (B) 45 x 10 cm
(C) 30 x 20 cm (D) 45 x 20 cm

133. Optimum sowing time of summer mungbean is:
(A) Mid March to Mid April
(B) First week of June to Last week of June
(C) Second fortnight of June to First fortnight of July
(D) Third week of July to Last week of July

134. Optimum spacing of summer mungbean is:
(A) 30 x 10 cm (B) 45 x 10 cm.
(C) 30 x 20 cm (D) 45 x 20 cm

135. Fertilizer requirement of mungbean is:
(A) 20-25 kg N + 45-50 kg P_2O_5 +40-50 kg K_2O/ha
(B) 50 kg N + 40 kg P_2O_5 + 40kg K_2O/ha
(C) 20 kg N + 20 kg P_2O_5 + 20 kg K_2O/ha
(D) 80 kg N + 60 kg P_2O_5 + 40 kg K_2O/ha

136. Optimum seed rate of kharif sown urdbean is
(A) 10-12 kg/ha (B) 12-15 kg/ha
(C) 15-17 kg/ha (D) 20-25 kg/ha

137. Optimum seed rate of summer sown urdbean is
(A) 10-12 kg/ha (B) 12-15 kg/ha
(C) 15-17 kg/ha (D) 20-25 kg/ha

138. Optimum sowing time of kharif sown urdbean is:
(A) Last week of May to First week of June
(B) First week of June - Last week of June
(C) Last week of June to First week of July
(D) Third week of July to Last week of July

139. Optimum sowing time of summer urdbean is:
(A) Last week of February to Last week of April
(B) First week of June - Last week of June
(C) Second fortnight of June to First fortnight of July
(D) Third week of July to Last week of July

140. Optimum spacing of kharif urdbean is:
(A) 30 x 10 cm (B) 45 x 10 cm
(C) 30 x 20 cm (D) 45 x 20 cm

141. Optimum spacing of summer urdbean is:
(B) 45 x 10 cm (A) 20-25 x 10-15 cm
(C) 30 x 20 cm (D) 45 x 20 cm

142. Optimum seed rate of bold seeded lentil is
(A) 25-30 kg/ha (B) 35-40 kg/ha
(C) 40-50 kg/ha (D) 50-60 kg/ha

143. Optimum seed rate of small seeded lentil is
(A) 25-30 kg/ha (B) 35-40 kg/ha
(C) 40-50 kg/ha (D) 50-60 kg/ha

144. Optimum spacing of lentil is:
(A) 25 x 10 cm (B) 30 x 10 cm
(C) 25 x 20 cm (D) 30 x 20 cm

145. Optimum sowing time of lentil under unirrigated conditions is:
(A) Mid September to Last September
(B) Last week of September to Mid October
(C) November to December
(D) Mid December to Mid January

146. Optimum sowing time of lentil under irrigated conditions is:
(A) Mid September to Last September
(B) Mid October to Mid November
(C) November to December
(D) Mid December to Mid January

147. Optimum seed rate of fieldpea is
(A) 70-80 kg/ha (B) 80-100 kg/ha
(C) 100-120 kg/ha (D) 120-130 kg/ha

148. Optimum seed rate of garden pea is
(A) 70-80 kg/ha
(B) 80-100 kg/ha
(C) 100-120 kg/ha
(D) 120-130 kg/ha

149. Optimum spacing of fieldpea is:
(A) 30 x 5 cm (B) 30 x 10 cm
(C) 25 x 10 cm (D) 30 x 20 cm

150. Optimum sowing time of fieldpea is:
(A) Mid September to Last September
(B) Second week of October to Fourth week of October
(C) November to December
(D) Mid December to Mid January

151. Optimum sowing time of garden pea is:
(A) Mid September to Last September
(B) First week of October to Second week of October
(C) Mid October to Mid November
(D) Mid December to Mid January

152. In plains French bean are generally grown in:
(A) Kharif season
(B) Rabi season
(C) Summer season
(D) None of these

153. In hills French bean are generally grown in:
(A) Kharif season
(B) Rabi season
(C) Summer season
(D) None of these

154. Optimum seed rate of French bean is
(A) 70-80 kg/ha (B) 80-100 kg/ha
(C) 100-120 kg/ha (D) 120-130 kg/ha

155. Optimum spacing of French bean is:
(A) 30 x 5 cm (B) 30 x 10 cm
(C) 25 x 10 cm (D) 40 x 10 cm

156. Optimum sowing time of French bean in plains of north India is:
(A) Mid September to Last September
(B) Second week of October to Fourth week of October
(C) Mid October to Mid November
(D) Mid December to Mid January

157. Optimum sowing time of French bean in hills of Eastern part of India is:
(A) Mid July to Mid August
(B) First week of August to Mid September
(C) Mid September to Last week of September
(D) Last week of September to Second week of October

158. Fertilizer requirement of French bean is:
(A)100-120 kg N+60 kg P_2O_5+40kg K_2O/ha
(B) 50 kg N + 40 kg P_2O_5 + 40 kg K_2O/ha
(C) 120 kg N + 120 kg P_2O_5 + 80 kg K_2O/ha
(D) 80 kg N + 60 kg P_2O_5 + 40 kg K_2O/ha

159. Optimum seed rate of cowpea for grain purposes is:
(A) 20-25 kg/ha (B) 35-45 kg/ha
(C) 35-40 kg/ha (D) 40-45 kg/ha

160. Optimum seed rate of cowpea for fodder purposes is:
(A) 20-25 kg/ha (B) 35-45 kg/ha
(C) 35-40 kg/ha (D) 40-45 kg/ha

161. Optimum seed rate of cowpea for green manure purposes is:
(A) 20-25 kg/ha (B) 35-45 kg/ha
(C) 35-40 kg/ha (D) 40-45 kg/ha

162. Optimum seed rate of cowpea for summer season is:
(A) 20-25 kg/ha (B) 35-45 kg/ha
(C) 35-40 kg/ha (D) 40-45 kg/ha

163. Optimum sowing time of cowpea in north India during kharif season is:
(A) Last week of May to First week of June
(B) First week of June to Last week of June
(C) First week of July to Second week of July
(D) Third week of July to Last week of July

164. Optimum sowing time of cowpea in north India during summer season is:
(A) Third week of February to Second week of March
(B) First week of March to Last week of March
(C) First week of April to Second week of April
(D) Third week of April -Last week of April

165. Optimum spacing of kharif season cowpea is:
(A) 30-45 x 8-10 cm
(B) 30-45 x 10-15 cm
(C) 45-60 x 8-10 cm
(D) 45-60 x 10-15 cm

166. Optimum sowing time of cowpea in south India during rabi season is:
(A) Last week of August to First week of September
(B) First week of September to Last week of September
(C) First week of September to First week of October
(D) Third week of October to Last week of October

167. Optimum spacing of summer season cowpea is:
(A) 25-30 x 8-10 cm
(B) 30-45 x 10-15 cm
(C) 45-60 x 8-10 cm
(D) 45-60 x 10-15 cm

168. Optimum seed rate of oat for fodder purposes is
(A) 60-70 kg/ha (B) 70-80 kg/ha
(C) 80-100 kg/ha (D) 100-120 kg/ha

169. Optimum seed rate of oat for seed production is:
(A) 60-65 kg/ha (B) 70-80 kg/ha
(C) 80-100 kg/ha (D) 100-120 kg/ha

170. Optimum sowing time of oats is:
(A) Last week of August to First week of September
(B) First week of September to Last week of September
(C) First week of September to First week of October
(D) Mid of October to Last week of October

171. Fertilizer requirement of oats is:
(A) 100-120 kg N + 50-60 kg P_2O_5+40 kg K_2O/ha
(B) 50 kg N + 40 kg P_2O_5 + 40kg K_2O/ha
(C) 20 kg N + 20 kg P_2O_5 + 20 kg K_2O/ha
(D) 80 kg N + 60 kg P_2O_5 + 40 kg K_2O/ha

172. Fodder can be obtained from oats in 2 cuttings is:
(A) 350-450 q/ha (B) 500-550 q/ha
(C) 550-600 q/ha (D) 600-650 q/ha

173. Fodder can be obtained from oats in 3-4 cuttings is:
(A) 350-450 q/ha (B) 500-550 q/ha
(C) 550-600 q/ha (D) 600-650 q/ha

174. Optimum seed rate of lucern sowing by Howard methods is:
(A) 8-10 kg/ha (B) 10-12 kg/ha
(C) 12-15 kg/ha (D) 15-20 kg/ha

175. Optimum seed rate of lucern sowing by broadcast methods is:
(A) 8-10 kg/ha
(B) 10-12 kg/ha
(C) 12-15 kg/ha
(D) 15-20 kg/ha

176. Optimum seed rate of lucern for seed production is:
(A) 8-10 kg/ha
(B) 10-12 kg/ha
(C) 12-15 kg/ha
(D) 15-20 kg/ha

177. Optimum sowing time of lucern for fodder purpose is:
(A) Last week of August to First week of September
(B) First week of September to Last week of September
(C) Last week of September to last week of October
(D) First week of November to Second week of November

178. Optimum sowing time of lucern for seed production purpose is:
(A) Last week of August to First week of September
(B) First week of September to Last week of September
(C) Last week of September to last week of October
(D) First week of November to Second week of November

179. Optimum seed rate of small seeded sorghum is:
(A) 20-25 kg/ha (B) 30-45 kg/ha
(C) 50-60 kg/ha (D) 60-70 kg/ha

180. Optimum seed rate of bold seeded sorghum is:
(A) 20-25 kg/ha (B) 30-45 kg/ha
(C) 50-60 kg/ha (D) 60-70 kg/ha

181. Optimum seed rate of fodder maize is:
(A) 20-25 kg/ha (B) 30-45 kg/ha
(C) 50-60 kg/ha (D) 60-70 kg/ha

182. Fertilizer requirement of fodder maize is:
(A) 120-140 kg N + 50-60 kg P_2O_5 +40-50 kg K_2O/ha
(B) 150 kg N + 40 kg P_2O_5 + 40kg K_2O/ha
(C) 120 kg N + 20 kg P_2O_5 + 20 kg K_2O/ha
(D) 80 kg N + 60 kg P_2O_5 + 40 kg K_2O/ha

183. Optimum sowing time of teosinte during summer season is:
(A) January to February
(B) February to March
(C) March to April
(D) May to June

184. Optimum sowing time of teosinte during winter season is:
(A) May - June (B) June - July
(C) July - August (D) August-September

185. Optimum seed rate of teosinte for broadcast method is:
(A) 35-40 kg/ha (B) 40-50 kg/ha
(C) 50-60 kg/ha (D) 60-70 kg/ha

186. Optimum seed rate of teosinte for line sowing method is:
(A) 35-40 kg/ha
(B) 40-50 kg/ha
(C) 50-60 kg/ha
(D) 60-70 kg/ha

187. Fertilizer requirement of teosinte is:
(A) 120-140 kg N + 50-60 kg P_2O_5 +40-50 kg K_2O/ha
(B) 150 kg N + 40 kg P_2O_5 + 40kg K_2O/ha
(C) 120 kg N + 20 kg P_2O_5 + 20 kg K_2O/ha
(D) 60 kg N + 50 kg P_2O_5 + 40 kg K_2O/ha

188. Optimum seed rate of cluster bean is:
(A) 20 kg/ha (B) 30 kg/ha
(C) 40 kg/ha (D) 50 kg/ha

189. Optimum seed rate of Rhodes grass by line sowing method is:
(A) 8-10 kg/ha (B) 10-12 kg/ha
(C) 12-15 kg/ha (D) 15-20 kg/ha

190. Optimum seed rate of Rhodes grass sowing by broadcast method is:
(A) 8-10 kg/ha (B) 10-12 kg/ha
(C) 12-15 kg/ha (D) 15-20 kg/ha

191. Optimum seed rate of guinea grass is:
(A) 8-10 kg/ha (B) 10-12 kg/ha
(C) 12-15 kg/ha (D) 15-20 kg/ha

192. Optimum seed rate of sudan grass for fodder purposes is:
(A) 8-10 kg/ha (B) 10-12 kg/ha
(C) 12-15 kg/ha (D) 15-20 kg/ha

193. Optimum seed rate of sudan grass for seed production purposes is:
(A) 6-8 kg/ha (B) 10-12 kg/ha
(C) 12-15 kg/ha (D) 15-20 kg/ha

194. Optimum seed rate of napier grass is:
(A) 12000-15000 sets/ha
(B) 15000-20000 sets/ha
(C) 20000-25000 sets/ha
(D) 25000-30000 sets/ha

195. Optimum sowing time of Rhodes grass is:
(A) May - June (B) June - July
(C) July - August (D) August-September

196. Optimum sowing time of guinea grass is:
(A) May - June (B) June - July
(C) July - August (D) August -September

197. Optimum sowing time of sudan grass is:
(A) Last week of February
(B) First fortnight of March
(C) Second fortnight of March
(D) First fortnight of April

198. Optimum sowing time of para grass is:
(A) May - June (B) June - July
(C) July - August (D) August -September

199. Optimum sowing time of napier grass is:
(A) May - June
(B) June - July
(C) July - August
(D) August- September

200. Fertilizer requirement of Rhodes grass is:
(A) 50 kg N + 40 kg P_2O_5 + 30 kg K_2O/ha
(B) 60 kg N + 40 kg P_2O_5 + 30 kg K_2O/ha
(C) 50 kg N + 20 kg P_2O_5 + 20 kg K_2O/ha
(D) 60 kg N + 50 kg P_2O_5 + 40 kg K_2O/ha

201. Fertilizer requirement of guinea grass is:
(A) 150 kg N + 40 kg P_2O_5 + 30 kg K_2O/ha
(B) 150-300 kg N + 40-50 kg P_2O_5 + 30-40 kg K_2O/ha
(C) 150 kg N + 20 kg P_2O_5 + 20 kg K_2O/ha
(D) 160 kg N + 50 kg P_2O_5 + 40 kg K_2O/ha

202. Fertilizer requirement of napier grass is:
(A) 120-150 kg N + 60-70 kg P_2O_5 + 40-50 kg K_2O/ha
(B) 150-300 kg N + 40-50 kg P_2O_5 + 30-40 kg K_2O/ha
(C) 150 kg N + 20 kg P_2O_5 + 20 kg K_2O/ha
(D) 160 kg N + 50 kg P_2O_5 + 40 kg K_2O/ha

203. Optimum sowing time of dinanath grass is:
(A) May - June
(B) June - July
(C) July - August
(D) August-September

204. Fertilizer requirement of dinanath grass is:
(A) 50 kg N + 40 kg P_2O_5 + 30 kg K_2O/ha
(B) 100-120 kg N + 40-50 kg P_2O_5 + 30-35 kg K_2O/ha
(C) 50 kg N + 20 kg P_2O_5 + 20 kg K_2O/ha
(D) 60 kg N + 50 kg P_2O_5 + 40 kg K_2O/ha

205. Optimum seed rate of dinanath grass is:
(A) 6-8 kg/ha
(B) 10-12 kg/ha
(C) 12-15 kg/ha
(D) 15-18 kg/ha

206. Seed treatment of potato should be done with:
(A) Areton @ 0.2%
(B) Mencozeb @ 0.2 %
(C) Decoleton @ 0.2 %
(D) All of these

207. Optimum sowing time of early potato is:
(A) Last week of August to First week of September
(B) First week of September to Last week of September
(C) Last week of September to First week of October
(D) First week of November to Second week of November

208. Fertilizer requirement of potato is:
(A) 50 kg N + 40 kg P_2O_5 + 30 kg K_2O/ha
(B) 90 kg N + 60 kg P_2O_5 + 120 kg K_2O/ha
(C) 50 kg N + 20 kg P_2O_5 + 20 kg K_2O/ha
(D) 60 kg N + 50 kg P_2O_5 + 40 kg K_2O/ha

209. Optimum sowing time of main crop of potato is:
(A) Last week of August to First week of September
(B) First week of September to Last week of September
(C) Last week of September to last week of October
(D) Last week of October to First week of November

210. Optimum sowing time of autumn planted sugarcane is:
(A) June - July
(B) April - May
(C) October - November
(D) February - March

211. Optimum sowing time of spring planted sugarcane is:
(A) June - July
(B) October-November
(C) February - March
(D) April - May

212. Optimum sowing time of adhsali planted sugar-cane is:
(A) June - July
(B) October- November
(C) February - March
(D) April - May

213. Tying, wrapping and propping are important agronomic practices of:
(A) Potato (B) Tobacco
(C) Sugarcane (D) Jute

214. Fertilizer requirement of sugarcane in North India is:
(A) 150-200 kg N + 70-80 kg P_2O_5 + 80-90 kg K_2O/ha
(B) 90 kg N + 60 kg P_2O_5 + 120 kg K_2O/ha
(C) 180-200 kg N + 30-90 kg P_2O_5 + 75-90 kg K_2O/ha
(D) 60 kg N + 50 kg P_2O_5 + 40 kg K_2O/ha

215. Optimum seed rate of sugarbeet is:
(A) 4-5 kg/ha
(B) 5-6 kg/ha
(C) 6-7 kg/ha
(D) 7-8 kg/ha

216. Fertilizer requirement of sugarcane in South India is:
(A) 150-200 kg N + 50-80 kg P_2O_5 + 40-90 kg K_2O/ha
(B) 90 kg N + 60 kg P_2O_5 + 120 kg K_2O/ha
(C) 180-200 kg N + 80-90 kg P_2O_5 + 75-90 kg K_2O/ha
(D) 60 kg N + 50 kg P_2O_5 + 40 kg K_2O/ha

217. Optimum sowing time of sugarbeet is:
(A) September to October
(B) Second week of October to Second week of November
(C) Second week of November to Last week of November
(D) First week of December to Second week of December

218. Optimum sowing time of sugarbeet for seed production in Himachal Pradesh is:
(A) Last week of March
(B) Second week of April
(C) Last week of April
(D) First week of May

219. Optimum sowing time of desi cotton in North India is:
(A) Last week of March-First week of April
(B) Mid of April to Mid of May
(C) Mid of May to Mid of June
(D) July to August

220. Optimum sowing time of American cotton in North India is:
(A) Last week of March-First week of April
(B) Mid of April to Mid of May
(C) Mid of May to Mid of June
(D) July to August

221. Optimum sowing time of desi cotton in South India is:
(A) Last week of March to First week of April
(B) Mid of April to Mid of May
(C) Mid of May to Mid of June
(D) July to August

222. Optimum sowing time of American cotton in Andhra Pradesh is:
(A) Last week of March to First week of April
(B) Mid of April to Mid of May
(C) Mid of May to Mid of June
(D) First week of August

223. Optimum sowing time of hybrid cotton in South India is:
(A) Last week of March-First week of April
(B) Mid of April to Mid of May
(C) Mid of May to Mid of June
(D) First week of August

224. Fertilizer requirement of Sugarbeet is:
(A) 120-130 kg N + 60-80 kg P_2O_5 + 40-60 kg K_2O/ha
(B) 90 kg N + 60 kg P_2O_5 + 120 kg K_2O/ha
(C) 180-200 kg N + 30-90 kg P_2O_5 + 75-90 kg K_2O/ha
(D) 60 kg N + 50 kg P_2O_5 + 40 kg K_2O/ha

225. Optimum spacing of sugarbeet is:
(A) 30 x 20-25 cm (B) 40 x 20-25 cm
(C) 30 x 30 cm (D) 40 x 30 cm

226. Optimum spacing of desi cotton in North India is:
(A) 60 x 25-30 cm
(B) 60 x 30-45 cm
(C) 60 x 15-30 cm
(D) 90 x 45 cm

227. Optimum spacing of American cotton in North India is:
(A) 60 x 25-30 cm
(B) 60 x 30-45 cm
(C) 60 x 15-30 cm
(D) 90 x 45 cm

228. Optimum spacing of desi cotton in South India is:
(A) 60 x 25-30 cm
(B) 60 x 30-45 cm
(C) 60 x 15-30 cm
(D) 90 x 45 cm

229. Optimum spacing of American cotton in South India is:
(A) 60 x 25-30 cm (B) 60 x 30-45 cm
(C) 60 x 15-30 cm (D) 90 x 45 cm

230. Optimum spacing of hybrid cotton in South India is:
(A) 60 x 25-30 cm (B) 60 x 30-45 cm
(C) 60 x 15-30 cm (D) 90 x 60 cm

231. Fertilizer requirement of hybrid cotton in Maharastra is:
(A) 80 kg N + 40 kg P_2O_5 + 40 kg K_2O/ha
(B) 140 kg N + 120 kg P_2O_5 + 120 kg K_2O/ha
(C) 80 kg N + 50 kg P_2O_5 + 40 kg K_2O/ha
(D) 70 kg N + 20 kg P_2O_5 + 20 kg K_2O/ha

232. Fertilizer requirement of American cotton is:
(A) 80 kg N + 40 kg P_2O_5 + 40 kg K_2O/ha
(B) 140 kg N + 120 kg P_2O_5 + 120 kg K_2O/ha
(C) 80 kg N + 50 kg P_2O_5 + 40 kg K_2O/ha
(D) 70 kg N + 20 kg P_2O_5 + 20 kg K_2O/ha

233. Fertilizer requirement of hybrid cotton in Gujarat is:
(A) 80 kg N + 40 kg P_2O_5 + 40 kg K_2O/ha
(B) 140 kg N + 120 kg P_2O_5 + 120 kg K_2O/ha
(C) 80 kg N + 50 kg P_2O_5 + 40 kg K_2O/ha
(D) 70 kg N + 20 kg P_2O_5 + 20 kg K_2O/ha

234. Fertilizer requirement of cotton in North India is:
(A) 80 kg N + 40 kg P_2O_5 + 40 kg K_2O/ha
(B) 140 kg N + 120 kg P_2O_5 + 120 kg K_2O/ha
(C) 80 kg N + 50 kg P_2O_5 + 40 kg K_2O/ha
(D) 70 kg N + 20 kg P_2O_5 + 20 kg K_2O/ha

235. Fertilizer requirement of hybrid cotton in South India is:
(A) 80 kg N + 40 kg P_2O_5 + 40 kg K_2O/ha
(B) 140 kg N + 120 kg P_2O_5 + 120 kg K_2O/ha
(C) 120 kg N + 60 kg P_2O_5 + 60 kg K_2O/ha
(D) 70 kg N + 20 kg P_2O_5 + 20 kg K_2O/ha

236. Fertilizer requirement of desi cotton is:
(A) 80 kg N + 40 kg P_2O_5 + 40 kg K_2O/ha
(B) 140 kg N + 120 kg P_2O_5 + 120 kg K_2O/ha
(C) 80 kg N + 50 kg P_2O_5 + 40 kg K_2O/ha
(D) 40 kg N + 20 kg P_2O_5 + 20 kg K_2O/ha

237. Seed rate of sunhemp for green manuring is:
(A) 15-20 kg/ha (B) 20-25 kg/ha
(C) 55-60 kg/ha (D) 30-40 kg/ha

238. Seed rate of sunhemp for fibre purposes is:
(A) 15-20 kg/ha (B) 20-25 kg/ha
(C) 55-60 kg/ha (D) 30-40 kg/ha

239. Seed rate of sunhemp for seed production is:
(A) 15-20 kg/ha
(B) 20-25 kg/ha
(C) 55-60 kg/ha
(D) 30-40 kg/ha

240. Optimum sowing time of sunhemp for summer season is:
(A) April to June
(B) June to July
(C) Last week of June to First week of July
(D) September to October

241. Optimum sowing time of sunhemp for kharif season is:
(A) April to June
(B) June to July
(C) Last week of June to First week of July
(D) September to October

242. Optimum sowing time of sunhemp for Rabi season is:
(A) April to June
(B) June to July
(C) Last week of June to First week of July
(D) September to October

243. Optimum sowing time of sunhemp for fibre and seed production is:
(A) April to June
(B) June to July
(C) Last week of June to First week of July
(D) September to October

244. Optimum spacing of sunhemp for fibre and seed production is:
(A) 15-20 x 10-15 cm
(B) 20-25 x 10-15 cm
(C) 15-20 x 15-20 cm
(D) 20-25 c 15-20 cm

245. Pikka tobacco mainly grown in:
(A) Orissa (B) Andhra Pradesh
(C) West Bengal (D) Tamilnadu

246. Natu, Lanka and Burley tobacco mainly grown in:
(A) Orissa (B) Andhra Pradesh
(C) West Bengal (D) Tamilnadu

247. Cigar filler tobacco mainly grown in:
(A) Orissa (B) Andhra Pradesh
(C) West Bengal (D) Tamilnadu

248. Cigar wrapper tobacco mainly grown in:
(A) Orissa (B) Andhra Pradesh
(C) West Bengal (D) Tamilnadu

249. HDBRG tobacco mainly grown in:
(A) Orissa (B) Andhra Pradesh
(C) West Bengal (D) Tamilnadu

250. Motihari and Jati tobacco mainly grown in:
(A) Orissa
(B) Andhra Pradesh
(C) West Bengal
(D) Tamilnadu

251. HDBRG tobacco also known as:
(A) White burley (B) Pikka
(C) Motihari (D) Jati

252. Optimum seed rate of Motihari tobacco grown in West Bengal is:
(A) 5 kg/ha (B) 6 kg/ha
(C) 7 kg/ha (D) 8 kg/ha

253. Optimum seed rate of Bidi and chewing tobacco grown in Gujarat is:
(A) 5 kg/ha (B) 6 kg/ha
(C) 7 kg/ha (D) 8 kg/ha

254. Fertilizer requirement of Cigarette tobacco grown in Karnataka is:
(A) 50 kg N + 80 kg P_2O_5 + 80 kg K_2O/ha
(B) 180 kg N + 150 kg P_2O_5 + 250 kg K_2O/ha
(C) 200 kg N + 150 kg P_2O_5 + 200 kg K_2O/ha
(D) 120 kg N + 50 kg P_2O_5 + 75 kg K_2O/ha

255. Fertilizer requirement of bidi tobacco grown in Gujarat is:
(A) 50 kg N + 80 kg P_2O_5 + 80 kg K_2O/ha
(B) 180 kg N + 150 kg P_2O_5 + 250 kg K_2O/ha
(C) 200 kg N+150 kg P_2O_5 + 200 kg K_2O/ha
(D) 120 kg N + 50 kg P_2O_5 + 75 kg K_2O/ha

256. Fertilizer requirement of Chewing tobacco grown in Gujarat is:
(A) 50 kg N + 80 kg P_2O_5 + 80 kg K_2O/ha
(B) 180 kg N+150 kg P_2O_5 + 250 kg K_2O/ha
(C) 200 kg N + 150 kg P_2O_5 + 200 kg K_2O/ha
(D) 120 kg N + 50 kg P_2O_5 + 75 kg K_2O/ha

257. Fertilizer requirement of chewing tobacco grown in West Bengal is:
(A) 50 kg N + 80 kg P_2O_5 + 80 kg K_2O/ha
(B) 180 kg N + 150 kg P_2O_5 + 250 kg K_2O/ha
(C) 200 kg N + 150 kg P_2O_5 + 200 kg K_2O/ha
(D) 120 kg N + 50 kg P_2O_5 + 75 kg K_2O/ha

258. Fertilizer requirement of jati tobacco grown in West Bengal is:
(A) 50 kg N + 80 kg P_2O_5 + 80 kg K_2O/ha
(B) 112 kg N + 50 kg P_2O_5 + 75 kg K_2O/ha
(C) 200 kg N+150 kg P_2O_5 + 200 kg K_2O/ha
(D) 120 kg N + 50 kg P_2O_5 + 75 kg K_2O/ha

259. Fertilizer requirement of cigar and wrapper tobacco grown in West Bengal is:
(A) 50 kg N + 80 kg P_2O_5 + 80 kg K_2O/ha
(B) 180 kg N+150 kg P_2O_5 + 250 kg K_2O/ha
(C) 200 kg N+150 kg P_2O_5 + 200 kg K_2O/ha
(D) 120 kg N+112 kg P_2O_5+22.4 kg K_2O/ha

260. Fertilizer requirement of chewing tobacco grown in Tamilnadu is:
(A) 50 kg N + 80 kg P_2O_5 + 80 kg K_2O/ha
(B) 180 kg N+150 kg P_2O_5 + 250 kg K_2O/ha
(C) 75-100 kg N + 100 kg P_2O_5/ha
(D) 120 kg N + 50 kg P_2O_5 + 75 kg K_2O/ha

261. Fertilizer requirement of cigar filler tobacco grown in Tamilnadu is:
(A) 50 kg N + 80 kg P_2O_5 + 80 kg K_2O/ha
(B) 180 kg N+150 kg P_2O_5 + 250 kg K_2O/ha
(C) 200 kg N+150 kg P_2O_5 + 200 kg K_2O/ha
(D) 40 kg N + 50 kg P_2O_5 /ha

262. Optimum spacing of FCV tobacco grown in Andhra Pradesh is:
(A) 70 x 50 cm (B) 90 x 90 cm
(C) 60 x 60 cm (D) 60 x 75 cm

263. Optimum spacing of bidi tobacco grown in Gujarat is:
(A) 70 x 50 cm (B) 90 x 90 cm
(C) 60 x 60 cm (D) 60 x 75 cm

264. Optimum spacing of rustica tobacco grown in Gujarat is:
(A) 70 x 50 cm (B) 90 x 90 cm
(C) 60 x 60 cm (D) 60 x 75 cm

265. Optimum spacing of chewing tobacco grown in Gujarat is:

(A) 70 x 50 cm (B) 90 x 90 cm
(C) 60 x 60 cm (D) 105 x 90 cm

266. Optimum spacing of chewing tobacco grown in Bihar is:

(A) 70 x 50 cm (B) 90 x 90 cm
(C) 60 x 60 cm (D) 105 x 90 cm

267. Optimum spacing of Motihari tobacco grown in West Bengal is:

(A) 45 x 60 cm (B) 90 x 90 cm
(C) 45 x 90 cm (D) 75 x 75 cm

268. Optimum spacing of Jati tobacco grown in West Bengal is:

(A) 45 x 60 cm (B) 90 x 90 cm
(C) 45 x 90 cm (D) 75 x 75 cm

269. Optimum spacing of Cigar wrapper tobacco grown in West Bengal is:

(A) 45 x 60 cm (B) 90 x 90 cm
(C) 45 x 90 cm (D) 75 x 75 cm

270. Optimum spacing of chewing tobacco grown in Tamilnadu is:

(A) 45 x 60 cm (B) 90 x 90 cm
(C) 45 x 90 cm (D) 75 x 75 cm

271. Optimum sowing time of FCV tobacco grown in Andhra Pradesh is:

(A) Mid Oct to Mid November
(B) Mid August to Last week of September
(C) Last week of October to First week of November
(D) September

272. Optimum sowing time of bidi tobacco grown in Gujarat is:

(A) Mid Oct to Mid November
(B) Mid August to Last week of September
(C) Last week of October to First week of November
(D) September

273. Optimum sowing time of rustica tobacco grown in Gujarat is:

(A) Mid Oct to Mid November
(B) Mid August to Last week of September
(C) Last week of October to First week of November
(D) September

274. Optimum sowing time of chewing tobacco grown in Gujarat is:

(A) Mid Oct to Mid November
(B) Mid August to Last week of September
(C) Last week of October to First week of November
(D) September

275. Optimum sowing time of chewing tobacco grown in Bihar is:

(A) Third week of September to Second week of October
(B) Mid November to Mid December
(C) Last week of October to First week of November
(D) Third week of October to First week of November

276. Optimum sowing time of Motihari tobacco grown in West Bengal is:

(A) Third week of September to Second week of October
(B) Mid November to Mid December
(C) Last week of October to First week of November
(D) Third week of October to First week of November

277. Optimum sowing time of Jati tobacco grown in West Bengal is:

(A) Third week of September to Second week of October
(B) Mid November to Mid December
(C) Last week of October to First week of November
(D) Third week of October to First week of November

278. Optimum sowing time of Cigar wrapper tobacco grown in West Bengal is:

(A) Third week of September to Second week of October
(B) Mid November to Mid December
(C) Last week of October to First week of November
(D) Third week of October to First week of November

279. Optimum sowing time of chewing tobacco grown in Tamilnadu is:

(A) Third week of September to Second week of October
(B) Mid November to Mid December
(C) Last week of October to First week of November
(D) Last week of November to Second week of December

280. Harvesting method of bidi tobacco is:
(A) Whole plant
(B) Leaf wise
(C) Gugro and Lilvo
(D) All of these

281. Harvesting method of chewing tobacco is:
(A) Whole plant (C) Gugro and Lilvo
(B) Leaf wise (D) All of these

282. Harvesting method of FCV tobacco is:
(A) Whole plant (B) Gugro and Lilvo
(C) Leaf wise (D) All of these

283. Depog method of nursery preparation of rice developed from:
(A) Phillipines (Manila)
(B) India
(C) China
(D) Pakistan

284. In south India, optimum transplanting time of rabi rice is:
(A) October-November
(B) November-December
(C) December-January
(D) January-February

285. Fertilizer requirement of hybrid rice is:
(A) 100 kg N + 75 kg P_2O_5 + 60 kg K_2O + 25 kg Zn /ha
(B) 180 kg N+150 kg P_2O_5 + 250 kg K_2O/ha
(C) 200 kg N+150 kg P_2O_5 + 200 kg K_2O/ha
(D) 40 kg N + 50 kg P_2O_5 /ha

286. The optimum seed rate of hybrid rice is:
(A) 15-20 kg/ha (B) 20-25 kg/ha
(C) 25-30 kg/ha (D) 30-35 kg/ha

287. Normal sowing time of wheat at low hills areas is:
(A) First week of October to Second week of November
(B) Last week of October to Second week of November
(C) Second week of November to last week of November
(D) Last week of November to First week of December

288. Normal sowing time of wheat at high hills areas is:
(A) First week of October to Second week of November
(B) Last week of October to Second week of November
(C) Second week of November to last week of November
(D) Last week of November to First week of December

289. Without ploughing sowing of wheat is done by:
(A) Zero-tilled seeddrill
(B) Seed drill
(C) Multicrop planter
(D) None of these

290. The suitable time of planting for spring planted sugarcane is:
(A) June-July
(B) Sept-October
(C) February-March
(D) April-May

291. Method more suitable for fertilizer application in transplanted paddy is:
(A) Broad casting
(B) Combine drilling
(C) Plough sole placement
(D) Band placement

292. Optimum sowing time of *Corchorus olitorius* is:
(A) First week of April to last week of April
(B) Mid of February to last week of March
(C) Last week of March to first week of April
(D) Last week of April to first week of May

293. Optimum sowing time of *Corchorus capsularis* is:
(A) First week of April to last week of April
(B) Mid of February to last week of March
(C) Last week of March to first week of April
(D) Last week of April to first week of May

294. Seed rate of Corchorus capsularis is:
(A) 4 kg/ha (B) 5 kg/ha
(C) 6 kg/ha (D) 7 kg/ha

295. Seed rate of *Corchorus olitorius* is:
(A) 4 kg/ha
(B) 5 kg/ha
(C) 6 kg/ha
(D) 7 kg/ha

296. Optimum fertilizer requirement of safed jute is
(A) 50 kg N + 20 kg P_2O_5 + 20 kg K_2O/ha
(B) 60 kg N + 30 kg P_2O_5 + 30 kg K_2O/ha
(C) 50 kg N + 30 kg P_2O_5 + 20 kg K_2O/ha
(D) 60 kg N + 20 kg P_2O_5 + 30 kg K_2O/ha

297. Harvesting of jute should be done at:
(A) 100-120 DAS (B) 120-150 DAS
(C) 150-160 DAS (D) 160-170 DAS

298. Optimum fertilizer requirement of Tosa jute is
(A) 50 kg N + 20 kg P_2O_5 + 20 kg K_2O/ha
(B) 60 kg N + 30 kg P_2O_5 + 30 kg K_2O/ha
(C) 50 kg N + 30 kg P_2O_5 + 20 kg K_2O/ha
(D) 60 kg N + 20 kg P_2O_5 + 30 kg K_2O/ha

299. Under intensive cropping system jute can be harvested at:
(A) 115-125 DAS (B) 120-150 DAS
(C) 150-160 DAS (D) 160-170 DAS

300. Seed rate of mustard per hectare is:
(A) 1-2 kg (B) 3-4 kg
(C) 5-6 kg (D) 7-8 kg

301. The row spacing of bunch type variety of groundnut is:
(A) 10 cm (B) 20 cm
(C) 30 cm (D) 50 cm

302. Which dose of gypsum is recommended for groundnut is:
(A) 50 kg/ha (B) 100 kg/ha
(B) 150 kg/ha (D) 200 kg/ha

303. How much three budded sugarcane sets are required for sugarcane planting:
(A) 40000/ha (B) 20000/ha
(C) 10000/ha (D) None of these

304. Autumn planted sugarcane is planted in the month of:
(A) July (B) October
(C) February (D) April

305. Spring planted sugarcane is planted in the month of:
(A) July (B) October
(C) February (D) April

306. The average productivity of sugarcane in India is about:
(A) 800 q/ha (B) 700 q/ha
(C) 500 q/ha (D) 400 q/ha

307. The sugar percentage in sugarcane juice is:
(A) 5-6% (B) 7-8%
(C) 8-9% (D) 10-11%

308. Which one is the correct plant to plant spacing of early pigeonpea:
(A) 45 x 10 cm (B) 60 x 15 cm
(C) 90 x 20 cm (D) 120 x 25 cm

309. The recommended dose of nitrogen for pigeonpea is:
(A) 20 kg/ha (B) 40 kg/ha
(C) 60 kg/ha (D) 80 kg/ha

310. The protein content in gram seed is about:
(A) 10% (B) 15%
(C) 20% (D) 25%

311. At which stage irrigation is not given in gram:
(A) Growth stage
(B) Pod formation stage
(C) Flowering stage
(D) Branching stage

312. The seed rate of Kabuli variety of gram is:
(A) 50-60 kg/ha (B) 80-90 kg/ha
(C) 100-125 kg/ha (D) 150-200 kg/ha

313. Mustard crop planted at a plant spacing of 50 x 20 cm will have ----- plants/ha:
(A) 75000 (B) 100000
(C) 125000 (D) 15000

314. Sorghum was planted at a plant spacing of 12.5 cm with a population of 1,60,000 plants/ha its row spacing will be:
(A) 30 cm (B) 40 cm
(C) 5 cm (D) 50 cm

315. The seed rate of cotton would be ------ kg/ha, if the planting geometry is 75 x 50 cm, and test-weight, germination and purity percentages are 140g, 90 and 95, respectively:
(A) 2.17 (B) 3.37
(C) 4.37 (D) 5.21

316. If the germination and purity percentage and test weight of seed are 95, 90 and 46 g, respectively and the wheat crop was planted at 20 x 3 cm. The seed rate of wheat per hectare will be:
(A) 50.67 kg (B) 60.12 kg
(C) 79.67 kg (D) 89.67 kg

317. The seed rate of potato tubers would be ---- q/ha, if the crop spacing, average weight of tubers and rottage loss of tubers during storage are 60 x 25 cm, 35 g/kg and 5 per cent, respectively:

(A) 20.5 (B) 22.5
(C) 24.5 (D) 26.5

318. The quantity of urea needed for supplying 69 kg nitrogen to sorghum crop will be:

(A) 100 kg (B) 125 kg
(C) 150 kg (D) 175 kg

319. The application rates of nitrogen and phosphorus are 64 and 46 kg/ha to an experimental crop of wheat as basal, for that the quantity of urea and diammonium phosphate will be:

(A) 50 kg urea, 100 kg DAP
(B) 75 kg urea, 75 kg DAP
(C) 75 kg urea, 100 kg DAP
(D) 100 kg urea, 100 kg DAP

320. To assure adequate population in rice, the number of hills per square metre area should be maintained at:

(A) 50 (B) 80
(C) 40 (D) 60

321. To supply 100 kg nitrogen and 50 kg potassium, the quantity of ammonium chloride and muriate of potash will be:

(A) 299.12 kg AC, 73.3 kg MoP
(B) 384.61 kg AC, 73.3 kg MoP
(C) 209.12 kg AC, 83.3 kg MoP
(D) 384.61 kg AC, 83.3 kg MoP

322. Potato tuber yield will be -------- q/ha, if crop was spaced at 60 x 25 cm, and average number of tubers/plant and average weight of tubers are 14 and 3 g, respectively:

(A) 286.17 (B) 317.29
(C) 326.67 (D) 356.87

323. If the average panicle density per square metre is 260, the average number of filled grains per panicle and test weight of rice are 136 and 20 g, respectively. The rice crop will yield:

(A) 5.98 t/ha
(B) 7.08 t/ha
(C) 9.12 t/ha
(D) 10.46 t/ha

324. The yield of wheat grain will be --------t/ha from the following observations, spacing is 20 x 3 cm, effective tillers/plant is 3, number of grains/ear is 32, 1000 grains weight is 40 g:

(A) 5.8 t/ha (B) 6.2 t/ha
(C) 6.4 t/ha (D) 6.8 t/ha

325. Topping of safflower plants is beneficial:

(A) To reduce lodging
(B) To promote branching and flowering
(C) To reduce water loss
(D) To promote early branching

326. The area needed to raise seedlings by dapog method sufficient for transplanting one hectare area is:

(A) 250-300 sqm (B) 25-30 sqm
(C) 40-50 sqm (D) 400-500 sqm

327. Dapog seedlings would be ready for transplanting within:

(A) 8-10 days (B) 18-20 days
(C) 21-24 days (D) 11-14 days

328. The seed rate of bold seeded varieties of rice to raise seedlings enough for one hectare is:

(A) 40-45 kg (B) 20-25 kg
(C) 30-35 kg (D) 50-55 kg

329. Under late sown conditions area of nursery, sufficient for transplating one hectare rice field is:

(A) 800-1000 sqm (B) 500-600 sqm
(C) 1000-1200 sqm (D) 600-700 sqm

330. 'Aman' (Kharif) rice is sown in:

(A) April-May (B) June-July
(C) Nov-Dec (D) May-June

331. In Orissa summer or spring season rice is known as:

(A) Dalua (B) Boro
(C) Aman (D) Aus

332. In parts of eastern region and peninsular India two or three crops of rice are taken in a year because of:

(A) Favourable mean temperature
(B) High rainfall
(C) Soils having good water retention capacity
(D) Prolonged sunshine

333. Anaerobic environment in paddy soil is responsible for gaseous loss of fertilizer nitrogen by:

(A) Denitrification (B) Volatilization
(C) Nitrification (D) Ammonification

334. Denitrification losses of nitrogen in paddy field can be prevented by incorporating ammonical nitrogen into the:

(A) Oxidized zone (B) Reduced zone
(C) Calcified zone (D) Hydronized zone

335. Recommendation of nitrogen (N), phosphorus (P_2O_5) and potassium (K_2O) for high yielding varieties of rice in UP is:

(A) 120+60+60 (B) 120+60+45
(C) 100+60+60 (D) 120+60+30

336. The technique, used for minimizing the nitrogen losses in paddy field under Indian conditions comprises:

(A) Use of pre-incubated urea
(B) Use of neem seed cake
(C) Use of mud ball
(D) All of these

337. The zinc deficiency in rice cause which disease:

(A) Bacterial blight (B) Sheath blight
(C) Bacterial leaf streak
(D) Khaira

338. The seed of wheat should be treated with which fungicide against seed-borne diseases:

(A) Vitavex (B) Thiram
(C) Both of these (D) Any of these

339. Dibbling (sowing) of wheat requires how much seed :

(A) 15-20 kg/ha (B) 25-30 kg/ha
(C) 30-40 kg/ha (D) 20-25 kg/ha

340. A crop of wheat yielding 50 quintals of grain per hectare removes how much nutrients from soil:

(A) 80-100 kg N + 50-60 kg P_2O_5 + 100-120 kg K_2O/ha
(B) 100-150 kg N + 70-80 kg P_2O_5 + 125-150 kg K_2O/ha
(C) 150-180 kg N + 90-100 kg P_2O_5 + 160-180 kg K_2O/ha
(D) 160-200 kg N + 80-100 kg P_2O_5 + 150-170 kg K_2O/ha

341. The major part of nitrogen uptake by the maize is over by the:

(A) Knee high stage
(B) Tasselling stage
(C) Silking stage
(D) Grain filling stage

342. The maize crop should be planted at the distance of:

(A) 30 x 20-25 cm
(B) 60 x 20-25 cm
(C) 30 x 10-15 cm
(D) 60 x 10-15 cm

343. The seed rate for composite maize varies from:

(A) 18-20 kg/ha (B) 20-25 kg/ha
(C) 28-30 kg/ha (D) 30-35 kg/ha

344. Under high rainfall conditions maize should be planted:

(A) In narrow furrows
(B) On a flat bed
(C) On the side of ridges
(D) In widely spaced furrows

345. Hybrid maize should be fertilized with:

(A) 120 + 30 + 15 kg N + P_2O_5+K_2O/ha
(B) 80 + 40 + 20 kg N + P_2O_5+K_2O/ha
(C) 120 + 60 + 40 kg N + P_2O_5+K_2O/ha
(D) 120 + 30 + 30 kg N + P_2O_5+K_2O/ha

346. Seed rate of sorghum is:

(A) 12-15 kg/ha (B) 10-12 kg/ha
(C) 8-10 kg/ha (D) 15-20 kg/ha

347. To attain maximum yield of sorghum per hectare, plant population should be maintained at:

(A) 150000 (B) 160000
(C) 175000 (D) 125000

348. The best time for sowing of chickpea in northern India is:

(A) Second fortnight of October
(B) First fortnight of October
(C) First fortnight of November
(D) Second fortnight of September

349. The starter dose of nitrogen for chickpea is recommended as:

(A) 20 kg/ha
(B) 10 kg/ha
(C) 30 kg/ha
(D) 40 kg/ha

350. The optimum sowing time for lentil in N.W. India is a:
(A) Middle of October
(B) Last week of October
(C) First week of October
(D) Last week of September

351. The recommended optimum seed rate for lentil, in general, varies from
(A) 20-30 kg/ha (B) 40-50 kg/ha
(C) 30-40 kg/ha (D) 10-20 kg/ha

352. Lentil under late sowing situations should be seeded with:
(A) Enhaned seed rate
(B) Reduced seed rate
(C) Widened row space
(D) Widebed plant space

353. A lentil crop grown with improved package of practices may record the yield upto:
(A) 1.0-1.5 t/ha (B) 3.0-3.5 t/ha
(C) 1.5-2.0 t/ha (D) 2.0-2.5 t/ha

354. Nitrogen is applied into transplanted rice in the proportion of:
(A) 50% at basal + nil at tillering + 50% at panicle initiation stage
(B) Nil at basal + 50% at tillering + 50% at panicle initiation stage
(C) 25% at basal + 50% at tillering + 25% at panicle initiation stage
(D) 50% at basal + 25% at tillering + 25% at panicle initiation stage

355. In north Indian conditions fieldpea should be sown in:
(A) First fortnight of November
(B) Second fortnight of September
(C) First fortnight of October
(D) Second fortnight of October

356. In phosphorus & potassium deficient soils, fieldpea crop should be fertilized with:
(A) 30-40 + 20-30 kg P_2O_5 + K_2O/ha
(B) 50-60 + 20-30 kg P_2O_5 + K_2O/ha
(C) 60-70 + 30-40 kg P_2O_5 + K_2O/ha
(D) 50-60 + 30-40 kg P_2O_5 + K_2O/ha

357. The main crop of potato is planted from:
(A) 25th September to 10th October
(B) 15th October to 25th October
(C) 25th September to 10th November
(D) March – April

358. The optimum seed rate for late maturing pigeonpea is:
(A) 12-15 kg/ha (B) 20-25 kg/ha
(C) 8-10 kg/ha (D) 18-20 kg/ha

359. Zinc deficiency in arhar (pigeonpea) can be rectified by spraying of:
(A) 5 kg $ZnSO_4$ + 2.5 kg lime/ha
(B) 20 kg $ZnSO_4$ + 2.5 kg lime/ha
(C) 5 kg $ZnSO_4$ + 5.0 kg lime/ha
(D) 10 kg $ZnSO_4$ + 5.0 kg lime/ha

360. Two hand weeding have been adequate to control weeds in pigeonpea reasonably well:
(A) 25-30 and 45-60 days after sowing
(B) 40-45 and 60-65 days after sowing
(C) 15-20 and 35-40 days after sowing
(D) 30-35 and 60-65 days after sowing

361. Seed rate for transplanting of bajra is:
(A) 1 kg/ha (B) 6 kg/ha
(C) 2 kg/ha (D) 4 kg/ha

362. The ideal stage of harvest of jute for fibre purpose is:
(A) One month old
(B) Flowering
(C) Small pod stage
(D) Mature pod stage

363. Mungbean cultivated with the improved package of improved package of production technology may yield:
(A) 22-25 q/ha
(B) 12-15 q/ha
(C) 18-20 q/ha
(D) 8-10 q/ha

364. Which one of the following sequences is being used in a combination seed treatment:
(A) Rhizobium, fungicide and insecticide
(B) Rhizobium, insecticide and fungicide
(C) Fungicide, insecticide and Rhizobium
(D) Rodenticide, Rhizobium and fungicide

365. Fertilizer application in maize should be completed before:
(A) Tasseling
(B) Silking
(C) Ripening
(D) Knee high

366. The recommended dose of fertilizers for urdbean is:

(A) 15-20+20-30+8-10 kg N+P_2O_5+ K_2O/ha

(B) 15-20 + 50-60+30-40 kg N + P_2O_5 + K_2O/ha, respectively

(C) 40-80+50-60+30-40 kg N+P_2O_5+K_2O/ha

(D) 15-20 + 0+30-40 kg N+P_2O_5+K_2O/ha

367. Trench method of sugarcane planting is used in:

(A) North India (B) Western India

(C) Coastal India (D) All of these

368. Sugarbeen matures in:

(A) April-May

(B) June-July

(C) September-October

(D) January-February

369. Application of nitrogen in cowpea at the time of planting is known as:

(A) Basic dose (B) Synergistic dose

(C) Starter dose (D) Additional dose

370. The bacteria responsible for nitrogen fixation in soybean is:

(A) *Rhizobium japonicum*

(B) *Rhizobium phasioli*

(C) *Rhizobium leguminosarum*

(D) *Rhizobium glycicum*

371. The optimum time or kharif soybean planting is:

(A) First fortnight of June

(B) First fortnight of July

(C) Second fortnight of June

(D) Second fortnight of July

372. The usually recommended plant to plant distance for soybean is:

(A) 2-3 cm (B) 4-5 cm

(C) 5-7 cm (D) 8-10 cm

373. The optimum seed rate of soybean for normal planting is:

(A) 50-60 kg/ha (B) 60-70 kg/ha

(C) 70-80 kg/ha (D) 80-90 kg/ha

374. Normally recommended N + P_2O_5+K_2O dose for soybean is:

(A) 20-30 + 70-80 + 50-60 kg/ha

(B) 20-30 + 100-120 + 50-60 kg/ha

(C) 50-60 + 70-80 + 50-60 kg/ha

(D) 50-60 + 70-80 + 80-100 kg/ha

375. For spring crop of soybean seed rate should be --------------- than that of kharif planting:

(A) More (B) Less

(C) Similar (D) Double

376. The most suitable soil for groundnut is:

(A) Sandy & sandy loam

(B) Clayey loam

(C) Silty loam & clay loam

(D) Loamy clay

377. Deep ploughing of soil in a groundnut field should normally be avoided because it:

(A) Makes harvesting difficult

(B) Makes pegging difficult in upper layers

(C) Delays germination

(D) Delays pegging and pod development

378. The recommended row to row spacing for bunch type of groundnut is:

(A) 20-30 cm (B) 30-40 cm

(C) 40-50 cm (D) 45-60 cm

379. The recommended row spacing for spreading type of groundnut is:

(A) 25-30 cm (B) 35-40

(C) 45-60 cm (D) 55-70 cm

380. Cultivation of duram wheat is primarily confined to:

(A) Central India

(B) Southern India

(C) Both (A) and (B)

(D) Eastern India

381. The seed normally required for planting of sesamum is:

(A) 2-3 kg/ha

(B) 3-4 kg/ha

(C) 4-5 kg/ha

(D) 5-6 kg/ha

382. For nitrogen and phosphorus application in groundnut ---------------- fertilizers are normally preferred:

(A) Urea and single super phosphate

(B) Ammonium sulphate and single super phosphate

(C) Urea and diammonium phosphate

(D) Ammonium sulphate and diammonium phosphate

383. Ammonium sulphate is preferred for nitrogen application in groundnut because it also provides:
(A) Calcium (B) Magnesium
(C) Sulphur (D) Boron

384. Sesamum cake contains :
(A) 6.0-6.2 + 2.0-2.2+1.0-1.2% N+ P_2O_5+K_2O
(B) 2.0-2.2 + 6.0-6.2+1.0-1.2% N+ P_2O_5+K_2O
(C) 6.0-6.2 + 1.0-1.2+2.0-2.2% N + P_2O_5+K_2O
(D) 2.0-2.2 + 1.0-1.2+6.0-6.2% N + P_2O_5+K_2O

385. The recommended planting geometry for sesame is:
(A) 25 x 15 cm (B) 30 x 15 cm
(C) 30 x 10 cm (D) 45 x 15 cm

386. The recommended optimum depth of sowing of castor is:
(A) 4 cm (B) 6 cm
(C) 8 cm (D) 10 cm

387. The optimum time for planting of castor varies between:
(A) 20th May to 5th June
(B) 5th June to 20th June
(C) 20th June to 5th July
(D) 5th July to 20th July

388. The first GM potato developed at CPRI for increasing protein content in tubers consists of genes from:
(A) Chickpea (B) Pigeonpea
(C) Field pea (D) Grain amaranthus

389. Irrigated crop of castor should normally be planted at a row to row and plant to plant distance is:
(A) 45 x 45 cm (B) 60 x 45 cm
(C) 60 x 60 cm (D) 90 x 45cm

390. The panicle initiation stage in rice plants comes:
(A) Immediate after transplanting
(B) After maximum tillering stage
(C) After boot leaf stage
(D) At any time after transplanting

391. Under irrigated conditions, nitrogen should be splitted in castor as:
(A) Half basal plus half at 60 DAS
(B) Half plus half at 30 DAS
(C) Three-fourth as basal and one-fourth at 60 DAS
(D) Three-fourth as basal and one-fourth at 30 DAS

392. In maize plants:
(A) Silk appears first
(B) Tassels appears first
(C) Both the appears at the same time
(D) None of these

393. Thining of rapeseed and mustard is done to maintain a plant to plant distance of:
(A) 5-10 cm (B) 10-15 cm
(C) 15-20 cm (D) 20-25 cm

394. Irrigated rapeseed and mustard is recommended to be fertilized with
(A) 60-90 + 60 + 40 kg N+P_2O_5 + K_2O/ha
(B) 90-120 + 60 + 40 kg N + P_2O_5 + K_2O/ha
(C) 60-90+ 90 + 40 kg N + P_2O_5 + K_2O/ha
(D) 90-120 + 90 + 40 kg N + P_2O_5 + K_2O/ha

395. The mustard crop is regarded ready for harvest when its pods turn in colour:
(A) Greenish-yellow
(B) Greenish-black
(C) Yellowish-green
(D) Yellowish-brown

396. The recommended row to row spacing for linseed planting is:
(A) 20 cm (B) 30 cm
(C) 40 cm (D) 50 cm

397. Linseed has been recommended to be fertilizes with ----- kg of P_2O_5/ha to raise a good crop:
(A) 10 (B) 20
(C) 30 (D) 40

398. Seedrate of hybrid maize is:
(A) 18-20 kg/ha (B) 20-25 kg/ha
(C) 25-30 kg/ha (D) 10-15 kg/ha

399. Mat type nursery is related to:
(A) Tobacco crop
(B) Paddy crop
(C) Onion crop
(D) Brinjal crop

400. The recommended row spacing normally maintained for safflower planting is:
(A) 25 cm (B) 35 cm
(C) 45 cm (D) 55 cm

401. Nitrogen application of safflower influences:
(A) Seed yield
(B) Seed composition
(C) Both of these
(D) None of these

402. Irrigated safflower is recommended to be fertilized with:
(A) 80 + 60 + 20 kg N + P_2O_5 + K_2O/ha
(B) 80 + 40 + 20 kg N + P_2O_5 + K_2O/ha
(C) 40 + 20 + 20 kg N + P_2O_5 + K_2O/ha
(D) 40 + 40 + 20 kg N + P_2O_5 + K_2O/ha

403. Topping of safflower plants is beneficial:
(A) To reduce the lodging
(B) To reduce the water losses
(C) To promote branching and flowering
(D) To reduce losses of seed and fruit

404. The top portion of sugarcane should be selected for seed purposes because bud tissues are rich in:
(A) Sucrose (B) Glucose
(C) Galactose (D) Lactose

405. The recommended seed rate of sunflower varies from:
(A) 6-8 kg/ha (B) 8-10 kg/ha
(C) 10-12 kg/ha (D) 12-14 kg/ha

406. The sunflower crop is recommended to be planted at:
(A) 45 x 20 cm (B) 60 x 20 cm
(C) 45 x 30 cm (D) 60 x 30 cm

407. Urea is a:
(A) Inorganic fertilizer
(B) Organic fertilizer
(C) Ammonical fertilizer
(D) Nitrogen fertilizer

408. What is the effect of potassium fertilizer?
(A) It improves grain quality
(B) It makes the plant more resistant to diseases
(C) It makes plants susceptible to disease
(D) It causes lodging

409. Which of the following roots are of permanent type on sugarcane cane setts:
(A) Sett roots (B) Shoot roots
(C) Prop roots (D) All of these

410. Nitrogen is taken in urea as a form of:
(A) Chloride (B) Nitrate
(C) Oxide (D) Amide

411. Nitrogen is taken in ammonium fertilizers in the form of:
(A) Ammonium (B) Nitrate
(C) Oxide (D) All of these

412. Which of the following can be used as foliar spray?
(A) Single Super Phosphate
(B) Calcium Ammonium Nitrate
(C) Urea
(D) Ammonium sulphate

413. Which of the following is not compound fertilizer?
(A) DAP (B) SSP
(C) Urea (D) CAN

414. Maximum amount of P_2O_5 is found in:
(A) Triple Super Phosphate
(B) DAP
(C) Single Super Phosphate
(D) DCP

415. Which of the following pair is matched correctly?

Fertilizer	Nutrient content
(A) Urea	36.0% N
(B) Diammonium phosphate	20.0% P_2O_5
(C) Ammonium sulphate	20.0% N
(D) Muriate of potash	45.0% K_2O

416. In dwarf wheat, sowing depth is related to:
(A) Humidity
(B) Length of coleoptiles
(C) Temperature
(D) Irrigation

417. Which of the following is water soluble ?
(A) Di ammonium phosphate
(B) Di calcium phosphate
(C) Rock phosphate
(D) Rhemania phosphate

418. Acid equivalent of ammonium sulphate is:
(A) 100 kg
(B) 110 kg
(C) 120 kg
(D) 130 kg

419. Which of the following is citrate soluble ?
(A) Di ammonium phosphate
(B) Single Super phosphate
(C) Rock phosphate
(D) Rhenania phosphate

420. Which of the following is water and citrate soluble?
(A) Di ammonium phosphate
(B) Single super phosphate
(C) Rock phosphate
(D) Rhemania phosphate

421. In tobacco, the basic aim of topping and desuckering is to:
(A) Reduce the plant height
(B) Encourage branching
(C) Divert energy and nutrients from flower heads to leaves
(D) Protect the plants against lodging

422. Delinting of cotton seed may be done with:
(A) Sulphuric acid
(B) Citric acid
(C) Sodium hydroxide
(D) Ammonium chloride

423. S.S.P. is the source of mainly:
(A) S (B) Ca
(C) P_2O_5 (D) All of these

424. Deficiency symptoms of calcium on plants are first appear at:
(A) Lower leaves
(B) Middle leaves
(C) Terminal leaves
(D) All leaves

425. Azotobactor fixes atmospheric nitrogen by the process of:
(A) Symbiosis (B) Non-symbiosis
(C) Both of these (D) None of these

426. The fertilizer containing nitrogen in amide form is:
(A) Urea
(B) CAN
(C) Ammonium sulphate
(D) Ammonium phosphate

427. Which of the following element causes general yellowing and poor growth of plant?
(A) Calcium (B) Boron
(C) Nitrogen (D) Phosphorus

428. Which group of microbes increases phosphate solubility?
(A) Rhizobium (B) Clostridium
(C) Azotobactor (D) Pseudomonas

429. Which one of the following plant nutrient is useful in increasing resistance to disease and insect-pest?
(A) Calcium (B) Phosphorus
(C) Nitrogen (D) Potassium

430. Whip tail symptom in cauliflower is due to the deficiency of:
(A) Zinc (B) Iron
(C) Molybdenum (D) Copper

431. Hard fruit of citrus is due to the deficiency of:
(A) B (B) Ca
(C) Fe (D) Mn

432. Crop logging is a method of:
(A) Soil fertility evaluation
(B) Plant analysis for assessing requirement of nutrients for crop production
(C) Assessing crop damage
(D) Testing suitability of fertilizers

433. Delinting of cotton helps in:
(A) Grading the seed
(B) Killing of hibernating boll worm larvae and disease pathogens
(C) Quicker germination
(D) All of these

434. Interculture in ground nut is avoided at:
(A) Flowering stage (B) Seedling stage
(C) Pegging stage (D) Maturity stage

435. The optimum plant population for cotton has been estimated at:
(A) 25000-50000 plants/ha
(B) 30000-60000 plants/ha
(C) 50000-80000 plants/ha
(D) 80000-100000 plants/ha

436. The optimum sowing time of cotton in northern region of India is:
(A) First fortnight of May
(B) Second fortnight of May
(C) First fortnight of June
(D) Second fortnight of June

437. How many root slips or stem cuttings are required for planting of one hectare of land in Napier grass:
(A) 18800 (B) 27800
(C) 33800 (D) 40800

438. The crown root in wheat appear:
(A) Above soil surface
(B) Below soil surface but above seed
(C) Below soil surface and below seed
(D) Below seed

439. Tossa jute is to be sown at a row spacing of:
(A) 15 (B) 20
(C) 25 (D) 30

440. For fibre purposes, jute crop may be harvested at:
(A) After pod formation
(B) After flowering
(C) Before flowering
(D) Before pod formation

441. Retting is a process by which:
(A) The poor fibre stalk is sorted-out
(B) The fibres in the bark get loosened and separated from woolly stalk
(C) The good quality fibres are sorted out
(D) Fibres are differentiated into various retted groups

442. Green manuring of sunhamp adds approximately -------- kg N/ha to the soil after decomposition:
(A) 20-40 (B) 40-60
(C) 60-80 (D) 80-100

443. For a fibre crop of sunhemp, recommended seed rate for one hectare is:
(A) 5-10 kg/ha (B) 10-15 kg/ha
(C) 15-20 kg/ha (D) 20-25 kg/ha

444. Sunhemp should be planted at a distance to raise a fibre crop:
(A) 20 x 7-10 cm (B) 25 x 5-7
(C) 30 x 7-10 cm (D) 30 x 5-7

445. Seed rate for sunflower crop is:
(A) 5-6 kg/ha (B) 8-10 kg/ha
(C) 15-18 kg/ha (D) 20-22 kg/ha

446. Sugarcane cut for seed-setts should essentially have:
(A) One bud (B) Two buds
(C) Three buds (D) Four buds

447. The average yield of sugarcane in north India is low as compared to that in the southern parts mainly due to:
(A) Climatic conditions
(B) Varietal status
(C) Level of management
(D) Soil fertility status

448. Which portion of sugarcane –stalk should normally be used for seed purpose:
(A) Top one-fourth portion
(B) Top one-third to half portion
(C) Bottom one-fourth portion
(D) Bottom one-third to half portion

449. The number of setts needed for planting one hectare of sugarcane is:
(A) 25000-30000
(B) 35000-40000
(C) 45000-50000
(D) 55000-60000

450. Late planting of autumn sugarcane reduces the germination due to:
(A) Short day
(B) Frost
(C) Low temperature
(D) High humidity

451. The planting of sugarcane by trench method:
(A) Improves germination
(B) Reduces lodging
(C) Improves tillering
(D) Saves seed setts

452. The seed rate is required for sugar beet planting is:
(A) 6-8 kg/ha (B) 8-10 kg/ha
(C) 10-12 kg/ha (D) 12-14 kg/ha

453. The recommended spacings for sugarbeet planting are:
(A) 30 x 15 cm (B) 50 x 15 cm
(C) 30 x 25 cm (D) 50 x 25 cm

454. Seed rate of sugarcane is:
(A) 50 q/ha (B) 75 q/ha
(C) 100 q/ha (D) 125 q/ha

455. Set treatment of sugarcane is done by Agallol at the rate of:
(A) 0.5% (B) 0.25%
(C) 0.75% (D) 1.0 %

456. In hybrid rice, seedlings would be ready for transplanting:
(A) 11-14 days
(B) 15-18 days
(C) 17-20 days
(D) 25- 30 days

457. Topping consists in removal of the terminal bud with or without some of the small leaves practiced in:
(A) Chickpea (B) Pearlmillet
(C) Cotton (D) Tobacco

458. In tobacco, the correct chronological order of practices is:
(A) Topping – desuckering – priming
(B) Desuckering – topping – priming
(C) Priming – topping – desuckering
(D) Desuckering – priming – topping

459. Seedrate for timely sown wheat is:
(A) 75 kg/ha (B) 100 kg/ha
(C) 125 kg/ha (D) 150 kg/ha

460. Seedrate for late sown wheat is:
(A) 75 kg/ha (B) 100 kg/ha
(C) 125 kg/ha (D) 150 kg/ha

461. Recommended dose of N + P_2O_5 +K_2O kg/ha for sugarbeet is:
(A) 120 + 80 + 80 (B) 90 + 60 + 40
(C) 120 + 60 + 40 (D) 90 + 80 + 60

462. The time of planting of main crop of potato in northern plains is:
(A) 10-15 October
(B) 15-25 October
(C) 10-15 November
(D) 15-25 November

463. The recommended seed rate of main crop of potato in northern plains is:
(A) 15-20 q/ha
(B) 20-25 q/ha
(C) 25-30 q/ha
(D) 30-35 q/ha

464. A good crop of potato yielding about 400 q/ha on an average removes N + P_2O_5 +K_2O kg/ha:
(A) 170-180 + 100 + 250
(B) 170-180+125 + 250
(C) 170-180 + 50 + 125
(D) 170-180 + 25 + 100

465. The rainfall at the time of maturity of tobacco is undesirable as it causes:
(A) Lodging
(B) Leaves to get rotten
(C) Leaves to get shattered
(D) Washing down of gums & resins from leaf

466. The optimum time for sowing nursery of tobacco (*Nicotiana rustica*) :
(A) First fortnight of August
(B) Second fortnight of August
(C) First fortnight of September
(D) Second fortnight of September

467. The optimum time for sowing nursery of tobacco (*Nicotiana tabacum*) :
(A) First fortnight of August
(B) Second fortnight of August
(C) First fortnight of September
(D) Second fortnight of September

468. The seed rate of tobacco (*Nicotiana tabacum*) to raise seedling for one hectare varies from:
(A) 5.0 kg/ha (B) 6.0 kg/ha
(C) 7.0 kg/ha (D) 8.0 kg/ha

469. The seed rate of tobacco (*Nicotiana rustica*) to raise seedling for one hectare varies from:
(A) 5.0 kg/ha (B) 6.0 kg/ha
(C) 7.0 kg/ha (D) 8.0 kg/ha

470. Excess chloride in tobacco:
(A) Inhibits the leaf burn and reduces leaf storage capacity
(B) Reduces the leaf size and aroma
(C) Increases the leaf storage capacity & leaf burn
(D) Increases the leaf size and aroma

471. Topping in tobacco means a process of:
(A) Removal of the buds in the axil of leaves
(B) Removal of leaves
(C) Removal of terminal buds
(D) Burning of leaves

472. Seedrate of sorghum crop is:
(A) 8-10 kg/ha (B) 12-15 kg/ha
(C) 18-20 kg/ha (D) 20-22 kg/ha

473. Optimum row to row spacing for sorghum crop is:
(A) 30 cm (B) 40 cm
(C) 45 cm (D) 50 cm

474. Sorghum plants show stunted growth, leaf top and leaf margins turn dark to light purple in colour due to deficiency of:
(A) Nitrogen (B) Phosphorus
(C) Iron (D) Zinc

475. Curing process is essential in which crop:
(A) Chickpea (B) Tobacco
(C) Jute (D) Cotton

476. Optimum time for sowing of gram is:
(A) First fortnight of October
(B) Second fortnight of October
(C) First fortnight of November
(D) Second fortnight of November

477. Best sowing time of gram in Tarai region is:
(A) First fortnight of October
(B) Second fortnight of October
(C) First fortnight of November
(D) Second fortnight of November

478. Best sowing time of gram in Bundelkhand region is:
(A) First fortnight of October
(B) Second fortnight of October
(C) First fortnight of November
(D) Second fortnight of November

479. Seed rate for lentil crop is:
(A) 20-25 kg/ha (B) 40-50 kg/ha
(C) 50-55 kg/ha (D) 60-65 kg/ha

480. Retting process is essential in which crop:
(A) Chickpea (B) Tobacco
(C) Jute (D) Cotton

481. The main aim of topping and desuckering in tobacco is:
(A) To reduce the plant height
(B) To divert energy and nutrients from flower head to leaves
(C) To increase the horizontal branching of plant
(D) To protect the plants against lodging

482. Maize crop is grown successfully in the soil pH at:
(A) 4-5.5 (B) 5.5-7.5
(C) 7.5-8.5 (D) 8.5-9.5

483. Which suckericide is mostly used in tobacco:
(A) Suckerout (B) Auxin
(C) Isoproturon (D) Endosulphan

484. Which of the following is not a tolerant crop for boron concentration?
(A) Palm (B) Sugarbeet
(C) Onion (D) Lemon

485. Exchangeable cation percentage (ECP) is equal to:
(A) $\frac{\text{Exchangeable cation}}{\text{CEC}} \times 100$
(B) $\frac{\text{Exchangeable sodium}}{\text{CEC}} \times 100$
(C) $\frac{\text{Exchangeable cation}}{\text{CEC}}$
(D) None of these

486. In wet bed method of nursery of rice, seeding would be ready for transplanting at an age of:
(A) 11-14 days (B) 15-20 days
(C) 20-25 days (D) 25-30 days

487. In rice, Dapog seedlings would be ready for transplanting:
(A) 11-14 days (B) 15-18 days
(C) 17-20 days (D) 22-25 days

488. Seed rate for line sowing of white jute is:
(A) 4-5 kg/ha (B) 6-8 kg/ha
(C) 9-10 kg/ha (D) 11-12 kg/ha

489. The ideal stage of harvesting of jute crop is:
(A) 100-110 days (B) 135-140 days
(C) 150-160 days (D) 165-170 days

490. Exchangeable sodium percentage (ESP) is equal to:
(A) $\frac{\text{Exchangeable cation Na}}{\text{CEC}} \times 100$
(B) $\frac{\text{Exchangeable sodium}}{\text{CEC}} \times 100$
(C) $\frac{Na^{+}}{\sqrt{Ca^{+2} + Mg^{+2}}}$
(D) None of these

491. Compost is:
(A) Bulky manure
(B) Concentrate manure
(C) Green manure (D) Biofertiliers

492. The inherent capacity of soil to produce crop is called:
(A) Soil productivity
(B) Soil fertility
(C) Productivity and fertility both
(D) None of these

493. Kjeldahl method is used to determine:
(A) Total nitrogen in the soil
(B) Total phosphorus in the soil
(C) Total potash in the soil
(D) Total organic matter in the soil

494. To determine organic matter of the soil, the content of organic carbon is multiplied by the factor:
(A) 1.824 (B) 1.800
(C) 1.724 (D) 1.834

495. Seed rate for American cotton varieties is:
(A) 10-18 kg/ha (B) 15-25 kg/ha
(C) 30-35 kg/ha (D) 35-40 kg/ha

496. Optimum time for sowing of cotton in North India is:
(A) First fortnight of May
(B) Second fortnight of May
(C) First fortnight of June
(D) Second fortnight of une

497. Seed treatment by conc. H_2SO_4 is done in cotton for:
(A) To protect seed from soil borne fungus
(B) To protect seed from insect
(C) To delint the seeds
(D) None of these

498. Requirement of nitrogen for hybrid cotton is:
(A) 60-75 kg/ha (B) 80-100 kg/ha
(C) 100-120 kg/ha (D) 120-150 kg/ha

499. Seed rate for rapeseed and mustard crop is:
(A) 2-3 kg/ha (B) 5-6 kg/ha
(C) 8-10 kg/ha (D) 12-14 kg/ha

500. Which crop required maximum amount of nitrogen?
(A) Wheat (B) Sugarcane
(C) Potato (D) Banana

501. Potassium uptake is highest in:
(A) Cereals (B) Oilseeds
(C) Pulses (D) Grasses

502. Long term effect of urea application is:
(A) Soil becomes neutral
(B) Soil becomes acidic
(C) Soil becomes saline
(D) None of these

503. Split dose of nitrogen application is recommended in:
(A) Clay soil (B) Sandy soil
(C) Saline soil (D) Alkaline soil

504. Diara lands are:
(A) Areas flooded by sea
(B) Areas created by sand storms
(C) Areas on both sides of canal
(D) Areas located on either side of river which get flooded every year

505. Soil fertility is reduced mainly due to:
(A) Poor drainage
(B) Over irrigation
(C) Continuous cropping
(D) Imbalanced use of fertilizers

506. Which one contains highest N?
(A) Dried blood
(B) Raw bone meal
(C) Fish manure
(D) Dry activatedsludge

507. Which of the following elements is not essential for plants but proves to be beneficial for some plants?
(A) Copper (B) Boron
(C) Sodium (D) Iron

508. Dapog method of raising seedlings is related to:
(A) Jute (B) Sugarcane
(C) Paddy (D) Tobacco

509. Seed rate required for palma rosa crop is:
(A) 4 kg/ha (B) 8 kg/ha
(C) 10 kg/ha (D) 14 kg/ha

510. Seed rate of lemon grass is
(A) 5-10 kg/ha
(B) 10-15 kg/ha
(C) 15-20 kg/ha
(D) 20-25 kg/ha

511. The best sowing time of soybean is:
(A) Ist week of June
(B) Ist week of July
(C) IIrd week of July
(D) Ist week of August

512. Bacteria responsible for oxidation of sulphur is:
(A) Rhizobium (B) Pseudomonas
(C) Bacillus (D) Thiobacillus

513. Seed treatment is done to control:
(A) Soil borne disease
(B) Air borne disease
(C) Seed borne disease
(D) None of these

514. Rock phosphate is used in:
(A) Neutral soil (B) Acidic soil
(C) Alkaline soil (D) Saline soil

515. Salt tolerant crop is:
(A) Cowpea (B) Pea
(C) Sugarbeet (B) Muskmelon

516. Temporary roots found in maize is known as:
(A) Tap root (B) Crown root
(C) Seminal root (D) None of these

517. Ratio of volume of air and solids of soil is known as:
(A) Porosity (B) Void ratio
(C) Bulk density (D) Particle density

518. What will be the seed rate of rape seed when spacing is 30 x 10 cm?
(A) 5 kg/ha (B) 10 kg/ha
(C) 15 kg/ha (D) 20 kg/ha

519. Organic matter can be calculated by multiplying organic carbon with:
(A) 1.72 (B) 1.82
(C) 1.92 (D) 2.00

520. Sugarcane is planted in furrows because of:
(A) To avoid water losses
(B) To reduce soil borne diseases
(C) It reduces lodging
(D) All of these

521. Average yield of sugarcane adhsali crop is:
(A) 1000-1200 q/ha
(B) 400-500 q/ha
(C) 800-900 q/ha
(D) 100-200 q/ha

522. White bud in maize is caused by deficiency of:
(A) N (B) P
(C) Zn (D) Cl

523. Deficiency symptoms of phosphorus is first appears on:
(A) Stem (B) Roots
(C) Upper leaves (D) Lower leaves

524. For transplanting of one hectare area of rice how much area is sufficient for nursery raising:
(A) 100m^2 (B) 500 m^2
(C) 1000 m^2 (D) 1500 m^2

525. Which one is a parameter used to determine cotton quality?
(A) Staple length (B) Fibre fineness
(C) Fibre strength (D) All of these

526. Source to sink ratio in pigeonpea crop is:
(A) 15-20 % (B) 20-30 %
(C) 35-50 % (D) >50 %

527. Inverse yield nitrogen law is given by:
(A) Leibig (B) Wilcox
(C) Mitscherlich (D) None of these

528. Recovery of sugar in sugarcane crop in India is:
(A) 6-8 % (B) 10 %
(C) 10-15 % (D) 4-6 %

529. Example of stem nodulating legume is:
(A) *Sesbania aculeate*
(B) *Crotalaria juncea*
(C) *Aeschynomene ajraspora*
(D) *Phaseolus trilobus*

530. Apatite contains the plant nutrient:
(A) Phosphorus (B) Nitrogen
(C) Potassium (D) Boron

531. Which one of the following operations is associated with the process of nipping in gram:
(A) Treating the seed with Rhizobium culture
(B) Tying the branched to avoid lodging
(C) Picking green leaves for vegeatable purposes
(D) Plucking the apical buds to promote branches

532. What is the seed rate of pearlmillet ?
(A) 8-10 kg/ha (B) 10-12 kg/ha
(C) 10-20 kg/ha (D) 4-5 kg/ha

533. Which of the plant growth regulator is responsible for breaking dormancy?
(A) Cytokinin (B) Auxin
(C) Gibberellic acid (D) 2, 4-D

534. Azotobacter is useful in fixing nitrogen from air in:
(A) Non-leguminous crops

(B) Leguminous crops
(C) Soybean
(D) None of these

535. Seedrate of American cotton is:
(A) 20 kg/ha (B) 30 kg/ha
(C) 35 kg/ha (D) 40 kg/ha

536. How many days old paddy seedling is used for SRI technology?
(A) 8-12 days
(B) 12-15 days
(C) 25-30 days
(D) 30-35 days

537. How much nursery area is required for transplanting one hectare of paddy crops?
(A) 1000 m^2 (B) 750 m^2
(C) 500 m^2 (D) 250-400 m^2

538. Single cut fodder sorghum crop is harvested at:
(A) 25 % Flowering (B) 50 % Flowering
(C) 75 % Flowering (D) Maturity

539. Napier crop is harvested at:
(A) Ground level
(B) 5-10 cm above the ground level
(C) 10-15 cm above the ground level
(D) None of these

540. Digestibility of dinanath grass is:
(A) 25 % (B) 50 %
(C) 75 % (D) 100 %

541. Seed rate of forage cowpea is:
(A) 20-25 kg/ha (B) 30-35 kg/ha
(C) 40-45 kg/ha (D) 50-55 kg/ha

542. Digestibility of Lucerne crops is:
(A) 62 % (B) 72 %
(C) 82 % (D) 92 %

543. Seed rate for oat crop is:
(A) 50 kg/ha (B) 100 kg/ha
(C) 150 kg/ha (D) 200 kg/ha

544. Protein content in oat crop is maximum at:
(A) Flowering stage
(B) Milk stage
(C) Maturing stage
(D) Equal at all stages

545. Which one is not a part of compost?
(A) Dung (B) Urea
(C) Urine (D) Crop residue

546. Now-a-day, NPK ratio for wheat production is:
(A) 2 : 1 : 1 (B) 4 : 2 : 1.
(C) 8 : 2.5 : 1 (D) 4 : 1 : 1

547. Papery white in maize is the deficiency symptom of:
(A) Sulphur (B) Sodium
(C) Iron (D) Potash

548. Harvesting of wheat crop is suggested at seed moisture content:
(A) 10-12 % (B) 15 %
(C) 20 % (D) 22 %

549. Which is the highest digestible protein fodder in the following?
(A) Lucerne (B) Cowpea
(C) Clusterbean (D) Berseem

550. Optimum seed rate of pea var. Arkil is:
(A) 50-60 kg/ha (B) 75-100 kg/ha
(C) 100-125 kg/ha (D) 140-150 kg/ha

551. Optimum sowing time of French bean in plains is:
(A) June (B) July
(C) October (D) December

552. In which crop earthing is not required?
(A) Sugarcane (B) Sunflower
(C) Potato (D) Rice

553. The recommended optimum seed rate of summer season mungbean is:
(A) 10-15 kg/ha (B) 15-20 kg/ha
(C) 20-25 kg/ha (D) 25-30 kg/ha

554. The recommended optimum seed rate of rabi season urdbean is:
(A) 10-15 kg/ha (B) 15-20 kg/ha
(C) 20-25 kg/ha (D) 25-30 kg/ha

555. Boro (summer season) rice is sown in:
(A) March-April (B) June – July
(C) Nov-Dec (D) May-June

556. Ashu (winter season) rice is sown in:
(A) March-April (B) June – July
(C) Nov-Dec (D) May-June

557. What is the seed rate of lentil?
(A) 20-25 kg/ha (B) 30-35 kg/ha
(C) 40-50 kg/ha (D) None of these

558. What is the optimum seed rate for lentil under utera cultivation?
(A) 30-35 kg/ha (B) 40-45 kg/ha
(C) 55-75 kg/ha (D) 75-100 kg/ha

559. Specific heat of dry mineral soil is:
(A) 1.0 calories/g (B) 0.40 calories/g
(C) 0.5 calories/g (D) 0.22 calories/g

560. Specific heat of clay soil is:
(A) 1.0 calories/g (B) 0.40 calories/g
(C) 0.5 calories/g (D) 0.22 calories/g

561. Specific heat of ice is:
(A) 1.0 calories/g (B) 0.40 calories/g
(C) 0.5 calories/g (D) 0.22 calories/g

562. Specific heat of Humus is:
(A) 1.0 calories/g (B) 0.40 calories/g
(C) 0.5 calories/g (D) 0.22 calories/g

563. How much calories is received by earth daily in winter season?
(A) 50 calories/g (B) 100 calories/g
(C) 500 calories/g (D) All of these

564. Optimum temperature range for most of the plants and animals is:
(A) 80 – 100 ^{0}F (B) 10 – 30 ^{0} F
(C) 120 – 140 ^{0} F (D) None of these

565. Which of the following cation is found under reduced conditions?
(A) Fe^{+++} (B) Mn^{+++}
(C) Fe^{++} (D) All of these

566. Which of the following gas has the highest concentration in soil?
(A) CO_2 (B) O_2
(C) S_2 (D) N_2

567. What is the CO_2 concentration in atmosphere?
(A) 21 % (B) 78.5 %
(C) 0.03 % (D) 0.3 %

568. Which of the following sequence of heat conductivity of different soil is correct?
(A) Sand > loam > clay > peat
(B) Peat > clay > loam > sand
(C) Clay > peat > loam > sand
(D) None of these

569. Which gas is released from paddy field?
(A) H_2S (B) CO_2
(C) NH_3 (D) CH_4

570. How much biomass in quintal per hectare is obtained from green manure of the sunnhemp and dhaincha crops?
(A) 300-325 (B) 225-275
(C) 200-225 (D) 325-350

571. Rouging is done for:
(A) Weed control
(B) Disease control
(C) Removing of off type plants
(D) None of these

572. Double Zero rapeseed & mustard varieties refer to:
(A) Low content of erusic acid
(B) Low content of glycosinolate acid
(C) Low content of both erusic and glycosinolates
(D) None of these

573. How much Kg of sugarcane seed is required for sowing one hectare?
(A) 70000-75000 Kg (B) 55000-60000 Kg
(C) 80000-85000 Kg (D) 50000-55000 Kg

574. Which one is the frost sensitive crop?
(A) Rice (B) Potato
(C) Wheat (D) Oat

575. Proper seed storage of wheat requires optimum seed moisture:
(A) 0-5 % (B) 5 – 10 %
(C) 10-14 % (D) 14-18%

576. Topping of tobacco is to be done at:
(A) At vegetative stage
(B) At before flowering stage
(C) After flowering stage
(D) At priming stage

577. How many ratoon of sugarcane are taken economically?
(A) One (B) Two
(C) Three (D) Four

578. N requirement for cotton crop is:
(A) 30-40 kg/ha (B) 40-60 kg/ha
(C) 80-100 kg/ha (D) 120-130 kg/ha

579. Crude protein content in forage maize crop is:
(A) 6.5 % (B) 10 %
(C) 12 % (D) None of these

580. Optimum sowing time for Napier grass is:
(A) January (B) February
(C) March (D) April

581. Optimum sowing time for forage maize is:
(A) January (B) May
(C) March (D) April

582. Digestibility of maize silage:
(A) 25% (B) 50 %
(C) 60 % (D) 75 %

583. Seed rate of poppy is:
(A) 2-3 kg/ha (B) 5-6 kg/ha
(C) 10-12 kg/ha (D) 15-16 kg/ha

584. Optimum harvesting time of poppy is:
(A) Jan-Feb (B) March-April
(C) May-June (D) Sep-Oct

585. If one plant species has a negative effect on other species the effect is known as:
(A) Allelopathic (B) Antagonistic
(C) Symbiotic (D) Mutalistic

586. FIRBS refers to:
(A) Sowing of crops without tillage
(B) Sowing of crops at raised beds
(C) A technique of weed control
(D) Allelopathic study

587. Optimum sowing time of Palma rosa grass is:
(A) Jan-Feb (B) March-April
(C) July-Aug (D) Sept-Oct

588. Optimum spacing for palma rosa grass is:
(A) 60 x 40 cm (B) 40 x 40 cm
(C) 80 x 50 cm (D) 100 x 100 cm

589. Nitrogen requirement for Palma rosa grass is:
(A) 40 kg/ha (B) 60 kg/ha
(C) 100 kg/ha (D) 150 kg/ha

590. Nitrogen requirement for Citrunella grass is:
(A) 40 kg/ha (B) 80 kg/ha
(C) 100 kg/ha (D) 150 kg/ha

591. Geranium oil smells as:
(A) Rose (B) Lemon
(C) Lavender (D) None of these

592. Geranium is propagated through:
(A) Seed (B) Cutting
(C) Pudding (D) All of these

593. What is the optimum seed rate of gram:
(A) 50-60 kg/ha (B) 60-70 kg/ha
(C) 70-80 kg/ha (D) 80-90 kg/ha

594. What is the optimum seed rate for dwarf pea variety:
(A) 80 kg/ha (B) 125 kg/ha
(C) 40 kg/ha (D) 100 kg/ha

595. What is the N:P:K requirement for lentil?
(A) 20 : 40 : 30 (B) 40 : 80 : 30
(C) 60 : 40 : 30 (D) All of these

596. What is the N:P:K requirement for gram?
(A) 20 : 60 : 30 (B) 40 : 80 : 30
(C) 60 : 40 : 30 (D) All of these

597. What is the optimum sowing time of pea at high hills:
(A) Sept-Oct (B) Oct-Nov
(C) March-April (D) Nov-Dec

598. Khaira disease in rice is due to:
(A) Seed borne fungus
(B) Nutrient deficiency/disorder
(C) Soil borne disease
(D) None of these

599. The series containing micronutrients is:
(A) Fe, Mn, Zn, Cu, Mo, B
(B) Mg, N, P, K, S, Fe
(C) Ca, Mg, P, K, B, Fe
(D) S, N, P, K, B, Al

600. The nitrogen fertilizer containing amino-nitrogen is:
(A) Ammonium sulphate
(B) Urea
(C) Ammonium sulphate nitrate
(D) Calcium ammonium nitrate

601. The phosphate fertilizer containing gypsum is:
(A) Diammonium phosphate
(B) Urea phosphate
(C) Nitrophosphate
(D) Single super phosphate

602. How much nitrogen could be saved in paddy crop if preceding crop was greengram?
(A) 20-25 kg/ha (B) 30-40 kg/ha
(C) 45-50 kg/ha (D) 50-60 kg/ha

603. The optimum time of harvesting of paddy crop is:
(A) 10-15 days after ear emergence
(B) 15-25 days after ear emergence
(C) 30-35 days after ear emergence
(D) 40-50 days after ear emergence

604. Severe problem in gram cultivation is:
(A) High fertilizer requirement
(B) Weed
(C) Rust
(D) Pod borer

605. In toria-wheat rotation, sowing of toria should be done during:
(A) First fortnight of Septeber
(B) Second fortnight of September
(C) First fortnight of October
(D) Second fortnight of October

606. How much seed per hectare is needed for berseem?
(A) 18-22 Kg (B) 25-30 Kg
(C) 30-35 Kg (D) 15-20 Kg

607. Recommended fertilizer requirement (NPK) for Zaid maize is:
(A) 100 : 50 : 50 kg/ha
(B) 30 : 40 : 40 kg/ha
(C) 120 : 60 : 60 kg/ha
(D) 40 : 20 : 20 kg/ha

608. Tying and wrapping is a essential process of sugarcane for:
(A) To avoid lodging of plant
(B) To decrease incidence of pest & diseases
(C) To improve nutrients uptake
(D) To decrease water requirement

609. Thinning should be done at:
(A) Before germination of seed
(B) After germination of seed
(C) At vegetative growth stages of crop
(D) At flowering stage of crop

610. In trench method of sugarcane planting, the depth of trench is:
(A) 10-15 cm (B) 15-20 cm
(C) 20-25 cm (D) 25-30 cm

611. The rice seedlings raised by depog method become ready for transplanting on:
(A) 21st days after sowing in nursery
(B) 18th days after sowing in nursery
(C) 15th days after sowing in nursery
(D) 12th days after sowing in nursery

612. Blind hoeing is a recommended practice for:
(A) Muskmelon (B) Sugarcane
(C) Sugarbeet (D) Tobacco

613. Recommended age of paddy seedlings for transplanting of medium duration paddy in sodic land on:
(A) 20-25 days after sowing in nursery
(B) 25-30 days after sowing in nursery
(C) 30-35 days after sowing in nursery
(D) 40-45 days after sowing in nursery

614. Urea can be safely used as foliar spray in wheat crop at a concentration (%) of:
(A) 2-3 (B) 4-5
(C) 6-7 (D) 9-10

615. Nipping should be done at:
(A) Before flowering
(B) After flowering
(C) Before harvesting
(D) After harvesting

616. The purpose of dehauling in potato seed crop is:
(A) Better quality of seed tuber
(B) Higher yield
(C) Higher protein content
(D) Higher starch content

617. Fertilizer to be used in acidic soil is:
(A) Urea
(B) Ammonium sulphate
(C) Ammonium chloride
(D) Sodium nitrate

618. Harmful substance in urea is:
(A) Uric acid (B) Biurete
(C) Amide (D) Cyanide

619. Legume crop in which usually no symbiotic nitrogen fixation occurs in the plain region:
(A) Soybean (B) Moong
(C) French bean (D) Field pea

620. The crop in which 'crop logging' is used:
(A) Corn (B) Soyabean
(C) Sugarcane (D) Pearlmillet

621. The cultivation practice is not useful for soybean:
(A) Sowing in the morning
(B) Sowing in the evening
(C) Sowing in the summer weather
(D) Planting just after sowing

622. Recommended crop for grey and black areas of the district is:
(A) Potato (B) Dwarf wheat
(C) Tobacco (D) Vegetable pea

623. The recommended nitrogen dose for hybrid cotton is:
(A) 90-120 kgN/ha (B) 120-150 kgN/ha
(C) 150-180 kgN/ha (D) 180-210 kgN/ha

624. How much seed of legume crop should be treated with one packet of rhizobium culture?

(A) 5 kg (B) 10 kg
(C) 15 kg (D) 20 kg

625. How much seed of legume crop should be treated with one packet of phosphate solubilizing bacteria?

(A) 5 kg (B) 10 kg
(C) 15 kg (D) 20 kg

626. Which one of the following is not correctly matched?

Crop	Seed rate (kg/ha)
(A) Wheat	100
(B) Urd	20
(C) Maize	18
(D) Mustard	20

627. Basic aim of nipping is:

(A) To increase growth of plants
(B) To increase nutrients uptake
(C) To increase number of branchs plant
(D) None of these

628. Basic aim of gap filling is:

(A) To increase flowering
(B) To maintain optimum plant population
(C) To increase growth of plants
(D) To increase nutrient uptake

629. Which one of the following statement is correct?

(A) Blue green algae fix 20-30 kgN/ha in a rice field
(B) Blue green algae fix 40-50 kgN/ha in a rice field
(C) Blue green algae fix 8-10 kgN/ha in a rice field
(D) Blue green algae fix 60-70 kgN/ha in a rice field

630. Which of the following seed rate of TPS is recommended for planting of one hectare area:

(A) 150 g (B) 200 g
(C) 250 g (D) 450 g

631. During early stage of growth, paddy takes nitrogen in the form of:

(A) Ammonia (B) Nitrate
(C) Nitrite (D) None of these

632. Gap filling in short duration crops should be done at:

(A) Upto 10 days after sowing
(B) Upto 20 days after sowing
(C) Upto 30 days after sowing
(D) Upto harvesting

633. Topping and desuckering operations are essential for increasing yield and improving quality of leaf is:

(A) Tobacco (B) Potato
(C) Jute (D) Sugarcane

634. Consider the following statements:

Assertion (A): Ground nut is not taken as a summer crop in North India

Reason (R): At the time of sowing the soil temperature should not remains very low

Select the correct answer using the code given below:

(A) A is true but R is false
(B) A is false but R is true
(C) Both A and R are true and R is the correct explanation of A
(D) Both A and R are true but R is not correct explanation of A

635. In bunch type of groundnut, earthing should be completed by:

(A) 30 days of sowing
(B) 45 days of sowing
(C) 60 days of sowing
(D) 75 days of sowing

636. Greening of potato results in:

(A) Increase in nutritional quality
(B) Decrease in nutritional quality
(C) Increase in disease resistance
(D) Decrease in disease resistance

637. The optimum plant population per hectare of sorghum is:

(A) 50000 plants
(B) 100000 plants
(C) 150000 plants
(D) 200000 plants

638. Strategy for sustainable crop production depends upon:

(A) Integrated plant nutrients management
(B) Use of high doses of chemical fertilizers
(C) Use of high doses of organic manure
(D) Increasing area under irrigation

639. In tobacco cultivation 'Priming' refers to:
(A) Removal of top of the leaves
(B) Removal of suckers
(C) Selective harvesting of leaves
(D) Curing of leaves

640. Which one is not correctly matched:

Crop	Maturity period in UP (days)
(A) Early pigeonpea	140
(B) Toria	120
(C) Safflower	170
(D) Summer moong	65

641. To encourage maximum number of mother shoots in sugarcane, it should be planted by:
(A) Ring method
(B) Flat method
(C) Furrow method
(D) Trench method

642. Nitrogen is applied into transplanted rice in the proportion of:
(A) 50% at basal + nil at tillering + 50% at panicle emergence stage
(B) Nil at basal + 50% at tillering + 50% at panicle emergence stage
(C) 25% at basal + 50% at tillering + 25% at panicle emergence stage
(D) 50% at basal + 25% at tillering + 25% at panicle emergence stage

643. Interculture in groundnut is avoided after:
(A) Flowering stage (B) Seedling stage
(C) Pegging stage (D) None of these

644. Maximum yield of mustard is obtained at a plant geometry of:
(A) 45 x 20 cm (B) 60 x 30 cm
(C) 90 x 30 cm (D) 105 x 30 cm

645. Application of gypsum is required more for:
(A) Paddy (B) Berseem
(C) Lucerne (D) Groundnut

646. Urea super granule is efficient source of nitrogen for waterlogged paddy because:
(A) Its leaching losses are minimum
(B) Its fixation in soil is low
(C) Its ammonia volatilization losses are low
(D) Denitrification losses from it are minimum

647. Gap filling is a essential process to:
(A) Wheat (B) Oilseed crops only
(C) Pulse crops (D) All crops

648. Thinning is a essential to:
(A) Bajra (B) Wheat
(C) Maize (D) All crops

649. Recommended seed rate per hectare of dhaincha for green manuring in sodic soil is:
(A) 20 kg (B) 40 kg
(C) 60 kg (D) 80 kg

650. The most efficient use of potassium is achieved by:
(A) Broadcasting at the sowing time
(B) Top dressing after 21 days of sowing
(C) Basal placement at the time of sowing
(D) Foliar spray

651. Nipping is a process to recommends in:
(A) Chickpea (B) Pigeonpea
(C) Fieldpea (D) Maize

652. Nipping is a process generally done in:
(A) Fieldpea (B) Chickpea
(C) Pigeonpea (D) Cowpea

653. Which group of microbes increases phosphorus solubility:
(A) Azatobactor (B) Pseudomonas
(C) Clastridium (D) Rhizobium

654. Wheat variety suitable for sodic soil is:
(A) K-68 (B) K 8027
(C) HP 1102 (D) Raj 3077

655. The most efficient use of nitrogen is achieved by:
(A) Broadcasting at the sowing time
(B) Top dressing after 21 days of sowing
(C) Basal placement at the time of sowing
(D) Foliar spray

656. Suitable age of paddy seedlings for transplanting in sodic soil:
(A) 20-25 days (B) 25-30 days
(C) 30-35 days (D) 40-45 days

657. Urea nitrogen become completely available to plants:
(A) 24 hours after its application
(B) 48 hours after its application
(C) 72 hours after its application
(D) Immediately after its application

658. Removal of undesirable auxiliary buds at initial stage from flowering plants in cotton is known as:
(A) Clipping (B) Notching
(C) Budding (D) Topping

659. The most efficient use of phosphorus is achieved by:
(A) Broadcasting at the sowing time
(B) Top dressing after 21 days of sowing
(C) Basal placement at the time of sowing
(D) Foliar spray

660. Berseem is a fodder crop and is recommended for sowing in the months of:
(A) March-April
(B) June-July
(C) October-November
(D) January-February

661. Richest source of nitrogen is:
(A) Dried blood
(B) Raw bone meal
(C) Fish manure
(D) Dry activated sludge

662. Beseem seeds are inoculated with:
(A) Azolla culture
(B) Azotobactor culture
(C) Rhizobium culture
(D) None of these

663. Winter maize is recommended to be sown on slopes of ridges:
(A) Southern (B) Northern
(C) Eastern (D) Western

664. Land area having definite boundary and contributing runoff to a single point is known as:
(A) Watershed
(B) Catchment area
(C) Command area
(D) A field unit

665. Rotational intensity of Maize-barley-late potato - mungbean (first year) and rice-potato-wheat-blackgram (Second year) calculated as:
(A) 800% (B) 600%
(C) 400% (D) 200%

666. The unit of crop growth rate :
(A) g/m^2 (leaf area) / day
(B) g/g/day
(C) g/m^2 (land area) / day
(D) g/day

667. Crop intensification in both time and space dimension observed in:
(A) Multiple cropping
(B) Double cropping
(C) Intercropping
(D) Intensive cropping

668. An intercropping can be beneficial if it has LER:
(A) = 1.0 (B) < 1.0
(C) > 1.0 (D) Zero

669. Which one of the following indices is used to assess the intercropping system:
(A) Leaf area index
(B) Land equivalent ratio
(C) Land utilization index
(D) Multiple cropping

670. Study of agronomy dealing with the relationship of yield to the quantity of an added/available fertilizer element is called:
(A) Soil fertility (B) Agrobiology
(C) Agrostology (D) Agrology

671. Agronomic efficiency is expressed as:
(A) Kg/kg (B) Kg/ha
(C) Kg/ha/day (D) Kg/day

672. Growing of annual crops in between the perennial crop is:
(A) Relay cropping (B) Intercropping
(C) Alley cropping (D) Multiple cropping

673. Annidation is:
(A) Legume effect
(B) Allelopathy
(C) Competitive interaction
(D) Complementary interaction

674. Which one of the following planting geometry is recommended to optimizing system productivity in intercropping system in semi-arid and sub-humid ecosystem?
(A) Normal planting of intercrop between two rows of normal planted main crop
(B) Skip row planting
(C) Three rows of main crop and two rows of intercrops
(D) Four rows of main crop and one row of intercrop

675. The formula LER-1/LER X value of combined intercrop represents:
(A) MAI (B) LER
(C) LAI (D) IER

676. If L_1 and L_2 are leaf area and W_1 and W_2 are dry weight of leaves at time t_1 and t_2 respectively, than $[(W_2 - W_1)(\text{Log } L_2 - \log L_1)]/[(t_2 - t_1)(L_2 - L_1)]$ is:
(A) RGR (B) NAR
(C) LAR (D) CGR

677. If on a 20 acres farm, crops are raised on 15 acres in rabi, 15 acres in kharif and 20 acres in zaid in a year, what shall be the cropping intensity:
(A) 150% (B) 250%
(C) 300% (D) 400%

678. Productivity of an intercrop per unit area of ground compared with that expected from sole crop sown in the same proportions in termed as:
(A) Land equivalent ratio
(B) Competitive ratio
(C) Land equivalent coefficient
(D) Crop performance ratio

679. Which one of the following crop rotation is the best for maintaining soil fertility?
(A) Maize-toria-wheat
(B) Paddy-wheat-cowpea
(C) Paddy-potato-greengram
(D) Soybean-wheat-greengram

680. The ratio between gross and net cropped area is called an index of :
(A) Cropping intensity
(B) Multiple cropping
(C) High intensity cropping
(D) None of these

681. Repitive growing of same sole crop on the same field is known as:
(A) Monocropping
(B) Sequential cropping
(C) Relay cropping
(D) Multiple cropping

ANSWER SHEET
8. Agronomy Practices

1. (D)	2. (A)	3. (B)	4. (C)	5. (A)	6. (C)	7. (D)	8. (A)	9. (B)	10. (C)
11. (A)	12. (C)	13. (B)	14. (C)	15. (A)	16. (A)	17. (B)	18. (D)	19. (A)	20. (B)
21. (B)	22. (C)	23. (B)	24. (A)	25. (A)	26. (B)	27. (B)	28. (A)	29. (B)	30. (D)
31. (B)	32. (A)	33. (B)	34. (D)	35. (A)	36. (B)	37. (C)	38. (D)	39. (A)	40. (C)
41. (D)	42. (A)	43. (A)	44. (A)	45. (D)	46. (C)	47. (A)	48. (B)	49. (B)	50. (B)
51. (B)	52. (B)	53. (A)	54. (D)	55. (A)	56. (A)	57. (B)	58. (C)	59. (C)	60. (D)
61. (A)	62. (B)	63. (A)	64. (C)	65. (D)	66. (D)	67. (B)	68. (B)	69. (B)	70. (B)
71. (A)	72. (A)	73. (B)	74. (C)	75. (C)	76. (B)	77. (D)	78. (C)	79. (C)	80. (A)
81. (B)	82. (A)	83. (B)	84. (C)	85. (D)	86. (D)	87. (B)	88. (D)	89. (A)	90. (D)
91. (A)	92. (D)	93. (C)	94. (A)	95. (D)	96. (C)	97. (A)	98. (B)	99. (A)	100. (B)
101. (C)	102. (A)	103. (B)	104. (C)	105. (B)	106. (C)	107. (C)	108. (B)	109. (A)	110. (A)
111. (C)	112. (C)	113. (D)	114. (A)	115. (C)	116. (C)	117. (D)	118. (B)	119. (C)	120. (D)
121. (C)	122. (A)	123. (C)	124. (B)	125. (A)	126. (C)	127. (B)	128. (B)	129. (D)	130. (C)
131. (B)	132. (B)	133. (A)	134. (A)	135. (A)	136. (B)	137. (D)	138. (C)	139. (A)	140. (B)
141. (A)	142. (B)	143. (A)	144. (A)	145. (B)	146. (B)	147. (B)	148. (C)	149. (A)	150. (B)
151. (B)	152. (B)	153. (A)	154. (B)	155. (D)	156. (C)	157. (B)	158. (A)	159. (A)	160. (B)
161. (C)	162. (D)	163. (C)	164. (A)	165. (A)	166. (C)	167. (A)	168. (C)	169. (A)	170. (D)
171. (A)	172. (A)	173. (B)	174. (B)	175. (C)	176. (A)	177. (C)	178. (D)	179. (A)	180. (C)
181. (C)	182. (A)	183. (C)	184. (C)	185. (C)	186. (A)	187. (D)	188. (C)	189. (A)	190. (C)
191. (A)	192. (B)	193. (A)	194. (A)	195. (B)	196. (B)	197. (A)	198. (B)	199. (B)	200. (B)

201. **(B)** 202. **(A)** 203. **(B)** 204. **(B)** 205. **(D)** 206. **(D)** 207. **(C)** 208. **(B)** 209. **(D)** 210. **(C)**

211. **(C)** 212. **(A)** 213. **(C)** 214. **(A)** 215. **(B)** 216. **(C)** 217. **(B)** 218. **(B)** 219. **(B)** 220. **(C)**

221. **(D)** 222. **(D)** 223. **(D)** 224. **(A)** 225. **(B)** 226. **(A)** 227. **(B)** 228. **(C)** 229. **(D)** 230. **(D)**

231. **(A)** 232. **(D)** 233. **(B)** 234. **(C)** 235. **(C)** 236. **(D)** 237. **(C)** 238. **(B)** 239. **(A)** 240. **(A)**

241. **(B)** 242. **(D)** 243. **(C)** 244. **(A)** 245. **(A)** 246. **(B)** 247. **(D)** 248. **(C)** 249. **(B)** 250. **(C)**

251. **(A)** 252. **(B)** 253. **(A)** 254. **(A)** 255. **(B)** 256. **(C)** 257. **(D)** 258. **(B)** 259. **(D)** 260. **(C)**

261. **(D)** 262. **(A)** 263. **(B)** 264. **(C)** 265. **(C)** 266. **(D)** 267. **(A)** 268. **(B)** 269. **(C)** 270. **(D)**

271. **(A)** 272. **(B)** 273. **(C)** 274. **(D)** 275. **(A)** 276. **(B)** 277. **(C)** 278. **(D)** 279. **(D)** 280. **(D)**

281. **(D)** 282. **(C)** 283. **(A)** 284. **(B)** 285. **(A)** 286. **(B)** 287. **(B)** 288. **(A)** 289. **(A)** 290. **(C)**

291. **(A)** 292. **(A)** 293. **(B)** 294. **(B)** 295. **(A)** 296. **(B)** 297. **(A)** 298. **(A)** 299. **(A)** 300. **(C)**

301. **(C)** 302. **(D)** 303. **(A)** 304. **(B)** 305. **(C)** 306. **(B)** 307. **(C)** 308. **(B)** 309. **(A)** 310. **(C)**

311. **(C)** 312. **(C)** 313. **(B)** 314. **(D)** 315. **(C)** 316. **(D)** 317. **(C)** 318. **(C)** 319. **(D)** 320. **(A)**

321. **(D)** 322. **(C)** 323. **(B)** 324. **(C)** 325. **(B)** 326. **(B)** 327. **(D)** 328. **(A)** 329. **(A)** 330. **(B)**

331. **(A)** 332. **(A)** 333. **(A)** 334. **(B)** 335. **(A)** 336. **(D)** 337. **(D)** 338. **(C)** 339. **(B)** 340. **(B)**

341. **(B)** 342. **(B)** 343. **(A)** 344. **(C)** 345. **(C)** 346. **(A)** 347. **(A)** 348. **(A)** 349. **(A)** 350. **(A)**

351. **(C)** 352. **(A)** 353. **(D)** 354. **(D)** 355. **(D)** 356. **(C)** 357. **(B)** 358. **(A)** 359. **(A)** 360. **(A)**

361. **(A)** 362. **(C)** 363. **(B)** 364. **(C)** 365. **(A)** 366. **(B)** 367. **(C)** 368. **(A)** 369. **(C)** 370. **(A)**

371. **(B)** 372. **(B)** 373. **(C)** 374. **(A)** 375. **(A)** 376. **(A)** 377. **(A)** 378. **(B)** 379. **(C)** 380. **(C)**

381. **(B)** 382. **(B)** 383. **(C)** 384. **(A)** 385. **(D)** 386. **(C)** 387. **(C)** 388. **(D)** 389. **(B)** 390. **(B)**

391. **(A)** 392. **(B)** 393. **(B)** 394. **(A)** 395. **(D)** 396. **(B)** 397. **(D)** 398. **(B)** 399. **(B)** 400. **(B)**

401. **(C)** 402. **(D)** 403. **(C)** 404. **(B)** 405. **(B)** 406. **(B)** 407. **(A)** 408. **(B)** 409. **(B)** 410. **(D)**

411. **(A)** 412. **(C)** 413. **(C)** 414. **(A)** 415. **(C)** 416. **(B)** 417. **(A)** 418. **(B)** 419. **(D)** 420. **(C)**

421. **(C)** 422. **(A)** 423. **(D)** 424. **(C)** 425. **(B)** 426. **(A)** 427. **(C)** 428. **(D)** 429. **(D)** 430. **(C)**

431. **(A)** 432. **(B)** 433. **(D)** 434. **(C)** 435. **(C)** 436. **(A)** 437. **(B)** 438. **(B)** 439. **(B)** 440. **(C)**

441. **(B)** 442. **(B)** 443. **(D)** 444. **(C)** 445. **(B)** 446. **(C)** 447. **(A)** 448. **(B)** 449. **(C)** 450. **(C)**

451. **(B)** 452. **(B)** 453. **(B)** 454. **(B)** 455. **(A)** 456. **(D)** 457. **(D)** 458. **(A)** 459. **(B)** 460. **(C)**

461. **(A)** 462. **(B)** 463. **(A)** 464. **(B)** 465. **(D)** 466. **(B)** 467. **(D)** 468. **(A)** 469. **(B)** 470. **(A)**

471. **(C)** 472. **(B)** 473. **(C)** 474. **(B)** 475. **(B)** 476. **(B)** 477. **(D)** 478. **(A)** 479. **(B)** 480. **(C)**

481. **(B)** 482. **(B)** 483. **(A)** 484. **(D)** 485. **(A)** 486. **(C)** 487. **(A)** 488. **(B)** 489. **(B)** 490. **(A)**

491. **(A)** 492. **(A)** 493. **(A)** 494. **(C)** 495. **(B)** 496. **(A)** 497. **(C)** 498. **(D)** 499. **(B)** 500. **(B)**

501. **(B)** 502. **(B)** 503. **(B)** 504. **(D)** 505. **(C)** 506. **(A)** 507. **(C)** 508. **(C)** 509. **(A)** 510. **(B)**

511. **(C)** 512. **(C)** 513. **(C)** 514. **(B)** 515. **(C)** 516. **(C)** 517. **(B)** 518. **(A)** 519. **(A)** 520. **(C)**

521. **(A)** 522. **(C)** 523. **(D)** 524. **(C)** 525. **(D)** 526. **(A)** 527. **(B)** 528. **(A)** 529. **(C)** 530. **(A)**

531. **(D)** 532. **(D)** 533. **(C)** 534. **(A)** 535. **(A)** 536. **(A)** 537. **(A)** 538. **(C)** 539. **(C)** 540. **(B)**

541. **(C)** 542. **(B)** 543. **(B)** 544. **(A)** 545. **(B)** 546. **(C)** 547. **(C)** 548. **(C)** 549. **(D)** 550. **(C)**

551. **(C)** 552. **(D)** 553. **(C)** 554. **(C)** 555. **(A)** 556. **(C)** 557. **(C)** 558. **(D)** 559. **(D)** 560. **(D)**

561. **(C)** 562. **(B)** 563. **(C)** 564. **(A)** 565. **(C)** 566. **(D)** 567. **(C)** 568. **(A)** 569. **(D)** 570. **(A)**

571. **(C)** 572. **(C)** 573. **(A)** 574. **(B)** 575. **(C)** 576. **(B)** 577. **(A)** 578. **(C)** 579. **(A)** 580. **(B)**

581. **(B)** 582. **(C)** 583. **(A)** 584. **(B)** 585. **(A)** 586. **(B)** 587. **(C)** 588. **(A)** 589. **(B)** 590. **(B)**

591. **(A)** 592. **(B)** 593. **(D)** 594. **(B)** 595. **(A)** 596. **(A)** 597. **(C)** 598. **(B)** 599. **(A)** 600. **(A)**

601. **(D)** 602. **(C)** 603. **(C)** 604. **(D)** 605. **(A)** 606. **(B)** 607. **(A)** 608. **(A)** 609. **(B)** 610. **(A)**

611. **(C)** 612. **(B)** 613. **(B)** 614. **(A)** 615. **(A)** 616. **(A)** 617. **(D)** 618. **(B)** 619. **(C)** 620. **(C)**

621. **(C)** 622. **(C)** 623. **(B)** 624. **(B)** 625. **(B)** 626. **(D)** 627. **(C)** 628. **(B)** 629. **(A)** 630. **(A)**

631. **(A)** 632. **(A)** 633. **(A)** 634. **(D)** 635. **(B)** 636. **(B)** 637. **(C)** 638. **(A)** 639. **(C)** 640. **(B)**

641. **(D)** 642. **(D)** 643. **(C)** 644. **(A)** 645. **(D)** 646. **(C)** 647. **(D)** 648. **(D)** 649. **(C)** 650. **(C)**
651. **(A)** 652. **(B)** 653. **(B)** 654. **(D)** 655. **(D)** 656. **(D)** 657. **(A)** 658. **(A)** 659. **(C)** 660. **(C)**
661. **(A)** 662. **(C)** 663. **(A)** 664. **(B)** 665. **(C)** 666. **(C)** 667. **(C)** 668. **(C)** 669. **(B)** 670. **(A)**
671. **(A)** 672. **(C)** 673. **(D)** 674. **(B)** 675. **(A)** 676. **(B)** 677. **(B)** 678. **(A)** 679. **(D)** 680. **(A)**
681. **(A)**

9 NUTRIENT MANAGEMENT

1. Fertilizer schedule includes:
(A) Dose only
(B) Source only
(C) Time of application
(D) Dose, source, time and method of application

2. N, P, K proportion in balanced fertilization is:
(A) 4 : 1 : 1 (B) 2 : 1 : 1
(C) 6 : 2 : 1 (D) 6 : 3 : 1

3. Name of the trace element is:
(A) Sulphur (B) Zinc
(C) Calcium (D) Potassium

4. Soil sickness is caused due to:
(A) Accumulation of autotoxins
(B) N-deficiency
(C) Accumulation of phosphates
(D) Zinc deficiency

5. Nitrogen use efficiency is lower in:
(A) Wheat (B) Maize
(C) Paddy (D) Vegetable

6. Phosphorus use efficiency in most of the crops is:
(A) 10-15% (B) 20-25%
(C) 30-35% (D) 40-45%

7. The cheaper source of sulphur is:
(A) Single super phosphate
(B) Ammonium sulphate
(C) Elemental sulphur
(D) Gypsum

8. Nitrogen as starter solution is applied in:
(A) Paddy
(B) Wheat
(C) Vegetable crops
(D) Mustard

9. FYM is applied to the field:
(A) One day before sowing
(B) 7 days before sowing
(C) 15 days before sowing
(D) 30 days before sowing

10. The nutrient responsible for development of root is:
(A) Nitrogen (B) Sulphur
(C) Zinc (D) Phosphorus

11. Bio fertilizer supplied nitrogen to the extent of:
(A) 10 kg/ha (B) 20 kg/ha
(C) 40 kg/ha (D) 60 kg/ha

12. The soil pH for rice cultivation is:
(A) 5.5-6.5 (B) 8.5-9.5
(C) 4.5-5.5 (D) 8.0-8.5

13. The north-east alluvium of Uttar Pradesh is:
(A) Light textured
(B) Heavy textured
(C) High calcareous
(D) Non calcareous

14. In alluvial soils of India, deficiency of the following nutrient is most wide spread:
(A) Iron (B) Zinc
(C) Copper (D) Manganese

15. Among major plant nutrients, the deficiency of the following nutrients is much common in soils of India:
(A) Nitrogen
(B) Potassium
(C) Nitrogen and Potassium both
(D) None of these

16. Clay fraction of Indo-Gangetic alluvium is dominated by the mineral:
(A) Vermiculite (B) Illite
(C) Kaolinite (D) Chlorite

17. Black cotton soils of India have originated from the rocks:
(A) Sand stones (B) Shales
(C) Lime stones (D) Deccan trap

18. The CEC Cmol (p+) kg^{-1} of black cotton soils of India ranges between:
(A) 15 to 25 (B) 5 to 20
(C) 35 to 50 (D) None of these

19. The clay minerals predominant in black cotton soils of India is/are:
(A) Montmorillonite (B) Beidellite
(C) Both (A) & (B) (D) None of these

20. Argillopedoturbation in black cotton soils of India is the result of:
(A) Slickensides -
(B) Shrinking and swelling of clays
(C) Gilgai
(D) None of these

21. SiO_2/R_2O_2 of the clay fraction of black cotton soils usually lies between:
(A) 1.5 to 2.0 (B) 2.0 to 2.5
(C) 2.5 to 3.0 (D) 3.0 to 3.5

22. Black cotton soils of India are poor in:
(A) Nitrogen (B) Phosphorus
(C) Both (A) & (B) (D) None of these

23. Red soil of India occupy an area of nearly:
(A) 2000 km^2 (B) 3000 km^2
(C) 4000 km^2 (D) 5000 km^2

24. Red soils of India are:
(A) Calcareous
(B) Rich in soluble salts
(C) Both (A) & (B)
(D) None of these

25. The CEC Cmol (p+) kg^{-1} red soils is:
(A) < 20 (B) 20 to 30
(C) 30 to 40 (D) >40

26. Red soils have the dominance of the following clay mineral:
(A) Illite (B) Kaolinite
(C) Smectite (D) Vermiculite

27. The colour of the red soils is due to:
(A) High content of iron oxide
(B) Wide diffusion of iron
(C) Both of these
(D) None of these

28. The laterites of India occupy nearly an area of:
(A) 50000 km^2 (B) 12500 km^2
(C) 250000 km^2 (D) 500000 km^2

29. Laterite soils have developed as:
(A) in situ formation
(B) Sedimentary formations
(C) Both (A) & (B)
(D) None of these

30. The SiO_2/R_2O_2 ratio of laterite clays varied between:
(A) 1.00 to 1.25 (B) 1.40 to 2.00
(C) 2.00 to 3.00 (D) > 3.00

31. Laterite soils are characterized by:
(A) Compact, honey combed/vesicular sub-soil mass
(B) Pallid zone/Plinthite above the water table
(C) Both (A) & (B)
(D) None of these

32. Laterites have a pH of:
(A) 4.5 to 6.0 (B) 6.0 to 7.0
(C) 7.0 to 8.0 (D) > 8.0

33. Laterite soils in India are commonly found in:
(A) Uttar Pradesh
(B) Assam
(C) Both of these
(D) None of these

34. The essentiality of nitrogen for higher plants was first established in the year:
(A) 1801 (B) 1802
(C) 1803 (D) 1804

35. The number of aminoacids forming the building-blocks of plant proteins are:
(A) 190 (B) 21
(C) 23 (D) 25

36. Nitrogen is an essential constituent of the following compound in plants:
(A) Chlorophyll (B) Carotenes
(C) Xanthine (D) Xanthocyanin

37. The essential nature of nitrogen to higher plants was first established by:
(A) JustusVon Leibig
(B) Sommer and Lipman
(C) Warrington
(D) Theodore de Saussure

38. Nitrogen is an essential constituent of the plant growth substance:
(A) Gibberellins (B) Abscisin II
(C) Cytokinin (D) Calins

39. High N supply to plants influences shoot/root ratio in plants by:
(A) Increasing the ratio
(B) Decreasing the ratio
(C) No alteration in the ratio
(D) None of these

40. Excess N supply to plants leads to following changes in their carbohydrate accumulation:
(A) Higher accumulation
(B) Depressed accumulation
(C) No change in accumulation
(D) None of these

41. The susceptibility of crops alters with excessive N-supply to plants as:
(A) Lodging increases
(B) Lodging decreases
(C) Lodging remains unchanged
(D) None of these

42. Rice plants prefer absorption of fertilizer nitrogen in the from of:
(A) NO_3^- (B) NH_4^+
(C) Amino acid (D) N_2

43. Deficiency symptoms of nitrogen in plants appear at first in:
(A) Lower older leaves
(B) Middle leaves
(C) Top young leaves
(D) All of these

44. Excessive N supply to plants promotes:
(A) Mildew and rusts
(B) Root-rot
(C) Bacterial blight
(D) Leaf mosaic

45. Under low N supply, tillering in plants:
(A) Decreases (B) Increases
(C) Does not alter (D) None of these

46. The critical value of N percentage for potato (in potato leaves of 40-45 days old) is:
(A) 1.5 (B) 3.0
(C) 4.5 (D) 6.0

47. The critical value of N percentage for sugarcane (in 3rd to 6th sugarcane leaf from the top) is:
(A) 1.0 (B) 1.5
(C) 2.0 (D) 2.5

48. Good 'Indicator plant' for N deficiency is:
(A) Rice (B) Wheat
(C) Maize (D) Sunflower

49. The essential nature of phosphorus for higher plants was first established experimentally by:
(A) Arnon (B) E. Gris
(C) Ville (D) Johonson and Stout

50. The essentiality of phosphorus for higher plants was reported at first in the year:
(A) 1829 (B) 1839
(C) 1849 (D) 1859

51. The phosphorus in plant seeds is mainly stored as:
(A) Phospholipids (B) Nucleic acids
(C) Inorganic –P (D) Phytin

52. The phosphorus in cell membranes is mainly present as:
(A) Phospholipids (B) ATP
(C) NADP (D) Nucleoproteins

53. Phosphorus is a structural component of the cell organelles:
(A) Cell membranes
(B) Cell membranes and chloroplast
(C) Cell membrane, chloroplast and mitochondria
(D) None of these

54. As a constituent of coenzyme NAD and NADP, phosphorus plays an important role in:
(A) Energy transformation
(B) Hydrolytic reactions
(C) Decarboxylation
(D) Oxidation-reduction reactions

55. Phosphorus supply to plants promotes growth of:
(A) Fibrous roots
(B) Lateral roots
(C) Fibrous as well as lateral roots
(D) None of these

56. With an ample phosphorus supply to crops:
(A) Maturity period is delayed
(B) Maturity period is early
(C) Maturity period is unchanged
(D) None of these

57. Phosphorus supply to crops influences the fungal disease incidences by:
(A) Increasing disease incidence
(B) Decreasing disease incidence
(C) No change in disease incidence
(D) None of these

58. In legumes, phosphorus supply improves crop yield and protein content of crop through:
(A) Higher utilization of fertilizer – N
(B) Higher N – fixation
(C) Both (A) & (B)
(D) None of these

59. An ample supply of phosphorus to plants improves the availability of the following nutrients to plants:
(A) Molybdenum
(B) Zinc
(C) Both Molybdenum and zinc
(D) None of these

60. In plants, the dominant form of available inorganic phosphorus in soil below pH 6.7 exists in the inionic forms:
(A) $HPO_4^=$ (B) $H_2PO_4^-$
(C) $PO_4^=$ (D) None of these

61. In organic forms, plants can directly absorb organic phosphorus in the form:
(A) Phytin (B) Phospholipids
(C) Both (A) & (B) (D) None of these

62. The optimum soil pH range for phosphorus availability to crops is:
(A) 5.0 to 6.0 (B) 6.0 to 7.0
(C) 7.0 to 8.0 (D) None of these

63. Phosphorus deficiency in plants appears at first in:
(A) Older leaves (B) Younger leaves
(C) Both (A) & (B) (D) None of these

64. The critical value of P in phosphorus deficient leaves of maize plants before silking stage is:
(A) 0.25 per cent (B) 0.35 per cent
(C) 0.45 per cent (D) 0.55 per cent

65. Indicator plant for P deficiency indication is:
(A) Maize
(B) Tomato
(C) Maize and tomato both
(D) None of these

66. Potassium was experimentally established as an element essential to higher plants during the year:
(A) 1839 (B) 1876
(C) 1886 (D) 1896

67. The essentiality of potassium to higher plants was at first reported by:
(A) Russel (B) Sprengel
(C) Steinberg (D) Arnon

68. The potassium is almost entirely present as K^+ in the:
(A) Cytoplasm (B) Vacuole
(C) Both (A) & (B) (D) None of these

69. Within the cell, potassium helps to maintain:
(A) Permeability of cellular membranes
(B) Hydration of protoplasm
(C) Both (A) & (B)
(D) None of these

70. Potassium supply to plants has a favourable effect on:
(A) Protein synthesis
(B) Respiration
(C) Both (A) & (B)
(D) None of these

71. Potassium supply to plants decreases crop lodging through strengthening stalk by:
(A) Promoting cellulose and lignin deposition on walls
(B) Improving turgidity of cells
(C) Increasing parenchymatous tissues
(D) Maintaining electrical charge within the cell

72. Potassium deficiency leads to increase the severity of:
(A) Cotton wilt (B) Potato rot
(C) Both (A) & (B) (D) None of these

73. Plants utilize potassium in the from of:
(A) K_2 (B) K^+
(C) Both (A) & (B)
(D) None of these

74. In general, potassium deficiency influences crop maturity by:
(A) Delaying maturity period
(B) Preceding maturity period
(C) Not changing maturity period
(D) None of these

75. Potassium deficiency symptoms in plants appear at first from:
(A) Lower most leaves
(B) Upper most leaves
(C) Middle leaves
(D) Scattered throughout the entire leaves

76. Deficiency of K in plants leads to an accumulation of:
(A) Methionine (B) Putrescine
(C) Ornithine (D) Isoleucine

77. In general, the shoot/root ratio of K-deficient crops is around:
(A) 1.0 (B) 2.0
(C) 3.0 (D) 4.0

78. Potassium is a structural component of:
(A) Nucleus (B) Mitochondria
(C) Chloroplast (D) None of these

79. Potassium deficiency in plants is characterized by:
(A) Leaf scorch (B) Tip burn
(C) Both (A) & (B) (D) None of these

80. The critical value of K percentage in petioles of deficient potato plants is:
(A) 4.0 (B) 5.0
(C) 6.0 (D) 7.0

81. The critical value for K deficiency in maize is:
(A) 0.5 per cent (B) 1.0 per cent
(C) 1.5 per cent (D) 2.0 per cent

82. Indicator plant for monitoring K deficiency is:
(A) Mustard (B) Pea
(C) Sunflower (D) Potato

83. Major part of dry weight of plant is derived from elements obtained from:
(A) Soil (B) Air
(C) Water
(D) Decomposing organic matter

84. Major function of phosphorus is in the formation of:
(A) Cell membranes (B) Cell wall
(C) Enzymes (D) Carbohydrates

85. Minerals are absorbed by the plants through:
(A) Pressure flow (B) Diffusion
(C) Active transport (D) Osmosis

86. Magnesium is required for:
(A) DNA replication
(B) RNA synthesis
(C) Stomatal movements
(D) Synthesis of chlorophyll

87. Potassium is employed in:
(A) Chlorophyll synthesis
(B) Protein synthesis
(C) Cementing cells
(D) Opening and closing of stomata

88. Calcium is a component of:
(A) Middle lamella
(B) Primary walls
(C) Secondary wall
(D) Chlorophyll

89. Chlorosis results from deficiency of:
(A) Chlorine (B) Sodium
(C) Calcium (D) Iron

90. Both nitrogen and sulphur are required by plants for:
(A) Chlorophyll synthesis
(B) Enzymes
(C) Cell wall
(D) Stomatal movements

91. Micronutrients mainly function as:
(A) Osmotic constituents of cell sap
(B) Cofactors of enzymes
(C) Constituents of chlorophyll
(C) Components of important biochemicals

92. Critical elements are:
(A) Na, K and Ca (B) N, P and Mg
(C) N, P and K (D) Mn, B and Mo

93. Macronutrients are elements that:
(A) Play major role in plant nutrition
(B) Are require in large quantities in plants
(C) Form large molecules in plants
(D) None of these

94. Minerals are absorbed in the form of:
(A) Molecules (B) Ions
(C) Compounds (D) Mixtures

95. Micronutrients/trace elements:
(A) Can be removed from the plant without any effect on it
(B) Play a minor role in plant nutrition
(C) Form very small molecules in plants
(D) Are present/required in small quantities in plants

96. In plant mineral nutrition, elements are called macro or micro elements depending upon their:
(A) Relative presence in plant ash
(B) Relative importance in plant growth
(C) Relative amount required in plants
(D) Relative availability in soil

97. Tracer elements are those that are:
(A) Required in small amount by the plant
(B) Radioactive and can be used to investigate a process in the plant
(C) First to be traced in the plant
(D) Present in traces in the soil solution

98. The absorption of minerals due to difference in the electropotential gradient without use of energy is:
(A) Active absorption
(B) Passive absorption
(C) Osmotic absorption (D) None of these

99. Passive absorption of mineral salts is not dependent on:
(A) Diffusion (B) Osmosis
(C) Donnan equilibrium
(D) Ionic exchange

100. Active absorption of minerals salts makes use of:
(A) Ionic exchange (B) Carriers
(C) Phosphorylation of elements
(D) None of these

101. Which of the following does not operate in the carrier concept of active absorption of minerals:
(A) Electron transport
(B) Contractile protein
(C) Protein synthesis
(D) Ascending water stream

102. Absorbed minerals are transported to other parts of the plants by:
(A) Diffusion (B) Imbibition
(C) Ion exchange (D) Water stream

103. A primary deficiency is caused by insufficient absorption of:
(A) Magnesium (B) Manganese
(C) Calcium (D) Potassium

104. Down word passage of mineral, from upper soil layers to lower soil strata is called:
(A) Percolation (B) Leaching
(C) Weathering (D) Runway

105. A number of minerals like Ca, Mg and K are held over the surface of clay particles because the latter are:
(A) Negatively charged
(B) Positively charged
(C) Neutral
(D) Having both positive and negative residual valencies

106. Soil nitrate is more likely to leach than ammonium due to its:
(A) Small size
(B) Negative charge
(C) Being useless to soil flora
(D) Abundance

107. The largest reservoir of nitrogen is in:
(A) Rocks (B) Sea
(C) Air (D) Soil

108. The enzymes required for nitrogen fixation is:
(A) Nitrogen deaminase
(B) Nitrodioxidase
(C) Amino acid decarboxylase
(D) Nitrogenase

109. Hydroponies is growing plants in:
(A) Laboratory (B) Phytotron
(C) Liquid culture medium
(D) Solid culture medium

110. Aeroponics is growing plants in:
(A) Air (B) Satellites
(C) Other planets and space
(D) Stands with liquid spray of their roots

111. Nodule formation of legume roots is reduced during deficiency of:
(A) Sulphur (B) Boron
(C) Nitrogen (D) Both (A) and (B)

112. Little leaf/leaf rosetting is deficiency symptoms of:
(A) Zn (B) Mn
(C) Fe (D) B

113. Internal cork of apple/browning of cauliflower is deficiency symptoms of:
(A) Mo (B) B
(C) Cu (D) Zn

114. Exanthema is due to deficiency of:
(A) B (B) Mo
(C) Mn (D) Cu

115. Whiptail of crucifers is due to deficiency of:
(A) Mo (B) Cu
(C) Zn (D) P

116. A trace element is the one which is:
(A) Radioactive (B) Heavy isotope
(C) Required in minute quantity
(D) Enters the protoplasm alongwith an other elements

117. Common free ion in the cell is:
(A) Iron (B) Potassium
(C) Sulphate (D) Phosphate

118. Which of the following are the most abundant elements in the living cells:
(A) C, H, S and P (B) C, O, N and P
(C) C, H, O and N (D) C, H, Mg and Ca

119. A fertilizer has a formula of three figures 19-9-9. They stand for percentage of:
(A) N, Ca and Mg (B) Mg, P and K
(C) Ca, N and Fe (D) N, P and K

120. The raw material most used by the plants is:
(A) O_2 (B) CO_2
(C) N_2 (D) K

121. Mineral salts which are absorbed by the root from the soil are in the form of:
(A) Very dilute solution
(B) Dilute solution
(C) Concentrated solution
(D) Very concentrated solution

122. Minerals absorbed by roots move to the leaf through:
(A) Xylem (B) Phloem
(C) Sieve tube (D) None of these

123. Which of the following is macronutrient:
(A) Ca (B) Mo
(C) Mn (D) Zn

124. Which of the following is a micronutrient:
(A) Mg (B) Zn
(C) Ca (D) P

125. Plants absorb minerals salts from the soil solution through:
(A) Semipermeable membrane into the cytoplasm
(B) Perforation at the apex of root hairs only
(C) The cell wall which is semipermeable
(D) None of these

126. Which of the following is not an essential micronutrient:
(A) Boron (B) Nickel
(C) Manganese (D) Molybdenum

127. Which one of the following elements is not required by plants for their healthy normal growth:
(A) Calcium (B) Magnesium
(C) Lead (D) Iron

128. Which one of the following non mineral elements can obtain directly from the air:
(A) Hydrogen (B) Carbon
(C) Nitrogen (D) Phosphorus

129. Which of the following elements is necessary for translocation of sugars in plants:
(A) Iron (B) Molybdenum
(C) Manganese (D) Boron

130. Which of the following is present in the core of chlorophyll molecule:
(A) Fe (B) Mg
(C) K (D) Mn

131. Which was not an essential element for plants:
(A) Iron (B) Potassium
(C) Zinc (D) Sodium

132. Which one is an inorganic nutrient:
(A) Cellulose (B) Vitamin
(C) Calcium (D) Protein

133. Chlorosis in plants occurs due to:
(A) High sunlight intensity
(B) Low sunlight intensity
(C) Absorption of yellow pigments from the soil
(D) Deficiency of Mg and Fe in the soil

134. Most of the plants obtain nitrogen from the soil in the form of:
(A) Free nitrogen gas (B) Nitric acid
(C) Nitrites (D) Nitrates

135. Mg and Fe are needed for plants in the:
(A) Energy transfer
(B) Synthesis of chlorophyll pigment in the leaves
(C) Stomatal opening
(D) Translocation of carbohydrates

136. Carbon becomes available to crop plant in the form of:
(A) Amino acids (B) Carbonates
(C) Carbon dioxide (D) Elemental carbon

137. On the basis of symptoms of chlorosis in leaves a student inferred that this Ws due to deficiency of nitrogen. This inference could be correct only if we assume that yellowing of leaves appeared first in:
(A) Young leaves
(B) Old leaves
(C) Young leaves followed by mature leaves
(D) Mature leaves followed by young leaves

138. Plants require two metallic compounds (minerals) for chlorophyll synthesis. They are:
(A) Fe and Ca (B) Fe and Mg
(C) Cu and Ca (D) Ca and K

139. If a plant shows interveinal chlorosis of the older leaves and later this reaches to younger leaves, the symptoms is due to lack of:
(A) Magnesium (B) Calcium
(C) Iron (D) Copper

140. When chlorophyll is burnt which one is obtained:
(A) Ca (B) Na
(C) Mg (D) Mn

141. Importance of microelements was recognized late due to:
(A) Their toxicity
(B) Presence as contaminants in macro nutrients
(C) Their absence in plant ash
(D) Leakage from roots

142. Major nitrogen fixation is carried out by:
(A) Lighting
(B) Chemical industries
(C) Symbiotic bacteria
(D) Leaching

143. Which one does not contribute to nitrogen availability in the soil:
(A) Decay of organic matter
(B) Electric storms
(C) Azotobactor
(D) Liming

144. Which one plays an independent role in energy metabolisms:
(A) Calcium (B) Sodium
(C) Sulphur (D) Phosphorus

145. In an experiment a plant was dried, crushed and heated strongly in a crucible for long. The residue contained:
(A) Oxides and carbonates of about ten elements
(B) Oxides and carbonates of three elements
(C) Carbon only
(D) Starch and related compounds

146. Rotation of crops is useful for:
(A) Increasing mineral quantity of soil
(B) Increasing fertility of soil
(C) High protein content
(D) Denitrification of crops

147. Deficiency of Molybdenum causes:
(A) Poor development of vasculature
(B) Bending of leaf tips
(C) Yellowing of leaves
(D) Mottling and marginal necrosis of leaves

148. In plants a common symptom caused by deficiency of P, K, Ca and Mg is:
(A) Bending of leaf tips
(B) Formation of anthocynin
(C) Poor development of vasculature
(D) Appearance of dead necrotic cells

149. Active transport from outside to inside of molecule across a membrane requires:
(A) Cyclic AMP (B) Acetyl choline
(C) ATP (D) Phloroglucinol

150. Yellowing of leaves or chlorosis is the first deficiency symptoms of which element:
(A) Calcium (B) Nitrogen
(C) Phosphorus (D) Chlorine

151. Premature leaf fall is due to deficiency of:
(A) Phosphorus (B) Nitrogen
(C) Iron (D) Calcium

152. Active uptake of minerals requires:
(A) Respiration
(B) Expenditure of energy
(C) Root metabolism
(D) Photophosphorylation

153. Ammonium sulphate is:
(A) Insecticide (B) Weedicide
(C) Fertilizer (D) Toxic chemical

154. Which element is needed in large amount:
(A) Nitrogen (B) Sulphur
(C) Phosphorus (D) Calcium

155. Phytotron is meant for:
(A) Controlled irradiation
(B) Induction of mutations
(C) Controlled humidity
(D) Growing plants under controlled environment

156. A nonessential element is:
(A) Calcium (B) Barium
(C) Potassium (D) Magnesium

157. Nonessential element taking part in plant growth is:
(A) Magnesium (B) Calcium
(C) Cobalt (D) Chlorine

158. Which is the role of molybdenum:
(A) Nitrogen fixation
(B) Flower induction
(C) Chromosome contraction
(D) Carbon assimilation

159. Which one of the following is required for nitrate reduction to nitrite:
(A) Fe (B) B
(C) Zn (D) Mo

160. An essential element is that which:
(A) Is found in plant ash
(B) Is available in soil
(C) Improves health of plants
(D) Is irreplaceable and indispensable for growth of plants

161. Micronutrients are:
(A) Less important than macronutrients
(B) As important as macronutrients
(C) Having a minor role in plant nutrition
(D) Omitted from culture medium without any detrimental effect

162. A micronutrient is the one which is:
(A) More important than any major element
(B) Less important than major elements
(C) Needed in small quantity but is as important as a major element
(D) Found in small quantities in the soil

163. Framework elements are:
(A) C, H and O
(B) Cu, Co and Fe
(C) Mg, Cu and Fe
(D) Mn, Ca and N

164. Which of the following is a macronutrient:
(A) Mo (B) Mg
(C) Mn (D) Zn

165. Which one of the following is a micronutrient:
(A) Mg (B) Ca
(C) Mn/Mo (D) P

166. Plastocyanin contains:
(A) Mo (B) Fe
(C) Cu (D) Zn

167. Deficiency of Magnesium causes:
(A) Necrosis (B) Plasmolysis
(C) Hydrolysis (D) Chlorosis

168. A sulphur containing amino acid is:
(A) Methionine (B) Asparagine
(C) Serine (D) Proline

169. Plant ash indicates:
(A) Organic matter of plant
(B) Mineral salts absorbed by plants
(C) Both mineral salts and organic matter
(D) Silica absorbed by plants

170. Active ion uptake means:
(A) Ions moves passively
(B) Ions move freely
(C) There is expenditure of energy
(D) Ions are active

171. Iron deficiency results in:
(A) Leaf tip necrosis
(B) Small leaves
(C) Decreased protein synthesis
(D) Intervenial chlorosis appearing first in young leaves

172. A trace element, also required for plant growth, and in the form of radioactive isotope useful in cancer therapy is:
(A) Iron (B) Calcium
(C) Cobalt (D) Sodium

173. Theory suggesting that carbon dioxide produced in respiration helps in mineral absorption is called:
(A) Carbonic acid exchange theory
(B) Contact exchange theory
(C) Active mineral absorption
(D) Donnan equilibrium

174. Phosphorus & nitrogen ions generally depleted in soil because they usually occur as:
(A) Nautral ions
(B) Negatively charged ions
(C) Positively charged ions
(D) Both positively and negatively charged but disproportionate mixture

175. Application of organic materials with C : N ratio wider than 20 to 30 : 1 leads to ------- of soil N:
(A) Nitrification (B) Ammonification
(C) Mineralization (D) Immobilization

176. NPK denotes:
(A) Nitrogen, Protein and Kinetin
(B) Nitrogen, Protein and Potassium
(C) Nitrogen, Potassium and Kinetin
(D) Nitrogen, Phosphorus & Potassium

177. Which one is a component of ferrodoxin:
(A) Zn (B) Mn
(C) Cu (D) Fe

178. Which one is an essential mineral, not constituents of any enzymes but stimulates the activity of many enzymes:
(A) Zn (B) Mn
(C) K (D) Mg

179. Potassium is involved in:
(A) Photosynthesis
(B) Promoting many enzymatic activities that regulate plant processes
(C) Providing reddish pigmentation to fruits
(D) Formation of vascular cambium

180. Which can function as carrier in active ion absorption:
(A) Cytochrome (B) Ferredoxin
(C) Lecithin (D) Plastoquinone

181. Stem and root tips die in deficiency of:
(A) Nitrogen (B) Phosphorus
(C) Calcium (D) Sulphur

182. Pungent principle, sinigrin of crucifers is a:
(A) Glycoside having sulphur
(B) Glycoside having cyanide
(C) Glycoside having special amino acids
(D) Tannin

183. Phosphorus brings about:
(A) Retardation of protein synthesis
(B) Healthy root growth
(C) fruit ripening
(D) retarded plant growth

184. Zinc is essential for:
(A) Stomatal closing
(B) Stomatal opening
(C) Biosynthesis of IAA
(D) Oxidation of carbohydrates

185. Brown heart disease is due to deficiency of:
(A) Iron (B) Boron
(C) Potassium (D) Molybdenum

186. Scientist credited with development of formula for balanced solution of mineral element is:
(A) Knopp (B) Went
(C) Hill (D) Emerson

187. A crop plant which can grow in nitrogen deficient soils without external supply of it is:
(A) *Allium sativum*
(B) *Gossypium herbaccum*
(C) *Cajanus cajan*
(D) *Helianthus annus*

188. Nitrogen fixing bacteria were discovered by:
(A) Koch (B) Winogradsky
(C) Pasteur (D) Lister

189. Ions are absorbed by the plants through:
(A) Molecular diffusion
(B) Carriers and pumps
(C) Difference in water potential
(D) Difference in DPD

190. Mineral absorption occurs:
(A) If soil solution is hypotonic
(B) If soil solution is hypertonic
(C) Independent of water potential
(D) Independent of water absorption

191. Whcih is not a micronutrient:
(A) P (B) Cu
(C) Zn (D) Mo

192. Essential macronutrients are:
(A) Manufactured during photosynthesis
(B) Produced by growth hormones
(C) Absorbed from soil
(D) Produced by enzymes

193. Active and passive transports across cell membrane differ in:
(A) Passive transport is nonselective
(B) Passive transport is along the concentration gradient while active transport is due to metabolic energy
(C) Active transport is more rapid
(D) Passive transport is confined to anions while active transport is confined to cations

194. Major role of phosphorus in plant metabolism is:
(A) O_2- evolution during photosynthesis
(B) CO_2 evolution during respiration
(C) Create anaerobic conditions
(D) Generate metabolic energy

195. Nitrogen fixing enzyme found in root nodules is:
(A) Nitrogen esterase (B) Nitrogenase
(C) Nitrase (D) Nitrosomonas

196. Sulphur is absorbed by plants as:
(A) SO_3 from soil (B) SO_2 from air
(C) SO_2 from soil (D) Both (B) and (C)

197. Mineral absorption is mostly:
(A) Physical process
(B) Chemical process
(C) Active process
(D) Passive process

198. For sugarcane 'N' fertilizer application has to be completed within ------ days after planting for better sugar content in juice:
(A) 60 (B) 40
(C) 30 (D) 80

199. Which one group is of micro-elements:
(A) Mn, Cu, Ca (B) Na, K, B
(C) Na, Ni, Cl (D) Cu, Mo, Zn

200. Element playing important role in nitrogen fixation is:
(A) Mn (B) Mo
(C) Zn (D) Cu

201. Which one is not a micronutrient:
(A) Mo (B) B
(C) Mg (D) Zn

202. Culturing of plants in water/soil-less medium is:
(A) Hydroponics (B) Sand culture
(C) Aeroponics (D) Soil-less culture

203. Which is not a micronutrient:
(A) Cu (B) Mn
(C) B (D) Ca

204. Molybdenum is involved in plant metabolism in:
(A) Translocation of solutes
(B) Nitrate reduction
(C) Tryptophan synthesis
(D) Ascorbic acid synthesis

205. Element responsible for maintaining turgor is:
(A) K (B) Na
(C) Ca (D) Hg

206. Element required is least quantity is:
(A) Zn (B) Mn
(C) Mo (D) Cl

207. Role of inorganic nutrients in plant growth and metabolism was first discovered by:
(A) Knop (B) Woodland
(C) DeSassure (D) Steward

208. Organic compounds invariably contain:
(A) Phosphorus (B) Sulphur
(C) Magnesium (D) Carbon

209. Element which is a component of proteinaceous materials and protoplasm is:
(A) Potassium (B) Nitrogen
(C) Magnesium (D) Oxygen

210. Element essential for photolysis of water is:
(A) Nitrogen (B) Oxygen
(C) Chlorine (D) Carbon

211. Which is absent in plant ash:
(A) Mineral elements
(B) Nitrogen
(C) Essential elements
(D) Trace elements

212. Sink is related to:
(A) Phytochrome (B) Enzyme
(C) Stomata
(D) Transport of minerals

213. Bidirectional translocation of mineral takes place through:
(A) Parenchyma (B) Cambium
(C) Xylem (D) Phloem

214. Fall of immature leaves is due to deficiency of:
(A) Potassium (B) Sodium
(C) Phosphorus (D) Zinc

215. Enzyme catalysed reactions can be inhibited by:
(A) Mg^{2+} (B) Cu^{2+}
(C) Hg^{2+} (D) Zn^{2+}

216. Permeability of protoplasm is accelerated by:
(A) Ca (B) Na
(C) P (D) K

217. Exothema is a result of deficiency of:
(A) Cu (B) Ca
(C) P (D) Na

218. Movement of minerals in plant roots through diffusion is:
(A) Endosmosis
(B) Osmosis
(C) Active absorption
(D) Passive absorption

219. Plants require Fe and Mg for:
(A) Synthesis of chlorophyll
(B) Opening and closing of stomata
(C) Energy transfer during photosynthesis and respiration
(D) Translocation of carbohydrate

220. Nitrogen is an important constituent of:
(A) Polyphosphates (B) Proteins
(B) Carbohydrates (D) Lipids

221. Sand culture is a component of:
(A) Soil-less cultivation
(B) Green house effect
(C) Photorespiration
(D) None of the above

222. Inorganic nutrients are present in the soil as:
(A) Molecules
(B) Atoms
(C) Electrically charged ions
(D) Colloids

223. Clay and organic matter of the soil have negative charges. They attract positively charged ions like:
(A) Ca^{2+} (B) Mg^{2+}
(C) K^{+} (D) All the above

224. Nitrogen absorbed by plants is:
(A) Converted to nitrate
(B) Reduced to ammonia
(C) Changed to nitrite
(D) Combined with oxygen

225. Micronutrient absorbed by foliage is:
(A) Zinc (B) Phosphorus
(C) Aluminium (D) Boron

226. Plants use zinc as:
(A) Zn (B) Zn^{2+}
(C) $ZnSO_4$ (D) $Zn(NO_3)_2$

227. Plants growing in urea sprayed but Mg deficient soil will show:
(A) Deep green foliage
(B) Early flowering
(C) Yellowing of leaves
(D) Loss of pigmentation in petals

228. Pedology is a study of:
(A) Locomotion of animals
(B) Rocks
(C) Soil
(D) Crop diseases

229. When is a man a part of biotic community:
(A) Always
(B) Only when he lives in big cities
(C) Only when he lives in forest
(D) Never

230. Laterite soils is rich in:
(A) Calcium (B) Magnesium
(C) Iron (D) Nitrogen

231. Amount of water available to the plants in the soil is called as:
(A) Holard (E+C)
(B) Chresard
(C) Echard
(D) None of these

232. Humus is defined as:
(A) Physical texture of soil
(B) Chemical composition of soil
(C) Decomposed organic matter in soil
(D) None of these

233. Almost decomposed organic matter in which original form is lost is in:
(A) Duff (B) Litter
(C) Leaf mould (D) None of these

234. Edaphic factors are related to:
(A) Soil (B) Man
(C) Animals (D) Temperature

235. Which one of the following is not an edaphic factor:
(A) Mineral matter (B) Organic matter
(C) Rainfall (D) Soil water

236. The kind of soil water most useful to plant is:
(A) Hygroscopic water
(B) Gravitational water
(C) Capillary water
(D) None of these

237. The soil with poorest water holding capacity is:
(A) Clay (B) Loam
(C) Sandy (D) None of these

238. The organic matter of soil has the value:
(A) 25% (B) 50%
(C) 40% (D) 10%

239. The size of clay particles is:
(A) Less than 0.002 mm
(B) More than 0.002 mm
(C) 0.2 mm to 2 mm
(D) None of these

240. Which of the following has smallest of soil particles:
(A) Loam soil (B) Sand
(C) Clay (D) Silt

241. Major constituent of soil is:
(A) Organic matter
(B) Inorganic matter
(C) Soil air (D) Soil water

242. Formation of organic matter takes place by:
(A) Chemical weathering
(B) Soil micro-organisms
(C) Herbivores
(D) Soil water

243. Formation of soil takes place by:
(A) Weathering (B) Pedogenesis
(C) Melting (D) Both (A) & (B)

244. Soil erosion is more where:
(A) Rainfall is high
(B) No rainfall
(C) Low rainfall (D) None of these

245. Solid matter from land washed annually into oceans is:
(A) 330 million tones
(B) 3300 million tones
(C) 3.3 million tones
(D) 33 million tones

246. Land mass occupies about:
(A) 14 million tones
(B) 140 million tones
(C) 1400 million tones
(D) 14000 million tones

247. Arable land constitutes:
(A) 36.6% (B) 30%
(C) 22% (D) 11%

248. Soil forms the surface layer in percentage of land:
(A) 80% (B) 60%
(C) 50% (D) 30%

249. Soil transported by air is:
(A) Alluvial (B) Colluvial
(C) Glacial (D) Aeolian

250. River water deposits:
(A) Loamy soil (B) Alluvial soil
(C) laterite soil (D) Sandy soil

251. Alluvial soils are found in:
(A) Northern India (B) Southern India
(C) Gangetic Yamuna plains
(D) Eastern India

252. Ao layer of the soil is made of:
(A) Decaying litter
(B) Rocky matter
(C) Soil mixed with organic matter
(D) Soil rich in inorganic matter

253. Development of soil from rocky matter is:
(A) Pedogenesis (B) Pedology
(C) Edaphic factors (D) Edaphic climax

254. Soil particles arranged in a series of increasing size is:
(A) Silt → Sand → Clay
(B) Sand → Silt → Clay
(C) Clay → Sand → Silt
(D) Clay → Silt → Sand

255. Soil best suited for plant growth is:
(A) Clay (B) Loam
(C) Sandy (D) Gravel

256. Least porous soil is:
(A) Clay soil (B) Sandy soil
(C) Loam soil (D) Gravelly soil

257. Good soil is the one which:
(A) Allows water to percolate slowly
(B) Allows water to pass quickly
(C) Allows limited amount of water into it
(D) Holds whole of water entering it

258. Deep black soil is productive due to high proportion of:
(A) Sand and Zinc
(B) Gravel and Calcium
(C) Clay and Humus
(D) Silt and Earthworm

259. Humus is good for plant growth because:
(A) It improves physical condition of soil
(B) It makes the soil porous
(C) It increases water holding and aeration of soil
(D) All the above

260. Humus is important for plant growth as it is:
(A) Made of dead organic matter
(B) Derived from leaves
(C) Rich in nutrients and increases water holding capacity of soil
(D) Partially decomposed

261. Humus is example of:
(A) A fertilizer
(B) Component of soil structure
(C) Organic colloids
(D) Crystalloids

262. Soil fertility can be maintained without addition of nutrients due to:
(A) Floods
(B) Crop residue
(C) Activity of microorganisms
(D) Favourable temperature and water

263. Fertility of soil is measured by its ability to:
(A) Retain nutrients
(B) Hold organic materials
(C) Hold water
(D) Support life

264. A fertility of agricultural soil appears deep coloured at the surface as compared to soil one metre down. The reason for colour of top soil is:
(A) More moisture
(B) Rich in organic matter
(C) Rich in Iron, calcium and magnesium
(D) Recent formation

265. Water holding capacity is highest in:
(A) Sandy soil (B) Silt soil
(C) Clay soil (D) Loam soil

266. Soil water available to root is:
(A) Surface water
(B) Hygroscopic water
(C) Gravitational water
(D) Capillary water

267. Water held tightly by soil particles as thin film over their surface is:
(A) Hygroscopic water
(B) Capillary water
(C) Chemical water
(D) Gravitational water

268. Which governs the solubility and availability of nutrients in the soil:
(A) Temperature (B) pH
(C) Aeration (D) Porosity

269. Water-logging commonly occurs in:
(A) Sandy soil (B) Gravelly soil
(C) Clayey soil (D) Loam soil

270. Soil conservation is:
(A) Conversion of sterile soil in fertile one
(B) Aeration of soil
(B) Erosion of soil
(D) Protecting against loss

271. Removal of top fertile soil by wind or water is:
(A) Siltation
(B) Soil erosion
(C) Weathering of soil
(D) Leaching

272. Soil erosion is prevented by:
(A) Afforestation
(B) Increased bird population
(C) Overgrazing
(D) Removal of vegetation

273. Terracing helps in soil conservation in:
(A) Plains (B) Deserts
(C) Hilly areas (D) Wet areas

274. Sheet erosion is due to:
(A) Fast running rivers
(B) Wind
(C) Heavy rains
(D) Occasional rains

275. Maximum productivity (on global basis) is shown by:
(A) Forests (B) Aquatic ecosystem
(C) Deserts (D) Grasslands

276. Which one is biotic in origin:
(A) Texture and porosity of soil
(B) Rainfall
(C) Carbon dioxide content of soil
(D) light intensity

277. Eroded soils are:
(A) Devoid of plant nutrients
(B) Richer in plant nutrients
(C) Unaltered in plant nutrients
(D) Fit for agriculture

278. Soil conservation is achieved through:
(A) Restricted human activities
(B) Good plant cover
(C) Wind screen
(D) Ploughing

279. Soil fertility is reduced by:
(A) Crop rotation
(B) Nitrogen fixing bacteria
(C) Decaying organic matter
(D) Intensive agriculture

280. Type of soil in Gangetic basin is:
(A) Eolian (B) Colluvial
(C) Residual (D) Alluvial

281. Soil particles determine its:
(A) Texture (B) Field capacity
(C) Water holding capacity
(D) Soil flora

282. Soil showing permanently wilted plants has:
(A) Gravitational water
(B) Capillary water
(C) Hygroscopic water
(D) Ground water

283. Water is a resource:
(A) Nondegradable nonmaintainable
(B) Degradable maintainable
(C) Renewable
(D) Non renewable

284. Main cause of soil erosion is:
(A) Afforestation (B) Less rains
(C) Thinning of ozone layer
(D) Deforestation

285. Largest amount of fresh water is found in:
(A) Lakes and streams
(B) Underground
(C) Polar ice and glaciers
(D) Rivers

286. Guano is the major source of:
(A) Nitrogen (B) Phosphorus
(C) Sulphur (D) Zinc

287. Sequence of humification and mineralization:
(A) Dead organic matter → Litter → Duff → Humus → Minerals
(B) Humus → Minerals → Litter → Duff
(C) Mineral → Humus → Litter → Duff
(D) Dead organic matter → Duff → Litter → Minerals → Humus

288. Soil is composed of:
(A) Mineral + Water +Air
(B) Mineral+Organic matter + Water
(C) Mineral+Organic matter + Air + Water
(D) Organic matter + water

289. Common type of soil found in India is:
(A) Sandy
(B) Alluvial and black soils
(C) Red soil
(D) Both (B) and (C)

290. Plants die in water-logged soils due to:
(A) Nutrient leaching
(B) Stoppage of root respiration
(C) Dilution of soil nutrients
(D) Dilution of cell sap

291. In hilly areas, erosion is maintained by:
(A) Terracing (B) Manuring
(C) Ploughing (D) Mixed cropping

292. Edaphology is connected with:
(A) Plant and biosphere
(B) Soil and living microorganisms
(C) Animals and ecosystem
(D) Soil and biosphere

293. Top soil is darker and:
(A) Contains more Na and Mg
(B) Is drier than subsoil
(C) Contains more organic matter
(D) Is wetter than subsoil

294. Soil salinity is measured by:
(A) Porometer (B) Potometer
(C) Calorimeter
(D) Conductivity meter

295. Wind velocity at 1 m height as compared to ground level is:
(A) Same (B) 2.5 times
(C) 10 times (D) 1.5 times

296. Soil covered by peaty fibrous layer is:
(A) Glei (B) Podsol
(C) Laterite (D) Chernozem

297. Reservoir, lakes, ponds, rivers and canals contain percentage of fresh water:
(A) 77.2% (B) 22.4%
(C) 0.36% (D) 30.3%

298. Red soils of India are:
(A) Laterite soils
(B) Deficient in lime, Mg, P and K
(C) Rich in iron and organic matter
(D) All the above

299. Macropores of soil take part in:
(A) Holding air
(B) Holding air & percolation of water
(C) Percolation of water
(D) Holding of capillary water

300. Soil strength for engineering purpose is determined by the characteristics of:
(A) Compressibility
(B) Compatibility
(C) Bearing resistance
(D) All of these

301. Type of water vapour movement occurs in soils is:
(A) Internal (B) External
(C) Both of these (D) None of these

302. Capillary movement of water is complemented by:
(A) Stem elongation
(B) Root extension
(C) Leaf orientation
(D) None of these

303. When a plant is under soil moisture stress, the roots:
(A) Shrink in size
(B) Expand in size
(C) Remain unaffected
(D) Both a and b

304. The positive and high value of redox potential indicates:
(A) Reducing conditions
(B) Strong oxidizing conditions
(C) Equally reducing &oxidizing conditions
(D) None of these

305. The concept of soil as a natural body was first developed by:
(A) C.F. Marbut (B) K.D. Glinka
(C) V.V. Dakuchaev
(D) E.W. Hillgard

306. In a waterlogged soil the concentration of which gas is high:
(A) Ethane (B) Methane
(C) Carbon dioxide (D) Carbon monoxide

307. A soil absorbs how much amount of incoming solar radiation:
(A) 5% (B) 10%
(C) 15% (D) 20%

308. The global average of solar radiation reaching the earth is:
(A) 20% (B) 30%
(C) 40% (D) 50%

309. Humus colloids are composed basically of:
(A) Carbon (B) Nitrogen
(C) Helium (D) Ozone

310. The most important silicate clays are known as:
(A) Photosilicates (B) Phyllosilicates
(C) Autosilicates (D) Simply silicates

311. Non-humic substances are less complex and are------- to microbial attack than those belonging to humic group:
(A) More resistant (B) Less resistant
(C) More susceptible (D) Less susceptible

312. Humus enhances the mineral breakdown and increases the availability of some:
(A) Anions (B) Cations
(C) Neutral ions (D) Nne of these

313. Which of the following is harmful to the higher plants:
(A) Phenol carboxylic acid
(B) Vanillic acid
(C) Dihydroxystearic acid
(D) All of these

314. Soil organic matter has many remarkable effects and one of them is given below, identify the same:
(A) Encouraged granulation
(B) Increased plasticity
(C) Enhanced cohesion
(D) Low cation exchange

315. The C:N ratio in the organic matter of the furrow slice (upper 15 cm) of arable soils commonly ranges from:
(A) 3:1 to 4:1 (B) 6:1 to 7:1
(C) 18:1 to 19:1 (D) 8:1 to 15:1

316. Which of the following rotations is likely to leave soil richer in organic matter:
(A) Continuous sorghum
(B) Continuous maize
(C) Continuous oats
(D) Maize-Oats-Clovers

317. Ammonium and potassium ions are apparently just the right size to fit into the cavities between are crystal units of:
(A) Kaolinite (B) Montmorillonite
(C) Vermiculite (D) None of these

318. Ammonia present in the soil can be lost in significant quantities from:
(A) Acid soils (B) Alkaline soils
(C) Both of these (D) None of these

319. Etridiazol and nitrapyrin ------- oxidation of ammonia to nitrates:
(A) Accelerates (B) Maintains
(C) Inhibits (D) No effect

320. In soils where the negative charge is pH dependent, liming increases CEC which results in an increased potassium by the soil colloids and ----- in solution:
(A) Increases (B) Decreases
(C) Little effect (D) None of these

321. The soils of which zones can supply adequate potassium for many years even under irrigations:
(A) Temperate (B) Humid
(C) Arid (D) Subtropical

322. The tropical soils high in hydrous oxides of iron and aluminium embody ---- to torrential rains:
(A) Susceptibility (B) Resistance
(C) None of these (D) Both of these

323. The prime objective of conservation tillage is:
(A) Removal of plant residues
(B) Complete retention of plant residues
(C) Partial retention of plant residues
(D) None of these

324. In no-tillage systems, the surface soil layers have:
(A) Higher bulk density
(B) Lower bulk density
(C) Higher total pore space
(D) Lower water infiltration

325. Conservation tillage tends to encourage:
(A) Higher microbial population
(B) Lower number of earthworms
(C) Reduces soil fauna
(D) None of these

326. Wind laden with soil particles embodied the abrasive action which is:
(A) Greatly increased
(B) Greatly decreased
(C) Absolutely unaffected
(D) None of these

327. The diameter of most erodible soil particles is:
(A) 0.2 mm (B) 0.4 mm
(C) 0.3 mm (D) 0.1 mm

328. Cost-sharing assistance for soil conservation has been in practice for many years in:
(A) India (B) England
(C) USA (D) China

329. Nitrogen fertilization in apples favours growth, but its excess:
(A) Hastens maturity
(B) Delays maturity
(C) Improves colour (D) Improves quality

330. On the basis of productivity and fertility, the fertile and productive lakes have been classified as:
(A) Eutrotrophic (B) Oligotrophic
(C) Dystrophic (D) None of these

331. The idea of a map showing the use of land was mooted in 1919 by:
(A) Robert S. Plat (B) A.J. Hunt
(C) Carl O. Sauer (D) Charles K. Paul

332. Biogeochemical nutrient cycles have led to recognize that ----- is very important:
(A) Balanced fertilization
(B) Over fertilization
(C) Under fertilization
(D) None of these

333. --------, nonessential elements in the ecosystem may accumulate at undesirable levels:
(A) Normally (B) Occasionally
(C) Mostly (D) Invariably

334. The soil without a well developed soil profile characteristics are known as azonal soil and include:
(A) Planosols (B) Latosols
(C) Lithosols (D) None of these

335. Geological distribution of clay minerals is of considerable importance, southern are predominantly found:
(A) Illite (B) Kaolinite
(C) Montmorillonite (D) None of these

336. Greasewood (vegetation) is found in desert soils where ------ problem prevails:
(A) Salinity (B) Acidity
(C) Alkalinity (D) Neutrality

337. The nodule in ------ plants tend to be large, pinkish, fleshy, singly or in clusters about the tap roots or the first formed lateral roots:
(A) Biennials (B) Perennials
(C) Annuals (D) All of these

338. The identification of haemoglobin in legume root nodules has been made by Kubo (1931) through its characteristic ------ band maxima of 575 mμ and 540 mμ and by comparing with haemoglobin of the horse:
(A) Reflection (B) Deflection
(C) Absorption (D) Adsorption

339. Nodule haemoglobin appeared to have the typical haemoglobin characteristic of taking up and giving off:
(A) Carbon dioxide (B) Hydrogen
(C) Nitrogen (gas) (D) Oxygen

340. The molecular nitrogen in a root nodule is reduced to ------ form which it changes rapidly to amino acids:
(A) Ammonium (B) Ammonia
(C) Nitrate (D) Nitrite

341. The humus myth was effectively deposed by --------- who presented a paper at a prominent scientific meeting:
(A) Justus Senebier
(B) Jean Young
(C) Justus von Liebig
(D) Justus Billright

342. Rothamsted Agricultural Experiment Station, England was founded by:
(A) J.B. Lawes
(B) J.H. Gilbert
(C) J.B. Lawes and J.H. Gilbert
(D) Lord Hastings

343. Some specialists have identified ------ factors that influence crop growth:
(A) 32 (B) 42
(C) 52 (D) 22

344. It has been suggested that the -------- of sunlight is generally most satisfactory for plant growth:
(A) Full spectrum
(B) Half spectrum
(C) Little more than full spectrum
(D) None of these

345. The maximum carbon dioxide concentration at normal atmospheric levels occur in relatively low light intensities of -------- foot candles for some plants:
(A) 2500-5000 (B) 6000-7500
(C) 1000-2000 (D) 2150-2450

346. Tomato plants grown on soils with low bulk densities of tended to be:
(A) High in sugar
(B) Low in protein
(C) High in anthrocyanin
(D) Showed no effect

347. -------in 1909 developed an equation that related growth to the supply of plant nutrients:
(A) V.J. Kilmer (B) W.J. Spillman
(C) J.W. Friedrich (D) E.A.Mitscherlich

348. The principle similar to that developed by E.A. Mitscherlich was also independently developed several years later by:
(A) F. Steenbjerg (B) W.J. Spillman
(C) M.E. Summer (D) J.G. Shannon

349. Based on dark metabolism, glucose consumption for protein production is about ----- per cent when nitrogen is provided as nitrate rather than as ammonium:
(A) 20 (B) 30
(C) 40 (D) 50

350. The sequence in which amino acids are linked in proteins depends on:
(A) Climatic control
(B) Agronomic control
(C) Genetic control
(D) None of these

351. The tendency of the young upper leaves to remain as the lower leaves yellow or die is an indication of nitrogen -------- in the plant:
(A) Mobility (B) Immobility
(C) Stability (D) Degradability

352. Build up of soluble sugars and amino acids due to inadequate potassium, low molecular weight compounds encourage the activity of:
(A) Plant hormones
(B) Plant Pathogens
(C) None of these (D) Both of these

353. Blossem-end rot in tomato has been found to be caused by the deficiency of:
(A) Nitrogen (B) Phosphorus
(C) Calcium (D) Sulphur

354. Affinity of the enzyme RuDP carboxylase for carbon dioxide is enhanced by:
(A) Sulphur (B) Phosphorus
(C) Nitrogen (D) Magnesium

355. The sufficiency range of iron is normally between how much ppm:
(A) 10-40 ppm (B) 50-250 ppm
(C) 300-400 ppm (D) None of these

356. The presence of ectotrophic mycorrhia enhances the uptake of nutrients in general, but particularly of:
(A) Molybdenum (B) Calcium
(C) Magnesium (D) Phosphorus

357. The quantity of nutrients that can come in direct contact with the plant roots is the amount in a volume of soil -------- the volume of roots:
(A) Equal to (B) Less than
(C) More than (D) No relation with

358. -------ion effect represent the effect of absorbed ion on the release of another from the surface of a colloid:
(A) Supplementary (B) Complementary
(C) Coordinated (D) Integrated

359. The movement of nutrients, ions by mass flow is reduced at low temperatures because of:
(A) Low respiration
(B) Low photosynthesis
(C) Low transpiration
(D) None of these

360. Organic matter content under well drained conditions is higher under:
(A) Field crops (B) Fruit crops
(C) Grassland (D) Forest covers

361. The ------ oxidation of ammonia has been termed as nitrification:
(A) Physical (B) Chemical
(C) Biological (D) Physico-chemical

362. ----- is considered to be most important of the soil organisms bringing about the conversion of NH_4^+ to NO_2^-:
(A) Nitrosomonas (B) Nitrrobacter
(C) Azotobacter (D) Azospirillum

363. The rapid increase in denitrification at the elevated soil temperatures suggests that which microorganisms plays a major role in denitrification:
(A) Chemophilic (B) Photophilic
(C) Thermophilic (D) Hydrophilic

364. ------ of the NH_4^+ in a region of alkaline microsite helps to restore the conditions suitable for the normal conversion of NO_2^- to NO_3^-:
(A) Diffusion (B) Dilution
(C) Both of these (D) None of these

365. Application of urea, aqua-ammonia, anhydrous ammonia and diammonium phosphate form alkaline microsite at the point of application, and in rice -------- encourage the taking up of nitrite:
(A) High NH_4^+concentration
(B) High pH
(C) Both of these
(D) None of these

366. The losses of ammonia usually become greater with:
(A) Increase in soil pH
(B) Decrease in soil pH
(C) Decrease in calcium
(D) Decrease in temperature

367. The highest percentage content of nitrogen among the commonly used nitrogenous fertilizers is embodied by:
(A) Urea
(B) Urea sulphate
(C) Ammonium nitrate
(D) Anhydrous ammonia

368. Permissible biuret level, in most of urea fertilizers is:
(A) 5 % (B) 4 %
(C) 3 % (D) 2 %

369. In the fertilized soil with soil solution concentration of one ppm of phosphorus, mass flow contribution to phosphorus nutrition of crops could approach:
(A) 0 % (B) 40 %
(C) 30 % (D) 20 %

370. The quantity of solid phase phosphorus that acts as reserve is often referred to as:
(A) Intensity factor
(B) Quantity factor
(C) Quality factor
(D) None of these

371. Phosphatase enzymes play a major role in -------- of organic phosphates in soil:
(A) Mineralization
(B) Immobilization
(C) Mass transportation
(D) None of these

372. The mineralization of organic phosphorus has been studied in relation to --- ratio:
(A) C/N/S (B) C/N/K
(C) C/N/P (D) C/N/Ca

373. The surfaces on which phosphate ions enter into sorption reaction are:
(A) Surfaces of constant charge
(B) Surfaces of variable charge
(C) Both of these (D) None of these

374. The mineral calcite develops a negative charge due to ----- tendency for Ca^{++} ions than CO_3^{--}ions to go into solution:
(A) Lower (B) Greater
(C) Equal (D) None of these

375. Clays such as kaolinite will larger quantities of phosphorus with -------- SiO_2/R_2O_3 ratio:
(A) Low (B) High
(C) Equal (D) None of these

376. Ammorphous aluminosilicate minerals like allophone have ------ negative charge which is partly or entirely balanced by complex aluminium cations:
(A) Small (B) Zero
(C) Large (D) None of these

377. A minor portion of the phosphorus sorption capacity of soils originates from :
(A) Sodium bicarbonate
(B) Sodium carbonate
(C) Calcium oxide
(D) Calcium carbonate

378. The maximum phosphorus availability in most of the soils is in the pH range:
(A) 4.0-4.5 (B) 5.0-5.5
(C) 6.0-6.5 (D) 7.0-7.5

379. In a predominantly biological system, mineralization of phosphorus from organic matter or crop residues is dependent on soil -------- activity:
(A) Chemical (B) Physical
(C) Physico-chemical (D) Biological

380. ------ generally augments the utilization of phosphorus which is water soluble:
(A) Broadcast
(B) Broadcast and incorporation
(C) Band placement
(D) None of these

381. Rock phosphate is treated with --------- to break the apatite bond and to render the contained phosphate more soluble:
(A) Acid alone (B) Alkali alone
(C) Heat alone (D) Either acid or heat

382. Various potassic fertilizers embody nearly an equal effectiveness in --------- situations requiring its supplemental application:

(A) Climatic (B) Edaphic
(C) Cropping (D) Economic

383. The concentration of which ppm of SO_2^- ion in the soil or solution cultures have proved adequate for the growth of many plant species:

(A) 1-2 (B) 3-5
(C) 6-8 (D) 9-11

384. Proper ---- nourishment for cabbage, potato and sugar beet etc., is essential since these are known to have it high requirement:

(A) Sulphur (B) Manganese
(C) Zinc (D) Calcium

385. The essentiality of copper for higher plants was first recognized in:

(A) 1911 (B) 1921
(C) 1931 (D) 1941

386. "Reclamation disease" due to copper deficiency occurs in plants grown on newly reclaimed acid:

(A) Histosols (B) Alfisols
(C) Aridosols (D) Ultisols

387. Among small grain crops which has an exceptional tolerance to low levels of soil copper supply:

(A) Barley (B) Wheat
(C) Rye (D) Oats

388. Repeated use of Bordeaux mixture, sewage sludge, pig and poultry manures and/or mine wastes may cause toxicity by:

(A) Iron (B) Aluminium
(C) Zinc (D) Copper

389. The accessibility and utilization of soil applied copper can be enhanced by:

(A) Reducing particle size
(B) Reducing number of granules
(C) Reducing droplets per unit soil valume
(D) None of these

390. Each unit increase in pH brings about a decrease in solubility of Fe^{++} of how many folds:

(A) 50 (B) 75
(C) 100 (D) 150

391. The total soluble iron concentration becomes sufficient to supply iron adequately to roots only at pH:

(A) 2 (B) 3
(C) 4 (D) 5

392. It has been observed that the iron chelated by:

(A) 25 % (B) 50 %
(C) 75 % (D) 100 %

393. Low total iron content of soil is ------- the basis for plant deficiencies:

(A) Commonly (B) Often
(C) Seldom (D) Mostly

394. Iron deficiency usually occurs in soil in the pH range of:

(A) 5.4-6.5 (B) 7.4-8.5
(C) 9.0-9.5 (D) 9.6-10.1

395. There can actually be a build up of toxic concentrations of Fe^{++} in the soil solution of:

(A) Aridosol (B) Histosol
(C) Latosol (D) Vertisol

396. For each unit increase in pH, the concentration of Mn^- decrease by how many folds:

(A) 50 (B) 100
(C) 150 (D) 200

397. Silicon is the ------ most abundant element in the earth's crust:

(A) Third (B) Fourth
(C) Sixth (D) Second

398. Properties of the silicon-deficient Hawaii soils include:

(A) High total aluminium content
(B) High base saturation
(C) High pH (D) High total silicon

399. It is generally accepted that the Gramineae members contain how many times the silicon concentration normally found in legumes and other dicotyledonous plants:

(A) 5-10 (B) 10-20
(C) 20-30 (D) 30-40

400. Superphosphate containing -------- selenium may provide enough to the plants in selenium deficient areas:

(A) 10 ppm or more (B) 20 ppm or more
(C) 30 ppm or more (D) 40 ppm or more

401. In the room and pillar pattern of mining potassium ------ rooms are mined out and the pillars of ore are left for support:
(A) Hexagonal (B) Spherical
(C) Rectangular (D) Square

402. ------ is probably the most important growth limiting factor in acid soils especially those having pH below 5.0-5.5:
(A) Nitrogen (B) Calcium
(C) Potassium (D) Aluminium

403. At pH ----------- H+ ion toxicity is increased by the removal of ions:
(A) 4.5 or less (B) 5.5 or less
(C) 6.0 or less (D) 5.0 or less

404. If an acid soil is limed to the pH level ------ it enhances nitrification:
(A) 3.5-4.5 (B) 4.5-5.5
(C) 5.5-6.5 (D) 6.5-7.5

405. When the phosphorus is either band placed or broadcast, the relative efficiency is related to :
(A) Its soil status
(B) Its applied quantum
(C) Both of these
(D) None of these

406. Crops during any season are generally able to recover only ------- of the added fertilizer phosphorus:
(A) Less than 25%
(B) Less than 30%
(C) Less than 35%
(D) None of these

407. Potassium salts are:
(A) More mobile than nitrates
(B) Less mobile than nitrates
(C) Less mobile than phosphorus
(D) As mobile as nitrates

408. Broadcast potassium, is ------ than banded:
(A) More efficient
(B) As efficient as
(C) Less efficient
(D) None of these

409. Winter application of phosphorus is followed in:
(A) Warm areas (B) Cool areas
(C) Both of these (D) None of these

410. Potassium is utilized more efficiently if applied:
(A) Before/at planting
(B) Side dressed
(C) Top dressed
(D) None of these

411. Tillage of the soil ------ rate of disappearance of organic matter:
(A) Reduces (B) Increases
(C) Maintains (D) Has no effect

412. Deficiency of nitrogen in maize causes:
(A) Yellowing of leaves from tip to base in V shaped pattern
(B) Chlorotic spots at base of leaves near the margins
(C) Yellow streaking of leaf edges
(D) Whitish-yellow spots on leaves

413. If urdbean is being grown for the first time in a field, the seed should be inoculated with:
(A) *Azospirillum* (B) *Rhizobium*
(C) PSB (D) *Clostridium*

414. Who gave the criteria of essentiality of plant nutrients:
(A) Arnon and Stout (1939)
(B) Arnon and Stout (1941)
(C) Leibig (1949)
(D) None of these

415. When O.M. undergoes enzymatic oxidation, it librates:
(A) CO_2 (B) Water
(C) Energy (D) All of these

416. Humus is a:
(A) Amorphous (B) Colloidal
(C) Both (A) & (B) (D) None of these

417. What percent of humic group present in Humus?
(A) 60-80% (B) 50-60%
(C) 30-40% (D) 20-30%

418. Humic group contains:
(A) Polypherals (B) Polyquenones
(C) Both (A) & (B) (D) None of these

419. Which of the following is less resistant, light in colour, suitable in acid and alkali and less in molecular weight?
(A) Fulvic acid (B) Humic acid
(C) Humin acid (D) All of these

420. Humic acid is soluble in:
(A) Acid & alkali (B) Alkali
(C) Acid (D) None of these

421. Non humic group contains:
(A) Polysaccharide (B) Polyquenones
(C) Polyphenols (D) Both (B) and (C)

422. Bacteria responsible for nitrification process is:
(A) Nitrosomonas (B) Nitrobactor
(C) Rhizobium (D) Both (A) & (B)

423. Biuret content in urea should not exceed -- ------ according to fertilizer control order, 1957:
(A) 4.0% (B) 1.5%
(C) 3.0% (D) 2.5%

424. Gaseous loss of N takes place in the form of:
(A) NO_3 (B) N_2
(C) N_2O (D) Both (B) & (C)

425. Cross inoculation group of clover is:
(A) *Rhizobium meliloti*
(B) *Rhizobium trifoli*
(C) *Rhizobium lupini*
(D) *Rhizobium phaseoli*

426. Which is generally used for correcting soil acidity:
(A) Gypsum (B) Lime
(C) Iron pyrite (D) Zinc

427. Which is the slow release urea?
(A) Sulphur coated urea
(B) Shellack coated urea
(C) Neem coated urea
(D) All of these

428. Indian soils generally low in:
(A) Organic matter (B) Phosphorus
(C) Potassium (D) None of these

429. Under submerged soils, urea must be placed at:
(A) Oxidized zone (B) Reduced zone
(C) Root zone (D) All of these

430. Super granulated urea release slow N because of:
(A) Higher weight
(B) Big size
(C) More surface area in contact with soil
(D) Less surface area in contact with soil

431. Which phosphatic fertilizer contains phosphoric acid but not soluble in water or citric acid?
(A) SSP
(B) Di calcium phosphate
(C) Rock phosphate
(D) DAP

432. Application of MOP is not suitable for:
(A) Sugarcane (B) Sugarbeet
(C) Tobacco (D) All of these

433. The entry of inorganic materials into xylem occurs at:
(A) Leaves (B) Stem
(C) Root tips (D) None of these

434. Nitrate assimilation in plant is mainly carried out at:
(A) Toot (B) Stem
(C) Branches (D) Leaves

435. Conversion of organic form to inorganic form is known as:
(A) Mineralization
(B) Immobilization
(C) Volatilization (D) None of these

436. Potassium is rich in:
(A) Albite (B) Biotite
(C) Silkinite (D) All the above

437. Alkali soil can be reclaimed by the application of:
(A) Gypsum (B) Basic slag
(C) Lime (D) Calcite

438. Acid soils can be reclaimed by:
(A) $CaCO_3$ (B) H_2SO_4
(C) $CaSO_4.2H_2O$ (D) HNO_3

439. Denitrification is more in:
(A) Waterlogged soils
(B) Well drained soils
(C) Heavy soils (D) Light soils

440. Which of the following fertilizer has maximum plant nutrient?
(A) DAP (B) TSP
(C) MOP
(D) Ammonium nitrate

441. Dolomite is:
(a) $CaCO_3$ (B) $MgSO_4$
(C) $Ca(OH)_2$ (D) $MgCO_3CaCO_3$

442. The process of formation of nitrogen and nitrous oxide gases from ammonical fertilizers in soil is known as:
(A) Ammonification
(B) Nitrification
(C) Denitrification
(D) Mineralization

443. Blue green algae fix nitrogen mainly in:
(A) Rice field (B) Cereals field
(C) Sugarcane field (D) Legumes field

444. Organic carbon content in the organic matter is:
(A) 32% (B) 18%
(C) 68% (D) 58%

445. The acidity of H+ in aqueous phase of soils is called as:
(A) Active acidity
(B) Reserve acidity
(C) Permanent acidity
(D) None of these

446. Soil is originated from word 'Solum'. The word 'Solum' is a:
(A) Latin word
(B) Greek word
(C) Spanish word
(D) English word

447. Edaphology refers the study of soil related to:
(A) Crop production
(B) Origin of soil
(C) Classification of soil
(D) None of these

448. Soil is a:
(A) Dynamic body
(B) Static body
(C) Rock
(D) None of these

449. Soil contains minerals approximately:
(A) 25 % (B) 45 %
(C) 5 % (D) 10 %

450. Soil contains organic matter approximately:
(A) 2 % (B) 4 %
(C) 6 % (D) 8 %

451. Soil contains porosity approximately:
(A) 25 % (B) 35 %
(C) 50 % (D) 60 %

452. Soil contains water approximately:
(A) 75 % (B) 25 %
(C) 35 % (D) 45 %

453. Sand particles contain highest amount of:
(A) Quartz (B) Silica
(C) Iron (D) All of them

454. Fine particles of soil contain higher amount of:
(A) Quartz (B) Silica
(C) Iron (D) All of them

455. Different identified soil layers are known as:
(A) Horizon (B) Solum
(C) Profile (D) All of these

456. A horizon is known as:
(A) Organic (B) Elluvial
(C) Accumulation (D) All of these

457. O horizon is known as:
(A) Organic (B) Elluvial
(C) Accumulation (D) All of these

458. Illuvial horizon is a:
(A) O horizon (B) A horizon
(C) B horizon (D) None of these

459. The colour of different soil layer varies due to presence of:
(A) Minerals (B) Organic matter
(C) Minerals and organic matter
(D) None of these

460. A and B horizons are collectively known as:
(A) Horizon (B) Solum
(C) Profile (D) All of these

461. Sial is a:
(A) Upper layer of earth's crust
(B) Lower of earth's crust
(C) Upper layer of atmosphere
(D) Lower layer of atmosphere

462. Sial is made of mainly by:
(A) Al silicate (B) Mg silicate
(C) Fe- silicate (D) All of them

463. Sima is situated just:
(A) Upper of Sial (B) Lower of Sial
(C) Middle of soil (D) None of these

464. Sima is made of:
(A) Al silicate (B) Mg silicate
(C) Fe- silicate (D) All of these

465. Oxy anions are generally:
(A) Lithophilic elements
(B) Chaleophilic elements
(C) Biophilic elements
(D) Siderophilic elements

466. N, O, He and Xe elements are grouped into:
(A) Lithophilic elements
(B) Chaleophilic elements
(C) Biophilic elements
(D) Atmosphilic elements

467. Chalophilic elements make covalent bonds with:
(A) Sulphide (B) Carbonate
(C) Oxide (D) Hydroxide

468. Igneous rocks are made of:
(A) Lava (B) Lime
(C) Marble (D) All of them

469. Silica percentage in acidic igneous rocks is:
(A) 55-60 % (B) 60-65 %
(C) 65-85 % (D) All of these

470. Silica percentage in basic rocks is:
(A) 55-60 % (B) 60-65 %
(C) 45-55 % (D) All of these

471. Granite is a:
(A) Acidic (B) Basic
(C) Saline (D) None of these

472. Basalt is a:
(A) Acidic (B) Basic
(C) Saline (D) None of these

473. Sedimetory rocks are:
(A) Layered rocks
(B) Layer less rocks
(C) Metamorphic rocks
(D) All of these

474. Gypsum is a example of:
(A) Igneous rock
(B) Sedimentary rocks
(C) Metamorphic rocks
(D) All of these

475. Regolith is known as:
(A) Elluvial + Illuvial layers
(B) Solum + Parent material
(C) O+A+B+C layers of soil profile
(D) All of them

476. The list of rocks is given below:

Rocks	Metamorphic rocks
1. Sand stone	i. Quartzite
2. Shale	ii. Slate
3. Limestone	iii Marble
4. Granite	iv Gneiss

Which combinations is correct:
(A) 1.i 2. ii 3. iii 4. iv
(B) 1. i 2. iii 3. ii 4. iv
(C) 1. iv 2. ii 3. iii 4. i
(D) 1. i 2. iv 3.iii 4. ii

477. Celluvial is the result of:
(A) Water (B) Air
(C) Gravity (D) None of these

478. Indo-Gangetic plains were developed by:
(A) Alluvial materials
(B) Aeolian materials
(C) Colluvial materials
(D) None of these

479. Sand dunes are the result of:
(A) Alluvial materials
(B) Aeolian materials
(C) Colluvial materials
(D) None of these

480. Peat and muck are the example of:
(A) Cumulose materials
(B) Residual materials
(C) Transported materials
(D) All of these

481. The plant roots are potent agent of weathering?
(A) Physical (B) Chemical
(C) Biological (D) None of these

482. Hematite is changed into limonite by the process of:
(A) Hydrolysis (B) Hydration
(C) Reduction (D) Carbonation

483. The colour of hematite is:
(A) Blue (B) Red
(C) Black (D) Yellow

484. Limonite is the source of:
(A) N (B) P
(C) Fe (D) Zn

485. Dolomite is a source of following elements?
(A) Ca (B) Mg
(C) Ca and Mg (D) K

486. Apatite is a source of following elements?
(A) Ca (B) Mg
(C) P (D) K

487. Which of the following mineral does not belong to primary mineral?
(A) Quartz (B) Orthoclase
(C) Biotite (D) Hematite

488. Size of particle of primary minerals is:
(A) ≥ 2μ (B) < 2 μ
(C) ≤ 2 μ (D) All of these

489. According to Mohr's scale, few minerals are arranged in order of its decreasing hardness. Which of the following is a correct order?
(A) Diamond > Corundum>Topaz>Talc
(B) Diamond > Topaz > Corundum>Talc
(C) Talc > Topaz > Corundum > Diamond
(D) None of these

490. Eluviation refers to:
(A) Percolation of minerals from upper layers to lower layers
(B) Deposition of percolated materials from upper to lower layers
(C) Erosion of soil
(D) None of these

491. Illuviation refers to:
(A) Percolation of minerals from upper layers to lower layers
(B) Deposition of percolated materials from upper to lower layers
(C) Erosion of soil
(D) None of these

492. Which of the following is the example of reduction process:
(A) $4FeO + O_2 \rightarrow 2Fe_2O_2$
(B) $2Fe_2O_3 + 2HOH \rightarrow 2Fe_2O_3.3H_2O$
(C) $2Fe_2O_3 \rightarrow O_2 + 4FeO$
(D) All of these

493. Shape of clay particles is:
(A) Angular (B) Smooth
(C) Platy (D) All of these

494. Humification helps in:
(A) Formation of Ao layer
(B) Decomposition of organic matter
(C) Synthesis of new organic materials
(D) All of these

495. Podzolization refers to:
(A) Decomposition of humus and sesquioxides in lower soil layers.
(B) Leaching of humus and sesquioxides from upper soil layers
(C) This process occurs in humid temperate regions
(D) All of these

496. Laterization refers to:
(A) Fe and Al remain in upper layers
(B) Leaching of Si to lower layers
(C) This process occurs in humid tropical regions
(D) All of these

497. Classification refers to:
(A) This process in low rainfall areas
(B) It happens due to presence of higher amount of bases in parent materials
(C) $CaCO_3$ is accumulated in soil profile
(D) All of these

498. Solanization refers to:
(A) Accumulation of salts in soil profile
(B) High water table and poor drainage enhance this process
(C) The structure of such soils are very poor
(D) All of these

499. Which soil particle has highest surface areas?
(A) Sand (B) Silt
(C) Clay (D) All of these

500. Size of coarse sand particles is:
(A) 2 – 0.2 mm (B) 0.2 – 0.02 mm
(C) 0.02 – 0.002 mm (D) <0.002 mm

501. Size of fine sand particles is:
(A) 2 – 0.2 mm (B) 0.2 – 0.02 mm
(C) 0.02 – 0.002 mm (D) <0.002 mm

502. Size of silt particles is:
(A) 2 – 0.2 mm (B) 0.2 – 0.02 mm
(C) 0.02 – 0.002 mm (D) <0.002 mm

503. Size of clay particles is:
(A) 2 – 0.2 mm (B) 0.2 – 0.02 mm
(C) 0.02-0.002 mm (D) <0.002 mm

504. Whih is not a soil forming factor?
(A) Climate (B) Topography
(C) Lithosphere (D) All of these

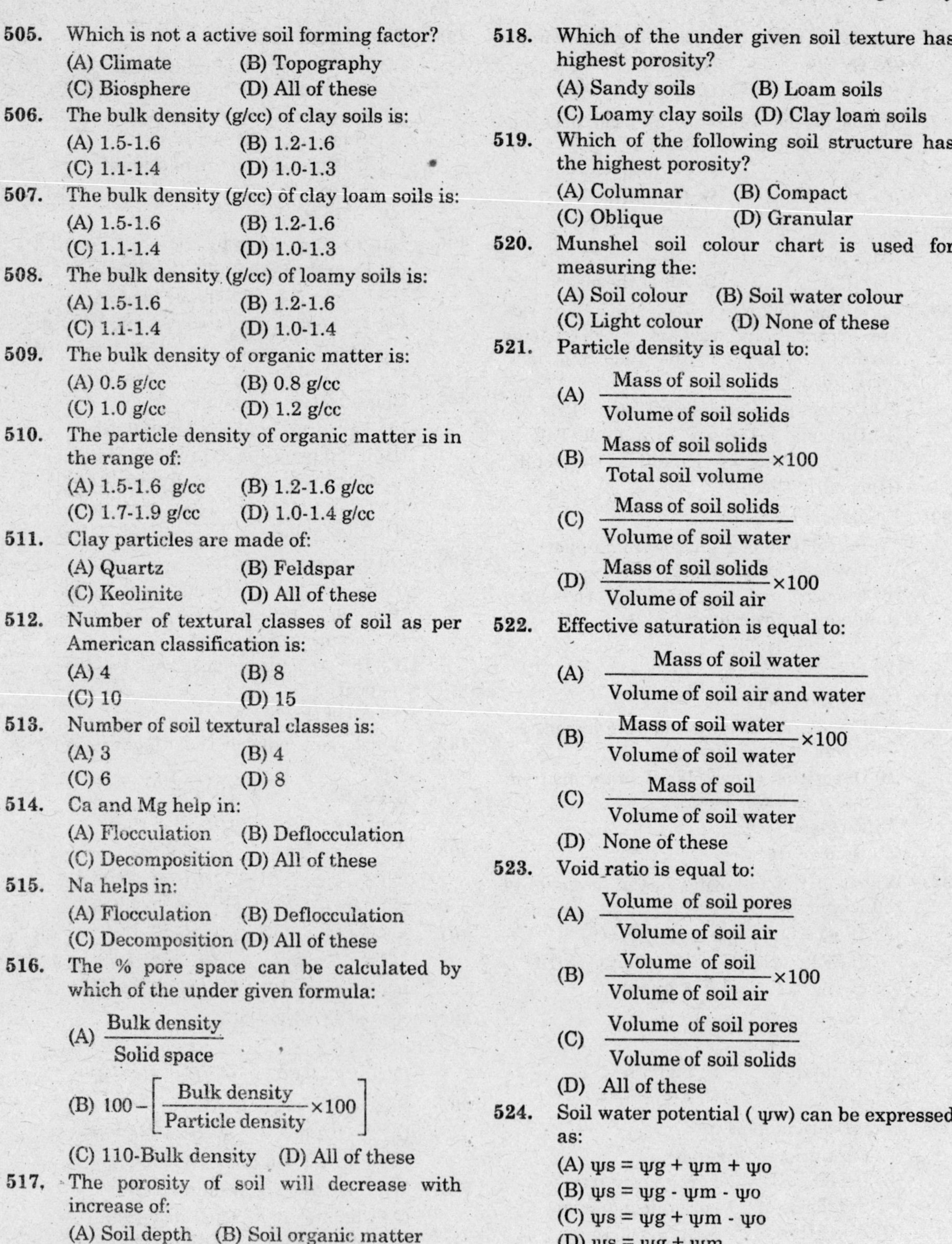

505. Which is not a active soil forming factor?
(A) Climate (B) Topography
(C) Biosphere (D) All of these

506. The bulk density (g/cc) of clay soils is:
(A) 1.5-1.6 (B) 1.2-1.6
(C) 1.1-1.4 (D) 1.0-1.3

507. The bulk density (g/cc) of clay loam soils is:
(A) 1.5-1.6 (B) 1.2-1.6
(C) 1.1-1.4 (D) 1.0-1.3

508. The bulk density (g/cc) of loamy soils is:
(A) 1.5-1.6 (B) 1.2-1.6
(C) 1.1-1.4 (D) 1.0-1.4

509. The bulk density of organic matter is:
(A) 0.5 g/cc (B) 0.8 g/cc
(C) 1.0 g/cc (D) 1.2 g/cc

510. The particle density of organic matter is in the range of:
(A) 1.5-1.6 g/cc (B) 1.2-1.6 g/cc
(C) 1.7-1.9 g/cc (D) 1.0-1.4 g/cc

511. Clay particles are made of:
(A) Quartz (B) Feldspar
(C) Keolinite (D) All of these

512. Number of textural classes of soil as per American classification is:
(A) 4 (B) 8
(C) 10 (D) 15

513. Number of soil textural classes is:
(A) 3 (B) 4
(C) 6 (D) 8

514. Ca and Mg help in:
(A) Flocculation (B) Deflocculation
(C) Decomposition (D) All of these

515. Na helps in:
(A) Flocculation (B) Deflocculation
(C) Decomposition (D) All of these

516. The % pore space can be calculated by which of the under given formula:
(A) $\frac{\text{Bulk density}}{\text{Solid space}}$
(B) $100-\left[\frac{\text{Bulk density}}{\text{Particle density}}\times 100\right]$
(C) 110-Bulk density (D) All of these

517. The porosity of soil will decrease with increase of:
(A) Soil depth (B) Soil organic matter
(C) Soil texture (D) All of these

518. Which of the under given soil texture has highest porosity?
(A) Sandy soils (B) Loam soils
(C) Loamy clay soils (D) Clay loam soils

519. Which of the following soil structure has the highest porosity?
(A) Columnar (B) Compact
(C) Oblique (D) Granular

520. Munshel soil colour chart is used for measuring the:
(A) Soil colour (B) Soil water colour
(C) Light colour (D) None of these

521. Particle density is equal to:
(A) $\frac{\text{Mass of soil solids}}{\text{Volume of soil solids}}$
(B) $\frac{\text{Mass of soil solids}}{\text{Total soil volume}}\times 100$
(C) $\frac{\text{Mass of soil solids}}{\text{Volume of soil water}}$
(D) $\frac{\text{Mass of soil solids}}{\text{Volume of soil air}}\times 100$

522. Effective saturation is equal to:
(A) $\frac{\text{Mass of soil water}}{\text{Volume of soil air and water}}$
(B) $\frac{\text{Mass of soil water}}{\text{Volume of soil water}}\times 100$
(C) $\frac{\text{Mass of soil}}{\text{Volume of soil water}}$
(D) None of these

523. Void ratio is equal to:
(A) $\frac{\text{Volume of soil pores}}{\text{Volume of soil air}}$
(B) $\frac{\text{Volume of soil}}{\text{Volume of soil air}}\times 100$
(C) $\frac{\text{Volume of soil pores}}{\text{Volume of soil solids}}$
(D) All of these

524. Soil water potential (ψw) can be expressed as:
(A) $\psi s = \psi g + \psi m + \psi o$
(B) $\psi s = \psi g - \psi m - \psi o$
(C) $\psi s = \psi g + \psi m - \psi o$
(D) $\psi s = \psi g + \psi m$

525. Loam soil has:

(A) Equal amount of sand and clay

(B) More sand and less clay

(C) Less sand and more clay

(D) None of these

526. The productivity level of different soils is given below on the basis of soil colour. Which of the following arrangement is correct?

(A) Black>Rust brown>Yellow>Grey

(B) Grey>Yellow>Rust brown>Black

(C) Black>Yellow>Rust brown>Grey

(D) All of these

527. Which of the soil has the highest infiltration rate?

(A) Loam soil (B) Clay soil

(C) Silt loam (D) Clay loam

528. Very slow infiltration rate will be called if the infiltration rate is:

(A) < 0.1 inch/hr (B) 0.1-0.5 inch/hr

(C) 0.5-1.0 inch/hr (D) > 1.0 inch/hr

529. The water is absorbed by soil due to the force of:

(A) Adhesion force (B) Cohesion force

(C) Suction force (D) None of these

530. The water particles are attracted to other water particle due to force of:

(A) Adhesion force (B) Cohesion force

(C) Suction force (D) All of these

531. pF is the unit of:

(A) Force (B) Distance

(C) Velocity (D) Weight

532. Hygroscopic coefficient is equal to:

(A) 1/3 bar (B) 5 bar

(C) 31 bar (D) None of these

533. Soil metrix is also known as:

(A) Soil solids (B) Soil water

(C) Soil air (D) Soil pores

534. The relationship between the soil water content on weight and volume basis can be expressed as:

(A) $Pv = \frac{Pw}{Bd}$ (B) $Pv = Pw \times Bd$

(C) $Pv = \frac{Bd}{Pw}$ (D) $Pw = Pv \cdot Bd$

535. Which of the following gas is not found in reduced conditions?

(A) CH_4 (B) N_2

(C) H_2S (D) O_2

536. At which concentration of soil CO_2, the roots will die?

(A) 5 % (B) 10 %

(C) 1 % (D) None of these

537. At which concentration of soil O_2, the root growth may be completely checked?

(A) > 15 % O_2 (B) 10 -15% O_2

(C) < 5 % O_2 (D) All of these

538. With increasing soil depth, concentration of CO_2 will be?

(A) Increased (B) Decreased

(C) Remained same (D) All of above

539. Which of the following is not correct statement?

(A) Soil has higher CO_2 concentration than atmospheric CO_2

(B) Soil has higher N_2 concentration than atmospheric N_2

(C) Soil has lower O_2 concentration than atmospheric O_2

(D) Soil has lower CO_2 concentration than atmospheric CO_2

540. 1 bar is equal to:

(A) pF of 3 (B) pF of 2

(C) pF of 4 (D) None of these

541. Osmotic potential is also known as:

(A) Solute potential

(B) Pressure potential

(C) Metrix potential (D) None of these

542. Pascal is a unit of:

(A) Water potential (B) Work

(C) Distance (D) None of these

543. The curve indicating available water against soil water tension is called:

(A) Soil water curve

(B) Soil water tension curve

(C) Soil tension curve

(D) Water tension curve

544. Which of the following water is available water is available to plant communities?

(A) Hygroscopic water

(B) Gravitational water

(C) Capillary water

(D) All of these

545. Moisture equivalent is equal to:
(A) 1/3 bar (B) 15 bar
(C) 0.3 - 1 bar (D) 1/3 – 31 bar

546. Which of the following is not responsible for soil water loss?
(A) Evaporation (B) Transpiration
(C) Percolation
(D) Hydraulic conductivity

547. When all soil macrospores are water filled, the water movement in soil is called:
(A) Unsaturated flow
(B) Saturated flow
(C) Run-off
(D) None of these

548. How much heat is required for vaporization of 1 g water at 20⁰C?
(A) 500 calories (B) 540 calories
(C) 560 calories (D) None of these

549. Saturated flow can be expressed by following law?
(A) Darcy's law (B) Newton's law
(C) William's law (D) None of these

550. Under Darcy law, water flow is inversely related to:
(A) Hydraulic conductivity
(B) Hydraulic heat
(C) Flow length
(D) Cross section area

551. Which of the following is not used for measuring soil water:
(A) Tensiometer
(B) Zypsum block
(C) Pressure bomb
(D) Neutron meter

552. Tensiometer can measure soil water up to soil moisture tension of:
(A) 0.0-0.85 bar (B) 0.0-0.50 bar
(C) -0.33-15.0 bar (D) All of these

553. Gypsum block can measure the soil water potential up to:
(A) 1-15 bar (B) 15-31 bar
(C) 31-1000 bar (D) All of these

554. Which of the following is the source of neutrons in neutron moisture meter?
(A) Radium (B) Thorium
(C) P^{32} (D) H^3

555. Which of the following can not prevent the soil water loss?
(A) Mulch (B) Hexadecanol
(C) Wind breakers (D) Deforestation

556. Which of the following gas has the highest concentration in atmosphere?
(A) CO_2 (B) O_2
(C) S_2 (D) N_2

557. Rhizobium bacteria is:
(A) Symbiotic bacteria
(B) Nonsymbiotic bacteria
(C) Anaerobic bacteria
(D) Autotropic bacteria

558. For calculating value of P, P_2O_5 is multiplied by:
(A) 2.29 (B) 0.437
(C) 1.20 (D) 3.00

559. Which one of the following has organic form of sulphur:
(A) Purine (B) Cysteine
(C) RNA (D) Phytin

560. If the soil moisture increases, the heat conductivity of soil will?
(A) Increase (B) Decrease
(C) Remain same (D) None of these

561. Which of the following can not diffuse across the semi permeable membrane?
(A) Crystalloid (B) Colloids
(C) Solid particles (D) None of these

562. The pF value at PWP is equal to:
(A) 4.2 (B) 5.2
(C) 3.2 (D) None of these

563. Which of the soil type has the lowest PWP?
(A) Sandy loam (B) Loam
(C) Clay loam (D) Clay

564. Water movement through soil columns is called:
(A) Permeability (B) Percolation
(C) Surface runoff (D) None of these

565. Water available in between hygroscopic coefficient and field capacity is known as:
(A) Capillary water
(B) Gravitational water
(C) Hygroscopic water
(D) None of these

566. Which of the soil type has the highest field capacity?
(A) Loam (B) Silt loam
(C) Clay loam (D) Clay

567. Sticky point refers to:
(A) Level of soil moisture at which water is completely absorbed by soil colloids
(B) Water status of soil
(C) Water level at which irrigation is applied
(D) None of these

568. Colloid is a Greek word that is made of:
(A) Kolla + Eidos (B) Colla + Oid
(C) Co + Lloid (D) None of these

569. Critical moisture point is equal to:
(A) Field capacity
(B) Permanent wilting point
(C) Sticky point
(D) None of these

570. Colloidal particles have:
(A) Positive charge
(B) Negative charge
(C) Both of these
(D) None of these

571. Zeta potential refers to:
(A) Magnitude of electric charges of colloidal particles
(B) Size of colloidal particles
(C) Movement of colloidal particles
(D) None of these

572. Browning movement can be seen in:
(A) Colloidal particles
(B) Crystalloid particles
(C) Urea
(D) None of these

573. Which of the following is not a colloidal material?
(A) Starch (B) Zelatin
(C) Fog (D) Sugar

574. What is the diameter of colloidal particles:
(A) 0-4 mμ (B) 5-100 mμ
(C) 101-200 mμ (D) None of these

575. Stem nodulation occurs in green manue crop of:
(A) *Sesbania aculeate*
(B) *Sesbania cannabina*
(C) *Crotolaria juncea*
(D) *Aeschynomene afraspera*

576. The uptake of which element is affected least under waterlogged conditions:
(A) N (B) P
(C) K (D) Ca

577. The uptake of which element is affected most under waterlogged conditions:
(A) N (B) P
(C) K (D) Ca

578. Which of the following harmful gas is produced under waterlogged conditions?
(A) Ethane (B) H_2S
(C) Methane (D) All of these

579. Which of the following disease incidence is increased under waterlogged conditions?
(A) Red rot of sugarcane
(B) Boll shedding of cotton
(C) Wilt of indigo
(D) All of these

580. Hexadecandole is a chemical used in soil particularly for:
(A) To increase evaporation
(B) To reduce evaporation
(C) To control runoff
(D) All of these

581. Higher salt concentration in soil solution increases osmotic pressure of solution but the available water will be:
(A) Increased (B) Decreased
(C) Remain same (D) None of these

582. What is the water tension of saturated flow of soil water?
(A) 1/3 bar (B) 15 bar
(C) 0 bar (D) None of these

583. Soil water flow is governed by:
(A) Conductivity of the medium
(B) Driving force of flow
(B) Proportionality factor
(D) All of these

584. Which of the following soil has the highest rate of saturated soil water flow?
(A) Sandy soil (B) Loamy soil
(C) Clay (D) None of these

585. Saturated soil water flow is inversely related to:
(A) Viscosity of liquid
(B) Pressure difference
(C) Radius of tube (D) All of these

586. According to Darcy's law, flow velocity of any liquid is directly proportional to:
(A) Hydraulic conductivity
(B) Length of porous medium
(C) Viscosity of liquid
(D) All of these

587. Unsaturated water flow can be understood by:
(A) Darcy's law (B) Poiseuille law
(C) Capillary principle
(D) All of these

588. Soil water is lost by:
(A) Evaporation (B) Transpiration
(C) Runoff (D) All of these

589. Which of the following elements become toxic to plants under waterlogged conditions?
(A) Fe and Mn (B) B and Mo
(C) N and P (D) N and K

590. Which of the following cation is not found under reduced conditions:
(A) NO_3 (B) Fe^{++}
(C) Mn_{++} (D) NH_4^+

591. CO_2 concentration in the atmosphere is:
(A) 1 % (B) 2 %
(C) 3 % (D) 0.03 %

592. Which of the following clay mineral absorbs less water?
(A) Kaolinite (B) Montmorillonite
(C) Vermiculite (D) None of these

593. Which of the following ion may cause more water absorption on soil colloids?
(A) Na^+ (B) Ca^{++}
(C) H^+ (D) All of these

594. Attraction force between two water molecules is known as:
(A) Cohesion (B) Adhesion
(C) Electromagnetic force
(D) None of these

595. Attraction force between two different molecules is known as:
(A) Cohesion (B) Adhesion
(C) Electromagnetic force
(D) None of these

596. Micelles or micro cells are:
(A) Micro clay colloids particles
(B) Micro-cells
(C) Micro organisms
(D) None of these

597. CO_2 concentration in soil is about:
(A) 0.03 % (B) 21.0 %
(C) 0.2-5.5 % (D) 78 %

598. The soil colloids of humid region have:
(A) Ca^{++} - H^+ complexes
(B) Ca^{++} - Mg^{++} complexes
(C) Na^+ - Ca^{++} complexes
(D) None of these

599. At what pH value, phosphate availability is the highest in the soil:
(A) 5.5 (B) 6.5
(C) 7.5 (D) 8.5

600. Which of the following methods is used to measure soil moisture?
(A) Tensiometer
(B) Neutron moisture meter
(C) Zypsum block
(D) All of these

601. The source of fast neutrons in neutron moisture meter?
(A) H^{3+} (B) P^{32}
(C) Actinium + Baryllium
(D) All of these

602. Average O_2 concentration in soil is about:
(A) 0.03 % (B) 21.0 %
(C) 15.5 % (D) 78 %

603. Which of the following soil type has higher per cent of CO_2?
(A) Sandy soil (B) Loam soil
(C) Clay soil (D) None of these

604. With increasing soil depth, CO_2 concentration will be:
(A) Increased (B) Decreased
(C) No changed (D) Unknown

605. In comparison to fallow land, CO_2 concentration in cropped land will be:
(A) Less (B) More
(C) Same (D) Unknown

606. Which of the following is a kaolinite?
(A) $Al_4Si_4O_{10}(OH)_8$
(B) $Al_4Si_8O_{20}(OH)_4$
(C) $Al_2Si_4O_{10}(OH)_2K$
(D) None of these

607. Higher silica-sesquioxide ratio of soil indicates that soil has:
(A) Higher CEC
(B) Higher WHC
(C) Higher soil fertility
(D) All of these

608. Which of the following minerals is important for formation of silicate minerals?
(A) Feldspars (B) Mica
(C) Amphibols (D) All of these

609. Which of the following mineral has the highest anion exchange capacity?
(A) Kaolinite (B) Montmorillonite
(C) Illite (D) Chlorite

610. Which of the following mineral has the largest crystal size?
(A) Kaolinite (B) Montmorillonite
(C) Illite (D) Vermiculite

611. Which of the clay mineral has higher amount of K?
(A) Kaolinite (B) Montmorillonite
(C) Illite (D) Vermiculite

612. Which of the following mineral has the highest CEC?
(A) Kaolinite (B) Montmorillonite
(C) Illite (D) Vermiculite

613. Which of the following is a 2 : 2 type mineral?
(A) Chlorite (B) Illite
(C) Kaolinite (D) None of these

614. Which of the following is a 1 : 1 type clay mineral?
(A) Kaolinite (B) Halloysite
(C) Anauxite (D) All of these

615. Which of the following is a 2 : 1 type expanding mineral?
(A) Vermiculite (B) Montmorillonite
(C) Beidellite (D) All of these

616. Which of the following is a 2 : 1 type non expanding mineral?
(A) Vermiculite (B) Montmorillonite
(C) Illite (D) None of these

617. The soil colloids of tropical and subtropical regions have more:
(A) Ca-H complexes
(B) Ca-Mg complexes
(C) Na-Ca complexes
(D) None of these

618. The soil colloids of arid region have more:
(A) Ca-H complexes
(B) Ca-Mg complexes
(C) Na-Ca complexes
(D) None of these

619. The chemical composition of clay is:
(A) Fixed (B) Dynamic
(C) Loose (D) None of these

620. Which of the following element is found in tetrahedral layer of clay crystals?
(A) Se^{4+} (B) Al^{3+}
(C) Fe^{3+} (D) Zn^{2+}

621. Which of the following element is found in Octahedral layer of clay crystals?
(A) Se^{4+} (B) Al^{3+}
(C) Fe^{3+} (D) Zn^{2+}

622. Which of the following is a montmorillonite?
(A) $Al_4Si_4O_{10}(OH)_8$ (B) $Al_4Si_8O_{20}(OH)_4$
(C) $Al_2Si_4O_{10}(OH)_2K$ (D) None of these

623. List A has soil groups and B main clay minerals as below:

List A	List B
1. Alluvial soil	i. Illite
2. Black soil	ii. Montmorillonite
3. Red soil	iii. Kaolinite and Illite
4. Hilly soils	iv. Illite and chlorite

Which of the pair is correctly matched?
(A) 1. i. 2. ii. 3. iii. 4. iv.
(B) 1. i. 2. iv. 3. iii. 4. ii.
(C) 1. ii. 2. iii. 3. iv. 4. iii.
(D) 1. iv. 2 ii. 3. iii. 4. i.

624. List A has soil orders and B main clay minerals as below:

List A	List B
1. Mollisol	i. Montmorillonite & Illite
2. Vertisol	ii. Montmorillonite
3. Inceptisol	iii.Illite
4. Alfisol	iv.Kaolinite & Montmorillonite

Which of the pair is correctly matched?
(A) 1. i. 2. ii. 3. iii. 4. iv.
(B) 1. i. 2. iv. 3. iii. 4. ii.
(C) 1. ii. 2. iii. 3. iv. 4. iii.
(D) 1. iv. 2. ii. 3. iii. 4.i.

625. Base saturated soils are those which have fully saturated soil colloids with:
(A) Ca^{2+} (B) Mg^{2+}
(C) H^+ (D) $Al(OH)^+_2$

626. Arid soils have higher soil pH because of:
(A) Higher cations on soil colloids
(B) Lower cations on soil colloids
(C) Higher anions on soil colloids
(D) All of these

627. If the base saturation of a soil is more the pH of soil will be:
(A) More (B) Low
(C) Normal (D) None of these

628. Which of the soil has higher CEC?
(A) Alluvial soil (B) Black soil
(C) Laterite soil (D) Red soil

629. In general, which of the following cation has higher concentration in soil?
(A) Ca^{2+} (B) Mg^{2+}
(C) H^+ (D) Na^+

630. Which of the following anion has higher concentration in soil?
(A) SH^- (B) Cl^-
(C) NO_3^- (D) $H_2PO_4^-$

631. Base saturation of soil can be calculated by which of the under given formula?
(A) $\frac{\mathbf{100 \times total\ exchangeable\ cations}}{\mathbf{CEC}}$
(B) $\frac{100 \times \text{total exchangeable anions}}{\text{CEC}}$
(C) $\frac{100 \times \text{CEC}}{\text{Total exchangeable cation}}$
(D) All of these

632. Ion exchange is a:
(A) Static process
(B) Reversible process
(C) Only cations can be exchanged
(D) Only anions can be exchanged

633. Which of the soil has higher silica sesquioxide ratio?
(A) Black soil (B) Red soil
(C) Laterite soil (D) Sandy soil

634. Loss of NH_3 will be more at pH:
(A) <6.0 (B) <7.0
(C) <8.0 (D) None of these

635. In which form of nitrogen is lost at <6 soil pH:
(A) N_2 (B) N_2O
(C) NO (D) None of these

636. Volatilization is a process in which:
(A) Loss of nitrogen in form of N_2 & N_2O
(B) Conversion of NH_4-N into NO_3-N
(C) Conversion of NO_3-N into NH_4-N
(D) None of these

637. Denitifying bacterias are:
(A) Aerobic (B) Anaerobic
(C) Both of these (D) None of these

638. Denytrifying bacteria belongs to genera:
(A) Pseudomonas (B) Acromobactor
(C) Bacillus (D) All of these

639. Which of the following genera of bacteria is found mainly in soil?
(A) Pseudomonas and Achromobacter
(B) Thiobacillus (C) Vibrio
(D) All of these

640. Denitrifying bacteria uses NO_3 as an electron acceptor in absence of:
(A) CO_2 (B) O_2
(C) H_2O (D) None of these

641. Optimum temperature for denitrification is:
(A) <25^{0}C (B) >25^{0}C
(C) 25-30 ^{0}C (D) >30^{0}C

642. The organic residues with wider C:N ratios as compared to narrow C:N ratios are decomposed at:
(A) Faster rate (B) Slow rate
(C) Equal rate
(D) Faster rate followed by slow rate

643. Loss of nitrogen under neutral soil pH occurs mainly in the form of:
(A) NO_2 (B) NO_3
(C) N_2O (D) NO

644. The red colour of legume nodules is due to the presence of:
(A) Haemoglobin (B) Leghaemoglobin
(C) RBC (D) None of these

645. Symbiotic N fixation governed by the group of bacteria is:
(A) Rhizobia (B) Azotobacter
(C) Azospirillum (D) All of these

646. The main enzyme of non symbiotic N fixation is:
(A) Nitrogenase (B) Hydrogenase
(C) Isomerase (D) All of these

647. Nitrogen fixation is a:
(A) Endothermic process
(B) Exothermic process
(C) Hypothermic process
(D) None of these

648. Azotobacter are active at soil pH of:
(A) >6.5-7.0 (B) 7.0-7.5
(C) 7.5-8.0 (D) 5.5-6.0

649. Average quantum of N fixed by non-symbiotic process is:
(A) 10-15 kg/ha (B) 25-30 kg/ha
(C) 50-60 kg/ha (D) None of these

650. The maximum decomposition shall be at soil WHC of:
(A) <30 % (B) 40-60 %
(C) 60-80 % (D) >80 %

651. The rate of decomposition is the lowest in:
(A) Sugars & proteins (B) Hemi cellulose
(C) Cellulose (D) Lignin

652. What is the end product of aerobic decomposition of sugars?
(A) CO_2 & H_2O (B) Pyruvic acid
(C) Glucose (D) None of these

653. What is the end product of anaerobic decomposition of sugars?
(A) Alcohols and acetones
(B) CO_2 and H_2O
(C) Starch (D) None of these

654. Salts in the xylem ducts of the root are carried upward with:
(A) Photosynthesis
(B) Transpiration stream
(C) Respiration
(D) Guttation

655. Conversion of proteins and organic N into amino acids is known as:
(A) Amonification (B) Amminization
(C) Nitrification (D) None of these

656. Conversion of amino acids into NH_3 and ammonium compounds are known as:
(A) Amonification (B) Amminization
(C) Nitrification (D) None of these

657. Ammonification rate will be higher at:
(A) 7-8 pH, 30-40 ^{0}C and 60-80 % WHC
(B) 4-5 pH, 30-40 ^{0}C and 60-80 % WHC
(C) 7-8 pH, 20-25 ^{0}C and >80 % WHC
(D) None of these

658. Nitrification is a:
(A) Aerobic process
(B) Anaerobic process
(C) Both of these
(D) None of these

659. Which of the following statement is true?
(A) Nitrosomonas produces nitrite and nitrobacter produces nitrate
(B) Conversion rate of nirite to nitrate is faster than ammonium to nitrite
(C) Nitrosomonas oxidizes 35-40 mole NH_4, while nitrobacter 70-100 mole nitrite
(D) All of these

660. Nitrification rate is equal to:
(A) $\frac{\textbf{Quantity of NO}_3}{\textbf{Time}}$
(B) $\frac{\text{Quantity of NO}_2}{\text{Time}}$
(C) $\frac{\text{Quantity of NO}_2\text{ \& NO}_3}{\text{Time}}$
(D) None of these

661. Optimum soil pH for nitrification is:
(A) 5.5-6.0 (B) 6.5-8.0
(C) 8.5-9.0 (D) >9.5

662. Which of the following is harmful for nitrification process?
(A) Hyposulphite (B) Bacteria
(C) Presence of Mo (D) None of these

663. Conversion of nitrate into nitrite and ammonia is called:
(A) Dinitrification (B) Assimilation
(C) Nitrification (D) None of these

664. Conversion of NO_3 into N_2 and N_2O is called:
(A) Denitrification (B) Nitrification
(C) Mineralization (D) None of these

665. What is the percentage of total exchangeable cations of CEC in normal fertile soils?
(A) 50-60 % (B) 60-70 %
(C) 70-80 % (D) 80-90 %

666. What is the percent of exchangeable H^+ in normal fertile soils?
(A) <10 % (B) <20 %
(C) <30 % (D) <40 %

667. Ca-clay has soil pH:
(A) 6.5-7.5 (B) 4.5-5.5
(C) 8.0-8.5 (D) None of these

668. Which of the clay mineral has highest anion exchange capacity?
(A) Kaolinite (B) Montmorillonite
(C) Vermiculite (D) All of these

669. In general, the range of soil pH is:
(A) 5-7 (B) 2-5
(C) 4-10 (D) 7-12

670. Active acidity of soil is due to presence of:
(A) H^+ (B) Al^{3+} and H^+
(C) $Al(OH)^{+2}$ (D) All of these

671. Residual acidity of soil is due to presence of:
(A) H^+ (B) Al^{3+} and H^+
(C) $Al(OH)^{+2}$ (D) All of these

672. Which of the acidity is the highest in soils?
(A) Active (B) Exchangeable
(C) Residual (D) All of these

673. The availability of Ca and Mg to plants in humid region is reduced due to:
(A) Leaching of Ca and Mg
(B) Precipitation of Ca and Mg
(C) Increase in soil pH
(D) All of these

674. The availability of Ca and Mg to plants in arid region is reduced due to:
(A) Leaching of Ca and Mg
(B) Precipitation of Ca and Mg
(C) Increase in soil pH
(D) None of these

675. The highest availability of NO_3 –N to plants is at soil pH:
(A) 4.0-5.0 (B) 6.5-7.5
(C) 8.0-8.5 (D) None of these

676. Bacteria responsible for N fixation in soybean is:
(A) *R. phaseoli* (B) *R. glycicum*
(C) *R. japonicum* (D) *R. leguminoserum*

677. At which pH value, both ions H_2PO_4 and HPO_4 $^-$ are at equal concentration in soil solution?
(A) 5.0 (B) 6.0
(C) 7.0 (D) 8.0

678. S availability is highest at soil pH of:
(A) 5.0-6.0 (B) 7.0-8.0
(C) 6.0-8.0 (D) 8.0-9.0

679. Zinc availability is highest at soil pH of:
(A) 4.0 (B) 4.5
(C) 5.0 (D) 5.5

680. Exchangeable acidity of soil is due to presence of:
(A) H^+ (B) Al^{3+} and H^+
(C) $Al(OH)^{+2}$ (D) All of these

681. Which of the following has the highest capacity of N fixation?
(A) Rhizobia (B) Azotobacter
(C) Beijerinckia (D) Azospirillum

682. The rate of symbiotic N fixation is higher than non-symbiotic N fixation to the tune of:
(A) 2 times (B) 5/2 times
(C) 3 times (D) 4 times

683. Higher C:N ratio enhances the:
(A) N fixation (B) N reduction
(C) Denitrification (D) All of these

684. The optimum pH for rhizobium activity is:
(A) 4.5-5.5 (B) 5.5-6.0
(C) 5.5-7.0 (D) 7.5-8.5

685. P content in Phytin is:
(A) 16 % (B) 26 %
(C) 36 % (D) 46 %

686. Nucleoprotein contains P:
(A) 7-9 % (B) 10-16 %
(C) 1-5 % (D) 20-25 %

687. Phospholipids contains P:
(A) 7-9 % (B) 13-14 %
(C) 16-21 % (D) 23-28 %

688. S reduction is mainly caused by:
(A) Desulphovibrio (B) Azotobacter
(C) Azospirillum (D) None of these

689. Mo is required for the activity of:
(A) Azotobacter (B) Legume bacteria
(C) Vibrio (D) None of these

690. B is required mainly by:
(A) Azotobacter
(B) Azospirillum
(C) Legume bacteria
(D) None of these

691. Humus is a:
(A) Part of organic matter
(B) Mixture of complex compounds
(C) Amorphous and colloidy
(D) All of these

692. Which of the following is correct for humus?
(A) It contains negative charge
(B) It is acidic
(C) Insoluble in water
(D) All of these

693. Which of the following pair is correct?
(A) Humus : C-55-60 %
(B) Humus: N -3-6 %
(C) Humus: O_2 – 35-40 %
(D) All of these

694. C : N ratio of humus is:
(A) 10 : 1 (B) 20 : 1
(C) 30 : 1 (D) 40 : 1

695. CEC of humus is:
(A) 150-300 C mol/kg
(B) 50-100 C mol/kg
(C) 250-300 C mol/kg
(D) All of these

696. Average C : N ratio of arable soil is:
(A) 10-12 : 1 (B) 20 : 1
(C) 80 : 1 (D) None of these

697. C : N ratio of arid soils in comparison to other arable soils is always:
(A) Higher (B) Lower
(C) Same (D) None of these

698. C : N ratio of hilly soils is:
(A) 9.2-13.7 : 1 (B) 6.4-11.4 : 1
(C) 20 : 1 (D) None of these

699. C : N ratio of legume fodder is:
(A) 10 : 1 (B) 25 : 1
(C) 80 : 1 (D) 225 : 1

700. C : N ratio of cereal straw is:
(A) 10 : 1 (B) 25 : 1
(C) 80 : 1 (D) 225 : 1

701. C : N ratio of saw dust is:
(A) 10 : 1 (B) 25 : 1
(C) 80 : 1 (D) 225 : 1

702. C : N ratio of FYM is:
(A) 20-30 : 1 (B) 10-12 : 1
(C) 80 : 1 (D) None of these

703. Accoding to modern classification, soils have been divided into:
(A) Order-Sub order-Great soil group-Sub group
(B) Order-soil group-family-series
(C) Order-family-series-types
(D) All of these

704. Soil orders have been divided into:
(A) Zonal – Interzonal - Azonal soils
(B) Zonal – Azonal – Series
(C) Zonal – Interzonal – Series – Azonal
(D) None of these

705. Which of the following soil type is not a interzonal soil?
(A) Hydromorphic soil
(B) Halomorphic
(C) Calcimorphic soil
(D) Alluvial soil

706. Calcimorphic soils have:
(A) Higher lime content
(B) Higher Na content
(C) Low lime content
(D) None of these

707. Azonal soils have three following great soil groups:
(A) Lithosol – Regosol – Alluvial
(B) Podzol – Planosol – Chernozems
(C) Latosol – Podzol – Lithosol
(D) None of these

708. Which is correct for Regosols?
(A) Soil profiles are not well developed
(B) Soils are made of marls, sand dunes and ash
(C) Parent materials affect the characters of soil
(D) All of these

709. Chernozems are like:
(A) Black cotton soils
(B) Red soils
(C) Brown soils
(D) Alluvial soils

710. Soil classification 7th approximation was developed in year:
(A) 1960 in America
(B) 1950 in America
(C) 1960 in UK
(D) 1960 in Russia

711. Epipedons are:
(A) Diagnostic surface horizons
(B) Diagnostic subsurface horizons
(C) Bed rocks
(D) None of these

712. Number of soil epipedons are:
(A) 6 (B) 10
(C) 16 (D) 21

713. As per new soil classification – 7th approximation, soil has been classified into following 6 categories
(A) Order – Suborder - Great soil group – Family - Series
(B) Order – Soil group – Family – Series – Type - Phase
(C) Order –Soil group – Family – Type – Series - Phase
(D) None of these

714. Number of soil orders are:
(A) 10 (B) 11
(C) 12 (D) 13

715. Which of the following soil order covers the highest area in the world?
(A) Vertisol (B) Entisol
(C) Aridisol (D) Alfisol

716. Usar was derived from 'Ustra word. Ustra is a word of:
(A) Latin (B) Greek
(C) Sanskrit (D) Bangla

717. Which of the following state has highest area under saline and alkaline soils:
(A) UP (B) Punjab
(C) Gujarat (D) Rajasthan

718. Finger and Toe disease in crucifer is caused in:
(A) Acidic soil (B) Neutral soil
(C) Alkaline soil (D) Saline soil

719. Total value of pH and POH is equal to:
(A) 7 (B) 10
(C) 14 (D) 21

720. pH word was first used by:
(A) Sorenson (B) Scofield
(C) Brady (D) Tisdale

721. The soil pH value of humid region is always:
(A) <7.0 (B) >7.0
(C) 7.0 (D) None of these

722. The fungi are highly active at soil pH of:
(A) Acidic (B) Neutral
(C) Alkaline (D) Saline

723. The highest activities of actinomycetes is found at soil pH of:
(A) Acidic
(B) Neutral to slight alkaline
(C) Alkaline
(D) All of these

724. Mo availability is higher under:
(A) Acidic condition (B) Neutral
(C) Alkaline (D) Saline

725. Fe availability is higher at soil pH of:
(A) 5.5 – 7.0 (B) 7.0 – 8.5
(C) 3.5 – 5.5 (D) None of these

726. B availability is highest at soil pH of:
(A) 3.5-5.5 (B) 5.5-8.0
(C) 8.0-8.5 (D) 9.0-9.5

727. Which of the following word is used for saline and alkaline soils?
(A) Rakar (B) Usar
(C) Kaller (D) All of these

728. Saline soil is also known as:
(A) White alkali soil
(B) Black alkali soil
(C) Solonetz soil
(D) All of these

729. Alkali soil is also known as:
(A) White alkali soil
(B) Black alkali soil
(C) Usar soil
(D) All of these

730. Which of the following is the characteristics of saline soils?
(A) EC>4 ds/m, ESP<15, pH<8.5
(B) EC>4 ds/m, ESP>15, pH<8.5
(C) EC<4 ds/m, ESP>15, pH>8.5
(D) None of these

731. Which of the following is the characteristics of alkali soils?
(A) EC>4 ds/m, ESP<15, pH<8.5
(B) EC>4 ds/m, ESP>15, pH<8.5
(C) EC>4 ds/m, ESP>15, pH>8.5
(D) None of these

732. Which of the following is the characteristics of saline alkali soils?
(A) EC>4 ds/m, ESP<15, pH<8.5
(B) EC>4 ds/m, ESP>15, pH<8.5
(C) EC<4 ds/m, ESP<15, pH>8.5
(D) None of these

733. Which of the anion is not found in saline soils?
(A) CO_3^- (B) Cl^-
(C) SO_4^- (D) HCO_3^-

734. Saline and alkaline soils are mainly found in:
(A) Arid and semiarid region
(B) Humid region
(C) Temperate region
(D) All of these

735. Which of the following chemical is used for reclamation of saline and alkaline soils?
(A) Zypsum (B) Phosphozypsum
(C) El.Sulphur (D) All of these

736. Saline and alkaline soils were classified by:
(A) D. Sigmoid (B) Arnon
(C) Wilcox (D) None of these

737. Which of the following process takes place in the development of saline and alkaline soils?
(A) Salinazation (B) Alkalization
(C) Salodization (D) All of these

738. At which salt concentration in soil solution, the water absorption by plants is reduced?
(A) 0.5 % (B) 1.0 %
(C) 1.5 % (D) 2.0 %

739. The water absorption by plants is completely checked at salt concentration in soil solution:
(A) 10 % (B) 20 %
(C) 30 % (D) 40 %

740. Which of the following is more toxic to roots?
(A) Nacl (B) Na_2SO_4
(C) Na_2CO_3 (D) $NaHCO_3$

741. The critical limit of boron in soil is:
(A) 0.7 ppm (B) 1.0 ppm
(C) 1.5 ppm (D) 2.0 ppm

742. Which of the following method is used for pH determination?
(A) Hydrogen electrode method
(B) Glass electrode method
(C) Calorimetric method
(D) All of these

743. Which of the following type of soil has the largest area?
(A) Alluvial soil
(B) Black cotton soil
(C) Red soil
(D) Laterite and lateritic soil

744. Old alluvial plains are known as:
(A) Bangar (B) Khadar
(C) Regur (D) None of these

745. New alluvial plains are known as:
(A) Bangar (B) Khadar
(C) Regur (D) None of these

746. Alluvial soils are generally deficient in:
(A) N (B) Humus
(C) P (D) All of these

747. Black cotton soils are black due to presence of:
(A) Titaniferrous magnetite
(B) $MgCO_3$
(C) $CaCO_3$
(D) All of these

748. Karail is a:
(A) Type of black soils
(B) Black soils of lower lands of both banks of Ganga in UP
(C) These soils have high lime and Mg
(D) All of these

749. Which of the following is correct for black cotton soils?
(A) Excess in lime and K_2O
(B) Deficient in N, P_2O_5 and organic matter
(C) Excess in Fe, $CaCO_3$ and $MgCO_3$
(D) All of these

750. Which of the following is correct for red soils?
(A) Made of gneiss, schist & granite rocks
(B) Deficient in N, humus, P and lime
(D) Excess of Fe and Al oxides
(D) All of these

751. Which of the following is correct for laterite soils?
(A) Lower layers ar made of Al & Fe oxides
(B) Deficient in P, K and lime
(C) Generally rich in organic matter
(D) All of these

752. Rakar is a:
(A) Red soil (B) Black soil
(C) Alluvial soil (D) None of these

753. Which of the following is correct for Tarai soils?
(A) Deficient in Zn
(B) High organic matter
(C) High soil fertility
(D) All of these

754. Arid and desert soils are characterized by:
(A) High evaporation
(B) Low organic matter
(C) High Ca CO_3
(D) All of these

755. Efflorescent deposits are found in:
(A) Saline and alkaline soils
(B) Alluvial soils
(C) Black soils (D) None of these

756. Which of the following is correct for peat and marshy soils?
(A) Low pH
(B) High organic matter
(C) Blue colour due to Fe^{++}
(D) All of these

757. Dhakar soils are çharacterized as:
(A) Good for rice soils
(B) Found in eastern UP
(C) Light acidic soils
(D) All of these

758. Saline and alkali soils are characterized as:
(A) Excess of Na_2CO_3 and NaCl
(B) Absence of $CaSO_4$
(C) Hard rock at subsoil
(D) All of these

759. In general, Indian soils have been classified into:
(A) 4 types (B) 6 type
(C) 8 types (D) None of these

760. Which of the following order of Indian soil has highest area:
(A) Alfisols (B) Entisols
(C) Inceptisols (D) Vertisols

761. Who classified Indian soils into 25 great soil groups?
(A) Govindarajan (1965) and Dutta Vishwash (1968)
(B) FAO (1965)
(C) ICAR (1937)
(D) Brady (1972)

762. Black soils are classified in order:
(A) Vertisols (B) Alfisols
(C) Spodosol (D) Histosols

763. Red soils are classified in order:
(A) Vertisols (B) Alfisols
(C) Spodosol (D) Histosols

764. Which of the following epipedon is present in mollisols?
(A) Mollic (B) Umbric
(C) Ochric (D) Histic

765. Which of the following is correct for histosols?
(A) They are bog and half bog soils
(B) They are developed under waterlogged conditions
(C) They contain 20-30% organic carbon
(D) All of these

766. Soil monoliths are:
(A) Permanent soil profile records
(B) Temporary soil profile records
(C) Types of soil classification
(D) None of these

767. Soil is divided on the basis of land capability into:
(A) 4 classes (B) 6 classes
(C) 8 classes (D) 10 classes

768. Land capability classification is based on:
(A) Utility of land
(B) Soil erosion
(C) Techniques to protect soil
(D) All of these

769. Which of the class is unfit for agricultural uses:

(A) Class I (B) Class IV
(C) Class VIII (D) Class VI

770. The categories of soil classification and its number are given in 'A' and 'B' list below:

A	B
1. Orders	i. 54
2. Suborders	ii. 11
3. Greatgroups	iii. 254
4. Subgroups	iv. 1200

Which of the pair is matched correctly:

(A) 1. i 2. ii 3. iii 4. iv
(B) 1. iii 2. ii 3. iv 4. i
(C) 1. ii 2. i 3. iii 4. iv
(D) 1. iv 2. ii 3. i 4. iii

771. List 'A' and 'B' represent the land capability clas-ses and colour on soil maps, respectively below:

A	B
1. Class I	i. Light green
2. Class II	ii. Yellow
3. Class III	iii. Red
4. Class IV	iv. Blue

Which of the pair is matched correctly:

(A) 1. i 2. ii 3. iii 4. iv
(B) 1. iii 2. ii 3. iv 4. i
(C) 1. ii 2. i 3. iii 4. iv
(D) 1. iv 2. ii 3. i 4. iii

772. Which soil has the highest cation exchange capacity:

(A) Loam (B) Loamy sand
(C) Sandy loam (D) Clay loam

773. Na per cent in list A and irrigation water quality groups in B are given below:

A	B
1. <20 % Na	i. Excellent
2. 20-40 % Na	ii. Good
3. 40-60 % Na	iii. Permissible
4. >60 % Na	iv. Unsuitable

Which of the pair is matched correctly:

(A) 1. i 2. ii 3. iii 4. iv
(B) 1. iii 2. ii 3. iv 4. i
(C) 1. ii 2. i 3. iii 4. iv
(D) 1. iv 2. ii 3. i 4. iii

774. List 'A' has SAR values and 'B' the quality of irrigation water below:

A	B
1. 0-10	i. Low Na
2. 10-18	ii. Medium Na
3. 18-26	iii. High Na
4.>26	iv. Very high Na

Which of the pair is matched correctly:

(A) 1. i 2. ii 3. iii 4. iv
(B) 1. iii 2. ii 3. iv 4. i
(C) 1. ii 2. i 3. iii 4. iv
(D) 1. iv 2. ii 3. i 4. iii

775. Which of the following is matched correct:

A	B
1. 0-250 micro mhos/cm	i. Low saline water
2. 250-500 micro mhos/cm	ii.Moderate saline water
3. 250-750 micro mhos/cm	iii.Moderate to high saline water
4. 750-2250 micro mhos/cm	iv. High saline water

Which of the pair is matched correctly:

(A) 1. i 2. ii 3. iii 4. iv
(B) 1. iii 2. ii 3. iv 4. i
(C) 1. ii 2. i 3. iii 4. iv
(D) 1. iv 2. ii 3. i 4. Iii

776. Which of the biological agent is used for amelioration of saline and alkali soils?

(A) FYM
(B) Molasses
(C) *Argemone mexicana*
(D) All of these

777. Which of the following grass is tolerant to alkali soils?

(A) Nandi grass (B) Rhodes grass
(C) Dinanth grass (D) All of these

778. Irrigation water is classified on the basis of Na per cent by:

(A) Wilcox
(B) Staff Richards
(C) Thorne and Peterson
(D) Kanwar

779. Gypsum contains Ca and S:

(A) 23 % Ca and 18 % S
(B) 12 % Ca and 12 % S
(C) 40 % Ca and 12 % S
(D) 40 % Ca and 20 % S

780. Pyrite contains S:
(A) 22-24 % (B) 12-18 %
(C) 28-30 % (D) 30-32 %

781. 1 ds/m EC is equal to:
(A) 640 ppm salt concentration
(D) 64 ppm salt concentration
(C) 6.4 gm/litre
(D) None of these

782. Sodium percentage is equal to:

(A) $\frac{Na^+}{Ca^{++} + Mg^{++} + K^+ + Na^+} \times 100$

(B) $\frac{Na^+ \times 100}{Ca^{++} + Mg^{++}}$

(C) $\frac{\sqrt{Na^+} + 100}{Ca^{++} + Mg^{++}}$

(D) None of these

783. SAR is equal to:

(A) $\frac{Na^+}{\frac{\sqrt{Ca^{++} + Mg^{++}}}{2}}$

(B) $\frac{Na^+}{\sqrt{Ca^{++} + Mg^{++}}}$

(C) $\frac{Na^+}{\sqrt{Ca^{++} + Mg^{++} + K^+ + Na^+}}$

(D) None of these

784. Lime requirement for acid soils was worked out by:
(A) Shoemaker etal (1961)
(B) Kanwar (1972)
(C) Arnon (1961)
(D) Biswas (1965

785. Over liming may cause:
(A) P availability reduced
(B) Fe, Mn, Cu and Zn availability reduced
(C) Adverse effect on soil particles
(D) All of these

786. Which of the following can be used for improvement of acid soils?
(A) NH_4^+ (B) NO_3^-
(C) Ca^{++} (D) Mg^{++}

787. Generally acid soils have pH range of:
(A) <4.0 (B) 4.0-6.8
(C) 6.5-7.5 (D) >7.5

788. Acid soils generally found in:
(A) Humid region
(B) Arid region
(C) Temperate region
(D) Tundra region

789. Which of the crop is sensitive to boron concentration in irrigation water?
(A) Grape (B) Apple
(C) Lemon (D) Sugarbeet

790. The safe limit of K^+ ions in irrigation water is:
(A) <5 me/l (B) 5-10 me/l
(C) 10-20 me/l (D) >20 me/l

791. The safe limit of NO_3^- ions in irrigation water is:
(A) 2.5-5.0 me/l (B) 5-10 me/l
(C) 10-15 me/l (D) >15 me/l

792. If residual sodium carbonate (RSC) is more than 2.50 me/l, the quality of irrigation water shall be:
(A) Safe (B) Marginal
(C) Unsafe (D) All of these

793. The good quality of irrigation water should have HCO_3^- quantity in water:
(A) Less than Ca^{++} and Mg^{++} quantity
(B) More than Ca^{++} and Mg^{++} quantity
(C) More than Ca^{++} quantity
(D) More than Mg^{++} quantity

794. Which of the following can be used for improvement of acid soils?
(A) Marl (B) Chalk
(C) Quick lime (D) All of these

795. Average P in soils is:
(A) 0.02 – 0.5 % (B) 0.5 – 1.5 %
(C) 1.6 – 2.2 % (D) 2.0-2.5 %

796. Plant contains N:
(A) 1-4 % of total dry weight
(B) 0.5-0.8% of total dry weight
(C) 5-7 % of total dry weight
(D) None of these

797. Excess N causes:
(A) Soft stem (B) High growth
(C) Late maturity (D) All of these

798. N deficiency causes:
(A) Chlorosis
(B) Stunted growth
(C) Reduced tillering
(D) All of these

799. Average N per cent in arable soils is:
(A) 0.02 – 0.10 % (B) 0.2 – 0.3 %
(C) 0.5 – 0.7 % (D) 1.0 - 1.5 %

800. Na is essential for:
(A) Green algae (B) Sugar beet
(C) Diatoms (D) Sugarcane

801. Va is essential for:
(A) Green algae (B) Sugar beet
(C) Rice (D) Wheat

802. Role of Mo is:
(A) Sugar synthesis
(B) NH_3 and NO_3 reduction
(C) N fixation by free living bacteria
(D) All of these

803. Role of Cu is:
(A) Synthesis of Indole acetic acid
(B) Control respiration
(C) Synthesis of chlorophyll
(D) All of these

804. Role of Zn is:
(A) Control disintegration of RNA
(B) Water absorption
(C) Sugar translocation
(D) All of these

805. Role of B is:
(A) Translocation of sugars
(B) Pollen formation
(C) N absorption
(D) All of these

806. Which of the following element was included recently in the list of essential plant nutrients?
(A) Co (B) Na
(C) Ni (D) Se

807. Plant requires micronutrients:
(A) 1 ppm (B) < 1ppm
(C) > 1 ppm (D) None of these

808. Auxin synthesis is catalysed by element:
(A) N (B) P
(C) Mo (D) S

809. Which of the following element helps in fat and lipid metabolism?
(A) N (B) P
(C) K (D) Zn

810. Drought resistance is increased by element:
(A) N (B) P
(C) K (D) Mo

811. Which of the following element is required for oil synthesis?
(A) N (B) P
(C) K (D) S

812. Alkali soils are deficient in:
(A) Ca (B) N
(C) N and Ca (D) None of these

813. Soil fertility refers to:
(A) Capacity of soil to provide all essential elements in balanced amounts to plants
(B) Availability of N, P and K elements
(C) High crop productivity
(D) None of these

814. Soil productivity refers to:
(A) Capacity of soil to produce high economic yield
(B) Capacity of soil to provide essential plant nutrients to plant
(C) Capacity of soil to produce high soil organisms
(D) All of these

815. Log phase of growth indicates:
(A) Slow growth rate
(B) It is observed at early plant growth stage
(C) Decreasing growth rate with time
(D) All of these

816. Log plant growth phase refers to:
(A) Maximum growth rate
(B) Increasing growth rate with time
(C) Maximum dry matter accumulation
(D) All of these

817. Who established the relationship between crop yield and growth factor?
(A) Wilcox
(B) Mitscherlich
(C) Brown
(D) None of these

818. The Baule unit refers to:
(A) Unit of any growth factor that provides 50% yield of expected yield
(B) It was proposed by German Scientist, Baule
(C) It varies with inputs
(B) All of these

819. Total essential plant nutrients are:
(A) 15 (B) 16
(C) 17 (D) 18

820. Plant nutrient contributes the total plant dry matter approximately:
(A) 2-5 % (B) 5-10 %
(C) 12-16 % (D) 17-19 %

821. Inverse nitrogen yield concept was given by:
(A) O.W. Wilcox (B) Mitscherlish
(C) Spillman (D) Tisdale

822. The optimum temperature range for most of the crops in:
(A) 10-25 ^{0}C (B) 15-40 ^{0}C
(C) 35-45 ^{0}C (D) > 45 ^{0}C

823. Mitscherlich's equation is:
(A) dy/dx=(A-Y)C
(B) Y = K/n
(C) Y=CX – A
(D) None of these

824. The baule units in lbs/acre for N, P_2O_5 and K_2O, respectively are:
(A) 218, 45, 70 (B) 223, 45, 76
(C) 255, 50, 75 (D) 280, 55, 80

825. Which of the following crop is not tolerant to acid soils?
(A) Berseem (B) Potato
(C) Tea (D) Rice

826. Lime requirement of sandy soil in comparison to clay soil at a specific soil pH will be:
(A) More (B) Less
(C) Equal (D) None of these

827. How much dry soil is produced by earthworms annually per acre?
(A) 10 tonnes (B) 15 tonnes
(C) 20 tonnes (D) None of these

828. What will be the effect of earthworms on C/N ratio of soil?
(A) Increase (B) Decrease
(C) Same (D) All of these

829. Among soil organisms, earthworms account nearly:
(A) 50-75 % weight of total soil organisms
(B) 75-85 % weight of total soil micro organisms
(C) > 85 % weight of total soil micro organisms
(D) None of these

830. Earthworms increase the highest availability of:
(A) K (B) Mg
(C) N (D) P

831. Number of earthworms may be present per m^2 area of arable land is:
(A) 30-300 (B) 500-800
(C) 1000-12000 (D) None of these

832. At which soil pH, the activity of earthworms is highest?
(A) <4.5 (B) 4.5-5.5
(C) 5.5-6.6 (D) 7.0-8.5

833. Which of the following reduces the population of earthworms:
(A) Drought (B) Frost
(C) Cold (D) All of these

834. Pasture lands and non-ploughed soils have earthworms population in comparison to ploughed soils:
(A) 6.5-11 % higher (B) 12-15 % higher
(C) 16-18 % higher (D) > 20 % higher

835. Which of the following is not a soil microorganisms?
(A) Nematodes (B) Protozoans
(C) Rotifers (D) Sludge and snails

836. Rotifers are:
(A) Parasite (B) Saprophytes
(C) Predators (D) None of these

837. Higher population of rotifers are available in:
(A) Dry region (B) Temperate region
(C) Humid region (D) None of these

838. Mycorrhiza is a
(A) Association of fungi with roots of higher plants
(B) Association of fungi and bacteria
(C) Association of bacteria & higher plants
(D) None of these

839. Maximum decomposition shall occur at temperature:
(A) 20-25^{0}C (B) 25-30^{0}C
(C) 30-40^0 C (D) > 40^0 C

840. If the C/N ratio is high in organic matter, the rate of decomposition will be:
(A) Increased (B) Reduced
(C) Unchanged (D) None of these

841. Which of the following is found highest in plant body?
(A) Sugar and Starch (B) Hemicellulose
(C) Cellulose (D) Lignin

842. Peat soils may contains organic matter:
(A) 20-95% (B) 0.1-5 %
(C) 5-10 % (D) >10 %

843. Mineral soils may contains organic matter:
(A) 1-5 % (B) 5-10 %
(C) 10-15 % (D) >20 %

844. Decomposition of organic matter in submerged soil is carried out by:
(A) Bacteria (B) Actinomycetes
(C) Fungi (D) Earthworm

845. The carbohydrates produced in leaves can not be translcated to different growing parts of plant in the absence of:
(A) Manganese (B) Boron
(C) Zinc (D) Iron

846. Optimum soil pH for nitrogen fixing bacteria is:
(A) 4.5-5.5 (B) 5.5-6.5
(C) 6.5-8.0 (D) >8.0

847. Mycorrhiza is generally found in:
(A) N and P deficient soils
(B) P and K deficient soils
(C) Ca deficient soils
(D) All of these

848. Highest activity of actinomycetes is found at soil pH:
(A) 4.5-5.5 (B) 5.5-6.5
(C) 6.5-8.0 (D) >8.0

849. Which of the following is not a fungi?
(A) Nocardia (B) Mucor
(C) Rhizopus (D) Yeast

850. Which of the following is not a actinomycetes?
(A) Nocardia (B) Mucor
(C) Micromonospora (D) Streptomycetes

851. Potato scab disease is caused by:
(A) Bacteria (B) Fungi
(C) Actinomycetes
(D) Algae

852. Sweet potato pox is caused by:
(A) Bacteria
(B) Fungi
(C) Actinomycetes
(D) Algae

853. Which of the following is present highest in soil?
(A) Bacteria (B) Fungi
(C) Actinomycetes (D) Algae

854. Which of the following type of bacteria is the largest in size?
(A) Bacillus (B) Coccus
(C) Spirillum (D) None of these

855. Average P in plants is:
(A) 0.05-1.0 % (B) 1 – 3 %
(C) 3 – 5 % (D) 4 – 5 %

856. Highest P is found in:
(A) Leaf (B) Stem
(C) Seed (D) Root

857. Organic P is available in form of:
(A) Phytin (B) Nucleoproteins
(C) Phospholipids (D) None of these

858. Which of the following P ion is available in acidic conditions?
(A) $H_2PO_4^-$ (B) HPO_4^{--}
(C) PO_4^{---} (D) PO_3^{---}

859. Which of the following P ion is absorbed by plants?
(A) $H_2PO_4^-$ (B) HPO_4^{--}
(C) PO_4^{---} (D) PO_3^{---}

860. The lower leaves of maize plant develop violet green colour due to deficiency of:
(A) N (B) P
(C) K (D) Zn

861. K % in plant is:
(A) Nearly 2% of total dry weight
(B) <1 % of total dry weight
(C) >2% of total dry weight
(D) None of these

862. Highest K content is found in:
(A) Leaf (D) Stem
(C) Seed (D) Root

863. Yellow white mottling of cotton is caused by:
(A) K deficiency (D) N deficiency
(C) S deficiency (D) Zn deficiency

864. Necrosis on lower leaf margin is caused by:
(A) K deficiency (D) N deficiency
(C) S deficiency (D) Zn deficiency

865. Tea yellow is caused by:
(A) K deficiency (B) N deficiency
(C) S deficiency (D) P deficiency

866. Interveinal chlorosis on lower leaves is caused by:
(A) Mg (B) Ca
(C) P (D) Cl

867. Average Mg content in plant is:
(A) 0.5-1.5 % (B) 1.7-1.9 %
(C) 2.0-2.2 % (D) 2.5-2.7 %

868. Average S content in plant is:
(A) 0.01-0.1 % (B) 1.7-1.9 %
(C) 2.0-2.2 % (D) 2.5-2.7 %

869. Papery white is an important deficiency symptoms of:
(A) Fe (B) N
(C) Cu (D) Mo

870. Marsh spots in pea is caused by the deficiency of:
(A) N (B) Mg
(C) Mn (D) Cu

871. If the EC of soil is >4 ds m^{-1}, the soil is rated as:
(A) Normal (B) Injurious
(C) Sensitive (D) None of these

872. If the soil nutrient index is <1.67, the soil is rated as:
(A) Low fertile (B) Medium fertile
(C) High fertile (D) None of these

873. If the available P in soil test value is >24.6 kg P/ha, the soil is rated as:
(A) Low in available P
(B) Medium in available P
(C) High in available P
(D) None of these

874. Which of the following cakes is not edible?
(A) Castor cake (B) Mustard cake
(C) Til cake (D) Groundnut cake

875. If the available P in soil test value is 10.0-24.6 kg P/ha, the soil is rated as:
(A) Low in available P
(B) Medium in available P
(C) High in available P
(D) None of these

876. If the available P in soil test value is <10.0 kg P/ha, the soil is rated as:
(A) Low in available P
(B) Medium in available P
(C) High in available P
(D) None of these

877. If the available N in soil test value is >280 kg N/ha, the soil is rated as:
(A) Low in available N
(B) Medium in available N
(C) High in available N
(D) None of these

878. If the available N in soil test value is 280-560 kg N/ha, the soil is rated as:
(A) Low in available N
(B) Medium in available N
(C) High in available N
(D) None of these

879. If the available N in soil test value is >560 kg N/ha, the soil is rated as:
(A) Low in available N
(B) Medium in available N
(C) High in available N
(D) None of these

880. Metabolites play a significant role in ion uptake and process is known as:
(A) Passive uptake
(B) Active uptake
(C) Mass flow
(D) All of these

881. Ion uptake caused by electro-chemical potential is known as:
(A) Passive uptake (B) Active uptake
(C) Mass flow (D) All of these

882. Which of the following is not responsible for transport of nutrient ions from soil solution to root surface?
(A) Root inception (B) Mass flow
(C) Diffusion (D) Soil solution

883. Which of the following is a B rich mineral in soil?
(A) Tourmaline (B) Tenorite
(C) Powellite (D) None of these

884. Which of the following is a chelating material?
(A) EDTA (B) DPTA
(C) NTA (D) All of these

885. Which of the following is correct for chelating agent?
(A) It is a organic material
(B) It combines or attracts the cations
(C) It increases the availability of chelated nutrients
(A) All of these

886. Khaira disease of rice is caused by:
(A) K deficiency (B) N deficiency
(C) Zn deficiency (D) P deficiency

887. Little leaf or mottle leaf is caused by:
(A) K deficiency (B) Zn deficiency
(C) S deficiency (D) P deficiency

888. White bud of maize is the deficiency symptoms of:
(A) Zn deficiency (B) N deficiency
(C) S deficiency (D) Mg deficiency

889. Yellow spot of citrus is caused by:
(A) K deficiency (B) N deficiency
(C) Mo deficiency (D) P deficiency

890. Which of the following diseases are caused by B deficiency:
(A) Heart rot of sugar beet
(B) Yellow tip of alfalfa
(C) Brown heart of turnip
(D) All of these

891. Average values of N, P and K in tank silt is:
(A) 3, 0.3 and 0.3 %
(B) 0.1, 0.2 and 0.3 %
(C) 1.2, 0.5 and 0.5 %
(D) None of these

892. Which of the following is a concentrated organic manure?
(A) Dried bone meal
(B) Oilseed cakes
(C) Guano
(D) All of these

893. Optimum N : P : K ratio for cereal crop production is:
(A) 4 : 2 : 1 (B) 2 : 1 : 1
(C) 2 : 1 : 2 (D) 2 : 2 : 2

894. Optimum N : P : K ratio for pulses is:
(A) 4 : 2 : 1 (B) 2 : 1 : 1
(C) 2 : 1 : 2 (D) 2 : 2 : 2

895. Optimum N : P : K ratio for root crops is:
(A) 4 : 2 : 1 (B) 2 : 1 : 1
(C) 2 : 1 : 2 (D) 2 : 2 : 2

896. Muriate of potash is known as:
(A) KCl (B) K_2SO_4
(C) K_2CO_3 (D) None of these

897. S content in single super phosphate is:
(A) 12-16 % (B) 8-12 %
(C) 16-20 % (D) 11-12 %

898. Ca content in single super phosphate is:
(A) 18-21 % (B) 15-18 %
(C) 12-15 % (D) 11-12 %

899. Which of the following formula is correct?
(A) $P_2O_5 = P \times 2.29$
(B) $P_2O_5 = P \times 2.09$
(C) $P_2O_5 = P \times 1.29$
(D) $P_2O_5 = P \times 0.43$

900. Which clay mineral is rich in potash:
(A) Montmorillonite (B) Kaolinite
(C) Illite (D) Chlorite

901. Pelo phos phosphate contains P_2O_5:
(A) 16-17 % (B) 16-17 %
(C) 21-23 % (D) 24-26 %

902. Heavy application of potassium fertilizers often lead to reduced absorption of:
(A) Mg and Ca (B) B and Mn
(C) N, P and K (D) Zn and Cu

903. P_2O_5 in basic slage is:
(A) 14-18 % (B) 21-25 %
(C) 34-39 % (D) 41-42 %

904. The most important potential contaminant of food produced on sewage sludge amended soil is:
(A) Chromium (B) Iron
(C) Zinc (D) Cadmium

905. P_2O_5 in double super phosphate is:
(A) 16 % (B) 32 %
(C) 18 % (D) 46 %

906. The quantity of SSP (16% P_2O_5) required for application to one acre of rice field at a dose of 40 kg P_2O_5 ha^{-1} is:
(A) 200 kg (B) 250 kg
(C) 300 kg (D) 400 kg

907. Which of the following elements is not considered as fertilizer nutrient:
(A) C (B) N
(C) Ca (D) Zn

908. Amongst oil cakes, the highest nitrogen content is in:
(A) Caster cake (B) Neem cake
(C) Groundnut cake (D) Coconut cake

909. Maximum residual acidity is associate with the continuous soil application of:
(A) Urea (B) Ammonium nitrate
(C) Zinc (D) Ammonium sulphate

910. Sulpho nitrate is known as:
(A) Ammonium sulphate
(B) Calcium Ammonium Nitrate
(C) Ammonium Nitrate
(D) Ammonium Sulphate Nitrate

911. Which of the following nutrient should be applied by placement method?
(A) N (B) P
(C) K (D) P and K

912. Incomplete manure provide:
(A) Only one nutrient
(B) Only one or two nutrients
(C) Most of the plant nutrients
(D) None of these

913. Complete manure provide:
(A) Only one nutrient
(B) Only one or two nutrients
(C) Most of the plant nutrients
(D) None of these

914. Sodium nitrate is a:
(A) Alkaline fertilizer
(B) Acidic fertilizer
(C) Neutral fertilizer
(D) None of these

915. Ammonium sulphate is a:
(A) Alkaline fertilizer
(B) Acidic fertilizer
(C) Neutral fertilizer
(D) None of these

916. Which of the following element is not found in stimulative & prophylactic manures?
(A) N (B) B
(C) Mn (D) Mo

917. In foliar spray, the solution should not contain the biurate concentration?
(A) < 1 % (B) > 1 %
(C) 3 % (D) 4 %

918. Sona is a trade name of:
(A) Urea
(B) Calcium Ammonium Nitrate
(C) Ammonium Chloride
(D) Sulphate of Potash

919. Flame photometer is used for the estimation of:
(A) Available K (B) Available P
(C) Available N (D) Available Ca

920. Walkley and Black method is used for estimation of:
(A) N (B) P
(C) K (D) S

921. A value technique is used for estimation of:
(A) Available K in soil
(B) Available P in soil
(C) Available N in soil
(D) Available Ca in soil

922. Cunninghmella plague method is used for testing the soil value of:
(A) N (B) P
(C) K (D) S

923. Which of the following nutrient is mobile in soil?
(A) N (B) P
(C) K (D) S

924. List A and B indicate the nutrient and indictor plants that shows deficiency of respective nutrient respectively below:

List A	List B
1. N	i. Cauliflower
2. P	ii. Toria
3. K	iii. Potato
4. B	iv. Sunflower

Which of the pair is correctly matched

(A) 1. i	2. ii	3. iii	4. iv
(B) 1. iii	2. ii	3. iv	4.i
(C) 1. ii	2. i	3. iii	4.iv
(D) 1. iv	2. ii	3. i	4. iii

925. N-lignin contains N %:
(A) 18 (B) 20
(C) 22 (D) 25

926. Which of the following is a slow release N fertilizer?
(A) IBDU (B) CDU
(C) N lignin (D) All of these

927. Phosphatic fertilizers and its P_2O_5 content are given below:

Phosphatic fertilizers	P_2O_5 content (%)
1. Steamed bonemeal	i 22
2. Rock phosphate	ii. 20-40
3. Udaipur rock phosphate	iii. 30-32
4. Massori rock phospahte	iv. 10-20

Which of the pair is correctly matched
(A) 1. i 2. ii 3. iii 4. iv
(B) 1. iii 2. ii 3. iv 4.i
(C) 1. ii 2. i 3. iii 4. iv
(D) 1. iv 2. ii 3. i 4. iii

928. Which of the following is a nitrification inhibitor?
(A) N serve (B) AM
(C) Thiourea (D) All of these

929. Which of the following is correct for Ankur?
(A) Liquid N fertilizer
(B) Contains 32 % N
(C) Useful for rice crop
(D) All of these

930. Aqua ammonium contains ammonia:
(A) 18 % (B) 20 %
(C) 22.4 % (D) 25 %

931. Anhydrous ammonia contain ammonia:
(A) >90 % (B) <90 %
(C) >99% (D) None of these

932. Which of the following is a source of Mo?
(A) Sodium molybdate
(B) Ammonium molybdate
(C) Molybdenum sulphate
(D) All of these

933. Copper sulphate contains Cu:
(A) 15 % (B) 20 %
(C) 25 % (D) 30 %

934. Nitrophosphate contains N, P and K respectively:
(A) 20 : 20 : 0 (B) 20 : 20 : 20
(C) 20 : 45 : 0 (D) 20 : 40 : 40

935. Nitrophosphate potash grade I contains N, P and K respectively:
(A) 18 : 18 : 9 (B) 18 : 18 : 0
(C) 17 : 17 : 17 (D) 20 : 20 : 20

936. Nitrophosphate potash grade II contains N, P and K respectively:
(A) 18 : 18 : 9 (B) 18 : 18 : 18
(C) 15 : 15 : 15 (D) 20 : 20 : 20

937. Zinc sulphate contains S:
(A) 15 % (B) 20 %
(C) 14.6 % (D) 17.8 %

938. Borax contains Boron:
(A) 5 % (B) 10 %
(C) 11 % (D) 13 %

939. Boric acid contain Boron:
(A) 15 % (B) 17 %
(C) 18 % (D) 19 %

940. Cu deficiency may be corrected by application of:
(A) Copper sulphate (B) Cupperous oxide
(C) Cupric oxide (D) All of these

941. Wood ash is generally:
(A) Acidic (B) Basic
(C) Neutral (D) None of these

942. KCl is not suitable for:
(A) Wheat (B) Rice
(C) Gram (D) Tobacco

943. K_2O in KCl is:
(A) 60 % (B) 42%
(C) 50 % (D) 51 %

944. K_2O in potassium carbonate is:
(A) 21 % (B) 46 %
(C) 65 % (D) 54 %

945. Which of the following is known as salt petre of nitre?
(A) KCl (B) KNO_3
(C) K_2SO_4 (D) None of these

946. KNO_3 contains:
(A) 12-13 % N and 44 % K_2O
(B) 12-13 % N and 36 % K_2O
(C) 18 % N and 48 % K_2O
(D) 12-13 % N and 48 % K_2O

947. Which of the following is a basic fertilizer?
(A) KCl
(B) Single super phosphate
(C) KNO_3
(D) None of these

948. Which of the following element is found maximum in ash?
(A) N (B) P
(C) K (D) All of these

949. K_2O in ash is approximately?
(A) 0.5-12 % (B) 17-18 %
(C) 25-28 % (D) 28-30%

950. S in K_2SO_4 is:
(A) 12 % (B) 21%
(C) 17.6 % (D) 20.6 %

951. Complete complex fertilizers contains:
(A) Only N (B) Only P
(C) Only K (D) N, P and K

952. Complex fertilizers generally contain total nutrients:
(A) >20 % (B) <20 %
(C) 10-15 % (D) 5-10 %

953. P in nitrophosphate is:
(A) Water soluble
(B) Citrate soluble
(C) Water and Citrate soluble
(D) None of these

954. Which of the following is a nitro phosphate fertilizer?
(A) Suphla
(B) Ammonium chloride
(C) Super phosphate
(D) Urea

955. Which of the following is a complex fertilizer?
(A) Urea
(B) Muriate of potash
(C) Mono Ammonium Phosphate
(D) None of these

956. Which of the following element is mainly found in guano?
(A) N (B) P
(C) N and P (D) K

957. N content in Guano is found in form of:
(A) NO_3^- (B) NH_4^+
(C) N_2O (D) N_2

958. Guano is a:
(A) Fish by product
(B) Beat of sea birds
(C) Mixture of cakes
(D) Inorganic fertilizer

959. Average N, P and K content of poultry manure are:
(A) 3.8, 3.5 & 1.7 % (B) 1.2, 0.5 & 0.7 %
(C) 12, 7 & 3% (D) 1.2, 5 & 7.0 %

960. N content in fishmeal is:
(A) 4-10 % (B) 1-2 %
(C) 0.1-1.5 % (D) >4-1.0 %

961. Pressmud contains mainly:
(A) Lime and P (B) N and P
(C) P and K (D) N

962. Which of the crop is used for green manuring in saline-alkaline soils?
(A) Berseem (B) Dhaincha
(C) Sunnhemp (D) Mung

963. Phosphate absorber bio fertilizer is:
(A) VAM (B) BGA
(C) FYM (D) Azolla

964. Which of the following is the phosphate mobilizing bio-fertilizers?
(A) *Bacillus polymyza*
(B) *Pseudomonas strata*
(C) *Aspergillus niger* (D) All of these

965. Average C : N ratio of digested gobar gas slurry is:
(A) 10-12 (B) 15-16
(C) 20-23 (D) 25-30

966. The average values of different parameters of vermicompost are given below:

1.N	i. 0.56 %
2.P	ii. 1.48 %
3.K	iii. 0.36 %
4.C :N ratio	iv. 34-35

Which of the pair is correctly matched

(A) 1.i	2.ii	3. iii	4. iv
(B) 1.iii	2.ii	3. iv	4. i
(C) 1.ii	2. i	3. iii	4. iv
(D) 1. iv	2. ii	3. i	4. iii

967. Average values (%) of N, P_2O_5 and K_2O in sludge are:
(A) 3.5, 2.5 & 0.5 (B) 0.5, 0.25 & 0.75
(C) 2.5, 0.5 & 2.5 (D) None of these

968. The values of different parameters of compost are given below:

Parameters	Average values
1. N	i. 2.0-2.4 %
2. P	ii. 1.0-1.2 %
3. K	iii. 3.5-3.6 %
4. C :N ratio	iv. 15-17

Which of the pair is correctly matched

(A) 1. i 2. ii 3. iii 4. iv
(B) 1. iii 2. ii 3. iv 4. i
(C) 1. ii 2. i 3. iii 4. iv
(D) None of these

969. Which of the following crop is used for green manuring?

(A) Sunnhemp (B) Dhaincha
(C) Indigo (D) All of these

970. Which of the following bacteria fixes nitrogen in non leguminous crops?

(A) Azotobacter (B) Azospirillum
(C) Beijerinckia (D) All of these

971. Dicyandiamide and S.T. (2-Silphathiazole) are:

(A) Nitrification inhibitors
(B) N slow release fertilizers
(C) High grade N fertilizer
(B) None of these

972. Low grade fertilizers have total percentage of N, P and K:

(A) 14 (B) >14
(C) <14 (D) None of these

973. Medium grade fertilizers have total percentage of N, P and K:

(A) 15-25 (B) >35
(C) <15 (D) None of these

974. High grade fertilizers have total percentage of N, P and K:

(A) 15-25 (B) >35
(C) <15 (D) None of these

975. Concentrated fertilizers have total percentage of N, P and K:

(A) 15-25 (B) >35
(C) <15 (D) None of these

976. Trench method of preparing FYM was developed by:

(A) Dr. C.N. Acharya
(B) Dr. Panjab Singh
(C) B.N. Ghildyal
(D) None of these

977. Which of the following micronutrient is found highest in FYM:

(A) Cu (B) Zn
(C) Mn (D) B

978. Sunflower act as an indicator plant to diagnose the deficiency of:

(A) Boron (B) Zinc
(C) Nitrogen (D) Phosphorus

979. Average N content in FYM nearly on dry weight basis:

(A) 0.2 % (B) 0.6 %
(C) 0.9 % (D) 1.0 %

980. Organic carbon content in FYM is :

(A) 5 % (B) 8 %
(C) 11 % (D) 15 %

981. Digested gobar gas slurry contains N % on dry weight basis nearly:

(A) 1.8-1.9 %
(B) 1.51-1.55 %
(C) 2.11-2.24 %
(D) None of these

982. Conditioners are mixed with mixed fertilizers because:

(A) To improve physical condition
(B) Avoid clouding
(C) To avoid formation of volatile compounds
(D) All of these

983. Fillers are mixed in fertilizers just to:

(A) Improve physical condition of fertilizers
(B) Increase weight of fertilizers
(C) Improve chemical composition of fertilizers
(D) None of these

984. Fertilizer control order was announced in year:

(A) 1947 (B) 1957
(C) 1965 (D) 1970

985. Which of the following is not a bulky organic manure?

(A) FYM (B) Compost
(C) Green manure (D) Fish manure

986. India's share in soil organic carbon stock of the world is only:

(A) 3 % (B) 0.3 %
(C) 0.03 % (D) 0.003 %

987. Sizeable amount of nitrogen could be escaped by volatilization is:
(A) Ammonia (B) Nitrate
(C) Nitrite (D) Amide

988. Sizeable amount of nitrogen could be escaped by denitrification is:
(A) Nitrous oxide or chemical nitrogen
(B) Nitrate
(C) Nitrite (D) Amide

989. Sizeable amount of nitrogen could be escaped by underground water is:
(A) Ammonia (B) Nitrate
(C) Nitrite (D) Amide

990. NPK removal ratio of crops is:
(A) 4 : 1 : 4 (B) 1 : 1 : 5
(C) 5 : 1 : 5 (D) 5 : 5 : 1

991. During 1998-99, consumption of N : P_2O_5 : K_2O ratio was:
(A) 1.33: 4.10 : 11.32
(B) 4.10: 11.32: 1.33
(C) 11.32 : 4.10 : 11.33
(D) 11.32 : 4.10 : 1.33

992. Since last 25 years the Indian soils are experiencing on an average net negative balance of:
(A) 8-10 mt/annum
(B) 10-12 mt/annum
(C) 12-14 mt/annum
(D) 14-16 mt/annum

993. Arrangement of soil sparticle is called:
(A) Soil texture (B) Soil structure
(C) Bulk density (D) Particle density

994. With increase in organic matter, the density will:
(A) Decrease (B) Increase
(C) Not change (D) Unknown

995. The weight per unit volume of the solid particle is called:
(A) Bulk density
(D) Particle density
(C) Soil porosity
(D) Soil Consistency

996. The soil consisting 40% or more of clay particles are classes as:
(A) Silt loam (B) Silt
(C) Clat loam (D) Clay

997. Red soil develops from the parent material of:
(A) Granite (B) Schist
(C) Both of these (D) None of these

998. Quartz, feldspar and mica are:
(A) Primary mineral
(B) Secondary mineral
(C) Tertiary mineral
(D) None of these

999. Dolomite, gypsum, apatite and hematite are:
(A) Primary mineral
(B) Secondary mineral
(C) Tertiary mineral
(D) None of these

1000. Calcium, magnesium and sulphur plant nutrients are called:
(A) Primary (B) Secondary
(C) Micro (D) Trace

1001. Non essential but useful plant nutrients are:
(A) Cobalt & Iodine
(B) Strontium & Flourine
(C) Vanadium & Bromine
(D) All of these

1002. The inherent capacity of soil to supply plant nutrients in adequate amount and suitable proportion is called:
(A) Fertility index (B) Soil fertility
(C) Soil productivity (D) None of these

1003. Under humid condition potassium fixation more in silicate clay of:
(A) Illite (B) Monmorillonite
(C) Kaolinite (D) None of these

1004. First stage in the evaluation of saline soils is:
(A) Solonchek (B) Solonetz
(C) Solodi (D) None of these

1005. The major organic cementing agent in soil aggregate formation is:
(A) Fats
(B) Proteins and protein derivatives
(C) Polysaccharides
(D) Organic acids

1006. Facultative free living 'N' fixing bacterium:
(A) Rhizobium (B) Azotobacter
(C) Klebsiella (D) None of these

1007. Cyst formation is characteristics feature of:
(A) *Acetobacter* (B) *Arthrobacter*
(C) *Azotobacter* (D) None of these

1008. Bacteria having especially high rate of respiration:
(A) *Rhizobium* (B) *Azotobacter*
(C) *E. coli* (D) *Acetobacter*

1009. "Ray Fungi" is the name given to:
(A) Rhizopus
(B) Yeast
(C) Actinomycetes
(D) Mycoplasma

1010. Actinomycetes are:
(A) Gram +ve, aerobic
(B) Gram +ve, anaerobic
(C) Gram –ve, aerobic
(D) Gram –ve, anaerobic

1011. Example for microaerophillic N_2 fixer:
(A) *Rhizobium* (B) *Azotobacter*
(C) *E. coli* (D) *Frankia*

1012. Phosphorus solublizing microorganism:
(A) VAM (B) Mucor
(C) Rhizopus (D) *E. coli*

1013. Liquid N_2 has a temperature:
(A) -196^0C (B) -186^0C
(C) -176^0C (D) -172^0C

1014. Low pH of media inhibits the growth of:
(A) Bacteria (B) Moulds
(C) Both of above (D) None of these

1015. Which of the following genera is most likely to contain organisms capable of surviving high temperature?
(A) Clostridium (B) Penicillium
(C) Torula (D) Pseudomonas

1016. Which of the following is/are Bio-insecticides?
(A) *Bacillus thuringiensis*
(B) *Bacilllus popilliae*
(C) Both (A) and (B)
(D) None of these

1017. Funaric acid is produced by:
(A) *Aspergillus terrus*
(B) *Rhizopus nigricans*
(C) *Aspergillus niger*
(D) *Asergillus fumigatus*

1018. ATP needed for N_2 fixation:
(A) 12 (B) 16
(C) 28 (D) 32

1019. Factor which does not affect legume root nodulation?
(A) Temperature (B) Light
(C) Combined N_2 (D) H^+ ion conc.

1020. Microaerophillic N_2 fixing bacteria:
(A) *Azotobacter* (B) *Azospirillium*
(C) *Clostridium* (D) *Enterobacter*

1021. Bacteria which can ferment sugar faster than yeast is:
(A) *Clostridium* (B) *Zygomonas*
(C) *Thiobacillus* (D) *Bacillus*

1022. In leghaemoglobin the heme portion is specified by:
(A) Plantgenes (B) Bacterial genes
(C) Both of above (D) None of these

1023. Leghaemoglobin is structurally and functionally similar to:
(A) Haemoglobin (B) Actin
(C) Myosin (D) Myogolobin

1024. Anaerobic free living 'N' fixing bacterium:
(A) Azotobacter (B) Acetobacter
(C) Kelbseilla (D) Clostridium

1025. Microorganism which are associated with non legumes:
(A) Anabaena (B) Klebseilla
(C) Both of these (D) None of these

1026. Bacteria are commonly seen in soils such as:
(A) Acidic (B) Neutral
(C) Alkaline (D) Saline

1027. Nitrate reductase enzyme present in:
(A) Cytoplasm (B) Chloroplast
(C) Mitochondria (D) Nucleus

1028. Microbes involved in fibre retting:
(A) *Micrococcus sp.*
(B) *Bacillus subtilis*
(C) *Clostridium butrycium*
(D) *Lactobacillus sp.*

1029. Microbe involved in tobacco curing:
(A) *Micrococcus sp.*
(B) *Bacillus subtilis*
(C) *Lactobacillus sp.*
(D) *Aspergillus niger*

1030. Urea degrading bacteria:
(A) *Bacillus pasturi*
(B) *Clostridium pasteurianum*
(C) *A. niger*
(D) *Micrococcus sp.*

1031. 'P' solubilizing bacteria:
(A) *Bacillus megatherium*
(B) *Clostridium pasteurianum*
(C) *A. niger*
(D) *Micrococcus sp.*

1032. The size of microbe as:
(A) Viriod<Virus<Bacteria<Fungi
(B) Virus<Viriod<Bacteria<Fungi
(C) Viriod<Virus<Fungi<Bacteria
(D) Virus<Viriod<Fungi<Bacteria

1033. Type of relationship between *Acetobacter diazotrophicus* and sugarcane is:
(A) Symbiotic (B) Associative
(C) Endophytic (D) Free living

1034. Example of VAM fungi:
(A) Pisoletecis (B) Sclerotenia
(C) Glomus (D) Trichoderma

1035. For calculating value of K, K_2O is multiplied by:
(A) 2.29 (B) 0.437
(C) 1.20 (D) 3.00

1036. For calculating value of S, SO_4 is multiplied by:
(A) 2.29 (B) 0.437
(C) 1.20 (D) 3.00

1037. For calculating value of Ca, CaO is multiplied by:
(A) 1.40 (B) 3.00
(C) 1.286 (D) 4.43

1038. For calculating value of Ca, $Ca(OH)_2$ is multiplied by:
(A) 1.40 (B) 3.00
(C) 1.286 (D) 4.43

1039. For calculating value of N, NH_4 is multiplied by:
(A) 1.40 (B) 3.00
(C) 1.286 (D) 4.43

1040. For calculating value of N, NO_3 is multiplied by:
(A) 1.40 (B) 3.00
(C) 1.286 (D) 4.43

1041. For calculating value of organic carbon, organic matter is multiplied by:
(A) 1.724 (B) 3.00
(C) 1.286 (D) 4.43

1042. Which of the following crops cause maximum reduction in soil alkalinity:
(A) Cowpea (B) Maize
(C) Paddy (D) Wheat

1043. Which one of the following is not a micronutrient?
(A) Cu (B) B
(C) S (D) Mn

1044. Choose the micronutrient:
(A) Magnesium (B) Zinc
(C) Phosphorus (D) Potassium

1045. How much urea contains nitrogen:
(A) 16% (B) 26%
(C) 36% (D) 46%

1046. How much calcium ammonium nitrate contains nitrogen:
(A) 16% (B) 26%
(C) 25% (D) 46%

1047. How much ammonium liquid contains nitrogen:
(A) 16% (B) 20-25%
(C) 36% (D) 46%

1048. How much ammonium nitrate contains nitrogen:
(A) 16% (B) 26%
(C) 33-34% (D) 46%

1049. How much sodium nitrate contains nitrogen:
(A) 16% (B) 26%
(C) 36% (D) 46%

1050. How much ammonium chloride contains nitrogen:
(A) 24-26% (B) 26%
(C) 36% (D) 46%

1051. How much Ammonium sulphate nitrate contains nitrogen:
(A) 16% (B) 20-21%
(C) 36% (D) 46%

1052. How much Calcium cynamide contains nitrogen:
(A) 16% (B) 21%
(C) 36% (D) 46%

1053. How much Ammonium anhydrous contains nitrogen:
(A) 16% (B) 26%
(C) 36% (D) 82%

1054. How much oxamide contains nitrogen:
(A) 16% (B) 26%
(C) 31.8% (D) 46%

1055. How much ammonium phosphate contains nitrogen:
(A) 16-20% (B) 26%
(C) 36% (D) 46%

1056. How much ammonium sulphate nitrate contains nitrogen:
(A) 16% (B) 26%
(C) 36% (D) 46%

1057. How much ammonium phosphate contains phosphorus:
(A) 16% (B) 20%
(C) 36% (D) 46%

1058. How much single super phosphate contains phosphorus:
(A) 16-18% (B) 26%
(C) 36% (D) 48%

1059. How much dicalcium phosphate contains phosphorus:
(A) 16% (B) 26%
(C) 32-34% (D) 46%

1060. How much triple super phosphate contains phosphorus:
(A) 16% (B) 26%
(C) 36% (D) 46-48%

1061. How much rhenania phosphate contains phosphorus:
(A) 16% (B) 25%
(C) 36% (D) 46%

1062. How much pelophos contains phosphorus:
(A) 17% (B) 26%
(C) 36% (D) 46%

1063. How much rock phosphate (imported) contains phosphorus:
(A) 16% (B) 26%
(C) 30-40% (D) 46%

1064. How much mussorie rock phosphate contains phosphorus:
(A) 16% (B) 23-24%
(C) 36% (D) 46%

1065. How much Jhabua rock phosphate contains phosphorus:
(A) 16% (B) 26%
(C) 31-38% (D) 46%

1066. How much basic slage contains phosphorus:
(A) 14-18% (B) 26%
(C) 31-38% (D) 46%

1067. How much bone meal contains phosphorus:
(A) 16% (B) 20-25%
(C) 36% (D) 46%

1068. How much stream bone meal contains phosphorus:
(A) 16% (B) 22%
(C) 36% (D) 46%

1069. The element which does not enter into permanent organic combination in plants is:
(A) Nitrogen (B) Zinc
(C) Magnesium (D) Potassium

1070. How much sulphate of potash contains potash:
(A) 16% (B) 26%
(C) 48-52% (D) 46%

1071. The soil carried in saltation consists of particle size:
(A) 0.1 to 0.5 mm (B) 0.1 to 0.2 mm
(C) 0.5 to 1.0 mm (D) None of these

1072. How much potassium magnesium sulphate contains potassium:
(A) 16% (B) 25-30%
(C) 36% (D) 46%

1073. How much potassium magnesium contains magnesium:
(A) 12% (B) 26%
(C) 36% (D) 46%

1074. How much bone meal contains nitrogen:
(A) 1-2% (B) 3-4%
(C) 4-5% (D) 5-6%

1075. How much nitrogen content in.coal ash:
(A) 0.7% (B) 0.4%
(C) 0.5% (D) 1.0%

1076. How much phosphorus content in coal ash:
(A) 0.7% (B) 0.4%
(C) 0.5% (D) 1.0%

1077. How much potassium content in coal ash:
(A) 0.7% (B) 0.4%
(C) 0.5% (D) 1.0%

1078. How much nitrogen content in house hold ash
(A) 0.5-1.9% (B) 0.4%
(C) 0.5% (D) 1.0%

1079. How much phosphorus content in house hold ash:
(A) 0.7% (B) 1.6-4.2%
(C) 0.5% (D) 1.0%

1080. How much potassium content in house hold ash:
(A) 0.7% (B) 0.4%
(C) 2.3-12.0% (D) 1.0%

1081. How much nitrogen content in cane trash ash:
(A) 0.1-0.2% (B) 0.4%
(C) 0.5% (D) 1.0%

1082. How much phosphorus content in cane trash ash:
(A) 0.7% (B) 0.8-1.3%
(C) 0.5% (D) 1.0%

1083. How much potassium content in cane trash ash:
(A) 0.7% (B) 0.4%
(C) 1.5-3.1% (D) 1.0%

1084. How much nitrogen content in babool wood ash:
(A) 0.1-0.2% (B) 0.4%
(C) 0.5% (D) 1.0%

1085. How much phosphorus content in babool wood ash:
(A) 0.7% (B) 2.5-3.0%
(C) 0.5% (D) 1.0%

1086. How much potassium content in babool wood ash:
(A) 0.7% (B) 0.4%
(C) 3.5-4.5% (D) 1.0%

1087. How much nitrogen content in casuarinas wood ash:
(A) Traces (B) 1.5%
(C) 0.5% (D) 1.0%

1088. How much nitrogen content in Eucalyptus wood ash:
(A) Traces (B) 1.5%
(C) 0.5% (D) 1.0%

1089. How much nitrogen content in Tobacco wood ash:
(A) Traces (B) 1.5%
(C) 0.5% (D) 1.0%

1090. How much phosphorus content in Eucalyptus wood ash:
(A) Traces (B) 1.5%
(C) 5.9% (D) 1.0%

1091. How much phosphorus content in tobacco wood ash:
(A) 0.7% (B) 2.6%
(C) 5.9% (D) 1.0%

1092. How much potassium content in casuarina wood ash:
(A) 14.0% (B) 23.8%
(C) 36.0% (D) 40.0%

1093. How much potassium content in Eucalyptus wood ash:
(A) 14.0% (B) 23.8%
(C) 36.0% (D) 40.0%

1094. How much potassium content in tobacco wood ash:
(A) 0.7% (B) 23.8%
(C) 36.0% (D) 40.0%

1095. How much nitrogen content in dung of cow and bull:
(A) 0.40% (B) 1.00%
(C) 0.75% (D) 1.35%

1096. How much nitrogen content in urine of cow and bull:
(A) 0.40% (B) 1.00%
(C) 0.75% (D) 1.35%

1097. How much nitrogen content in dung of sheep and goat:
(A) 0.40% (B) 1.00%
(C) 0.75% (D) 1.35%

1098. How much nitrogen content in urine of sheep and goat:
(A) 0.40% (B) 1.00%
(C) 0.75% (D) 1.35%

1099. How much nitrogen content in dung of horse:
(A) 0.40% (B) 0.55%
(C) 0.75% (D) 1.35%

1100. How much nitrogen content in urine of horse:
(A) 0.40% (B) 0.55%
(C) 0.75% (D) 1.35%

1101. How much nitrogen content in dung of pigs:
(A) 0.40% (B) 0.55%
(C) 0.75% (D) 1.35%

1102. How much nitrogen content in urine of pigs:
(A) 0.40% (B) 0.35%
(C) 0.75% (D) 1.35%

1103. How much phosphorus content in dung of cow and bull:
(A) 0.20% (B) Trace
(C) 0.30% (D) 0.50%

1104. How much phosphorus content in urine of cow and bull:
(A) 0.20% (B) Trace
(C) 0.30% (D) 0.50%

1105. How much phosphorus content in dung of sheep and goat:
(A) 0.20% (B) Trace
(C) 0.30% (D) 0.50%

1106. How much phosphorus content in urine of sheep and goat:
(A) 0.20% (B) Trace
(C) 0.30% (D) 0.50%

1107. How much phosphorus content in dung of Horse:
(A) 0.20% (B) Trace
(C) 0.30% (D) 0.50%

1108. How much phosphorus content in urine of Horse:
(A) 0.20% (B) Trace
(C) 0.30% (D) 0.50%

1109. How much phosphorus content in dung of pigs:
(A) 0.20% (B) Trace
(C) 0.30% (D) 0.50%

1110. How much phosphorus content in urine of pigs:
(A) 0.10% (B) Trace
(C) 0.30% (D) 0.50%

1111. How much potassium content in dung of cow and bull:
(A) 0.10% (B) 0.45 %
(C) 1.35% (D) 2.10%

1112. How much potassium content in urine of cow and bull:
(A) 0.10% (B) 0.45 %
(C) 1.35% (D) 2.10%

1113. How much potassium content in dung of sheep and goat:
(A) 0.10% (B) 0.45 %
(C) 1.35% (D) 2.10%

1114. How much potassium content in urine of sheep and goat:
(A) 0.10% (B) 0.45 %
(C) 1.25% (D) 2.10%

1115. How much potassium content in dung of Horse:
(A) 0.40% (B) 0.45 %
(C) 1.25% (D) 2.10%

1116. How much potassium content in urine of Horse:
(A) 0.10% (B) 0.45 %
(C) 1.25% (D) 2.10%

1117. How much potassium content in dung of pigs:
(A) 0.40% (B) 0.45 %
(C) 1.25% (D) 2.10%

1118. How much potassium content in urine of pigs:
(A) 0.10% (B) 0.45 %
(C) 1.25% (D) 2.10%

1119. How much nitrogen is fixed in the soil by *Crotolaria juncea*?
(A) 75-85 kg/ha (B) 70-78 kg/ha
(C) 55-65 kg/ha (D) 45-52 kg/ha

1120. How much nitrogen is fixed in the soil by *Sesbania aculeata*?
(A) 75-85 kg/ha (B) 70-78 kg/ha
(C) 55-65 kg/ha (D) 45-52 kg/ha

1121. How much nitrogen is fixed in the soil by *Phaseolus tribulus*?
(A) 75-85 kg/ha (B) 70-78 kg/ha
(C) 55-65 kg/ha (D) 45-52 kg/ha

1122. How much nitrogen is fixed in the soil by *Vigna radiata*?
(A) 75-85 kg/ha (B) 70-78 kg/ha
(C) 55-65 kg/ha (D) 45-52 kg/ha

1123. How much nitrogen is fixed in the soil by *Vigna mungo*?
(A) 42-53 kg/ha (B) 55-58 kg/ha
(C) 60-67 kg/ha (D) 120-125 kg/ha

1124. How much nitrogen is fixed in the soil by *Vigna sinensis*?
(A) 42-53 kg/ha (B) 55-58 kg/ha
(C) 60-67 kg/ha (D) 120-125 kg/ha

1125. How much nitrogen is fixed in the soil by *Cyamopsis tetragonoloba*?
(A) 42-53 kg/ha (B) 55-58 kg/ha
(C) 60-67 kg/ha (D) 120-125 kg/ha

1126. How much nitrogen is fixed in the soil by *Melilotus alba*?
(A) 42-53 kg/ha (B) 55-58 kg/ha
(C) 60-67 kg/ha (D) 120-125 kg/ha

1127. How much nitrogen is fixed in the soil by *Trifolium alexandrinum*?
(A) 42-53 kg/ha (B) 58-64 kg/ha
(C) 62-75 kg/ha (D) 120-125 kg/ha

1128. How much nitrogen is fixed in the soil by *Lathyrus sativus*?
(A) 42-53 kg/ha (B) 58-64 kg/ha
(C) 62-75 kg/ha (D) 120-125 kg/ha

1129. How much nitrogen contains in castor cake?
(A) 4.3% (B) 2.5%
(C) 5.2% (D) 3.0%

1130. How much nitrogen contains in mahua cake?
(A) 4.3% (B) 2.5%
(C) 5.2% (D) 3.0%

1131. How much nitrogen contains in neem cake?
(A) 4.3% (B) 2.5%
(C) 5.2% (D) 3.0%

1132. How much nitrogen contains in coconut cake?
(A) 4.3% (B) 2.5%
(C) 5.2% (D) 3.0%

1133. How much nitrogen contains in cotton seed cake?
(A) 6.4% (B) 7.3%
(C) 4.9% (D) 5.2%

1134. How much nitrogen contains in groundnut cake?
(A) 6.4% (B) 7.3%
(C) 4.9% (D) 5.2%

1135. How much nitrogen contains in linseed cake?
(A) 6.4% (B) 7.3%
(C) 4.9% (D) 5.2%

1136. How much nitrogen contains in mustard cake?
(A) 6.4% (B) 7.3%
(C) 4.9% (D) 5.2%

1137. How much nitrogen contains in safflower cake?
(A) 6.4% (B) 7.9%
(C) 4.9% (D) 5.2%

1138. How much nitrogen contains in sesame cake?
(A) 6.4% (B) 7.3%
(C) 6.2% (D) 5.2%

1139. How much potassium contains in castor cake?
(A) 1.3% (B) 1.4%
(C) 1.8% (D) 2.2%

1140. How much potassium contains in mahua cake?
(A) 1.3% (B) 1.4%
(C) 1.8% (D) 2.2%

1141. How much potassium contains in neem cake?
(A) 1.3% (B) 1.4%
(C) 1.8% (D) 2.2%

1142. How much potassium contains in coconut cake?
(A) 1.3% (B) 1.4%
(C) 1.8% (D) 2.2%

1143. How much potassium contains in cotton seed cake?
(A) 1.3% (B) 1.4%
(C) 1.8% (D) 2.2%

1144. How much potassium contains in groundnut cake?
(A) 1.3% (B) 1.4%
(C) 1.8% (D) 2.2%

1145. How much potassium contains in linseed cake?
(A) 1.2% (B) 1.3%
(C) 1.9% (D) 2.0%

1146. How much potassium contains in mustard cake?
(A) 1.2% (B) 1.3%
(C) 1.9% (D) 2.0%

1147. How much potassium contains in safflower cake?
(A) 1.2% (B) 1.3%
(C) 1.9% (D) 2.0%

1148. How much potassium contains in sesame cake?
(A) 1.2% (B) 1.3%
(C) 1.9% (D) 2.0%

1149. How much phosphorus contains in castor cake?
(A) 0.8% (B) 1.0%
(C) 1.8% (D) 1.9%

1150. How much phosphorus contains in mahua cake?
(A) 0.8% (B) 1.0%
(C) 1.8% (D) 1.9%

1151. How much phosphorus contains in neem cake?
(A) 0.8% (B) 1.0%
(C) 1.8% (D) 1.9%

1152. How much phosphorus contains in coconut cake?
(A) 0.8% (B) 1.0%
(C) 1.8% (D) 1.9%

1153. How much phosphorus contains in cotton seed cake?
(A) 1.4% (B) 1.5%
(C) 1.8% (D) 2.9%

1154. How much phosphorus contains in groundnut cake?
(A) 1.4% (B) 1.5%
(C) 1.8% (D) 2.2%

1155. How much phosphorus contains in linseed cake?
(A) 1.4% (B) 1.5%
(C) 1.8% (D) 2.2%

1156. How much phosphorus contains in mustard cake?
(A) 1.4% (B) 1.5%
(C) 1.8% (D) 2.2%

1157. How much phosphorus contains in safflower cake?
(A) 1.4% (B) 1.5%
(C) 1.8% (D) 2.2%

1158. How much phosphorus contains in sesame cake?
(A) 1.4% (B) 1.5%
(C) 1.8% (D) 2.0%

1159. How much nitrogen contains in blood meal?
(A) 10-12% (B) 4-10%
(C) 13% (D) 15.2%

1160. How much nitrogen contains in fish meal?
(A) 10-12% (B) 4-10%
(C) 13% (D) 15.2%

1161. How much nitrogen contains in horn and hoof meal?
(A) 10-12% (B) 4-10%
(C) 13% (D) 15.2%

1162. How much phosphorus contains in blood meal?
(A) 1-2% (B) 3-9%
(C) 10% (D) 15.2%

1163. How much phosphorus contains in fish meal?
(A) 1-2% (B) 3-9%
(C) 10% (D) 15.2%

1164. How much potassium contains in fish meal?
(A) 1-2% (B) 3-9%
(C) 0.3-1.5% (D) 15.2%

1165. How much nitrogen contains in farm yard manure:
(A) 0.5-0.6% (B) 0.50%
(C) 1.0-1.5% (D) 1.5%

1166. How much nitrogen contains in rural compost:
(A) 0.5-0.6% (B) 0.50%
(C) 1.0-1.5% (D) 1.5%

1167. How much nitrogen contains in urban compost:
(A) 0.5-0.6% (B) 0.50%
(C) 1.0-1.5% (D) 1.5%

1168. How much nitrogen contains in sewage and sludge:
(A) 0.5-0.6% (B) 0.50%
(C) 1.0-1.5% (D) 1.5%

1169. How much nitrogen contains in Human urine:
(A) 0.60% (B) 1.1-1.6%
(C) 1.50% (D) 3.0%

1170. How much nitrogen contains in bird beat:
(A) 0.60% (B) 1.1-1.6%
(C) 1.50% (D) 3.0%

1171. How much nitrogen contains in molasses:
(A) 0.60% (B) 1.1-1.6%
(C) 1.50% (D) 3.0%

1172. How much nitrogen contains in vermicompost:
(A) 0.60% (B) 1.1-1.6%
(C) 1.50% (D) 3.0%

1173. How much phosphorus contains in farm yard manure:
(A) 0.25% (B) 0.40%
(C) 1.0% (D) 1.0-1.25%

1174. How much phosphorus contains in rural compost:
(A) 0.25% (B) 0.40%
(C) 1.0% (D) 1.0-1.25%

1175. How much phosphorus contains in urban compost:
(A) 0.25% (B) 0.40%
(C) 1.0% (D) 1.0-1.25%

1176. How much phosphorus contains in sewage and sludge:
(A) 0.25% (B) 0.40%
(C) 1.0% (D) 1.0-1.25%

1177. How much phosphorus contains in Human urine:
(A) 0.1% (B) 0.40%
(C) 1.0% (D) 1.0-1.25%

1178. How much phosphorus contains in bird beat:
(A) 0.25% (B) 0.40%
(C) 1.0% (D) 1.0-1.20%

1179. How much phosphorus contains in molasses:
(A) 0.25% (B) 0.40%
(C) 1.0% (D) 3.0%

1180. How much phosphorus contains in vermicompost:
(A) 0.25% (B) 0.40%
(C) 1.0% (D) 1.25%

1181. How much potassium contains in farm yard manure:
(A) 0.5% (B) 0.50-1.0%
(C) 0.8% (D) 1.5%

1182. How much potassium contains in rural compost:
(A) 0.5% (B) 0.50-1.0%
(C) 0.8% (D) 1.5%

1183. How much potassium contains in urban compost:
(A) 0.5% (B) 0.50-1.0%
(C) 0.8% (D) 1.5%

1184. How much potassium contains in sewage and sludge:
(A) 0.5% (B) 0.50-1.0%
(C) 0.8% (D) 1.5%

1185. How much potassium contains in Human urine:
(A) 0.80% (B) 1.1-1.6%
(C) 1.50% (D) 3.0%

1186. How much potassium contains in bird beat:
(A) 0.5% (B) 0.50-1.0%
(C) 0.8% (D) 6.1-8.5%

1187. How much potassium contains in vermicompost:
(A) 0.60% (B) 1.1-1.6%
(C) 1.50% (D) 3.0%

1188. How much sulphur contains in gypsum (Ag grade)?
(A) 12% (B) 13%
(C) 15% (D) 16%

1189. How much sulphur contains in phosphogypsum?
(A) 12% (B) 13%
(C) 15% (D) 16%

1190. How much sulphur contains in pyrite (Ag grade)?
(A) 12% (B) 13%
(C) 15% (D) 22%

1191. How much sulphur contains in copper sulphate?
(A) 12% (B) 13%
(C) 15% (D) 16%

1192. How much sulphur contains in ferrus sulphate?
(A) 12% (B) 13%
(C) 15% (D) 16%

1193. How much sulphur contains in magnesium sulphate?
(A) 12% (B) 13%
(C) 15% (D) 16%

1194. How much sulphur contains in zinc sulphate?
(A) 12%% (B) 13%
(C) 15% (D) 11-16%

1195. How much calcium contains in gypsum (Ag grade)?
(A) 21% (B) 13%
(C) 15% (D) 16-19%

1196. How much calcium contains in phopshogypsum?
(A) 21% (B) 13%
(C) 15% (D) 16-19%

1197. How much magnesium contains in magnesium sulphate?
(A) 12% (B) 13%
(C) 15% (D) 16%

1198. How much boron contains in borax?
(A) 10.5% (B) 13%
(C) 15% (D) 16%

1199. How much boron contains in solabar?
(A) 10% (B) 13%
(C) 19% (D) 26%

1200. How much copper contains in copper sulphate?
(A) 20% (B) 22%
(C) 24% (D) 26%

1201. How much iron contains in ferrus sulphate?
(A) 12% (B) 14%
(C) 14% (D) 16%

1202. How much mangnese contains in magnesium sulphate?
(A) 20% (B) 30.5%
(C) 35% (D) 40%

1203. How much zinc contains in zinc sulphate?
(A) 21 or 33% (B) 22%
(C) 24% (D) 26%

1204. How much molybdenum contains in sodium molybdate?
(A) 17-39% (B) 30-45%
(C) 24-39% (D) 26-50%

1205. How much zinc contains in zinc chelate?
(A) 10% (B) 12%
(C) 14% (D) 16%

1206. How much molybdenum contains in ammonium molybdate?
(A) 39% (B) 45%
(C) 54% (D) 60%

1207. The nitrogen is removed by grain of paddy:
(A) 66 kg/ha (B) 39 kg/ha
(C) 180 kg/ha (D) 75 kg/ha

1208. The nitrogen is removed by straw of paddy:
(A) 66 kg/ha (B) 39 kg/ha
(C) 180 kg/ha (D) 75 kg/ha

1209. The nitrogen is removed by grain of wheat:
(A) 66 kg/ha (B) 39 kg/ha
(C) 180 kg/ha (D) 75 kg/ha

1210. The nitrogen is removed by straw of wheat:
(A) 66 kg/ha (B) 39 kg/ha
(C) 180 kg/ha (D) 75 kg/ha

1211. The nitrogen is removed by grain of maize:
(A) 45 kg/ha (B) 70kg/ha
(C) 80 kg/ha (D) 95 kg/ha

1212. The nitrogen is removed by straw of maize:
(A) 45 kg/ha (B) 70kg/ha
(C) 80 kg/ha (D) 95 kg/ha

1213. The nitrogen is removed by grain of sorghum:
(A) 45 kg/ha (B) 70kg/ha
(C) 80 kg/ha (D) 95 kg/ha

1214. The nitrogen is removed by straw of sorghum:
(A) 45 kg/ha (B) 70kg/ha
(C) 80 kg/ha (D) 95 kg/ha

1215. The nitrogen is removed by grain of barley:
(A) 45 kg/ha (B) 70kg/ha
(C) 80 kg/ha (D) 95 kg/ha

1216. The nitrogen is removed by straw of barley:
(A) 45 kg/ha (B) 70kg/ha
(C) 80 kg/ha (D) 95 kg/ha

1217. The nitrogen is removed by mustard:
(A) 45 kg/ha (B) 70kg/ha
(C) 95 kg/ha (D) 150 kg/ha

1218. The nitrogen is removed by groundnut:
(A) 45 kg/ha (B) 70kg/ha
(C) 95 kg/ha (D) 150 kg/ha

1219. The nitrogen is removed by sugarcane:
(A) 45 kg/ha (B) 70kg/ha
(C) 95 kg/ha (D) 150 kg/ha

1220. The nitrogen is removed by tobacco in light soisl:
(A) 35 kg/ha (B) 70kg/ha
(C) 95 kg/ha (D) 150 kg/ha

1221. The nitrogen is removed by tobacco in heavy soils:
(A) 48 kg/ha (B) 70kg/ha
(C) 95 kg/ha (D) 150 kg/ha

1222. The nitrogen is removed by potato:
(A) 45 kg/ha (B) 70kg/ha
(C) 95 kg/ha (D) 150 kg/ha

1223. The nitrogen is removed by cotton seed:
(A) 45 kg/ha (B) 70kg/ha
(C) 80 kg/ha (D) 150 kg/ha

1224. The nitrogen is removed by jute:
(A) 45 kg/ha (B) 70kg/ha
(C) 95 kg/ha (D) 150 kg/ha

1225. The nitrogen is removed by grain of soybean:
(A) 45 kg/ha (B) 70kg/ha
(C) 95 kg/ha (D) 210 kg/ha

1226. The phosphorus is removed by grain of paddy:
(A) 15 kg/ha (B) 24 kg/ha
(C) 35 kg/ha (D) 70 kg/ha

1227. The phosphorus is removed by straw of paddy:
(A) 15 kg/ha (B) 24 kg/ha
(C) 35 kg/ha (D) 70 kg/ha

1228. The phosphorus is removed by grain of wheat:
(A) 15 kg/ha (B) 24 kg/ha
(C) 35 kg/ha (D) 70 kg/ha

1229. The phosphorus is removed by straw of wheat:
(A) 15 kg/ha (B) 24 kg/ha
(C) 35 kg/ha (D) 70 kg/ha

1230. The phosphorus is removed by grain of maize:
(A) 27 kg/ha (B) 30 kg/ha
(C) 41 kg/ha (D) 48 kg/ha

1231. The phosphorus is removed by straw of maize:
(A) 27 kg/ha (B) 30 kg/ha
(C) 41 kg/ha (D) 48 kg/ha

1232. The phosphorus is removed by grain of sorghum:
(A) 27 kg/ha (B) 30 kg/ha
(C) 41 kg/ha (D) 48 kg/ha

1233. The phosphorus is removed by straw of sorghum:
(A) 27 kg/ha (B) 30kg/ha
(C) 41 kg/ha (D) 48 kg/ha

1234. The phosphorus is removed by grain of barley:
(A) 22 kg/ha (B) 25 kg/ha
(C) 38 kg/ha (D) 48 kg/ha

1235. The phosphorus is removed by straw of barley:
(A) 22 kg/ha (B) 25 kg/ha
(C) 38 kg/ha (D) 48 kg/ha

1236. The phosphorus is removed by mustard:
(A) 22 kg/ha (B) 25kg/ha
(C) 38 kg/ha (D) 48 kg/ha

1237. The phosphorus is removed by groundnut:
(A) 22 kg/ha (B) 25 kg/ha
(C) 38 kg/ha (D) 48 kg/ha

1238. The phosphorus is removed by sugarcane:
(A) 45 kg/ha (B) 58 kg/ha
(C) 65 kg/ha (D) 190 kg/ha

1239. The phosphorus is removed by tobacco in light soils:
(A) 45 kg/ha (B) 58 kg/ha
(C) 65 kg/ha (D) 190 kg/ha

1240. The phosphorus is removed by tobacco in heavy soils:
(A) 45 kg/ha (B) 58 kg/ha
(C) 65 kg/ha (D) 190 kg/ha

1241. The phosphorus is removed by potato:
(A) 45 kg/ha (B) 58 kg/ha
(C) 65 kg/ha (D) 190 kg/ha

1242. The phosphorus is removed by cotton seed:
(A) 30 kg/ha (B) 35 kg/ha
(C) 60 kg/ha (D) 150 kg/ha

1243. The phosphorus is removed by jute:
(A) 30 kg/ha (B) 35 kg/ha
(C) 60 kg/ha (D) 150 kg/ha

1244. The phosphorus is removed by grain of soybean:
(A) 30 kg/ha (B) 35 kg/ha
(C) 60 kg/ha (D) 150 kg/ha

1245. The potassium is removed by grain of paddy:
(A) 18 kg/ha (B) 52 kg/ha
(C) 96 kg/ha (D) 98 kg/ha

1246. The potassium is removed by straw of paddy:
(A) 18 kg/ha (B) 52 kg/ha
(C) 96 kg/ha (D) 98 kg/ha

1247. The potassium is removed by grain of wheat:
(A) 18 kg/ha (B) 52 kg/ha
(C) 96 kg/ha (D) 98 kg/ha

1248. The potassium is removed by straw of wheat:
(A) 18 kg/ha (B) 52 kg/ha
(C) 96 kg/ha (D) 98 kg/ha

1249. The potassium is removed by grain of maize:
(A) 20 kg/ha (B) 35 kg/ha
(C) 90 kg/ha (D) 105 kg/ha

1250. The potassium is removed by straw of maize:
(A) 20 kg/ha (B) 35 kg/ha
(C) 90 kg/ha (D) 105 kg/ha

1251. The potassium is removed by grain of sorghum:
(A) 20 kg/ha (B) 35 kg/ha
(C) 90 kg/ha (D) 105 kg/ha

1252. The potassium is removed by straw of sorghum:
(A) 20 kg/ha (B) 35 kg/ha
(C) 90 kg/ha (D) 105 kg/ha

1253. The potassium is removed by grain of barley:
(A) 30 kg/ha (B) 75 kg/ha
(C) 89 kg/ha (D) 110 kg/ha

1254. The potassium is removed by straw of barley:
(A) 30 kg/ha (B) 75 kg/ha
(C) 89 kg/ha (D) 110 kg/ha

1255. The potassium is removed by mustard:
(A) 30 kg/ha (B) 75 kg/ha
(C) 89 kg/ha (D) 110 kg/ha

1256. The potassium is removed by groundnut:
(A) 30 kg/ha (B) 75 kg/ha
(C) 89 kg/ha (D) 110 kg/ha

1257. The potassium is removed by sugarcane:
(A) 60 kg/ha (B) 70 kg/ha
(C) 210 kg/ha (D) 265 kg/ha

1258. The potassium is removed by tobacco in light soils:
(A) 60 kg/ha (B) 70 kg/ha
(C) 210 kg/ha (D) 265 kg/ha

1259. The potassium is removed by tobacco in heavy soils:
(A) 60 kg/ha (B) 70 kg/ha
(C) 210 kg/ha (D) 265 kg/ha

1260. The potassium is removed by potato:
(A) 60 kg/ha (B) 70 kg/ha
(C) 210 kg/ha (D) 265 kg/ha

1261. The potassium is removed by cotton seed:
(A) 22 kg/ha (B) 30 kg/ha
(C) 85 kg/ha (D) 150 kg/ha

1262. The potassium is removed by jute:
(A) 22 kg/ha (B) 30 kg/ha
(C) 85 kg/ha (D) 150 kg/ha

1263. The potassium is removed by grain of soybean:
(A) 22 kg/ha (B) 30 kg/ha
(C) 85 kg/ha (D) 150 kg/ha

1264. Which rhizobium species is suitable for pea and lentil?
(A) *Rhizobium leguminosorum*
(B) *Rhizobium phaseoli*
(C) *Rhizobium meliloti*
(D) *Rhizobium japonicum*

1265. Which rhizobium species is suitable for arhar, mung, urd and lobia?
(A) *Rhizobium leguminosorum*
(B) *Rhizobium phaseoli*
(C) *Rhizobium meliloti*
(D) *Rhizobium japonicum*

1266. Which rhizobium species is suitable for lucern?
(A) *Rhizobium leguminosorum*
(B) *Rhizobium phaseoli*
(C) *Rhizobium meliloti*
(D) *Rhizobium japonicum*

1267. Which rhizobium species is suitable for soyabean?
(A) *Rhizobium leguminosorum*
(B) *Rhizobium phaseoli*
(C) *Rhizobium meliloti*
(D) *Rhizobium japonicum*

1268. Which rhizobium species is suitable for lupin?
(A) *Rhizobium lupine*
(B) Rhizobium trifoli
(C) *Rhizobium meliloti*
(D) *Rhizobium japonicum*

1269. Which rhizobium species is suitable for berseem?
(A) *Rhizobium lupine*
(B) Rhizobium trifoli
(C) *Rhizobium meliloti*
(D) *Rhizobium japonicum*

1270. The plants with little growth rate and continuing accumulation of nutrient element in the plants can be defined as:
(A) Severe deficiency range
(B) Toxic range
(C) Moderate deficiency range
(D) Sufficiency range

1271. Nitrate formation in soil is carriedout by the microbe named?
(A) Azatobactor
(B) Azospirillum
(C) Nitrobactor
(D) Nitrosomonas

1272. Organic matter in soil is an index of:
(A) Available nitrogen
(B) Available phosphorus
(C) Available potassium
(D) Available zinc

1273. $CaSO_4 + 2H_2O \rightarrow CaSO_4.2H_2O$ is an example of the weathering process is known as:
(A) Hydration (B) Hydrolysis
(C) Oxidation (D) Reduction

1274. Rock phosphate can be safely used in:
(A) Saline soil (B) Alkali soil
(C) Calcareous soil (D) Acid soil

1275. Nitrate inhibitors inhibit the formation of:
(A) NH_4^+ (B) NO_2^-
(C) NO_3^- (D) NH_3

1276. Silt and clay content can be determined with the help of:
(A) Tensiometer (B) Thermometer
(C) Hydrometer (D) Osmometer

1277. Organism whose population in soil is maximum:
(A) Bacteria (B) Virus
(C) Earthworm (D) Fungi

1278. Fraction of humus solubilized by acid or alkali is known as:
(A) Humic acid (B) Fulvic acid
(C) Humin (D) None of these

1279. Easily oxidized rocks contain:
(A) Fe (B) K
(C) Mg (D) $CaCO_3$

1280. Soil organisms obtaining energy from the oxidation of mineral constituents are known as:
(A) Autotrophic (B) Heterotrophic
(C) Facultative (D) Aerobic

1281. Soils with high silica/sesquioxide ratio are called:
(A) Siliceous (B) Sesquioxide
(C) Both (A) & (B) (D) None of these

1282. Soils with low silica/sesquioxide ratio are called:
(A) Siliceous (B) Sesquioxide
(C) Both (A) & (B) (D) None of these

1283. Due to presence of iron and aluminium, the colour of soil is:
(A) Red to yellow (B) Yellow to red
(C) Red to Brown (D) Brown to red

1284. The processes of replacement of one atom by another atom of similar size, in a crystal lattice of a soil clay without disrupting or changing the crystal structure of the mineral is termed as:
(A) Ion exchange
(B) Percentage base saturation
(C) Isomorphic substitution
(D) Polymorphism

1285. Optimum temperature for nitrification is:
(A) 18^0C (B) 22^0C
(C) 30^0C (D) 35^0C

1286. Aggregates are more stable in soils dominated by:
(A) Kaolinite (B) Montmorillonite
(C) Vermiculite (D) Chlorite

1287. Heavy soils refer to:
(A) High density soils
(B) Soils dominated by high density particles
(C) Soils which are difficult to work
(D) Soils rich in nutrients

1288. Bulk density of $1m^3$ soil weighing 1.4 mg is:
(A) 1.33 mg/m_3 (B) 1.44 mg/m_3
(C) 1.50 mg/m_3 (D) 1.40 mg/m_3

1289. Mechanical analysis of soil makes use of:
(A) Darcy's law (B) Stoke's law
(C) Schofield's law (D) Mendel's law

1290. Number of master horizons are:
(A) 3 (B) 4
(C) 5 (D) 6

1291. Exchange of O_2 and CO_2 between soil pores and atmospheric air refers to:
(A) Soil-air capacity
(B) Soil-oxygen diffusion
(C) Soil-air permeability
(D) Soil aeration

1292. Most porous soil structure is:
(A) Granular (B) Crumb
(C) Prismatic (D) Platy

1293. Liming of soil is done to reclaim:
(A) Saline soil (B) Sodic soil
(C) Acidic soil (D) Saline-sodic soil

1294. Regolith refers to:
(A) A horizon (B) A+B horizon
(C) A+B+C horizon (D) Solum

1295. Carbon exchange capacity is found maximum in:
(A) Montmorillonite (B) Kaolinite
(C) Vermiculite (D) Humus

1296. Di-calcium phosphate is written as:
(A) $Ca(H_2PO_4)_2$ (B) $CaHPO_4$
(C) $Ca_3(PO_4)_2$ (D) None of these

1297. Hydraulic conductivity of soil is directly proportional to:
(A) Radius of capillary pores
(B) Capillary porosity
(C) Total porosity
(D) Water holding capacity

1298. Application of urea super granules (USG) is beneficial in:
(A) Water logged soil (B) Saline soils
(C) Both (A) & (B) (D) None of these

1299. Humic acid is a fraction of humus, which is:
(A) Soluble in alkali and acid
(B) Soluble in alkali and insoluble in acid
(C) Insoluble in alkali and soluble in acid
(D) Insoluble in both in alkali and acid

1300. Azolla may fix atmospheric nitrogen in the soil upto:
(A) 251 kg/ha (B) 400 kg/ha
(C) 451 kg/ha (D) None of these

1301. Azolla is nitrogen fixing:
(A) Bacteria (B) Fern
(C) Fungi (D) Algae

1302. The capacity of a soil to resist appreciable change in pH value is called:
(A) Anion-exchange capacity
(B) Buffering capacity
(C) Percentage base saturation
(D) Cation-exchange capacity

1303. The process of loosening for separating fibre from plant is known as:
(A) Curing (B) Retting
(C) Wetting (D) Netting

1304. Azolla is used for nitrogen fixation in:
(A) Wheat crop (B) Gram crop
(C) Paddy crop (D) Arhar crop

1305. Tick the correct group of plant nutrients which acts as energy exchange from the following:
(A) K, Na, Ca and Mg
(B) C, N, P and S
(C) N and O
(D) Fe, Mn, Mo, Cu, B and Zn

1306. Which gas is released from paddy fields:
(A) CH_4 (B) H_2S
(C) CO_2 (D) NH_3

1307. Peat soil are generally characterized as:
(A) Black, heavy and alkaline soil
(B) Red, light and acidic soil
(C) Black, light and alkaline soil
(D) Black, heavy and neutral soil

1308. Solubility of rock phosphate can be improved by:
(A) Nitrosomonas
(B) Nitrobactor
(C) Azotobactor
(D) *Bacillus polymexa*

1309. The principal way in which P and K ions move from the soil to the root of field crops:
(A) Root interception (B) Mass flow
(C) Diffusion (D) None of these

1310. Biuret content in urea should not exceed in foliar spray on wheat crop more than the following:
(A) 0.5% (B) 1.0%
(C) 1.5% (D) 2.0%

1311. Which element causes dispersion of soil particles:
(A) Na (B) K
(C) Ca (D) Mg

1312. VA micorrhiza promotes mainly the plant uptake of:
(A) Nitrogen absorption
(B) Phosphorus absorption
(C) Potassium absorption
(D) Molybdenum absorption

1313. Immobilization of sulphur takes place when the 'S content' of organic matter is less than:

(A) 0.60% (B) 0.45%
(C) 0.30% (D) 0.15%

1314. Read the following statements and mark your answer according to the code given below:

I. Soil acidity increases with exchangeable Al^{++}
II. Soil acidity increases with exchangeable H
III. Soil alkalinity increases with exchangeable calcium
IV. Soil alkalinity increases with exchangeable aluminium

Answer code
(A) I and III are correct
(B) I and II are correct
(C) II and III are correct
(D) III and IV are correct

1315. Consider the following statement:

Assertion (A): Alkali soils are poor in nitrogen
Reason (R): Nitrogen is leached from alkali soil

Select the correct answer from the code given below:
(A) A is true but R is wrong
(B) R is true but A is false
(C) Both A and R are correct
(D) Both A and R are false

1316. Carbon dioxide content in soil air is:

(A) 0.03% (B) 0.003%
(C) 0.25% (D) 0..50%

1317. How much is the porosity percentage if the particle density of soil is 2.6 and bulk density is 1.3:

(A) 35% (B) 40%
(C) 50% (D) 55%

1318. Potassium content is maximum in:

(A) Feldspar (B) Pyroxine
(C) Amphibole (D) Olivine

1319. Which element has greater significance in biological nitrogen fixation:

(A) Nickel (B) Sulphur
(C) Potassium (D) Molybdenum

1320. Read the following statements and mark your answer according to the code given below:

I. Organic matter improves the structure of clayey soil
II. Organic matter improves the structure of sandy soil
III. Sandy soils are heavy in texture
IV. Clayey soils are lighter in texture

Answer code
(A) I and II are correct
(B) I and III are correct
(C) I and IV are correct
(D) III and IV are correct

1321. Read the following statements and mark your answer according to the code given below:

I. Alfisols are lighter in texture
II. Vertisols are higher in texture
III. Entisols have low potassium content
IV. Ultisols have low phosphorus fixation

Answer code
(A) I and II are correct
(B) I and III are correct
(C) I and IV are correct
(D) III and IV are correct

1322. Chemical formula of single super phosphate is:

(A) $Ca(H_2PO_4)_2$
(B) $CaHPO_4$
(C) $Ca_3(PO_4)_2$
(D) $Ca(H_2PO_4)_2.CaSO_4.2H_2O$

1323. The most concentrated fertilizer used for nutrient supply is:

(A) Urea
(B) DAP
(C) Anhydrous ammonia
(D) SSP

1324. Iron containing enzyme responsible for the biological nitrogen fixation is:

(A) Nitrate reductase
(B) Nitrite reductase
(C) Nitrogenase
(D) Peroxidase

1325. Decomposition of organic matter in submerged soil is carried out by:

(A) Bacteria (B) Actinomycetes
(C) Fungi (D) Earthworm

1326. The highest nitrogen content is found in:
(A) Farm Yard Manure
(B) Vermicompost
(C) Poultry manure
(D) Biogas slurry

1327. Excess uptake of which element is known as luxury consumption:
(A) N (B) P
(C) K (D) Zn

1328. Urea hydrolysis in the soil produces:
(A) Uric acid
(B) Ammonium carbonate
(C) Ammonium carbonate
(D) Amide

1329. Sodic soils are characterized by:
(A) High soluble salt concentration and low pH
(B) High soluble salt concentration
(C) Low pH and high exchangeable sodium
(D) High exchangeable sodium and high pH

1330. Which of the following is not correctly matched:
(A) Magnesium Macro nutrient
(B) Sandy loam Soil texture
(C) Rhizobium Symbiotic N_2-fixation
(D) Tensiometer Soil porosity

1331. Iron deficiency is prominent in:
(A) Neutral soil
(B) Calcareous soil
(C) Acidic soil
(D) Waterlogged soil

1332. Direct application of rock phosphate as phosphorus source for crops is recommended in:
(A) Acid soils
(B) Neutral soils
(C) Alkaline soils
(D) Saline soils

1333. Soils of which soil order have no diagnostic horizon:
(A) Mollisols (B) Entisols
(C) Andisols (D) Eridisols

1334. Organic matter content in organic soils in the range of:
(A) 10-15% (B) 16-20%
(C) 20-25% (D) Above 25%

1335. Consider the following statements:
Assertion (A): Nitrification inhibitors are the chemicals used to control nitrogen transformation in the soils
Reason (R): They promote the activities of nitrifying bacteria
Select the correct answer using the code given below:
(A) A is true but R is false
(B) A is false but R is true
(C) Both A and R are true and R is the correct explanation of A
(D) Both A and R are true but R is not the correct explanation of A

1336. Integrated Plant Nutrients Management pertains with:
(A) Use of organic manures
(B) Use of chemical fertilizers
(C) Use of biofertilizers
(D) All of these

1337. For foliar spray, zinc sulphate fertilizer is dissolved in water at the rate of:
(A) 0.25% (B) 0.50%
(C) 0.75% (D) 1.0%

1338. Phosphate containing mineral is:
(A) Dolomite (B) Apatite
(C) Marble (D) Pyrites

1339. Which organism is responsible for phosphorus availability to plants?
(A) Azotobacter (B) Nitrosomonas
(C) Mycorrhiza (D) Earthworms

1340. With increasing soil pH, base saturation % is:
(A) Decrease
(B) Increase
(C) Increases upto pH 7 and then decline
(D) Decrease upto pH 7 and then decline

1341. Zinc sulphate should not be mixed with:
(A) Urea (B) Muriate of potash
(C) DAP (D) FYM

1342. Which clay mineral is rich in potash?
(A) Montmorillonite (B) Kaolinite
(C) Illite (D) Chlorite

1343. Which form of nitrogen is absorbed by paddy under waterlogged conditions:
(A) N_2 (B) NH_4 ion
(C) NO_2 ion (D) Nitrate ion

1344. Which soil has the highest cation exchange capacity?
(A) Loam (B) Loamy sand
(C) Sandy loam (D) Clay loam

1345. The carbohydrates produced in leaves cannot be translocated to different growing parts of plants in the absence of:
(A) Manganese (B) Boron
(C) Zinc (D) Iron

1346. Application of potash increases:
(A) Disease resistance in plants
(B) Resistance for water logging
(C) Frost resistance in plants
(D) None of these

1347. Ammonia is converted into nitrate by:
(A) Fungus (B) Algae
(C) Earthworms (D) Bacteria

1348. Electrical conductivity of saturation extract of saline-sodic soils in mmhos per cm is:
(A) 2.0
(B) Between 2.0 to 3.0
(C) 3.5
(D) More than 4.0

1349. When organic matter is added to the soil:
(A) Bulk density is increased
(B) Bulk density is decreased
(C) Bulk density does not change
(D) None of these

1350. The availability of molybdenum is more at:
(A) pH 5.0 (B) pH 7.0
(C) pH 7.5 (D) pH 9.0

1351. Azotobacter fixes atmospheric nitrogen:
(A) Symbiotically
(B) Non-symbiotically
(C) Both symbiotically & non-symbiotically
(D) None of these

1352. The soil is well flocculated if it has more of:
(A) Sodium (B) Zinc
(C) Potassium (D) Calcium

1353. Black cotton soil is rich in:
(A) Montmorillonite (B) Kaolinite
(C) Illite (D) Chlorite

1354. Which one of the following microorganisms is involved in nitrification in the soil:
(A) Pseudomonas (B) Micro-coccus
(C) Thio-bacillus (D) Nitrosomonas

1355. Montmorillonite clay has silica and alumina layers respectively in the following proportion:
(A) 2 : 1 (B) 1 : 2
(C) 3 : 2 (D) 2 : 3

1356. Bactria responsible for converting nitrite into nitrate is called:
(A) Pseudomonas
(B) Nitrobacter
(C) Heterosomonas
(D) Rhizobium

1357. Sulphur is needed for which group of the following amino acids:
(A) Lysine and glutamine
(B) Methionine and cystine
(C) Tryptophane and cystine
(D) None of these

1358. Energy metabolism in plants is performed by the element:
(A) Potassium (B) Magnesium
(C) Phosphorus (D) Iron

1359. Brayer and others discovered which of the essential elements?
(A) Copper (B) Boron
(C) Chlorine (D) Molybdenum

1360. Potassic fertilizer containing maximum quantity of potash is:
(A) MOP
(B) Sulphate of potash
(C) Potassium nitrate
(D) Potassium ammonium nitrate

1361. Which one of the following pairs is commonly matched:
(A) Peat soil : Calcareous
(B) Alkali soil : pH 9
(C) Black soil : Rich in kaolinite
(D) Laterite soil : pH 9

1362. Nitrogen is a major nutrient because:
(A) It is available in soil more
(B) It is constituent of protein
(C) It is available in atmosphere
(D) It is required by plants more

1363. Peat soils are generally:
(A) Black, heavy and acidic
(B) Red, light and acidic
(C) Black, light and basic
(D) Black, heavy and neutral

1364. With too much increase in nitrogen application, suagr content in juice is:
(A) Decreased (B) Increased
(C) Remains constant (D) None of these

1365. Soil amendments used to reclaim usar (barren) land is:
(A) Calcium sulphate
(B) Sodium sulphate
(C) Potassium sulphate
(D) Calcium carbonate

1366. The salt affected soils are divided in how many classes depending upon electrical conductivity and sodium contents of their solution extract:
(A) Two (B) Three
(C) Four (D) Five

1367. Which microorganisms is present in large quantities in the soil?
(A) Algae (B) Bacteria
(C) Fungi (D) Protozoa

1368. Propounding of the essentiality of the following elements for plants was made in different years:
(A) Nitrogen (B) Manganese
(C) Zinc (D) Copper
Which of the following chronological order is correct in respect of year of propounding:
(A) 2, 3, 1, 4 (B) 3, 2, 4, 1
(C) 1, 3, 2, 4 (D) 1, 2, 3, 4

1369. Scorching or burning on margins of bottom leaves manifests the deficiency of:
(A) Nitrogen (B) Zinc
(C) Copper (D) Potassium

1370. Browning in cauliflower can be corrected by use of:
(A) Ammonium sulphate
(B) Zinc sulphate
(C) Boric acid (D) Urea

1371. Non – essential but beneficial element to plants is:
(A) Calcium (B) Sodium
(C) Potassium (D) Magnesiumm

1372. Carbon is an essential element for plant growth. Plant take it from:
(A) Soil (B) Water
(C) Air (D) Light

1373. Hydrogen is an essential element for plant growth. Plant take it from:
(A) Soil (B) Water
(C) Air (D) Light

1374. Oxygen is an essential element for plant growth. Plant take it from:
(A) Soil (B) Water
(C) Air (D) Light

1375. Near the canal banks soils become:
(A) Acidic (B) Neutral
(C) Alkaline (D) Well drained

1376. True density of most mineral soils is usually:
(A) 1.35 g/cm^3
(B) 1.65 g/cm^3
(C) 2.65 g/cm^3
(D) 3.50 g/cm^3

1377. Each unit change in pH represents a change in activity of H^+ or OH ions by:
(A) One fold
(B) Ten fold
(C) Hundred fold
(D) Thousand fold

1378. Which of the following is micronutrients:
(A) Calcium (B) Zinc
(C) Nitrogen (D) Phosphorus

1379. Deficiency of which of the following element causes general yellowing and poor growth of the plant?
(A) Phosphorus (B) Nitrogen
(C) Calcium (D) boron

1380. Most effective phosphatic fertilizer for crops in acidic soil is:
(A) DAP (B) Rock phosphate
(C) SSP (D) None of these

1381. The availability of zinc is reduced when zinc sulphate is mixed with:
(A) Nitrogenous fertilizers
(B) Phosphatic fertilizers
(C) Potassium fertilizers
(D) None of these

1382. Kadar and Bangar are the soil types of belonging to:
(A) Black soil
(B) Laterite soil
(C) Alluvial soil
(D) Arid soil

1383. Laterite soil is formed by the presence of:
(A) Transportation
(B) Weathering
(C) Leaching
(D) Deposition

1384. This soil is silt and loamy in valley areas and coarse grained in upper slopes. It is called:
(A) Alluvial soil (B) Khadar soil
(C) Forest soil (D) Arid soil

1385. Select the most appropriate combination to indicates the proportion of important relief features in India:
(A) Plains 43%, Plateau 25%, Mountains 32%
(B) Plains 30%, Plateau 42%, Mountains 28%
(C) Plains 43%, Plateau 27%, Mountains 30%
(D) None of these

1386. The general structural formula of montmorrilonite is:
(A) $X_{0.8}(Al_{0.3}Si_{7.7})Al_{2.6}Fe^{3+}{}_{0.9}Mg_{0.5})O_{20}OH)_4nH_2O$
(B) $K_2Al_2Si_6Al_4O_{20}(OH)_4$
(C) $K_2Al_2Si_6(Fe^{2+},Mg)_6O_{20}(OH)_4$
(D) $Si_4Al_4O_{10}OH_8$

1387. The structural formula of muscovite is:
(A) $X_{0.8}(Al_{0.3}Si_{7.7})Al_{2.6}Fe^{3+}{}_{0.9}Mg_{0.5})O_{20}(OH)_4nH_2O$
(B) $K_2Al_2Si_6Al_4O_{20}(OH)_4$
(C) $K_2Al_2Si_6(Fe^{2+}, Mg)_6O_{20}(OH)_4$
(D) $Si_4Al_4O_{10}OH_8$

1388. The structural formula of biotite is:
(A) $X_{0.8}(Al_{0.3}Si_{7.7})Al_{2.6}Fe^{3+}{}_{0.9}Mg_{0.5})O_{20}(OH)_4nH_2O$
(B) $K_2Al_2Si_6Al_4O_{20}(OH)_4$
(C) $K_2Al_2Si_6(Fe^{2+},Mg)_6O_{20}(OH)_4$
(D) $Si_4Al_4O_{10}OH_8$

1389. The structural formula of kaolinite is:
(A) $X_{0.8}(Al_{0.3}Si_{7.7})Al_{2.6}Fe^{3+}{}_{0.9}Mg_{0.5})O_{20}(OH)_4nH_2O$
(B) $K_2Al_2Si_6Al_4O_{20}(OH)_4$
(C) $K_2Al_2Si_6(Fe^{2+},Mg)_6O_{20}(OH)_4$
(D) $Si_4Al_4O_{10}OH_8$

1390. The minerals of montmorillonite or smectic isomorphism series are freely expandable layer silicate are:
(A) Beidellite
(B) Monmorillonite
(C) Saponite
(D) All of these

1391. Which cation has low absorption capacity on clay:
(A) H (B) Ca
(C) Fe (D) Na

1392. Silicate minerls, phyllo-silicates or clay minerals are:
(A) Primary minerals
(B) Secondary minerals
(C) Tertiary minerals
(D) Accessory minerals

1393. Apatite, pyroxins, Tourmaline and Magnetite are:
(A) Primary minerals
(B) Secondary minerals
(C) Tertiary minerals
(D) Accessory minerals

1394. Feldspar are alluminosilicates of:
(A) Potassium
(B) Sodium
(C) Calcium
(D) All of these

1395. As the inter-layer potassium in subjected to depletion through chemical equilibrium with the soil solution, the exchangeable ions and inter-layer water is gained and leads to another mineral according to following relation:
(A) Micas → Illite → Vermiculite → Montmorillonite
(B) Micas → Vermiculite → Illite → Montmorillonite
(C) Montmorillonite → Illite→ Vermiculite→ Micas
(D) Vermiculite → Micas → Illite→ Montmorillonite

1396. Maximum number of bacteria are available in:
(A) Horizon A (B) Solum
(C) Horizon B (D) Horizon C

1397. Guano is not seen in the excreta of:
(A) Scals (B) Bats
(C) Sea birds (D) Whales

1398. Nitrogen bases do not contain:
(A) Phosphorus
(B) Hydrogen
(C) Nitrogen
(D) Carbon

1399. Nitrogen fixation in blue green algae is done by the specialized structures known as:

(A) Akinete (B) Hormogonium
(C) Nanocyst (D) Heterocys

1400. Vermicomposting is a type of composting process involving:

(A) Earthworms
(B) Wooden ash
(C) Plant material
(D) Animal manure

1401. How much phosphorus content in casuarina wood ash:

(A) Traces (B) 1.5%
(C) 5.9% (D) 1.0%

1402. Non-expanding 1:1 clay mineral is:

(A) Chlorite (B) Illite
(C) Vermiculite (D) Kaolinite

1403. One of the following sets of element has been termed as beneficial for plant growth:

(A) S, Na, Se, V (B) Cl, Si, Co, Al
(C) B, V, Si, Na (D) Si, Na, Se, Co

1404. Which cation has higher aggregation capacity in soil:

(A) Al^{3+} (B) Ca^{2+}
(C) Mg^{2+} (D) K^+

1405. DRIS system of fertility evaluation is:

(A) Diagnosis and Research Information System
(B) Development and Research Integrated system
(C) Diagnosis and Recommendation Integrated System
(D) Development & Research Information System

1406. Which one is correctly matches?

Plant nutrient	Common extractants
(A) K+	KCl
(B) $H_2PO_4^-$	NH_4OAC
(C) NH_4^+	$NaHCO_3$
(D) SO_4^-	$Ca(H_2PO_4)_2$

1407. Which one is not correctly matched?

Element	Deficiency disease
(A) N	Khaira is paddy
(B) Mn	Grey speck of oat
(C) B	Brown heart in sugarcane
(D) Mo	Whiptail of Brassica

1408. Most resistant mineral to weathering is:

(A) Quartz (B) Apatite
(C) Fluospar (D) Corundum

1409. Number of soil orders recognized in the taxonomy are:

(A) 10 (B) 11
(C) 12 (D) 13

1410. Soil colours are generally due to:

(A) Iron oxides (B) Humus
(C) Clays
(D) Iron oxide and humus

1411. More weathering causing mineral is:

(A) Quartz (B) Apatite
(C) Fluospar (D) Corundum

1412. *Axorhizobium candidas* forms nodules on the roots of:

(A) Legume (B) Non-leguminous trees
(C) Cereals (D) Sugarbeet

1413. Complete oxidation of a gram of a dry organic matter gives rise to:

(A) 0.4 Kcal energy
(B) 2.0 Kcal energy
(C) 4.0 Kcal energy
(D) 6.0 Kcal energy

1414. Group of salts formed in sodic soils are:

(A) Sodium carbonate and Sodium sulphate
(B) Sodium chloride and Sodium carbonate
(C) Sodium carbonate and Sodium bicarbonate
(D) Sodium chloride and Sodium sulphate

1415. Read the following statements and mark your answer according to these code given below:

I. Alkali soils are structureless
II. Saline soils are poorly drained
III. Alkali soils are poorly drained
IV.Saline soils are devoid of organic matter

Answer code

(A) I and III are correct
(B) II and III are correct
(C) I and IV are correct
(D) III and IV are correct

1416. Nitrogen fixation in soil by Rhizobium is increased by:

(A) P (B) Ca
(C) Mg (D) K

1417. Organic matter content of most Indian soils is:
(A) Less than 1.0%
(B) Between 1.0 to 5.0%
(C) Between 6.0 to 10.0%
(D) More than 10.0 %

1418. Highest share of fertilizer consumption in India is by:
(A) Rice (B) Wheat
(C) Maize (D) Potato

1419. Soil with less than 10 kg available P/ha are classified as:
(A) Low (B) Medium
(C) High (D) Extreme high

1420. Soil with 10-24.6 kg available P/ha are classified as:
(A) Low (B) Medium
(C) High (D) Extreme high

1421. Soil with more than 24.6 kg available P/ha are classified as:
(A) Low (B) Medium
(C) High (D) Extreme high

1422. Soil with less than 280 kg available K/ha are classified as:
(A) Low (B) Medium
(C) High (D) Extreme high

1423. Soil with 280-569 kg available K/ha are classified as:
(A) Low (B) Medium
(C) High (D) Extreme high

1424. Soil with more than 560 kg available K/ha are classified as:
(A) Low (B) Medium
(C) High (D) Extreme high

1425. Soil with less than 28 kg available N/ha are classified as:
(A) Low (B) Medium
(C) High (D) Extreme high

1426. Soil with 28-56 kg available N/ha are classified as:
(A) Low (B) Medium
(C) High (D) Extreme high

1427. Soil with more than 56 kg available N/ha are classified as:
(A) Low (B) Medium
(C) High (D) Extreme high

1428. Soil with less than 0.5 % organic carbon in the soil are classified as:
(A) Low (B) Medium
(C) High (D) Extreme high

1429. Soil with 0.5 to 0.75% organic carbon in the soil are classified as:
(A) Low (B) Medium
(C) High (D) Extreme high

1430. Soil with more than 0.75% organic carbon is the soil are classified as:
(A) Low (B) Medium
(C) High (D) Extreme high

1431. Soil with less than 18 kg available S/ha are classified as:
(A) Low (B) Medium
(C) High (D) Extreme high

1432. Soil with 18 to 36 kg available S/ha are classified as:
(A) Low (B) Medium
(C) High (D) Extreme high

1433. Soil with more than 36 kg available S/ha are classified as:
(A) Low (B) Medium
(C) High (D) Extreme high

1434. General sufficiency or optimum range of N in plants is:
(A) 2.0-5.0% (B) 0.2-0.5%
(C) 1.0-5.0% (D) 0.1-1.0%

1435. General sufficiency or optimum range of P in plants is:
(A) 2.0-5.0% (B) 0.2-0.5%
(C) 1.0-5.0% (D) 0.1-1.0%

1436. General sufficiency or optimum range of K in plants is:
(A) 2.0-5.0% (B) 0.2-0.5%
(C) 1.0-5.0% (D) 0.1-1.0%

1437. General sufficiency or optimum range of Ca in plants is:
(A) 2.0-5.0% (B) 0.2-0.5%
(C) 1.0-5.0% (D) 0.1-1.0%

1438. General sufficiency or optimum range of S in plants is:
(A) 0.1-0.4% (B) 0.2-0.5%
(C) 1.0-5.0% (D) 0.1-1.0%

1439. General sufficiency or optimum range of Mg in plants is:
(A) 2.0-5.0% (B) 0.1-0.3%
(C) 1.0-5.0% (D) 0.1-1.0%

1440. General sufficiency or optimum range of Zn in plants is:
(A) 20-100 μg/g (B) 50-250 μg/g
(C) 20-300 μg/g (D) 5-20 μg/g

1441. General sufficiency or optimum range of Fe in plants is:
(A) 20-100 μg/g (B) 50-250 μg/g
(C) 20-300 μg/g (D) 5-20 μg/g

1442. General sufficiency or optimum range of Mn in plants is:
(A) 20-100 μg/g (B) 50-250 μg/g
(C) 20-300 μg/g (D) 5-20 μg/g

1443. General sufficiency or optimum range of Cu in plants is:
(A) 20-100 μg/g (B) 50-250 μg/g
(C) 20-300 μg/g (D) 5-20 μg/g

1444. General sufficiency or optimum range of B in plants is:
(A) 10-100 μg/g (B) 50-250 μg/g
(C) 20-300 μg/g (D) 5-20 μg/g

1445. General sufficiency or optimum range of Mo in plants is:
(A) 20-100 μg/g (B) 0.1-0.5 μg/g
(C) 20-300 μg/g (D) 5-20 μg/g

1446. General sufficiency or optimum range of Cl in plants is:
(A) 20-100 μg/g (B) 50-250 μg/g
(C) 2000-20000 μg/g (D) 5-20 μg/g

1447. The clay minerals having CEC 200 meq/ 100 g belong to:
(A) Illite (B) Kaolinite
(C) Vermiculite (D) Montmorillonite

1448. Availability of which element is decreased under waterlogged conditions?
(A) Fe (B) Zn
(C) P (D) Mn

1449. Saline-alkali soils are generally found in:
(A) Arid region (B) Humid region
(C) Temperate region
(D) Flood region

1450. Acidic soils are generally found in:
(A) Arid region (B) Humid region
(C) Temperate region
(D) Flood region

1451. Phosphate Solubilizing Bacteria (PSB) improves soil fertility by:
(A) Fixing atmospheric nitrogen
(B) Fixing atmospheric phosphorus
(C) Releasing fixed phosphorus in the soil
(D) All of these

1452. Soils most suitable for agriculture are those whose pH value in between:
(A) 5 to 6 (B) 6 to 7
(C) 7 to 8 (D) 8 to 9

1453. Which element is essential for sugar translocation in sugarcane:
(A) P (B) K
(C) B (D) Zn

1454. Which element is now essential for plant growth:
(A) Na (B) Si
(C) Ni (D) None of these

1455. Phosphorus is called macronutrient because it is:
(A) Available in soil in relatively large amount
(B) Available in plant in relatively in large amount
(C) Used by plants in relatively large amount
(D) Found in grains in relatively large amount

1456. Seed inoculation in pulses is done with:
(A) Azotobacter (B) Rhizobium
(C) Azospirillum (D) Phosphotica

1457. Which elements are useful in energy storage, transfer and bonding?
(A) NPK (B) NSP
(C) NKS (D) None of these

1458. Which of the following elements play an important role in nitrogen metabolism:
(A) Magnesium (B) Manganese
(C) Molybdenum (D) Iron

1459. The leaching loss of nitrogen (N) is more in the form of:
(A) Ammonia (B) Nitrate
(C) Nitrogen (D) Water soluble

1460. Which is not a natural factor affecting sol fertility?
(A) Topography (B) Soilage
(C) Air (D) Parent material

1461. How much percentage of soluble salts are present in alkali soils?
(A) 1-2% (B) 0.1-0.2%
(C) 1.5-2.5% (D) 2-3%

1462. At what stage more nutrients are essentially required in cereal crops:
(A) Growth stage of plants
(B) Formation of leaves
(C) Panicle initation (D) Maturity of crop

1463. Which is not an organic matter in the following?
(A) Crude protein (B) Carbohydrate
(C) Vitamins (D) Hormones

1464. White colour of soil is due to the process of?
(A) Silica (B) Iron
(C) Manganese oxide (D) All of these

1465. Which of the complex fertilizer?
(A) Ammonium sulphate nitrite
(B) Single surface phosphate
(C) Ammonium phosphate
(D) Calcium ammonium nitrate

1466. Iron is an important component of:
(A) Siroheme (B) Ferredoxin
(C) Cytochromoes (D) All of these

1467. Teh pyrite is mostly found in:
(A) Bihar (B) Rajasthan
(C) Andhra Pradesh (D) Maharashtra

1468. The gypsum is mainly found in:
(A) Bihar (B) Rajasthan
(C) Andhra Pradesh (D) Maharashtra

1469. Plant micronutrient is:
(A) Carbon (B) Boron
(C) Magnesium (D) Sulphur

1470. Biofertilizers are:
(A) Organic manure (B) Green Manure
(C) Culture of Microorganism
(D) None of these

1471. For waterlogged rice cultivation, thé fertilizer that may be available is:
(A) Sodium nitratte (B) Urea
(C) Ammonium chloride
(D) None of these

1472. Chemical formula of pyrite is:
(A) FeS (B) CuS
(C) MnS (D) FeS_2

1473. Soil fertility is reduced due to:
(A) Over irrigation
(B) Poor drainage
(C) Imbalanced use of fertilizers
(D) Continuous cropping

1474. Percentage of P_2O_5 in Pelofas fertilizer is:
(A) 12 (B) 16
(C) 11 (D) 18

1475. Which one of the following is not a primary nutrient ?
(A) S (B) N
(C) K (D) P

1476. Which group of plant nutrients involves in N_2 fixation?
(A) P, S. Mo (B) P, S, Co
(C) P, Mo, Co (D) All of these

1477. What culture should be given priority in grundnut cultivation?
(A) Mycorrhiza
(B) Rhizobium
(C) Phosphobacteria
(D) Azospirillum

1478. Sulphur coated urea contains N:
(A) 21% (B) 33%
(C) 40% (D) 26%

1479. Granite is a .... rock :
(A) Metamorphic (B) Igneous
(C) Sedimentary (D) None of these

1480. Empirical formula of

$$100-\left[\frac{\text{Bulk density}}{\text{Particle density}}\times 100\right]$$ is:

(A) Bulk density (B) Percent pore space
(C) Pore space (D) Particle density

1481. Empirical formula of $1-\frac{\text{Bulk density}}{\text{Particle density}}$ is:
(A) Bulk density
(B) Percent pore space
(C) Pore space
(D) Particle density

1482. The % pore space in a soil having a bulk density of 1.16 mg m^{-3} and particle density of 2.62 mg m^{-3} is:
(A) 30.4 (B) 44.2
(C) 55.7 (D) 80.2

1483. Among the following the set of essential nutrient elements, which are absorbed as anions by plants is:

(A) Boron, Chlorine, Copper, Iron, Manganese

(B) Copper, Iron, Phosphorus, Sulphur, Zinc

(C) Potassium, manganese, Molubdenum, Phosphorus, Zinc

(D) Boron, Chhloride, Molybdenum, Phoshorus, Sulphur

1484. A fertilizer which supplies three essential plant nutrients is:

(A) DAP (B) MOP

(C) SSP (D) SOP

1485. The Fe-chelate suitable for application to crops grown on acid soils is:

(A) Fe-EDTA (B) Fe-EDDHA

(C) Fe-DTPA (D) Fe-HEDTA

1486. Which one of the following fertilizers contain water spluble P:

(A) SSP (B) DCP

(C) MAP (D) Both (A) and (C)

1487. Texture of soil can be changed by:

(A) Use of fertilizers

(B) Use of pesticides

(C) Use of tillage operations

(D) None of these

1488. The slow growing species of legume root nodule bacteria are included in bacterial genus:

(A) Rhizobium

(B) Brady- rhizobium

(C) Sino-rhizobium

(D) Azo-rhizobium

1489. Pollen viability of wheat is related to supply of:

(A) Zinc (B) Molybdenum

(C) Boron (D) Potash

1490. Soil structure is improved by application of:

(A) Urea

(B) Super phosphate

(C) Muriate of potash

(D) Zinc sulphate

1491. Hydraulic conductivity of soils varies directly with:

(A) Capillary porosity

(B) Non-capillary porosity

(C) Total porosity

(D) Water holding capacity

1492. The principal way in which P and K ions move from the soil to the root of field crops:

(A) Root interception

(B) Mass flow

(C) Diffusion

(D) None of these

1493. Immobilization of sulphur take place when the 'S' content of organic matter is less than:

(A) 0.45% (B) 0.60%

(C) 0.30% (D) 0.15%

1494. Organic carbon is a measure of:

(A) Available N in soil

(B) Available P in soil

(C) Available K in soil

(D) Available Mg in soil

1495. Soil structure concerns with:

(A) Particle size distribution

(B) Arrangment of soil particles

(C) Aggregation soil particles

(D) Water Holding Capacity

1496. Per cent content of sulphur in single super phosphate is:

(A) 8 (B) 12

(C) 16 (D) 20

1497. Hydaulic conductivity of soil is very low in case of:

(A) Acid soils (B) Saline soils

(C) Loam soils (D) Alkali soil

1498. in a good quality manure C:N ratio should not be more than:

(A) 10 (B) 20

(C) 35 (D) 50

1499. Approximate weight of surface 14 cm soil of one hectare field is:

(A) 1×10^6 kg (B) 1.5×10^6 kg

(C) 2×10^6 kg (D) 2.24×10^6 kg

1500. Concentration of which element is highest in soil:

(A) O_2 (B) Fe

(C) Al (D) None of these

1501. Wheat crop yielding 50 quintals of grain per hectare removes N, P_2O_5 and K_2O kg/ha, respectively:
(A) 100-150, 70-80 and 125-150
(B) 90-100, 60-70 and 100-110
(C) 120, 40 and 80
(D) 80, 30-40 and 20

1502. The N fertilizer use efficiency in rice can be increased by using:
(A) S-coated urea
(B) Urea super granules
(C) Blue Green Algae
(D) Both (A) and (B)

1503. The nitrogen losses in rice can be reduced by placing NH_4 fertilizer in:
(A) Oxidized zone
(B) Reduced zone
(C) Both (A) and (B)
(D) None of these

1504. At pH 4.0 the predominant ionic form of phosphorus present is:
(A) HPO_4 (B) $H_2PO_4^-$
(C) PO_4^{3-} (D) None of these

1505. Which crop of the following has the double symbiotic relationship with nitrogen fixing bacteria:
(A) *Phaseolus vulgaris*
(B) *Glycine max*
(C) *Sesbania rostrata*
(D) None of these

1506. Which one of the following increases phosphate solubility:
(A) Clostridium (B) Pseudomonas
(C) Azotobacter (D) Rhizobium

1507. Permanent soil property is:
(A) Fertility status (B) pH
(C) Soil texture (D) Soil structure

1508. The N-fertilizer use efficiency in rice can be increased by using:
(A) BGA
(B) Urea super granules
(C) S-coated urea
(D) Both (B) and (C)

1509. The continuous use of straight fertilizers leads to decline in poroduction due to:
(A) Sharp fluctuation in soil
(B) Drastic depletion of micronutrients
(C) Loss of soil life form
(D) Crop response to fertilizers slow down over years

1510. With too much increase in nitrogen application, sugar content in juice is:
(A) Increased (B) Decreased
(C) Remain constant
(D) None of these

1511. Which growth stage of sugarcane is critical for irrigation:
(A) Germination
(B) Grand growth phase
(C) Formative stage
(D) Ripening stage

1512. The actual K content (%) of a feritilizer sample having 10% K_2O will be:
(A) 0.63 (B) 0.73
(C) 0.83 (D) 0.93

1513. The general value of the proportionality factor C in Mitscherlich's equation is:
(A) 0.122 (B) 0.314
(C) 0.301 (D) 0.601

1514. The principal way in which P and K ions moves from the soil to the root of field crops is:
(A) Root interception
(B) Mass flow
(C) Diffusion
(D) None of these

1515. Transformation of ammonia salts into nitrate is called as:
(A) Ammonification
(B) Nitrification
(C) Nitrogenation
(D) Mineralization

1516. The process of K fixation in the soil is:
(A) Chemical
(B) Mechanical
(C) Biological
(D) Both chemical and biological

1517. Wheat crop yield 50 q of grain per hectare removes N, P_2O_5 and K_2O kg/ha respectively:
(A) 100-150, 70-80 and 125-150
(B) 90-100, 60-70 and 100-110
(C) 120-40-80
(D) 8-30/40-20

1518. Which one of the following is an example of positive interaction:

(A) Lime × P (B) P × Si
(C) Mo × P (D) Na × K

1519. The N-fertilizer use efficiency in rice can be increased by using:

(A) BGA
(B) Urea super granules
(C) S-coated urea
(D) Both (B) and (C)

1520. The nitrogen losses in rice can be reduced by placing ammonium fertilizers:

(A) Oxadized zone
(B) Reduced zone
(C) Both (A) and (B)
(D) None of these

1521. Highest hydraulic conductivity is recorded in:

(A) Clay loam soil
(B) Loam soil
(C) Silty clay loam soil
(D) Sandy loam soil

1522. Which one of the following fertilizer schedule is recommended to optimize the productivity in intercropping system?

(A) Full recommended dose of both the main crop and intercrop
(B) Full recommended dose of main crop and half of intercrop
(C) Half recommended dose of main crop and intercrop
(D) Full recommended dose of main crop only

1523. The biuret percent limit of urea for foliar application is:

(A) 0.25 (B) 0.75
(C) 1.25 (D) 1.75

1524. The phosphate fertilizer recommended for perennial crops in acid soil is:

(A) SSP (B) TSP
(C) Ammonium sulphate
(D) Rock phosphate

1525. Economic optimum dose of a nutrient is calculated if the yield response is:

(A) Quadratic (B) Exponential
(C) Linear (D) None of the above

1526. If C : N and N : P (ratios) of a soil are 10 : 1 each, its C : N : P ratio would be:

(A) 100 : 100 : 1 (B) 100 : 10 : 1
(C) 10 : 10 : 1 (D) None of the above

1527. Humic acid is a fraction of humus, which is:

(A) Soluble in alkali and acid
(B) Souble in alkali and insoluble in acid
(C) Insoluble in alkali and soluble in acid
(D) Insoluble in both in alkali and acid

1528. Soil structure is improved by application of:

(A) Urea
(B) Super phosphate
(C) Muriate of Potash
(D) Zinc sulphate

1529. Each unit change in pH represents a change in activity of H^+ or OH^- ions by:

(A) One fold (B) Ten fold
(C) Hundred fold (D) Thousand fold

1530. Fresh organic residues as addition to a standing crop leads to :

(A) Denitrification loss of soil nitrogen
(B) Volatilization loss of nitrogen
(C) Decomposition of ammonia in soil
(D) Immobilization of available nitrogen in soil

1531. Which of the following micronutrients are most widely deficient in Indian soils?

(A) Zinc and Boron
(B) Zinc and Iron
(C) Zinc and Menganese
(D) Zinc and Copper

1532. Acid soils, as per USDA are those which in 1 : 1 soil water suspension, have pH less than:

(A) 5.5 (B) 6.0
(C) 6.5 (D) 7.0

1533. The equivalent acidity of $(NH_4)_3$ SO_4 fertilizer is:

(A) 100 (B) 110
(C) 115 (D) 118

1534. The pH of 0.1 N NaOH solution would theorectically by:

(A) 14 (B) 13
(C) 12 (D) 11

1535. The microorganism, constituting the largest share in the total microbial population in soil is:
(A) Nematodes (B) Fungi
(C) Bacteria (D) Actinomycetes

1536. One Baule unit of P (lb P_2O_5/acre) corresponds to:
(A) 50 (B) 55
(C) 45 (D) 40

1537. Irrigation water is not suitable for irrigation if it contains boron (ppm) greater than:
(A) 1.00 (B) 1.25
(C) 1.50 (D) 2.50

1538. The value of `K' (lb/acre) used in the Wilcox's inverse yield N law is:
(A) 218 (B) 318
(C) 418 (D) 518

1539. An irrigation water with EC 1.0 dS/m would have total soluble salt concentration (mg/litre or ppm):
(A) 32 (B) 64
(C) 320 (D) 640

1540. The indicator plant for B deficiency in soil is:
(A) Soybean (B) Rice
(C) Mustard (D) Sunflower

1541. The diameter (in mm) of the soil particles which are carried away due to saltation mode of soil erosion is:
(A) 0.02-0.2 (B) 0.01-0.1
(C) 0.1-0.5 (D) 0.5 - 1.0

1542. A highly benefitted and essential element for rice along with other macro and micro nutrients is:
(A) CO (B) V
(C) Ni (D) Si

1543. The most common growth limiting factor in acid soil is:
(A) Sodium (B) Calcium
(C) Aluminium (D) Potassium

1544. NaCl decomposed compost have C : N ratio of:
(A) 10-15 : 1 (B) 20-25 : 1
(C) 30-25 : 1 (D) 40-45 : 1

1545. Rock phosphates are generally used in soils which are:
(A) Strongly acidic
(B) Strongly alkaline
(C) Neutral soils
(D) Saline soils

1546. Browning of curd of cauliflower is due to the deficiency of:
(A) Iron (B) Manganese
(C) Boron (D) Copper

1547. Percent calcium content in single superphosphate is:
(A) 12 (B) 16
(C) 20 (D) 24

1548. Biuvet content of urea should more appropriately be:
(A) <0.5% (B) <1.0%
(C) <2.0% (D) <3.0%

1549. Movement of P and K from soil to the rest surface take place due to:
(A) Root interception
(B) Diffusion
(C) Mass flow
(D) Osmesis

1550. What is the chemical formula of slaked lime?
(A) $CaCO_3$ (B) CaO
(C) $CaMg\ (CO_3)_2$ (D) $Ca\ (OH)_2$

1551. Biosuper is:
(A) Rock phosphate + S-oxidizing bacteria
(B) Rock phsphate
(C) Rock phosphate + S + Zn
(D) Rock phosphate + S + Thiobacillus

1552. The optimum temperature for nitrification is:
(A) 10-20°C (B) 20-25°C
(C) 25-30°C (D) 30-35°C

1553. Which one of the following is a vertisol?
(A) Black cotton soil
(B) Red sandy loam soils
(C) Sandy loam sodic soil
(D) Sub mountaine (tarai) soil

1554. Which of the following is commonly referred to as muriate of Potash?
(A) Potassium nitrate
(B) Potassium chloride
(C) Potassium sulphate
(D) Potassium silicate

1555. For applying 100 kg of nitrogen, how much urea would one use?
(A) 45 Kg (B) 111 Kg
(C) 222 Kg (D) 333 Kg

1556. The crop highly sensitive to Mo deficiency:
(A) Cauliflower (B) Tomato
(C) Carrot (D) Wheat

1557. Which of the following is not a chelate compound?
(A) EDTA (B) CDMA
(C) DTPA (D) EDDHA

1558. In the process of nitrogen mineralization, the first step is called:
(A) Immobilization
(B) Ammonification
(C) Aminization
(D) Nitrification

1559. Which species of Rhizobium is suitable for fenugreek?
(A) Meliloti (B) Leguminosarum
(C) Phaseoli (D) Trifoli

1560. Which of the following in Na liking crop?
(A) Chickpea (B) Maize
(C) Soybean (D) Sugarbeet

1561. Bacteria *Thiobacillus thioxidans* is involved in :
(A) Reduction of nitrogen to ammonia
(B) Phosphorus solubilization
(C) Oxidation of S in soil
(D) Conversion of urea to ammonia

1562. In cotton, Zn deficiency is known as :
(A) White bud (B) Little leaf
(C) Mottle leaf (D) Fern leaf

1563. Which of the following nutrients is involved in sugar and starch formation, lipid metabolism and N-fixation?
(A) Nitrogen (B) Phosphorus
(C) Potassium (D) Sulphur

1564. Van Bemmel factors for computing soil organic matter from soil organic carbon is:
(A) 6.250 (B) 2.292
(C) 1.724 (D) 0.895

1565. The half life of ^{32}P is:
(A) 13.8 days (B) 14.3 days
(C) 14.8 days (D) 27.6 days

1566. Which of the following is an essential nutrient for higher plants?
(A) CO (B) Na
(C) Na (D) Ni

1567. Akochi diseae in paddy is caused by:
(A) Reduction of sulphate to sulphide
(B) Oxidation of S to SO_4
(C) Transformation of H_2S into SO_4
(D) Transformation of SO_2 into SO_4

1568. Which group of microorganisms is most active at the terminal stage of composting?
(A) Actinomycetes (B) Bacteria
(C) Fungi (D) Protozoa

1569. At the end point of titration for determination of organic carbon, colour changes from:
(A) Blue to red (B) Red to green
(C) Green to blue (D) Blue to green

1570. The first micronutrient reported deficient in Indian soil is:
(A) Zn (B) Mn
(C) Fe (D) Mo

1571. Who is known as father of agricultural chemistry?
(A) Jathrotull (B) Justis Von Liebig
(C) J.H. Gilbert (D) J.B. Lawes

1572. In which crop, responses to Si application has been established widely?
(A) Onion (B) Rice
(C) Potato (D) Chickpea

1573. As per land capability classification, the classes of soils suitable for cultivation of crops are:
(A) Classes I-IV
(B) Classes V-VI
(C) Classes VII-VIII
(D) All of the above

1574. Two molecules that have the same chemical formula but are mirror images of each other are called:
(A) Enantimoers (B) Isomers
(C) Isotopes (D) Mirrormers

1575. The concept of nutrient mobility in soils was first proposed by:
(A) R.H. Bray (B) M.L. Jackson
(C) E. Mitscherlich (D) E.W. Russell

1576. Which micro nutrient is usually deficient in calcarious soils?

(A) Mn (B) Mo
(C) Cu (D) B

1577. The C-axis spacing of smectite group of silicate clays is:

(A) 0.66 nM (B) 0.72 nM
(C) 1.01 nM (D) 1.40 nM

1578. The saturation extract of a soil has EC 2.0 dS/m. What would be its osmotic potential (in bars)?

(A) -0.18 (B) -0.36
(C) -0.72 (D) -1.08

1579. By increasing the electrolytic concentration, the thickness of diffuse double layer:

(A) Increase (B) Decreases
(C) Remain unaffected
(D) Depends on pH

1580. If the nitrate formation in a soil were independent of the amount of ammoniacal N, in that case the reaction would be considered as:

(A) Zero order (B) First order
(C) Second order
(D) Either first or second order

1581. The first N fertilizer manufactured in India was:

(A) Ammonium chloride
(B) Ammonium sulphate
(C) Urea
(D) CAN

1582. Which one is a major component of organic P in soils P:

(A) Nucleic acids (B) Phospholipids
(C) Inositol phosphate
(D) ATP

1583. Which position at the layer silicate has the maximum specificity for K^+ ion?

(A) e-position (B) p-position
(C) i-position
(D) Both e- and p- position

1584. Which micronutrient is fixed on the same sites of the soil colloids as a phosphate?

(A) Boron (B) Copper
(C) Zinc (D) Molybdenum

1585. Which micronutrient is most strongly bound to organic matter?

(A) Fe (B) Cu
(C) Zn (D) B

1586. The molarity of concentrated H_2SO_4 (AR grade) is:

(A) 10 M (B) 12 M
(C) 16 M (D) 18 M

1587. Which one is not a pedogenic process?

(A) Humification (B) Calcification
(C) Laterization (D) Ferruginization

1588. Division of the factors of soil formation into active and passive groups was proposed by:

(A) H. Jenny (B) C.F. Marbut
(C) J.S. Joffe (D) V.V. Dokuchaiev

1589. A subsoil layer of high bulk density which is brittle when moist and very hard when dry, is known as:

(A) Eragipan (B) Duripan
(C) Durinodes (D) Plinthite

1590. Ultisols are comparable with alfisols except for having:

(A) High base saturation of the exchange complex
(B) Low base saturation of the exchange complex
(C) Uniform soil profile
(D) High water table

1591. Total number of agro-ecological sub-regions (AESR) in India delineated by NBSS & LUP is:

(A) 21 (B) 42
(C) 60 (D) 78

1592. Dehydrogenase activity is a good index of biological activity of soil because it plays role in:

(A) Carbon metabolism
(B) Respiration
(C) Synthesis of cell macro molecules
(D) Cell division

1593. Association of two microorganisms in which one is benefitted while other remains unaffected is known as:

(A) Commensalism
(B) Protocooperation
(C) Ammensalism
(D) Competition

1594. The presence of `Hartig net' is a characteristic of:

(A) Endomycorrhizae
(B) Ectomycorrhizae
(C) Arbuscular micorrhizae
(D) Ericold mycorrhizae

1595. In soil taxonomy, great groups are defined largely by:

(A) Soil foundation process
(B) Presence or absence of dignostic horizon
(C) Thermal regime
(D) Moisture regime

1596. Rhizosphere is dominated by:

(A) Gram positive spore forming bacteria
(B) Gram negative spore forming bacteria
(C) Gram negative non-spore forming bacteria
(D) Gram positive non-spore forming bacteria

1597. The toxicity of oxygen for strict anaerobic is due to the production of :

(A) Hydroxyl radical
(B) Superoxide radical
(C) Hydrogen peroxide
(D) Cyamide

1598. The specificity between legume host and rhizobium spp. is govern by:

(A) Flavonoids
(B) Tryptophan
(C) Polysaccharides
(D) Indole acitic acid

1599. Some microorganisms produce organic ligands to chelate iron in the soil and make it available to the plants. These organic ligands are known as:

(A) Porphyrines
(B) Siderophores
(C) Chromophores
(D) Ferrodoxins

1600. Soil test based fertilizer recommendation for targeted yield of crops was first introduced in India by:

(A) J.S. Kanwar
(B) N.P. Dutta
(C) B. Ramamoorthy
(D) R.V. Tamhane

1601. TDR is a method of monitoring:

(A) Soil wetness
(B) Vapour pressure
(C) Salt concentration
(D) Solar radiation

1602. Which one is not a process of chemical weathring?

(A) Hydration (B) Hydrolysis
(C) Precipitation (D) Carbonation

1603. In the Jenny's factors of soil formation, factor `b' refers to :

(A) Biological factor
(B) Biochemical factor
(C) Base exchange
(D) Biosphere

1604. Sediments deposited in fresh water lakes are called:

(A) Delta (B) Aeolin
(C) Lacustrine (D) Outwash

1605. The number of diagnostic surface horizons (epipedons) included in soil taxonomy is:

(A) 6 (B) 9
(C) 12 (D) 15

1606. Argillic horizon is rich in:

(A) Fe and Al-oxides
(B) Sodium
(C) $CaCO_3$
(D) Silicate clay

1607. Phosphate fixation is minimum of soil pH between:

(A) 8 and 9 (B) 7 and 8
(C) 6 and 7 (D) 5 and 6

1608. Chlorite are the minerals belonging to:

(A) Tectosilicates (B) Cyclosilicates
(C) Indosilicates (D) Phyllosilicates

1609. Intensity of blue colour in phosphorus determination is measured at a wavelength of:

(A) 420 nm (B) 470 nm
(C) 540 nm (D) 660 nm

1610. Montmorillomite minerals have external surface area (m^2/g) of:

(A) 80-150 (B) 150-180
(C) 180-250 (D) 250-300

1611. Which of the following is not a slow-release fertilizer?

(A) N-serve (B) LCU
(C) SCU (D) NCCU

1612. Capillary porosity is given by the formula:
(A) 1-Field capacity:
(B) Field capacity x Bulk density
(C) Bulk density x Particle density
(D) $\left(1-\frac{\text{Bulk density}}{\text{Field capacity}}\right)\times 100$

1613. Azomethine -H is used in determination of:
(A) Molybdenum (B) Sulphur
(C) Boron (D) Calcium

1614. The first compound formed on urea hydrolysis in:
(A) Ammonium carbonate
(B) Ammonium caramate
(C) Anhydrous ammonia
(D) Ammonium hydroxide

1615. In the universal soil loss equation, factor `S' stands for:
(A) Slope length
(B) Slope gradient
(C) Soil texture
(D) Soil erodibility

1616. The process of bringing bulk soil sample to the desired quality is called:
(A) Reduction (B) Minimization
(C) Quartering (D) Partitioning

1617. Double rings infilterometer is used for:
(A) Maximizing the vertical and horizontal flows
(B) Reducing the vertical flow
(C) Reducing the upward flow
(D) Minimizing the horizontal flow

1618. Which is not a resource conservation technology (RCT)?
(A) Zero tillage (B) Puddling
(C) Laser levelling (D) Bad planting

1619. *Aspergillus awamori* is a :
(A) K-mobilizer
(B) Organic matter decomposer
(C) S - oxidizer
(D) P-solubilizer

1620. In India, the work on `biogas' was first carried out by:
(A) J.K. Mukherjee (B) S.V. Desai
(C) N.P. Datta (D) N.N. Goswami

1621. Cfu is a unit for measuring:
(A) Soil pollution
(B) Metal contamination
(C) Soil quality
(D) Microbial population

1622. Acid sulphate soils are formed mainly due to:
(A) Reduction of sulphates
(B) Oxidation of sulphides
(C) Evolution of hydrogen sulphide
(D) Rice cultivation using ammonium sulphide

1623. Find out the most appropriate order of organic soil nitrogen in order of their ease of mineralization :
(A) Urea> amino acids > proteins > nucleic acids > amino sugars > humified N
(B) Urea > amino sugars > proteins > amino acids > humified N> nucleic acids
(C) Urea > nucleic acids > proteins > amino acids > humified N > amino sugards
(D) Urea > amino acids > nucleic acids > humified N > proteins > amino sugars

1624. Number of recognised land capability classes given by soil conservation service of USA is:
(A) IV (B) VI
(C) VIII (D) X

1625. If total soil water potential at a point A in soil is -150 cm and at point B is -200 cm, then:
(A) Water will flew from point A to B
(B) Water will flow from point B to A
(C) Water will be no flow of water between points of A and B
(D) More information is needed to draw conclusion on direction of flow of water

1626. *Rhizobium japonicum* fixes N in symbiotic relationship with :
(A) Pea group (B) Lupin group
(C) Soybean group
(D) Phaseolus group

1627. If 2 mL of 1 N $K_2Cr_2O_4$ solution was consumed for oxidation of organic C present in 1 gram of a soil, the organic matter percentage of the soil is:
(A) 0.3 (B) 0.6
(C) 1.0 (D) 1.5

1628. Eriochrome Black-T is used for determination of:

(A) Lime requirement
(B) Gypsum requirement
(C) Organic - C
(D) Boron determinations

1629. Anion exchange capacity is highest in:

(A) Vermiculite
(B) Montomorillonite
(C) Kaolinite
(D) Illite

1630. DRIS approach was given by:

(A) Bray (B) Beaufils
(C) Jackson (D) Nemath

1631. Amount of micronutrients required by various crops can be arranged in an order:

(A) Fe > Mn > Zn > Cu > Mo
(B) Mn > Zn > Fe > Cu > Mo
(C) Fe > Zn > Mn > Cu > Mo
(D) Mo > Cu > Zn > Mn > Fe

1632. Mitscherlich related the potential yield of a crop to the amount of a given nutrient with the following equation:

$$\text{Log}(A - Y) = \log A - Cb$$

In the equation `b' denotes

(A) Proportionality constant
(B) Potetial yield
(C) Amount of given form of nutrient present in the soil
(D) Growth factor specific to a crop when all other factors are not limiting

1633. Which one of the following organisms is an obligate chemoautotraph:

(A) Frankia (B) Nitrosomonas
(C) Rhizobium (D) Anabaena

1634. Which one of the following minerals would weather most rapidly:

(A) Muscovite (B) Albite
(C) Quartz (D) Microline

1635. BIOLOG is use for determining:

(A) Genetic fingerprints of soil microorganisms
(B) Taxonomic changes of soil microorganisms
(C) Metabolic fingerprints of soil bacteria
(D) All of the above

1636. Carbon -14 (^{14}C) is used for C dating. What is the half life of ^{14}C?

(A) 8370 years (B) 5730 years
(C) 1620 years (D) 13 years

1637. Prokaryotes in soil include the following groups of organisms:

(A) Fungi, Actinomycetes, Bacteria, Algae
(B) Bacteria, Archaea, Actinomycetes, BGA
(C) Protozoa, Bacteria, Fungi, Archaea
(D) Algae, Protozoa, Fungi, Bacteria

1638. There are number of fundamental assumptions upon which the validity of Stoke's law is based. Which one of the assumptions given below is not correct?

(A) The particles must be large in comparison to liquid molecules
(B) Practice must not be right and smooth
(C) There must be no shippig between the particle and the liquids
(D) The velocity of the fall must not exceed a certain critical values so that the viscosity of the liquid remains the only resistance to the fall of the particles

1639. Suppose that 100 kg of green manure containing 40% C and 2% N is decomposed by bacteria with C to N ratio of 5 : 1 and C assimilation efficiency of 5%. What will be the quantity of N mineralized or immobilized from the addition of green manuring?

(A) 0.4 Kg N Immobilized
(B) 1.6 Kg N Mineralized
(C) 2 Kg N Immobilized
(D) 2 Kg N Mineralized

1640. Which of the following processes explains the formation of secondary crystalline colloids from the mica and feldspar present in the parent rock?

(A) Hydrolysis (B) Solution
(C) Coagulation (D) Carbonation

1641. When a mixture of rhizobium, Azotobacter and Vibrio was applied to rhizosphere, fixation of atmospheric nitrogen was increased. It was due to the activity of :

(A) All the above
(B) Azospirillum and Vibrio
(C) Azotobacter and Vibrio
(D) Azotobacter and Azospirillum

1642. An elements in soil will be considered as a trace element if its concentration is:
(A) 1-2 wt % (B) > 0.1 wt %
(C) > 3 wt % (D) < 1.0 wt %

1643. Which one of the following can be estimated by Flame Photometer?
(A) Sodium and Potassium
(B) Cadmium and Cobalt
(C) Chlorine and Bramine
(D) Mercury and Arsenic

1644. Walkley and Black rapid titration method is used for the determination of:
(A) Organic carbon content of soil
(B) Nitrate content of soil
(C) Phosphorus content of soil
(D) Fluoride content of soil

1645. In biological systems sulphur is largely bound in:
(A) Lipids (B) Proteins
(C) Nucleic acid
(D) Proteins and nucleic acid

1646. pH of 0.01 M HNO_3 is:
(A) 0.1 (B) 1
(C) 2 (D) 10

1647. Lightening in the atmosphere produces:
(A) NO (B) CO_2
(C) CO (D) NH_3

1648. pOH of 0.001 M solution of HCl is :
(A) 0.1 (B) 1
(C) 10 (D) 11

1649. Which one of the following category of earth worms is most suitable for vermicomposting?
(A) Epigeic (B) Anecic
(C) Endogeic (D) All the above

1650. Carbon dioxide evolved from soil mainly comes from:
(A) Microbial respiration
(B) Root respiration
(C) Small animal respiration
(D) All of the above

1651. Which one of the following is the most predominant element in a majority of igneous rocks?
(A) Al (B) Fe
(C) O (D) Si

1652. Which one of the following is used to determine total organic matter by Walkley and Black method ?
(A) KOH and H_2SO_4
(B) $Na_2S_2O_3$ and H_2SO_4
(C) $K_2Cr_2O_7$ and H_2SO_4
(D) HNO_3 and H_2SO_4

1653. The pOH of a 0.1 M solution of HCl :
(A) 12 (B) 10
(C) 11 (D) 13

1654. Under anaerobic conditions denitrifying pseudomonas changes:
(A) Nitriate to molecular nitrogen
(B) Nitrate to ammonia
(C) Nitrate to Nitrite
(D) Nitrite to Nitrate

1655. Aerobic bacteria in most active in:
(A) Moist and saturated soil
(B) Moist and unsaturated soil
(C) Alluvial soil
(D) Permafrost

1656. Andisols are formed by:
(A) Biological activity
(B) Erosion by wind
(C) Erosion by waves
(D) Volcanoes

1657. Hydraulic conductivity is a function of:
(A) Medium alone
(B) Fluid alone
(C) Either medium or fluid
(D) Both fluid and medium

1658. What is OH^- ion concentration of HCl whose pH is 3 ?
(A) -3 (B) 3
(C) 10^{-3} (D) 10^{-11}

1659. Particles which have maximum ability to attract and hold K^+, Ca^{++} and NH_4^+ ions on their surface are:
(A) Clay (B) Sand
(C) Loam (D) Loamy sand

1660. Laterite soil contains more of:
(A) Iron and Aluminium
(B) Magnesium and Boron
(C) Manganese and Silicate
(D) Potassium and Lead

1661. Peaty soil is found more in:
(A) Kerala (B) Uttar Pradesh
(C) Maharashtra (D) Gujarat

1662. Brown forest soil is also known as:
(A) Entisols (B) Alfisols
(C) Spodosols (D) Mollisols

1663. The soil type which is good for agriculture is:
(A) Podzole (B) Latosols
(C) Serpent soil (D) Solanchak

1664. Laterite represents:
(A) Regolith soil (B) Glacial soil
(C) Black cotton soil
(D) Red soil

1665. The mineral, most resistant to chemical weathering:
(A) Olivine (B) Quartz
(C) K-feldspar (D) Biotite

1666. Removal of top fertile soil by water is called:
(A) Leaching (B) Siltation
(C) Weathering of soil
(D) Soil erosion

1667. Indian monsoon period is from:
(A) October to November
(B) December to February
(C) June to September
(D) March to May

1668. Normality of 0.25 M phosphoric acid is:
(A) 0.25 (B) 0.50
(C) 0.75 (D) 2.50

1669. *Azolla pinnata* is a :
(A) Blue green algae (B) Green algae
(C) Red algar (D) Ferns

1670. Which of the following material has the highest hydraulic conductivity?
(A) Clay (B) Sand stone
(C) Lime stone (D) Quartzite

1671. Which of the following mixture of gases is called biogas:
(A) CO_2, CH_4, NH_3, H_2S, H_2O (Vapour)
(B) CO, CH_4, NH_3, H_2S, H_2O (Vapour)
(C) CO_2, CH_4, N_2O, NH_3, H_2O (Vapour)
(D) CO_2, NO_x, H_2O, CH_4

1672. The chemical used in a fermenter with molasses as a substrate is:
(A) Diammonium sulphate
(B) Diammonium phosphate
(C) Diammonium nitrate
(D) Diammonium chloride

1673. permanent wilting coefficient represents .... form of water:
(A) Hygroscopic
(B) Runaway
(C) Capillary
(D) Chemical combined

1674. "Law of restitution" was proposed by:
(A) Olsen (B) Das
(C) Liebig (D) Bray

1675. Stem nodulating *Rhizobium* spp. is:
(A) Sino-rhizobium
(B) Bradyrhizobium
(C) Azo-rhizobium
(D) All of these

1676. Who is considered as father of soil testing:
(A) Colwell (B) Troug
(C) Parker (D) Puri

1677. The value of Von Bemmelen factor is :
(A) 1.724 (B) 1.200
(C) 2.290 (D) 2.000

1678. Who developed the DTPA tests for available soil micronutrients:
(A) Kurtz LT (B) Lindsay WL
(C) Larsen (D) Jackson ML

1679. The tension of the water at the upper plastic limit is equivalent to pF value:
(A) 1.5 (B) 0.5
(C) 2.0 (D) 2.8-3.3

1680. Thermal conductivity follows the order:
(A) Sand>loam>clay>peat
(B) Sand >Peat > loam > clay
(C) Loam > Sand > Peat > Clay
(D) Peat > Sand > loam > Clay

1681. Nutrient index between 1.5 and 2.5 in catogorized as:
(A) Low (B) Medium
(C) High (D) None of these

1682. Hartig net in soil produce by:
(A) Rhizobium
(B) Ectomycorrhiza
(C) Endomycorrhiza
(D) None of these

1683. Soil order characterized by excess of weathering with dominant iron and aluminium oxide clays, common in tropics is:
(A) Entisols (B) Mollisols
(C) Alfisols (D) Oxisols

1684. Water stable aggregates in soil increase by addition of:
(A) N-fertilizer (B) P-fertilizer
(C) K-fertilizer (D) Micronutrients

1685. Fertilizer gradient approach was given by:
(A) B.V. Mehta (B) B. Ramamoorthy
(C) A.B. Ghosh (D) B. Murthy

1686. Which of the following nutrients is lost in the greatest amount by allowing exposure of manure pile to rain:
(A) N (B) P
(C) Zn (D) Ca

1687. Nitrogen deficiency in plants leads to :
(A) Chlorosis (B) Excessive growth
(C) Profuse flowering
(D) Dark green colour

1688. Soil productivity takes into accounts the following :
(A) Soil structure (B) Soil moisture
(C) Soil fertility
(D) Soil fertility and productivity

1689. Ammonium sulphate is :
(A) Straight fertilizer
(B) Binary fertilizer
(C) Tertiary fertilizer
(D) Mixed fertilizer

1690. Gypsum Equivalent in iron sulphate is :
(A) 1.24 (B) 1.78
(C) 1.62 (D) 0.77

1691. Peat soils are:
(A) Acidic (B) Saline
(C) Alkaline (D) Saline-alkaline

1692. Essentiality of which element was established by "Arnon and Stout"?
(A) Nitrogen (B) Magnesium
(C) Chlorine (D) Molybdenum

1693. Tailor fertilizer for legume is:
(A) MOP (B) DAP
(C) SSP (D) NPK

1694. The average fertilizer use ratio of NPK in India is a about:
(A) 4 : 6 : 2 : 1 (B) 7 : 3 : 3 : 1
(C) 4 : 2 : 1 (D) 6 : 3 : 1

1695. The permanent property of soil is:
(A) Texture (B) Structure
(C) Aggregation (D) pH

1696. Government of India started subsidy on fertilizers with effect from:
(A) Nov. 1, 1966 (B) Nov. 1, 1969
(C) Nov. 1, 1977 (D) Oct 2, 1988

1697. The capacity of a soil to resist appreciable change in pH value is called:
(A) Percent base saturation
(B) CEC
(C) Buffering capacity
(D) Anion exchange capacity

1698. Mechanical analysis of soil makes use of:
(A) Darcy's law (B) Stoke's law
(C) Schefield's law
(D) Ohm's law

1699. The process of replacement of one atom by another atom of similar size, in a crystal lattice of a soil clay without disrupting or changing the crystal structure of the mineral is termed as :
(A) Ion exchange (B) Isomerism
(C) Isomorphic substitution
(D) Polymorphism

1700. The element which does not enter into permanent organic combination in plants is:
(A) Nitrogen (B) Phosphorus
(C) Magnesium (D) Potassium

1701. The plateau with little growth rate and continuing accumulation of nutrient element in the plants can be defined as:
(A) Severe deficiency range
(B) Toxic range
(C) Moderate deficiency range
(D) Sufficiency range

1702. The major organic cementing agent in soil aggregate formation is:
(A) Lipids
(B) Protein and protein derivatives
(C) Polysaccharides
(D) Organic acid

1703. Biuret content in urea should not exceed ..... according to fertilizer control order, 1957:

(A) 1.0% (B) 1.5%
(C) 2.0% (D) 2.5%

1704. For which types of fertilizer in India is fully dependents on imports:

(A) N-fertilizer (B) P-fertilizer
(C) K-fertilizer (D) None of these

1705. Remote sensing helps in studying:

(A) Cropped area
(B) Soil characters
(C) Underground water
(D) All of the above

1706. Amongst oil cakes the highest nitrogen content is in:

(A) Castor cake
(B) Neem cake
(C) Groundnut cake
(D) Coconut cake

1707. Maximum residual acidity is associated with the continuous soil application of:

(A) Urea
(B) Ammonium nitrate
(C) Ammonium sulphate
(D) CAN

1708. The most important potential contaminant of food produced on sevage sludge ammended soil is:

(A) Chromium (B) Lead
(C) Zinc (D) Cadmium

1709. Which of the following phytohormone is involved in nutrient mobilization?

(A) Auxin (B) Gibberellin
(C) ABA (D) Cytokinin

1710. Heavy application of potassium fertilizers after lead to reduced absorption of:

(A) Mg and Ca (B) B and Mn
(C) N and P (D) Zn and Cu

1711. If a plant shows extensive interveinal chlorosis of the older leaves and late this reaches the younger leaves, the symptoms is due to lack of :

(A) Phosphorus (B) Magnesium
(C) Copper (D) Calcium

1712. The concept of Q/1 relationship was developed by:

(A) Mitschlrlich (B) Beckett
(C) Schofield (D) Martin

1713. The organic residues with wider C : N ratios as compared to narrow C : N ratios are decomposed at:

(A) Faster rate
(B) Slow rate
(C) Equal rate
(D) Faster rate followed by slow rate

1714. Application of organic materials with C : N ratio wider than 20 to 30 : 1 leads to .... of soil N:

(A) Nitrification (B) Fixation
(C) Immobilization (D) Mineralization

1715. `A' value concept was given by:

(A) Sorenson (B) Backett
(C) Schofield (D) Fried and Dean

1716. The process of formation of nitrogen and nitrous oxidate gasses from ammonical fertilizers in soil is known as:

(A) Ammonification
(B) Nitrification
(C) Denitrification
(D) Mineralization

ANSWER SHEET
9. Nutrient Management

---

1. (D)	2. (B)	3. (B)	4. (A)	5. (C)	6. (A)	7. (D)	8. (C)	9. (D)	10. (D)
11. (B)	12. (A)	13. (D)	14. (B)	15. (A)	16. (B)	17. (D)	18. (C)	19. (C)	20. (B)
21. (D)	22. (D)	23. (A)	24. (D)	25. (A)	26. (B)	27. (B)	28. (B)	29. (C)	30. (A)
31. (C)	32. (A)	33. (B)	34. (D)	35. (B)	36. (A)	37. (D)	38. (C)	39. (A)	40. (B)
41. (A)	42. (B)	43. (A)	44. (A)	45. (A)	46. (C)	47. (A)	48. (C)	49. (C)	50. (B)

51. (D)	52. (A)	53. (C)	54. (D)	55. (C)	56. (B)	57. (B)	58. (B)	59. (A)	60. (A)
61. (B)	62. (B)	63. (A)	64. (A)	65. (C)	66. (A)	67. (B)	68. (C)	69. (C)	70. (A)
71. (A)	72. (C)	73. (B)	74. (B)	75. (A)	76. (B)	77. (A)	78. (D)	79. (C)	80. (D)
81. (B)	82. (D)	83. (B)	84. (A)	85. (C)	86. (D)	87. (D)	88. (A)	89. (D)	90. (B)
91. (B)	92. (C)	93. (B)	94. (B)	95. (D)	96. (C)	97. (B)	98. (B)	99. (B)	100. (B)
101. (D)	102. (D)	103. (D)	104. (B)	105. (A)	106. (B)	107. (C)	108. (D)	109. (C)	110. (D)
111. (D)	112. (A)	113. (D)	114. (D)	115. (A)	116. (B)	117. (C)	118. (C)	119. (D)	120. (B)
121. (A)	122. (A)	123. (A)	124. (B)	125. (A)	126. (B)	127. (C)	128. (B)	129. (D)	130. (B)
131. (D)	132. (C)	133. (D)	134. (D)	135. (B)	136. (C)	137. (B)	138. (B)	139. (A)	140. (C)
141. (B)	142. (C)	143. (C)	144. (D)	145. (A)	146. (B)	147. (D)	148. (D)	149. (C)	150. (B)
151. (B)	152. (B)	153. (C)	154. (A)	155. (D)	156. (B)	157. (C)	158. (A)	159. (D)	160. (D)
161. (B)	162. (C)	163. (A)	164. (B)	165. (C)	166. (C)	167. (D)	168. (A)	169. (B)	170. (C)
171. (D)	172. (C)	173. (A)	174. (B)	175. (D)	176. (D)	177. (B)	178. (C)	179. (B)	180. (C)
181. (C)	182. (A)	183. (B)	184. (C)	185. (B)	186. (A)	187. (C)	188. (B)	189. (B)	190. (D)
191. (A)	192. (C)	193. (B)	194. (D)	195. (B)	196. (D)	197. (C)	198. (A)	199. (D)	200. (B)
201. (C)	202. (A)	203. (D)	204. (B)	205. (A)	206. (C)	207. (A)	208. (D)	209. (B)	210. (C)
211. (B)	212. (D)	213. (D)	214. (C)	215. (C)	216. (D)	217. (A)	218. (D)	219. (A)	220. (B)
221. (A)	222. (C)	223. (D)	224. (B)	225. (A)	226. (B)	227. (C)	228. (C)	229. (A)	230. (C)
231. (B)	232. (C)	233. (C)	234. (A)	235. (C)	236. (C)	237. (C)	238. (D)	239. (A)	240. (C)
241. (B)	242. (B)	243. (D)	244. (A)	245. (B)	246. (D)	247. (D)	248. (A)	249. (D)	250. (B)
251. (C)	252. (A)	253. (A)	254. (D)	255. (B)	256. (A)	257. (A)	258. (C)	259. (D)	260. (C)
261. (C)	262. (C)	263. (D)	264. (B)	265. (C)	266. (D)	267. (A)	268. (B)	269. (C)	270. (D)
271. (B)	272. (A)	273. (C)	274. (C)	275. (B)	276. (C)	277. (A)	278. (B)	279. (D)	280. (D)
281. (A)	282. (C)	283. (B)	284. (D)	285. (C)	286. (B)	287. (A)	288. (C)	289. (D)	290. (B)
291. (A)	292. (B)	293. (C)	294. (D)	295. (B)	296. (D)	297. (C)	298. (D)	299. (B)	300. (D)
301. (C)	302. (B)	303. (A)	304. (B)	305. (C)	306. (B)	307. (B)	308. (D)	309. (A)	310. (B)
311. (B)	312. (B)	313. (C)	314. (A)	315. (D)	316. (D)	317. (C)	318. (B)	319. (C)	320. (B)
321. (C)	322. (B)	323. (C)	324. (A)	325. (A)	326. (A)	327. (D)	328. (C)	329. (B)	330. (A)
331. (A)	332. (A)	333. (A)	334. (C)	335. (B)	336. (C)	337. (C)	338. (C)	339. (D)	340. (B)
341. (C)	342. (C)	343. (C)	344. (A)	345. (C)	346. (C)	347. (D)	348. (B)	349. (D)	350. (C)
351. (A)	352. (B)	353. (C)	354. (D)	355. (B)	356. (D)	357. (A)	358. (B)	359. (C)	360. (C)
361. (C)	362. (A)	363. (C)	364. (C)	365. (C)	366. (A)	367. (D)	368. (D)	369. (D)	370. (B)
371. (A)	372. (C)	373. (C)	374. (B)	375. (A)	376. (C)	377. (D)	378. (C)	379. (D)	380. (C)
381. (D)	382. (D)	383. (B)	384. (D)	385. (C)	386. (A)	387. (C)	388. (D)	389. (A)	390. (C)
391. (B)	392. (D)	393. (D)	394. (B)	395. (C)	396. (B)	397. (D)	398. (A)	399. (B)	400. (B)
401. (C)	402. (D)	403. (A)	404. (C)	405. (C)	406. (A)	407. (B)	408. (C)	409. (A)	410. (A)
411. (B)	412. (A)	413. (B)	414. (A)	415. (D)	416. (C)	417. (A)	418. (C)	419. (A)	420. (D)
421. (A)	422. (D)	423. (B)	424. (D)	425. (B)	426. (B)	427. (D)	428. (A)	429. (B)	430. (D)
431. (C)	432. (C)	433. (C)	434. (D)	435. (A)	436. (C)	437. (C)	438. (A)	439. (C)	440. (A)
441. (D)	442. (C)	443. (A)	444. (D)	445. (A)	446. (A)	447. (A)	448. (A)	449. (B)	450. (B)
451. (C)	452. (B)	453. (A)	454. (B)	455. (A)	456. (B)	457. (A)	458. (C)	459. (C)	460. (B)
461. (A)	462. (A)	463. (B)	464. (B)	465. (A)	466. (D)	467. (A)	468. (A)	469. (C)	470. (C)

471. (A) 472. (B) 473. (A) 474. (B) 475. (C) 476. (A) 477. (C) 478. (A) 479. (B) 480. (A)
481. (C) 482. (B) 483. (B) 484. (C) 485. (C) 486. (C) 487. (D) 488. (A) 489. (A) 490. (A)
491. (B) 492. (C) 493. (C) 494. (D) 495. (D) 496. (D) 497. (D) 498. (D) 499. (C) 500. (A)
501. (B) 502. (C) 503. (D) 504. (C) 505. (B) 506. (D) 507. (C) 508. (B) 509. (A) 510. (B)
511. (C) 512. (C) 513. (C) 514. (A) 515. (B) 516. (B) 517. (A) 518. (D) 519. (D) 520. (A)
521. (A) 522. (A) 523. (D) 524. (A) 525. (A) 526. (A) 527. (A) 528. (A) 529. (A) 530. (B)
531. (A) 532. (C) 533. (A) 534. (B) 535. (D) 536. (C) 537. (C) 538. (B) 539. (D) 540. (A)
541. (A) 542. (A) 543. (B) 544. (C) 545. (C) 546. (D) 547. (B) 548. (B) 549. (A) 550. (C)
551. (C) 552. (A) 553. (A) 554. (A) 555. (D) 556. (D) 557. (A) 558. (A) 559. (B) 560. (A)
561. (A) 562. (A) 563. (A) 564. (B) 565. (A) 566. (D) 567. (A) 568. (A) 569. (B) 570. (B)
571. (A) 572. (A) 573. (D) 574. (B) 575. (D) 576. (B) 577. (C) 578. (D) 579. (D) 580. (B)
581. (B) 582. (C) 583. (D) 584. (A) 585. (A) 586. (A) 587. (C) 588. (D) 589. (A) 590. (A)
591. (D) 592. (A) 593. (A) 594. (A) 595. (B) 596. (A) 597. (C) 598. (A) 599. (B) 600. (D)
601. (D) 602. (C) 603. (C) 604. (A) 605. (B) 606. (A) 607. (D) 608. (D) 609. (A) 610. (A)
611. (C) 612. (D) 613. (A) 614. (D) 615. (D) 616. (C) 617. (B) 618. (C) 619. (B) 620. (A)
621. (B) 622. (B) 623. (A) 624. (A) 625. (A) 626. (A) 627. (A) 628. (B) 629. (A) 630. (A)
631. (A) 632. (B) 633. (A) 634. (C) 635. (C) 636. (A) 637. (B) 638. (D) 639. (A) 640. (B)
641. (C) 642. (B) 643. (C) 644. (B) 645. (A) 646. (A) 647. (A) 648. (B) 649. (B) 650. (C)
651. (D) 652. (A) 653. (A) 654. (B) 655. (B) 656. (A) 657. (A) 658. (A) 659. (D) 660. (A)
661. (B) 662. (A) 663. (B) 664. (A) 665. (D) 666. (B) 667. (A) 668. (A) 669. (C) 670. (A)
671. (C) 672. (C) 673. (A) 674. (B) 675. (B) 676. (C) 677. (C) 678. (C) 679. (D) 680. (B)
681. (A) 682. (B) 683. (A) 684. (C) 685. (B) 686. (A) 687. (B) 688. (A) 689. (A) 690. (C)
691. (D) 692. (D) 693. (D) 694. (A) 695. (A) 696. (A) 697. (B) 698. (B) 699. (B) 700. (C)
701. (D) 702. (A) 703. (A) 704. (A) 705. (D) 706. (A) 707. (A) 708. (D) 709. (A) 710. (A)
711. (A) 712. (A) 713. (A) 714. (B) 715. (C) 716. (C) 717. (A) 718. (A) 719. (C) 720. (A)
721. (A) 722. (A) 723. (B) 724. (C) 725. (A) 726. (B) 727. (D) 728. (A) 729. (B) 730. (A)
731. (C) 732. (B) 733. (A) 734. (A) 735. (D) 736. (A) 737. (D) 738. (A) 739. (C) 740. (C)
741. (D) 742. (D) 743. (A) 744. (A) 745. (B) 746. (D) 747. (A) 748. (D) 749. (D) 750. (D)
751. (D) 752. (A) 753. (D) 754. (D) 755. (A) 756. (D) 757. (D) 758. (D) 759. (C) 760. (C)
761. (A) 762. (A) 763. (B) 764. (A) 765. (D) 766. (A) 767. (C) 768. (D) 769. (C) 770. (C)
771. (A) 772. (D) 773. (A) 774. (A) 775. (A) 776. (D) 777. (B) 778. (A) 779. (A) 780. (A)
781. (A) 782. (A) 783. (A) 784. (A) 785. (D) 786. (A) 787. (B) 788. (A) 789. (A) 790. (C)
791. (A) 792. (C) 793. (A) 794. (D) 795. (A) 796. (A) 797. (D) 798. (D) 799. (A) 800. (B)
801. (A) 802. (D) 803. (D) 804. (D) 805. (D) 806. (B) 807. (B) 808. (A) 809. (B) 810. (C)
811. (D) 812. (C) 813. (A) 814. (A) 815. (D) 816. (D) 817. (B) 818. (B) 819. (C) 820. (B)
821. (A) 822. (B) 823. (A) 824. (B) 825. (A) 826. (B) 827. (B) 828. (B) 829. (A) 830. (A)
831. (A) 832. (C) 833. (D) 834. (A) 835. (A) 836. (A) 837. (C) 838. (A) 839. (C) 840. (B)
841. (C) 842. (A) 843. (A) 844. (A) 845. (B) 846. (C) 847. (A) 848. (C) 849. (A) 850. (B)
851. (C) 852. (C) 853. (A) 854. (C) 855. (A) 856. (C) 857. (D) 858. (B) 859. (A) 860. (B)
861. (A) 862. (A) 863. (A) 864. (A) 865. (C) 866. (A) 867. (A) 868. (A) 869. (A) 870. (C)
871. (A) 872. (A) 873. (C) 874. (A) 875. (B) 876. (A) 877. (A) 878. (B) 879. (C) 880. (B)
881. (A) 882. (D) 883. (A) 884. (D) 885. (A) 886. (C) 887. (D) 888. (A) 889. (C) 890. (D)

891. (A) 892. (D) 893. (A) 894. (B) 895. (C) 896. (A) 897. (D) 898. (A) 899. (A) 900. (C)

901. (A) 902. (D) 903. (D) 904. (D) 905. (A) 906. (B) 907. (A) 908. (C) 909. (D) 910. (D)

911. (D) 912. (B) 913. (C) 914. (A) 915. (B) 916. (A) 917. (B) 918. (B) 919. (A) 920. (A)

921. (B) 922. (B) 923. (A) 924. (A) 925. (A) 926. (D) 927. (A) 928. (D) 929. (D) 930. (C)

931. (C) 932. (D) 933. (C) 934. (A) 935. (A) 936. (C) 937. (D) 938. (C) 939. (B) 940. (D)

941. (B) 942. (D) 943. (A) 944. (C) 945. (B) 946. (A) 947. (C) 948. (B) 949. (A) 950. (C)

951. (D) 952. (A) 953. (C) 954. (A) 955. (C) 956. (C) 957. (B) 958. (B) 959. (A) 960. (A)

961. (A) 962. (B) 963. (A) 964. (D) 965. (C) 966. (A) 967. (A) 968. (A) 969. (D) 970. (D)

971. (A) 972. (C) 973. (A) 974. (B) 975. (B) 976. (A) 977. (C) 978. (A) 979. (C) 980. (C)

981. (A) 982. (D) 983. (B) 984. (B) 985. (D) 986. (A) 987. (A) 988. (A) 989. (B) 990. (C)

991. (D) 992. (A) 993. (B) 994. (A) 995. (D) 996. (D) 997. (C) 998. (A) 999. (B) 1000. (B)

1001. (D) 1002. (B) 1003. (A) 1004. (A) 1005. (C) 1006. (C) 1007. (C) 1008. (B) 1009. (C) 1010. (A)

1011. (D) 1012. (A) 1013. (A) 1014. (A) 1015. (A) 1016. (C) 1017. (B) 1018. (B) 1019. (B) 1020. (B)

1021. (B) 1022. (B) 1023. (A) 1024. (D) 1025. (C) 1026. (B) 1027. (A) 1028. (C) 1029. (A) 1030. (A)

1031. (A) 1032. (A) 1033. (C) 1034. (C) 1035. (C) 1036. (D) 1037. (A) 1038. (B) 1039. (C) 1040. (D)

1041. (A) 1042. (C) 1043. (C) 1044. (B) 1045. (D) 1046. (C) 1047. (B) 1048. (C) 1049. (A) 1050. (A)

1051. (B) 1052. (B) 1053. (D) 1054. (C) 1055. (A) 1056. (B) 1057. (B) 1058. (A) 1059. (C) 1060. (D)

1061. (B) 1062. (A) 1063. (C) 1064. (B) 1065. (C) 1066. (A) 1067. (B) 1068. (B) 1069. (D) 1070. (C)

1071. (A) 1072. (B) 1073. (A) 1074. (B) 1075. (A) 1076. (B) 1077. (C) 1078. (A) 1079. (B) 1080. (C)

1081. (A) 1082. (B) 1083. (C) 1084. (A) 1085. (B) 1086. (A) 1087. (A) 1088. (A) 1089. (A) 1090. (C)

1091. (B) 1092. (A) 1093. (B) 1094. (C) 1095. (A) 1096. (B) 1097. (C) 1098. (D) 1099. (B) 1100. (D)

1101. (B) 1102. (A) 1103. (A) 1104. (B) 1105. (D) 1106. (B) 1107. (C) 1108. (B) 1109. (D) 1110. (A)

1111. (A) 1112. (C) 1113. (B) 1114. (D) 1115. (A) 1116. (C) 1117. (A) 1118. (B) 1119. (A) 1120. (B)

1121. (C) 1122. (D) 1123. (A) 1124. (B) 1125. (C) 1126. (D) 1127. (C) 1128. (B) 1129. (A) 1130. (B)

1131. (C) 1132. (D) 1133. (A) 1134. (B) 1135. (C) 1136. (D) 1137. (B) 1138. (C) 1139. (A) 1140. (C)

1141. (B) 1142. (C) 1143. (D) 1144. (A) 1145. (B) 1146. (A) 1147. (C) 1148. (A) 1149. (C) 1150. (A)

1151. (B) 1152. (D) 1153. (D) 1154. (B) 1155. (A) 1156. (C) 1157. (D) 1158. (D) 1159. (A) 1160. (B)

1161. (C) 1162. (A) 1163. (B) 1164. (C) 1165. (A) 1166. (B) 1167. (C) 1168. (D) 1169. (A) 1170. (B)

1171. (D) 1172. (C) 1173. (A) 1174. (B) 1175. (C) 1176. (C) 1177. (A) 1178. (D) 1179. (D) 1180. (D)

1181. (B) 1182. (C) 1183. (D) 1184. (A) 1185. (A) 1186. (D) 1187. (C) 1188. (B) 1189. (D) 1190. (D)

1191. (B) 1192. (A) 1193. (C) 1194. (D) 1195. (D) 1196. (A) 1197. (D) 1198. (A) 1199. (C) 1200. (C)

1201. (A) 1202. (B) 1203. (A) 1204. (A) 1205. (B) 1206. (C) 1207. (A) 1208. (B) 1209. (C) 1210. (D)

1211. (D) 1212. (A) 1213. (C) 1214. (A) 1215. (C) 1216. (A) 1217. (B) 1218. (C) 1219. (D) 1220. (A)

1221. (A) 1222. (D) 1223. (C) 1224. (B) 1225. (D) 1226. (B) 1227. (A) 1228. (D) 1229. (C) 1230. (C)

1231. (B) 1232. (D) 1233. (A) 1234. (C) 1235. (A) 1236. (D) 1237. (B) 1238. (D) 1239. (A) 1240. (B)

1241. (C) 1242. (B) 1243. (A) 1244. (C) 1245. (A) 1246. (C) 1247. (B) 1248. (D) 1249. (C) 1250. (A)

1251. (B) 1252. (D) 1253. (A) 1254. (B) 1255. (C) 1256. (D) 1257. (D) 1258. (A) 1259. (B) 1260. (C)

1261. (B) 1262. (A) 1263. (C) 1264. (A) 1265. (B) 1266. (C) 1267. (D) 1268. (A) 1269. (B) 1270. (D)

1271. (C) 1272. (A) 1273. (A) 1274. (D) 1275. (B) 1276. (C) 1277. (A) 1278. (C) 1279. (B) 1280. (B)

1281. (A) 1282. (B) 1283. (A) 1284. (C) 1285. (C) 1286. (D) 1287. (B) 1288. (D) 1289. (B) 1290. (C)

1291. (D) 1292. (B) 1293. (C) 1294. (C) 1295. (D) 1296. (B) 1297. (A) 1298. (A) 1299. (B) 1300. (D)

1301. (B) 1302. (B) 1303. (B) 1304. (C) 1305. (B) 1306. (A) 1307. (A) 1308. (D) 1309. (C) 1310. (A)

1311. **(A)** 1312. **(B)** 1313. **(C)** 1314. **(B)** 1315. **(A)** 1316. **(C)** 1317. **(C)** 1318. **(A)** 1319. **(D)** 1320. **(D)**
1321. **(A)** 1322. **(A)** 1323. **(B)** 1324. **(C)** 1325. **(A)** 1326. **(C)** 1327. **(C)** 1328. **(B)** 1329. **(D)** 1330. **(D)**
1331. **(B)** 1332. **(A)** 1333. **(B)** 1334. **(A)** 1335. **(D)** 1336. **(D)** 1337. **(A)** 1338. **(B)** 1339. **(C)** 1340. **(B)**
1341. **(A)** 1342. **(C)** 1343. **(B)** 1344. **(D)** 1345. **(B)** 1346. **(A)** 1347. **(D)** 1348. **(D)** 1349. **(B)** 1350. **(D)**
1351. **(B)** 1352. **(D)** 1353. **(A)** 1354. **(D)** 1355. **(A)** 1356. **(B)** 1357. **(C)** 1358. **(C)** 1359. **(C)** 1360. **(A)**
1361. **(B)** 1362. **(D)** 1363. **(A)** 1364. **(A)** 1365. **(A)** 1366. **(B)** 1367. **(B)** 1368. **(D)** 1369. **(D)** 1370. **(C)**
1371. **(B)** 1372. **(C)** 1373. **(B)** 1374. **(C)** 1375. **(C)** 1376. **(C)** 1377. **(B)** 1378. **(B)** 1379. **(B)** 1380. **(B)**
1381. **(B)** 1382. **(C)** 1383. **(C)** 1384. **(C)** 1385. **(C)** 1386. **(A)** 1387. **(B)** 1388. **(C)** 1389. **(D)** 1390. **(D)**
1391. **(D)** 1392. **(B)** 1393. **(D)** 1394. **(D)** 1395. **(A)** 1396. **(B)** 1397. **(D)** 1398. **(A)** 1399. **(D)** 1400. **(A)**
1401. **(B)** 1402. **(D)** 1403. **(C)** 1404. **(A)** 1405. **(C)** 1406. **(A)** 1407. **(A)** 1408. **(A)** 1409. **(A)** 1410. **(D)**
1411. **(C)** 1412. **(B)** 1413. **(B)** 1414. **(C)** 1415. **(C)** 1416. **(A)** 1417. **(A)** 1418. **(B)** 1419. **(A)** 1420. **(B)**
1421. **(C)** 1422. **(A)** 1423. **(B)** 1424. **(C)** 1425. **(A)** 1426. **(B)** 1427. **(C)** 1428. **(A)** 1429. **(B)** 1430. **(C)**
1431. **(A)** 1432. **(B)** 1433. **(C)** 1434. **(A)** 1435. **(B)** 1436. **(C)** 1437. **(D)** 1438. **(A)** 1439. **(B)** 1440. **(A)**
1441. **(B)** 1442. **(C)** 1443. **(D)** 1444. **(A)** 1445. **(B)** 1446. **(C)** 1447. **(A)** 1448. **(C)** 1449. **(A)** 1450. **(B)**
1451. **(C)** 1452. **(B)** 1453. **(B)** 1454. **(A)** 1455. **(C)** 1456. **(B)** 1457. **(C)** 1458. **(C)** 1459. **(B)** 1460. **(C)**
1461. **(B)** 1462. **(A)** 1463. **(D)** 1464. **(C)** 1465. **(C)** 1466. **(D)** 1467. **(D)** 1468. **(B)** 1469. **(B)** 1470. **(C)**
1471. **(D)** 1472. **(D)** 1473. **(C)** 1474. **(C)** 1475. **(A)** 1476. **(A)** 1477. **(C)** 1478. **(B)** 1479. **(C)** 1480. **(B)**
1481. **(C)** 1482. **(C)** 1483. **(D)** 1484. **(C)** 1485. **(D)** 1486. **(D)** 1487. **(D)** 1488. **(B)** 1489. **(C)** 1490. **(B)**
1491. **(C)** 1492. **(C)** 1493. **(D)** 1494. **(A)** 1495. **(B)** 1496. **(B)** 1497. **(D)** 1498. **(A)** 1499. **(D)** 1500. **(A)**
1501. **(A)** 1502. **(D)** 1503. **(B)** 1504. **(B)** 1505. **(C)** 1506. **(B)** 1507. **(C)** 1508. **(D)** 1509. **(B)** 1510. **(B)**
1511. **(C)** 1512. **(C)** 1513. **(A)** 1514. **(C)** 1515. **(B)** 1516. **(C)** 1517. **(A)** 1518. **(B)** 1519. **(D)** 1520. **(B)**
1521. **(D)** 1522. **(D)** 1523. **(A)** 1524. **(D)** 1525. **(A)** 1526. **(B)** 1527. **(B)** 1528. **(B)** 1529. **(B)** 1530. **(A)**
1531. **(B)** 1532. **(A)** 1533. **(B)** 1534. **(B)** 1535. **(C)** 1536. **(C)** 1537. **(D)** 1538. **(C)** 1539. **(D)** 1540. **(D)**
1541. **(C)** 1542. **(D)** 1543. **(C)** 1544. **(A)** 1545. **(A)** 1546. **(C)** 1547. **(A)** 1548. **(A)** 1549. **(C)** 1550. **(D)**
1551. **(D)** 1552. **(D)** 1553. **(A)** 1554. **(B)** 1555. **(C)** 1556. **(A)** 1557. **(B)** 1558. **(B)** 1559. **(A)** 1560. **(D)**
1561. **(C)** 1562. **(B)** 1563. **(C)** 1564. **(C)** 1565. **(B)** 1566. **(D)** 1567. **(C)** 1568. **(B)** 1569. **(B)** 1570. **(A)**
1571. **(B)** 1572. **(B)** 1573. **(A)** 1574. **(B)** 1575. **(A)** 1576. **(D)** 1577. **(A)** 1578. **(C)** 1579. **(B)** 1580. **(A)**
1581. **(B)** 1582. **(C)** 1583. **(C)** 1584. **(D)** 1585. **(B)** 1586. **(D)** 1587. **(B)** 1588. **(C)** 1589. **(A)** 1590. **(B)**
1591. **(C)** 1592. **(B)** 1593. **(A)** 1594. **(B)** 1595. **(B)** 1596. **(B)** 1597. **(B)** 1598. **(A)** 1599. **(B)** 1600. **(C)**
1601. **(A)** 1602. **(C)** 1603. **(D)** 1604. **(C)** 1605. **(B)** 1606. **(D)** 1607. **(C)** 1608. **(D)** 1609. **(D)** 1610. **(A)**
1611. **(A)** 1612. **(D)** 1613. **(C)** 1614. **(B)** 1615. **(B)** 1616. **(A)** 1617. **(D)** 1618. **(B)** 1619. **(D)** 1620. **(B)**
1621. **(D)** 1622. **(B)** 1623. **(B)** 1624. **(C)** 1625. **(A)** 1626. **(C)** 1627. **(A)** 1628. **(B)** 1629. **(C)** 1630. **(B)**
1631. **(A)** 1632. **(D)** 1633. **(B)** 1634. **(C)** 1635. **(D)** 1636. **(B)** 1637. **(B)** 1638. **(B)** 1639. **(B)** 1640. **(A)**
1641. **(D)** 1642. **(B)** 1643. **(A)** 1644. **(A)** 1645. **(B)** 1646. **(C)** 1647. **(A)** 1648. **(D)** 1649. **(A)** 1650. **(A)**
1651. **(C)** 1652. **(C)** 1653. **(D)** 1654. **(A)** 1655. **(B)** 1656. **(D)** 1657. **(D)** 1658. **(D)** 1659. **(A)** 1660. **(A)**
1661. **(A)** 1662. **(A)** 1663. **(B)** 1664. **(A)** 1665. **(B)** 1666. **(D)** 1667. **(C)** 1668. **(C)** 1669. **(D)** 1670. **(B)**
1671. **(A)** 1672. **(B)** 1673. **(A)** 1674. **(C)** 1675. **(C)** 1676. **(B)** 1677. **(A)** 1678. **(B)** 1679. **(B)** 1680. **(A)**
1681. **(B)** 1682. **(B)** 1683. **(D)** 1684. **(B)** 1685. **(B)** 1686. **(A)** 1687. **(A)** 1688. **(D)** 1689. **(A)** 1690. **(C)**
1691. **(A)** 1692. **(D)** 1693. **(D)** 1694. **(B)** 1695. **(A)** 1696. **(C)** 1697. **(C)** 1698. **(B)** 1699. **(C)** 1700. **(D)**
1701. **(D)** 1702. **(C)** 1703. **(B)** 1704. **(C)** 1705. **(D)** 1706. **(C)** 1707. **(C)** 1708. **(D)** 1709. **(A)** 1710. **(D)**
1711. **(B)** 1712. **(B)** 1713. **(B)** 1714. **(B)** 1715. **(D)** 1716. **(C)**

10 WEED MANAGEMENT

1. The practice of following a systematic, rotational sequence of herbicide in the same field is called:
   (A) Herbicide rotation
   (B) Crop rotation
   (C) Herbicide sequence
   (D) None of these
2. The basic aim of formulating herbicide are:
   (A) Ease of application
   (B) Increase efficiencies on the target plants
   (C) Both of these
   (D) None of these
3. Sprayable concentration may be done in the form of:
   (A) Emulsifiable concentration
   (B) Wettable powder
   (C) Dry flowable (D) All of these
4. The physically homogenous mixture of herbicide concentrate and the carrier is called:
   (A) Herbicide solution
   (B) Herbicide emulsion
   (C) Herbicide suspension
   (D) All of these
5. The fine particles of the wettable powder are dispersed in a suitable carrier is called:
   (A) Herbicide solution
   (B) Herbicide emulsion
   (C) Herbicide suspension
   (D) All of these
6. Herbicide concentration dispersed in carrier (and vice-versa), where each component maintain its original identity as minute globules is called:
   (A) Herbicide solution
   (B) Herbicide emulsion
   (C) Herbicide suspension
   (D) All of these
7. Herbicide formulations may be in the form of:
   (A) EC & SC (B) WP & DF
   (C) FL & G (D) All of these
8. The movement of a herbicide from the surface into the plant body in a limited zone is called:
   (A) Absorption (B) Adsorption
   (C) Translocation (D) None of these
9. The transfer of a herbicide, food, or other material from one part of the plant to another is called:
   (A) Absorption (B) Adsorption
   (C) Translocation (D) None of these
10. The molecules with electro positive and electro negative regions and they tend to ionize is called:
    (A) Polar compound (B) Apolar compound
    (C) Molecules (D) All of these
11. Molecules with no surface distribution of charge are said to be :
    (A) Polar compound
    (B) Apolar compound
    (C) Molecules (D) All of these
12. Substance capable of antagonizing specific herbicides phytotoxicity to plant is called:
    (A) Safner (B) Antidote
    (C) Antagonists (D) All of these
13. The herbicide safners are presently used:
    (A) Benoxacor & Dichlormid
    (B) Flurazole & Fluxofenim
    (C) Oxabetrinil & R 29148
    (D) All of these
14. Sprayable bioagent, presently the fungal spores of their extracts, applied like herbicides to kill the existing growth of weeds is called:
    (A) Bioherbicide (B) Mycoherbicide
    (C) Both of these (D) None of these

15. Commercial name of bioherbicides are:
(A) VELGO & ABG 5003
(B) BIOMEL & DEVINE
(C) CASST
(D) All of these

16. Name of plant pathogen of VELGO bioherbicide is:
(A) *Colletrotrichum cocodes*
(B) *Cercospora rodmanii*
(C) *Phytophthora palmivora*
(D) *Alternaria cassiae*

17. Name of plant pathogen of ABG 5003 bioherbicide is:
(A) *Colletrotrichum cocodes*
(B) *Cercospora rodmanii*
(C) *Phytophthora palmivora*
(D) *Alternaria cassiae*

18. Name of plant pathogen of DEVINE bioherbicide is:
(A) *Colletrotrichum cocodes*
(B) *Cercospora rodmanii*
(C) *Phytophthora palmivora*
(D) *Alternaria cassiae*

19. Name of plant pathogen of CASST bioherbicide is:
(A) *Colletrotrichum cocodes*
(B) *Cercospora rodmanii*
(C) *Phytophthora palmivora*
(D) *Alternaria cassiae*

20. Name of plant pathogen of BIOMEL bioherbicide is:
(A) *Colletrotrichum glocosporioidest sp malvae*
(B) *Cercospora rodmanii*
(C) *Phytophthora palmivora*
(D) *Alternaria cassiae*

21. VELGO bioherbicide effectively controlled
(A) *Abutilon theophorcii*
(B) *Aeschynomone virginica*
(C) Both of these
(D) None of these

22. ABG 5003 bioherbicide effectively controlled
(A) *Abutilon theophorcii*
(B) *Morrenia odorata*
(C) *Echhornia crassipus*
(D) *Malva pustilla*

23. BIOMEL bioherbicide effectively controlled
(A) *Abutilon theophorcii*
(B) *Morrenia odorata*
(C) *Echhornia crassipus*
(D) *Malva pustilla*

24. DEVINE bioherbicide effectively controlled
(A) *Abutilon theophorcii*
(B) *Morrenia odorata*
(C) *Echhornia crassipus*
(D) *Malva pustilla*

25. CASST bioherbicide effectively controlled
(A) *Cassia obtisifolia*
(B) *Morrenia odorata*
(C) *Echhornia crassipus*
(D) *Malva pustilla*

26. The name of natural bio-toxin in plant metabolites of *Streptomyces spp* is:
(A) Anisomycin (B) Biolophos
(C) Cercosporium (D) Tob toxin

27. The name of natural bio-toxin in plant metabolites of *Streptomyces hygrocopicus* is:
(A) Anisomycin (B) Biolophos
(C) Cercosporium (D) Tob toxin

28. The name of natural bio-toxin in plant metabolites of *Cercospora spp* is:
(A) Anisomycin (B) Biolophos
(C) Cercosporium (D) Tob toxin

29. The name of natural bio-toxin in plant metabolites of *Pseudomonas tabaci* is:
(A) Anisomycin (B) Biolophos
(C) Cercosporium (D) Tob toxin

30. The name of natural bio-toxin in plant metabolites of *Alternaria alternata* is:
(A) Anisomycin
(B) Biolophos
(C) Cercosporium
(D) Teu toxin

31. Chemical capable of improving the herbicidal effect in some way is known as:
(A) Adjuvants (B) Safners
(C) Antidotes (D) All of these

32. The adjuvants are:
(A) Triton & APSA 80
(B) Agrocer & Active 80
(C) AG – F & Luffmul
(D) All of these

33. The important kind of adjuvants are:
(A) Surfactants and Stabilizing agents
(B) Solvants and Humicants
(C) Stickers and activator
(D) All of these

34. Materials which facilitated and accentuates the emulsifying, dispersing, spreading, wetting and other surface modifying proportion of herbicide formulation is called:
(A) Surfactants
(B) Stabilizers
(C) Herbicide extenders
(D) None of these

35. Chemical for obtaining stable emulsion and suspension in the spray tank, individually known as emulsifier and dispersant, respectively known as:
(A) Surfactants
(B) Stabilizers
(C) Herbicide extenders
(D) None of these

36. Dietholate applied at the rate of:
(A) 0.50 – 0.75 kg/ha
(B) 0.75 – 1.0 kg/ha
(C) 1.0 – 1.5 kg/ha
(D) 1.5 – 2.0 kg/ha

37. The non-phytotoxic, specific microbial activity inhibitors delay the biodegradation of soil applied herbicides and thus maintain then weed control activity for some extended period, same time by a week or more is called:
(A) Surfactants
(B) Stabilizers
(C) Herbicide extenders
(D) None of these

38. Dietholate herbicide extender extends the soil residual life of:
(A) EPTC (B) Butylate
(C) Both of these (D) None of these

39. Coupling of herbicide molecules to some plant constituent, with consequent herbicide deactivation is called:
(A) Conjugation
(B) Coupling agents
(C) Compatibility agent
(D) All of these

40. Solvants used to solublize parent herbicide compound to form either its soluble concentrate or emulsifiable concentrate is called:
(A) Conjugation
(B) Coupling agents
(C) Compatibility agent
(D) All of these

41. Chemical used as aid to keep two or more kinds of toxicant molecules together is called:
(A) Conjugation
(B) Coupling agents
(C) Compatibility agent
(D) All of these

42. Adjuvant Triton (0.1%) used with:
(A) Metsulfuron methyl
(B) Atrazine
(C) Glyphosate
(D) Various herbicides

43. Adjuvant APSA (0.1 %) used with:
(A) Metsulfuron methyl
(B) Atrazine
(C) Glyphosate
(D) Various herbicides

44. Adjuvant Agrocer 02 (0.1%) used with:
(A) Metsulfuron methyl
(B) Atrazine
(C) Glyphosate
(D) Various herbicides

45. Adjuvant AG – F (0.5 %) used with:
(A) Metsulfuron methyl
(B) Atrazine
(C) Glyphosate
(D) Various herbicides

46. Adjuvant Laffmul DA (0.2 %) used with:
(A) Metsulfuron methyl
(B) Atrazine
(C) Glyphosate
(D) Various herbicides

47. Adjuvant Castor oil (0.2 %) used with:
(A) Metsulfuron methyl
(B) Atrazine
(C) Glyphosate
(D) Various herbicides

48. The measurement of the potency and stimulus by a chemical by means of reach in which it produces in living matter is called:
(A) Bioassey
(B) Spray drift
(C) Fate of herbicide
(D) None of these

49. The movement of air borne spray particles from the intended area of application to non-target area:
(A) Bioassey (B) Spray drift
(C) Fate of herbicide (D) None of these

50. The diluting the herbicide in soil by its transfer or decomposition is known as:
(A) Bioassey
(B) Spray drift
(C) Fate of herbicide
(D) None of these

51. A herbicide that kills the target plant by way of inducing excessive production of singlet oxygen in its green tissue cells is called:
(A) Bioassey
(B) Lesser herbicide
(C) Fate of herbicide
(D) None of these

52. The tendency of a compound to undergo reaction is called:
(A) Bioassey
(B) Spray drift
(C) Redox potential
(D) None of these

53. The redox potential is measured by:
(A) Potentiometer (B) Potometer
(C) Tensiometer (D) None of these

54. Weeds that has become an integral part of a crop ecosystem is called:
(A) Satellite weed
(B) Allien weed
(C) Facultative weed
(D) Obligate weed

55. The ratio of ED 50 of test crop to ED 50 of associated weeds is called:
(A) Selectivity index
(B) Crop index
(C) Both of these (D) None of these

56. Chemical used to kill tree is called:
(A) Silvicide (B) Arboricide
(C) Both of these (D) None of these

57. Grain yield due to weeding as percentage of yield of the weed free crop is called:
(A) Weed Competition Index
(B) Crop-Weed Competition
(C) Weed Control Efficiency
(D) None of these

58. The complete elimination of all live plants, plant parts and seeds from an area is called:
(A) Weed eradication
(B) Weed prevention
(C) Weed control
(D) None of these

59. Chemical employed to keep a herbicide moist longer on the leaf surface is called:
(A) Humicant (B) Hygroscopic agent
(C) Both of these (D) None of these

60. A component which causes the leaves or foliage to drop from the plant is called:
(A) Defoliant (B) Defoliator
(C) Both of these (D) None of these

61. The time needed for half the pesticide to disappear from the soil is known as:
(A) Half time (B) Full time
(C) Both of these (D) None of these

62. All measure which deny the entry and establishment of weeds in an area is called:
(A) Weed eradication
(B) Weed prevention
(C) Weed control
(D) None of these

63. Probable origin of *Convolvulus arvensis* is:
(A) Eurasia (B) Europe
(C) Tropical America
(D) Central America

64. Probable origin of *Cirsium arvense* is:
(A) Eurasia (B) Europe
(C) Tropical America
(D) Central America

65. Probable origin of *Cyperus rotundus* is:
(A) Eurasia (B) Europe
(C) Tropical America
(D) Central America

66. Probable origin of *Euphorbia esula* is:
(A) Eurasia (B) Europe
(C) Tropical America
(D) Central America

67. Probable origin of *Eichhornia crassipes* is:
(A) Eurasia (B) Europe
(C) Tropical America
(D) Central America

68. Probable origin of *Lantana camara* is:
(A) Eurasia (B) Europe
(C) Tropical America
(D) Central America

69. Probable origin of *Opuntia spp.* is:
(A) Eurasia (B) Europe
(C) South America
(D) Central America

70. Probable origin of *Orobanche spp* is:
(A) Eurasia
(B) Europe
(C) South America
(D) Central America

71. Probable origin of *Pluchea spp.* is:
(A) Eurasia
(B) Europe
(C) South-West USA
(D) Central America

72. Probable origin of *Sorghum helepense* is:
(A) Eurasia (B) South Europe
(C) Tropical America
(D) Central America

73. Probable origin of *Tribulus terrestris* is:
(A) Eurasia (B) Africa
(C) Tropical America
(D) Central America

74. Which weed species produces highest number of seeds per plant:
(A) *Amaranthus spp.*
(B) *Chenopodium album*
(C) *Trianthema spp.*
(D) *Cuscuta spp.*

75. State of seeds in which these are alive but not ready to germinate or sprout is called:
(A) Dormancy
(B) Enforced dormancy
(C) Innate dormancy
(D) Induced dormancy

76. The dormancy in weed seeds is due to their placement deeper than 5 cm, resulting usually from tillage of the field is called:
(A) Dormancy
(B) Enforced dormancy
(C) Innate dormancy
(D) Induced dormancy

77. The dormancy is a genetically controlled character and it is a feature of specific weed seeds is called:
(A) Dormancy
(B) Enforced dormancy
(C) Innate dormancy
(D) Induced dormancy

78. The dormancy results from some sudden physiological change in otherwise non-dormant weed seeds is called:
(A) Dormancy
(B) Enforced dormancy
(C) Innate dormancy
(D) Induced dormancy

79. The important herbicide in phenol group is/are:
(A) DNBP (B) Dicamba
(C) Fenac (D) Amitrole

80. The important herbicide in phenoxyalkanoic acid group is/are:
(A) 2,4-D (B) MCPA
(C) Macroprop (D) All of these

81. The important herbicide in phenylalkanoic acid group is/are:
(A) DNBP (B) Dicamba
(C) Fenac (D) Amitrole

82. The important herbicide in Chlorobenzoic and anisic acid group is/are:
(A) DNBP (B) Dicamba
(C) Fenac (D) Amitrole

83. The important herbicide in aliphatic acid group is/are:
(A) TCA (B) Dalapan
(C) Both of these (D) None of these

84. The important herbicide in Triazines group is/are:
(A) Simazine (B) Atrazine
(C) Prometron (D) All of these

85. The important herbicide in Triazoles group is/are:
(A) DNBP (B) Dicamba
(C) Fenac (D) Amitrole

86. The important herbicide in Diazoles group is/are:
(A) Difenzoquat (B) Methozole
(C) Both of these (D) None of these

87. The important herbicide in Diazines group is/are:
(A) MH (B) Pyrazon
(C) Bentazone (D) All of these

88. The important herbicide in Puridines group is/are:
(A) DNBP (B) Dicamba
(C) Picloram (D) Amitrole

89. The important herbicide in Quaternary ammonium compound group is/are:
(A) Diquat (B) Paraquat
(C) Both of these (D) None of these

90. The important herbicide in Uracils group is/are:
(A) Bromosil (B) Terbacil
(C) Both of these (D) None of these

91. The important herbicide in substitutes ureas group is/are:
(A) Diuran (B) Isoproturon
(C) Methabenzathiazuron
(D) All of these

92. The important herbicide in Carbanilates and carbomats group is/are:
(A) Thiobencarb (B) EPTC
(C) Barban (D) All of these

93. The important herbicide in Amides and anilides group is/are:
(A) Alachlor (B) Butachlor
(C) Metolachlor (D) All of these

94. The important herbicide in Anilines and toluidines group is/are:
(A) Fluchloralin (B) Pendimethalin
(C) Trifluralin (D) All of these

95. The important herbicide in Benonitriles group is/are:
(A) Bromoxynil (B) Ioxynil
(C) Dichlobenil (D) All of these

96. The important herbicide in Sulfonylureas group is/are:
(A) Metsulfuron (B) Chlorsulfuron
(C) Both of these (D) None of these

97. The important herbicide in organic arsenicals group is/are:
(A) MSMA (B) DSMA
(C) MAA (D) All of these

98. The important herbicide in miscellaneus group is/are:
(A) Glyphosate (B) Imazathapyr
(C) Fluazifop-p (D) All of these

99. Recommended dose of sulfosulfuron is:
(A) 15-20 g/ha (B) 25-50 g/ha
(C) 50-75 g/ha (D) 75-100 g/ha

100. Recommended dose of triasulfuron is:
(A) 15-20 g/ha (B) 25-50 g/ha
(C) 50-75 g/ha (D) 75-100 g/ha

101. Recommended dose of chlorimuron-ethyl is:
(A) 4-5 g/ha (B) 25-50 g/ha
(C) 50-75 g/ha (D) 75-100 g/ha

102. Recommended dose of acetachlor is:
(A) 0.5-1.0 kg/ha (B) 1.0-2.0 kg/ha
(C) 1.5-3.0 kg/ha (D) 3.0-4.0 kg/ha

103. Recommended dose of carfentrazone-ethyl is:
(A) 15-20 g/ha (B) 20-25 g/ha
(C) 25-40 g/ha (D) 50-60 g/ha

104. Recommended dose of chlorsulfuron is:
(A) 15-45 g/ha (B) 50-75 g/ha
(C) 75-100 g/ha (D) 100-125 g/ha

105. Recommended dose of cinmethylin is:
(A) 15-20 g/ha (B) 25-50 g/ha
(C) 50-75 g/ha (D) 75-100 g/ha

106. Recommended dose of cinsulfuron is:
(A) 10-80 g/ha (B) 100-120 g/ha
(C) 125-150 g/ha (D) 160-170 g/ha

107. Recommended dose of clodinafop-propargyl is:
(A) 15-20 g/ha (B) 40-90 g/ha
(C) 100-120 g/ha (D) 125-140 g/ha

108. Recommended dose of clomazone is:
(A) 150 g/ha (B) 200 g/ha
(C) 250 g/ha (D) 300 g/ha

109. Recommended dose of cyhalofop-butyl is:
(A) 15-20 g/ha (B) 25-50 g/ha
(C) 60-100 g/ha (D) 100-120 g/ha

110. Recommended dose of ethoxysulfuron is:
(A) 100-150 g/ha (B) 200-250 g/ha
(C) 250-300 g/ha (D) 300-400 g/ha

111. Recommended dose of phenoxaprop-p-ethyl is:
(A) 15-20 g/ha (B) 25-50 g/ha
(C) 60-120 g/ha (D) 125-150 g/ha

112. Recommended dose of flupropanate is:
(A) 1-2 kg/ha (B) 2-3 kg/ha
(C) 3-4 kg/ha (D) 4-5 kg/ha

113. Recommended dose of fluroxypur is:
(A) 100-200 g/ha (B) 200-300 g/ha
(C) 300-400 g/ha (D) 400-500 g/ha

114. Recommended dose of foramsulfuron is:
(A) 15-20 g/ha (B) 30-40 g/ha
(C) 50-75 g/ha (D) 75-100 g/ha

115. Recommended dose of glufosinate ammecium is:
(A) 100-200 g/ha (B) 200-300 g/ha
(C) 300-400 g/ha (D) 400-500 g/ha

116. Recommended dose of haloxyfop EE is:
(A) 200-250 g/ha (B) 250-400 g/ha
(C) 400-500 g/ha (D) 500-600 g/ha

117. Recommended dose of Imezethapyr is:
(A) 0.7-1.5 kg/ha (B) 1.5-2.0 g/ha
(C) 2.0-2.5 kg/ha (D) 2.5-3.0 kg/ha

118. Recommended dose of mesosulfuron is:
(A) 10-12 g/ha (B) 12-19 g/ha
(C) 20-25 g/ha (D) 25-30 g/ha

119. Recommended dose of metsulfuron-methyl is:
(A) 1-2 g/ha (B) 2-3 g/ha
(C) 3-8 g/ha (D) 8-10 g/ha

120. Recommended dose propaquizafop is:
(A) 15-20 g/ha (B) 25-50 g/ha
(C) 50-60 g/ha (D) 75-100 g/ha

121. Recommended dose of oxadiargyl is:
(A) 15-20 g/ha (B) 25-50 g/ha
(C) 50-150 g/ha (D) 150-200 g/ha

122. Recommended dose quizalofop-p is:
(A) 15-20 g/ha (B) 25-50 g/ha
(C) 50-75 g/ha (D) 75-100 g/ha

123. Recommended dose of pyrozosulfuron-ethyl is:
(A) 15-60 g/ha (B) 60-75 g/ha
(C) 75-100 g/ha (D) 100-125 g/ha

124. Recommended dose of sulfometuron-methyl is:
(A) 15-60 g/ha (B) 60-75 g/ha
(C) 75-100 g/ha (D) 50-200 g/ha

125. A herbicide treatment made any time before the crop is planted is called:
(A) Pre-plant (B) Pre-emergence
(C) Post-plant (D) None of these

126. Application of herbicides soon after sowing of a crop is:
(A) Pre-plant (B) Pre-emergence
(C) Post-plant (D) None of these

127. Application of herbicides made after emergence or standing crop plant or weeds is called:
(A) Pre-plant (B) Pre-emergence
(C) Post-plant (D) None of these

128. All application involving wiping, rubbing and smearing of herbicide into the target plant surface is called:
(A) Pre-plant
(B) Pre-emergence
(C) Post-plant
(D) Direct contact application

129. Which herbicide applied by soil injection method is:
(A) Ethylene (B) Carbon bi sulphide
(C) Vernolate (D) All of these

130. If the rate of application per hectare is 3.00 kg a.i, the quantity of simazine WP (80% a.i.) required to be sprayed in 0.20 hectare area would be:
(A) 0.50 kg (B) 0.75 kg
(C) 1.25 kg (D) 1.87 kg

131. Pre-emergence herbicides are applied in a restricted area over the crop row only is called:
(A) Pre-plant
(B) Band application
(C) Blanket application
(D) Direct contact application

132. Certain non-selective or partially selective herbicides can be used to control weed in the inter rows by directing the spray on weeds is called:
(A) Direct application
(B) Band application
(C) Blanket application
(D) Direct contact application

133. The herbicide containing carbon and hydrogen in their molecule are called:
(A) Arsenic (B) Acid
(C) Organic herbicide (D) Salt

134. Which one of thee following causes more wastages of herbicide by drift:
(A) High volume sprayer
(B) Ultra-low volume sprayer
(C) Hand sprayer
(D) Duster

135. Spot application practices used for treating patches of:
(A) Annual weeds
(B) Biennial weed
(C) Perennial weeds
(D) Noxious perennial weeds

136. Which one of the following microbial agents is being commercially exploited as biocontrol agent:
(A) *Penicillium notatum*
(B) *Bacillus subtilis*
(C) *Trichoderma viridiae*
(D) *Sclerotium rolfisii*

137. *Parthenium hysterophorus* (Congress grass) can be controlled by:
(A) *Chrysomella spp.*
(B) *Dactylopius tomentosus*
(C) *Zygogramma bicolarata*
(D) *Neochetina spp.*

138. Most common method of fertilizer application is:
(A) Spraying (B) Broadcast
(C) Rope-wick (D) None of these

139. By which method of application may be prevent crop phytotoxicity caused by spray drift that occurs with conventional spray application is:
(A) Spraying (B) Broadcast
(C) Rope-wick (D) None of these

140. Ideal method of controlling perennial weeds (*Cyperus rotundus*) in standing crop is:
(A) Spraying (B) Broadcast
(C) Rope-wick (D) None of these

141. Low dose of herbicides is recommended in:
(A) Light soil (B) Heavy soil
(C) Both of these (D) None of these

142. High dose of herbicides is recommended in:
(A) Light soil (B) Heavy soil
(C) Both of these (D) None of these

143. Optimum soil moisture is pre-requisite for which type of herbicides:
(A) Pre-emergence
(B) Post emergence
(C) Both of these
(D) None of these

144. At what time of herbicide spraying gave maximum benefit:
(A) Sunny day (B) Cloudy day
(C) Night (D) None of these

145. Untimely rain or heavy irrigation may accentuate:
(A) Crop injury (B) Bo effect
(C) Unknown (D) None of these

146. The spray tank should be atleast full:
(A) One-third (B) One-fourth
(C) Half (D) Two-third

147. Which type of spray nozzle should be used for spraying the herbicides:
(A) Flood jet nozzle
(B) Flat fan nozzle
(C) Both of these
(D) None of these

148. Most herbicides are formulated as:
(A) Wettable powder (WP)
(B) Emulsifiable concentration (EC)
(C) Both of these
(D) All of these

149. Under submerged condition which type of herbicides formulation is used:
(A) Wettable powder (WP)
(B) Emulsifiable concentration (EC)
(C) Granular formulation (G)
(D) All of these

150. Application of herbicides through irrigation water is called:
(A) Herbigation (B) Herbicultre
(C) Herbilogy (D) None of these

151. Herbigation method of herbicide application mainly used in:
(A) Maharastra (B) Gujarat
(C) Rajasthan (D) Punjab

152. Commercial product of herbicide can be computed by:

(A) $\text{Commercial product} = \frac{\text{Dose in kg a.i./ha}}{\text{\%a.i. in the product}} \times 100$

(B) $\text{Commercial product} = \frac{\text{Dose in kg a.i./ha}}{\text{\% a.i. in the product}}$

(C) $\text{Commercial product} = \frac{\text{Dose in kg/ha}}{\text{\% a.i. in the product}} \times 100$

(D) $\text{Commercial product} = \frac{\text{Dose in kg a.i./ha}}{\text{a.i. in the product}} \times 100$

153. Which one of the following is not narrow leaved weeds:

(A) *Cynodon dactylon*
(B) *Cyperus rotundus*
(C) *Setaria glauca*
(D) *Melilotus indica*

154. Most common herbicide used in maize is:

(A) Atrazine (B) Alachlor
(C) Simazine (D) Metribuzin

155. Which one of the following was introduced in the country to eradicate prickly pear (*Opuuntia dilleni*) weed:

(A) *Chrysomella spp.*
(B) *Dactylopius tomentosus*
(C) *Zygogramma bicolarata*
(D) *Neochetina spp.*

156. Herbicide used to clear the vegetation and as a substitute for tillage is:

(A) Glyphosate (B) Atrazine
(C) Simazine (D) Pendimethalin

157. Excellent control of barn yard grass by:

(A) Cyhalofop-butyl (B) Sulfosulfuron
(C) Isoproturon
(D) Metribuzin

158. Excellent control of *Phalaris minor* by:

(A) Cyhalofop-butyl
(B) Sulfosulfuron
(C) Isoproturon
(D) Metribuzin

159. When chickpea intercropped with mustard which herbicide to be spray for controlling of broad leaved weeds is:

(A) Pendimethalin
(B) Fluchloralin
(C) Both of these
(D) None of these

160. Pendimethalin should be applied in pulses at:

(A) Pre-emergence
(B) Pre-planting
(C) Post-emergence
(D) None of these

161. Fluchloralin should be applied in pulses at:

(A) Pre-emergence (B) Pre-planting
(C) Post-emergence (D) None of these

162. Which is not correctly match in the following statement:

(A) Flocculation : Setting of the soil particles
(B) Dispersion : Repellence of the soil particles
(C) Cohesion : Attraction of the soil particles
(D) Structure : Relaive distribution of the soil particles

163. Most critical weed competition period is upto ------- months after transplanting exist in sugarcane:

(A) 6 (B) 4
(C) 3 (D) 8

164. Most common herbicides used for weed control in sugarcane is/are:

(A) Simazine (B) Atrazine
(C) Alachlor (D) All of these

165. Relative weeds are:

(A) Plants of same crops but other variety in the field
(B) Plants of other crop in the crop field
(C) Seasonal weeds in the same field
(D) All of these

166. 'Atrazine' effectively control grasses in maize field without causing harm to the crop plants, as degradation of the chemical in maize plant occurs due to the presence of:

(A) R-Q enzyme (B) Amylases
(C) Aryl acyl amidase
(D) GSH enzyme

167. Which one of the following is mycoherbicide:

(A) Diquat
(B) Met sulphuron methyl
(C) Collego
(D) Both (A) and (B)

168. WP abbreviates as:
(A) Wettable Paste (B) Wettable Powder
(C) Water Paste (D) None of these

169. *Cyperus rotundus* is a:
(A) Perrenial (B) Rabi
(C) Kharif (D) Annual

170. *Cynodon dactylon* is a:
(A) Perrenial (B) Rabi
(C) Kharif (D) Annual

171. *Digera arvensis* is a:
(A) Perrenial (B) Rabi
(C) Kharif (D) Annual

172. *Euphorbia hirta* is a:
(A) Perrenial (B) Rabi
(C) Kharif (D) Annual

173. *Launea aspara* is a:
(A) Perrenial (B) Rabi
(C) Kharif (D) Annual

174. *Lantana camara* is a:
(A) Perrenial (B) Rabi
(C) Kharif (D) Annual

175. *Melilotus alba* is a:
(A) Perrenial (B) Rabi
(C) Kharif (D) Annual

176. *Argemone mexicana* is a:
(A) Perrenial (B) Rabi
(C) Kharif (D) Annual

177. *Asphodelus tenuifolius* is a:
(A) Perrenial (B) Rabi
(C) Kharif (D) Annual

178. *Avena fatua* is a:
(A) Perrenial (B) Rabi
(C) Kharif (D) Annual

179. *Convolvulus arvensis* is a:
(A) Perrenial (B) Rabi
(C) Kharif (D) Annual

180. *Chenopodium album* is a:
(A) Perrenial (B) Rabi
(C) Kharif (D) Annual

181. *Comelina benghalensis* is a:
(A) Perrenial (B) Rabi
(C) Kharif (D) Annual

182. Calculate weed control efficiency in terms of drymatter production of weeds. If weed drymatter production in mustard crop under weeded control (check) is 500 g/m^2 and weed drymatter accumulation in Pendimethalin treated plot in 20 g/m^2:
(A) 91% (B) 92%
(C) 93% (D) 94%

183. Calculate weed index, if the crop yield of sesame where minimum weed competition is 700 g/m^2 and yield of sesame for treatment for which weed index has to be calculated in 350 g/m^2:
(A) 80% (B) 70%
(C) 60% (D) 50%

184. Which herbicide residue for the longest period in the soil?
(A) Chlorprophan (B) Linuron
(C) Propachlor (D) Simazine

185. *Phyllanthus niruri* is a:
(A) Perrenial (B) Rabi
(C) Kharif (D) Annual

186. *Mimosa pudica* is a:
(A) Perrenial (B) Rabi
(C) Kharif (D) Annual

187. *Cuscuta* spp is a:
(A) Perrenial (B) Rabi
(C) Kharif (D) Annual

188. Weeder is the trade name of:
(A) 2,4,5-T (B) MCPA
(C) Simazine (D) Monuron

189. Methoxone and Agroxone-3 is the trade name of:
(A) 2,4,5-T (B) MCPA
(C) Simazine (D) Monuron

190. Geseprim is the trade name of:
(A) 2,4,5-T (B) MCPA
(C) Simazine (D) Monuron

191. Talvar is the trade name of:
(A) 2,4,5-T (B) MCPA
(C) Simazine (D) Monuron

192. Gramaxone is the trade name of:
(A) Paraquat (B) MCPA
(C) Simazine (D) Monuron

193. Ansar 8100 is the trade name of:
(A) 2,4,5-T (B) MCPA
(C) DSMA (D) Monuron

194. Tolkan is the trade name of:
(A) 2,4,5-T (B) Isoprotoron
(C) Simazine (D) Monuron

195. Lasso is the trade name of:
(A) 2,4,5-T (B) MCPA
(C) Simazine (D) Alachlor

196. Treflan is the trade name of:
(A) 2,4,5-T (B) Trifluralin
(C) Simazine (D) Monuron

197. Basaline is the trade name of:
(A) 2,4,5-T (B) MCPA
(C) Fluchloralin (D) Monuron

198. Stomp and herbidex is the trade name of:
(A) Pendimethalin (B) MCPA
(C) Simazine (D) Monuron

199. Pursuit is the trade name of:
(A) 2,4,5-T (B) MCPA
(C) Simazine (D) Imazathapyr

200. Reglone is the trade name of:
(A) Diquat (B) MCPA
(C) Simazine (D) Monuron

201. Ronstar is the trade name of:
(A) 2,4,5-T (B) MCPA
(C) Simazine (D) Oxadiazon

202. Dual is the trade name of:
(A) 2,4,5-T (B) Metolachor
(C) Simazine (D) Monuron

203. Goal is the trade name of:
(A) Oxyfluorfen (B) MCPA
(C) Simazine (D) Monuron

204. Targasuper is the trade name of:
(A) Quizalofop-ethyl
(B) MCPA
(C) Simazine (D) Monuron

205. Roundup and Glycil is the trade name of:
(A) 2,4,5-T (B) Glyphosate
(C) Simazine (D) Monuron

206. Klass is the trade name of:
(A) 2,4,5-T (B) MCPA
(C) Diuron (D) Monuron

207. Almix is the trade name of:
(A) Chlorimuron ethyl + Metsulfuron methyl
(B) MCPA
(C) Simazine (D) Monuron

208. Dowpon is the trade name of:
(A) 2,4,5-T (B) MCPA
(C) Dalapon (D) Monuron

209. Valour is the trade name of:
(A) 2,4,5-T (B) MCPA
(C) Imazethapyr + Pendimethalin
(D) Monuron

210. Focus and Gallant is the trade name of:
(A) 2,4,5-T (B) MCPA
(C) Simazine
(D) Haloxyfop-butyl

211. Sofit is the trade name of:
(A) Pretilachlor + safner
(B) MCPA
(C) Simazine (D) Monuron

212. Leader is the trade name of:
(A) 2,4,5-T (B) Sulfosulfuron
(C) Simazine (D) Monuron

213. Saturn is the trade name of:
(A) 2,4,5-T (B) MCPA
(C) Simazine (D) Thiobencarb

214. Yield is the trade name of:
(A) 2,4,5-T
(B) Methabenzthiazuron
(C) Simazine
(D) Monuron

215. Imparting resistance to normally herbicide susceptible crops to produce:
(A) Herbicide resistant crop
(B) Genetically modified crops
(C) Both of these
(D) None of these

216. A chemical in commercial product that is directly responsible for its herbicidal effect is known as:
(A) Active ingredient
(B) Acid equivalent
(C) Both in these
(D) None of these

217. The theoretical yield of parent acid from such a herbicide formulation is called:
(A) Active ingredient
(B) Acid equivalent
(C) Both in these
(D) None of these

218. Long term use of herbicide can lead to the development in weeds:
(A) Resistance (B) Persistence
(C) Both of these (D) None of these

219. In rice-wheat cropping system, which herbicide is more effective for control of *Eichinoclova colonum* is:

(A) Butachlor (B) Metholachlor
(C) Alachlor (D) Isoproturon

220. Prominent bioagent (*Puccinia chondrillina*) is effective against is:

(A) *Chondrilla juncea*
(B) *Cirsium arvense*
(C) *Cyperus rotundus*
(D) *Echinoclova spp.*

221. Prominent bioagent (*Septonia cirsii*) is effective against is:

(A) *Chondrilla juncea*
(B) *Cirsium arvense*
(C) *Cyperus rotundus*
(D) *Echinoclova spp.*

222. Prominent bioagent (*Bactra vendana*) is effective against is:

(A) *Chondrilla juncea*
(B) *Cirsium arvense*
(C) *Cyperus rotundus*
(D) *Echinoclova spp.*

223. Prominent bioagent (*Emmalocera spp.*) is effective against is:

(A) *Chondrilla juncea*
(B) *Cirsium arvense*
(C) *Cyperus rotundus*
(D) *Echinoclova spp.*

224. Prominent bioagent (*Sclerotinia spp*) is effective against is:

(A) *Chondrilla juncea*
(B) *Cirsium arvense*
(C) *Orobanche cernua*
(D) *Echinoclova spp.*

225. Prominent bioagent (*Zygogramma bicolorata*) is effective against is:

(A) *Chondrilla juncea*
(B) *Cirsium arvense*
(C) *Cyperus rotundus*
(D) *Parthenium hysterophorus*

226. Prominent bioagent (*Uromyces rumicis*) is effective against is:

(A) *Rumex spp*
(B) Cirsium arvense
(C) *Cyperus rotundus*
(D) *Echinoclova spp.*

227. Prominent bioagent (*Microplatinus lareynil*) is effective against is:

(A) *Chondrilla juncea*
(B) *Tribulus terrestris*
(C) *Cyperus rotundus*
(D) *Echinoclova spp.*

228. Prominent bioagent (*Crocidosema lantana*) is effective against is:

(A) *Chondrilla juncea*
(B) *Cirsium arvense*
(C) *Lantana camara*
(D) *Echinoclova spp.*

229. Prominent bioagent (*Leptospermum reoparium*) is effective against is:

(A) *Chondrilla juncea*
(B) Manuka weeds
(C) *Cyperus rotundus*
(D) *Echinoclova spp.*

230. Which species of fish consume large quantity of aquatic weeds:

(A) Chinese grass carp
(B) Carp
(C) Both of these
(D) None of these

231. Calculate weed control efficiency if, weed population in untreated plot of maize is 350 weed/m^2 and herbicide sprayed plot with simazine the weed population is 15 weeds/m^2:

(A) 95% (B) 95.7%
(C) 97% (D) 98%

232. How much area by the above material would be covered?

(A) 0.8 ha (B) 1.0 ha
(C) 1.2 ha (D) 1.4 ha

233. Weeds grow in highly acidic soils are called:

(A) Acidophills (B) Basiphills
(C) Neutrophills (D) None of these

234. Weeds grow in saline and alkali soils are called:

(A) Acidophills (B) Basiphills
(C) Neutrophills (D) None of these

235. Weeds grow in neutral soils are called:

(A) Acidophills (B) Basiphills
(C) Neutrophills (D) None of these

236. Those weed species that grow primarily in wild commodities but often escape to cultivated field, associated themselves closely with men's affairs is called:
(A) Alien weeds
(B) Native weeds
(C) Facultative weeds
(D) Noxicious weeds

237. How much herbicide having 70 per cent acid equivalent or active ingredient would you add in 500 litre of water to obtain a recommended rate of 0.5 kg a.i./ha?
(A) 0.71 kg
(B) 0.81 kg
(C) 0.85 kg
(D) 0.91 kg

238. The weeds occur only in the cultivated or otherwise disturbed land is called:
(A) Obligate weeds
(B) Native weeds
(C) Facultative weeds
(D) Noxicious weeds

239. Weed control efficency formula has been suggested by:
(A) Patil and Patil (1983)
(B) Gill and Kumar (1969)
(C) Misra (1968)
(D) Raju (1977)

240. Weed index formula has been suggested by:
(A) Patil and Patil (1983)
(B) Gill and Kumar (1969)
(C) Misra (1968)
(D) Raju (1977)

241. Weed density formula has been suggested by:
(A) Patil and Patil (1983)
(B) Gill and Kumar (1969)
(C) Misra (1968)
(D) Raju (1977)

242. Some weeds survive because of their similarity in morphology with the host crop is called:
(A) Mimicry
(B) Native weeds
(C) Facultative weeds
(D) Noxicious weeds

243. Criteria for successful bioagents are:
(A) Host specific
(B) Chiefly & easily available
(C) Hardy in nature
(D) All of these

244. Useful aspects of weeds are:
(A) Weeds helps in scientific discoveries and studies
(B) Weeds helps in reducing wind and water erosion
(C) Legume weeds improves fertility status of soil
(D) All of these

245. It is estimated that the herbicide market would grow at:
(A) 5 per cent per annum
(B) 10 per cent per annum
(C) 15 per cent per annum
(D) 20 per cent per annum

246. Highest consumption of herbicide in:
(A) Wheat (B) Rice
(C) Soybean (D) Pulses

247. Herbicide are being used in India:
(A) 15 mha (B) 20 mha
(C) 25 mha (D) 30 mha

248. Out of the total losses due to various pests, weeds account for nearly:
(A) One-third (B) Half
(C) Two-third (D) Three-fourth

249. The undesirable, injurious, unsightly & troublesome plant, which interfere with cultivated crops and affect human affairs is called:
(A) Weeds (B) Insect
(C) Crop (D) None of these

250. The weed seed mainly mixed with berseem is:
(A) *Chinchorium intybus*
(B) *Avena fatua*
(C) *Phalaris minor*
(D) Melilotus album

251. The weed seed is similar to mustard crop is:
(A) Chinchorium intybus
(B) Argemone mexicana
(C) Phalaris minor
(D) Melilotus album

252. The weed plant is similar to rice crop is:
(A) *Chinchorium intybus*
(B) *Avena fatua*
(C) *Phalaris minor*
(D) Oryza sativa var fatua

253. The weed plant is similar to wheat crop is:
((A) *Chinchorium intybus*
(B) *Avena fatua*
(C) *Phalaris minor*
(D) Melilotus album

254. The weed plant is similar to oat crop is:
(A) *Chinchorium intybus*
(B) *Avena fatua*
(C) *Phalaris minor*
(D) Melilotus album

255. The fibre obtained from weed is:
(A) *Sida cardifolia* (B) *Abutilon indicum*
(C) *Corchorus spp* (D) All of these

256. The industrial effluents absorb to heavy metal by:
(A) Water hyacinth
(B) Hydrilla
(C) Pimpernells
(D) All of these

257. The leaves of weed are used for preparing mats is:
(A) *Typha latifolia*
(B) *Saccharum munja*
(C) Both of these
(D) None of these

258. The natural agencies for dissemination of weeds is:
(A) Wind (B) Water
(C) Animals (D) All of these

259. Dissemination by *Xanthium strumarium* mainly by:
(A) Wind (B) Water
(C) Animals (D) All of these

260. Some weed seeds are dormant when they leave the parent plant is called :
(A) Innate dormancy
(B) Induced dormancy
(C) Inforced dormancy
(D) None of these

261. Seeds which ordinarily would germinate immediately if planted under favourable conditions may be thrown into dormancy by an unfavourable environment so that they will not germinate even when conditions because favourable is called:
(A) Innate dormancy
(B) Induced dormancy
(C) Inforced dormancy
(D) None of these

262. Seed may be prevented from germination by various environmental factors is known as:
(A) Innate dormancy
(B) Induced dormancy
(C) Inforced dormancy
(D) None of these

263. Weeds endanger biodiversity is:
(A) *Mimosa rubicaulis*
(B) *Mikania micrantha*
(C) Both of these
(D) All of these

264. Weeds are a great nuisance to fisheries, navigation, irrigation and hydroelectric projects are:
(A) Water hyacinth (B) *Salvinia molesta*
(C) Hydrilla (D) All of these

265. Management of an alien weeds is a great challenge to the weed research scientists in the country is:
(A) *Ipomea cernua*
(B) *Chenopodium album*
(C) *Avena fatua*
(D) *Convolvulus arvensis*

266. Herbicides are being used in approximately of the total cropped area is:
(A) 10 % (B) 15 %
(C) 20 % (D) 25 %

267. Recently resitance to isoproturon is developed in:
(A) *Phalaris minor* *(B) Avena fatua*
(C) *Chenopodium album* (D) None of these

268. New herbicide successfully controlled *Phalaris minor* is:
(A) Cladinafop (B) Fenoxaprop
(C) Sulfosulfuron (D) All of these

269. Which weedicide is used to kill weeds of broad leaves in legumes:
(A) 2,4-D (B) 2,4,5-T
(C) 2, 4-DB (D) None of these

270. Post emergence herbicide mentribuzin is recommended for control of broad leaves weeds in:
(A) Chickpea (B) Lentil
(C) Field pea (D) Lathyrus

271. Post emergence herbicide pursuit is recommended for control of broad leaves weeds in:
(A) Pigeonpea (B) Lentil
(C) Field pea (D) Black gram

272. Orobanche is a:
(A) Total root parasite
(B) Total stem parasite
(C) Semi root parasite
(D) Semi stem parasite

273. In pigeonpea + soybean intercropping following herbicide can be used for successful weed control:
(A) 2,4-D (B) Pendimethalin
(C) Isoproturon (D) Butachlor

274. The recommended herbicides for weed control in chickpea is:
(A) Basalin (B) Butachlor
(C) Alachlor (D) 2,4-D

275. Herbicide used for weed control in sorghum is:
(A) Atrazine at 0.5-1.0 kg a.i./ha
(B) Atrazine at 1.25-1.50 kg a.i./ha
(C) Basalin at 0.5-1.0 kg a.i./ha
(D) Basalin at 1.25-1.50 kg a.i./ha

276. Herbicide has been recommended for the control of weeds in mungbean:
(A) Basalin @ 1.0 kg a.i./ha
(B) 2,4-D @ 1 litre a.i./ha
(C) Simazine @ 1 kg a.i./ha
(D) Atrazine @ 1 kg a.i./ha

277. It has been found that the incorporation of basalin provides relatively more effective control of weeds in mungbean:
(A) Pre-planting incorporation
(B) Post-emergence (20 DAS)
(C) Pre-emergence
(D) Post emergence (40 DAS)

278. Crop yield losses occurred due to weeds nearly:
(A) 15 % (B) 22 %
(C) 29 % (D) 37 %

279. Crop yield losses occurred due to diseases nearly:
(A) 15 % (B) 22 %
(C) 29 % (D) 37 %

280. Crop yield losses occurred due to insects nearly:
(A) 15 % (B) 22 %
(C) 29 % (D) 37 %

281. Herbicides is used in India approximately of the total pesticides consumption is:
(A) 10% (B) 15 %
(C) 20 % (D) 25 %

282. Use of herbicides in wheat in India is:
(A) 4 % (B) 6 %
(C) 17 % (D) 57 %

283. Use of herbicides in rice in India is:
(A) 4 % (B) 6 %
(C) 17 % (D) 57 %

284. Use of herbicides in soybean in India is:
(A) 4 % (B) 6 %
(C) 17 % (D) 57 %

285. Use of herbicides in plantation crops in India is:
(A) 4 % (B) 6 %
(C) 17 % (D) 57 %

286. Weeds which complete their life cycle within one season or year is called:
(A) Annual (B) Binnial
(C) Perennial (D) None of these

287. Weeds which complete their life cycle in two years is called:
(A) Annual (B) Binnial
(C) Perennial (D) None of these

288. Weeds which complete their life cycle in more than two years is called:
(A) Annual (B) Binnial
(C) Perennial (D) None of these

289. Annual weeds is:
(A) *Daucos carota*
(B) Trianthema portulacastrum
(C) *Agropyron repens* (D) All of these

290. Binnial weeds is:
(A) *Daucos carota*
(B) Trianthema portulacastrum
(C) *Agropyron repens*
(D) All of these

291. Perennial weeds is:
(A) *Daucos carota*
(B) Trianthema portulacastrum
(C) *Agropyron repens*
(D) All of these

292. Monocot weed is:
(A) *Echinoclova colonum*
(B) Trianthema portulacastrum
(C) *Chenopodium album*
(D) Anagalis arvensis

293. Dicot weed is:
(A) *Echinoclova colonum*
(B) Trianthema portulacastrum
(C) *Avena sativa*
(D) Phalaris minor

294. Many weeds in India originated in some other part of the world is called:
(A) Native weeds
(B) Allien weeds
(C) Facultative weeds
(D) Obligate weeds

295. Weeds occur within the geographical limits of their origin is called:
(A) Native weeds
(B) Allien weeds
(C) Facultative weeds
(D) Obligate weeds

296. Weeds of crop land is capable of surviving in a wild community is called:
(A) Native weeds
(B) Allien weeds
(C) Facultative weeds
(D) Obligate weeds

297. Facultative weeds are also called:
(A) Apopytes (B) Symplasts
(C) Apoplasts (D) None of these

298. Alien weeds are:
(A) *Eichhornia crasiipus*
(B) *Lantana camara*
(C) *Argemone mexicana*
(D) All of these

299. Weeds that undesirable, troublesome and difficult to control is called:
(A) Native weeds (B) Noxious weeds
(C) Facultative weeds
(D) Obligate weeds

300. Noxious weeds whose seeds are difficult to separate, when once mixed with crop stand is called:
(A) Objectionable weeds
(B) Allien weeds
(C) Facultative weeds
(D) Obligate weeds

301. Difficult perennial weeds are also called:
(A) Native weeds
(B) Allien weeds
(C) Pernicious
(D) Obligate weeds

302. Which one is a troublesome weeds of agricultural fields?
(A) *Chenopodim album*
(B) *Cyperus rotundus*
(C) *Partthenium hysterophorus*
(D) *Eichhornia crassipus*

303. Congress grass is:
(A) *Cynodon dactylon*
(B) *Digitaria purpurea*
(C) *Cymbopogon citrates*
(D) *Parthenium hysterophorus*

304. Troublesome American water weed found in India is:
(A) *Eichhornia* *(B) Typha*
(C) *Cyprus* (D) *Trapa*

305. The herbicide applied in pulse crops is:
(A) Isoproturon (B) Pendimethalin
(C) 2, 4-D (D) Atrazine

306. Isoproturon herbicide is available in:
(A) Liquid form (B) Solid form
(C) Suspended form (D) Powder form

307. Pendimethalin in pulses is applied as:
(A) Pre-planting
(B) Pre-emergence
(C) Post emergence
(D) None of the above

308. Butachlor herbicide is applied in:
(A) Wheat crop (B) Pulse crop
(C) Paddy crop (D) Mustard crop

309. *Phalaris minor* is important weed of :
(A) Rice (B) Wheat
(C) Maize (D) Ground nut

310. *Echinocloa crussgalli* is a common weed of:
(A) Rice (B) Wheat
(C) Maize (D) Ground nut

311. Which weedicide and its correct dose for groundnut is:
(A) Isoproturon @ 1.5 kg/ha
(B) Glyphoset @ 4.00 lit/ha
(C) Pendimethalin @ 2.2 lit/ha
(D) Pendimethalin @ 3.3 lit/ha

312. The spray drift of 2,4-D (ester form) can cause considerable damage in:
(A) Maize (B) Rice
(C) Sorghum (D) Cotton

313. Dispersing agents reduce ---------between similar particles:
(A) Adhesion (B) Cohesion
(C) Tension (D) None of these

314. Addition of ammonium sulphate to a herbicidal spray solution enhances the ------- against *Agropyron repens*:
(A) Glyphosate (B) Amylase
(C) Reductase (D) Oxi-reductase

315. Uptake and translocation of alachlor -------- germinating and emerging seedlings is faster at higher day and night temperatures than at lower temperatures:
(A) Wheat (B) Castor
(C) Mustard (D) Snap beans

316. Propanil, a post-emergence herbicide is readily absorbed by the :
(A) Stem (B) Root
(C) Leaf (D) None of these

317. Once the phenoxy herbicide penetrates the cuticle, it moves inside the plant:
(A) Only apoplastically
(B) Only symplastically
(C) By both mechanisms
(D) None of these

318. Thiocarbamates viz., a soil-incorporated herbicide is readily absorbed by roots and coleoptile and--------- translocated:
(A) Only upward (B) Only downward
(C) Both (A) & (B) (D) None of these

319. It has been observed that propanil (a herbicide) interfred with ATP generation and electron transport in ----- mitochondria:
(A) Wheat (B) Barley
(C) Srghum (D) Greengram

320. Triazines and phenoxy acids are subject to ring cleavage, triazine, heterocyclic ring can be cleaved resulting in liberation of:
(A) Chlorine (B) Oxygen
(C) Hydrogen (D) Nitrogen

321. The accumulation of dinitramine in the plant tissue is believed to be:
(A) Significant (B) Negligible
(C) Unchanged (D) None of these

322. The -------- acids which undergo –β-oxidation are MCPB, 2,4-DB and 2,4,5-TB:
(A) Phenoxyacetic (B) Phenoxyoxalic
(C) Phenoxymaleic (D) Phenoxybutyric

323. Urea herbicides undergo conjugation with:
(A) Lipids (B) Carbohydrates
(C) Proteins (D) All of these

324. Glyphosate has been reported to be metabolized to aminomethylphosphoric acid and glyphosate in:
(A) Maize (B) Sugarcane
(C) Sunflower (D) Safflower

325. The ----- structure of the herbicide also has a great influence on its volatility through adsorption to soil:
(A) Biological (B) Chemical
(C) Physical (D) Biphysical

326. -------- comprises an important mechanism of herbicide detoxification in soil under field conditions especially surface incorporation subsequent to rainfall or irrigation:
(A) Thermodecomposition
(B) Biodecomposition
(C) Chemodecomposition
(D) Photodecomposition

327. Photodecomposition plays an important role in the loss of urea herbicides from the soil surface in which areas:
(A) Moist (B) Wet
(C) Dry (D) None of these

328. Most of the herbicides embody ------phytotoxicity at lower soil moisture content:
(A) Higher (B) Lower
(C) Equal (D) Unchanged

329. ---------- degradation plays a major role in the persistence and behaviour of herbicides in the soil:
(A) Microbial (B) Physical
(C) Chemical (D) Thermal

330. The organisms involved in herbicide detoxification include:
(A) Earthworms (B) Nematodes
(C) Algae (D) None of these

331. Triazine-tolerant broadleaf weed biotypes include:
(A) Lambsquarter (*Chenopodium album*)
(B) Morning glorry (*Convolvulus arvensis*)
(C) Sweet clover (*Melilotus indica*)
(D) None of these

332. If the weedicide kills the plants after reaching in to the system, the weedicide is a:
(A) Contact type
(B) Translocated type
(C) Selective type
(D) None-selective type

333. The herbicide that kills the contact part of the weeds is known as:
(A) Contact type
(B) Translocated type
(C) Selective type
(D) None-selective type

334. 2,4-D kills only:
(A) Broad leaved weeds
(B) Narrow leaved weeds
(C) Both of the above
(D) None of these

335. Prickly pear can be killed biologically by:
(A) *Dactylopius tomentosus*
(B) *Lemna minor*
(C) *Hydrilla*
(D) None of these

336. Oxyfluorfen has been found to inhibit apical bud growth in young ------- plants to better bush formation:
(A) Coffee (B) Tea
(C) Cacao (D) Coconut

337. Butylate stimulates germination in -------- seeds by enhancing gibberellic activity or by changing membrane or seed permeability:
(A) Potato (B) Pigeonpea
(C) Safflower (D) Barley

338. Application of propanil after ------ in rice causes prohibitive vegetative injury:
(A) DDT (B) Endosulfan
(C) Carbofuran (D) Sevin

339. Weed competition in rice is more in the :
(A) Transplanted crop
(B) Direct seeded crop
(C) Flooded crop (D) Crop sown in rows

340. The *Phalaris minor* can be controlled by using the weedicide called ---- at 0.75-1.25 kg a.i./ha depending upon soil type:
(A) 2,4-D (B) Isoproturon
(C) Basalin (D) Simazine

341. The rate of application of atrazine in maize under heavy soils varies from:
(A) 0.75 kg a.i./ha
(B) 1.25-1.50 kg a.i./ha
(C) 1.75-2.00 kg a.i./ha
(D) 1.00-1.25 kg a.i./ha

342. Soybean crop should be free from weed competition during first --------- days for its good productivity:
(A) 10-20 (B) 20-30
(C) 30-40 (D) 40-50

343. Fluchloralin can be used for weed control in soybean as:
(A) Pre-emergence
(B) Pre-plant incorporation
(C) Post-emergence
(D) Through herbigation

344. Use of -------- provides effective control of weeds in groundnut:
(A) 2,4-D (B) Simazine
(C) Isoprotoron (D) Tok E-25

345. The weeds can be controlled in linseed by the application of MCPB at:
(A) 0.5 kg a.i./ha (B) 1.5 kg a.i./ha
(C) 1.0 kg a.i./ha (D) 2.0 kg a.i./ha

346. Two hand weeding in safflower are required for its good growth at:
(A) 15 and 30 days stages
(B) 20 and 40 days stages
(C) 25 and 50 days stages
(D) 30 and 60 days stages

347. The annual broad leaf and grassy weeds in sugarbeet can be controlled effectively by use of:
(A) Basalin @ 1.0 kg a.i.ha
(B) 2,4-D @ 1.5 kg a.i./ha
(C) Butachlor @ 2.0 kg a.i./ha
(D) Pyramin @ 3.0 kg a.i./ha

348. 2,4-D is a herbicide:
(A) Selective (B) Non-selective
(C) Contact (D) None of these

349. One of the following herbicide is a selective herbicide:
(A) Simazine
(B) Bisulphide
(C) Pentachlorophenol
(D) Sodium chlorate

350. The critical period of weed control in upland rice varies between:
(A) 15-20 DAS
(B) 20-25 DAS
(C) 25-30 DAS
(D) 30-35 DAS

351. For mungbean the period between ------------ days after sowing has been regarded as critical for weed control:
(A) 0 and 30 (B) 5 and 25
(C) 15 and 40 (D) 25 and 45

352. *Parthenium hysterophorus* may be controlled biologically by:
(A) *Zygogramma biocolorata*
(B) *Delias hyperata*
(C) *Melanogromyra spp.*
(D) All of these

353. Which one is the variety of Radish?
(A) Pusa Deshi (B) Pusa Rasmi
(C) Pusa Chetaki (D) All of the above

354. Which weedicide is used to control weeds in rice?
(A) Anilophos (B) Butachlor
(C) Thiobencarb (D) All of these

355. *Phalaric minor* is cotrolled with:
(A) Butachlor (B) Isoproturon
(C) Pendimethalin (D) Fluchloralin

356. Paraquat is a weedicide of:
(A) Selective type
(B) Non selective type
(C) Fumigant
(D) Sterilent

357. 2,4-D is abbreviated as:
(A) 2,4 di phenolic acid
(B) 2,4 dichlorophenoxy acetic acid
(C) 2,4 di phenoxy acetic acid
(D) None of these

358. Integrated weed management means:
(A) Mechnical control
(B) Chemical control
(C) Biological control
(D) All of these

359. *Sorghum helepense* may be controlled biologically by:
(A) Geese (B) Hydrilla
(C) Verticillata (D) None of these

360. Which of the following tillage system is useful to control *Phalaris minor*?
(A) Reduced tillage(B) FIRBS
(C) Zero tillage (D) None of these

361. Which of the following weedicide is used to control weeds in cotton?
(A) Fluchloralin @ 0.75-1.0 kg a.i./ha
(B) Pendimethalin @ 1.0-1.5 kg ai/ha
(C) Diuron @ 0.5 kg ai/ha
(D) All of these

362. Match correctly from list A and B. A contains the weeds and B, its origin place below:

A	B
1. *Parthenium*	i. Central & North America
2. *Phalaris minor*	ii. Mediterranean region
3. Water ferm	iii. S.E. Brazil
4. Water hyacinth	iv. S. America

(A) I. i 2. ii 3. iv 4. iii
(B) I. iv 2. ii 3. i 4. iii
(C) I. i 2. ii 3. iii 4. iv
(D) I. ii 2. iv 3. iii 4. i

363. Which one is pre plant applied herbicide?
(A) Diquat (B) Fluchloralin
(C) Alachlor (D) Both (B) and (C)

364. Which one is pre emergence applied herbicide?
(A) Diquat (B) Fluchloralin
(C) Alachlor (D) Pendimethalin

365. Butachlor herbicide belongs to:
(A) Acetamide (B) Urea
(C) Uracils (D) Phenols

366. 2,4-D herbicide belongs to:
(A) Acetamide (B) Urea
(C) Uracils (D) Phenoxy

367. Atrazine herbicide belongs to:
(A) Amide (B) Urea
(C) Triazine (D) Phenols

368. Isoproturon herbicide belongs to:
(A) Amide (B) Pyridazinones
(C) Uracils (D) Phenols

369. Pendimethalin herbicide belongs to:
(A) Amide (B) Urea
(C) Denitroanilenes (D) Phenols

370. *Argimone mexicana* is a:
(A) Obligate weed
(B) Facultative weed
(C) Both (A) & (B) (D) None of these

371. *Convolvulus arvensis* is a:
(A) Obligate weed (B) Facultative weed
(C) Both (A) & (B) (D) None of these

372. *Opuntia spp* is a:
(A) Obligate weed (B) Facultative weed
(C) Both (A) & (B) (D) None of these

373. Chenopodium album is a:
(A) Obligate weed (B) Facultative weed
(C) Absolute weed (D) None of these

374. Phalaris minor is a:
(A) Obligate weed (B) Facultative weed
(C) Absolute (D) Mimicry

375. If farmer wants to apply Atrazine 80% @ 0.5 kg a.i./ha, the quantity of atrazine would be:
(A) 0.40 kg (B) 0.48 kg
(C) 0.52 kg (D) 0.62 kg

376. Which weedicide is used to kill broad leaves weeds in cereals:
(A) 2,4-D (B) 2,4,5-T
(C) 2,4,D B (D) All of these

377. Important monocot weed found in most of the wheat field is:
(A) *Cyperus rotundus* (B) *Phalaris minor*
(C) *Avena fatua* (D) *Cynodon dactylon*

378. The weed that causes Asthama is:
(A) Dommon grass weed
(B) Kunch weed
(C) Parthenium (D) Calotrapis

379. Puddling of rice crop is done for:
(A) Reduce water loss
(B) Weed control
(C) Both (A) & (B) (D) None of these

380. Glyphosat is mainly used as:
(A) Pre emergence (B) Post emergence
(C) Pre planting (D) None of these

381. Weedicide 2,4-D is not recommended in the crop:
(A) Sugarcane (B) Maize
(C) Moong (D) Potato

382. Glyphosate is a:
(A) Selective translocated herbicide
(B) Non selective translocated herbicide
(C) Non selective, non translocated herbicide
(D) Contact type herbicide

383. *Cyperus rotundus* is a:
(A) Annual grass
(B) Biennial grass
(C) Perennial grass
(D) Creeper

384. Triazines are mostly used in:
(A) Barley (B) Wheat
(C) Pulses (D) Maize

385. Motha belongs to the family:
(A) Gramineae (B) Malvaceae
(C) Cyperaceae (D) Liliaceae

386. Isoproturon resistance in *Phalaris minor* was reported by:
(A) Singh and Mishra
(B) Malik and Singh
(C) Singh and Singh
(D) Singh and Prasad

387. Doddar is a serious weed of:
(A) Til (B) Lucerne
(C) Oat (D) Berseem

388. *Anagalis arvensis* is a:
(A) Annual weed
(B) Biennial weed
(C) Perennial weed
(D) None of these

389. Which of the following is a selective contact herbicide:
(A) 2,4-D (B) Propanil
(C) Butachlor (D) Bromoxynil

390. For control of weeds in peas, herbicide used is:
(A) 2, 4-D (B) Dosanax
(C) MCPB (D) Atrazine

391. Butachlor is an ideal weedicide for:
(A) Pre-emergence application in direct seeded rice
(B) Soon after the transplanting of rice
(C) Before the transplantation of rice
(D) None of these

392. Doob grass belongs to the family of:
(A) Gramineae (B) Malvaceae
(C) Cyperaceae (D) Liliaceae

393. Which type of herbicides also known as Mycoherbicides:
(A) Bioherbicide
(B) Bioinsecticide
(C) Biopesticide
(D) None of these

394. The plant pathogen *Colletrotrichum cocodes* is used to control the weeds of :
(A) *Abutilon theophorsii*
(B) *Aeschynomene virginiea*
(C) *Echhornia crassipes*
(D) *Malva pustilla*

395. The plant pathogen *Colletrotrichum glocosporioidest sp. Aeschynomene* is used to control the weeds of :
(A) *Abutilon theophorsii*
(B) *Aeschynomene virginiea*
(C) *Echhornia crassipes*
(D) *Malva pustilla*

396. The plant pathogen *Cercospora rodmanii* is used to control the weeds of :
(A) *Abutilon theophorsii*
(B) *Aeschynomene virginiea*
(C) *Echhornia crassipes*
(D) *Malva pustilla*

397. The plant pathogen *Colletrotrichum glocosporioidest sp malvae* is used to control the weeds of :
(A) *Abutilon theophorsii*
(B) *Aeschynomene virginiea*
(C) *Echhornia crassipes*
(D) *Malva pustilla*

398. The plant pathogen *Phytophthora palmivora* is used to control the weeds of :
(A) *Morrenia odorata*
(B) *Aeschynomene virginiea*
(C) *Echhornia crassipes*
(D) *Malva pustilla*

399. The plant pathogen *Alternaria cassiae* is used to control the weeds of :
(A) *Abutilon theophorsii*
(B) *Cassia obtisifolia*
(C) *Echhornia crassipes*
(D) *Malva pustilla*

400. Application of herbicides to small patches of weeds leaving the weed free-gaps unsprayed. This is mostly practices in treating patches of noxious perennial weeds with potential herbicides:
(A) Protected spraying
(B) Rope Wick application
(C) Blanket application
(D) Spot application

401. Application of herbicides after the last cultivation in crops, for example after ridging in sugarcane or cotton, after earthing - up in potato, etc. Usually soil-acting herbicides are used, which will prevent further emergence of weeds is known as:
(A) Lay by application
(B) Rope Wick application
(C) Blanket application
(D) Band application

402. Technique is used in the application of non-selective herbicides particularly the translocative type (e.g. glyphosate) is known as:
(A) Protected spraying
(B) Rope Wick application
(C) Blanket application
(D) Band application

403. Non-selective herbicides can be employed to control weeds in widely spaced vegetables or ornamentals by covering the crop plants with plastic of metallic covers. This is practicable only in high value plants or crops is known as:
(A) Protected spraying
(B) Rope Wick application
(C) Blanket application
(D) Band application

404. DNBP herbicide are classified in:
(A) Phenol group
(B) Phenoxyalkanoic acids group
(C) Phenylalkanoic acid group
(D) Chlorobenzoic and anisic acids group

405. Application of herbicides over an entire area. In case of post-emergence applications this includes crop and weeds and hence only highly selective herbicides are suitable for such applications is known as:

(A) Protected spraying
(B) Rope Wick application
(C) Blanket application
(D) Band application

406. Isoproturon is available as 50% WP. If the recommended rate of application is 0.75 kg a.i./ha then the amount of commercial product required is:

(A) 1.0 kg/ha (B) 1.5 kg/ha
(C) 2.0 kg/ha (D) 2.5 kg/ha

407. Isoproturon is available as 75% WP. If the recommended rate of application is 0.75 kg a.i./ha then the amount of commercial product required is:

(A) 1.0 kg/ha (B) 1.5 kg/ha
(C) 2.0 kg/ha (D) 2.5 kg/ha

408. Paraquat is to be applied at 0.5 kg a.i./ha. The herbicide is available as Gramoxone (commercial name of paraquat) which contains 25% paraquat. The quantity of Gramoxone required is:

(A) 1.5 kg/ha (B) 2.0 kg/ha
(C) 2.5 kg/ha (D) 3.0 kg/ha

409. 2,4-D, MCPA, MCPB, 2,4-DB, Macoprop, Dichloroprop and Diclofop herbicides are classified in:

(A) Phenol group
(B) Phenoxyalkanoic acids group
(C) Phenylalkanoic acid group
(D) Chlorobenzoic and anisic acids group

410. Fenac herbicide is classified in:

(A) Phenol group
(B) Phenoxyalkanoic acids group
(C) Phenylalkanoic acid group
(D) Chlorobenzoic and anisic acids group

411. Acifluorfen, Dicamba herbicide are classified in:

(A) Phenol group
(B) Phenoxyalkanoic acids group
(C) Phenylalkanoic acid group
(D) Chlorobenzoic & anisic acids group

412. Endothal, DCPA and Naptalam herbicides are classified in:

(A) Phthalic & Phthalmic acids group
(B) Phenoxyalkanoic acids group
(C) Phenylalkanoic acid group
(D) Chlorobenzoic and anisic acids group

413. TCA and Dalapan herbicides are classified in:

(A) Alipatic acid group
(B) Triazines groups
(C) Traizoles group
(D) Diazoles group

414. Amitrole herbicide is classified in:

(A) Alipatic acid group
(B) Triazines groups
(C) Traizoles group
(D) Diazoles group

415. Triazine, Simazine, Atrazine, Propazine, Prometon, Desmetryne, Ametryn, Terbutryn, Prometryn, Hexazinone and Metribuzin herbicides are classified in:

(A) Alipatic acid group
(B) Triazines groups
(C) Traizoles group
(D) Diazoles group

416. Difenzoquat, and Methozole herbicides are classified in:

(A) Alipatic acid group
(B) Triazines groups
(C) Traizoles group
(D) Diazoles group

417. MH, Pyrazon and Bentazon herbicides are classified in:

(A) Diazines group
(B) Puridines groups
(C) Quaternary ammonium compounds group
(D) Uracils group

418. Picloram herbicide is classified in:

(A) Diazines group
(B) Puridines groups
(C) Quaternary ammonium compounds group
(D) Uracils group

419. Diquat and Paraquat herbicides are classified in:
(A) Diazines group
(B) Puridines groups
(C) Quaternary ammonium compounds group
(D) Uracils group

420. Bromocil and Terbacil herbicides are classified in:
(A) Diazines group
(B) Puridines groups
(C) Quaternary ammonium compounds group
(D) Uracils group

421. Fenuron, Monuron, Diuron, Linuron, Chloroxuron, Siduron, Norea, Methabenzthiazuron, Metoxuron and Isoproturon herbicides are classified in:
(A) Substitutes ureas group
(B) Carbanilates and carbomats groups
(C) Amides and Anilides group
(D) Anilines and toluidines group

422. Propham, Chloropham, Barban, Karbutilate, EPTC, Vernolate, Cycloate, Triallate, Phenmediphan and Thiobencarb herbicides are classified in:
(A) Substitutes ureas group
(B) Carbanilates & carbomats groups
(C) Amides and Anilides group
(D) Anilines and toluidines group

423. Diphenamide, Nepropomide, CDAA, Oryzalon, Bensulide, Propanil, Alachlor, Butachlor, Metolachlor and Propachlor herbicides are classified in:
(A) Substitutes ureas group
(B) Carbanilates and carbomats groups
(C) Amides and Anilides group
(D) Anilines and toluidines group

424. Fluchloralin, Ethafluralin, Trifluralin, Pendimethalin and Dinitramine herbicides are classified in:
(A) Substitutes ureas group
(B) Carbanilates and carbomats groups
(C) Amides and Anilides group
(D) Anilines and toluidines group

425. Bromoxynil, Ioxynil and Dichlobenil herbicides are classified in:
(A) Benonitriles group
(B) Other compounds groups
(C) Sulfonylureas group
(D) Organic arsenicals group

426. Nitrofen and Oxyflurfen herbicides are classified in:
(A) Benonitriles group
(B) Other compounds groups
(C) Sulfonylureas group
(D) Organic arsenicals group

427. Tribenuron, Thiofensulfuron, Metsulfuron and Chlorsulfuron herbicides are classified in:
(A) Benonitriles group
(B) Other compounds groups
(C) Sulfonylureas group
(D) Organic arsenicals group

428. MSMA, MAA and DSMA herbicides are classified in:
(A) Benonitriles group
(B) Other compounds groups
(C) Sulfonylureas group
(D) Organic arsenicals group

429. Acrolein, AMS, Benzoylprop (ethyl salt), Fluridone, Fusamine, Oxadiazon, Petrolium oils, Anilofos, Alamprop, Chlorimuron, Fluroxypyr, Glyphosate, Imidazolinones, Imazapyr, Imazathapyr, Fomesafen, Fluazifop-p, Haloxafop and Lactofenherbicides are classified in:
(A) Benonitriles group
(B) Miscellaneous herbicides groups
(C) Sulfonylureas group
(D) Organic arsenicals group

430. Application of herbicide in the form of dust not recommended due to:
(A) More hazards (B) Less hazards
(C) Drift hazards (D) None of these

431. *Pluchea lanceolata* is a weed of:
(A) Rabi (B) Kharif
(C) Zaid (D) Perennial

432. *Sachharum spontaneum* is a weed of:
(A) Rabi (B) Kharif
(C) Zaid (D) Perennial

433. *Solanum nigrum* is a weed of:
(A) Rabi (B) Kharif
(C) Zaid (D) Perennial

434. *Spergula arvensis* is a weed of:
(A) Rabi (B) Kharif
(C) Zaid (D) Perennial

435. *Sorghum helepense* is a weed of:
(A) Rabi (B) Kharif
(C) Zaid (D) Perennial

436. *Trianthema monogyna* is a weed of:
(A) Rabi (B) Kharif
(C) Zaid (D) Perennial

437. *Tribulus teristrus* is a weed of:
(A) Rabi (B) Kharif
(C) Zaid (D) Perennial

438. *Vicia sativa* is a weed of:
(A) Rabi (B) Kharif
(C) Zaid (D) Perennial

439. *Vicia hirsuta* is a weed of:
(A) Rabi (B) Kharif
(C) Zaid (D) Perennial

440. *Zizyphus rotandifolia* is a weed of:
(A) Rabi (B) Kharif
(C) Zaid (D) Perennial

441. *Celosia argentia* is a weed of:
(A) Rabi (B) Kharif
(C) Zaid (D) Perennial

442. *Medicago denticulata* is a weed of:
(A) Rabi (B) Kharif
(C) Zaid (D) Perennial

443. *Opuntia spp.* is a weed of:
(A) Rabi (B) Kharif
(C) Zaid (D) Perennial

444. *Orobanche aegyptiaca* is a weed of:
(A) Rabi (B) Kharif
(C) Zaid (D) Perennial

445. *Parthenium hysrophorus* is a weed of:
(A) Rabi (B) Kharif
(C) Zaid (D) Perennial

446. *Phylanthus niruri* is a weed of:
(A) Rabi (B) Kharif
(C) Zaid (D) Perennial

447. *Portulaca quadrifida* is a weed of:
(A) Rabi (B) Kharif
(C) Zaid (D) Perennial

448. *Portulaca oleracea* is a weed of:
(A) Rabi (B) Kharif
(C) Zaid (D) Perennial

449. *Amaranthese spinosus* is a weed of:
(A) Rabi (B) Kharif
(C) Zaid (D) Perennial

450. *Argemone maxicana* is a weed of:
(A) Rabi (B) Kharif
(C) Zaid (D) Perennial

451. *Chinchoreum intybus* is a weed of:
(A) Rabi (B) Kharif
(C) Zaid (D) Perennial

452. *Cuscuta refloxa* is a weed of:
(A) Rabi (B) Kharif
(C) Zaid (D) Perennial

453. *Datura fertuosa* is a weed of:
(A) Rabi (B) Kharif
(C) Zaid (D) Perennial

454. *Digera arvensis* is a weed of:
(A) Rabi (B) Kharif
(C) Zaid (D) Perennial

455. *Eclipta alba* is a weed of:
(A) Rabi (B) Kharif
(C) Zaid (D) Perennial

456. *Echhornia crassipus* is a weed of:
(A) Rabi (B) Kharif
(C) Zaid (D) Perennial

457. *Fumaria purviflora* is a weed of:
(A) Rabi (B) Kharif
(C) Zaid (D) Perennial

458. *Gynandropsis pentaphylla* is a weed of:
(A) Rabi (B) Kharif
(C) Zaid (D) Perennial

459. *Launcea aspera* is a weed of:
(A) Rabi (B) Kharif
(C) Zaid (D) Perennial

460. *Lathyrus sativa* is a weed of:
(A) Rabi (B) Kharif
(C) Zaid (D) Perennial

461. *Lathyrus aphaca* is a weed of:
(A) Rabi (B) Kharif
(C) Zaid (D) Perennial

462. *Launea splenifolia* is a weed of:
(A) Rabi (B) Kharif
(C) Zaid (D) Perennial

463. *Lolium tameolantum* is a weed of:
(A) Rabi (B) Kharif
(C) Zaid (D) Perennial

464. *Achranthus aspera* is a weed of:
(A) Rabi (B) Kharif
(C) Zaid (D) Perennial

465. *Ageratum conijoides* is a weed of:
(A) Rabi (B) Kharif
(C) Zaid (D) Perennial

466. *Ahlagi camelorum* is a weed of:
(A) Rabi (B) Kharif
(C) Zaid (D) Perennial

467. The chemical Butachlor is mostly used in agriculture is:
(A) Growth hormone
(B) Antitranspirants
(C) Post-emergence herbicide
(D) Pre-emergence herbicide

468. A herbicide formulation which is a clear solution turns milky while when mixed with water is:
(A) Soluble concentrate
(B) Suspension concentrate
(C) Aqueous concentrate
(D) Emulsifiable concentrate

469. Butachlor is recommended for control of weeds in:
(A) Maize
(B) Wheat
(C) Transplanted rice
(D) Sugarcane

470. Most critical period for weed competition in wheat is:
(A) 12-15 days after sowing
(B) 25-30 days after sowing
(C) 40-45 days after sowing
(D) All of these

471. Successful control of Kans (*Saccharum spontaneum*) is done by:
(A) Glyphosate 41/SL 2-3 litres/ha
(B) Glyphosate 41/SL 3-4 litre/ha
(C) Atrazine 50% w.p. 3 litre/ha
(D) Pendimethalin 30 E.C. 3.3 litre/ha

472. Isoproturon is a:
(A) Fungicide (B) Insecticide
(C) Herbicide (D) Nematicide

473. *Orobanche* sp. is an associated weed of:
(A) Potato (B) Sugarcane
(C) Tobacco (D) Berseem

474. Which one is not correctly matched:

Crop	Weed
(A) Wheat	*Phalaris minor*
(B) Paddy	Bathua
(C) Berseem	Kasni
(D) Pea	Piyazi

475. The herbicides containing carbon and hydrogen in their molecules are called:
(A) Arsenic
(B) Organic herbicide
(C) Salt
(D) Acid

476. Which of the following is now more suitable for control of *Phalaris minor* in wheat field:
(A) Sulfo-sulphuran
(B) Pendimethalin
(C) Isoproturon
(D) Methabenzthiazuron

477. Which of the following herbicide is more suitable for control of broad leave weeds in pulse crop as a pre-emergence application:
(A) Pendimethaline
(B) Methabenthiazuran
(C) Metribuzin
(D) Basalin

478. Which of the following herbicide is more suitable for control of broad leave weeds in pulse crop as a pre-plant application:
(A) Pendimethaline
(B) Methabenthiazuran
(C) Metribuzin
(D) Basalin

479. Which of the following herbicide is more suitable for control of broad leave weeds in urd crop as a standing crop:
(A) Pendimethaline
(B) Methabenthiazuran
(C) Metribuzin
(D) Basalin

480. Which of the following herbicide is more suitable for control of broad leave weeds in field pea crop as a standing crop:
(A) Pendimethaline
(B) Methabenthiazuran
(C) Metribuzin
(D) Basalin

481. Which one of the following is nonselective herbicide?

(A) Alachlor
(B) Butachlor
(C) Paraquat
(D) Atrazine

482. Which one of the following is not narrow leaved weeds?

(A) *Cynodon dactylon*
(B) *Cyperus rotundus*
(C) *Setaria glouca*
(D) *Melilotus indica*

483. Which of the following is not kharif season weed?

(A) *Chenopodium album*
(B) *Amarathus vividis*
(C) *Echinocloa colonum*
(D) Comelina benghalensis

484. Weedicide used for control of broadleaf weeds in pulse crops is:

(A) 2, 4-D (B) Basaline
(C) Atrazine (D) Butachor

485. Propanil herbicide is readily absorbed by:

(A) Stem (B) Root
(C) Leaf (D) None of these

486. Read the following and mark the correct answer:

Cropping system	Herbicide
(A) Potato + wheat	Pendimethalin
(B) Potato + mustard	Pendimethalin
(C) Chickpea+Linseed	Pendimethalin
(D) Sugarcane + Potato	Pendimethalin

487. Imperical formula of weed control efficiency:

(A) $\dfrac{\textbf{Dry matter weight of weeds in control plot – Dry matter weight of weeds in treated plot}}{\textbf{Dry matter weight of weeds in control plot}} \times \mathbf{100}$

(B) $\dfrac{\text{Dry matter weight of weeds in control plot – Dry matter weight of weeds in treated plot}}{\text{Dry matter weight of weeds in treated plot}} \times 100$

(C) $\dfrac{\text{Dry matter weight of weeds in control plot – Dry matter weight of weeds in treated plot}}{\text{Dry matter weight of weeds in control plot}}$

(D) $\dfrac{\text{Dry matter weight of weeds in control plot – Dry matter weight of weeds in treated plot}}{\text{Dry matter weight of weeds in treated plot}}$

488. Imperical formula of weed index:

(A) $\dfrac{\text{Yield from weed free plots – Yield from treated plots}}{\text{Yield from treated plots}} \times 100$

(B) $\dfrac{\textbf{Yield from weed free plots – Yield from treated plots}}{\textbf{Yield from weed free plots}} \times \mathbf{100}$

(C) $\dfrac{\text{Yield from weed free plots – Yield from treated plots}}{\text{Yield from weed free plots}}$

(D) $\dfrac{\text{Yield from weed free plots – Yield from treated plots}}{\text{Yield from treated plots}}$

489. Imperical formula of weed density:

(A) $\dfrac{\text{No. of individuals in all quadrates}}{\text{No. of all species in all quadrates}}$

(B) $\dfrac{\text{No. of all species in all quadrates}}{\text{No. of individuals in all quadrates}}$

(C) $\dfrac{\textbf{No. of individuals in all quadrates}}{\textbf{No. of all species in all quadrates}} \times \mathbf{100}$

(D) $\dfrac{\text{No. of all species in all quadrates}}{\text{No. of individuals in all quadrates}} \times 100$

490. Imperical formula of weed per cent occurrence:

(A) $\dfrac{\text{No. of occurrences of a species in a district}}{\text{Total no. of observations recorded in a district}}$

(B) $\dfrac{\text{No. of occurrences of a species in a district}}{\text{Total no. of observations recorded in a district}} \times 100$

(C) $\dfrac{\text{Total no. of observations recorded in a district}}{\text{No. of occurrences of a species in a district}} \times 100$

(D) $\dfrac{\textbf{No. of occurrences of a species in a district}}{\textbf{Total no. of observations recorded in a district}} \times \mathbf{100}$

491. Weed per cent occurrence formula has been suggested by:

(A) Patil and Patil (1983)
(B) Gill and Kumar (1969)
(C) Misra (1968)
(D) Raju (1977)

492. To findout the area sprayed by a full tank of sprayer, if the volume of the sprayer tank is 250 litres and rate of spray application is 150 litres/ha:

(A) 1.50 ha
(B) 1.67 ha
(C) 1.87 ha
(D) 2.00 ha

493. To findout the volume of alachlor 50% needed to spray one hectare. The rate of application of alachlor is 0.75 litres/ha:

(A) 1.0 kg (B) 1.5 kg
(C) 1.6 kg (D) 1.7 kg

494. To findout the quantity of Simazine wettable powder to be sprayed in one hectare area. The rate of application is 3.2 kg a.i./ha and active ingredient is 80% in simazine commercial product:

(A) 4.0 kg (B) 3.5 kg
(C) 3.0 kg (D) 2.5 kg

495. Express the concentration of solution in ppm (μg/ml), if 25 kg of 2,4-D is mixed with 800 litres of water :

(A) 25000 ppm (B) 24000 ppm
(C) 23000 ppm (D) 20000 ppm

Statement for linked Answer Question 506 to 509. Findout the quantity of commercial material and water needed for spraying of 50 hectare area of wheat crop with a solution of 6000 ppm of 2,4-D available in ester and amine forms. The active ingredient of ester formulation is 400 g/litres and amine formulation is 500 g/litre. The recommended doses are 0.25 kg/ha and 0.5kg/ha for ester and amine, respectively.

496. The quantity of commercial materail of 2,4-D in ester formulation is:

(A) 31.25 kg/ha (B) 30.25 kg/ha
(C) 29.25 kg/ha (D) 28.25 kg/ha

497. The quantity of commercial material of 2,4-D in amine formulation is:

(A) 30 kg/ha (B) 35 kg/ha
(C) 40 kg/ha (D) 50 kg/ha

498. The volume of water needed to make 6000 ppm solution of ester formulation is:

(A) 5008.33 litres (B) 5108.33 litres
(C) 5208.33 litres (D) 5308.33 litres

499. The volume of water needed to make 6000 ppm solution of amine formulation is:

(A) 7333.33 litres (B) 8333.33 litres
(C) 9333.33 litres (D) 10333.33 litres

Statement for linked Answer Question 510 to 514 Which one of the herbicide formulation is cheaper to apply in 20 hectare of wheat crop

(i) 2,4-D ester formulation with 400 g a.i./litres
(ii) 2,4-D amine salt with 500 g a.i./litre
The recommended dose of application are 0.30 kg a.i./ha ester and 0.6 kg a.i./ha amine formulation. The cost of commercial material of 2,4-D in ester formulation is Rs. 60/litre and that of amine salt is Rs. 40/ litre.

500. The quantity of commercial material of ester is needed for supply of 6.0 kg of active ingredient:

(A) 10 kg (B) 12 kg
(C) 15 kg (D) 20 kg

501. The quantity of commercial material of amine is needed for supply of 12 kg of active ingredient:

(A) 10 kg (B) 15 kg
(C) 20 kg (D) 25 kg

502. The cost of 15 kg of commercial material is 2,4-D is ester formulation would be:

(A) Rs. 900 (B) Rs. 800
(C) Rs. 700 (D) Rs. 600

503. The cost of 15 kg of commercial material of 2,4-D in amine formulation would be

(A) Rs. 900 (B) Rs. 800
(C) Rs. 700 (D) Rs. 600

504. Based on the cost analysis which formulation is cheaper:

(A) Ester (B) Amine
(C) Both formulation (D) None of these

505. Herbicide related to substituted urea group is:

(A) 2,4-D (B) Alachlor
(C) Atrazine (D) Isoprofuron

506. Herbicide which can be applied before planting/sowing:

(A) Alachlor (B) Atrazine
(C) Fluchloralin (D) Glyphosate

507. Which is the contact, non-selective and zero persistant herbicide in the soil?

(A) 2,4-D (B) Bromacil
(C) Triazine (D) Paraquat

508. Grassy weed in barley field can be effectively controlled by:

(A) Isoprofuron (B) Pendimethalin
(C) 2,4-D (D) Both (A) & (B)

509. Loranthus is a parasitic weed of:

(A) Rice (B) Tobacco
(C) Mango (D) Wheat

510. The following is an example of a crop bound weed:

(A) Nut sedge (B) Lantana
(C) Cuscuta (D) Eichornia

511. Paraquat kills plant by:

(A) Inhibiting respiration
(B) Inhibiting Pigment System - II
(C) Inhibiting Pigment System - I
(D) Inhibiting dark reaction

512. Glyphosate is a broad spectrum herbicide upon application to leaves, it is translocated to meristematic areas and underground/rhizome's by:

(A) Phloem (B) Xylem
(C) Border pits (D) Tracheids

513. Atrazine effectively control grasses in maize field without causing harm to the crop plants, as degradation of the chemical in maize plant occurs due to the presence of:

(A) R-Q enzyme
(B) Amylases
(C) Aryl aceyl amidase
(D) GSH enzyme

514. A broad leaved monocat weed is:

(A) *Stellaria indica*
(B) *Heliotropium indicum*
(C) *Melilotus indica*
(D) *Eichhornia crassipes*

515. A narrow leaf dicotyledonous weed is:

(A) *Sphenoclea zeylanica*
(B) *Spergula arvensis*
(C) *Conyza canadienrin*
(D) *Aspholatus tennifolius*

516. Which one of the following causes more wastages of herbicide by drift?

(A) High volume sprayer
(B) Ultra-low volume sprayer
(C) Hand sprayer
(D) Low volume sprayer

517. The resistance of *Phalaris minor* to Isoproturon is due to :

(A) Target site alteration
(B) Enhanced degradation
(C) Sequestration & compart mentation
(D) All of the above

518. An annual sedge found in rice is:

(A) *Cyperus rotundus*
(B) *Cyperus esculentus*
(C) *Cyperus iria*
(D) *Cyperus bulbosus*

519. The biological control of Parthenium using a Maxican beetle:

(A) *Zygogramma bicolorata*
(B) *Epiblema strenuana*
(C) *Agromyza lantanae*
(D) *Theola echion*

520. If dry weight of weeds from weedy check, hand weeded plots were 400 kg, 4 kg and 100 kg/ha, respectively. The weed control efficiency of metsulfuron will be:

(A) 60% (B) 75%
(C) 85% (D) 90%

521. Which one of the following weeds is used as vegetable:

(A) *Setaria glauca*
(B) *Vicia hirsuta*
(C) *Amaranthus edulis*
(D) *Amaranthus virids*

522. Which of the following herbicides has the longest persistancy:

(A) Dinosets (B) Chlorfenae
(C) Glyphosate (D) 2,4-D

523. Which of the following weed is used as vegetable (Sag)?

(A) *Chenopodium murale*
(B) *Convolvulus arvensis*
(C) *Avena sativa*
(D) *Chenopodium album*

524. The weed index (WI) value for weed free check treatment is:

(A) 0 (B) 1
(C) -1 (D) ±1

525. Which part of the leaf is present in wheat but not in the grassy weeds such as *Phalaris minor*, wild oats?

(A) Auricle
(B) Ligule
(C) Both auricle and ligule
(D) None of the above

526. Chenopodium album is a pollution indicator plant for:

(A) H_2S (B) SO_2
(C) Both H_2S and SO_2
(D) None of the above

527. Pondimethalin belongs to which group of herbicide:

(A) 2,4-dinitro phenols
(B) 2,6-dinitro phenols
(C) Nitrophenyl ethers
(D) Phenyl ureas

528. Phalaris minor has developed resistance against which herbicides?

(A) 2,4-D (B) Isoprotoron
(C) Sulfo sulfuron (D) Oxadiazon

529. Recommended herbicide for weed control in chickpea is:

(A) Fluchloralin
(B) Butachlor
(C) Isoproturon
(D) 2,4-D

530. What kind of parasite weed, the Orobanche is?

(A) Holo-root (B) Hemi-root
(C) Holo-stem (D) Hemi-stem

531. The broad spectrum systematic herbicide glyphosate kills the weeds of inhibiting the biosynthesis of:

(A) Phenylalamine (B) Alanine
(C) Glutamine (D) Cysteine

532. Which one of the following mycoherbicide?

(A) Paraquat (B) Diquat
(C) Bromacil (D) Collego

533. The triazin group of herbicides act by:

(A) Inhibiting root and shoot growth
(B) Developing albino leaves
(C) Disrupting membrane integrity
(D) Inhibiting photosynthesis

534. Which of these herbicide in used to control weeds in zero tillage?

(A) Pendimethalin
(B) Fluchloralin
(C) Paraquat
(D) Alachor

ANSWER SHEET
10. Weed Management

1. (A)	2. (C)	3. (D)	4. (A)	5. (C)	6. (B)	7. (D)	8. (A)	9. (C)	10. (A)
11. (B)	12. (D)	13. (D)	14. (C)	15. (D)	16. (A)	17. (B)	18. (C)	19. (D)	20. (A)
21. (C)	22. (C)	23. (D)	24. (B)	25. (A)	26. (A)	27. (B)	28. (C)	29. (D)	30. (D)
31. (A)	32. (D)	33. (D)	34. (A)	35. (B)	36. (B)	37. (C)	38. (C)	39. (A)	40. (B)
41. (C)	42. (A)	43. (B)	44. (B)	45. (C)	46. (D)	47. (C)	48. (A)	49. (B)	50. (C)
51. (B)	52. (C)	53. (A)	54. (A)	55. (A)	56. (C)	57. (A)	58. (A)	59. (C)	60. (C)
61. (A)	62. (B)	63. (A)	64. (B)	65. (A)	66. (B)	67. (C)	68. (D)	69. (C)	70. (B)
71. (C)	72. (B)	73. (B)	74. (A)	75. (A)	76. (B)	77. (C)	78. (D)	79. (A)	80. (D)
81. (C)	82. (B)	83. (C)	84. (D)	85. (D)	86. (C)	87. (D)	88. (C)	89. (C)	90. (C)
91. (D)	92. (D)	93. (D)	94. (D)	95. (D)	96. (C)	97. (D)	98. (D)	99. (B)	100. (A)
101. (A)	102. (C)	103. (B)	104. (A)	105. (D)	106. (A)	107. (B)	108. (B)	109. (C)	110. (A)
111. (C)	112. (B)	113. (B)	114. (B)	115. (C)	116. (A)	117. (A)	118. (B)	119. (C)	120. (C)
121. (C)	122. (B)	123. (A)	124. (D)	125. (A)	126. (B)	127. (C)	128. (D)	129. (D)	130. (B)
131. (B)	132. (A)	133. (C)	134. (B)	135. (D)	136. (C)	137. (C)	138. (A)	139. (C)	140. (C)
141. (A)	142. (B)	143. (A)	144. (A)	145. (A)	146. (A)	147. (C)	148. (C)	149. (C)	150. (A)

151. (B)	152. (A)	153. (D)	154. (C)	155. (B)	156. (A)	157. (A)	158. (B)	159. (C)	160. (A)
161. (B)	162. (D)	163. (B)	164. (D)	165. (A)	166. (D)	167. (C)	168. (B)	169. (A)	170. (A)
171. (C)	172. (C)	173. (C)	174. (A)	175. (B)	176. (B)	177. (B)	178. (B)	179. (B)	180. (B)
181. (C)	182. (A)	183. (D)	184. (B)	185. (C)	186. (B)	187. (A)	188. (A)	189. (B)	190. (C)
191. (D)	192. (A)	193. (C)	194. (B)	195. (D)	196. (B)	197. (C)	198. (A)	199. (D)	200. (A)
201. (D)	202. (B)	203. (A)	204. (A)	205. (B)	206. (C)	207. (A)	208. (C)	209. (C)	210. (D)
211. (A)	212. (B)	213. (D)	214. (B)	215. (C)	216. (A)	217. (B)	218. (A)	219. (A)	220. (A)
221. (B)	222. (C)	223. (D)	224. (C)	225. (D)	226. (A)	227. (B)	228. (C)	229. (B)	230. (A)
231. (B)	232. (B)	233. (A)	234. (B)	235. (C)	236. (C)	237. (A)	238. (A)	239. (A)	240. (B)
241. (C)	242.(A)	243.(D)	244.(D)	245.(B)	246.(A)	247.(B)	248.(A)	249.(A)	250.(A)
251.(B)	252.(D)	253.(C)	254.(B)	255.(D)	256.(A)	257.(C)	258.(D)	259.(C)	260.(A)
261.(B)	262.(C)	263.(C)	264.(D)	265.(A)	266.(A)	267.(A)	268.(D)	269.(C)	270.(C)
271.(D)	272.(A)	273.(B)	274.(A)	275.(A)	276.(A)	277.(A)	278.(D)	279.(B)	280.(C)
281.(B)	282. (D)	283. (C)	284. (A)	285. (B)	286. (A)	287. (B)	288. (C)	289. (B)	290. (A)
291. (C)	292. (A)	293. (B)	294. (B)	295. (A)	296. (D)	297. (A)	298. (D)	299. (B)	300. (A)
301. (C)	302. (C)	303. (D)	304. (A)	305. (B)	306. (B)	307. (B)	308. (C)	309. (B)	310. (A)
311. (D)	312. (D)	313. (B)	314. (A)	315. (D)	316. (C)	317. (C)	318. (C)	319. (D)	320. (B)
321. (B)	322. (D)	323. (C)	324. (A)	325. (B)	326. (D)	327. (C)	328. (B)	329. (A)	330. (C)
331. (A)	332. (B)	333. (A)	334. (A)	335. (C)	336. (B)	337. (D)	338. (C)	339. (B)	340. (B)
341. (B)	342. (C)	343. (B)	344. (D)	345. (A)	346. (B)	347. (D)	348. (A)	349. (A)	350. (C)
351. (B)	352. (A)	353. (D)	354. (B)	355. (B)	356. (B)	357. (B)	358. (D)	359. (C)	360. (C)
361. (B)	362. (B)	363. (D)	364. (D)	365. (A)	366. (D)	367. (C)	368. (B)	369. (C)	370. (B)
371. (A)	372. (B)	373. (C)	374. (D)	375. (D)	376. (D)	377. (B)	378. (C)	379. (C)	380. (C)
381. (B)	382. (B)	383. (C)	384. (D)	385. (C)	386. (B)	387. (B)	388. (A)	389. (B)	390. (C)
391. (B)	392. (A)	393. (A)	394. (A)	395. (B)	396. (C)	397. (D)	398. (B)	399. (B)	400. (D)
401. (A)	402. (B)	403. (A)	404. (A)	405. (C)	406. (B)	407. (A)	408. (B)	409. (B)	410. (C)
411. (D)	412. (A)	413. (A)	414. (C)	415. (B)	416. (D)	417. (A)	418. (B)	419. (C)	420. (D)
421. (A)	422. (B)	423. (C)	424. (D)	425. (A)	426. (B)	427. (C)	428. (D)	429. (B)	430. (C)
431. (D)	432. (D)	433. (A)	434. (A)	435. (B)	436. (B)	437. (B)	438. (A)	439. (A)	440. (D)
441. (B)	442. (A)	443. (D)	444. (A)	445. (D)	446. (B)	447. (B)	448. (A)	449. (B)	450. (A)
451. (A)	452. (D)	453. (D)	454. (B)	455. (B)	456. (D)	457. (A)	458. (B)	459. (B)	460. (A)
461. (A)	462. (A)	463. (A)	464. (D)	465. (A)	466. (C)	467. (C)	468. (D)	469. (C)	470. (B)
471. (B)	472. (C)	473. (C)	474. (B)	475. (B)	476. (A)	477. (A)	478. (D)	479. (B)	480. (C)
481. (C)	482. (D)	483. (A)	484. (B)	485. (C)	486. (C)	487. (A)	488. (B)	489. (C)	490. (D)
491. (D)	492. (B)	493. (C)	494. (A)	495. (C)	496. (A)	497. (D)	498. (C)	499. (B)	500. (C)
501. (B)	502. (A)	503. (D)	504. (B)	505. (D)	506. (C)	507. (D)	508. (A)	509. (C)	510. (C)
511. (C)	512. (A)	513. (D)	514. (D)	515. (D)	516. (B)	517. (A)	518. (A)	519. (A)	520. (B)
521. (D)	522. (C)	523. (D)	524. (A)	525. (A)	526. (A)	527. (B)	528. (B)	529. (A)	530. (A)
531. (A)	532. (D)	533. (D)	534. (C)						

11 IRRIGATION MANAGEMENT

1. Artificial moisture application of water to soil to escape stress of plant is called:
(A) Irrigation (B) Drainage
(C) Both of these (D) None of these

2. The main objectives of irrigation is/are:
(A) To make soil surface soft
(B) For proper seed germination
(C) Proper growth of plant
(D) All of these

3. Water available in the soil at standard crop growth- stages is called:
(A) Soil moisture constant
(B) Water holding capacity
(C) Field capacity
(D) None of these

4. When all the pores of the soil are filled with water, the soil said to be:
(A) Water holding capacity
(B) Saturation
(C) Both of these
(D) None of these

5. At maximum water holding capacity, water held at:
(A) Less than 1 bar
(B) 1 bar
(C) More than 1 bar
(D) None of these

6. Water stored in soil than field capacity is:
(A) Same (B) Double
(C) Triple (D) Any amount

7. The maximum amount of moisture that will be held in the pores of a saturated soil is known as:
(A) Water holding capacity
(B) Saturation
(C) Field capacity
(D) All of these

8. To constitute field capacity together with:
(A) Capillary water
(B) Hygroscopic water
(C) Both of these
(D) None of these

9. The moisture constant of a soil at field capacity is taken as:
(A) 25 % (B) 50 %
(C) 75 % (D) 100 %

10. Field capacity generally comes after irrigation:
(A) 10-12 hrs (B) 12-24 hrs
(C) 24-48 hrs (D) 48-72 hrs

11. At field capacity water held between:
(A) 0.1 to 0.3 bar (B) less than 0.1 bar
(C) 15 bar (D) 31 bars

12. The moisture level at which plants will no longer get enough moisture for their survival is called:
(A) Wilting point
(B) Filed capacity
(C) Saturation
(D) Water holding capacity

13. Wilting point in most of the soils is in the region of field capacity is:
(A) 25 % (B) 50 %
(C) 75 % (D) 100 %

14. Permanent wilting point is also known as:
(A) Wilting coefficient
(B) Hygroscopic coefficient
(C) Moisture equivalent
(D) None of these

15. Moisture held at permanent wilting point is:
(A) 0.1 to 0.3 bar
(B) less than 0.1 bar
(C) 15 bar
(D) 31 bars

16. Soil moisture between field capacity and permanent wilting point is referred as:
(A) Available soil moisture
(B) Field capacity
(C) Saturation
(D) Water holding capacity

17. Moisture storage capacity of fine sandy soil at field capacity is:
(A) 3-5 % (B) 5-15 %
(C) 12-18 % (D) 15-30 %

18. Moisture storage capacity of sandy loam soil at field capacity is:
(A) 3-5 % (B) 5-15 %
(C) 12-18 % (D) 15-30 %

19. Moisture storage capacity of silt loam soil at field capacity is:
(A) 3-5 % (B) 5-15 %
(C) 12-18 % (D) 15-30 %

20. Moisture storage capacity of clay loam soil at field capacity is:
(A) 3-5 % (B) 5-15 %
(C) 12-18 % (D) 15-30 %

21. Moisture storage capacity of clay soil at field capacity is:
(A) 5-15 % (B) 12-18 %
(C) 15-30 % (D) 25-40 %

22. Moisture storage capacity of fine sandy soïl at permanent wilting point is:
(A) 1-3 % (B) 3-8 %
(C) 6-10 % (D) 7-16 %

23. Moisture storage capacity of sandy loam soil at permanent wilting point is:
(A) 1-3 % (B) 3-8 %
(C) 6-10 % (D) 7-16 %

24. Moisture storage capacity of silt loam soil at permanent wilting point is:
(A) 1-3 % (B) 3-8 %
(C) 6-10 % (D) 7-16 %

25. Moisture storage capacity of clay loam soil at permanent wilting point is:
(A) 1-3 % (B) 3-8 %
(C) 6-10 % (D) 7-16 %

26. Moisture storage capacity of clay soil at permanent wilting point is:
(A) 3-8 % (B) 6-10 %
(C) 7-16 % (D) 12-20 %

27. Air dried soil at atmosphere contains tightly bound water with soil particles and water of macro and micro pores are completely replaced by air is called:
(A) Wilting coefficient
(B) Hygroscopic coefficient
(C) Moisture equivalent
(D) Sticky point

28. Hygroscopic coefficient is mainly depend upon:
(A) Colloidal phase of the soil
(B) Available soil moisture of the soil
(C) Field capacity of the soil
(D) None of these'

29. Clay soil contains more colloidal materials and hygroscopic water than:
(A) Sandy soil (B) Clay loam
(C) Sandy loam (D) All of these

30. The amount of water retained by a sample of initially saturated soil material after being subjected to a centrifugal force of 1000 times that of gravity for a definite period of time, usually half a hour is called:
(A) Wilting coefficient
(B) Hygroscopic coefficient
(C) Moisture equivalent
(D) Sticky point

31. Moisture equivalent is measures in laboratory, generally used in place of:
(A) Field capacity
(B) Permanent wilting point
(C) Temporary wilting point
(D) All of these

32. Moisture equivalent is also used for available moisture in:
(A) Micro capillary pores
(B) Macro capillary pores
(C) Both of these (D) None of these

33. In medium textured soil moisture content equivalent to:
(A) Field capacity
(B) Permanent wilting point
(C) Temporary wilting point
(D) All of these

34. In clay soil the field capacity is -------- than moisture equivalent:
(A) Less than (B) More than
(C) Same (D) Unknown

35. In sandy soil the field capacity is -------- moisture equivalent:

(A) Less than (B) More than
(C) Same (D) Unknown

36. The performance of soil moisture of hygroscopic water is:

(A) 31-10000 atm (B) 31 atm
(C) 15 atm (D) 0.3-31 atm

37. The performance of soil moisture of hygroscopic coefficient is:

(A) 31-10000 atm (B) 31 atm
(C) 15 atm (D) 0.3-31 atm

38. The performance of soil moisture of wilting coefficient is:

(A) 31-10000 atm (B) 31 atm
(C) 15 atm (D) 0.3-31 atm

39. The performance of soil moisture of capillary water is:

(A) 31-10000 atm (B) 31 atm
(C) 15 atm (D) 0.3-31 atm

40. The performance of soil moisture of moisture equivalent is:

(A) 0.3-1 atm (B) 0.3 atm
(C) 0.3-31 atm (D) less than 0.3 atm

41. The performance of soil moisture of field capacity is:

(A) 0.3-1 atm (B) 0.3 atm
(C) 0.3-31 atm (D) less than 0.3 atm

42. The performance of soil moisture of sticky point is:

(A) 0.3-1 atm (B) 0.3 atm
(C) 0.3-31 atm (D) less than 0.3 atm

43. The performance of soil moisture of gravita-tional water is:

(A) 0.3-1 atm (B) 0.3 atm
(C) less than 0.3 atm
(D) 0.3-31 atm

44. Moisture content of a soil at which it does not sticks with other materials is called:

(A) Hygroscopic coefficient
(B) Wilting coefficient
(C) Moisture equivalent
(D) Sticky point

45. Sticky point occurs at:

(A) 1/3 atm (B) ½ atm
(C) 1 atm (D) 2/3 atm

46. After the soil is completely saturated the excess infiltration that starts moving downwards due to the force of gravity to finally join the ground water reservoir is known as:

(A) Gravitational water
(B) Capillary water
(C) Hygroscopic water
(D) Crystalline water

47. Movement of gravitational water down the soil profile depends upon the:

(A) Permeability
(B) Infiltration
(C) Seepage
(D) None of these

48. Gravitational water is also called:

(A) Free water (B) Combined water
(C) Bound water (D) None of these

49. Gravitational water retained in the soil at:

(A) Below 1/3 atm
(B) at 1/3 atm
(C) Above 1/3 atm
(D) None of these

50. Gravitational water remain in the soil too long, serious damage may take place due to:

(A) Lack of O_2 (B) Lack of CO_2
(C) Excess of O_2 (D) Excess of CO_2

51. The amount of water will remain held within the soil pores due to capillary force is called:

(A) Gravitational water
(B) Capillary water
(C) Hygroscopic water
(D) Crystalline water

52. Capillary water behave according to:

(A) Law of gravity
(B) Law of capillary
(C) Law of atmosphere
(D) None of these

53. Capillary water retained in the soil at tension of:

(A) 0.3-1 atm
(B) 0.3 atm
(C) less than 0.3 atm
(D) 0.3-31 atm

54. Capillary water held at tension of 15-31 atm is known as:
(A) Intercapillary water
(B) Intra capillary water
(C) Both of these
(D) None of these

55. Availability of intercapillary water to the plants is:
(A) Never available
(B) Hardly available
(C) Not available
(D) More available

56. The part of the water which a dry soil will absorbed in order to reach a static state with the atmosphere water is known as:
(A) Gravitational water
(B) Capillary water
(C) Hygroscopic water
(D) Crystalline water

57. Hygroscopic water retained in the soil at tension of:
(A) 0.3-31 atm
(B) 31 atm
(C) less than 31 atm
(D) More than 31 atm

58. Hygroscopic water held by the force of molecular attraction and can be removed only by:
(A) Heating (B) Condensation
(C) Both of these (D) None of these

59. Availability of Hygroscopic water to the plants is:
(A) Never available (B) Hardly available
(C) Not available (D) More available

60. The water is held in the soil particles with the chemical forces is called:
(A) Gravitational water
(B) Capillary water
(C) Combined water
(D) Crystalline water

61. The water is present in intra spaces of minerals and binds with minerals is called:
(A) Gravitational water
(B) Capillary water
(C) Combined water
(D) Crystalline water

62. The water is present in soil in the form of vapour specially at higher temperature is known as:
(A) Vapour water
(B) Capillary water
(C) Combined water
(D) Crystalline water

63. Which form of water is not available to the plants is:
(A) Vapour water
(B) Crystalline water
(C) Combined water
(D) All of these

64. The quantity of water, which is needed for the maximum yield of crop is called:
(A) Water Requirement
(B) Irrigation Requirement
(C) Consumptive Use of water
(D) None of these

65. Quantity of surface and ground water absorbed by crops at transpired or use directly in the building of plant tissues together with that evaporated from the cropped area is called:
(A) Water Requirement
(B) Irrigation Requirement
(C) Consumptive Use of water
(D) None of these

66. Total water required of a crop, excluding of effective rainfall and soil moisture stored in the root zone or that contributed from the shallow ground water table is known as:
(A) Water Requirement
(B) Irrigation Requirement
(C) Consumptive Use of water
(D) None of these

67. Downward movement of water through saturated or nearly saturated soil in response to the force of gravity is known as:
(A) Percolation
(B) Seepage
(C) Infiltration
(D) Permeability

68. Lateral movement of water is called:
(A) Percolation (B) Seepage
(C) Infiltration (D) Permeability

69. The ratio of flow of water through a porus medium as it directly proportional to the cross- sectional area of the flow and the hydraulic gradient under which the flow moves is known as:

(A) Percolation (B) Seepage
(C) Infiltration (D) Permeability

70. Flow of ground water through a porus medium is controlled by:

(A) Law of gravity
(B) Law of capillary
(C) Law of atmosphere
(D) Darcy's law

71. The rate of flow of water through a unit cross-sectional area under a unit hydraulic gradient is known as:

(A) Coefficient of permeability
(B) Hygroscopic coefficient
(C) Wilting coefficient
(D) None of these

72. The ratio of irrigation water transpired by the crop on irrigation farm or project during their growth period over the water diverted from a river or other natural source into the farm or project canal or canals during the same period is known as:

(A) Irrigation application efficiency
(B) Project efficiency
(C) Water conveyance efficiency
(D) Water storage efficiency

73. The ratio of volume of irrigation water consumptively used in a specified period to the total volume of irrigation water that has been diverted from the capture system is called:

(A) Irrigation application efficiency
(B) Project efficiency
(C) Water conveyance efficiency
(D) Water storage efficiency

74. The ratio of water diverted to the farm to water diverted from the river or reservoir is called:

(A) Irrigation application efficiency
(B) Project efficiency
(C) Water conveyance efficiency
(D) Water storage efficiency

75. The ratio of water stored in the root zone during the irrigation to the water needed in the root zone prior to the irrigation is called:

(A) Irrigation application efficiency
(B) Project efficiency
(C) Water conveyance efficiency
(D) Water storage efficiency

76. In drought areas appears in a field which is not irrigated uniformly unless excess water is applied,' which is results in a waste of water is termed as:

(A) Irrigation application efficiency
(B) Water distribution efficiency
(C) Water conveyance efficiency
(D) Water storage efficiency

77. The proportional of normal consumptive use of water to the net amount of water depleted from root zone is known as:

(A) Consumptive use efficiency
(B) Project efficiency
(C) Water conveyance efficiency
(D) Water storage efficiency

78. The natural process by which water continue to change into the vapour from its free surface known as:

(A) Evaporation
(B) Transpiration
(C) Evapo-transpiration
(D) None of these

79. Evaporation is measured by:

(A) US Open evaporimeter
(B) Lysimeter
(C) Potometer
(D) Pressure membrane apparatus

80. The average annual potential evaporation in different agro-climatic zones ranges from:

(A) Less than 1000 mm per annum
(B) 1000-3500 mm per annum
(C) 3500-5000 mm per annum
(D) More than 5000 mm per annum

81. Water entering the soil from the ground surface is referred to:

(A) Evaporation
(B) Transpiration
(C) Evapo-transpiration
(D) None of these

82. The infiltration rate is less than 12.5 mm/hr categorized as:
(A) Low (B) Medium
(C) High (D) Very high

83. The infiltration rate is 12.5-25.0 mm/hr categorized as:
(A) Low (B) Medium
(C) High (D) Very high

84. The infiltration rate is more than 25.0 mm/hr categorized as:
(A) Low (B) Medium
(C) High (D) Very high

85. Water losses through combined effect of evaporation and transpiration is called:
(A) Evapo-transpiration
(B) Consumptive use
(C) Both of these
(D) None of these

86. Water held tightly by soil colloids through gravitational force is called:
(A) Capillary water
(B) Hygroscopic water
(C) Gravitational water
(D) Free water

87. Availability of gravitational water to the plant is:
(A) More (B) Less
(C) Limited (D) Unknown

88. Furrow length for irrigation should be kept under normal soil:
(A) Less than 100 meter
(B) 100-300 meter
(C) 500-1000 meter
(D) Morethan 1000 mtr

89. Depth of furrow for irrigation should be:
(A) 5-6 inches (B) 8-12 inches
(C) 12-15 inches (D) more than 15 in.

90. Water Use Efficiency highest in case of:
(A) Flood method (B) Border method
(C) Drip method (D) Sprinkler method

91. Among the following types of soil-water which one is the most useful for the plants:
(A) Capillary water
(B) Hygroscopic water
(C) Gravitational water
(D) Combined water

92. Near the canal banks soils become:
(A) Alkaline (B) Acidic
(C) Neutral (D) Well drained

93. The weight of a half mole of water is:
(A) 0.5 g (B) 1.0 g
(C) 9.0 g (D) 18.0

94. Water requirement of mentha crop is:
(A) High
(B) Less
(C) Same as other crop
(D) Drought crop

95. What will be the moisture content of soil at 50% ASM, if it has field capacity and permanent wilting point values of 20% and 8%, respectively:
(A) 18 % (B) 14 %
(C) 12 % (D) 10 %

96. A device for measuring percolation and leaching losses from a column of soil under controlled conditions is called:
(A) Lysimeter
(B) USDA Open pan evoporimeter
(C) Parshal flume
(D) Permeameter

97. A calibrated device for measuring the flow of water in open conditions is:
(A) Lysimeter
(B) USDA Open pan evoporimeter
(C) Parshal flume
(D) Permeameter

98. A device for measuring the permeability of soils or other materials is:
(A) Lysimeter
(B) USDA Open pan evoporimeter
(C) Parshal flume
(D) Permeameter

99. Instrument used to measure the humidity of the atmosphere is:
(A) Psychrometer
(B) USDA Open pan evoporimeter
(C) Parshal flume
(D) Permeameter

100. An instrument that measures the effective terrestrial radiation is:
(A) Lysimeter (B) Pyregeometer
(C) Parshal flume (D) Permeameter

101. A soil which has its interspaces or void spaces completely filled with water to the point where runoff occurs:

(A) Saturated soil
(B) Saturated deficit
(C) Saturation Extract
(D) Saturation index

102. The irrigation projects having culturable command area of more than 10 thousand hectare is:

(A) Major irrigation
(B) Medium irrigation
(C) Minor irrigation (D) None of these

103. The amount of water vapours required to bring non saturated air at a given temperature and pressure to the point of saturation is called:

(A) Saturated soil (B) Saturated deficit
(C) Saturation Extract
(D) Saturation index

104. The solution obtained by pressure or vacuum filtration of a soil paste that has been made up to a saturated condition by adding water while stirring is called:

(A) Saturated soil (B) Saturated deficit
(C) Saturation Extract
(D) Saturation index

105. An estimate of carbonate precipitation from irrigation water as a function of the degree of calcium carbonate saturation of the soil solution is called:

(A) Saturated soil
(B) Saturated deficit
(C) Saturation Extract
(D) Saturation index

106. The irrigation projects having culturable command area of less than 10 thousand but more than 1 thousand hectare is:

(A) Major irrigation
(B) Minor irrigation
(C) Medium irrigation
(D) None of these

107. The irrigation projects having culturable command area of less than 1 thousand hectare is

(A) Major irrigation
(B) Medium irrigation
(C) Minor irrigation
(D) None of these

108. Irrigation done through tanks, surface percolation, wells, tube wells and fluxial water is:

(A) Major irrigation
(B) Medium irrigation
(C) Minor irrigation
(D) None of these

109. The critical stages of rapeseed and mustard for irrigation is/are:

(A) Vegetative (B) Flowering
(C) Pod filling (D) All of these

110. The critical stages of cotton for irrigation is/are:

(A) Sympodial Branch formation
(B) Flowering
(C) Ball formation and ball bursting
(D) All of these

111. The critical stages of potato for irrigation is/are:

(A) Stolonization
(B) Tuber formation
(C) Tuber enlargement
(D) All of these

112. The critical stages of sugarcane for irrigation is/are:

(A) Seedling emergence
(B) Shoot elongation
(C) Tasseling
(D) All of these

113. The critical stages of linseed for irrigation is/are:

(A) Flower initiation
(B) Seed setting
(C) Both of these
(D) None of these

114. The critical stages of barley for irrigation is/are:

(A) Tillering (B) Milking
(C) Both of these (D) None of these

115. The critical stages of sorghum for irrigation is/are:

(A) Grain filling (B) Flower initiation
(C) Both of these (D) None of these

116. Optimum depth of irrigation water in tall wheat is:

(A) 5-8 cm (B) 4-8 cm
(C) 5-12 cm (D) 6-8 cm

117. Optimum depth of irrigation water in dwarf wheat is:
(A) 5-8 cm (B) 4-8 cm
(C) 5-12 cm (D) 6-8 cm

118. Optimum depth of irrigation water in rice is:
(A) 7-8 cm (B) 4-8 cm
(C) 5-12 cm (D) 6-8 cm

119. Optimum depth of irrigation water in sorghum is:
(A) 7-8 cm (B) 4-8 cm
(C) 5-12 cm (D) 5-8 cm

120. Optimum depth of irrigation water in maize is:
(A) 6-8 cm (B) 6-10 cm
(C) 3-8 cm (D) 5-8 cm

121. Optimum depth of irrigation water in sugarcane is:
(A) 5-8 cm (B) 6-10 cm
(C) 3-8 cm (D) 6-8 cm

122. Optimum depth of irrigation water in cotton is:
(A) 5-8 cm (B) 6-10 cm
(C) 3-8 cm (D) 6-8 cm

123. Optimum depth of irrigation water in rapeseed and mustard is:
(A) 5-8 cm (B) 6-10 cm
(C) 3-8 cm (D) 6-8 cm

124. Optimum depth of irrigation water in groundnut is:
(A) 5-8 cm (B) 3-6 cm
(C) 4-6 cm (D) 5-8 cm

125. Optimum depth of irrigation water in sesamum is:
(A) 5-8 cm (B) 3-6 cm
(C) 4-6 cm (D) 5-8 cm

126. Optimum depth of irrigation water in potato is:
(A) 5-8 cm (B) 3-6 cm
(C) 4-6 cm (D) 5-8 cm

127. Optimum depth of irrigation water in tobacco is:
(A) 5-8 cm (B) 3-6 cm
(C) 4-6 cm (D) 5-8 cm

128. Optimum depth of irrigation water in sunflower is:
(A) 5-8 cm (B) 3-6 cm
(C) 4-6 cm (D) 5-8 cm

129. Optimum depth of irrigation water in chickpea/ pigeonpea and lentil is:
(A) 6-8 cm (B) 5-6 cm
(C) 5-8 cm (D) 4-5 cm

130. Optimum depth of irrigation water in soybean is:
(A) 6-8 cm (B) 5-6 cm
(C) 5-8 cm (D) 4-5 cm

131. Optimum depth of irrigation water in green gram, black gram and cowpea is:
(A) 6-8 cm (B) 5-6 cm
(C) 5-8 cm (D) 4-5 cm

132. Optimum depth of irrigation water in finger millet is:
(A) 6-8 cm (B) 5-6 cm
(C) 5-8 cm (D) 4-5 cm

133. Optimum depth of irrigation water in berseem is:
(A) 6-8 cm (B) 5-6 cm
(C) 5-8 cm (D) 4-5 cm

134. Optimum depth of irrigation water in barley is:
(A) 4.5-8 cm (B) 5-6 cm
(C) 5-8 cm (D) 4-5 cm

135. Highest number of irrigations is applied in:
(A) Rice (B) Sugarcane
(C) Berseem (D) Sugarbeet

136. Water requirement of boro paddy is:
(A) 150-190 cm (B) 90 cm
(C) 100 cm (D) 300 cm

137. Water requirement of aman paddy is:
(A) 150-190 cm (B) 90 cm
(C) 100 cm (D) 300 cm

138. Water requirement of aus paddy is:
(A) 150-190 cm (B) 90 cm
(C) 100 cm (D) 300 cm

139. Which furrow method of irrigation is suitable where rate of infiltration is minimum :
(A) Flat furrow
(B) Alternate furrow
(C) Contour furrow
(D) Corrugated furrow

140. Under saline-alkali soil & water scarce areas, suitable furrow method of irrigation is:
(A) Flat furrow
(B) Alternate furrow
(C) Contour furrow
(D) Corrugated furrow

141. When furrow are prepared across the slope, the method of irrigation is called:
(A) Flat furrow
(B) Alternate furrow
(C) Contour furrow
(D) Corrugated furrow

142. When furrow are prepared to the same direction of slop, suitable furrow method of irrigation is called:
(A) Flat furrow
(B) Alternate furrow
(C) Sloppy furrow
(D) Corrugated furrow

143. Furrow method of irrigation is suitable in:
(A) Potato
(B) Groundnut
(C) Tobacco
(D) All of these

144. Which is the suitable irrigation method where root rot is a severe problem:
(A) Furrow method
(B) Basin method
(C) Ring method
(D) Surface flooding

145. Sprinkler irrigation is also called:
(A) Over head irrigation
(B) Trickle irrigation
(C) Under head irrigation
(D) Gravity irrigation

146. Surface irrigation method is also called:
(A) Over head irrigation
(B) Trickle irrigation
(C) Under head irrigation
(D) Gravity irrigation

147. The suitable sprinkler system of irrigation at low pressure (0.5-25 kg/cm2):
(A) Rotating head system
(B) Perforated pipe system
(C) Both of these
(D) None of these

148. Component of sprinkler system is:
(A) Pumping set or water lifting unit
(B) Pipe line
(C) Sprinkler head
(D) All of these

149. Under steep sloppy areas which irrigation system is more suitable:
(A) Furrow method
(B) Border method
(C) Sprinkler method
(D) Drip method

150. The most suitable method of irrigation in horticultural and vegetable crops is:
(A) Furrow method
(B) Border method
(C) Sprinkler method
(D) Drip method

151. Flood method of irrigation is suitable when:
(A) High water requiring crop
(B) Water is not a limiting factor
(C) Less erosion affected areas
(D) All of these

152. The practice of irrigation method is mainly adopted in India is:
(A) Surface (B) Sub surface
(C) Aerial (D) All of these

153. Border method of irrigation is suitable under:
(A) Permeability of soil should be less
(B) Soil erosion should be less
(C) Light slope should be one direction
(D) All of these

154. The optimum width of bed in border method of irrigation is:
(A) 3-5 meter (B) 3-10 meter
(C) 3-15 meter (D) 3-20 meter

155. Border method of irrigation should not be applied under:
(A) More infiltration
(B) Less infiltration
(C) Both of these
(D) None of these

156. Flood method of irrigation is more suitable in:
(A) Paddy (B) Sugarcane
(C) Berseem (D) All of these

157. Basin method of irrigation is more suitable in:

(A) Orchard (B) Field crops
(C) Fodder crops (D) Cash crops

158. Ring basin method of irrigation is more suitable in:

(A) Mango (B) Jackfruit
(C) Jamun (D) All of these

159. Water use efficiency is maximum in:

(A) Sprinkler system
(B) Drip irrigation
(C) Surface method
(D) None of these

160. Main feature of drip irrigation system is:

(A) Less infiltration
(B) Less runoff
(C) Less evaporation
(D) All of these

161. Under undulated soil and high infiltration rate, which method of irrigation is more suitable:

(A) Sprinkler (B) Drip
(C) Surface (D) Subsurface

162. When moisture movement is in two direction, which method of irrigation is more suitable:

(A) Sprinkler (B) Drip
(C) Surface (D) Subsurface

163. Removal of excess water form the soil or crop by artificial means to maintain water table and salt concentration in the soil to increase crop production is called:

(A) Irrigation (B) Drainage
(C) Percolation (D) None of these

164. Soils are most suitable for irrigation when ground water table is:

(A) Less than 1 meter
(B) 2 meter
(C) 1-2 meter
(D) More than 3 meter

165. Soils are good for irrigation when ground water table is:

(A) Less than 1 meter
(B) 2 meter
(C) 1-2 meter
(D) More than 3 meter

166. Soils are medium for irrigation when ground water table is:

(A) Less than 1 meter
(B) 2 meter
(C) 1-2 meter
(D) More than 3 meter

167. Soils are less suitable for irrigation when ground water table is:

(A) Less than 1 meter
(B) 2 meter
(C) 1-2 meter
(D) More than 3 meter

168. More waterlogged affected areas are:

(A) West Bengal (B) Punjab
(C) Haryana (D) All of these

169. Waterlogging is not a problem where soils are:

(A) Leveled (B) Undulating
(C) Saline and alkali (D) None of these

170. Removal of excess water from surface of the soil is called:

(A) Surface drainage
(B) Subsurface drainage
(D) Both of these (D) None of these

171. Removal of excess water from subsurface of the soil is called:

(A) Subsurface drainage
(B) Surface drainag
(D) Both of these (D) None of these

172. Electrical conductivity of low saline water is:

(A) Less than 250 mmhos/cm
(B) 250-750 mmhos/cm
(C) 750-2250 mmhos/cm
(D) More than 2250 mmhos/cm

173. Electrical conductivity of medium saline water is:

(A) Less than 250 mmhos/cm
(B) 250-750 mmhos/cm
(C) 750-2250 mmhos/cm
(D) More than 2250 mmhos/cm

174. Electrical conductivity of high saline water is:

(A) Less than 250 mmhos/cm
(B) 250-750 mmhos/cm
(C) 750-2250 mmhos/cm
(D) More than 2250 mmhos/cm

175. Electrical conductivity of very high saline water is:
(A) Less than 250 mmhos/cm
(B) 250-750 mmhos/cm
(C) 750-2250 mmhos/cm
(D) More than 2250 mmhos/cm

176. Water is not suitable for irrigation is:
(A) Low saline water
(B) Medium saline water
(C) High saline water
(D) Very high saline water

177. Sodium absorption ratio of low quality water is:
(A) Less than 10% (B) 10-18 %
(C) 18-26 % (D) More than 26 %

178. Sodium absorption ratio of medium quality water is:
(A) Less than 10% (B) 10-18 %
(C) 18-26 % (D) More than 26 %

179. Sodium absorption ratio of high quality water is:
(A) 10-18 % (B) Less than 10 %
(C) 18-26 % (D) More than 26 %

180. Sodium absorption ratio of very high quality water is:
(A) 10-18 % (B) Less than 10 %
(C) 18-26 % (D) More than 26 %

181. Low sodium absorption ratio containing water is suitable for:
(A) All crops (B) Few crops
(C) Not fit for irrigation
(D) None of these

182. Very high sodium absorption ratio containing water is:
(A) All crops (B) Few crops
(C) Not fit for irrigation
(D) None of these

183. Very high sodium absorption ratio containing water can be used for irrigation by:
(A) Use of gypsum (B) Limited use
(C) Both of these (D) None of these

184. Very high carbonate and bicarbonate containing water is:
(A) Fit for irrigation
(B) No effect on crops
(C) Not fit for irrigation
(D) None of these

185. Very high potassium and nitrate containing water is:
(A) Fit for irrigation
(B) No effect on crops
(C) Not fit for irrigation
(D) None of these

186. Very high florine containing water is:
(A) Fit for irrigation
(B) No effect on crops
(C) Not fit for irrigation
(D) None of these

187. Florine absorption is more in:
(A) Saline soil (B) Alkali soil
(C) Saline-alkali soil
(D) Acidic soil

188. Permissible limit of florine in drinking water is:
(A) 1-10 ppm (B) 10-20 ppm
(C) 20-30 ppm (D) 30-40 ppm

189. Irrigation water containing more than 7 ppm florine will affect in wheat:
(A) Germination (B) Growth
(C) Both of these (D) None of these

190. Irrigation water containing florine more than 2.5 meq/litre of water is:
(A) Fit for irrigation
(B) No effect on crops
(C) Not fit for irrigation
(D) None of these

191. The effect on soil due to salt containing water is:
(A) Nutrient fixation increased
(B) Nutrient fixation decreased
(C) No affect of nutrient fixation
(D) None of these

192. The effect on soil due to more calcium containing water is:
(A) Potassium absorption decreased
(B) Potassium absorption increased
(C) Calcium absorption decreased
(D) Calcium absorption increased

193. The effect on soil due to more magnesium containing water is:
(A) Potassium absorption decreased
(B) Potassium absorption increased
(C) Calcium absorption decreased
(D) Calcium absorption increased

194. The effect on soil due to more sodium containing water is:
(A) Salt defloculate
(B) Formation of crust layer
(C) Beneficial microorganism decreased
(D) All of these

195. Highly salt tolerant crop is:
(A) Barley
(B) Sugarbeet
(C) Taramira
(D) All of these

196. The effect on plant due to irrigation of salt containing water is:
(A) Harmful effect on cell differentiation
(B) Harmful effect of cell division
(C) Harmful effect on photosynthesis
(D) All of these

197. Plant show physiological drought when:
(A) Increase in salt concentration
(B) Decrease salt conecentration
(C) Both of these
(D) None of these

198. The optimum irrigation requirement of potato after sowing is:
(A) 25 cm (B) 45-60 cm
(C) 50 cm (D) 10 cm

199. The optimum irrigation requirement of maize after sowing is:
(A) 25 cm (B) 45-60 cm
(C) 50 cm (D) 10 cm

200. The optimum irrigation requirement of bajra after sowing is:
(A) 25 cm (B) 45 cm
(C) 50 cm (D) 15-20 cm

201. The optimum irrigation requirement of wheat after sowing is:
(A) 25 cm (B) 45 cm
(C) 50-60 cm (D) 10 cm

202. The optimum irrigation requirement of sorghum after sowing is:
(A) 25 cm (B) 45-50 cm
(C) 50 cm (D) 10 cm

203. The optimum irrigation requirement of sunflower after sowing is:
(A) 25 cm (B) 40-55 cm
(C) 50 cm (D) 10 cm

204. The optimum irrigation requirement of rapeseed and mustard after sowing is:
(A) 25-30 cm (B) 45 cm
(C) 50 cm (D) 10 cm

205. The optimum irrigation requirement of groundnut after sowing is:
(A) 25 cm (B) 45 cm
(C) 50-70 cm (D) 10 cm

206. The optimum irrigation requirement of berseem after sowing is:
(A) 250 cm (B) 450 cm
(C) 500 cm (D) 150-200 cm

207. The optimum irrigation requirement of chickpea after sowing is:
(A) 25 cm (B) 25-40 cm
(C) 50 cm (D) 10 cm

208. The optimum irrigation requirement of field pea after sowing is:
(A) 25 cm (B) 25-40 cm
(C) 50 cm (D) 10 cm

209. The optimum irrigation requirement of cotton after sowing is:
(A) 25 cm (B) 45-70 cm
(C) 50 cm (D) 10 cm

210. The optimum irrigation requirement of rice after sowing is:
(A) 125-135 cm (B) 45-70 cm
(C) 50-100 cm (D) 100 cm

211. The optimum irrigation requirement of tobacco after sowing is:
(A) 20-25 cm (B) 45-70 cm
(C) 50 cm (D) 10 cm

212. The optimum irrigation requirement of sugarcane after sowing is:
(A) 250 cm (B) 400 cm
(C) 200-300 cm (D) 100-150 cm

213. The optimum irrigation requirement of pigeonpea after sowing is:
(A) 25 cm (B) 45-70 cm
(C) 50 cm (D) 7-15 cm

214. The yield of marketable crop produced per unit of water used in evapotraspiration is known as:
(A) Water distribution efficiency
(B) Water use efficiency
(C) Water application efficiency
(D) Conveyence efficiency

215. At what stages, wheat crop should be irrigated if two irrigations are available?
(A) Crown root and tillering
(B) Tillering and flowering
(C) Flowering and milking
(D) Crown root and flowering

216. What would happen when pigeonpea crop is irrigated during flowering stage?
(A) More flowering
(B) Dropping of flowers and reduction in yield
(C) No favourable effect or adverse effect
(D) None of these

217. Which type of soil has higher water holding capacity?
(A) Sandy soil (B) Sandy loam soil
(C) Loam soil (D) Clay loam soil

218. If the level of mercury (Hg) suddenly drops, it indicates:
(A) Excess heat (B) Stormy
(C) Heavy rain (D) Clear weather

219. The application of water divided in strips along with contour is known as:
(A) Strip irrigation
(B) Contour irrigation
(C) Control irrigation
(D) Basin irrigation

220. Specific heat of soil water is:
(A) 1.0 calories/g (B) 0.40 calories/g
(C) 0.5 calories/g (D) 0.22 calories/g

221. If rainfall is P and potential Evapotranspiration is PET, the moisture deficit index (MDI) will be:
(A) $\frac{P-PET}{P}\times100$
(B) $\frac{P-PET}{PET}\times100$
(C) $\frac{PET}{P}\times100$ (D) $\frac{P\times100}{PET}$

222. pF is:
(A) Logarithm of water column height in cm
(B) Logarithm of H+
(C) Equal to pH
(D) None of these

223. Which irrigation system gives the highest irrigation efficiency:
(A) Surface method
(B) Subsurface method
(C) Sprinkler method
(D) Trickle method

224. Total surface water availability in the country:
(A) 1900 BCM (B) 1953 BCM
(C) 400 BCM (D) 650 BCM

225. Total utilizable surface water in India is:
(A) 600 BCM (B) 690 BCM
(C) 710 BCM (D) 1953 BCM

226. Total ground water resources of India are:
(A) 432 BCM (B) 532 BCM
(C) 350 BCM (D) 342 BCM

227. Total utilizable ground water resources of India are:
(A) 432 BCM (B) 342 BCM
(C) 400 BCM (D) 350 BCM

228. Total utilizable water resources of India are:
(A) 920 BCM (B) 1032 BCM
(C) 1142 BCM (D) 1500 BCM

229. Total utilized water resources of India are:
(A) 35% (B) 44%
(C) 51% (D) 60%

230. How much of total water utilization goes to irrigation sector in India?
(A) 2% (B) 3%
(C) 4% (D) 5%

231. Total irrigation potential created upto 1999-2000, is
(A) 92 mha (B) 94.73 mha
(C) 96 mha (D) 100 mha

232. Total utilized irrigation potential upto 1999-2000, is
(A) 82 mha (B) 84.7 mha
(C) 87 mha (D) 97 mha

233. Major irrigation project should have total cultural command area:
(A) 500 ha (B) 1000 ha
(C) 10000 ha (D) >10000 ha

234. Minor irrigation project should have total cultural command area:
(A) 2000 ha (B) <1000 ha
(C) 10000 ha (D) 2000 - 10000 ha

235. Medium irrigation project should have total cultural command area:
(A) 2000 ha (B) 1000 - 10000 ha
(C) >10000 ha (D) None of these

236. National water policy was adopted in year:
(A) 1975 (B) 1981
(C) 1987 (D) 1965

237. Average world wide irrigation efficiency is:
(A) 30% (B) 35%
(C) <40% (D) 75%

238. The sensitive stage to moisture stress in sorghum are:
(A) Flowering (B) Grain filling
(C) Both of these (D) None of these

239. The critical stages for irrigation in chickpea are
(A) Branching-preflowering
(B) Branching-pod development
(C) Pre-flowering and pod development
(D) Branching, pre-flowering and pod development

240. Method of irrigation more suitable for potato crop is:
(A) Furrow method
(B) Flooding
(C) Check basin (D) Ring method

241. Method of irrigation more suitable for fruit crop is:
(A) Surface method
(B) Furrow method
(C) Drip method
(D) Sprinkler method

242. Method of irrigation more suitable in undulated land is:
(A) Sprinkler method
(B) Drip method
(C) Check basin method
(D) Border strip method

243. Border method of sowing wheat crop saves:
(A) 25% seed only
(B) 25% fertilizer only
(C) 25% seed as well as 25% fertilizer
(D) None of the above

244. The most critical stage for irrigation in wheat is:
(A) CRI stage (B) Tillering stage
(C) Milky stage (D) Dough stage

245. Osmotic potential of pure water is:
(A) One (B) Zero
(C) Less than zero
(D) Between zero & one

246. A plant cell placed in water will:
(A) Swell up and become turgid
(B) Swell up and burst
(C) lose water and become flaccid
(D) Shrink and die

247. Passage of water across a selectively permeable membrane is:
(A) Active transport
(B) Pinocytosis
(C) Facilitated diffusion
(D) Osmosis

248. Seeds soaked in water imbibe the same because of:
(A) Adsorption (B) Higher O.P
(C) Lower O.P. (D) Pressure of vacuoles

249. The force which determines the flow of water from one cell to another is:
(A) T.P (B) D.P.D.
(C) O.P. (D) W.P.

250. DPD is abbreviated form of:
(A) Daily Photosynthetic Deficit
(B) Daily Phosphorus Deficit
(C) Daily Pressure Deficit
(D) Diffusion Pressure Deficit

251. Swelling of wooden frames during rains is caused by:
(A) Endosmosis (B) Imbibition
(C) Capillarity (D) Osmosis

252. Initial absorption of water by root hair is through:
(A) Osmosis (B) Endosmosis
(C) Desorption (D) Imbibition

253. Root hairs absorb water from soil when:
(A) Osmotic concentration is same in the two
(B) Solute concentration is higher in soilsolution
(C) Solute concentration is higher in root hairs
(D) Absorption is active

254. DPD is equal to:
(A) OP xTP (B) OP+TP
(C) OP-TP (D) TP-OP

255. The actual pressure with which placed in solution, the solution is:
(A) DPD (B) OP
(C) WP (D) Diffusion

256. When a cell is fully turgid which of the following will be zero:
(A) Turgor pressure/pressure potential
(B) Wall pressure
(C) Suction pressure/DPD/Water potential
(D) Osmotic pressure (Solute pressure)

257. Turgidity of cell is maintained by:
(A) Osmotic pressure
(B) Wall pressure
(C) Turgor pressure
(D) Diffusion pressure

258. Seed germination is accompanied by:
(A) Absorption of heat
(B) Starch synthesis
(C) Absorption of water
(D) Absorption of nutrient

259. Purple cabbage leaves do not pass out colour in cold water but do so in hot water because:
(A) Hot water enters the cells faster
(B) Pigment is not soluble in cold water
(C) Hot water destroys cell walls
(D) Hot water kills plasmalemma and makes it permeable

260. Osmotic pressure is maximum in:
(A) Hydrophytes (B) Halophytes
(C) Xerophytes (D) Mesophytes

261. When concentration of solutes is low in the soil, absorption of water is:
(A) Stopped (B) Increase
(C) Retarded (D) Remains normal

262. Excessive supply of chemical fertilizers often causes death of crop plants due to:
(A) Exosmosis (B) Endosmosis
(C) Imbibition (D) Turgidity

263. Supply of excess fertilizer and watering of a grass lawn causes browning of grass leaves due to:
(A) Decreased photosynthesis
(B) Waterlogging of soil
(C) Leaching of fertilizer lower soil strata
(D) Osmosis and death of root

264. During absorption of water by roots, the water potential of cell sap is lower than that of:
(A) Pure water and soil solution
(B) Neither pure water nor soil solution
(C) Pure water but higher than that of soil solution
(D) Soil solution but higher than that of pure water

265. Cotton fibre dipped in water absorb water through:
(A) Endosmosis (B) Exosmosis
(C) Capillarity (D) Imbibition

266. The term water potential was coined by:
(A) Sayre (B) Von Mohl
(C) Slater & Taylor (D) Lloyd

267. Water potential of a solution is depicted by:
(A) ψ_p (B) ψ_s
(C) ψ_x (D) $\Delta\psi$

268. For raising a healthy nursery of tobacco on highly leached soils, besides other nutrients, borax are ---g/m is also recommended
(A) 0.4-0.5 (B) 0.6-0.7
(C) 0.1-0.2 (D) 0.3-0.4

269. A basal dose of phosphorus at ---------- g/m² of tobacco nursery area has been recommended:
(A) 5.5-6.5 (B) 7.5-8.0
(C) 9.0-9.5 (D) 10.0-10.5

270. The most critical stage of rice for moisture stress is:
(A) Germination-tillering
(B) Tillering-flowering
(C) Panicle initiation-flowering
(D) Flowering-ripening

271. When only three irrigations are available, the wheat crop should be irrigated:
(A) CRI, tillering and dough
(B) CRI, flowering and milking
(C) CRI, tillering and milking
(D) CRI, flowering and late jointing

272. First irrigation to linseed is normally recommended after:
(A) 30-40 DAS (B) 40-50 DAS
(C) 20-30 DAS (D) 50-60 DAS

273. Sufficient soil moisture at which stage is essential for optimum productivity of urdbean:
(A) Branching and pod development
(B) Branching and flowering
(C) Pod development and maturity
(D) Flowering and pod development

274. The ideal moisture content in soybean seed at the time of threshing has been estimated at:
(A) 10-11% (B) 11-12%
(C) 13-14% (D) 14-15%

275. When only one irrigation is available, it should be given at which stage in safflower:
(A) 20 days (B) 30 days
(C) 40 days (D) 50 days

276. The identified critical stages for irrigation of safflower are:
(A) Rosette and stem elongation
(B) Rosette and flowering
(C) Flowering and grain filling
(D) Stem elongation and flowering

277. The most sensitive stages of sunflower to water stress lie between:
(A) Germination and seedling
(B) Germination and flowering
(C) Seedling and flowering
(D) Flowering and grain filling

278. Heavy rainfall or irrigation during fruiting periods of cotton may cause:
(A) Lodging of crop
(B) Shedding of flowers and young bolts
(B) Yellowing of young bolls
(D) Heavy incidence of boll worm

279. The water requirement of sugarcane is the maximum during which phase:
(A) Germination and formation phase
(B) Formation and grand growth phase
(C) Grand growth phase and maturity phase
(D) Formation phase and maturity phase

280. One of the following stages of rice is not critical in respect of irrigation:
(A) Seedling (B) Panicle initiation
(C) Flowering (D) Maturity

281. The first irrigation in wheat is normally recommended at:
(A) Crown root initiation
(B) Tillering stage
(C) Flowering
(D) Milking

282. The identified critical stages of irrigation in barley is:
(A) Tillering (B) Flowering
(C) Both of these (D) None of these

283. Water requirement of crop is equal to:
(A) CUW + Application losses
(B) CUW
(C) CUW + Application losses + water needed for special operations
(D) All of the above

284. Irrigation requirement is equal to:
(A) Water requirement – (Effective rainfall + Soil water contribution + Ground water contribution)
(B) Water requirement – CUW
(C) CUW
(D) None of these

285. CUW is equal to:
(A) ET
(B) ET + ≤ 1% for plant metabolite activities
(C) E + T
(D) ET + Effective rainfall

286. The amount of water consumptively used in 24 hours period is called:
(A) CUW
(B) Daily consumptive use
(C) Seasonal consumptive use
(D) All the above

287. Actual crop ET (Et crop) and reference crop ET (ETo) has the relation ship equal to:
(A) CT crop = Ki. ETo
(B) ETo = Kc. ET crop
(C) Kc = ETo/ET crop
(D) None of these

288. Crop ET can be estimated by:
(A) Lysimeter method
(B) Field Experiment method
(C) Soil water depletion method
(D) All the above

289. Reference crop ET was proposed by:
(A) Doorenbos and Pruitt
(B) Blaney –Criddle
(C) Tanner
(D) Sharma and Dastane

290. Weighing and non-weighing type lysimeter were suggested by:
(A) Harrold and Dreibelbis
(B) Slatyer and Mellroy
(C) Pelton
(D) None of these

291. PET can be calculated as:
(A) PET = Epan x crop factor
(B) $PET = \frac{Epan}{Crop\ factor}$
(C) PET = Epan (D) None

292. Duty of water refers to:
(A) Number of hectares of a crop successfully raised with a constant flow of one cusec of water throughout the growth period
(B) Total area irrigated
(C) Total cropped area irrigated
(D) None of these

293. The amount of water consumptively used by a crop during the entire growing season or crop period is called as:
(A) Daily consumptive use
(B) Seasonal consumptive use
(C) CUW (D) All of the above

294. Furrow irrigation method compared to other surface methods saves water to the tune of:
(A) 10% (B) 20%
(C) 30% (D) 40%

295. The furrow length of furrow irrigation methods varies with:
(A) Furrow slope
(B) Intake rate of soil
(C) Soil type
(D) All the above

296. Surge irrigation is useful mainly for:
(A) Reducing percolation loss
(B) Increased water use efficiency
(C) Infiltration increased
(D) All the above

297. 1 bar is equal to:
(A) 10 pascal (B) 10^3 pascal
(C) 10^5 pascal (D) 10^7 pascal

298. 1 pascal is equal to:
(A) 1 newton m^{-2}
(B) 10 Newton m^{-2}
(C) 100 Newton m^{-2}
(D) None of these

299. Total soil water potential is equal to:
(A) $\psi g + \psi p + \psi o$ (B) $\psi g + \psi p$
(C) $\psi g + \psi o$ (D) $\psi g + \psi p - \psi o$

300. 1 Joule is equal to:
(A) 10^6 ergs (B) 10^7 ergs
(C) 10^8 ergs (D) 10^{10} ergs

301. 1 Newton is equal to:
(A) 10^5 dynes (B) 10^6 dynes
(C) 10^7 dynes (D) 10^8 dynes

302. Which soil type holds maximum water?
(A) Clay soil (B) Loam soil
(C) Sandy soil (D) Sandy loam soil

303. Hydraulic conductivity is measured in:
(A) cm hr^{-1} (B) hr cm^{-1}
(C) cm (D) hr

304. Water movement in saturated soil is measured by:
(A) Brown's law (B) Poiseuille's law
(C) Newton's law (D) None of these

305. Poiseulle's law describes as:
(A) Volume of water flow is proportional to radius of the tube
(B) Volume of water flow is proportional to pressure difference
(C) Volume of water flow is inversely related to viscosity of liquid
(D) All of these

306. Darcy's law is related to:
(A) Water flow
(B) Infilterability
(C) Permeability
(D) Water potential

307. Soil water potential can be measured by:
(A) Tensiometer
(B) Neutron moisture meter
(C) Pressure plate apparatus
(D) All the above

308. Darcy's law states that:
(A) Velocity of water flow is directly proportional to the difference of hydraulic heads and flow length
(B) Velocity of water flow is directly proportional to the difference of hydraulic heads and inversely proportional to the flow length
(C) Both of the above
(D) None of these

309. Gypsum block was developed by:
(A) Scofield
(B) Bouyoucos & Mills
(C) Richards (D) None of these

310. Gypsum block measures the soil water tension in the range of:
(A) 0-10 atm (B) 1-20 atm
(C) 25-40 atm (D) None of these

311. Pressure plate apparatus was developed by:
(A) Scofield (B) Bouyoucos & Milk
(C) Richards (D) None of these

312. The source of fast neutron in neutron moisture meter is:
(A) Radium 226
(B) Beryllium
(C) Mixture of radium and beryllium
(D) None of these

313. Plant/cell water potential is equal to:
(A) ψs + ψm + ψp
(B) ψs + ψp
(C) ψs + ψm (D) ψs + ψs

314. Which one has the positive value?
(A) Osmotic potential
(B) Metric potential
(C) Turgor potential
(D) All of these

315. Range of soil water potential is:
(A) 0 to -5 bar (B) -0.1 to -20 bar
(C) -5 to -30 bar (D) -100 to -200 bar

316. Range of leaf water potential is:
(A) 0 to -5 bar (B) -0.1 to -20 bar
(C) -5 to -30 bar (D) -100 to -200 bar

317. Range of atmospheric water potential is:
(A) 0 to -5 bar (B) -0.1 to -20 bar
(C) -5 to -30 bar (D) -100 to -200 bar

318. Transpiration is maximum at:
(A) Morning
(B) Evening
(C) Afternoon at 2:0 PM
(D) Night

319. Transpiration ratio refers to:
(A) Volume of water transpired by a plant to produce a unit quantity of dry matter
(B) Production of unit quantity of dry matter to unit volume of water used
(C) Both of these
(D) None of these

320. Unit of transpiration ratio is:
(A) Dimension less (B) Kg
(C) Cm/hr
(D) Cm/gm

321. Which of the following shall be affected first under water stress conditions:
(A) Cell division and elongation
(B) Photosynthesis
(C) Transpiration
(D) Protein synthesis

322. If the plant water potential drops by -5 to -6 bar, the water stress will be known as:
(A) Mild (B) Moderate
(C) Severe (D) Very severe

323. If the plant water potential drops by -12 to -15 bar, the water stress will be known as:
(A) Mild (B) Moderate
(C) Severe (D) Very severe

324. Which of the following is accumulate water in plant body during water stress?
(A) Fats (B) Proteins
(C) Proline (D) Water

325. Discharge of water from trapezoidal weir is equal to:
(A) $186x10^{-4}LH^{3/2}$ (B) $184x10^{-4}LH^{3/2}$
(C) $138x10^{-4}LH^{5/2}$ (D) None of these

326. Discharge of water from V-notch is equal to:
(A) $186x10^{-4}LH^{3/2}$ (B) $184x10^{-4}LH^{3/2}$
(C) $138x10^{-4}LH^{5/2}$ (D) None of these

327. Discharge of water from rectangular weir is equal to:
(A) $186x10^{-4}LH^{3/2}$ (B) $184x10^{-4}LH^{3/2}$
(C) $138x10^{-4}LH^{5/2}$ (D) None of these

328. Discharge from an orifice is calculated as:
(A) $Q=61x10^{-5}A\sqrt{2gh}$
(B) $Q=6.1x10^{-5}A\sqrt{2gh}$
(C) $Q=184x10^{-4}LH^{3/2}$
(D) $Q=138x10^{-4}LH^{5/2}$

329. If the ratio of Ha/Hb of a parshall flume is <0.5, the flow will be:
(A) Free flow (B) Submerged flow
(C) Over flow (D) Under flow

330. Field water use efficiency is equal to:
(A) $\frac{Y}{WR}$ (B) $\frac{WR}{Y}$
(C) $\frac{Y}{ET}$ (D) $\frac{ET}{Y}$

331. Crop water use efficiency is equal to:
(A) $\frac{Y}{WR}$ (B) $\frac{WR}{Y}$
(C) $\frac{Y}{ET}$ (D) $\frac{ET}{Y}$

332. Optimum soil water regime for rice crop is:
(A) 50% ASW
(B) 75% ASW
(C) Submergence to 4 cm water height
(D) None of these

333. Optimum soil water regime required for higher yield of wheat is:
(A) 25% ASW (B) 50% ASW
(C) 65% ASW (D) 50-100 % ASW

334. Which one is not a critical growth stage of wheat for its scheduling irrigation?
(A) CRI
(B) Flowering
(C) Grand growth
(D) Milking and dough

335. Which of the following is a critical stage of sunflower for its scheduling irrigation?
(A) 4-6 leaf stage (B) Flowering
(C) Seed filling (D) All of these

336. Indicator plants can be used for:
(A) Scheduling irrigation
(B) Identifying nutritional deficiency symptoms
(C) Soil water variations
(D) All of these

337. Tensiometer was first used for scheduling irrigation by:
(A) Richards (B) Taylor
(C) Miller (D) All the above

338. IW/CPE ratio as a scheduling irrigation technique was suggested by:
(A) Richards
(B) Parihar & Co-workers
(C) Penman
(D) Blaney-Criddle

339. For higher productivity, the irrigation frequency should be:
(A) High (B) Medium
(C) Low (D) Very low

340. Most critical stage of rice for water need is:
(A) Seedling stage (B) Tillering stage
(C) Flowering stage
(D) Grain filling

341. Which growth stage of maize is most sensitive to water need?
(A) Seedling stage
(B) Knee high stage
(C) Tasselling stage
(D) Grain filling stage

342. Most important critical growth stage for irrigation of sorghum is:
(A) Flag leaf appearance stage.
(B) Flowering stage
(C) Grain filling stage
(D) All of these

343. Most important critical growth stage for irrigation of groundnut is:
(A) Fruiting stage
(B) Flowering stage
(C) Grain filling stage
(D) All of these

344. Ground water is considered good if EC (ds/m) is:
(A) <2 (B) 2
(C) >2 (D) 4

345. Safe irrigation water should contain boron concentration:
(A) <2 ppm (B) 2 ppm
(C) 3 ppm (D) 5 ppm

346. Most critical stage in wheat for irrigation:
(A) CRI (B) Flowering
(C) Milking (D) Soft dough

347. CRI stage in wheat comes at:
(A) 3-4 DAS (B) 20-25 DAS
(C) 40-45 DAS (D) 30-40 DAS

348. Wheat crop is irrigated at:
(A) 25% depletion of available soil moisture
(B) 35% depletion of available soil moisture
(C) 50% depletion of available soil moisture
(D) 75% depletion of available soil moisture

349. Cotton crop is irrigated at:
(A) 25% depletion of available soil moisture
(B) 35% depletion of available soil moisture
(C) 50% depletion of available soil moisture
(D) 75% depletion of available soil moisture

350. What is cumulative evaporation required for scheduling of irrigation at 0.75 (IW/CPE) with 5 cm of irrigations?
(A) 10 cm (B) 8.33 cm
(C) 6.66 cm (D) 6.25 cm

351. Water potential at permanent wilting point is:
(A) -0.3 bar (B) -15 bar
(C) -30 bar (D) -60 bar

352. Flooding method of irrigation is used when:
(A) Field is leveled
(B) Water is free
(C) Both (A) & (B)
(D) Water is limited

353. Sprinkler method of irrigation is used when:
(A) Field is leveled
(B) Water is free
(C) Both (A) and (B)
(D) Water is limited

354. Drip or trickle method of irrigation is used when:
(A) Field is leveled
(B) Water is free
(C) Both (A) and (B)
(D) Water is limited

355. Check basin method of irrigation is used where the land slope is:
(A) <1% (B) 1%
(C) >1% (D) 2%

356. Sprinkler method of irrigation is used when:
(A) Field is leveled
(B) Water is limited
(C) Field is undulated
(D) Both (B) and (C)

357. Size of check basin is depends upon:
(A) Land slope (B) Stream size
(C) Intake rate (D) All of these

358. Ideal slope for furrow method of irrigation is:
(A) 0.01% (B) 0.1%
(C) 1.0% (D) >1.0%

359. Sprinkler irrigation is suitable where the soil has:
(A) Hugh intake rate (B) Low intake rate
(C) Steeper slope (D) All of these

360. Spacing between the sprinklers at low wind velocity is:
(A) ½ of wetting diameter
(B) 2/3 of wetting diameter
(C) 3/4 of wetting diameter
(D) Equal to wetting diameter

361. Spacing between the sprinklers at high wind velocity is:
(A) ½ of wetting diameter
(B) 2/3 of wetting diameter
(C) 3/4 of wetting diameter
(D) Equal to wetting diameter

362. Sprinkler irrigation stopped at wind velocity of:
(A) <10 km/hr (B) >10 km/hr
(C) <15 km/hr (D) >15 km/hr

363. Entry of water into soil is known as:
(A) Infilteration
(B) Percolation
(C) Waterlogging
(D) Evaporation

364. If EC of drainage water is 8 mhos/cm and EC of irrigation water is 1 mhos/cm, the leaching requirement will be equal to:
(A) 10% (B) 11%
(C) 13% (D) 15%

365. Rice can be survive under water logged conditions due to:
(A) Absence of aerenchymatous tissues
(B) Presence of aerenchymatous tissues
(C) Rice plant likes water
(D) Presence of parenchymatous tissue

366. Which of the following crop requires more total water?
(A) Maize (B) Paddy
(C) Sugarcane (D) Wheat

367. Canopy temperature is measured by:
(A) Hygrometer
(B) Irrometer
(C) Infra red thermometer
(D) None of these

368. Alternate wetting and drying in soil change the:
(A) Soil texture (B) Soil structure
(C) Soil colour (D) Soil temperature

369. Percentage of water available at wilting point:
(A) 0% (B) 2%
(C) 10% (D) 20%

370. Moisture content at field capacity in comparison to percent of moisture content at saturation, is approximately:
(A) 100% (B) 50%
(C) 75% (D) 30%

371. Water left in soil after 24-72 hrs of irrigation is known as:
(A) Available water
(B) Field capacity
(C) PWP
(D) TWP

372. Name of the first phase of water erosion is:
(A) Sheet erosion (B) Rill erosion
(C) Gully erosion (D) Splash erosion

373. Which phase of soil erosion losses upper layer of fertile soil:
(A) Sheet erosion (B) Rill erosion
(C) Gully erosion (D) Splash erosion

374. Most critical stage for irrigation of mustard crop is:
(A) Germination
(B) Branching
(C) Flowering
(D) Pod formation

375. Quantity of water required to produce one kg of dry matter is:
(A) Consumptive use
(B) ET
(C) Water requirement
(D) Transpiration ratio

376. Maximum soil moisture is drawn from:
(A) Ist quarter of soil depth
(B) IInd quarter of soil depth
(C) IIIrd quarter of soil depth
(D) IVth quarter of soil depth

377. Wilting of pulses can be controlled by:
(A) Follow 3-4 year crop rotation
(B) Sowing of resistant varieties
(C) Intercropping with linseed
(D) All of these

378. Moisture equivalent generally used in place of:
(A) Field capacity
(B) Saturation
(C) Wilting point
(D) Sticky point

379. Field capacity of fine sandy soil is:
(A) 3-5 % (B) 5-15 %
(C) 12-18 % (D) 15-30 %

380. Field capacity of sandy loam soil is:
(A) 3-5 % (B) 5-15 %
(C) 12-18 % (D) 15-30 %

381. Field capacity of silt loam soil is:
(A) 3-5 % (B) 5-15 %
(C) 12-18 % (D) 15-30 %

382. Field capacity of clay loam soil is:
(A) 3-5% (B) 5-15 %
(C) 12-18% (D) 15-30 %

383. Field capacity of clay soil is:
(A) 3-5 % (B) 5-15 %
(C) 12-18 % (D) 25-40 %

384. Permanent wilting point of fine sandy soil is:
(A) 1-3 % (B) 3-8 %
(C) 6-10 % (D) 7-16 %

385. Permanent wilting point of sandy loam soil is:
(A) 1-3 % (B) 3-8 %
(C) 6-10 % (D) 7-16 %

386. Permanent wilting point of silt loam soil is:
(A) 1-3 % (B) 3-8 %
(C) 6-10 % (D) 7-16 %

387. Permanent wilting point of clay loam soil is:
(A) 1-3 % (B) 3-8 %
(C) 6-10 % (D) 7-16 %

388. Permanent wilting point of clay soil is:
(A) 1-3 % (B) 3-8 %
(C) 6-10 % (D) 12-20 %

389. The clay soil contains more colloidal metarials in hygroscopic water than:
(A) Sandy soil (B) Clay soil
(C) Silt soil (D) Clay loam soil

390. Water held at water holding capacity is:
(A) Less than 0.1 bar
(B) At 0.1 bar
(C) More than 0.1 bar
(D) None of these

391. When all the pores of the soil are filled with water, the soil said to be under saturation is called:
(A) Water holding capacity
(B) Field capacity
(C) Permanent wilting point
(D) Saturation

392. Water stored in the soil just double than field capacity is:
(A) Field capacity
(B) Water holding capacity
(C) Permanent wilting point
(D) Saturation

393. The tension of the water at the upper plastic limit is equivalent to pF value:
(A) 0.5 (B) 1.5
(C) 2.5 (D) 3.0

394. 1 bar is equal to:
(A) 0.9869 atm
(B) Weight of a 1020 cm water column
(C) 75.01 cm high mercury colimn
(D) All of these

395. Water held at wilting point is:
(A) Less than 0.1 bar
(B) 0.1 bar
(C) 15 bar
(D) None of these

396. Water stable aggregates in soil increase by addition of:
(A) N-fertilizers (B) P-fertilizer
(C) K-fertilizers (D) Micronutrients

397. Capillary and hygroscopic water together constitute:
(A) Water holding capacity
(B) Field capacity
(C) Permanent wilting point
(D) Saturation

398. In neutron moisture meter, source of fast neutrons is/are:
(A) Radium
(B) Americium-beryllium
(C) Both (A) and (B)
(D) None of these

399. Pressure membrane apparatus is used to measure matric potential moisture content relations at potential value as low as:
(A) 1 bar (B) 10 bar
(C) 100 bar (D) None of these

400. Height of rise of water in a capillary tube is directly proportional to:
(A) Surface tension of water
(B) Radius of capillary tube
(C) Viscosity of water
(D) Both (A) and (C)

401. The water holding capacity (WHC) of humus on a mass basis is .... times that of the silicate clays:
(A) 2 – 3 (B) 4 – 5
(C) 9 – 10 (D) 6 – 15

402. Available water holding capacity is maximum in:
(A) Clay loam (B) Sandy clay loam
(C) Silty clay loam (D) Loam

403. The moisture equivalent was introduced by:
(A) Briggd and McLane (1907)
(B) Richards and Weaver (1944)
(C) Whitney
(D) Mendel

404. Chemically composed water is known as:
(A) Combined water
(B) Free water
(C) Capillary water
(D) Hygroscopic water

405. The combined water held in the soil particles with the force of:
(A) Physical force
(B) Chemical force
(C) Biological force
(D) All of these

406. The water is required for special need is:
(A) Field preparation
(B) Nursery raising
(C) Puddling
(D) All of these

407. The down ward movement of water from the surface is called:
(A) Infiltration (B) Percolation
(C) Seepage (D) Permeability

408. From top 25% of effective root zone depth how much quantity of total water extraction take place by roots:
(A) 40% (B) 20%
(C) 30% (D) 10%

409. Haryali yojna is related to:
(A) Soil management
(B) Water management
(C) Crop management
(D) Air management

410. Permeability of surface layers in clay soil is varies in the range of:
(A) 0.8-1.2 mm/day
(B) 5-7 mm/day
(C) 7-8 mm/day
(D) More than 9 mm/day

411. Permeability of surface layers in sandy soil is varies in the range of:
(A) 0.8-1.2 mm/day
(B) 5-7 mm/day
(C) 7-8 mm/day
(D)Morethan 9 mm/ day

412. Most critical stage of irrigation requirement for ground nut is:
(A) Growth stage
(B) Flowering stage
(C) Pegging stage
(D) Pod development stage

413. Critical stages for irrigation requirement of potato is:
(A) Stoloniation
(B) Ruber formation
(C) Tuber enlargement
(D) All of these

414. Low infiltration rate is refer as:
(A) 2.5-12.5 mm/hr
(B) 12.5-25 mm/hr
(C) More than 25 mm/hr
(D) None of these

415. Medium infiltration rate is refer as:
(A) 2.5-12.5 mm/hr
(B) 12.5-25 mm/hr
(C) More than 25 mm/hr
(D) None of these

416. High infiltration rate is refer as:
(A) 2.5-12.5 mm/hr
(B) 12.5-25 mm/hr
(C) More than 25 mm/hr
(D) None of these

417. Most important critical growth stage for irrigation of cotton is:
(A) Boll formation stage
(B) Flowering stage
(C) Seedling stage
(D) All of these

418. In crop logging, during growing season, plant tissue is sampled every ------ days and analysed for N, sugar, moisture and weight of the young sheath tissue:
(A) 10
(B) 20
(C) 35
(D) 40

419. Osmotic pressure (atm) is equal to:
(A) 0.36 x EC mmhos/cm
(B) 36 x EC mmhos/cm
(C) 360 x EC mmhos/cm
(D) None of these

420. ppm is equal to:
(A) mg/l
(B) g/l
(C) l/l
(D) None of these

421. Transpiration is a:
(A) Evaporation
(B) Depletion of ground water
(C) Evaporation from plant
(D) Water loss from plant surfaces

422. The available water for plants in the soil is in between:

(A) Saturation and field capacity
(B) 1/3 atmospheric pressure and 15 atmospheric pressure
(C) Saturation and wilting point
(D) None of these

423. Which of the following soil types has high water holding capacity?

(A) Clayey soil (B) Loamy soil
(C) Sandy soil (D) Sandy loam soil

424. Water retained in loam soil between field capacity and wilting point is about:

(A) 5% (B) 10%
(C) 15% (D) 20%

425. In a saturated soil, tension with which water is held at the air-water interface is:

(A) Zero bar (B) 1/3 bar
(C) 15 bar (D) 31 bar

426. Tensiometer measures the:

(A) Total soil water potential
(B) Total soil moisture tension
(C) Metric potential
(D) Total soil moisture suction

427. Crop logging developed by CF Clement is carried out for sugarcane in:

(A) Brazil (B) Hawai
(C) New Zealand (D) India

428. The depth of water stored in root zone between field capacity and permanent wilting point is known as:

(A) Belt of soil water
(B) Gravitational water
(C) Available soil water
(D) Capillary water

429. The upper limit of available water in soil is determined at:

(A) 1/3 bar (B) ½ bar
(C) 2/3 bar (D) 1 bar

430. Leaching requirement can be estimated by:

(A) $LR = EC_{iw}/EC_{dw}$
(B) $LR = EC_{dw}/EC_{iw} \times 100$
(C) $LR = D_{iw}/D_{dw}$
(D) Both (B) and (C)

431. The rate of water absorption by plants is decreased due to deficiency of:

(A) Boron (B) Potassium
(C) Iron (D) Manganese

432. Main objective of puddling in paddy is to:

(A) Destroy insects – pests
(B) Create impermeable layer in the soil
(C) Mix the fertilizer in the soil
(D) None of these

433. Which are not the sources of fresh water?

(A) Running water, melting of ice
(B) Precipitation
(C) Lakes and ponds
(D) Municipal water

434. Paddy crop can be successfully grown under waterlogged conditions because:

(A) Paddy roots do not require O_2
(B) Paddy roots have capacity to absorb atmospheric O_2 directly
(C) Needed O_2 is supplied from the foliar parts of paddy to the roots
(D) Needed O_2 is supplied by the micro-organisms present in the soil

435. What will be the moisture content of soil at 50% ASM, if it has field capacity and permanent wilting point values of 20% and 8%, respectively:

(A) 18% (B) 14%
(C) 12% (D) 10%

436. Water use efficiency is the highest in case of:

(A) Flood irrigation (B) Border irrigation
(C) Sprinkler irrigation
(D) Drip irrigation

437. Among the following types of soil water which one is the most useful for the plants:

(A) Capillary water
(B) Hygroscopic water
(C) Gravitational water
(D) Combine water

438. If only four irrigations are available indicate critical stages of wheat, when there will be applied:

(A) CRI, jointing, flowering and dough stages
(B) CRI, tillering, flower & dough stages
(C) Tillering, jointing, milk & dough stage
(D) CRI, jointing, milking & dough stages

439. If only three irrigations are available indicate critical stages of wheat, when there will be applied:
(A) CRI, flowering and dough stages
(B) CRI, tillering and dough stages
(C) Tillering, jointing and milking stage
(D) CRI, jointing and dough stages

440. If only two irrigations are available indicate critical stages of wheat, when there will be applied:
(A) Flowering and dough stages
(B) CRI and flowering stages
(C) Tillering and milking stage
(D) CRI and dough stages

441. If only one irrigations are available indicate critical stages of wheat, when there will be applied:
(A) CRI stage (B) Tillering stage
(C) Jointing stage (D) Dough stage

442. Whenever we think of water shortage, what comes to our mind?
(A) Rajasthan (B) Gujarat
(C) Madhya Pradesh (D) Haryana

443. Name an important dam in Jammu & Kashmir and Orissa, respectively:
(A) Salal and Rihand
(B) Hirakud and Bhakra Nangal
(C) Salal and Hirakud
(D) Hirakud and Salal

444. Why are hydraulic structures constructed in ancient India?
(A) For beautification
(B) For bathing
(C) For washing clothes
(D) For storing water

445. Sardar sarovar project is associated with which river?
(A) Salty (B) Narmada
(C) Chambal (D) Damodar

446. Construction of multi-purpose projects has resulted in:
(A) Sharing of water
(B) Interstate water disputes
(C) Water sports
(D) Friendly relations

447. Which two dams are behind new social movement?
(A) Salal and Rihand
(B) Hirakud and Bhakra Nangal
(C) Narmada and Tehri
(D) Rihand and Chambal

448. Water scarcity in most cases is caused by:
(A) Over exploitation, excessive use, unequal access to water
(B) Shortage of rainfall
(C) Seasonal & annual variation of rainfall
(D) All of these

449. On which of the following rivers in Hirakud dam project located?
(A) Mahanadi (B) Satluj
(C) Beas (D) Narmada

450. Rivers of India have turned into toxic streams because:
(A) Dumping of domestic wastes
(B) Discharge of chemical effluents
(C) Industrialization and agriculture modernization
(D) All of these

451. Which one of the following reasons is not an appropriate reason for conserving water?
(A) To safeguard ourselves from health hazard
(B) To ensure food scarcity
(C) Continuation of our livelihood
(D) For the degradation of natural ecosystems

452. Assured irrigation due to multi-purpose projects has charged:
(A) Cropping pattern with farmers shifting to growing water intensive crops
(B) Growing of commercial crops and soil salinization
(C) Transformation of social landscapes
(D) Rise in social conflict due to increasing gap between the poor and rich farmers
(E) All of these

453. Which of the following causes are against the construction of multi-purpose projects?
(A) Fragmentation of river is bad for fish spawning
(B) Stream beds are rockier
(C) Production of hydro power
(D) Large-scale displacement of local communities

454. Are we the only ones chosen to make sacrifices for the nation? Who spoke these words?

(A) Protectors against Narmada (Sardar Sarovar Project)
(B) The survival of Rihand
(C) Farmers of Gujarat
(D) Industrialists and rich land owners

455. Which of the following is not a cause for a fall in underground water?

(A) Overdraft of ground water
(B) Use of tube wells for irrigation
(C) Rain water harvesting
(D) Overuse for domestic purpose in urban areas

456. Rooftop rain water harvesting helps in:

(A) Recharging ground water resources
(B) Rise in sea level
(C) Increased flow of river water
(D) Flooding of roads

457. Which one of the following combination of river dams is not correct?

(A) Salal dam on Chenab river
(B) Koyna on Krishna river
(C) Bhakra Nangal on Satluj river
(D) Tehri on Yamuna river

458. Which of the following are the natural resources of fresh water for the mankind?

(A) Ground water
(B) Precipitation
(C) Surface water
(D) All of these

459. Which of these statements is not an argument in favour of multipurpose river valley projects?

(A) They bring water to those areas which suffer from water scarcity
(B) They regulate water flow to help in controlling floods
(C) They lead to large-scale displacement and lose of livelihood
(D) They generate electricity for our industries and our homes

460. Multipurpose projects were referred to as 'Temples of Modern India' by Pt. Jawahar Lal Nehru because:

(A) Some important temples were built here
(B) City dwellers visit these temples
(C) They initiates all round development of the area
(D) None of these

461. Which of the following cause is not related with underground water depletion?

(A) Overdraft of water for irrigation in rural areas
(B) Excessive use of water in urban areas
(C) Decline in seepage due to cementation of spaces used by man
(D) Excessive evaporation of underground water

462. Water under the surface is known as:

(A) Surface water
(B) Underground water
(C) Both of these
(D) None of these

463. Scarcity of water can be both natural and mankind are:

(A) Uneven distribution of rainfall over space and time
(B) Seasonal variations in rainfall
(C) Over-exploitation of water for various activities
(D) Excessive and wasteful use of water
(E) All of these

464. Water resource is mismanaged by man is:

(A) Spilling of water
(B) Pipe leakages
(C) Dripping, overflowing of residential tanks
(D) All of these

465. Natural reason for causing water scarcity is:

(A) Uneven distribution of rainfall over space and time
(B) Seasonal variations in rainfall
(C) Both (A) and (B)
(D) None of these

466. What would be the repercussion of ground water depletion?

(A) Falling level of underground water
(B) Shortage of water for use
(C) Deteorating quality of water due to pollution
(D) All of these

467. Water storage structure in tribals of Rajasthan are:
(A) Johads
(B) Khadins
(C) Both (A) and (B)
(D) None of these

468. Roof top rain water harvesting was commonly practices to store drinking water in:
(A) Rajasthan (B) Madhya Pradesh
(C) Gujarat (D) Haryana

469. In hilly areas and in Western Himalayas, people have constructed diversion channels called:
(A) Guls
(B) Kuls
(C) Both (A) and (B)
(D) None of these

470. Rajaon Ki Bais or Sukhi Baoli (15th – 16th century) built during:
(A) Akbar reign
(B) Sikandar Lodi's reign
(C) Prthviraj Chauhan reign
(D) Tipu Sultan reign

471. Gadhak ki Baoli in Mehrauli complex in Delhi built in 14th centaury during the reign of:
(A) Akbar (B) Sikandar Lodi's
(C) Iltutmish (D) Gajnavi

472. Step well at Patan in Gujarat built 11th century also called:
(A) Raja Ki Vav (B) Rani Ki Vav
(C) Sena Ki Vav (D) Rajput Ki Vav

473. Plenty of 'Kunds' and "Tankas' in forts and houses in:
(A) Rajasthan
(B) Madhya Pradesh
(C) Both (A) and (B)
(D) None of these

474. Earth's surface is covered with water is:
(A) 70% (B) 71%
(C) 73% (D) 74%

475. World's water is contained in oceans as salty water:
(A) 97.5% (B) 96.5%
(C) 95.5% (D) 94.5%

476. World's water is fresh water in liquid state is:
(A) 1% (B) 1.5%
(C) 2% (D) 2.5%

477. India receives how much of the global precipitation is:
(A) 3% (B) 4%
(C) 5% (D) 6%

478. World's water in fresh and most of it occurs in frozen state as ice sheets of Arctic, Antarctic and Tundra regions, Ice caps in mountain and glaciers:
(A) 1.5% (B) 2.5%
(C) 3.5% (D) 4.5%

479. The total water resources of India are estimated at:
(A) 1897 km^2/annum (B) 1907 km^2/annum
(C) 1917 km^2/annum (D) 1927 km^2/annum

480. Dams, lakes and irrigation systems were built on a large scale during the reign of:
(A) Akbar reign
(B) Chandragupta Maurya reign
(C) Prthviraj Chauhan reign
(D) Tipu Sultan reign

481. Empirical formula of water requirement is:
(A) ET/CU + Application loses + special needs
(B) ET + CU + Application losses
(C) ET + CU
(D) ET + Special need

482. Empirical formula of permeability is:
(A) KA $\delta l/\delta h$ (B) KA δh
(C) KA $\delta h/\delta l$ (D) KA δl

483. Empirical formula of irrigation requirement is:
(A) ET/CU + Application loses – Effective rainfall
(B) ET + CU + Application losses + special needs – Effective rainfall
(C) ET + CU –Effective rainfall
(D) ET + Special need – Effective rainfall

484. The rate of flow of water through a unit cross-sectional area under a unit hydraulic gradient is called:
(A) Percolation
(B) Seepage
(C) Coefficient of permeability
(D) Infiltration

485. The permeability coefficient K is normally expressed in:

(A) cm/day (B) m/day
(C) m/hr (D) cm/hr

486. Postponing of first irrigation to 40-45 days after sowing is always preferable for cotton crop to:

(A) Promote sympodial branching
(B) Prevent excessive vegetative growth
(C) Promote early flowring
(D) Promote monopodial branches

487. The rate of flow of water through a porus medium is directly proportional to the cross-sectional area of the flow and the hydraulic gradient under which the flow moves is called:

(A) Permeability
(B) Seepage
(C) Coefficient of permeability
(D) Infiltration

488. At field capacity the water is held at:

(A) 0.033 MP
(B) 0.30 MPa
(C) 3.00 MPa
(D) 30.00 MPa

489. The ratio of irrigation water transpired by the crops on an irrigation farm or project during their growth period, over the water diverted from a river or other natural source in to the farm or project canal or canals during the same period of time is called:

(A) Irrigation application efficiency
(B) Project efficiency
(C) Water conveyance efficiency
(D) Water use efficiency

490. The ratio of volume of irrigation water consumptively used in a specified period to the total volume of irrigation water that has been diverted from the capture system is called:

(A) Irrigation application efficiency
(B) Project efficiency
(C) Water conveyance efficiency
(D) Water use efficiency

491. Most irrigation water that come from diversion from streams or reservoirs. Losses which occurred while conveying water were often excessive is known as:

(A) Irrigation application efficiency
(B) Project efficiency
(C) Water conveyance efficiency
(D) Water use efficiency

492. What proportion of the water delivered was beneficially used on the project, farm, or field is called:

(A) Irrigation application efficiency
(B) Project efficiency
(C) Water conveyance efficiency
(D) Water use efficiency

493. Empirical formula of Irrigation Application Efficiency is:

(A) $\frac{\text{Average depth of irrigation water contributing to target}}{\text{Average depth of irrigation water applied}} \times 100$

(B) $\frac{\text{Volume of irrigation water consumptively used}}{\text{Volume of irrigation water diverted to the project area}} \times 100$

(C) $\frac{\text{Water diverted to the farm}}{\text{Water diverted from the river or reservoir}} \times 100$

(D) $\frac{\text{Water beneficially used}}{\text{Water delivered}} \times 100$

494. Empirical formula of Project Efficiency is:

(A) $\frac{\text{Average depth of irrigation water contributing to target}}{\text{Average depth of irrigation water applied}} \times 100$

(B) $\frac{\text{Volume of irrigation water consumptively used}}{\text{Volume of irrigation water diverted to the project area}} \times 100$

(C) $\frac{\text{Water diverted to the farm}}{\text{Water diverted from the river or reservoir}} \times 100$

(D) $\frac{\text{Water beneficially used}}{\text{Water delivered}} \times 100$

495. Empirical formula of Water Conveyance Efficiency is:

(A) $\frac{\text{Average depth of irrigation water contributing to target}}{\text{Average depth of irrigation water applied}} \times 100$

(B) $\frac{\text{Volume of irrigation water consumptively used}}{\text{Volume of irrigation water diverted to the project area}} \times 100$

(C) $\frac{\text{Water diverted to the farm}}{\text{Water diverted from the river or reservoir}} \times 100$

(D) $\frac{\text{Water beneficially used}}{\text{Water delivered}} \times 100$

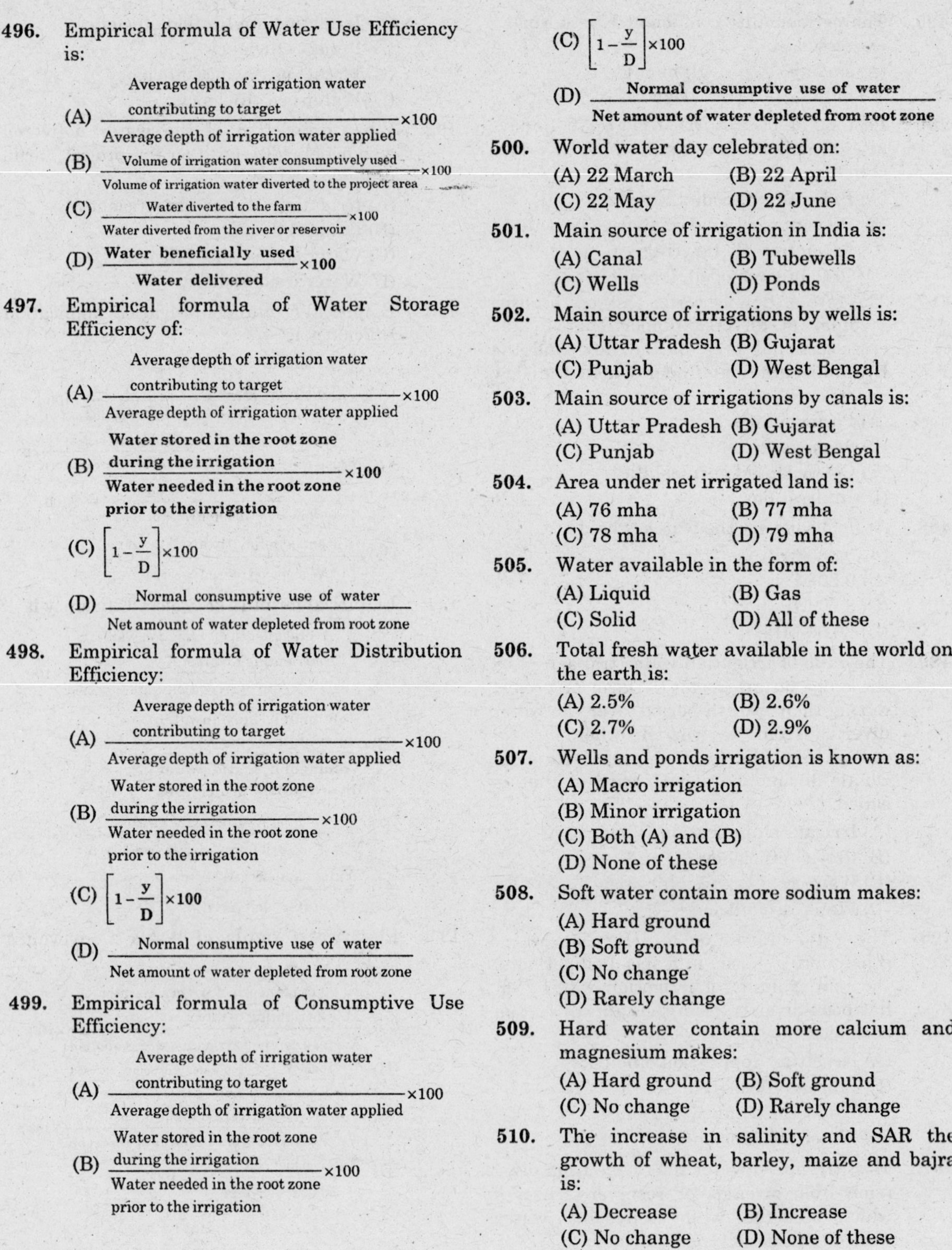

496. Empirical formula of Water Use Efficiency is:

(A) $\frac{\text{Average depth of irrigation water contributing to target}}{\text{Average depth of irrigation water applied}} \times 100$

(B) $\frac{\text{Volume of irrigation water consumptively used}}{\text{Volume of irrigation water diverted to the project area}} \times 100$

(C) $\frac{\text{Water diverted to the farm}}{\text{Water diverted from the river or reservoir}} \times 100$

(D) $\frac{\textbf{Water beneficially used}}{\textbf{Water delivered}} \times 100$

497. Empirical formula of Water Storage Efficiency of:

(A) $\frac{\text{Average depth of irrigation water contributing to target}}{\text{Average depth of irrigation water applied}} \times 100$

(B) $\frac{\textbf{Water stored in the root zone during the irrigation}}{\textbf{Water needed in the root zone prior to the irrigation}} \times 100$

(C) $\left[1 - \frac{y}{D}\right] \times 100$

(D) $\frac{\text{Normal consumptive use of water}}{\text{Net amount of water depleted from root zone}}$

498. Empirical formula of Water Distribution Efficiency:

(A) $\frac{\text{Average depth of irrigation water contributing to target}}{\text{Average depth of irrigation water applied}} \times 100$

(B) $\frac{\text{Water stored in the root zone during the irrigation}}{\text{Water needed in the root zone prior to the irrigation}} \times 100$

(C) $\left[1 - \frac{\mathbf{y}}{\mathbf{D}}\right] \times 100$

(D) $\frac{\text{Normal consumptive use of water}}{\text{Net amount of water depleted from root zone}}$

499. Empirical formula of Consumptive Use Efficiency:

(A) $\frac{\text{Average depth of irrigation water contributing to target}}{\text{Average depth of irrigation water applied}} \times 100$

(B) $\frac{\text{Water stored in the root zone during the irrigation}}{\text{Water needed in the root zone prior to the irrigation}} \times 100$

(C) $\left[1 - \frac{y}{D}\right] \times 100$

(D) $\frac{\textbf{Normal consumptive use of water}}{\textbf{Net amount of water depleted from root zone}}$

500. World water day celebrated on:
(A) 22 March (B) 22 April
(C) 22 May (D) 22 June

501. Main source of irrigation in India is:
(A) Canal (B) Tubewells
(C) Wells (D) Ponds

502. Main source of irrigations by wells is:
(A) Uttar Pradesh (B) Gujarat
(C) Punjab (D) West Bengal

503. Main source of irrigations by canals is:
(A) Uttar Pradesh (B) Gujarat
(C) Punjab (D) West Bengal

504. Area under net irrigated land is:
(A) 76 mha (B) 77 mha
(C) 78 mha (D) 79 mha

505. Water available in the form of:
(A) Liquid (B) Gas
(C) Solid (D) All of these

506. Total fresh water available in the world on the earth is:
(A) 2.5% (B) 2.6%
(C) 2.7% (D) 2.9%

507. Wells and ponds irrigation is known as:
(A) Macro irrigation
(B) Minor irrigation
(C) Both (A) and (B)
(D) None of these

508. Soft water contain more sodium makes:
(A) Hard ground
(B) Soft ground
(C) No change
(D) Rarely change

509. Hard water contain more calcium and magnesium makes:
(A) Hard ground (B) Soft ground
(C) No change (D) Rarely change

510. The increase in salinity and SAR the growth of wheat, barley, maize and bajra is:
(A) Decrease (B) Increase
(C) No change (D) None of these

511. Which water is good for house hold work:
(A) Hard water (B) Mild water
(C) Soft water (D) None of these

512. Excess salinity usually causes:
(A) Poor stand of crop
(B) Stunted growth
(C) Reduced yield
(D) All of these

513. Saline water have predominant salt is:
(A) Sodium carbonate
(B) Sodium chloride
(C) Magnesium sulphate
(D) Calcium chloride

514. The germination of seed is delayed and retarded but the cropping period is slightly reduced with the:
(A) Increase in salinity
(B) Decrease in salinity
(C) Increase in acidity
(D) Decrease in acidity

515. Salt content of irrigation water is measured as:
(A) pH
(B) Electrical conductivity
(C) SAR
(D) None of these

516. Electrical conductivity is expressed as:
(A) g/cc (B) g/cm^3
(C) mmhos/cm (D) mmhos/g

517. Electrical conductivity is suitable for crop growth is:
(A) Less than 0.5 mmhos
(B) 0.5 to 15 mmhos
(C) 15 to 30 mmhos
(D) More than 15 mmhos

518. When in irrigation water is high in salt, which method is only way to remove salt:
(A) Leaching (B) Scrapping
(C) Adding organic material
(D) Use of amendments

519. Soluble salt damage plants through an:
(A) Osmotic effects (B) Diffusion effect
(C) Endosmosis (D) Exosmosis

520. Water with an EC of 1.0 contains 640 ppm salt. When irrigating with 3 acre feet of this water, approximately how much of salt per acre are applied every year:
(A) 2.5 tons salt per acre
(B) 2.6 tons salt per acre
(C) 3.0 tons salt per acre
(D) 3.6 tons salt per acre

521. Sugerbeet are sensitive to salt as seedlings, but are more tolerant to salt at:
(A) Early stage (Б) Middle age
(C) Later stage (D) Maturity stage

522. Smaller seeds are sensitive to salts is:
(A) Less (B) Mild
(C) More (D) None of these

523. Water with an ED below 0.2 mmhos/cm can also cause:
(A) Problem (B) No problem
(C) Beneficial (D) Negative effect

524. Very low EC water dilutes and/or leaches calcium and makes soil aggregates susceptible to disintegration cause:
(A) Water infiltration
(B) Water percolation
(C) Water evaporation
(D) Water runoff

525. The presence of high concentration of lime in irrigation water can precipitate:
(A) Nitrogen (B) Phosphorus
(C) Potash (D) Calcium

526. Adding a calcium salt, such as gypsum or calcium chloride to the irrigation water and raising the EC to 0.2 to 0.3 mmhos/cm can prevent:
(A) Water percolation problem
(B) Water infiltration problem
(C) Water evaporation problem
(D) Water runoff problem

527. Irrigation water has an EC of 0.1 mmhos/cm. A grower needs to know how much gypsum (75% purity) to add to 1 acre foot of water to increase EC to 0.25 mmhos/cm:
(A) 265 lb/acre feet (B) 365 lb/acre feet
(C) 465 lb/acre feet (D) 565 lb/acre feet

528. Microirrigation systems and drip irrigation systems are more prone to plugging of calcium than are:
(A) Pivot sprinkler system
(B) Flat fan sprinkler system
(C) Simple sprinkler system
(D) None of these

529. For fruits and vegetable crops, the presence of lime residues reduces:

(A) Storage (B) Self life
(C) Marketability
(D) None of these

530. The presence of lime deposits in soil can reduce the solubility of plant nutrients:

(A) Phosphorus (B) Zinc
(C) Manganese (D) Iron
(E) All of these

531. Higher lime deposition potential, the higher the risk of:

(A) Lime deposition
(B) Lime deflocculating
(C) Lime utilization
(D) None of these

532. Sodium Absorption Ratio is increased when electrical conductivity is:

(A) Increased (B) Decreased
(C) Unchanged (D) None of these

533. The Sodium Absorption Ratio is related with:

(A) pH
(B) Exchangeable sodium percentage
(C) Infiltration
(D) Ground water

534. Which method reduces sodium hazard by reducing water SAR and increasing water EC:

(A) Nitrogen injection
(B) Phosphorus injection
(C) Calcium injection
(D) All of these

535. Calcium also prevent the formation of:

(A) Sodium chloride
(B) Sodium carbonate
(C) Magnesium carbonate
(D) Sodium bicarbonate

536. Residual sodium carbonates (bicarbonates + carbonates) minus the sum of the divalent cations are:

(A) Calcium
(B) Magnesium
(C) Calcium and magnesium
(D) Iron

537. Residual Sodium Carbonates is not excess in those water when EC is less than:

(A) 2 mmhos/cm (B) 3 mmhos/cm
(C) 4 mmhos/cm (D) 5 mmhos/cm

538. Water with an RSC above 1.0 usually require amendments to:

(A) Decrease RSC (B) Increase RSC
(C) Unchanged RSC (D) None of these

539. Crude fibre content is unaffected by P application but decreased with:

(A) Increased N application
(B) Decreased N application
(C) Increased P application
(D) Decreased P application

540. Irrigation efficiency of surface method at conveyance is:

(A) 60-70% (B) 50-60%
(C) 30-40% (D) 20-25%

541. Irrigation efficiency of surface method at application is:

(A) 60-70% (B) 50-60%
(C) 30-40% (D) 20-25%

542. Irrigation efficiency of surface method at surface water after evaporation is:

(A) 60-70% (B) 50-60%
(C) 30-40% (D) 20-25%

543. Irrigation efficiency of sprinkler method at conveyance is:

(A) 100% (B) 70-80%
(C) 30-40% (D) 20-25%

544. Irrigation efficiency of sprinkler method at application is:

(A) 100% (B) 70-80%
(C) 30-40% (D) 20-25%

545. Irrigation efficiency of sprinkler method at surface water after evaporation is:

(A) 100% (B) 70-80%
(C) 30-40% (D) 20-25%

546. Irrigation efficiency of drip method at conveyance is:

(A) 100% (B) 90%
(C) 30-40% (D) 15-30%

547. Irrigation efficiency of drip method at application is

(A) 100% (B) 90%
(C) 30-40% (D) 15-30%

548. Irrigation efficiency of drip method at surface water after evaporation is:

(A) 100% (B) 90%
(C) 30-40% (D) 15-30%

549. Prediction of that the share of water for agriculture will reduce by 2020 AD is:

(A) 78% (B) 73%
(C) 70% (D) 65%

550. Prediction of that the share of water for agriculture will reduce by 2025 AD is:

(A) 78% (B) 73%
(C) 70% (D) 65%

551. On domestic sector front, the present per capita annual water availability of 2100 m^3 is likely to decrease to about 1700 m^3 by:

(A) 2000 AD (B) 2025 AD
(C) 2030 AD (D) 2035 AD

552. In India, agriculture sector consumes around --------- of total water use is:

(A) 89 % (B) 80%
(C) 75% (D) 70%

553. Indian, domestic sector consumes around --------- of total water use is:

(A) 7% (B) 8 %
(C) 9% (D) 10%

554. Indian, industrial sector consumes around --------- of total water use is:

(A) 3% (B) 4 %
(C) 5% (D) 6%

555. India have how numbers of wells are:

(A) 10 million (B) 20 million
(C) 25 million (D) 30 million

556. A farmer applied 30 inches of water via sprinkler during the growing season. The irrigation water contains 10 mg nitrate-N/liter. Calculate nitrate – N applied in lb/acre;

(A) 68 lb/acre (B) 70 lb/acre
(C) 72 lb/acre (D) 74 lb/acre

557. Conversion of ppm nitrate (NO_3) into ppm nitrate – N (NH_3-N) is:

(A) 3.0 (B) 4.4
(C) 39.1 (D) 40.2

558. Conversion of ppm sulphate (SO_4) into ppm sulphate – N (SO_4-S) is:

(A) 3.0 (B) 4.4
(C) 39.1 (D) 40.2

559. Conversion of ppm potassium (K) into ppm potassium (K) is:

(A) 4.4 (B) 3.0
(C) 39.1 (D) 40.2

560. Water pH generally not a problem itself, but it is an indicator of other problems such as:

(A) Carbonates of sodium
(B) Bicarbonates of sodium
(C) Both (A) and (B)
(D) None of these

561. As irrigation water pH increases above 8.2, the potential for sodium problems:

(A) Increases (B) Decreases
(C) No changed (D) Unknown

562. Long term irrigation with high pH water may cause soil pH to:

(A) Increases (B) Decreases
(C) No changed (D) Unknown

563. For production of 1 kg dry matter, how much water is needed:

(A) 300-400 litres (B) 400-500 litres
(C) 500-600 litres (D) 600-700 litres

564. Very low EC water dilutes and leaches calcium and make soil aggregated susceptible to disintegration, causing problem in:

(A) Water infiltration
(B) Water percolation
(C) Water runoff (D) Water seepage

565. Loss of cell turgor and wilting cessation of cell enlargement, closure of stomata, reduction of photosynthesis and interference with many basic metabolic processes is accompanied by:

(A) Decreasing water content
(B) Increasing water content
(C) Increasing water pressure
(D) Decreasing water pressure

566. The apical portion of roots and stem contain water:

(A) More than 90% (B) 80 %
(C) 70 % (D) Less than 60%

567. The wood of large trees may contain water:

(A) More than 90%
(B) 80 %
(C) 70 %
(D) 50- 60%

568. The stem of barley, wheat and sorghum contain water:

(A) More than 90% (B) 80 %
(C) 60-70 % (D) Less than 50%

569. The freshly harvested grains of most crop contain water:

(A) More than 30% (B) 20-25 %
(C) 10-15 % (D) Less than 10%

570. The moisture content in the grains which determines that:

(A) Storage life (B) Viability
(C) Germinability (D) All of these

571. Water is needed mainly to meet the demands of evaporation, transpiration and metabolic needs of plants, all together is known as:

(A) Consumptive use
(B) Irrigation requirement
(C) Evapo-transpiration
(D) Water requirement

572. Soil salinity affects:

(A) Cell division
(B) Cell elongation
(C) Protein synthesis
(D) All of these

573. Soils with high exchangeable sodium soil become:

(A) Soft on drying
(B) Hard on drying
(C) Soft on wetting
(D) Hard on wetting

574. The harmful effects of poor quality of irrigation water can be minimized by adding of chemicals like:

(A) Sulphate
(B) Sulphuric acids
(C) Both (A) & (B)
(D) None of these

575. Water bloom is generally formed by the accumulation of:

(A) Bacteria (B) Bryophytes
(C) Green algae (D) Blue green algae

576. The term "Hydroponics" means:

(A) Soil less cultivation
(B) Growing of plants in pots
(C) Growing of aquatic plants
(D) Airless cultivation

577. When large amount of sewage is dumped into the river, the BOD value will be:

(A) Increased
(B) Slightly decreased
(C) Unchanged
(D) Decreased

578. Hardness in water is due to:

I. Iron II. Aluminium
III. Calcium IV. Magnesium
(A) I and II (B) III and IV
(C) I and III (D) II and III

579. Water requirement of rice crop is estimated as:

(A) 2000 mm (B) 2200 mm
(C) 600 mm (D) 1200 mm

Statement for linked Answer Question 580 to 582

Calculate the soil moisture content in per cent by weight (Pw), volumetric water content in per cent (Pv) and depth of water (cm) in soil sampled depth (Dw), with the following information

(i) Weight of alluminium tin (empty) =40 g.
(ii) Weight of moist soil + tin = 145 g
(iii) Weight of dry soil + tin = 135 g
(iv) Soil bulk density (BD) = 1.4 g/cm^3
(v) Sampling soil depth (Sd) = 30 cm

580. Soil moisture content in per cent by weight (Pw):

(A) 10.5 (B) 11.5
(C) 12.5 (D) 13.5

581. Soil moisture content in per cent by volumetric (Pv):

(A) 13.7 (B) 14.7
(C) 15.7 (D) 16.7

582. Depth of water (cm) is the soil (Dw):

(A) 2.41 (B) 3.41
(C) 4.41 (D) 5.41

583. Calculate the total depth of water in the 60 cm soil depth from teh following:

Depth of soil (cm)	FC (%)	PWD (%)	BD (g/cm^2)
0-15	23	8	1.30
15-30	24	9	1.35
30-45	25	10	1.40
45-60	26	11	1.50

(A) 11.0 (B) 11.5
(C) 12.0 (D) 12.5

Statement for linked Answer Question 584 to 587

Calculate the net depth of irrigation application, gross depth of water application, irrigation period with the help of following given information:

Soil moisture conent at F.C. = 30%
Soil moisture content at PWP = 15%
Soil bulk density = 1.5 g/cm^3
Effective root zone depth = 75 cm
Rate of evaporation = 0.6 cm/day
Available soil moisture depletion = 75%
Level for irrigation
Irrigation efficiency = 80%

584. Soil moisture content (cm) in the root zone:

(A) 16.88 (B) 15.88
(C) 14.88 (D) 13.88

585. Net depth of irrigation (cm) to be applied:

(A) 11.66 (B) 12.66
(C) 13.66 (D) 14.66

586. Gross depth (cm) of irrigation to be applied:

(A) 13.83 (B) 14.83
(C) 15.83 (D) 16.83

587. Irrigation period (days):

(A) 18 (B) 19
(C) 29 (D) 21

Statement for linked Answer Question 588 to 590

If soil water content in the root depth at the start of irrigation is 12 cm for a soil with field capacity of 22 cm. Peak period consumptive use rate is 9 mm/day. If the irrigation efficiency is 60%:

588. Gross irrigation requirement (cm):

(A) 16.67 (B) 17.00
(C) 17.67 (D) 18.00

589. Net irrigation requirement (cm) is:

(A) 9 (B) 10
(C) 11 (D) 12

590. Irrigation period (days) is:

(A) 9 (B) 10
(C) 11 (D) 12

Statement for linked Answer Question 591 to 594

A mustered crop was irrigated at 1W/CPE ratios of 0.50, 0.60, 0.75 and 1.0 with 6 cm depth of irrigation water.

591. CPE for 0.5 IW/CPE ratio is:

(A) 12 (B) 10
(C) 9 (D) 8

592. CPE for 0.6 1W/CPE ratio is:

(A) 12 (B) 10
(C) 9 (D) 8

593. CPE for 0.75 1W/CPE ratio is:

(A) 12 (B) 10
(C) 8 (D) 6

594. CPE for 1.00 1W/CPE ratio is:

(A) 12 (B) 10
(C) 8 (D) 6

Statement for linked Answer Question 595 to 598

It is proposed to schedule irrigation to wheat crop at CPE of 9 cm at 1W/CPE ratio of 0.5, 0.6, 0.75 and 0.8.

595. Depth of irrigation water for 0.5 1W/CPE ratio is:

(A) 4.5 (B) 4.7
(C) 4.9 (D) 5.1

596. Depth of irrigation water for 0.6 1W/CPE ratio is:

(A) 4.5 (B) 5.4
(C) 5.8 (D) 6.0

597. Depth of irrigation water for 0.75 1W/CPE ratio is:

(A) 4.55 (B) 5.45
(C) 6.75 (D) 6.90

598. Depth of irrigation water for 0.80 IW/CPE ratio is:

(A) 4.2 (B) 5.2
(C) 6.2 (D) 7.2

599. Which of the following are benefits of fertigation?

1. Increased nutrient absorption by plants
2. Reduction in fertilizer & chemical need
3. Reduced leaching to the water table

(A) 1,2 and 3 (B) 1 only
(C) 3 only (D) 2 only

Statement for linked Answer Question 600 to 602

Calculate the water conveyance, water storage and water application efficiencies using the following data

(i) Area irrigated in 8 hours = 1.5 ha
(ii) Size of canal head = 140 litres/second
(iii) Water delivered to the field = 80 litres/second
(iv) Deep of root zone soil = 1.8 m
(v) Deep percolation losses = Nil
(vi) Runoff loss from the field = 410 m^3

(vii) Available water holding capacity of the soil = 16 cm/m

(viii) Depth of water penetration at the head of the field = 1.6 m

(ix) Depth of water penetration as the tail end of the field = 1.0 m

(x) Irrigation was applied at 50% depletion of availabe moisture in root zone

600. Calculate water conveyance efficiency (%) is:

(A) 57.14 (B) 67.14
(C) 77.14 (D) 87.14

601. Calculate water application efficiency (%) is:

(A) 81.02 (B) 82.20
(C) 83.20 (D) 84.20

602. Calculate water storage efficiency used (%) is:

(A) 71.07 (B) 72.07
(C) 73.07 (D) 74.07

603. Evaporation is measured by which instrument:

(A) Baro meter
(B) Psychrometer
(C) Lysimeter
(D) Hygrometer

604. Which of the following methods of surface irrigation is considered as water and fertilizer efficient and ideal for several crops?

(A) Check basin and border ship irrigation
(B) Furrow irrigation
(C) Surge irrigation
(D) Flood irrigation

605. Water use efficiency is the highest in case of:

(A) Border irrigation
(B) Drip irrigation
(C) Sprinker irrigation
(D) Flood irrigation

606. The ratio between marketable crop yield and water used in evapotranspiration is known as:

(A) Field water use efficiency
(B) Economic irrigation efficiency
(C) Consumptive use efficiency
(D) Water use efficiency

607. Drip irrigation is useful in which region?

(A) Humid
(B) Dry
(C) High rainfall
(D) None of these

608. Empirical formula of gross irrigation requirement is:

(A) $\frac{\text{Net depth of irrigation}}{\text{Rate of evaporation}}$

(B) $\frac{\text{Net irrigation requiremnt}}{\text{Peak period consumptive use}}$

(C) $\frac{\text{Net irrigation requiremnt}}{\text{Irrigation efficiency}}$

(D) None of these

609. Empirical formula of irrigation period is:

(A) $\frac{\text{Net depth of irrigation}}{\text{Rate of evaporation}}$

(B) $\frac{\text{Net irrigation requiremnt}}{\text{Peak period consumptive use}}$

(C) Both (A) & (B)
(D) None of these

610. Empirical formula of net depth of irrigation to be applied:

(A) Available soil moisture in the root zone x Available soil moisture depletion
(B) Water content at field capacity — Water content at starting of irrigation
(C) Available soil moisture in the root zone x Available soil moisture depletion/ irrigation efficiency
(D) Field capacity - Permanent wilting point/ 100 x bulk density x soil depth

611. Empirical formula of net irrigation requirement is:

(A) Available soil moisture in the root zone x Available soil moisture depletion
(B) Water content at field capacity — Water content at starting of irrigation
(C) Available soil moisture in the root zone x Available soil moisture depletion/ irrigation efficiency
(D) Field capacity - Permanent wilting point/ 100 x bulk density x soil depth

612. Empirical formula of gross depth of irrigation to be applied is:

(A) Available soil moisture in the root zone x Available soil moisture depletion
(B) Water content at field capacity — Water content at starting of irrigation
(C) Available soil moisture in the root zone x Available soil moisture depletion/ irrigation efficiency
(D) Field capacity - Permanent wilting point/ 100 x bulk density x soil depth

613. Empirical formula of available soil moisture in the root zone is:

(A) Available soil moisture in the root zone x Available soil moisture depletion
(B) Water content at field capacity — Water content at starting of irrigation
(C) Available soil moisture in the root zone x Available soil moisture depletion/ irrigation efficiency
(D) Field capacity - Permanent wilting point/ 100 x bulk density x soil depth

614. 'Drip' irrigation is generally followed in the country:

(A) Israel (B) America
(C) Australia (D) Japan

615. In India, under canal net-work system of irrigation, generally, how much amount of water is allowed to run in small canal ?

(A) Less than 4 cusec
(B) Less than 10 cusec
(C) Between 15 to 20 cusec
(D) Between 25 to 30 cusec

616. Method of irrigation is:

(A) Surface irrigation
(B) Sprinkler irrigation
(C) Drip irrigation
(D) All of these

617. Swing-basket (Dhenkuli) is used for:

(A) Making furrow
(B) Lifting water from wells
(C) Destroying weeds
(D) Leveling of land

618. Which of the following is used maximum for lifting water from wells ?

(A) Persian wheel (Rahat)
(B) Swing basket (Bedi)
(C) Dhenkuli (D) Don

619. Which of the following is used for lifting water from 8-10 m depth ?

(A) Washer Rahat
(B) Hand pump
(C) Mayadas lift (D) None of these

620. In which implement. bullocks are not used for lifting water ?

(A) Buldev Balti
(B) Charsa
(C) Egyptian screw
(D) Rahat (Persian wheel)

Statement for Linked Answer Questions 621 and 622:

In an experimental setup, the discharge through a triangular notch is 0.0074 $m^3 s^{-1}$ at an operating head of 0.1 m. The coefficient of discharge for the notch is 0.7. If the required discharge is 0.1 $m^3 s^{-1}$,

621. The corresponding head in m is:

(A) 0.00015 (B) 0.035
(C) 0.283 (D) 67.13

622. The corresponding width of water surface in m is:

(A) 0.80 (B) 0.099
(C) 0.00043 (D) 0.00086

623. The pressure that does not have any measurable influence on the void ratio or shearing resistance of the soil mass is:

(A) Pore water pressure
(B) Intergranular pressure
(C) Capillary pressure
(D) Surcharge pressure

624. A trapezoidal canal, having a bottom width of 5.0 m and a side slope of 1 H : 1 V, is carrying a discharge of 20 $m^3 s^{-1}$. The critical depth, in m, is:

(A) 1.09 (B) 1.18
(C) 2.12 (D) 2.62

625. A border strip of 8 × 250 m is being irrigated by a border stream of 50 lps. The infiltration capacity of the soil is 25 mm h^{-1} (assumed to be constant throughout the period of irrigation). The average depth of the advancing sheet of water over the land is 70 mm. The time required to irrigate the border strip, in minutes, will be:

(A) 15-18 (B) 23-26
(C) 52-57 (D) 65-69

626. A regime channel carrying a discharge of 25 m^3 s^{-1} is designed using Lacey's regime theory. The side slope of the channel is ½ H : 1 V, and Lacey's silt factor is unity. The bottom width and depth of flow in the channel, in m, respectively are:

(A) 20.26, 1.38 (B) 20.26, 1.56
(C) 23.75, 1.56 (D) 32.78, 1.56

627. A 200 mm well fully penetrates a confined aquifer. After a long period of pumping at a rate of 1400 litres per minute, the drawdowns in the observation wells located at 25 m and 40 m from the pumping well are found to be 2.6 m and 1.9 m, respectively. The transmissivity of the aquifer in m^2day^{-1} is:

(A) 190 (B) 198
(C) 206 (D) 215

628. One ha cm of water is equal to:

(A) 100 tonnes water
(B) 100000 litres water
(C) 10 tonnes water
(D) Both (A) and (B)

629. Which one of the following water potential in soil is always positive:

(A) Osmotic potential
(B) Matric potential
(C) Gravitational potential
(D) None of these

630. Pan evaporation data recorded at a certain location over a period of one week are 4.0, 4.3, 4.6, 4.9, 5.12, 5.18 and 6.21 mm. If irrigation scheduling based on ratio of irrigation water (IW) to cumulative pan evaporation (CPE) is practiced, the depth of irrigation at an interval of a week for IW/CPE = 0.9 is:

(A) 3.60 (B) 4.41
(C) 5.59 (D) 30.88

631. A piston pump is driven by a 5 m diameter horizontal axis wind turbine for supplying water from a borehole with a total pump head of 10 m. The mean velocity of air is 18 km h^{-1} and the density of air is 1.29 kg m^{-3}. The actual power coefficient of the wind turbine is 0.30 and the overall pump efficiency is 60%. Neglecting the transmission losses, the expected pump discharge in L s^{-1} will be:

(A) 2.90 (B) 5.80
(C) 28.50 (D) 32.27

632. A 10 ha field has 1.2 m deep layer of sandy loam soil underlain by sandy soil up to a depth of 5 m. A pre-irrigation rainfall brings moisture content of the top 0.3 m layer to its field capacity. The moisture content of rest of the sandy loam layer remains at permanent wilting point. The volumetric moisture content at field capacity and permanent wilting point are 32 and 16%, respectively for the sandy loam soil. The field is irrigated with a stream size of 240 L s^{-1} for 24 hours. Considering the drainage from the sandy loam soil as deep percolation, application efficiency and deep percolation ratio in percent respectively are:

(A) 56.40 and 43.60
(B) 69.44 and 30.56
(C) 75.18 and 24.82
(D) 92.60 and 27.40

633. At field capacity the value of pF is:

(A) 0.5 (B) 2.5
(C) 4.0 (D) 6.0

634. At permanent wilting point (PWP) the pF value is:

(A) 1.5 (B) 2.5
(C) 4.2 (D) 5.0

635. If the gravimetric moisture content of soil is 15% and the bulk density is 1.4 g/cc, the volumetric moisture content is:

(A) 19.0 % (B) 21.0%
(C) 25.0% (D) 30.0%

636. The total volume of water on melting of ice:

(A) Increases (B) Remains constant
(C) Decreases (D) None of these

637. Heavy shedding of buds and balls occurs in cotton due to:

(A) Deficiency of nitrogen in the soil
(B) Deficiency of phosphorus in the soil
(C) Deficiency of magnesium in the soil
(D) Water stress at bud formation stage

638. The normal form of plant is maintained by:

(A) Water (B) Gases
(C) Solids (D) None of these

639. Flow of water in saturated soil is described by:
(A) Poiseuille's law(B) Darcy's law
(C) Fick/s law (D) Both (A) and (B)

640. Which growth stage of sugarcane is critical for irrigation:
(A) Germination
(B) Grand growth phase
(C) Formative stage
(D) Ripening stage

641. The use of tensiometer is confined upto metric potential of:
(A) -0.8 bar (B) -0.6 bar
(C) -0.4 bar (D) -0.2 bar

642. Which one of the following crops is most sensitive to both excess moisture and drought:
(A) Wheat (B) Maize
(C) Sunflower (D) Oat

643. Crop logging is used in:
(A) Sugarcane (B) Sugarbeet
(C) Maize (D) Bajra

644. Irrigation in gram are recommended at:
(A) Early flowering and maturity stage
(B) Late flowering and maturity stage
(C) Early flowering and pod formation stage
(D) Before flowering and pod formation stage

645. The downward movement f surface soil water is known as:
(A) Infiltration (B) Percolation
(C) Leaching (D) Washout

646. The neutron scattering method is not useful in estimating the moisture in:
(A) Acidic soils (B) Alluvial soils
(C) Lateritic soils (D) Organic soils

647. In furrow method of irrigation:
(A) Only 1/4th of the furrow is wetted
(B) Only 4/5th of the furrow is wetted
(C) Only ½ of the furrow is wetted
(D) Only 3/4th of the furrow is wetted

648. The upper limit of the soil moisture available for the plant growth is:
(A) PWP (15 bars)
(B) Field capacity (1/3 bars)
(C) Hygroscopic coefficient
(D) Matric suction

649. A device for measuring percolation and leaching osses from a column of soil under controlled conditions is known as:
(A) Inflitrometer (B) Odometer
(C) Psycrometer (D) Lysimeter

650. A calibrated device for measuring the flow of water in open channel is known as:
(A) V-notch (B) Parshall flume
(C) Watermeter (D) Lactometer

651. The instrument is used for measuring depth of water table is known as:
(A) Lysimeter (B) Psycrometer
(C) Piezometer (D) Evaporimeter

652. Water requirement (WR) includes the losses due to:
(A) Evapotranspiration and application of water
(B) Consumptive use of water and application of water
(C) Consumptive use, water required for special operation and other economically unavoidable losses of water
(D) ET and water required for special operations

653. The method of irrigation employed where surface is undulating:
(A) Flood irrigation
(B) Sprinkler irrigation
(C) Drip irrigation
(D) Both (A) and (C)

654. The most critical stage of maize from irrigation point of view is:
(A) Silking stage (B) Tasselling stage
(C) Boot stage (D) Dough stage

655. Moisture content at wilting point is highest in
(A) Clay loam (B) Silty clay loam
(C) Loam (D) Sandy loam

656. Darcy's law is applicable under unsaturated conditions if hydraulic conductivity is expressed as a function of
(A) Porosity
(B) Hydraulic gradient
(C) Hydraulic head
(D) Water content

657. Capillary fringe in soil refers to:
(A) Height of water rise in a capillary
(B) Saturated zone above the water table
(C) Water in capillary pores
(D) Depth of water table

658. At field capacity the value of pH is:
(A) 0.0 (B) 2.5
(C) 4.2 (D) 6.0

659. Water potential of pure water is:
(A) Zero (B) 1 Bar
(C) 100 Bar (D) 10^6 Bar

660. Capillary movement of water is complemented by:
(A) Root extension
(B) Epinasty
(C) Stem elongation
(D) Leaf orientation

661. pF value (4.18) indicates:
(A) Field capacity
(B) Wilting point
(C) Hygroscopic water
(D) Capillary water

662. Mechanism to improve water uptake by plants in dry farming areas:
(A) Lipid deposition on leaves
(B) Reduction in leaf area
(C) Development of awns
(D) Increase root growth and root shoot ratio

663. Relationship between water and fertilizer as production factors in crops is:
(A) Additive (B) Synergetic
(C) Antagonistic (D) None of these

664. The permanent wilting point (PWP) the pH value is:
(A) 0.0 (B) 2.5
(C) 4.2 (D) 6.0

665. PF is a measure of :
(A) Soil aeration (B) Soil moisture
(C) Soil erosion (D) Soil temperature

666. Which of the following method of irrigation is most widely used for orchard is:
(A) Drip (B) Sprinkler
(C) Flood (D) Furrow

667. The method of irrigation employed where surface is undulating is :
(A) Flood irrigation
(B) Sprinkler irrigation
(C) Drip irrigation
(D) Subsurface irrigation

668. The force of attraction binds the molecules of the same kind is:
(A) Adhesion (B) Matric force
(C) Cohesion (D) None of these

669. The upper limit of the soil moisture available for the plant growth is:
(A) PWP (15 bars)
(B) Field capacity (1/3 bars)
(C) Hygroscopic coefficient
(D) Matric suction

670. A device for measuring percolation and leaching losses from a column of soil under controlled conditions is known as:
(A) Infiltrometer (B) Evaporimeter
(C) Psycrometer (D) Lysimeter

671. A calibrated device for measuring the flow of water in open conditions is knwon as:
(A) V-notch (B) Parshall flume
(C) Water meter (D) None of these

672. The instrument used for measuring depth of watertable is known as :
(A) Lasimeter (B) Piezometer
(C) Colometer (D) Evaporimeter

673. Darcy's law is valid for:
(A) Only saturated flow
(B) Only unsaturated flow
(C) Both saturated and unsaturated flow
(D) None of the above

674. If the field capacity is 18% and permanent wilting point is 6%. The 50% of the available soil moisture will be:
(A) 6% (B) 8%
(C) 10.9% (D) 12%

675. Tensiometer can effectively measures the soil moisture content upto:
(A) 0.33 atm (B) 0.85 atm
(C) 1.50 atm (D) 5.00 atm

676. Flow of water in saturated soil is described by:
(A) Poiseuille's law
(B) Darcy's law
(C) Fick's law
(D) Both (A) and (B)

677. Consumptive use of water per kg of rice production approximately ranges between:
(A) 1000-2000 liters
(B) 1500-2000 liters
(C) 2000-5000 liters
(D) 5000-7000 liters

678. Heavy rainfall or irrigation during fruiting period of cotton may cause:
(A) Shadding of flowers and young bolls
(B) Lodging of crop
(C) Heavy incidence of bollworm
(D) Yellowing of young bolls

679. In general, irrigation is given to a crop when the available soil moisture in root zone is depleted by:
(A) 1-10% (B) 10-25%
(C) 50-75% (D) 80-90%

680. The sensitivity of tensiometer is only upto:
(A) 0.50 atm (B) 0.65 atm
(C) 0.85 atm (D) 1.00 atm

681. The small pores, or the bulk volume of small pores, which hold water in soils against a tension usually more than 60 cm of water:
(A) Capillary porosity
(B) Pore space
(C) Bulk density
(D) pF

682. The ratio between the irrigated area and the volume of water used in known as:
(A) Delta of water (B) Duty of water
(C) Base of water (D) Coarse of water

683. Which of the following crops has the highest percentage of irrigated area?
(A) Rice (B) Cotton
(C) Wheat (D) Sugarcane

684. Soil-moisture characteristics curve depicts relationship between:
(A) Soil moisture and tension
(B) Soil temperature and moisture
(C) Soil moisture and bulk density
(D) Soil texture and moisture

685. In furrow method of irrigation:
(A) Only 1/4th of the furrow in wetted
(B) Only 4/5th of the furrow in wetted
(C) Only 1/2 of the furrow in wetted
(D) Only 3/4th of the furrow in wetted

ANSWER SHEET
11. Irrigation Management

1. (A)	2. (D)	3. (A)	4. (C)	5. (A)	6. (B)	7. (C)	8. (C)	9. (D)	10. (D)
11. (A)	12. (A)	13. (B)	14. (A)	15. (C)	16. (A)	17. (A)	18. (B)	19. (C)	20. (D)
21. (D)	22. (A)	23. (B)	24. (C)	25. (D)	26. (D)	27. (B)	28. (A)	29. (D)	30. (C)
31. (A)	32. (A)	33. (A)	34. (A)	35. (B)	36. (A)	37. (B)	38. (C)	39. (D)	40. (A)
41. (B)	42. (B)	43. (C)	44. (D)	45. (A)	46. (A)	47. (A)	48. (A)	49. (A)	50. (A)
51. (B)	52. (B)	53. (D)	54. (A)	55. (B)	56. (C)	57. (D)	58. (A)	59. (C)	60. (C)
61. (D)	62. (A)	63. (D)	64. (A)	65. (C)	66. (B)	67. (A)	68. (B)	69. (D)	70. (D)
71. (A)	72. (A)	73. (B)	74. (C)	75. (D)	76. (B)	77. (A)	78. (A)	79. (A)	80. (B)
81. (B)	82. (A)	83. (B)	84. (C)	85. (C)	86. (C)	87. (C)	88. (B)	89. (B)	90. (C)
91. (A)	92. (A)	93. (C)	94. (A)	95. (B)	96. (A)	97. (C)	98. (D)	99. (A)	100. (B)
101. (A)	102. (A)	103. (B)	104. (C)	105. (D)	106. (C)	107. (C)	108. (C)	109. (D)	110. (D)
111. (D)	112. (D)	113. (C)	114. (C)	115. (C)	116. (A)	117. (B)	118. (C)	119. (D)	120. (A)
121. (B)	122. (C)	123. (D)	124. (A)	125. (A)	126. (B)	127. (C)	128. (D)	129. (A)	130. (B)

131. (C) 132. (C) 133. (D) 134. (A) 135. (A) 136. (A) 137. (B) 138. (C) 139. (A) 140. (B)
141. (C) 142. (C) 143. (D) 144. (A) 145. (A) 146. (D) 147. (B) 148. (D) 149. (C) 150. (D)
151. (D) 152. (A) 153. (D) 154. (C) 155. (C) 156. (D) 157. (A) 158. (D) 159. (B) 160. (D)
161. (B) 162. (B) 163. (B) 164. (D) 165. (B) 166. (C) 167. (A) 168. (A) 169. (B) 170. (A)
171. (A) 172. (A) 173. (B) 174. (C) 175. (D) 176. (D) 177. (A) 178. (B) 179. (C) 180. (D)
181. (A) 182. (C) 183. (C) 184. (C) 185. (C) 186. (C) 187. (C) 188. (A) 189. (C) 190. (C)
191. (A) 192. (A) 193. (C) 194. (D) 195. (D) 196. (D) 197. (A) 198. (B) 199. (B) 200. (D)
201. (C) 202. (B) 203. (B) 204. (A) 205. (C) 206. (D) 207. (B) 208. (B) 209. (B) 210. (A)
211. (A) 212. (C) 213. (D) 214. (B) 215. (D) 216. (B) 217. (D) 218. (B) 219. (A) 220. (A)
221. (B) 222. (A) 223. (D) 224. (B) 225. (B) 226. (A) 227. (B) 228. (B) 229. (B) 230. (C)
231. (D) 232. (B) 233. (D) 234. (B) 235. (B) 236. (B) 237. (C) 238. (C) 239. (C) 240. (A)
241. (C) 242. (A) 243. (C) 244. (A) 245. (B) 246. (A) 247. (D) 248. (A) 249. (B) 250. (D)
251. (B) 252. (D) 253. (C) 254. (C) 255. (A) 256. (C) 257. (C) 258. (C) 259. (D) 260. (B)
261. (B) 262. (A) 263. (D) 264. (A) 265. (C) 266. (C) 267. (B) 268. (C) 269. (B) 270. (B)
271. (B) 272. (A) 273. (D) 274. (C) 275. (B) 276. (C) 277. (D) 278. (B) 279. (B) 280. (D)
281. (A) 282. (C) 283. (A) 284. (A) 285. (B) 286. (B) 287. (A) 288. (D) 289. (A) 290. (A)
291. (A) 292. (A) 293. (B) 294. (C) 295. (D) 296. (D) 297. (C) 298. (A) 299. (A) 300. (B)
301. (A) 302. (A) 303. (A) 304. (B) 305. (D) 306. (A) 307. (D) 308. (B) 309. (B) 310. (B)
311. (C) 312. (C) 313. (A) 314. (C) 315. (B) 316. (C) 317. (D) 318. (C) 319. (A) 320. (A)
321. (A) 322. (A) 323. (B) 324. (C) 325. (C) 326. (A) 327. (C) 328. (A) 329. (A) 330. (A)
331. (C) 332. (C) 333. (D) 334. (C) 335. (D) 336. (D) 337. (A) 338. (B) 339. (A) 340. (C)
341. (A) 342. (A) 343. (A) 344. (A) 345. (D) 346. (A) 347. (B) 348. (A) 349. (C) 350. (C)
351. (B) 352. (C) 353. (D) 354. (D) 355. (A) 356. (D) 357. (D) 358. (B) 359. (C) 360. (B)
361. (A) 362. (D) 363. (A) 364. (C) 365. (B) 366. (C) 367. (C) 368. (C) 369. (C) 370. (C)
371. (B) 372. (D) 373. (D) 374. (C) 375. (A) 376. (A) 377. (D) 378. (A) 379. (A) 380. (B)
381. (C) 382. (D) 383. (D) 384. (A) 385. (B) 386. (C) 387. (D) 388. (D) 389. (A) 390. (A)
391. (A) 392. (B) 393. (D) 394. (A) 395. (C) 396. (B) 397. (B) 398. (C) 399. (C) 400. (A)
401. (B) 402. (A) 403. (A) 404. (A) 405. (B) 406. (D) 407. (A) 408. (A) 409. (B) 410. (A)
411. (C) 412. (D) 413. (D) 414. (A) 415. (B) 416. (C) 417. (A) 418. (C) 419. (A) 420. (A)
421. (D) 422. (B) 423. (A) 424. (A) 425. (A) 426. (B) 427. (B) 428. (C) 429. (A) 430. (A)
431. (B) 432. (B) 433. (D) 434. (C) 435. (B) 436. (D) 437. (A) 438. (A) 439. (A) 440. (B)
441. (A) 442. (A) 443. (C) 444. (D) 445. (B) 446. (B) 447. (C) 448. (D) 449. (A) 450. (D)
451. (D) 452. (E) 453. (C) 454. (B) 455. (C) 456. (C) 457. (D) 458. (D) 459. (C) 460. (C)
461. (D) 462. (B) 463. (E) 464. (D) 465. (C) 466. (D) 467. (C) 468. (A) 469. (C) 470. (B)
471. (C) 472. (B) 473. (C) 474. (B) 475. (B) 476. (A) 477. (B) 478. (C) 479. (A) 480. (B)
481. (A) 482. (C) 483. (B) 484. (C) 485. (B) 486. (B) 487. (A) 488. (A) 489. (A) 490. (B)
491. (C) 492. (D) 493. (A) 494. (B) 495. (C) 496. (D) 497. (B) 498. (C) 499. (D) 500. (A)
501. (A) 502. (B) 503. (C) 504. (A) 505. (D) 506. (B) 507. (B) 508. (A) 509. (B) 510. (A)
511. (C) 512. (D) 513. (B) 514. (A) 515. (B) 516. (C) 517. (B) 518. (A) 519. (A) 520. (B)
521. (C) 522. (C) 523. (A) 524. (A) 525. (B) 526. (B) 527. (B) 528. (A) 529. (C) 530. (E)
531. (A) 532. (A) 533. (B) 534. (C) 535. (D) 536. (C) 537. (B) 538. (A) 539. (A) 540. (A)

541. **(B)**	542. **(C)**	543. **(A)**	544. **(B)**	545. **(C)**	546. **(A)**	547. **(B)**	548. **(D)**	549. **(A)**	550. **(B)**
551. **(C)**	552. **(A)**	553. **(B)**	554. **(A)**	555. **(B)**	556. **(A)**	557. **(B)**	558. **(A)**	559. **(C)**	560. **(C)**
561. **(A)**	562. **(A)**	563. **(B)**	564. **(A)**	565. **(A)**	566. **(A)**	567. **(D)**	568. **(C)**	569. **(C)**	570. **(D)**
571. **(A)**	572. **(D)**	573. **(B)**	574. **(C)**	575. **(D)**	576. **(A)**	577. **(A)**	578. **(B)**	579. **(D)**	580. **(A)**
581. **(B)**	582. **(C)**	583. **(D)**	584. **(A)**	585. **(B)**	586. **(C)**	587. **(D)**	588. **(A)**	589. **(B)**	590. **(C)**
591. **(A)**	592. **(B)**	593. **(C)**	594. **(D)**	595. **(A)**	596. **(B)**	597. **(C)**	598. **(D)**	599. **(A)**	600. **(A)**
601. **(B)**	602. **(C)**	603. **(C)**	604. **(C)**	605. **(B)**	606. **(A)**	607. **(B)**	608. **(C)**	609. **(C)**	610. **(A)**
611. **(B)**	612. **(C)**	613. **(D)**	614. **(A)**	615. **(B)**	616. **(D)**	617. **(B)**	618. **(A)**	619. **(D)**	620. **(C)**
621. **(C)**	622. **(A)**	623. **(A)**	624. **(A)**	625. **(C)**	626. **(B)**	627. **(D)**	628. **(D)**	629. **(C)**	630. **(D)**
631. **(A)**	632. **(B)**	633. **(B)**	634. **(C)**	635. **(B)**	636. **(C)**	637. **(D)**	638. **(A)**	639. **(D)**	640. **(C)**
641. **(A)**	642. **(B)**	643. **(A)**	644. **(D)**	645. **(B)**	646. **(D)**	647. **(D)**	648. **(B)**	649. **(D)**	650. **(B)**
651. **(C)**	652. **(C)**	653. **(B)**	654. **(B)**	655. **(A)**	656. **(D)**	657. **(B)**	658. **(B)**	659. **(A)**	660. **(A)**
661. **(B)**	662. **(D)**	663. **(A)**	664. **(C)**	665. **(B)**	666. **(A)**	667. **(B)**	668. **(D)**	669. **(B)**	670. **(D)**
671. **(B)**	672. **(C)**	673. **(A)**	674. **(A)**	675. **(B)**	676. **(D)**	677. **(C)**	678. **(A)**	679. **(C)**	680. **(C)**
681. **(A)**	682. **(B)**	683. **(D)**	684. **(A)**	685. **(D)**					

12 DRYLAND MANAGEMENT

1. Which one is drought resistant crop?
(A) Paddy (B) Sorghum
(C) Chickpea (D) Wheat

2. The areas where drought has occurred in -------- per cent of years during the period are considered drought areas:
(A) 10 (B) 20
(C) 30 (D) 40

3. The area where drought has occurred in more than ------- per cent of years are called chronic drought areas:
(A) 30 (B) 40
(C) 50 (D) 60

4. The Central Arid Zone Research Institute, Jodhpur (Rajasthan) was established in:
(A) 1952 (B) 1957
(C) 1959 (D) 1961

5. The International Centre for Agricultural Research in Dry Areas (ICARDA) was established in:
(A) 1957 (B) 1967
(C) 1977 (D) 1987

6. Under rainfed conditions one of the suitable varieties of wheat is for central India:
(A) Mukta (B) Bijaga red
(C) VL-421 (D) K-65

7. Mulching is helpful in:
(A) Increasing soil fertility
(B) Improvement of soil structure
(C) Moisture conservation
(D) Better crops

8. Best suited soil for rainfed cotton crop is:
(A) Sandy loam (B) Clay loam
(C) Silty loam (D) Black soil

9. Physiological drought is caused by:
(A) Salinity & alkalinity
(B) Excess water
(C) Diseases (D) All of these

10. The chemical which induces the stomata closing is:
(A) PMA (B) 2,4-D
(C) Kaoline (D) Cycocil

11. Maximum wind soil erosion occurs in which state:
(A) Uttar Pradesh (B) West Bengal
(C) Rajasthan (D) Himachal Pradesh

12. Contour bunding and terracing are the measures of:
(A) Agronomical measures
(B) Mechanical measures
(C) Biological measures
(D) None of these

13. Under drought the sorghum plants synthesize Dhurin in:
(A) Roots (B) Shoots
(C) Leaves (D) None of these

14. Geological erosion is also known as:
(A) Accelerated erosion
(B) Natural erosion
(C) Soil erosion
(D) Water erosion

15. Wind breaks are planted in orchard towards:
(A) South-West
(B) North-West
(C) North-East
(D) East-South

16. Mulching is related to:
(A) Check percolation of moisture from the soil
(B) Check evaporation of moisture from the soil
(C) Check loss of any type of moisture from the soil
(D) None of these

17. Permits a crop to grow for a longer time in a given moisture stress conditions is called:
(A) Drought avoidance
(B) Drought resistance
(C) Drought escape (D) Drought prone

18. Relative ability to maintain growth and yield with little or no major set back under moisture stress conditions is called:
(A) Drought avoidance
(B) Drought resistance
(C) Drought escape
(D) Drought prone

19. The practice of crop production entirely with rain water received during the crop season or on conserved soil moisture in low rainfall (<750 mm) areas of arid and semi-arid climate and the crop may face mild to very severe moisture stress during their life cycle is called:
(A) Dry farming (B) Dry season
(C) Drought year (D) Rainfed farming

20. Growing of field crops entirely with rain water received during the crop season (rainfall usually > 750 mm) under humid or subhumid climate and the crop may face little or no moisture stress during their life cycle is called:
(A) Dry farming (B) Dry season
(C) Drought year (D) Rainfed farming

21. A period during which water deficiency occurs as stored water is used for evapotranspiration and actual evapotranspiration falls below potential evapotraspiration is called:
(A) Dry farming (B) Dry season
(C) Drought year (D) Drought prone

22. The year in which rainfall of particular place is short by more than twice the standard deviation is called:
(A) Dry farming (B) Dry season
(C) Drought year (D) Drought prone

23. A period of atleast 15 consecutive days, during which not a single wetting rain has fallen is called:
(A) Dry farming (B) Dry season
(C) Dry spell (D) Drought prone

24. The lack of resistance of a soil to soil erosion is called:
(A) Erodibility
(B) Erosion
(C) Erosion accelerated
(D) Erosion resistant

25. The transport away of the land surface by water, winds or other forces is called:
(A) Erodibility
(B) Erosion
(C) Erosion accelerated
(D) Erosion resistant

26. Abnormally rapid erosion in an environment disturbed by animal life, chiefly man is called:
(A) Erodibility
(B) Erosion
(C) Erosion accelerated
(D) Erosion resistant

27. A crop, because of its dense foliage, root system etc., provides effective protection against soil loss by erosion is called:
(A) Erodibility
(B) Erosion
(C) Erosion accelerated
(D) Erosion resistant crops

28. A crop, which are increased the soil losses by erosion is called:
(A) Erosion permitting crops
(B) Erosion
(C) Erosion accelerated
(D) Erosion resistant crops

29. Water erosion producing very small and numerous channels is called:
(A) Splash erosion (B) Sheet erosion
(C) Rill erosion (D) Gully erosion

30. A form of soil erosion resulting from soil splash under the impact of falling rain drops is called:
(A) Splash erosion (B) Sheet erosion
(C) Rill erosion (D) Gully erosion

31. Erosion of a fairly uniform layer of material from the land surface, often scarcely perceptible especially when caused by wind is called:
(A) Splash erosion (B) Sheet erosion
(C) Rill erosion (D) Gully erosion

32. Removal of soil by running water, with formation of channels, that cannot be smothered out completely by normal cultivation is called:
(A) Splash erosion (B) Sheet erosion
(C) Rill erosion (D) Gully erosion

33. Climate under crop canopy is called:
(A) Eco climate (B) Crop climate
(C) Both of these (D) None of these

34. The amount of water vapour in the air is called:
(A) Humidity (B) Frost
(C) Fog (D) Mist

35. The fraction of total precipitation which forms a part of the crop consumptive use or that becomes utilizable for crop production is called:
(A) Effective rainfall
(B) Effective irrigation
(C) Effective climate
(D) None of these

36. Controlled and uniform application of water to crop land in required amount at required time, with minimum cost to produce optimum yields without waste of water and adverse effect on soil in form of soil salinity and waterlogging problems is called:
(A) Effective rainfall
(B) Effective irrigation
(C) Effective climate
(D) None of these

37. A small body of water retained behind a small dam or held in a pit dug in the ground by water harvesting techniques is called:
(A) Farm pond
(B) Gully plug
(C) Both of these
(D) None of these

38. Leaf and canopy temperature can be measured by:
(A) Infra red thermometer
(B) Barometer
(C) Lysimeter
(D) Dry bulb thermometer

39. Organic substances which in minute amounts may participate in the control of growth processes is called:
(A) Growth regulator
(B) Growth retardant
(C) Growth substances
(D) None of these

40. A chemical that selectivity interferes with normal hormonal production of growth but without appreciable toxic effects is called:
(A) Growth regulator
(B) Growth retardant
(C) Growth substances
(D) None of these

41. A growth regulating chemical produced by a plant of a synthetic one with similar activity even at very low concentration is called:
(A) Growth regulator
(B) Growth retardant
(C) Growth substances
(D) None of these

42. Chemical such as auxins, gibberellins, kinins etc. are the example of:
(A) Growth regulator
(B) Growth retardant
(C) Growth substances
(D) None of these

43. A chemical such as CCC is the example of:
(A) Growth regulator
(B) Growth retardant
(C) Growth substances
(D) None of these

44. Reshaping of the field surface to a desired elevation and slope is called:
(A) Land forming (B) Land grading
(C) Land leveling (D) Land planning

45. The reshaping of the land surface to facilitate a more uniform application of irrigation water or for preventing soil erosion is called:
(A) Land forming (B) Land grading
(C) Land leveling (D) Land planning

46. Usually crop face little or no moisture stress during their life cycle in:
(A) Dry farming (B) Dry season
(C) Drought year (D) Rainfed farming

47. Usually crop face moisture stress during their life cycle in:
(A) Dry farming (B) Dry season
(C) Drought year (D) Rainfed farming

48. Atrazine used as an antitranspirant:
(A) Reduces the growyh of the crop
(B) Does not reflect light from plant leaf surface
(C) Affects the closure and opening of stomata
(D) Forms thin layer on eaf surface

49. Phenylmercuric acitate (PMA) is a chemical used in agricultural crops in order to:
(A) Increase CO_2 uptake
(B) Reduce repiration
(C) Reduce transpiration
(D) Decrease CO_2 uptake

50. Supplemental irrigation can be given to save the crop in:
(A) Dry farming (B) Dry season
(C) Drought year (D) Rainfed farming

51. No supplemental irrigation is required to save the crop in:
(A) Dry farming (B) Dry season
(C) Drought year (D) Rainfed farming

52. Drier part of Rajasthan, Gujarat, Maharashtra etc comes under:
(A) Dry farming (B) Dry season
(C) Drought year (D) Rainfed farming

53. Humid and sub humid part of North-eastern states comes under:
(A) Dry farming (B) Dry season
(C) Drought year (D) Rainfed farming

54. Measuring depth of water from precipitation is:
(A) Raingauge (B) Sunshine recorder
(C) Anemometer (D) None of these

55. The period of the year in which monthly rainfall exceeds 10 per cent of the monthly evapotrans-piration in:
(A) Rainy season
(B) Dry season
(C) Wet season
(D) None of these

56. A long, low ridge of earth with gently sloping sides and a shallow channel along the upper side, to control erosion by diverting surface runoff across the slope instead of permitting it to flow uninterrupted down to slope is called:
(A) Ridge terrace (B) Ridge planting
(C) Rotary tillage (D) None of these

57. Planting on ridges is called:
(A) Ridge terrace (B) Ridge planting
(C) Rotary tillage (D) None of these

58. A kind of drought which occurs during distinct annual periods of dry weather is called:
(A) Seasonal drought
(B) Recurrent drought
(C) Both of these
(D) None of these

59. A land or climate where the rainfall is greater many crops. Irrigation or dryland farming methods are commonly required for crop production is called:
(A) Arid (B) Semi arid
(C) Humid (D) Sub humid

60. The zone delineated by Thornthwaite's moisture indices varying from -20 to -40 is:
(A) Semi-arid (B) Arid
(C) Severe arid (D) Humid

61. The zone delineated by Thornthwaite's moisture indices varying from -60 to -80 is:
(A) Semi-arid (B) Arid
(C) Severe arid (D) Humid

62. The zone delineated by Thornthwaite's moisture indices varying from less than -80 is:
(A) Semi-arid (B) Arid
(C) Severe arid (D) Humid

63. The zone delineated by Thornthwaite's moisture indices varying from 20 to 50 is:
(A) Semi-arid (B) Arid
(C) Severe arid (D) Humid

64. The zone delineated by Thornthwaite's moisture indices varying from 50 to 100 is:
(A) Semi-arid (B) Humid
(C) Severe humid (D) Extreme Humid

65. The zone delineated by Thornthwaite's moisture indices varying from more than 100 is:
(A) Semi-arid (B) Humid
(C) Severe humid (D) Extreme Humid

66. Thornthwaite's moisture index can be calculated by:

(A) $MI = \frac{P - PE}{PE} \times 100$

(B) $MI = \frac{PE - P}{PE} \times 100$

(C) $MI = \frac{P - PE}{P} \times 100$

(D) $MI = \frac{P - PE}{PE}$

67. A dry soil layer at the surface may be created in-situ by loosening the surface soil and hastening the drying of the upper soil is known as:

(A) Soil mulch (B) Straw mulch
(C) Plastic mulch (D) Stuble mulch

68. Mulching with crop residue like straw, husk, dry leaves, twigs etc. helps to reduce surface runoff is known as:

(A) Soil mulch (B) Straw mulch
(C) Plastic mulch (D) Stuble mulch

69. Mulching with costly plastic films and paper is practices to reduce moisture losses from soil surface is called:

(A) Soil mulch
(B) Straw mulch
(C) Plastic mulch
(D) Stuble mulch

70. Which is mulching is generally practiced in high value crops is:

(A) Soil mulch
(B) Straw mulch
(C) Plastic mulch
(D) Stuble mulch

71. Which type of plastic is more effective in mulching:

(A) Black plastic
(B) White plastic
(C) Transparent plastic
(D) None of these

72. Plastic mulching is also act as:

(A) Reduces weed infestation
(B) Increases seed germination
(C) Both of these
(D) None of these

73. The amount of water needed to return the soil moisture status to the field capacity is called:

(A) Soil moisture deficit
(B) Soil moisture stress
(C) Soil moisture tension
(D) None of these

74. It is the sum of soil moisture tension and the osmotic pressure of the soil solution is called:

(A) Soil moisture deficit
(B) Soil moisture stress
(C) Soil moisture tension
(D) None of these

75. It is the force with which the water is held by soil against gravity is called:

(A) Soil moisture deficit
(B) Soil moisture stress
(C) Soil moisture tension
(D) None of these

76. Incorporation of vegetative mulch in a band in the soil for the purpose of harvesting rain water and conserving soil and water is called:

(A) Tillage action (B) Tillage depth
(C) Tillability (D) Vertical mulching

77. Conservation of rain water under unirrigated conditions, by collecting runoff of precipitation in order to supplement soil moisture in an adjacent area is called:

(A) Water harvesting
(B) Soil conservation
(C) Runoff harvesting (D) None of these

78. A combination of all management and land use methods which safe guard the soil against depletion or deterioration by natural or by man induced factors is called:

(A) Water harvesting
(B)Soil conservation
(C) Runoff harvesting
(D) None of these

79. The area above a given point on a stream that contributes water to the flow at that point is called:

(A) Water harvesting
(B) Soil conservation
(C) Runoff harvesting
(D) Watershed

80. An implement used for puddling the paddy field before transplanting is:
(A) Rotavotor
(B) Harrow
(C) Ridge & furrow makers
(D) Puddler

81. Water absorbing chemicals are:
(A) Grow soak (B) Agrolite-400
(C) Jalshakti (D) All of these

82. Bund construction across the slopes are known as:
(A) Gulley plugs (B) Terraces
(C) Field bunds (D) Contour bunds

83. Series of flat plateform constructed across a slopping land which give rise to the formation of a step or bench like feature is known as:
(A) Gulley plugs (B) Terraces
(C) Field bunds (D) Contour bunds

84. Small earthen barriers constructed in field is known as:
(A) Gulley plugs (B) Terraces
(C) Field bunds (D) Contour bunds

85. Construction of field bunds along the contour line is known as:
(A) Gulley plugs (B) Terraces
(C) Field bunds (D) Contour bunds

86. Over flow weir is also known as:
(A) Check dams (B) Field bunds
(C) Gully plugs (D) Terraces

87. Soil erosion can be caused by:
(A) High plant density
(B) Over grazing
(C) Terracing
(D) Strip cropping

88. Soil temperature can be regulated by use of straw mulch both in summer and winter season because:
(A) It reduces evaporation
(B) It lowers the soil temperature
(C) It keeps the temperature of soil high
(D) It provides insulating effect

89. Chemical which regulates opening and closing of stomata is:
(A) Kaoline (B) 2, 4-D
(C) PMA (D) Paraquat

90. Gully erosion of soil is mainly caused by:
(A) Stream formed by rain water
(B) Wind
(C) Terrace farming
(D) Sea waves

91. Which one of the following can lead to soil erosion?
(A) Contour ploughing
(B) Terrace farming
(C) Deforestation
(D) Mixed cropping

92. Soil erosion is the state of Punjab is due to which one of the following reasons?
(A) Excessive cultivation
(B) Irrigation
(C) Overgrazing
(D) Deforestation

93. In which one of the following states is the terrace cultivation practice?
(A) Punjab (B) Haryana
(C) Plains of Uttar Pradesh
(D) Uttarakhand

94. In India, first Dry Farming Research Station, manjri was started in:
(A) 1923 (B) 1933
(C) 1935 (D) 1938

95. Water harvesting is defined as:
(A) Application of water at the time of harvesting cops like colocasia, tapioca and yam
(B) Harvesting deep water from soils by pumping devices
(C) Unit of water applied to crops
(D) Collecting the excess runoff from rain on the farm in ponds and utilizing it later for agriculture

96. Area under dryland is:
(A) 33.5 mha (B) 34.5 mha
(C) 35.5 mha (D) 36.5 mha

97. Share of dryland area in total agriculture land is:
(A) 15.0% (B) 16.5%
(C) 17.0% (D) 18.5%

98. Share of total food grain production under dryland/rainfed area is:
(A) 35% (B) 44%
(C) 45% (D) 50%

99. Drought affected area in the country is:
(A) 107 mha (B) 108 mha
(C) 109 mha (D) 110 mha

100. When precipitation received less than 50% is called:
(A) Severe drought
(B) Medium drought
(C) Less drought
(D) Drought

101. Maximum evapo-transpiration occurs in:
(A) Jalgaon
(B) Ahmedabad
(C) Jalgaon and Ahmedabad
(D) None of these

102. World average rainfall is:
(A) 890 mm (B) 990 mm
(C) 1090 mm (D) 1190 mm

103. Average rainfall in India is:
(A) 950 mm (B) 1195 mm
(C) 1260 mm (D) 1350 mm

104. Indian agriculture is gamble of:
(A) Monsoon (B) Indian economy
(C) Politics (D) Social

105. Share of human population under rainfed agriculture is:
(A) 30% (B) 40%
(C) 50% (D) 60%

106. Per cent area under rainfed agriculture is:
(A) 60% (B) 65%
(C) 70% (D) 75%

107. Per cent livestock population under rainfed agriculture is:
(A) 60% (B) 65%
(C) 70% (D) 75%

108. Per cent under pulses and millets under rainfed conditions is:
(A) 87% (B) 77%
(C) 66% (D) 50%

109. Per cent under oilseeds under rainfed conditions is:
(A) 87% (B) 77%
(C) 66% (D) 50%

110. Per cent under cereals under rainfed conditions is:
(A) 87% (B) 77%
(C) 66% (D) 50%

111. Per cent under cotton under rainfed conditions is:
(A) 87% (B) 77%
(C) 66% (D) 50%

112. Area under rainfed condition in India is:
(A) 80 mha (B) 84 mha
(C) 52 mha (D) 50 mha

113. Area under irrigated condition in India is:
(A) 80 mha (B) 82 mha
(C) 52 mha (D) 50 mha

114. Wilting of plants occurs when:
(A) Lateral roots are removed
(B) Cortex is injured
(C) Phloem is blocked
(D) Xylem is blocked

115. Mulching is a process that helps mainly in:
(A) Improvement of soil
(B) Moisture conservation
(C) Weed control
(D) Soil drainage

116. Contour trenching is preferred to:
(i) Control of landslides
(ii) Conserve the mositure
(A) (i) is false and (ii) is true
(B) Both (i) and (ii) are false
(C) Both (i) and (ii) are true
(D) (i) is true and (ii) is false

117. How does the moisutre stress alter effect the cell?
(A) Cell division
(B) Cell expansion
(C) Cell mortality rate
(D) No effect on cell

Common Data for Questions 118 and 120

The hourly discharge observations at the mouth of a watershed due to 2 cm excess rainfall during 0 to 1 h and 3 cm excess rainfall during 1 to 2 h are given in the table below. Assume a constant base flow of 1 $m^3 s^{-1}$.

Time (h)	0	1	2	3	4	5	6
Discharge ($m s^{-1}$)	1	7	26	37	27	13	1

118. The area of the watershed, in km^2:
(A) 7.56 (B) 8.24
(C) 8.35 (D) 8.86

119. The peak of 1 h unit hydrograph in m^3s^{-1} for the watershed and its time of occurrence in h, respectively are:

(A) 6, 1 (B) 7, 2
(C) 8, 2 (D) 9, 1

120. Drought Prone Area Programme means:

(A) Development and use of regional food resources
(B) Conservation, development and use of land, water, cattle and human resources, especially in those areas where dry spell is found for a long period
(C) Providing nutrition for a long period in drought affected areas
(D) None of these

121. Phenylmercuric acitate (PMA) is a chemical used in agricultural crops in order to:

(A) Increase CO_2 uptake
(B) Reduce respiration
(C) Reduce transpiration
(D) Increase respiration

122. Contour bunds are effective when slope of the land is:

(A) Upto 6% (B) Upto 16%
(C) Upto 20% (D) Upto 50%

123. A film farming anti-transparent used in dryland agriculture is:

(A) Kaolin (B) Atrazine
(C) Mobileaf (D) PMA

124. In universal soil loss equation, what does K refer to ?

(A) Cropping management factor
(B) Slope gradient factor
(C) Soil erodibility factor
(D) Rainfall factor

125. Which of the following tillage operations is particularly beneficial under dry farming situations?

(A) Planking (B) Harrowing
(C) Dust mulching (D) Zero tillage

126. An imaginary line drawn on a map connecting the points of equal rainfall is called:

(A) Isotherm (B) Isobath
(C) Isohyt (D) Isobar

127. Watershed is area of:

(A) Biological system
(B) Physical system
(C) Social system
(D) All of those

128. Semi-arid climate has a moisture index of

(A) -80 to -100 (B) -40 to -80
(C) -33.3 to -66.7 (D) 0 to 20

129. Activities of which harmone increases with moisture stress?

(A) Gibberellic acid
(B) Indoleacetic acid
(C) Abscisic acid (D) Cytokinins

130. Drought avoidance mechanism is found in:

(A) Barley (B) Pearlmillet
(C) Sorghum (D) Maize

131. The size of micro-watershed is:

(A) 10000-20000 ha
(B) 1000-10000 ha
(C) 100-1000 ha (D) 10-100 ha

132. CAM system is prevalent in:

(A) Arid legumes (B) Cactus
(C) Pearlmillet (D) Sorghum

133. In India, First Dry Farming Research Station, Manjri was started in:

(A) 1923 (B) 1933
(C) 1934 (D) 1935

134. Under stress condition, which amino acid accumulated in crop plants:

(A) Methionine (B) Tryptophan
(C) Proline (D) Phenyl alanine

135. Headquarter of Rainfed Authority of India is located at:

(A) New Delhi (B) Bikaner
(C) Hyderabad (D) Kolkata

136. Sorghum crop is considered as camel crop because of :

(A) Deep root system
(B) Resistant to drought
(C) Shallow root system
(D) Nutrient exhaustiveness

137. Soil mulch is useful in:

(A) Improving aeration
(B) Improving drainage
(C) Minimize evaporation losses
(D) Removing weeds

ANSWER SHEET
12. Dryland Management

1. (B)	2. (B)	3. (B)	4. (C)	5. (C)	6. (A)	7. (C)	8. (D)	9. (A)	10. (A)
11. (C)	12. (B)	13. (A)	14. (B)	15. (A)	16. (B)	17. (A)	18. (B)	19. (A)	20. (D)
21. (B)	22. (C)	23. (C)	24. (A)	25. (B)	26. (C)	27. (D)	28. (A)	29. (C)	30. (A)
31. (B)	32. (D)	33. (A)	34. (A)	35. (A)	36. (B)	37. (A)	38. (A)	39. (A)	40. (B)
41. (C)	42. (A)	43. (B)	44. (B)	45. (C)	46. (D)	47. (A)	48. (C)	49. (C)	50. (A)
51. (D)	52. (A)	53. (D)	54. (A)	55. (A)	56. (A)	57. (B)	58. (A)	59. (B)	60. (A)
61. (B)	62. (C)	63. (D)	64. (C)	65. (D)	66. (A)	67. (A)	68. (B)	69. (C)	70. (C)
71. (A)	72. (C)	73. (A)	74. (B)	75. (C)	76. (D)	77. (A)	78. (B)	79. (D)	80. (D)
81. (D)	82. (A)	83. (B)	84. (C)	85. (D)	86. (A)	87. (B)	88. (D)	89. (C)	90. (A)
91. (C)	92. (C)	93. (D)	94. (A)	95. (D)	96. (B)	97. (C)	98. (B)	99. (C)	100. (A)
101. (C)	102. (B)	103. (B)	104. (A)	105. (B)	106. (B)	107. (A)	108. (A)	109. (B)	110. (D)
111. (C)	112. (B)	113. (C)	114. (D)	115. (B)	116. (A)	117. (A)	118. (A)	119. (C)	120. (B)
121. (C)	122. (A)	123. (C)	124. (C)	125. (C)	126. (D)	127. (D)	128. (C)	129. (C)	130. (C)
131. (C)	132. (B)	133. (A)	134. (C)	135. (A)	136. (B)	137. (C)			

13 AGROFORESTRY

1. Growing trees along with agricultural crops is known as:
   (A) Silviculture (B) Horticulture
   (C) Social forestry (D) Agroforestry
2. The sustainable land use systems involving trees combined with crops and/or animals on the same unit of land is known as:
   (A) Agroforestry (B) Silviculture
   (C) Social forestry (D) Horticulture
3. The suitable intercrop in *Populus deltoids* based agrisilvicultural system is:
   (A) Wheat (B) Rice
   (C) Maize (D) Ginger
4. Who initiated Van Mahotsava?
   (A) K.M. Munshi (B) Mahatma Gandhi
   (C) J.L. Nehru (D) A.K. Benerjee
5. Van Mahotsava is a festival of:
   (A) Wild life (B) Tree plantation
   (C) Timbers (D) All of these
6. Van Mahotsava was started in:
   (A) 1847 (B) 1850
   (C) 1947 (D) 1950
7. The practice of forestry in and around the farms/village land is called:
   (A) Social forestry (B) Community forestry
   (C) Farm forestry (D) All of these
8. The main criteria for good agroforestry design are:
   (A) Productivity (B) Sustainability
   (C) Adaptability (D) All of these
9. Based on the nature of components, Nair (1985) classified the agroforestry system into how many groups:
   (A) Four (B) Five
   (C) Six (D) Seven
10. Forest conservation act was enacted during:
    (A) 1880 (B) 1980
    (C) 1988 (D) 1998
11. The first independent National Forest Policy was enunciated during:
    (A) 1950 (B) 1952
    (C) 1954 (D) 1956
12. The first Forest Policy was enunciated during:
    (A) 1964 (B) 1874
    (C) 1884 (D) 1894
13. Declaration of national parks is covered in which section of wildlife protection act, 1972?
    (A) Section 25 (B) Section 35
    (C) Section 36 (D) Section 37
14. Conservation of ecological fragile eco-system and preservation of biological diversity in terms of flora and fauna was emphasized in:
    (A) Forest policy, 1894
    (B) Forest policy, 1952
    (C) Forest policy, 1988
    (D) None of these
15. Wildlife protection act was enacted during:
    (A) 1975 (B) 1980
    (C) 1972 (D) 1995
16. The project tiger was launched in the year:
    (A) 1972 (B) 1995
    (C) 1973 (D) 1980
17. India's first Forest Policy was enunciated during:
    (A) 1894 (B) 1927
    (C) 1952 (D) 1988
18. Wildlife protection act of 1972 was amended during the year:
    (A) 1988 (B) 1990
    (C) 1991 (D) 1980

19. In Wildlife Protection Act, 1972, Nilgiri Tahr comes under:
(A) Schedule I Part II
(B) Schedule I Part I
(C) Schedule II Part II
(D) Schedule III

20. World Forest Day is celebrated on:
(A) 15th June (B) 1st July
(C) 1st October (D) 21st March

21. The Environment (Protection) Act is attributed to the year:
(A) 1972 (B) 1980
(C) 1986 (D) 1988

22. The Indian Board of wildlife was constituted in the year:
(A) 1952 (B) 1972
(C) 1966 (D) 1994

23. The forests are brought on concurrent list by 42nd amendment to the constitution in the year:
(A) 1972 (B) 1976
(C) 1980 (D) 1988

24. Forest law is:
(A) A general law
(B) A special law
(C) An international law
(D) A constitutional law

25. Man And Biosphere (MAB) programmes was started during:
(A) 1971 (B) 1980
(C) 1982 (D) 1990

26. Sub chapter 6.2 of the Indian Forest Act, 1927 deals with the:
(A) Protected forests
(B) Penalities and procedures
(C) Duty on timber and other forest produce
(D) Reserved forests

27. National Forest Policy, 1988 envisages that --------- % of the total geographical area of India should be under forest:
(A) 22 (B) 33
(C) 40 (D) 40

28. Forest and wildlife were brought under concurrent list as per:
(A) 40th amendment
(B) 42nd amendment
(C) 44th amendment
(D) 48th amendment

29. Indian Grassland and Fodder Research Institute is located at:
(A) Jhansi (B) Kanpur
(C) Jaipur (D) Bhopal

30. National Research Centre for Agroforestry is located at:
(A) Jhansi (B) Kanpur
(C) Jaipur (D) Bhopal

31. ICRAF stands for:
(A) International Council for Research in Agro Forestry
(B) Indian Council for Research in Agro Forestry
(C) Indian Centre for Rainfall Assessment in Forest
(D) International Centre for Rainfall Assessment in Forest

32. Centre for International Forestry Research (CIFOR) is in:
(A) Nairobi, Kenya
(B) Vienna, Austria
(C) Bogor, Indonesia
(D) Ibadan, Nigeria

33. International Centre for Research on Agro Forestry is located at:
(A) Nairobi, Kenya
(B) Vienna, Austria
(C) Bogor, Indonesia
(D) Ibadan, Nigeria

34. Tropical Forest Research Institute is located at:
(A) Kanpur (B) Jabalpur
(C) Jhansi (D) Jodhpur

35. Forest Research Institute is located at:
(A) Jodhpur (B) Jhansi
(C) Dehradun (D) New Delhi

36. ICFRE is located at:
(A) New Delhi (B) Dehradun
(C) Jhansi (D) Ibanan

37. The literary meaning of agroforestry is:
(A) Agri + Forest
(B) Agri + Horti
(C) Agri + Fish
(D) Agri + Honey bee

38. The literary meaning of agri-horticulture is:
(A) Agri + Forest (B) Agri + Horti
(C) Agri + Fish (D) Agri + Honey bee

39. The literary meaning of agri-apiculture is:
(A) Agri + Forest (B) Agri + Horti
(C) Agri + Fish (D) Agri + Honey bee

40. The literary meaning of agri -pissiculture is:
(A) Crops + Forest (B) Crops + Horti
(C) Crops + Fish (D) Crops + Honey bee

41. The literary meaning of silvi-pastoral is:
(A) Forest + Pasture
(B) Forest + Horti
(C) Forest + fish (D) Forest + Honey bee

42. The literary meaning of silvi- horticulture is:
(A) Forest + Pasture
(B) Forest + Horti
(C) Forest + fish
(D) Forest + Honey bee

43. The literary meaning of silvi-pissiculture is:
(A) Forest + Pasture
(B) Forest + Horti
(C) Forest + fish
(D) Forest + Honey bee

44. The literary meaning of silvi-apiculture is:
(A) Forest + Pasture
(B) Forest + Horti
(C) Forest + fish
(D) Forest + Honey bee

45. The literary meaning of agri-silvi-pissiculture system is:
(A) Crops + Forest + Pasture
(B) Crops + Forest + Horti
(C) Crops + Forest + fish
(D) Crops + Forest + Honey bee

46. The literary meaning of agri-silvi-pastoral system is:
(A) Crops + Forest + Pasture
(B) Crops + Forest + Horti
(C) Crops + Forest + fish
(D) Crops + Forest + Honey bee

47. The literary meaning of agri-silvi-horticulture system is:
(A) Crops + Forest + Pasture
(B) Crops + Forest + Horti
(C) Crops + Forest + fish
(D) Crops + Forest + Honey bee

48. The literary meaning of agri-silvi-apiculture system is:
(A) Crops + Forest + Pasture
(B) Crops + Forest + Horti
(C) Crops + Forest + fish
(D) Crops + Forest + Honey bee

49. The literary meaning of alley cropping system is:
(A) Crops + Herb tree
(B) Trees around house + Vegetables
(C) Crops + Legume grasses + Grasses
(D) Crops + Forest + Honey bee

50. The literary meaning of home gardening system is:
(A) Crops + Herb tree
(B) Trees around house + Vegetables
(C) Crops + Legume grasses + Grasses
(D) Crops + Forest + Honey bee

51. The literary meaning of lay farming system is:
(A) Crops + Herb tree
(B) Trees around house + Vegetables
(C) Crops + Legume grasses + Grasses
(D) Crops + Forest + Honey bee

52. The literary meaning of homestead agroforestry system is:
(A) Crops + Herb tree
(B) Trees around house + Vegetables
(C) Crops + Legume grasses + Grasses
(D) Crops + Forest + Honey bee

53. The literary meaning of timfib system is:
(A) Crops + Herb tree
(B) Trees around house + Vegetables
(C) Crops + Legume grasses + Grasses
(D) Timber tree + fibre tree

54. The branch of forestry which deals with the determination of dimension, form and age and increment of single trees, stands or whole woods either standing or after feeling is called:
(A) Dendrology
(B) Forest management
(C) Dendrometry
(D) Forest protection

55. Management of forests for multiple uses involves which of the following uses:

(A) Timber (B) Recreation
(C) Water (D) All of these

56. The practical application of scientific, technical and economical principles of forestry is called:

(A) Forest management
(B) Forest mensuration
(C) Forest economics
(D) Forest legislation

57. Silviculture rotation is applicable to:

(A) Aesthetic
(B) Recreational purpose
(C) Both (A) and (B)
(D) None of these

58. Which type of rotation is useful for aesthetic and recreational purpose?

(A) Silvicultural rotation
(B) Physical rotation
(C) Technical rotation
(D) Rotation of maximum volume production

59. Which rotations controlling the financial returns?

(A) Silvicultural
(B) Physical rotation
(C) Both (A) and (B)
(D) None of these

60. Which type of rotation is most favourable for natural reproduction?

(A) Silvicultural rotation
(B) Physical rotation
(C) Technical rotation
(D) Financial rotation

61. Which social forestry practice is not included in the scope of social forestry?

(A) Farm forestry
(B) Extension forestry
(C) Recreational forestry
(D) Protection forestry

62. Social forestry include:

(A) Farm forestry
(B) Extension forestry
(C) Recreational forestry
(D) All of these

63. Which is false about type of rotation?

(A) Regular (B) Silvicultural
(C) Technical (D) Financial

64. The rotation in which a species retains satisfactory vigour of growth and reproduction on a given site is:

(A) Silvicultural rotation
(B) Financial rotation
(C) Technical rotation
(D) Rotation with maximum volume production

65. The rotation followed for sandal wood in south India is:

(A) Silvicultural rotation
(B) Financial rotation
(C) Technical rotation
(D) Physical rotation

66. Rotation of roadside and avenue trees is:

(A) Silvicultural rotation
(B) Financial rotation
(C) Physical rotation
(D) Rotation with maximum volume production

67. Tropical dry evergreen forests are found in:

(A) Carnatic coast
(B) Andaman and Nicobar island
(C) Coastal areas of Kerala
(D) North east India

68. Most suitable silvicultural system of deodar in hilly terrain for control of soil erosion is:

(A) Selection system
(B) Clear felling
(C) Coppice system (D) All of these

69. The silviculture of India trees was written by:

(A) L.S. Khanna (B) M.L. Chaudhary
(C) R.S. Singh (D) Benerjee

70. Mimocin is present in:

(A) Teak (B) Leucaena
(C) Neem (D) Acacia

71. Eucalyptus is a:

(A) Bad coppicer
(B) Moderately good coppicer
(C) Good coppicer
(D) Very bad coppicer

72. Non leguminous nitrogen fixing tree is:
(A) *Alunus nepalensis*
(B) *Leucaena leucocephala*
(C) *Acacia nilotica*
(D) *Calliandra callothyrsus*

73. A silviculture prescription:
(A) Changes in forest density
(B) Is a type of forest inventory
(C) Plans for pesticide application to control insects
(D) Usually contains cultural operations relating to objectives

74. A very complex community of associated trees, shrubs, other plants and animals is a/an:
(A) Arboretum (B) Forest
(C) Green house (D) Park

75. Removal of lower branches of a tree is called:
(A) Thinning (B) Cutting
(C) Prunning (D) Pinching

76. The prunned materials of sesbania in alley cropping can be utilized as:
(A) For feed
(B) For moisture conservation
(C) For organic matter
(D) All of these

77. Khejari tree in Rajasthan is used for:
(A) For feed to animals
(B) For vegetable
(C) Both (A) and (B)
(D) None of these

78. *Prosopis julifloa, Acacia nilotica, Casuarina equisetifolia* and *Tamerix articulate* can be successfully grown at pH of:
(A) Less than 7.0 (B) 8.2 – 9.0
(C) 9.1 – 10.0 (D) Morethan 10.0

79. *Terminalia arjuna, Albizzia lebbeck, Pongamia pinnata* and *Sesbania sesban* can be successfully grown at pH of:
(A) Lessthan 7.0 (B) 8.2 – 9.0
(C) 9.1 – 10.0 (D) Morethan 10.0

80. *Eucalyptus tereticornis, Parkinsonia auleata, Salvidora Persia* and *Capparis decidua* can be successfully grown at pH of:
(A) Lessthan 7.0 (B) 8.2 – 9.0
(C) 9.1 – 10.0 (D) Morethan 10.0

81. *Butea monosperma, Kijellea pinnata, Acacia auriculiformis* and *Moringa olifera* can be successfully grown at pH of:
(A) Less than 7.0 (B) 8.2 – 9.0
(C) 9.1 – 10.0 (D) Morethan 10.0

82. *Grevillia robusta, Azadiracta indica, Populus deltoids* and *Melia azadirach* can be successfully grown at pH of:
(A) Less than 7.0 (B) 8.2 – 9.0
(C) 9.1 – 10.0 (D) Morethan 10.0

83. *Acacia albida* is used for:
(A) Timber (B) Nitrogen fixation
(C) Fodder (D) All of these

84. *Albizia amara, A. lebbeck, A. procera* are used for:
(A) Shade (B) Nitrogen fixation
(C) Fodder (D) All of these

85. *Annona squamosa* and *Annona muricata* are used for:
(A) Fruits (B) Nitrogen fixation
(C) Fodder (D) All of these

86. *Areca catchu* is used for:
(A) Fuits
(B) Mulching
(C) Both (A) and (B)
(D) None of these

87. *Casuarina spp* is used for:
(A) Timber (B) Nitrogen fixation
(C) Energy (D) All of these

88. *Bombax insignis* is used for:
(A) Timber (B) Fodder
(C) Both (A) and (B)
(D) None of these

89. *Cassia siamea* is used for:
(A) Mulching (B) Nitrogen fixation
(C) Fodder (D) Soil conservation
(E) All of these

90. *Dalbergia sissoo* and *D. latifolia* are used for:
(A) Timber (B) Nitrogen fixation
(C) Fodder (D) All of these

91. *Gliricidia sepium* is a MPTs tree mainly used for:
(A) Timber (B) Nitrogen fixation
(C) Fodder (D) Mulching
(E) Fencing (F) All of these

92. *Genelina arborea* is used for:
(A) Timber (B) Soil amendment
(C) Fodder (D) All of these

93. *Moringa oleifera* is used for:
(A) Vegetables (B) Medicine
(C) Fodder (D) All of these

94. The product of form factor and total height of tree:
(A) Form - quotient
(B) Form – class
(C) Tree - height
(D) Form - height

95. International Institute of Tropical Forestry is situated at:
(A) Jodhpur (B) Dehradun
(C) Puerto Rica (D) Kenya

96. National Research Institute for Arid Horticulture is situated at:
(A) New Delhi (B) Jodhpur
(C) Jhansi (D) Dehradun

97. Central Arid Zone Research Institute is located at:
(A) Jodhpur (B) New Delhi
(C) Jhansi (D) Dehradun

98. National Institute on Subtropical Horticulture is located at:
(A) New Delhi (B) Bikaner
(C) Jhansi (D) Lucknow

99. *Psidium guajava* is used for:
(A) Fruits (B) Nitrogen fixation
(C) Fodder (D) All of these

100. *Sesbania grandiflora* is used for:
(A) Timber (B) Nitrogen fixation
(C) Vegetables (D) All of these

101. *Mallotus phillipnensis* is used for:
(A) Timber (B) Nitrogen fixation
(C) Fodder (D) All of these

102. *Grewia optiva* and *G. oppositifolia* are used for:
(A) Timber (B) Soil amendment
(C) Fodder (D) All of these

103. *Leucaena leucocephala* is used for:
(A) Mulching (B) Nitrogen fixation
(C) Fodder (D) Soil conservation
(E) All of these

104. "The Journal of Agroforestry Today" is published from:
(A) Kenya (B) Denmark
(C) Japan (D) Canada

105. "Rangeland and Agroforestry" published from:
(A) New Delhi (B) Jhansi
(C) Jodhpur (D) Dehradun

106. Alley cropping means:
(A) Growing of pastures in between two widely spaced rows of fast growing trees
(B) Growing of field crops in between two widely spaced rows of fast growing trees
(C) Growing of only short-duration crops in between two widely spaced rows of fast growing trees
(D) Growing of only fodder crops in between two widely spaced rows of fast growing trees

107. Alley cropping is also known as:
(A) Hedge row intercropping
(B) Avenue cropping
(C) Both (A) & (B) (D) Woody hedge

108. The type of forests in which Teak plants grow is:
(A) Alpine (B) Deciduous
(C) Evergreen (D) Scrub jungle

109. The process of cultivating wild species to suit human needs is called:
(A) Hybridization (B) Acclimatization
(C) Selection (D) Domestication

110. The practice of intensive exploitation of land capability by regular harvesting of such trees which produce numerous shoots from the stump after cutting is called:
(A) Coppice farming (B) Cover crop
(C) Crop competition (D) Crop canopy

111. A concept involving the farming of fast growing plants/trees for the purpose of providing biomass that can be used directly as fuel or converted in to other forms of fuel or energy products is called:
(A) Energy farming
(B) Organic farming
(C) Diversified farming
(D) Energy productivity ratio

112. A type of agro-forestry system in which trees are planted with food crops which serve as a nursery for the young tree seedlings is called:

(A) Taungya cultivation
(B) Shifting cultivation
(C) Both (A) and (B)
(D) None of these

113. The art of growing and reproducing a forest is called:

(A) Horticulture (B) Silviculture
(C) Alley cropping (D) Agroforestry

114. The land management system in which agriculture and forest crops are grown in combined form on same unit of the land is called:

(A) Conservation farming system
(B) Ley farming system
(C) Alley cropping system
(D) Agroforestry system

115. The scientific basis for the practice of silviculture is:

(A) Silvics
(B) Forest management
(C) Forest biology
(D) None of these

116. In India, common types of forest is:

(A) Tropical thorn forests
(B) Sal and Teak forests
(C) Tropical most deciduous forest
(D) Tropical dry deciduous forest

117. Rotation of arable crops with two or more years of sown pasture is called as:

(A) Mixed farming
(B) Mixed cropping
(C) Land use systems
(D) Lay farming

118. Integration of trees with grasses is known as:

(A) Aqua forestry
(B) Silvipasture
(C) Apiculture (D) Sericulture

119. The practice of managing rows of closely planted woody plants with annual crops planted in alleys in between hedges is called:

(A) Hedge row intercropping
(B) Woody hedgerows
(C) Conservation hedges
(D) Multipurpose tree gardens

120. The deliberated use of land for the production of crops including trees and fruit tree is called:

(A) Agri silvohorticulture
(B) Agripasture
(C) Agrisilvipasture
(D) Agrisilviculture

121. Integration of trees with pasture grasses and crops is called:

(A) Agri horticulture
(B) Agripasture
(C) Agrisilvipasture
(D) Agrisilviculture

122. Integration of horticultural tree with different types of crops is called:

(A) Agri horticulture
(B) Agripasture
(C) Agrisilvipasture
(D) Agrisilviculture

123. Integration of pasture grasses with crops is called:

(A) Agri horticulture
(B) Agripasture
(C) Agrisilvipasture
(D) Agrisilviculture

124. The practice of forestry for the simultaneous use of a forest area for two or more purpose is called:

(A) Multiple use forestry
(B) Mixed use forestry
(C) Multipurpose trees and shrubs
(D) Multispecies tree garden

125. The practice of forestry with the object of obtaining maximum gain per unit area is called:

(A) Mixed forestry
(B) Multiple use forestry
(C) Intensive forestry
(D) All of these

126. Home gardens found in tropics are one of the classic example of:

(A) Agrisilvicultural systems
(B) Silvopastoral systems
(C) Agrosilvopastoral systems
(D) Apiculture with trees

127. Hedge row intercropping is:
(A) Agrisilvicultural systems
(B) Silvopastoral systems
(C) Agrosilvopastoral systems
(D) None of these

128. Most common agroforestry system practiced in arid regions are:
(A) Agrisilvicultural systems
(B) Silvopastoral systems
(C) Agrisilvopastoral systems
(D) Aquaforestry

129. Which forestry practice is not included in the scope of social forestry:
(A) Farm forestry
(B) Extension forestry
(C) Recreational forestry
(D) Protection forestry

130. Which social forestry programme is purely managed by private efforts:
(A) Farm forestry
(B) Community forestry
(C) Urban forestry
(D) Silvopastoral

131. Shifting cultivation is most prevalent in:
(A) North-east India
(B) Eastern part of India
(C) Northern part of India
(D) Western part of India

132. In north-eastern hill region, shifting cultivation is called:
(A) Jhum (B) Podu
(C) Kumari (D) Slash and burn

133. Jhuming is the other name for:
(A) Shifting cultivation
(B) Taungya
(C) Both (A) & (B) (D) None of these

134. Cutting and burning of forests or woodlands to create field for agriculture or pasture for livestock, or for a variety of other purposes is called:
(A) Industrial agriculture
(B) Slash and Burn agriculture
(C) Bio-dynamic agriculture
(D) Community supported agriculture

135. In Kerala the most important agroforestry systems is:
(A) Alley cropping
(B) Home garden
(C) Lay farming
(D) Shifting cultivation

136. Kangeyam region in Tamilnadu, the famous silvipastoral system is:
(A) *Cenchrus ciliaris* & *Acacia nilotica*
(B) *Cenchrus ciliaris* & *Acacia leucophlea*
(C) *Eulaliopsis binata* & *Acacia nilotica*
(D) *Eulaliopsis binata* & *Acacia leucophlea*

137. Which among tree species are best suited for alley cropping?
(A) *Eucalyptus tereticornis*
(B) *Casuarina equisetifolia*
(C) *Cassia siamea*
(D) *Ailanthus excels*

138. Which among these factors has degraded the Indian grasslands the most?
(A) Adverse site factors
(B) Poor management
(C) Poor productivity
(D) Over grazing

139. Extensive fallow systems are:
(A) Shifting cultivation system
(B) Bush fallow or grass fallow systems
(C) Fallow system
(D) None of these

140. Intensive fallow systems are:
(A) Shifting cultivation system
(B) Bush fallow or grass fallow systems
(C) Fallow system
(D) None of these

141. Stingless honey bee is:
(A) *Melipona spp.* (B) *Apis dorsata*
(C) *Apis florae* (D) None of these

142. Sting is found only in:
(A) Queen (B) Drone
(C) Worker (D) All of these

143. Queen completes its life cycle in:
(A) 15-16 days (B) 18-20 days
(C) 23-24 days (D) 25-26 days

144. Drone completes its life cycle in:
(A) 15-16 days (B) 18-20 days
(C) 23-24 days (D) 25-26 days

145. Workers completes its life cycle in:
(A) 15-16 days (B) 18-20 days
(C) 23-24 days (D) 25-26 days

146. One honey bee hive contains nearly total honey bees:
(A) 100-500 (B) 1000-5000
(C) 30000-50000 (D) More than 50000

147. One honey bee hive has drones:
(A) 1 - 5 (B) 10 - 20
(C) 200 - 300 (D) More than 300

148. One honey bee hive contains nearly queen honey bees:
(A) 1 (B) 2
(C) 3 (D) 4

149. Which of the following honey bee is found in India?
(A) *Apis dorsata* (B) *Apis florae*
(C) *Apis indica* (D) All of these

150. Which of the following honey bee smallest?
(A) *Apis dorsata* (B) *Apis florae*
(C) *Apis indica* (D) *Melipona spp.*

151. Which of the following honey bee produce the sweetest honey?
(A) *Apis dorsata* (B) *Apis florae*
(C) *Apis indica* (D) *Melipona spp.*

152. Which of the following honey bee is largest one?
(A) *Apis dorsata* (B) *Apis florae*
(C) *Apis indica* (D) *Melipona spp.*

153. Air conditioning system is found only in:
(A) Honey bee (B) Silkworm
(C) Butterfly (D) None of these

154. Which of the following is most common silk insects in India?
(A) *Bombyx mori*
(B) *Antheraca puphia*
(C) *Antheraea ossima*
(D) *Phlocania ricini*

155. Which of the following is a moonga silk insects?
(A) *Bombyx mori*
(B) *Antheraca puphia*
(C) *Antheraea ossima*
(D) *Phlocania ricini*

156. Silk worm completes its life cycle in:
(A) 20-30 days (B) 40-50 days
(C) 50-60 days (D) 60-0 days

157. Silk production and its related aspects are studied under:
(A) Sericulture (B) Apiculture
(C) Floricultre (D) Pisciculture

158. Stifiling is a process in which:
(A) Cocoon are boiled to kill the insect
(B) Cocoons are stored at 4^0C for 15 days
(C) Adults are allowed to remove from cocoons
(D) None of these

159. Lac is a:
(A) Resinous matter
(B) Solid matter
(C) Liquid
(D) Gaseous matter

160. Lac is mainly used in:
(A) Gramophone industry
(B) Leather industry
(C) Printing industry
(D) Paper industry

161. Which of the following state is the largest producer of lac in India?
(A) Bihar (B) UP
(C) MP (D) AP

162. Lac is generally produced by:
(A) Female Lac insect
(B) Male Lac insect
(C) Both male and female Lac insects
(D) None of these

163. Which of the following is a raw lac?
(A) Eric lac (B) Stick lac
(C) Seed lac (D) Shell ac

164. Lac insects may be reared on following trees?
(A) *Butea monosperme*
(B) *Zizyphus spp.*
(C) *Schleichera spp* (D) All of these

165. Scientific name of lac insects is:
(A) *Kerria lacca* (B) *Apis dorsata*
(C) *Bombyx mori* (D) *Phlocania ricini*

166. Which of the following is the predatory insect of lac insect?
(A) *Crysopa spp.*
(B) *Eublema amabilis*
(C) *Holcocera pulvarea*
(D) All of these

167. European honey bee species is:
(A) *Apis mellifera* (B) *Apis dorsata*
(C) *Melipona spp* (D) *Apis florae*

168. Honey robber is also known as:
(A) *Acherontia styx*(B) Wax moth
(C) *Pebrine* (D) None of these

169. Honey contains sugar:
(A) 55% (B) 78%
(C) 91% (D) 99%

170. Which of the following is a natural enemies of honey bees?
(A) Wax moth (B) *Acherontia styx*
(C) *Vespa cincta* (D) All of these

171. Which is the origin place of silk?
(A) China (B) India
(C) Egypt (D) Malaysia

172. Mustard aphid mainly attacks on:
(A) Stem (B) Leaf
(C) Flower (D) All of these

173. Which of the following silk race is found in our country?
(A) Univoltine race
(B) Multi-voltine race
(C) Bi-voltine race
(D) None of these

174. The species of silk worm that complete only one life cycle in a year, is called:
(A) Univoltine race
(B) Multi-voltine race
(C) Bi-voltine race
(D) None of these

175. Central Silk Research Board is situated at:
(A) Bahrampur (WB)
(B) Cuttack (Orissa)
(C) Dehradune (Uttarakhand)
(D) Lucknow (UP)

176. Which of the country is the highest producer of lac?
(A) India (B) China
(C) Nepal (D) Bangladesh

177. How many times the female lac insect is larger than male lac insects?
(A) Two times (B) Three times
(C) Four times (D) Five times

178. Which of the following organ is found in female lac insects?
(A) Eyes (B) Legs
(C) Wings (D) Antennae

179. Adult lac male insect survive only for:
(A) 1-2 days (B) 3-4 days
(C) 5-7 days (D) More than 7 days

180. Which of the instar does not have eyes, legs and antennae?
(A) Ist (B) 2nd
(C) 3rd (D) 4rth

181. The colour of the lac insect is:
(A) Pink and Red (B) Green – red
(C) Blue – red (D) Black

182. Brood lac is a:
(A) The branch of a plant that contains lac cells
(B) Caterpiller of lac insect
(C) Pupa of lac insect
(D) None of these

183. The scientific management of forests for the continued output of products is called:
(A) Dendrology (B) Forestry
(C) Ecology
(D) Forest mensuration

184. Hoe, das and digging sticks are used in which of the following farming types?
(A) Subsistence farming
(B) Intensive subsistence farming
(C) Shifting cultivation
(D) Commercial farming

185. Practice of raising forest or fruit trees on the society land to reduce atmospheric pollution is known as:
(A) Farm forestry
(B) Silvipasture
(C) Agroforestry
(D) Social forestry

186. Lay farming refers:
(A) Growing of field crops in rotation with grasses and fodder legumes
(B) Growing trees in rotation with grasses and fodder legumes
(C) Growing of pastures in rotation with grasses and fodder legumes
(D) Growing of grasses and fodder legumes in forest areas

187. A rotation of arable crops requiring annual cultivation and artificial pastures occupying field for two years or longer is called:
(A) Lay farming
(B) Alley cropping
(C) Silvi pastoral system
(D) None of these

188. The practice of cultivating clearings scattered in the reservoir of natural vegetation and of abandoning them as soon as the soil is exhausted; and this includes the practice of shifting homesteads in order to fallow the cultivator's search for new fertile land is called:
(A) Row intercropping
(B) Shifting cultivation
(C) Strip cropping
(D) None of these

189. A type of agro-forestry system in which trees are planted with food crops which serve as a nursery for the young tree seedlings is called:
(A) Taungya cultivation
(B) Shifting cultivation
(C) Both (A) and (B)
(D) None of these

190. The practice of forestry with the object of raising flowering trees is called:
(A) Recreation forestry
(B) Urban forestry
(C) Social forestry (D) Farm forestry

191. Plantations grow as linear strips of lands on the sides of public roads, canals and railway lines are called:
(A) Linear strip plantations
(B) Commercial forestry
(C) Avenue plantations
(D) Urban forestry

192. In which agroforestry system, agricultural crops is of secondary importance?
(A) Silvipasture (B) Agrisilviculture
(C) Silviagriculture (D) Hortisilviculture

193. Tree planted in silvi-agriculture should be oriented in the:
(A) East-west direction
(B) North-south direction
(C) East-south direction
(D) West-south direction

194. Silvipasture means:
(A) Growing trees with grass
(B) Growing trees with annual crops
(C) Growing trees with fruit crops
(D) Growing trees with crops and horticultural crops

195. Agri-silviculture means:
(A) Growing trees with grass
(B) Growing trees with annual crops
(C) Growing trees with fruit crops
(D) Growing trees with crops and horticultural crops

196. Silvi-horticulture means:
(A) Growing trees with grass
(B) Growing trees with annual crops
(C) Growing trees with fruit crops
(D) Growing trees with crops and horticultural crops

197. Agri-silvi-horticulture means:
(A) Growing trees with grass
(B) Growing trees with annual crops
(C) Growing trees with fruit crops
(D) Growing trees with crops and horticultural crops

198. Agri -silvipasture means:
(A) Growing trees with grass and crops
(B) Growing trees with annual crops
(C) Growing trees with fruit crops
(D) Growing trees with crops and horticultural crops

199. The most prevalent agroforestry systems in high rainfall areas:
(A) Silvipasture
(B) Silviagriculture
(C) Horti-silvi-pasture
(D) Home gardens

200. The most prevalent agroforestry systems in low rainfall (arid) regions is:
(A) Agrisilvicultural system
(B) Silvopastoral system
(C) Agrosilvopastoral system
(D) All of these

201. In which agroforestry system, agricultural crops is of secondary importance?
(A) Silvipasture
(B) Agrisilviculture
(C) Silviagriculture
(D) Hortisilviculture

202. Identification of trees according to species is called:
(A) Forestry (B) Silviculture
(C) Ecology (D) Dendrology

203. National Wasteland Development Board came into existence during the period:
(A) 1970 (B) 1975
(C) 1980 (D) 1985

204. The India, plantation forestry was started in:
(A) 1842 (B) 1944
(C) 1947 (D) 1950

205. ICAR had launch All India Co-ordinated Research Project on Agroforestry in the country in:
(A) 1974 (B) 1979
(C) 1981 (D) 1983

206. First Indian Forest Act was drafted in the year:
(A) 1865 (B) 1848
(C) 1894 (D) 1927

207. Indian lilac is the common name for:
(A) *Azadiracta indica*
(B) *Tectonia grandis*
(C) *Casuarina equisetifolia*
(D) *Acacia nilotica*

208. Deforestation brings about:
(A) Soil erosion
(B) Weed control
(C) Decreases drought
(D) Increased sunlight

209. Deforestation is caused due to:
(A) Silviculture
(B) Construction of roads
(C) Rainfall
(D) Plantation of trees

210. Deforestation is the major causal agent of:
(A) Depletion of natural resources
(B) Environmental pollution
(C) Desertification of habitat
(D) Genetic erosion

211. Deforestation will decrease:
(A) Soil erosion
(B) Land slides
(C) Soil fertility
(D) Rainfall

212. When the use of land is directed towards satisfying basic needs and is managed by the owner is said to be:
(A) Commercial Agroforestry systems
(B) Intermediate Agroforestry systems
(C) Subsistence Agroforestry systems
(D) All of these

213. The practice of forestry in areas devoid of tree growth and other vegetation outsides the conventional forest areas is:
(A) Social forestry (B) Extension forestry
(C) Farm forestry (D) Afforestation

214. Homegarden agroforestry system is practices in:
(A) Humid tropical climates
(B) Semi arid lands
(C) High lands
(D) Sub humid low lands

215. A forest which has been reproduced through sprouting is called:
(A) Epicormic (B) Stocked
(C) Coppice (D) All of these

216. The forest products can be broadly classified into:
(A) Major Timber Products
(B) Non Timber Forest Products
(C) Both (A) and (B)
(D) Secondary Timber Products

217. Food, fuel, fodder and fibre obtained from the forests is its:
(A) Protective role
(B) Productive role
(C) Ameliorative role
(D) Developmental role

218. The closeness of the tree tops is called:
(A) Closed canopy (B) Open canopy
(C) Crown density (D) Carrying capacity

219. A method of harvesting that removes all of the trees in the stand is called:
(A) Clearing (B) Clear cutting
(C) Total cutting (D) Total removal

220. A cutting which removes trees of poor quality, undesirable species or those in poor condition is called:
(A) Improvement cutting
(B) Upgrading (C) Thinning
(D) Defect clearing

221. Darbergia sisso is:
(A) Frost hardy
(B) Moderately frost hardy
(C) Frost tender
(D) None of these

222. Shorea robusta is:
(A) Susceptible to drought
(B) Drought hardy
(C) Moderately drought hardy
(D) None of these

223. Deciduous trees are:
(A) *Melia azedarach*
(B) *Acacia catechu*
(C) *Ailanthus excelsa*
(D) All of these

224. Evergreen trees are:
(A) *Abies pindrow*
(B) *Michelia chamoaca*
(C) *Hopea parviflora*
(D) All of these

225. Light demanders trees for moist temperate forest are:
(A) *Pinus wallichiana*
(B) *Populus ciliate*
(C) Both (A) and (B)
(D) None of these

226. Shade demander trees for tropical moist deciduous forest are:
(A) *Syzygium cumini*
(B) *Schleichera oleosa*
(C) *Mallotus philinensis*
(D) All of these

227. Shade bearer trees for tropical wet evergreen forest are:
(A) *Artocarpus spp.*
(B) *Dipterocarpus* spp.
(C) *Calophyllum*
(D) *Mesua ferrea*

228. Frost tender tree species are:
(A) *Azadiracta indica*
(B) *Tectona grandis*
(C) *Terminalia arjuna*
(D) All of these

229. Frost hardy tree species are:
(A) *Acacia catechu*
(B) *Anogeissus pendula*
(C) *Toona ciliate*
(D) All of these

230. Drought hardy tree species are:
(A) *Acacia arabica*
(B) *Diospyros melanoxylon*
(C) *Syzygium cumini*
(D) All of these

231. The commercial bidi leaves are plucked from:
(A) *Diospyros melanoxylon*
(B) *Shorea robusta*
(C) *Pterosperma* spp
(D) *Gmelina arborea*

232. Indigenous populous tree species is/are:
(A) *Populus ciliate*
(B) *Populus robusta*
(C) *Populus deltoides*
(D) *Populus rubrapoiret*

233. Agar wood tree is/are:
(A) *Aquilaria malaccensis*
(B) *Teactona grandis*
(C) *Dalbergia sissoo*
(D) *Acacia catechu*

234. The silvicultural stem in which regeneration fellings are distributed over the whole felling series and the crop is always irregular is known as:
(A) Systems of diffused regeneration
(B) Accessory system
(C) System of concentrated regeneration
(D) Coppice system

235. Most suitable silvicultural system of deodar in hilly terrain for control of soil erosion is:
(A) Selection system
(B) Clear felling
(C) Coppice system
(D) All of these

236. Extensive planting of trees to increase forest cover is called:
(A) Deforestation (B) Agroforestry
(C) Social forestry (D) Afforestation

237. Forests control drought by:
(A) Preventing soil erosion
(B) Increasing oxygen
(C) Increasing humidity and rainfall
(D) Preventing floods

238. The silvicultural system in which regeneration fellings instead of being done uniformly all over the compartment are carried out in scattered group is defined as:

(A) Uniform systems
(B) Group systems
(D) Selection systems
(D) Shelter wood strip systems

239. Silvicultural systems in which clear felling are done in the form of strips which progress successively in one direction is known as:

(A) Alternate strip system
(B) Shelterwood strip system
(C) Clear strip system
(D) Clear-felling with standards

240. Which of the silvicultural system is used for the regeneration and management of high forests:

(A) Clear felling (B) Shelter wood
(C) Selection felling
(D) All of these

241. The silvicultural system followed for black wattle (*Acacia mearnsii*) in the Nilgiris is:

(A) Clear felling followed by natural regeneration
(B) Clear felling followed by artificial regeneration
(C) Uniform system
(D) Simple coppice system

242. Silvicultural systems prevalent during early part of 20th century was:

(A) Shelter wood system
(B) Clear felling systems
(C) Selection systems
(D) Coppice systems

243. Shading effect of trees can be overcome by:

(A) Canopy management
(B) Root management
(C) Water
(D) Fertilizer

244. Multistory arrangement reduces:

(A) Productivity (B) Nutrition
(C) Competition (D) All of these

245. The oldest known agroforestry systems is:

(A) Alley cropping (B) Taungya systems
(C) Home gardens (D) Shifting cultivation

246. Social forestry programme was launched in the:

(A) Fifth five year plan
(B) Sixth five year plan
(C) Seventh five year plan
(D) Eight five year plan

247. Carrying capacity is more related to:

(A) Agrisilviculture
(B) Improvement of grasslands
(C) Improvement of nutritive value of grasslands
(D) None of these

248. Social forestry was first coined by:

(A) Singh & Sharma
(B) Nair
(C) Brandis (D) Westoby

249. Social forestry was first adopted successfully in:

(A) Gujarat (B) Karnataka
(C) Uttar Pradesh (D) Rajasthan

250. Neem and mango are preferred for:

(A) Railway line plantation
(B) Canal bank plantation
(C) Tank bed plantation
(D) Avenue plantation

251. TCIP stands:

(A) Tree Cultivation Inducement Programme
(B) Tree Cultivation Intensive Programme
(C) Tree Cultivation Incentive Programme
(D) Tree Cultivation Income Programme

252. Component of social forestry is:

(A) Farm forestry
(B) Extension forestry
(C) Recreational forestry
(D) All of these

253. *Prosopis cineraria* and *Cenchrus ciliaris* is a good fodder tree with grass combination for:

(A) Semi-arid climate with red soil
(B) Temperate climate
(C) Arid climate with sandy soil
(D) Semi-arid black cotton soil

254. Tree competes with the neighbouring crops mainly for:

(A) Light (B) Moisture
(C) Nutrients (D) All of these

255. The fundamental attributes of all agroforestry systems are:
(A) Recover depleted soil nutrients
(B) Increase crop yields
(C) Production and protection
(D) Stabilize the soil structure

256. Planting of nectar producing trees on the boundary of the agricultural field is called:
(A) Aquaforestry
(B) Apiculture with sericulture
(C) Multipurpose wood lots
(D) Woody hedgerows

257. Agroforestry D&D covers entire ecological zone within a country is:
(A) Macro D&D (B) Micro D&D
(C) Both (A) & (B) (D) None of these

258. Agroforestry D&D covers within the chosen ecological zone within a country is:
(A) Macro D&D (B) Micro D&D
(C) Both (A) & (B) (D) None of these

259. Which is false about shelter belt:
(A) Single tree species
(B) Multiple tree species
(C) Used to arrest wind
(D) Prevent sand dune formation

260. The ratio of height and width in shelter belt is:
(A) 1:10 (B) 1:15
(C) 1:20 (D) 1:25

261. The cross sectional shape of any shelter belt is:
(A) Parallel (B) Neloid
(C) Cylinder (D) Pyramid

262. Taungya was first introduced by:
(A) Nair (B) Brandis
(C) Khanna (D) Clements

263. Taungya system was originated in:
(A) North Bengal (B) West Bengal
(C) Myanmar (D) China

264. Taungya is a word derived from:
(A) Nepali (B) Chinese
(C) Burmese (D) Bengali

265. In southern India, the taungya system is also called:
(A) Jhum (B) Kumari
(C) Podu (D) None of these

266. First taungya plantation was established in:
(A) West Bengal (B) North Bengal
(C) Rajasthan (D) Uttar Pradesh

267. The main aim of raising crops in departmental taungya is to:
(A) Reduce the weed growth
(B) To provide additional income
(C) Maximum utilization of site
(D) All of these

268. The queen of timber is:
(A) Shisham (B) Vagai
(C) Teak (D) Eucalyptus

269. Loyal timber of the poor is:
(A) Acacia (B) Prosopis
(C) Bamboo (D) Teak

270. Shisham is also called:
(A) Rose wood (B) Soft wood
(C) Hard wood (D) None of these

271. Energy plantation is established for:
(A) Pulp wood production
(B) Fuel wood production
(C) Soil conservation
(D) Wind break

272. Based on temporal arrangement of agro-forestry system, 'home garden' is the example of:
(A) Coincident (B) Concurrent
(C) Interpoleted (D) Alley cropping

273. Home gardening is highly suitable for:
(A) Humid/sub humid region
(B) Arid/sub arid region
(C) Tropical region
(D) Temperate region

274. The home garden is characterized by:
(A) 2-3 vertical strata
(B) 3-4 vertical strata
(C) 3-5 vertical strata
(D) 5-6 vertical strata

275. In the manufacture of Braile paper which tree species is used:
(A) *Eucalyptus globules*
(B) *Eucalyptus tereticornis*
(C) *Eucalyptus camandulensis*
(D) *Eucalyptus urophylla*

276. Indigofera tinctoria is used for:
(A) Rasin industry
(B) Dye industry
(C) Tannin industry
(D) Gum industry

277. The major tree species are most suitable for planting on tank bed is:
(A) *Acacia leucophloea*
(B) *Prosopis juliflora*
(C) *Acacia nilotica*
(D) *Eucalyptus* spp.

278. Heliotropic type of branching in trees should be preferred for:
(A) Roadside planting
(B) Canal planting
(C) Railway planting
(D) Wasteland planting

279. Which type of trees are suitable for noise abatement?
(A) Deciduous trees
(B) Evergreen trees
(C) Semi evergreen trees
(D) All of these

280. Cutting of large trees which throw shade on the young trees is called:
(A) Thinning (B) Selection cutting
(C) Shade tree cutting
(D) Release cutting

281. Removing trees damaged by fire, insects or disease is called:
(A) Salvage cut (B) Clean cutting
(C) Clear cut (D) Sanitation cut

282. Most suitable tree species for reclamation of saline-sodic soils is:
(A) *Prosopis juliflora*
(B) *Albizia lebbeck*
(C) *Azadiracta indica*
(D) *Casuarina equisetifolia*

283. Most suitable tree species for reclamation of sandy soils is:
(A) *Acacia species*
(B) *Casuarina equisetifolia*
(C) *Melia dubia*
(D) *Azadiracta indica*

284. "Block wood" is obtained from:
(A) Acacia (B) Pongamia
(C) Dalbergia (D) Albizia

285. As the ability to regenerate by shoot or root suckers, or a forest so established is called:
(A) Coppice farming
(B) Cover crop
(C) Crop competition
(D) Crop canopy

286. Generally shelterbelt assumes the shape of:
(A) Quadrangle (B) Rectangle
(C) Triangle (D) trapezoidal

287. Characteristics of multipurpose tree suitable for agroforestry is:
(A) High adaptability
(B) Nitrogen fixing
(C) Coppicing ability
(D) All of these

288. Basic set of elements or components of agroforestry are:
(A) Tree (B) Herbs
(C) Animals (D) All of these

289. The National Afforestation programme is being implemented through a which tier structure:
(A) 2 (B) 3
(C) 4 (D) 5

290. Wind breaks are:
(A) Strips of trees
(B) Line of trees
(C) Both (A) and (B)
(D) None of these

291. The average size of tree patta land could be:
(A) 5.0 (B) 4.0
(C) 3.0 (D) 2.0

292. Which is false about classification of agro-forestry based on arrangement of components in respect of time:
(A) Coincident (B) Mixed dense
(C) Concomitant (D) Interploted

293. The tree species having xerophytic chara-cteristics are suitable for:
(A) Humid Bengal Assam Basin
(B) Arid zones
(C) Sub humid areas
(D) Humid to semi arid areas

294. In North India agroforestry the combination of which crop is common:

(A) Rice with poplar
(B) Rice with eucalyptus
(C) Rice with melia
(D) Rice with shisham

295. The combination of wheat with eucalyptus was common in:

(A) Uttar Pradesh (B) Punjab
(C) Maharashtra (D) Bihar

296. The important tree species for shape in tea estates is:

(A) *Grevillea robusta*
(B) *Eucalyptus* spp.
(C) *Melia dubia*
(D) All of these

297. Which tree species is preferred for water logged areas:

(A) Coconut (B) Eucalyptus
(C) Neem (D) Casuarina spp.

298. Agroforestry practice of growing *Albizia lebbek* with sorghum is called:

(A) Xeromorphic system
(B) Hydromorphic system
(C) Mesophytic system
(D) None of these

299. Which is not correct about an advantages of agroforestry system?

(A) Maximum production
(B) Improve soil fertility
(C) Maximize adverse impact on climatic factors
(D) Aid industrial growth

300. In order to get maximum net return from agroforestry system, we have to:

(A) Choose suitable tree species
(B) Choose high yielding agricultural crops
(C) Plant clonally propagated tree seedlings
(D) Adopt suitable tree crop combination

301. Which is false about shelterbelt:

(A) Single species
(B) Multiple tree species
(C) Used to arrest wind
(D) Prevalent sand dune formation

302. Which is false about planting of forest trees in agroforestry system:

(A) Reduces temperature
(B) Reduces evaporation
(C) Increase irrigation requirement
(D) Increase humidity

303. The important practices in silvicultural operation in agroforestry is:

(A) Pruning (B) Thinning
(C) Weeding (D) Climber cutting

304. Protein bank is a sub system of:

(A) Agri -silviculture
(B) Agri - pastoral
(C) Agri -horticulture
(D) Agri-silvi-pastoral

305. Which tree species is called green gold is?

(A) Sal (B) Sandal
(C) Bamboo (D) Teak

306. The system in which woody component is combined with agricultural crop is known as:

(A) Agri -silviculture
(B) Agri - pastoral
(C) Agri -horticulture
(D) Agri-silvi-pastoral

307. Essential attributes of agroforestry systems are:

(A) Adaptability (B) Productivity
(C) Sustainability (D) All of these

308. Shelterbelt should posses a typical:

(A) Triangular shape
(B) Linear belt
(C) Square shape
(D) Rectangular shape

309. A common example of the zonal pattern is:

(A) Hedgerow intercropping
(B) Home gardens
(C) Silvopastoral systems
(D) Agrosilvopastoral system

310. Commercial agroforestry system is used:

(A) Production of perennial cash crops
(B) Satisfying basic needs
(C) Sale of the production of output
(D) All of these

311. Shading effect of trees can be overcome by:

(A) Canopy management
(B) Root management
(C) Water management
(D) Nutrient management

312. Which classification of agroforestry is more important and commonly adopted?
(A) Socioeconomic (B) Structural
(C) Floristic (D) Ecological

313. The broad and well known classification of agroforestry systems was given by:
(A) Chundawat (B) Gautam
(C) Nair (D) Sharma

314. The term social forestry first time used by:
(A) Shah (B) Brandis
(C) Nair (D) Jack Westoby

315. Social forestry was first recommended by:
(A) National Commission on Agriculture
(B) Van Mahotsava
(C) Ninth Commonwealth Forestry Congress
(D) All of these

316. The first state in India who created a separate social forestry wing in the forest department was:
(A) Gujarat (B) Maharashtra
(C) Tamilnadu (D) Kerala

317. Social Forestry Project Phase – I was launched in:
(A) 1981 (B) 1985
(C) 1988 (D) 1992

318. Social Forestry Project Phase – II was launched in:
(A) 1981 (B) 1985
(C) 1988 (D) 1992

319. Major thrust to social forestry concept was given by:
(A) NCA report
(B) Forest Conservation Report
(C) National Forest Policy
(D) Indian Forest Act

320. Which is not a factor in which the agroforestry systems are classified?
(A) Floristic
(B) Function
(C) Agro climatic zones
(D) None of these

321. The Diagnosis and Design (D&D) survey in agroforestry was given by:
(A) P.K.R. Nair (B) Dwivedi
(C) Sharma (D) Singh

322. Forest products contribution to world gross domestic product (GDP) is:
(A) 1.0 % (B) 1.7 %
(C) 2.0 % (D) 2.7 %

323. When an immediate supply of a single necessary factor falls below the combined demand of the plant is called:
(A) Allelopathy
(B) Allelospoly
(C) Both (A) & (B)
(D) None of these

324. The direct or indirect of one plant upon another through the production of chemical inhibitors is called:
(A) Allelopathy
(B) Allelospoly
(C) Both (A) and (B)
(D) None of these

325. Allelopathy denotes:
(A) Positive interaction
(B) Negative interaction
(C) Both (A) and (B)
(D) None of these

326. Allelopathy is the best example for:
(A) Association (B) Ammensalism
(C) Symbiosis (D) Photo cooperation

327. Allelochemicals produced by plants are:
(A) Primary plant metabolites
(B) Secondary plant metabolites
(C) Both (A) and (B)
(D) None of these

328. Allelochemicals are released by means of:
(A) Volatilization (B) Leaching
(C) Exudation (D) All of these

329. Examples of allelochemicals:
(A) Gallic acid (B) Vanilic acid
(C) Syringic acid (D) All of these

330. Strips of trees planted to protect fields, homes, canal from wind is referred to:
(A) Shelter belts (B) Wind breaks
(C) Alley cropping (D) All of these

331. Characteristics of multipurpose tree (MPTs) suitable for agroforestry is:
(A) High adaptability
(B) Nitrogen fixing
(C) Coppicing ability
(D) All of these

332. Suitable tree species for MPTs are:
(A) Casuarina (B) Acacia
(C) Gliricidia (D) All of these

333. Suitable tree species for nitrogen fixing tree (NFTs) is/are:
(A) Teak (B) Leucaena
(C) Neem (D) All of these

334. Suitable herbs for live fencing is:
(A) Euphorbia (B) Gliricidia
(C) Acacia (D) Sesbania

335. Nitrogen fixing ability of Casuarina is attributed to:
(A) Azotobacter (B) Frankia
(C) Rhizobium (D) Azospirilum

336. Benefit sharing concept is a core objective in
(A) Joint Forest Management
(B) Sustainable Forest Management
(C) Community Forestry
(D) Social Forestry

337. Joint Forest Management concept was first implemented at:
(A) Ahmedabad (B) Arabari
(C) Mumbai (D) Pune

338. In Joint Forest Management programme, the unit of management is:
(A) Village (B) Block
(C) Taluka (D) District

339. In Joint Forest Management programme, the objectives are to be decided at:
(A) Division level (B) District level
(C) Taluka level (D) Village level

340. Protection and reforestation of degraded forests with the involvement of local people is:
(A) Joint Forest Management
(B) Sustainable Forest Management
(C) Community Forestry
(D) Social Forestry

341. Silviculture systems followed for Bamboo is:
(A) Clear felling systems
(B) Shelter wood systems
(C) Group selection systems
(D) Culm selection system

342. Silviculture system for road side plantation is:
(A) Simple coppice systems
(B) Shelter wood systems
(C) Clear strip systems
(D) None of these

343. Chipko movement is an example of forest conservation through:
(A) Tehri-Garhwal district
(B) Uttarakhand
(C) Public awareness and participation
(D) Political issue

344. Chipko movement is related to:
(A) Forest conservation
(B) Soil conservation
(C) Water conservation
(D) Wetland conservation

345. Chipko movement was launched for protection of:
(A) Forests (B) Grasslands
(C) Wetlands (D) Live stocks

346. Conservation of species in its natural habitat is:
(A) In-situ (B) Ex-situ
(C) In-vitro (D) Both (B) and (C)

347. Development of botanical garden is what type of conservation:
(A) Ex-situ (B) A common
(C) An easy (D) In-situ

348. An agroforestry system suitable for small orchards prone to wind damage is:
(A) Taungya (B) Home garden
(C) Windbreak (D) Shifting cultivation

349. A well designed windbreak can offer protection upto how many times of its height towards leeward side:
(A) 10 (B) 20
(C) 30 (D) 40

350. Swidden is a synonym for one of the following agroforestry sytem. Identify the correct one:
(A) Taungya (B) Home garden
(C) Perkland system
(D) Shifting cultivation

ANSWER SHEET
13. Agroforestry

1. (D) 2. (A) 3. (A) 4. (A) 5. (B) 6. (D) 7. (D) 8. (D) 9. (A) 10. (B)
11. (B) 12. (D) 13. (B) 14. (C) 15. (C) 16. (C) 17. (A) 18. (C) 19. (B) 20. (D)
21. (C) 22. (A) 23. (B) 24. (D) 25. (A) 26. (D) 27. (B) 28. (B) 29. (A) 30. (A)
31. (A) 32. (C) 33. (A) 34. (B) 35. (C) 36. (B) 37. (A) 38. (B) 39. (D) 40. (C)
41. (A) 42. (B) 43. (C) 44. (D) 45. (C) 46. (A) 47. (B) 48. (D) 49. (A) 50. (B)
51. (C) 52. (B) 53. (D) 54. (C) 55. (D) 56. (A) 57. (C) 58. (A) 59. (C) 60. (A)
61. (D) 62. (D) 63. (A) 64. (A) 65. (D) 66. (C) 67. (A) 68. (A) 69. (D) 70. (B)
71. (C) 72. (A) 73. (D) 74. (B) 75. (C) 76. (D) 77. (C) 78. (D) 79. (C) 80. (C)
81. (B) 82. (B) 83. (D) 84. (D) 85. (A) 86. (C) 87. (D) 88. (C) 89. (E) 90. (D)
91. (F) 92. (D) 93. (D) 94. (D) 95. (C) 96. (B) 97. (A) 98. (D) 99. (A) 100. (D)
101. (B) 102. (D) 103. (E) 104. (A) 105. (B) 106. (C) 107. (C) 108. (B) 109. (D) 110. (A)
111. (A) 112. (A) 113. (B) 114. (D) 115. (A) 116. (D) 117. (D) 118. (B) 119. (A) 120. (A)
121. (C) 122. (A) 123. (B) 124. (A) 125. (C) 126. (C) 127. (A) 128. (A) 129. (C) 130. (A)
131. (A) 132. (A) 133. (A) 134. (B) 135. (B) 136. (B) 137. (C) 138. (D) 139. (A) 140. (B)
141. (A) 142. (C) 143. (A) 144. (C) 145. (B) 146. (C) 147. (C) 148. (A) 149. (D) 150. (D)
151. (B) 152. (A) 153. (A) 154. (A) 155. (B) 156. (C) 157. (A) 158. (A) 159. (A) 160. (A)
161. (A) 162. (A) 163. (A) 164. (D) 165. (A) 166. (D) 167. (A) 168. (A) 169. (B) 170. (D)
171. (A) 172. (D) 173. (B) 174. (A) 175. (A) 176. (A) 177. (B) 178. (D) 179. (B) 180. (B)
181. (A) 182. (A) 183. (B) 184. (C) 185. (D) 186. (A) 187. (A) 188. (B) 189. (A) 190. (A)
191. (A) 192. (C) 193. (A) 194. (A) 195. (B) 196. (C) 197. (D) 198. (A) 199. (D) 200. (A)
201. (C) 202. (D) 203. (D) 204. (A) 205. (D) 206. (A) 207. (A) 208. (A) 209. (B) 210. (C)
211. (D) 212. (A) 213. (B) 214. (A) 215. (C) 216. (C) 217. (B) 218. (C) 219. (A) 220. (A)
221. (A) 222. (A) 223. (D) 224. (D) 225. (C) 226. (D) 227. (A) 228. (D) 229. (D) 230. (D)
231. (A) 232. (A) 233. (A) 234. (A) 235. (A) 236. (D) 237. (C) 238. (B) 239. (C) 240. (D)
241. (A) 242. (B) 243. (A) 244. (C) 245. (D) 246. (A) 247. (B) 248. (D) 249. (A) 250. (D)
251. (C) 252. (D) 253. (C) 254. (D) 255. (C) 256. (B) 257. (A) 258. (B) 259. (B) 260. (A)
261. (B) 262. (B) 263. (C) 264. (C) 265. (B) 266. (B) 267. (D) 268. (C) 269. (B) 270. (A)
271. (B) 272. (C) 273. (A) 274. (A) 275. (A) 276. (B) 277. (C) 278. (A) 279. (B) 280. (D)
281. (A) 282. (A) 283. (B) 284. (A) 285. (A) 286. (C) 287. (D) 288. (D) 289. (A) 290. (A)
291. (D) 292. (B) 293. (B) 294. (A) 295. (B) 296. (A) 297. (B) 298. (A) 299. (C) 300. (D)
301. (A) 302. (C) 303. (A) 304. (B) 305. (C) 306. (A) 307. (D) 308. (A) 309. (A) 310. (C)
311. (C) 312. (B) 313. (C) 314. D) 315. (A) 316. (A) 317. (A) 318. (C) 319. (A) 320. (C)
321. (A) 322. (B) 323. (B) 324. (A) 325. (C) 326. (C) 327. (B) 328. (D) 329. (D) 330. (B)
331. (D) 332. (D) 333. (B) 334. (A) 335. (B) 336. (A) 337. (B) 338. (A) 339. (B) 340. (A)
341. (D) 342. (C) 343. (A) 344. (A) 345. (A) 346. (A) 347. (A) 348. (C) 349. (B) 350. (D)

14 PLANT PROTECTION MEASURES

1. Pest is:
(A) Organism feeding on and damaging others
(B) Fungal pathogen attacking plants and animals
(C) Bacterial pathogen attacking plants and animals
(D) Viral pathogen attacking organisms

2. Agricultural chemicals denote:
(A) Pesticides
(B) Pesticides and fertilizers
(C) Fertilizers and growth regulators
(D) Pesticides, fertilizers and growth regulators

3. Growth regulators produce:
(A) Stimulation of growth
(B) Retardation of growth
(C) Some cause stimulation while others bring about retardation
(D) Shedding of leaves

4. Loss of agricultural productivity due to pests and diseases is:
(A) 1% (B) 30%
(C) 5% (D) 15%

5. Household pesticide sprays are:
(A) Finit (Flit)
(B) Finit and Baygon
(C) Finit, Baygon, Mosquito coils & Cakes
(D) DDT

6. Cabamates include:
(A) Fungicides, herbicides and insecticides
(B) Insecticides only
(C) Herbicides, insecticides & nematicides
(D) Insecticides, nematicide & rotendicides

7. The first organic pesticide to be used commercially was:
(A) Bordeaux mixture
(B) Burgandy mixture
(C) DDT (D) 2,4-D

8. Bordeaux mixture is:
(A) Organochlorine
(B) Antibiotic
(C) Organophosphate
(D) Inorganic fungicide

9. The most famous pesticide of world is:
(A) BHC (B) DDT
(C) Baygon (D) Flit

10. In India, the most commonly used pesticide BHC is:
(A) Carbamate (B) Triazine
(C) Organochlorine (D) Antibiotic

11. Annual loss due to pests, diseases and weeds annually is estimated nearly:
(A) >50000 crores (B) <50000 crores
(C) > 1.0 crores (D) < 1.0 crores

12. India is using biopesticides presently:
(A) 1% of total pesticide
(B) 5 % to total pesticide
(C) 10 % of total pesticide
(D) 15 % of total pesticide

13. Average biopesticide use globally is:
(A) 5 % of total pesticide
(B) 10 % of total pesticide
(C) 12 % of total pesticide
(D) 15 % of total pesticide

14. Total pesticide consumption in India during 2006-07 is:
(A) 35.65 million tonnes
(B) 37.95 million tonnes
(C) 39.26 million tonnes
(D) 42.56 million tonnes

15. Directorate of Plant Protection Quarantine and Storage was started in:
(A) 1907 (B) 1914
(C) 1921 (D) 1946

16. Destructive Insects and Pest Act was launched in India in:
(A) 1907 (B) 1914
(C) 1921 (D) 1946

17. Which of the airport does not have plant quarantine station?
(A) Lucknow (B) Kolkata
(C) New Delhi (D) Madras

18. Tobacco cultivar A.C. Change is resistant to:
(A) Leaf curl (B) Black rot
(C) Damping off (D) TMV

19. Horizontal resistance is:
(A) Resistance against one race of pathogen
(B) Resistance against all the races of pathogen
(C) Resistance against lodging
(D) None of these

20. White grub is an important insect of:
(A) Sugarcane (B) Potato
(C) Ground nut (D) Mustard

21. Army worm is a serious pest of:
(A) Wheat (B) Chickpea
(C) Rice (D) Groundnut

22. At which stage, Gandhi bug damages the rice crop:
(A) Tillering stage
(B) Panicle initiation stage
(C) Milking stage
(D) Maturity stage

23. Yellow stem borer of rice is:
(A) *Scirpophaga innotata*
(B) *Scirpophaga incertulas*
(C) *Orseolia oryzae*
(D) *Dicladispa armigera*

24. 'Hopper burn' is caused by:
(A) *Leptocorisa varicornis*
(B) *Bemisia tabaci*
(C) *Spodoptera litura*
(D) *Nilaparvata lugens*

25. 'Dead heart' is caused by:
(A) Paddy stem borer
(B) Termite
(C) Rice gandhi bug
(D) Paddy stem weevil

26. 'Silver shoot' is caused by:
(A) Grass hopper (B) Stem borer
(C) White fly (D) Gall midge

27. 'Chaffy appearance' in the paddy is caused on which part of plant:
(A) Roots (B) Tillers
(C) Flowers (D) Ears

28. Rice Gandhi bug attacks appear on the plant at:
(A) Seedling stage (B) Flowering stage
(C) Milky stage (D) Tillering stage

29. Female of paddy hispa lays eggs:
(A) Inside the roots
(B) Inside the stem
(C) Inside the leaves
(D) Inside the flowers

30. Systemic granules are applied @ 1.5 kg a.i.:
(A) Per hectare (B) Per acre
(C) Per 100 m^2 (D) Per 1000 m^2

31. *Odontotermis obesus* is controlled by the application of:
(A) Phosphamidon (B) Aluminium chloride
(C) Chlorpyriphos (D) Brestanol

32. Nuptial flight is found in:
(A) Termite (B) Grasshopper
(C) White fly (D) Gujhia weevil

33. Which one of the following is called as 'Army worm'?
(A) Leptocorisa (B) Mythimna
(C) Spodoptera (D) Tanymecus

34. Wheat shoot fly attacks the plants at:
(A) Seedling stage (B) Flowering stage
(C) Milky stage (D) Maturity stage

35. The wheat seeds are treated with chlorpyriphos 20 EC against termite @:
(A) 3 to 4 ml/kg (B) 15-20 ml/kg
(C) 25 to 30 ml/kg (D) 35 to 40 ml/kg

36. Eggs of *Atherigona naqvi* are laid in the:
(A) Buds (B) Flowers
(C) Inside the leaves
(D) At the base of stem and under side of leaves

37. The field rates are controlled by:
(A) Zinc chloride (B) Zinc phosphide
(C) Zinc oxide (D) Copper oxide

38. Which one of the following caste of termite is sterile:
(A) King (B) Queen
(C) Soldiers (D) Winged termite

39. White grubs damage:
(A) Roots (B) Stems
(C) Leaves (D) Flowers

40. Which one of the following is 'June Beetle':
(A) *Holotrichia consanguinea*
(B) *Xanthopimia sp.*
(C) *Rodolia sp.* (D) *Coccinella sp.*

41. Sugarcane shoot borer damages:
(A) Roots (B) Stem
(C) Leaves (D) All of these

42. 'Bunchy top' in sugarcane is caused by:
(A) Root borer (B) Pyrilla
(C) White fly (D) Scale insects

43. 'Anal-tufts' are found in the:
(A) Adult pyrilla
(B) Nymphs of pyrilla
(C) Nymphs and adults both
(D) None of these

44. *Beauvania brassiana* is effective against control of:
(A) Termite (B) Pod borer
(C) Grasshopper (D) Grubs

45. There is a pupal stage in:
(A) Pyrilla (B) Termite
(C) White fly (D) Jassids

46. The female of which insect has a tuft of hairs at its anal end?
(A) Sugarcane top borer
(B) Sugarcane leaf hopper
(C) Sugarcane root borer
(D) Sugarcane white grubs

47. *Epiricanea melanoleuca* is a parasite of pyrilla:
(A) Nymphs only
(B) Adults only
(C) Nymphs and adult both
(D) Eggs

48. Amrasca biguttula is:
(A) Jassid (B) Leaf roller
(C) Whitefly (D) Stem weevil

49. 'Hibernacula' is found in:
(A) Flower beetle (B) Stem weevil
(C) Spotted boll worm
(D) Pink boll worm

50. Trycho card is very effective against control of:
(A) Termite (B) Pod borer
(C) Grasshopper (D) Cutworm

51. Spotted boll worm of cotton pupaes is:
(A) Inside the boll
(B) Out of the boll but naked
(C) Out of the boll in a coccon
(D) Inside the seeds

52. Which one of the following is cotton leaf roller?
(A) *Oxycarenus laetus*
(B) *Sylepta derogata*
(C) *Dysdercus cingulatus*
(D) *Pempherulus affinis*

53. The larvae of *Sylepta derogata* are:
(A) Greenish bluish
(B) Pinkish
(C) Blackish (D) Yelowish

54. Nymphs of red cotton bug also damage:
(A) Wheat (B) Sugracane
(C) Gram (D) Lady's finger

55. Which one of the following insects come from store to infest crop in field?
(A) Red cotton bug (B) Pink boll worm
(C) Leaf roller (D) Spotted boll worm

56. In which of the following insects a greenish strip is found in forewings?
(A) *Earia vitella*
(B) *Earias insulana*
(C) *Oxycarenus leatus*
(D) *Pectinophora gossypii*

57. Which one of the following lays stalked eggs?
(A) *Bemisia tabaci*
(B) *Aphis gossypii*
(C) *Myllocerms sp*
(D) *Pempherulus affinis*

58. Larvae of *Helicoverpa armigera* damage:
(A) Pods (B) Roots
(C) Stem (D) Buds

59. Which one of the following is a cut worm:
(A) Plume moth (B) Pod borer
(C) *Agrotis sp* (D) *Heliothis sp.*

60. Seed treatment of *Beauvania brassiana* @ 10 g/kg is effective against control of:
(A) Termite (B) Pod borer
(C) Grasshopper (D) Cutworm

61. Against *Helicoverpa armigera*, NPV is applied @:
(A) 250 LE/ha (B) 100 LE/ha
(C) 50 LE/ha (D) 10 LE/ha

62. The number of organophosphates synthesized is:
(A) 1000 (B) 100
(C) 10000 (D) 100000

63. The mode of action of insecticides is through the attack on:
(A) Respiratory system
(B) Nervous system
(C) Muscular system
(D) Both (A) and (B)

64. One of earliest pesticides employed by humans was:
(A) Nicotine
(B) Margosa (Neem) leaves
(C) Pyrethrum
(D) DDT

65. Which of the following is a rat species?
(A) *Rattus rattus* (B) *Rattus norvegicus*
(C) *Mus musculus* (D) All of these

66. Brown or Norvey rat is:
(A) *Rattus rattus* (B) *Rattus norvegicus*
(C) *Mus musculus* (D) None of these

67. Miticides is used to control of:
(A) Mites (B) Rats
(C) Insects (D) Nematodes

68. Which of the following poison is used to control of rats?
(A) Zinc phosphide
(B) Barium carbonate
(C) Arsenium oxide
(D) All of these

69. An example of pesticide treadmill is:
(A) Forty times increase in requirement for cotton
(B) Development of mosquito resistance to DDT, BHC, malathion and fenitrothion
(C) Decreasing use of pesticides due to introduction of biological control
(D) Decreasing use of pesticides due to decrease in the population of pests

70. DDT concentration increases in the human body because it is soluble in:
(A) Water (B) Vitamins
(C) Fat (D) Protein

71. Which of the fumigants can be used against control of rats?
(A) Sulphos (B) Cynogas dust
(C) Hydrocynic acid
(D) All of these

72. DDT and 2,4-D were introduced:
(A) Towards end of World War II
(B) Towards beginning of World War II
(C) During World War I
(D) Just after the World War I

73. Bordeaux mixture is named after:
(A) Scientist
(B) Chemical composition
(C) Country
(D) University

74. Insecticides kill:
(A) Harmful insects
(B) Both harmful and useful insects
(C) Specific insects
(D) Only plant pests

75. Which is organochlorine:
(A) Malathion (B) Parathion
(C) Aldicarb (D) DDT

76. Which one is an organophosphate:
(A) BHC (B) Aldrin
(C) Fenitrothion (D) Endosulphan

77. A carbamate pesticide is:
(A) Propoxur (B) Simazine
(C) Atrazine
(D) Tetraethyl pyrophosphate

78. Urea derivative pesticides belong to the category of:
(A) Organochlorines
(B) Organophosphates
(C) Carbamates
(D) Triazines

79. Carbamates employed as insecticide are:
(A) Phenylcarbamates
(B) Thiocabamates
(C) Propoxur (D) Dithiocarbamates

80. Carbamates used as fungicides are:
(A) Thiocarbamates
(B) Dithiocarbamates
(C) Carbofuron
(D) Carbofuron and Aldicarb

81. The pesticides having very low biodegradation and strong affinity for fatty tissues are:
(A) Organochlorines
(B) Organophosphates
(C) Triazines
(D) Pyrethroids

82. Pesticide ingredient methyl isocyanate which killed thousands in Bhopal gas tragedy is intermediate of:
(A) Mustard gas (B) Nitrous acid
(C) Carbaryl
(D) Carbon tetrachloride

83. The most common pesticide used in India is:
(A) DDT (B) Endrin
(C) BHC (D) Baygon

84. Pesticides comprise:
(A) Herbicides and insecticides
(B) Insecticides only
(C) Insecticides, nematicides and rodenticides
(D) Fungicides, herbicides, insecticides, nematicides and rotendicides

85. Herbicide carbamates are:
(A) Dithiocarbamates
(B) Thiocarbamates
(C) Phenyl carbamates
(D) Both (B) and (C)

86. Household pesticides consist of:
(A) Finit, Baygon, Mosquito repellent cakes and coils
(B) Finit and Baygon
(C) Finit/Flit
(D) DDT

87. DDT is:
(A) Carbamate
(B) Organophosphate
(C) Organochlorine
(D) Triazine

88. First commercial pesticides was:
(A) Burgandy mixture
(B) Bordeaux mixture
(C) DDT (D) 2,4-D

89. Mode of pesticides action is similar in:
(A) DDT and parathion
(B) Organochlorine and organophosphates
(C) Organochlorine and carbamates
(D) Organophosphates and carbamates

90. BHC is:
(A) Benzene hexachloride
(B) Benzine heptachloride
(C) Benzene hydrochloride
(D) Bromide herbicide

91. Herbicides prevent:
(A) Transpiration
(B) Water absorption
(C) Translocation of salts
(D) Photolysis of water

92. Pesticides generally attacks:
(A) Muscular system
(B) Respiratory system
(C) Nervous system
(D) Circulatory system

93. Pests resistant to insecticides have:
(A) Enzymes for metabolizing toxins
(B) Ability to store toxin in fat
(C) Less permeable cuticles
(D) All the above

94. Most effective pesticides are:
(A) Carbamates
(B) Organochlorine
(C) Organophosphates
(D) All the above

95. DDT content in Indian population has touched:
(A) 1-12 ppm (B) 11-30 ppm
(C) 25-50 ppm (D) 48-98 ppm

96. Which one is not matched:
(A) Organochlorine – DDT
(B) Pyrethroid – Atrazine
(C) Organophosphate – Malathion
(D) Carbamate - Carbofuran

97. Pesticides with very low biodegradation but strong affinity for fatty tissues are:
(A) Organochlorines
(B) Organophosphates
(C) Pyrethrines (D) Triazines

98. Bordeaux mixture was formulated by:
(A) Leeuwehoek (B) Tillet
(C) De Bary (D) Millardet

99. Baygon contains:
(A) Malathion (B) Carbofuran
(C) Propoxur (D) Aldicarb

100. Which ingredient killed hundreds of people in Bhopal gas tragedy:
(A) Nitrous acid (B) Mustard gas
(C) Carbon tetrachloride
(D) Methyl isocyanate

101. DDT is:
(A) Biodegradable (B) Non-biodegradable
(C) Antibiotic (D) Growth regulator

102. Which is a fungicide:
(A) 2,4-D (B) DDT
(C) Bordeaux mixture
(D) BHC

103. Zinc phosphide is a pesticide used for:
(A) Spraying
(B) Fumigation
(C) Killing of weevils
(D) Killing rodents

104. Rotenone is:
(A) A bioherbicide
(B) A natural hormone
(C) An insect hormone
(D) A natural insecticide

105. 2,4-D is an effective
(A) Weedicide (B) Pesticide
(C) Fungicide (D) Rodenticide

106. Organophosphate inhibit:
(A) Cholinesterases (B) Enterokinase
(C) Urease (D) Carbohydrases

107. Fungicide Bordeaux mixture is:
(A) Magnesium hydroxide
(B) Copper sulphate + Sodium hydroxide
(C) Copper sulphate + Calcium hydroxide
(D) Magnesium sulphate + Calcium hydroxide

108. Which of the following is cholinesterase inhibitor:
(A) Endosulphan (B) Aldrin
(C) BHC (D) Malathion

109. Bordeaux mixture is a:
(A) Organic fungicide
(B) Inorganic fungicide
(C) Organic insecticide
(D) Inorganic insecticide

110. Which is major component of Bordeaux mixture:
(A) Copper sulphate
(B) Calcium chloride
(C) Sodium chloride
(D) Magnesium sulphate

111. Ethylene dibromide is:
(A) Stomach poison
(B) Contact poison
(C) Fumigant
(D) Biological control

112. A widely but periodically occupying plant disease is:
(A) Andemic (B) Epidemic
(C) Sporodic (D) Epiphytotic

113. Insecticides generally attack:
(A) Respiratory system
(B) Nervous system
(C) Muscular system
(D) Circulatory system

114. Carbamates used as fungicides are:
(A) Thiocarbamates
(B) Dithiocarbamates
(C) Carbofuran
(D) Carbofuran and Aldicarb

115. Biopesticides include:
(A) Bioherbicides
(B) Bioinsecticides
(C) Bioherbicides and bioinsecticides
(D) Bioherbicides, bioinsecticdes and biofertilizers

116. A transgenic crop that is resistant to hornworm larvae is:
(A) Sunflower (B) Tomato
(C) Tobacco (D) Sugarcane

117. A major pest eradicated through release of sterile males is:
(A) Screw worm (B) Aphids
(C) Ladybug (D) Praying mantis

118. Pheromone are chemicals that are used in animals for:
(A) Sex attractants
(B) Sending alarm signals
(C) Marking trails
(D) All the above

119. Rotenone is a:
(A) Natural herbicide
(B) Bioherbicides
(C) Natural insecticide
(D) Insect hormone

120. A natural insecticide obtained from plants is:
(A) Azadirachtin (B) Rotenone
(C) Nicotine, pyrethrum and cinerin
(D) All the above

121. Cochineal insect has been used in checking the wild growth of:
(A) Opuntia (B) Eichhornia
(C) Aphids (D) Screw worm

122. Pheromone traps have:
(A) Sterile males
(B) Male pheromones
(C) Female pheromones
(D) Sterile females

123. In confusion technique, the pheromone coated paper strips are thrown over an area in order to:
(A) Attract insects and kill them
(B) Repel insects from an area
(C) Confuse males so that they are unable to locate females
(D) Confuse females so that they are unable to locate males

124. Juvenile hormone causes:
(A) Early sexual maturity
(B) Late sexual maturity
(C) Completion of larval stages
(D) Development of giant larva without ability to undergo maturation

125. Pyrethrin is a common ingredient of:
(A) Mosquito coils (B) Fly sprays
(C) Mosquito mats (D) All of these

126. Thurioside is:
(A) Insecticide (B) Fungicide
(C) Antibiotic (D) Weedicide

127. Antifeedant property occurs in:
(A) Nicotine (B) Azadirachtin
(C) Rotenone (D) Cinerin

128. Ladybug is a predator of:
(A) Moths (B) Beetles
(C) Bacteria (D) Aphids

129. Which one is an effective plant insecticide:
(A) Pyrethrin (B) Cinerin
(C) Nicotine (D) All of these

130. Pyrethrin is extracted from:
(A) *Azadiracta indica*
(B) *Helianthus annus*
(C) *Poa indica*
(D) *Chrysanthemum cinerarifolium*

131. The earliest pesticide was:
(A) Margosa (B) Pyrethrum
(C) Nicotine (D) DDT

132. First bioinsecticide developed on commercial scale was:
(A) Quinone (B) DDT
(C) Organophosphate
(D) Sporeine

133. Sporoneine was developed in:
(A) Germany (B) France
(C) USA (D) Russia

134. The phenomenon of using a predator for controlling a pest is:
(A) Biological control
(B) Genetic engineering
(C) Artificial control
(D) Confusion technique

135. Which is a microbial insecticide:
(A) Bacillus thuringensis
(B) *B. subtilis*
(C) *B. polymixa*
(D) *B. brexis*

136. IPM (Integrated Pest Management) involves :
(A) Biological control
(B) Pesticides
(C) Confusion technique
(D) All above

137. *Corpora allata* is removed from a nymph. It will:
(A) Remain nymph for life
(B) Become adult
(C) Change to next nymph and remain in that state for ever
(D) Die after some time

138. To insects ccdysone takes part in:
(A) Growth and development
(B) Maturation into adult
(C) Moulting till pupa is formed
(D) Secretion of cuticle

139. Thurioside is proteinaceous toxin obtained from:
(A) Biofertilier
(B) Green manure
(C) Bacterial origin
(D) Farmyard manure

140. Latest trend in plant disease control is:
(A) Chemical control
(B) Biological control
(C) Good manure and fertilizer
(D) Breeding for disease resistance

141. A third generation pesticide is:
(A) Pheromone
(B) Weedicide
(C) Juvenile hormone analogue
(D) Sterilogastrone

142. Genus *Radopholus* of nematode was first described in 1893 on the basis of a male specimen obtained from necrotic roots of -------- growing in Fiji Islands:
(A) Sugarbeet (B) Sugarcane
(C) Banana (D) Coconut

143. The nematode *Radopholus similes* has shown the maximum numbers at a soil depth of:
(A) 5-15 m (B) 20-30 m
(C) 32-42 m (D) 1-4 m

144. Stem borer can be controlled b:
(A) Putting mixture of chloroform and carbon bisulphide into the holes
(B) Spray of monocrotophos
(C) Insect may be killed with the help of wire
(D) All of these

145. Red spider can be controlled by:
(A) Spray of sulphur
(B) Spray of zinc
(C) Spray of iron
(D) Spray of 2,4-D

146. Flying poxes are:
(A) Bats (B) Porcupine
(C) Parrot (D) None of these

147. Larvae are preserved in:
(A) 70% alcohol solution
(B) 5 % salt solution
(C) 80 % sugar solution
(D) 10% sulphuric acid solution

148. It has been estimated that insects had existed on this earth for at least:
(A) 100 million (B) 150 million
(C) 200 million (D) 250 million

149. Stored grain pests can be killed by ---- hours exposure to 51.5-54.5 ^{0}C:
(A) 2 (B) 3
(C) 5 (D) 7

150. Most of the insects cannot survive the temperatures above:
(A) 29.9 ^{0}C (B) 39.9 ^{0}C
(C) 49.9 ^{0}C (D) 59.9 ^{0}C

151. Ultra – Low Volume (ULV) or Low Volume Concentration (LVC) refers to the use of spray volume (liquid) less than :
(A) 5 litres (B) 7 litres
(C) 9 litres (D) 11 litres

152. Malathion is a insecticide:
(A) Slow fumigant
(B) Contact plus systemic
(C) Contact plus stomach
(D) Systemic

153. Metasystox – R is a -------- insecticide:
(A) Sytemic
(B) Fumigant
(C) Contact
(D) Contact plus stomach

154. Phorate is a ---------- insecticide:
(A) Systemic
(B) Contact plus systemic
(C) Contact
(D) Contact plus stomach

155. Phosdrin is a ----------insecticides:
(A) Slow fumigant
(B) Systemic
(C) Systemic plus contact
(D) Contact

156. A minimum period of ----- years is to be given before assessing the effectiveness of an important biocontrol agent:
(A) 2 (B) 3
(D) 4 (D) 5

157. The queen termite produces --------- eggs per second:
(A) 40 (B) 50
(C) 60 (D) 70

158. Honey bee queen produces ---------- thousand eggs per day:
(A) 0.5-1.5 (B) 2-3
(C) 4.0-4.5 (D) 5.0-5.5

159. Malathion, parathion and fumitrothion belong to group of pesticides:
(A) Triazines (B) Carbamates
(C) Pyrethroids (D) Organophosphates

160. Which of the following is not a ingredient of universal antidote for insecticide poisoning?
(A) Charcoal (B) Tannic acid
(C) Milk of magnesia (D) All of these

161. Insect requires relative humidity for its better growth and development?
(A) 30-40 % (B) 40-50%
(C) 50-60 % (D) 60-80 %

162. Most of the insects become inactive at temperature?
(A) $<40^{0}$F (B) $>40^{0}$F
(C) 40^{0}F (D) 50^{0}F

163. Dead heart in rice is the symptom of insect damage:
(A) Stem borer (B) Leaf roller
(C) Grass hopper (D) Gandhi bug

164. Pod borer of pulses can be controlled by:
(A) Spray of Endosulphan 35 EC @ 1.5 l/ha
(B) Spray of Monocrotophos @ 0.75 l/ha
(C) Spray of Quinolphos 25 EC @ 1.25 l/ha
(D) All of these

165. Grasshoppers lay eggs:
(A) On soil surface (B) Below soil surface
(C) In nest (D) All of these

166. The colour of insects is:
(A) Red (B) Green and yellow
(C) Voilet (D) Pink

167. Most of the insects become active at temperature?
(A) $<140^{0}$F (B) $>140^{0}$F
(C) 80^{0}F (D) 100^{0}F

168. Insects may be killed by using mainly:
(A) Low frequency waves
(B) High frequency waves
(C) Ultrasound waves
(D) None of these

169. *Pseudoapliyeus utilis* is a parasite insect of:
(A) Mealy bug
(B) Sugarcane weevil
(C) Cottony cushion scale
(D) None of these

170. *Cytrohinus mundulus* is a predator insect of:
(A) Mealy bug
(B) Sugarcane weevil
(C) Cottony cushion scale
(D) None of these

171. *Aphelinus mali* is predator insects of:
(A) Wooly aphid (B) Potato tuber moth
(C) Fruit fly (D) None of these

172. Milky disease in Japani beetle is caused by:
(A) Protozoa (B) Bacteria
(C) Fungi (D) Virus

173. Insects can be trapped by using the light?
(A) Infrared light
(B) Ultra violet light
(C) Far infrared light
(D) None of these

174. Insects may be attracted maximum at light intensity of:
(A) 365 Å (B) 400Å
(C) 500 Å (D) More than 600 Å

175. Fruit fly can be biologically controlled by:
(A) *Opius vandenboschi*
(B) *Rodolia cardinalis*
(C) *Aphelinus mali*
(D) None of these

176. Insects were controlled by parasitic viruses first of all in:
(A) USA (B) Canada
(C) China (D) India

177. *Cocobacillis acridiorum* bacteria can be used against:
(A) Locust (B) Housefly
(C) Grasshopper (D) Mites

178. Japanese beetle is controlled by bacteria:
(A) *Bacillus popilliae*
(B) *Bacillis lentimarbus*
(C) Both (A) and (B)
(D) None of these

179. Levelling of insect should indicate the information:
(A) Name of insect and its host plant
(B) Date and place of collection
(C) Name of collector
(D) All of these

180. Potato crop is mainly damaged by:
(A) Crow (B) Wild boar
(C) Parrot (D) None of these

181. Acaricides are used against:
(A) Beetles (B) Mites
(C) Moths (D) Grubs

182. Sesamin acts as a synergist if it is mixed with:
(A) Pyrethrum (B) Lindane
(C) Endosulphan (D) Monocrotophos

183. The dark yellow triangle on insecticide container indicates:
(A) Highly poisonous
(B) Poisonous
(C) Medium poisonous
(D) Less poisonous

184. Dark green triable on insecticide container indicates:
(A) Highly poisonous
(B) Poisonous
(C) Medium poisonous
(D) Less poisonous

185. Which of the following poison is used in killing bottle?
(A) Potassium cynide
(B) Hydrocynic acid
(C) H_2S (D) None of these

186. Snails and sludge are:
(A) Phytophagous (B) Polyphagous
(C) Saprophytes (D) None of these

187. Yellowing of flour is caused by:
(A) Rice weevil
(B) Red rust flour beetle
(C) Khapra beetle
(D) All of these

188. Which stage of pulse weevil is harmful?
(A) Grub only (B) Adult only
(C) Grub and adult(D) None of these

189. Khapra beetle is also known as:
(A) Pai (B) Grain moth
(C) Red surahi (D) All of these

190. Bihar hairy caterpillar can be controlled by:
(A) Chlorpyriphos (B) Endosulphan
(C) Nuvan (D) All of these

191. Honey dew secretion is found in:
(A) Grasshopper (B) Pyrilla
(C) White fly (D) None of these

192. Sugar cane shoot borer completes its life cycle in:
(A) 34-44 days (B) 24-26 days
(C) 14-16 days (D) 44-54 days

193. Pyrilla can be controlled by spraying of:
(A) Endosulphan (B) Rogor
(C) Phosphemidon (D) All of these

194. Red pumpkin beetle is a harmful insects of:
(A) Tomato (B) Cucurbits
(C) Rice (D) Potato

195. Gandhi bug can be controlled by:
(A) Thiodan 4% dust @ 20 kg/ha
(B) Malathion 5% dust @ 30 kg/ha
(C) Carbaryl 10% dust @ 20 kg/ha
(D) All of these

196. Which of the following is a chemosterilants?
(A) Tepa (B) Matepa
(C) Colchison (D) All of these

197. Which of the following is a systemic poison?
(A) Lindane (B) Thiodan
(C) Aldrin (D) All of these

198. *Bacillus thuringiensis* is used against:
(A) Larvae of lapidoptera
(B) Beetle (C) Cockroach
(D) Grasshopper

199. Ovoviviparous insects can:
(A) Adult can yield nymph either directly or from eggs
(B) Adult produces only larvae
(C) Adult can produce nymphs for eggs only
(D) All of these

200. Ovoviviparous insect is:
(A) Grasshopper (B) Mustard aphid
(C) Fruit fly (D) Shoot borer

201. Which of the following is a natural enemy of mustard aphid as:
(A) *Cocciniella septumpunctata*
(B) *Chrysopa spp.*
(C) *Syrphid maggot*
(D) All of these

202. Gram caterpillar is harmful to:
(A) Chickpea (B) Sunflower
(C) Tomato (D) All of these

203. Scientific name of gram caterpiller is:
(A) *Heliothis armigera*
(B) *Athalia spp*
(C) *Nezara viridula*
(D) None of these

204. Scientific name of mustard saw fly is:
(A) *Athalia proxima*
(B) *Bagrada cruciferaum*
(C) *Trogogerma granarium*
(D) None of these

205. Which of the following stage of mealy bug is harmful?
(A) Larvae only
(B) Female adult only
(C) Male adult only
(D) Larvae and female adult

206. Caterpillar of rice weevil is called as:
(A) Nymph (B) Larvae
(C) Maggot (D) Grub

207. Most harmful stage of khapra beetle is:
(A) Grub (B) Adult
(C) Pupa (D) All of these

208. Which of the following stage of lesser grain moth is harmful to grains?
(A) Grub only (B) Adult only
(C) Grub & adult (D) Nymph only

209. Maximum harm occurs due to presence of pulse weevil during:
(A) July – September
(B) September – December
(C) January – February
(D) None of these

210. Red rust flour remain most active during:
(A) Winter season (B) Summer season
(C) Rainy season (D) All season

211. Grain moth is also known as:
(A) Grain and Flour caterpillar
(B) Potato tuber moth
(C) Rice weevil (D) Cut worm

212. Which of the life stage of grain moth is harmful?
(A) Egg and grub
(B) Caterpillar
(C) Nymph and adult
(D) Adult

213. Khpara beetle may be controlled at:
(A) $<16\ \%\ O_2$ (B) $>16\ O_2$
(C) $<21\ \%\ O_2$ (D) $>21\ \%\ O_2$

214. The life cycle of pulse weevil is:
(A) 15-20 days (B) 25-28 days
(C) 30-35 days (D) 40-42 days

215. Net like structures are formed in flour due to presence of:
(A) Khapra beetle (B) Red rust flour beetle
(C) Rice weevil (D) None of these

216. Following insecticides may be used to control of mealy bug is:
(A) Phosphomedon 100 EC
(B) Malathion 50 EC
(C) Thiodan 35 EC
(D) All of these

217. Mango leaf hopper is most active in the month of:
(A) February – November
(B) December – January
(C) September – October
(D) All of these

218. *Sitotroga cerealella* is a scientific name of:
(A) Grain moth (B) Pulse beetle
(C) Rice weevil (D) None of these

219. Which of the following part of grains is damaged by khapra beetle?
(A) Seed (B) Endosperm
(C) Embryo (D) Cotyledons

220. Azospirillum biofertilizer induce resistance in sorghum against pest of:
(A) Shoot fly (B) Head borer
(C) Cut worm (D) None of these

221. Which of the preventive measure should be adopted for safe storage?
(A) Fumigation of godowns with E.D.C.T.
(B) Bags should be treated with 0.1% malathion solution
(C) Put E.D.B. ampules 3.0 ml/q grain
(D) All of these

222. Heat spots in stored grains are:
(A) Favourable to insect multiplication due to optimum temperature
(B) Harmful to insects
(C) Spots of low temperature and high humidity
(D) None of these

223. Which of the storage pest may survive at ≤ 10 % moisture in stored grains?
(A) Rice weevil (B) Khapra beetle
(C) Pulse beetle (D) All of these

224. The process of which insect has many life stage, is called as:
(A) Monocious (B) Diocious
(C) Polymorphism (D) None of these

225. Mustard aphid can be controlled by spray of:
(A) Phosphomidon 100 EC @ 250 ml/ha
(B) Rogor 30 EC @ 1000 ml/ha
(C) Thiodan 35 EC @ 1250 ml/ha
(D) All of these

226. Painted mustard bug can be controlled by spraying of:
(A) Malathion @ 1.25 l/ha
(B) Dimecron @ 250 ml/ha
(C) Rogor @ 1.00 l/ha (D) All of these

227. Cut worm generally attack on plants at:
(A) Day only (B) Day and night
(C) Night only (D) None of these

228. Highest production of mustard aphid is found in the month of:
(A) November – December
(B) January – February
(C) February – March
(D) March – April

229. Which of the following is a parasite insect of Gurdaspur borer?
(A) *Trichogramme spp*
(B) *Apanteles flavipes*
(C) *Sturmiopsis interens*
(D) All of these

230. Dead heart in sugarcane is caused by:
(A) Top borer (B) Pyrilla
(C) Gandhi bug (D) Shoot borer

231. Sugarcane crop is damaged severely by shoot borer in the month of:
(A) April – June
(B) July – September
(C) October – December
(D) January –March

232. Termite control in house building is possible by using:
(A) 25% Tri chlorobenzene in fuel oil
(B) 5% penta chlorophenol
(C) 1 % Chlordane
(D) All of these

233. Cactus can be controlled by which insect:
(A) *Cactoblastic cactorium*
(B) *Coccinellids*
(C) *Apanteles*
(D) None of these

234. The virus of Bunchy top disease of banana is transmitted by:
(A) Fruit fly (B) Pentalonia
(C) Thrips (D) None of these

235. Plant tolerance refers to:
(A) Plant kills insects
(B) Plant dies after insect attack
(C) Plant does not show any side effect of insect incidence
(D) None of these

236. Antibiosis refers to:
(A) Ability of plant to suppress the reproduction potential of the insect
(B) Production of antibodies by insect
(C) Insecticide to kill insects
(D) None of these

237. Phytoecdysone is a:
(A) Plant origin hormone similar to moulting hormone of insect
(B) Insect moulting hormone
(C) Plant steroids to kill insects
(D) None of these

238. The pheromone of honey bee that suppresses sex differentiation among workers is known as:
(A) 9- Keto-trans-2 decenoic acid
(B) Gyplure
(C) Dolphene
(D) None of these

239. Which of the following is a repellent for mosquitoes?
(A) Dolphene (B) Citronella oil
(C) Camphor oil (D) All of these

240. Baygon acts as antifedant against:
(A) Cotton ball weevil
(B) Moth worm
(C) Cut worm
(D) None of these

241. Which of the following is a protoplasmic poison?
(A) Nitrophenols
(B) Lead arsinate
(C) Sodium tetra borate
(D) All of these

242. Which of the following is a fumigants?
(A) Methyl bromide (B) HCN
(C) EDCT (D) All of these

243. Which of the following is a plant origin?
(A) Nicotine sulphate
(B) Pyrethrins
(C) Ryania
(D) All of these

244. Which of the following is a stomach poison?
(A) Arsenic salt
(B) Flourine compounds
(C) Chlorinated hydrocarbons
(D) All of these

245. Rogor is also known as:
(A) Metasystox (B) Dimethoate
(C) Phosdrin (D) None of these

246. Endosulphan is also known as:
(A) Lindane (B) Thiodan
(C) Aldrin (D) BHC

247. Sterility method as a synergist if it is mixed with?
(A) Pyrethrum (B) Lindane
(C) Endosulphan (D) None of these

248. The chemicals that make insects sterile are known as:
(A) Systemic poison (B) Chemosterilants
(C) Contact poison (D) Pheromones

249. DDVP is known as:
(A) Nuvan (B) Malathoate
(C) Thiodan (D) Sulfec

250. DDT is a:
(A) Fumigant (B) Contact poison
(C) Stomach poison (D) All of these

251. Polyhedrosis virus can control:
(A) *Heliothis armigera*
(B) *Heliothus virescence*
(C) *Alfalfa caterpillar*
(D) All of these

252. Which of the radio active material was used for γ rays in sterile male technique?
(A) Ra^{232} (B) P^{32}
(C) Co^{60} (D) Ba^{232}

253. Which of the insect was first of all controlled by knipling techniques?
(A) Screw worm fly
(B) Mealy bug
(C) Mango hopper
(D) Mustard aphid

254. Cannibalism is a process in which insects:
(A) Fight and kill each other
(B) Protect each other
(C) Attract for copulation
(D) None of these

255. Stored grains can be saved from insect damage if the grain moisture content is:
(A) $< 10\%$ (B) $> 10\%$
(C) 10 % (D) None of these

256. The grasshopper always move towards:
(A) South-east direction
(B) South-west direction
(C) North-east direction
(D) North-west direction

257. Which of the following is the optimum temperature for insect growth and development?
(A) 5-10^{0}C (B) 13-40^{0}C
(C) 42-55^{0}C (D) 15-20^{0}C

258. Which of the following is the antidote of insecticide poisoning?
(A) Atropine sulphate
(B) Diacetyl monohexane
(C) Pyridine aldoxymet
(D) All of these

259. Apposition refers to:
(A) Mosaic vision
(B) It is found in diurnal insects
(C) Separate image is made in each omatidium
(D) All of these

260. On pH scale, insect blood is a:
(A) Acidic (B) Basic
(C) Neutral (D) None of these

261. Typical or primitive type of mouth parts are found in:
(A) Grubs (B) Honey bee
(C) Bugs (D) Mosquitoes

262. Trachea and tracheoles are found in:
(A) RBC (B) WBC
(C) Blood (D) All of these

263. Insects can live in:
(A) Water only (B) Land only
(C) Air only (D) All of these

264. Exoskeleton of insects is made of:
(A) Hemi-cellulose (B) Cellulose
(C) Chitin (D) None of these

265. Which is the most important pest of cauliflower is?
(A) Sawfly (B) Aphids
(C) Cabbage borer (D) Flea beetle

266. Which is the most important pest of onion is?
(A) Thrips (B) Maggot
(C) Flea beetle (D) Tobacco caterpiller

267. Which is the most important pest of bitter guard is:
(A) Red Pumpkin beetle
(B) Aphid
(C) Cutworm (D) Fruitfly

268. Fruit fly is pest of bottle guard and damaged fruits so, fruit fly can be cotrolled by:
(A) By staking
(B) BY seed treatment
(C) Bt spray of Aldrin dust
(D) By spray of 0.2 % Sevin solution at flowering stage

269. Which is the most important pest of carrot is:
(A) Cut worm (B) Wire worm
(C) White grub (D) All of these

270. Sweet potato aphinx attacks mainly on:
(A) Leaves (B) Stem
(C) Roots (D) All of these

271. Insects hide at low temperature is known as:
(A) Aestivation (B) Hibernation
(C) Natural safety (D) None of these

272. Insects hide at high temperature is known as:
(A) Aestivation (B) Hibernation
(C) Natural safety (D) None of these

273. Which of the following rat species is found in agricultural farms only:
(A) *Bandicota bengalensis*
(B) *Nesokia indica*
(C) *Tetera indica*
(D) All of these

274. Red triangle on insecticide container indicates:
(A) Highly poisonous
(B) Poisonous
(C) Medium poisonous
(D) Less poisonous

275. Relaxing insects of insect means:
(A) To collect dead insects in a special bottle for establishing its bodies
(B) To kill the insects
(C) To mount the insects
(B) None of these

276. Which of the following insects are insect parasites?
(A) *Trichogramma spp.*
(B) *Microbacon spp.*
(C) *Sarcophaga spp.*
(D) All of these

277. Which of the following is a nocturnal insect?
(A) Butterfly (B) Beetle
(C) Moth (D) Honey bee

278. Dragon fly belongs to order:
(A) Odonota (B) Orthoptera
(C) Isoptera (D) Hemipetra

279. Termites belong to order:
(A) Odonota (B) Hemiptera
(C) Isoptera (D) None of these

280. Simple metamorphosis refers to:
(A) Simple change in size from Nymph to adult Nymph are almost similar to adult
(B) Egg – Nymph – Adult stages are found
(C) It is found only in exopterygota insects
(D) All of these

281. Which of the following is known as incomplete metamorphosis?
(A) A metamorphosis
(B) Complete metamorphosis
(C) Simple metamorphosis
(D) None of these

282. Mciller's organ is a:
(A) Sense organ
(B) It is a group of small many sense organs
(C) It is found in Tympanum
(D) All of these

283. Complete metamorphosis is found in:
(A) Exopterygota (B) Endopterygota
(C) Apterygota (D) None of these

284. Mustard aphid has a mouth organs:
(A) Sucking type
(B) Piercing and sucking type
(C) Siphoning type (D) Chewing type

285. Which of the following stage of mustard aphid is harmful to crop plants?
(A) Nymph (B) Egg
(C) Adult (D) Nymph and Adult

286. Which stage of mustard saw fly is:
(A) Egg (B) Grubs
(C) Adult (D) Grub and Adult

287. Harmful stage of painted mustard bug is:
(A) Nymph (B) Egg
(C) Adult (D) Nymph and Adult

288. Harmful stage of red pumpkin beetle is:
(A) Nymph (B) grubs
(C) Adult (D) Grubs and Adult

289. Which of the following is a natural enemy of mustard aphid is:
(A) *Cocciniella septumpunctata*
(B) *Chrysopa spp.*
(C) *Syrphid maggot*
(D) All of these

290. Harmful stage of mango leaf hopper is:
(A) Nymph (B) Egg
(C) Adult (D) Nymph and Adult

291. Mango leaf roller commonly known as:
(A) Mango chanpa (B) Lassi
(C) Tella (D) All of these

292. Mango leaf hopper has mouth parts:
(A) Piercing type
(B) Sucking type
(C) Piercing and sucking type
(D) Chewing type

293. Scientific name of mango hopper is:
(A) *Amritodus atkinsoni*
(B) *Athalia spp*
(C) *Heliothis spp*
(D) None of these

294. Empusa fungus can be used against:
(A) Housefly (B) Housefly and Locust
(C) Moth (D) Grubs

295. Dimethoate is very effective against:
(A) Housefly (B) Fruit fly
(C) Aphids (D) All of these

296. Which of the following is a respiratory poison?
(A) Hydrogen sulphide
(B) Carbon monoxide
(C) Phosphin (D) All of these

297. Which of the following form of BHC is most toxic?
(A) α BHC (B) β BHC
(C) γ BHC (D) δ BHC

298. Which of the following is a physical repellent?
(A) Water barriers
(B) Tar
(C) Oil bands around tree trunks
(D) All of these

299. Which of the following is not attractant?
(A) Pheromones (B) Sound
(C) Light (D) Chemosterilants

300. Nerve poisons are:
(A) Chlorinated hydrocarbons
(B) Organo phosphates
(C) Plant origin insecticides
(D) All of these

301. Which of the following is a contact poison?
(A) Chlorinated hydrocarbons
(B) Organo phosphate
(C) Carbamates (D) All of these

302. BHC refers to:
(A) Benzene Hexa Chloride
(B) Benzene Hepta chloride
(C) Benzene Hema Chloride
(D) None of these

303. DDT refers to:
(A) Dichloro-diphenyl-trichloro ethane
(B) Dichloro-dinitro-tetra chloride
(C) Diphenyl-dichloride-trichloro ethane
(D) None of these

304. Pheromone is released by:
(A) Male insects
(B) Female insects
(C) Male and female insects
(D) None of these

305. Antifeedants are substances that:
(A) Kill the insects
(B) Repell the insects
(C) Repell insects without killing them
(D) None of these

306. Which of the following is a predator insects?
(A) Trichogramma (B) Coccinellids
(C) Apanteles (D) None of these

307. Which of the following is a predator insects?
(A) Trichogramma (B) Coccinellids
(C) Apanteles (D) None of these

308. Termites are very active during:
(A) Day (B) Night
(C) Day & Night (D) None of these

309. Six segmented head is found in:
(A) Myriapods (B) Arthopoda
(C) Annelida (D) None of these

310. Homeostasis refers to:
(A) Stability of population in a ecosystem
(B) Change in insect population
(C) Increase of insect population
(D) None of these

311. Gyplure is a:
(A) Synthetic compound related to natural compounds of gypsy moth
(B) Natural compound of gypsy moth
(C) Natural sex hormone of fruit fly
(D) None of these

312. Which of the following is not a locust species?
(A) *Locusta migraioria*
(B) *Pantanga succinata*
(C) *Schistocerca gregarta*
(D) *Cactoblastic spp.*

313. Which stage of pink boll worm is harmful to plants?
(A) Larvae only (B) Larvae and Adult
(C) Adult only (D) None of these

314. The only plant that remained untouched by *Schistocerca gregarta* (Desert locust):
(A) Mango (B) Ber
(C) Neem (D) None

315. Lantana can be controlled by:
(A) Lantana bug (B) Scale insect
(C) Fruit fly (D) None of these

316. Sexone is a:
(A) Insecticide (B) Weedicide
(C) Pesticide (D) Hormone mimics

317. Which of the following is a repellent for mosquitoes?
(A) Dolphene (B) Citronella oil
(C) Camphore oil (D) All of these

318. Monophagy refers to:
(A) Insect feeding on simple plant species
(B) Insect feeding on large variety of plants
(C) Insect feeding on few plants
(D) None of these

319. Oligophagy refers to:
(A) Insect feeding on simple plant spp.
(B) Insect feeding on large number of plants
(C) Insect feeding on few plants
(D) None of these

320. Which of the following stage of termite is harmful to plants?
(A) Nymph (B) Adult
(C) Pupae (D) Nymph and adult

321. Termite caste refers to:
(A) Sexual differentiation
(B) Physical differentiation
(C) Difference in mouth organs
(D) Difference in colour

322. Which of the following is the sterile caste of termite?
(A) Queen (B) Workers
(C) Soldiers (D) Workers & soldiers

323. Termite can be controlled by use of:
(A) BHC (B) Aldrin
(C) Chlordane (D) All of these

324. Gandhi bud is active in south India during:
(A) June – October
(B) November – December
(C) February – March
(D) May - June

325. Gandhi bud is active in north India during:
(A) June – October
(B) November – December
(C) February – March
(D) May – June

326. Pyrilla is a severe insect of:
(A) Rice (B) Sugarcane
(C) Maize (D) Wheat

327. Gurdaspur borer is a pest of:
(A) Rice (B) Sugarcane
(C) Maize (D) Wheat

328. Fruit fly lays eggs:
(A) On leaves (B) Inside fruits
(C) In soil (D) All of these

329. Which of the stage of pyrilla is more harmful:
(A) Adult (B) Nymph
(C) Pupa (D) Adult and Nymph

330. Which of the stage of fruit fly is more harmful:
(A) Adult (B) Nymph
(C) Maggot (D) Adult and Nymph

331. Gandhi bug has mouth organs:
(A) Piercing and sucking type
(B) Sponging type
(C) Siphoning type
(D) All of these

332. Which of the polyphagous insect?
(A) Termite (B) Bihar hairy caterpillar
(C) Locust (D) All of these

333. Fruit fly is harmful to:
(A) Papaya (B) Guava
(C) Cucurbits (D) All of these

334. Bacterial disease can be controlled by use of chemicals:
(A) Antibiotics (B) Viricides
(C) Fungicides (D) Kelthane

335. The Kresek occurs in early stage of plant growth of rice in:
(A) BLB (B) Bacterial leaf streak
(C) Tungro virus (D) False smut

336. Tungro virus of rice is transmitted by:
(A) Stem borer (B) Leaf roller
(C) Green leaf hopper
(D) Gandhi bug

337. Dead heart and white head damage to rice is caused by:
(A) Gall midge (B) Leaf roller
(C) Army worm (D) Stem borer

338. Which one of the following disease not induce floral abnormalities:
(A) Downey mildew of mustard
(B) Green ear of bajra
(C) White rust of crucifers
(D) None of these

339. Which one of the following pairs is not correctly matched:
(A) Paddy gall fly - Silver leaf
(B) Rice bug - Repugnant smell
(C) Rice stem borer - Black ears
(D) Rice zig-zag leaf hopper - Orange leaf

340. Which one of the following disease of pigeonpea is transmitted by Eryophid mite:
(A) Phytophthora stem rot
(B) Cercospora leaf spot
(C) Yellow mosaic
(D) Sterility mosaic

341. The most serious pest of Bengal gram is:
(A) Pod borer (B) Cut worm
(C) Aphid (D) Stem borer

342. Silvery shoot or onion leaf symptom of rice tillers is caused by:
(A) Rice hispa (B) Gall midge
(C) Stem borer (D) Gandhi bug

343. Grassy stunt virus desease of rice is transmitted by:
(A) Green leaf hopper
(B) Brown plant hopper
(C) White fly
(D) Gandhi bug

344. 'Akiochi' disease in rice is due to the toxicity caused by:
(A) Lead (B) Iron
(C) Phosphorus (D) Hydrogen sulphide

345. *Epricania melanoleuca* is a parasitoid effective against:
(A) Sugarcane scale
(B) Rice mealy bug
(C) Sugarcane pyrilla
(D) Rice leag hopper

346. Which one of the following types of resistance is present in a crop variety showing high degree of resistance to a pathogen but becoming susceptible to it ate large scale cultivation for 4-5 years or so:
(A) Vertical resistance
(B) Horizontal resistance
(C) Durable resistance
(D) None of these

347. Which one of the following is the characteristic of stripe rust of wheat:
(A) Brown, elongated ruptured pustules on stem and leaf sheath
(B) Yellow, small, spherical to oval ruptured uredo-pustules in rows of leaf
(C) White, spherical to oval, ruptured uredo-pustules on leaf
(D) Black teliopustules, scattered on under surface of the leaf

348. Soil sickness is caused by:
(A) Pea (B) Jowar
(C) Linseed (D) Both (B) and (A)

349. Studies of plant diseases is called:
(A) Virology (B) Plant Pathology
(C) Mycology (D) None of these

350. Who studies the plant diseases:
(A) Virologist
(B) Pathologist
(C) Plant Pathologist
(D) Mycologist

351. Plant Pathology word derived from three Greek words. Which of the following is not related to it?
(A) Phyton (B) Pathos
(C) Logus (D) Logy

352. Downey mildew fungus was discovered by:
(A) Anton de Bery (C) Luis Pasteur
(C) Morren (D) Tillet

353. The main source of primary inoculum of bacterial blight of cotton is:
(A) Seed (B) Soil
(C) Air (D) Water

354. Kresek phase is found in the rice disease:
(A) Bacterial leaf streak
(B) Bacterial blight
(C) Blast (D) Sheath blight

355. Which factor does not favour the rapid spread of bacterial blight of rice:
(A) Rainy weather
(B) Strong wind
(C) High humidity
(D) Temperature below 20⁰C

356. The symptoms only appear on leaf in case of:
(A) Angular leaf spot of cotton
(B) Bacterial leaf streak of rice
(C) Blast of rice
(D) Brown spot of rice

357. In the genus *Fusarium*, conidia are produced in:
(A) Pycnidia (B) Acervuli
(C) Sporodocia (D) Synnemata

358. Wilt disease can be controlled by eliminating soil borne infection through:
(A) Crop rotation
(B) Seed treatment
(C) Spraying of fungicide
(D) None of these

359. Wilt pathogen is mainly:
(A) Soil borne (B) Seed borne
(C) Air borne (D) None of these

360. The size of pesticide molecules, if all other conditions are constant, greatly influences it is:

(A) Absorption (B) Solubilization
(C) Adsorption (D) Volatilization

361. Some pesticides with positively charged groups are also absorbed by:

(A) Illite clays
(B) Kaolinite clays
(C) Monmorillonite clays
(D) Silicate clays

362. In general, herbicides are ---- than either fungicides or insecticides:

(A) More mobile (B) Less mobile
(C) Equally mobile (D) More immobile

363. Some degradation products of pesticides are even ------ than the original pesticides.

(A) Less toxic (B) More toxic
(C) Non-toxic (D) Equally toxic

364. The total number of bacteria frequently ----- after fumigation:

(A) Increases
(B) Decreases
(C) Remains constant
(D) None of these

365. The process of ammonification is often ---- by the use of pesticides:

(A) Harmful (B) Benefitted
(C) Not affected (D) None of these

366. The principle that in modern times "the distribution of plant species on the earth is not uniform" was established by:

(A) Charles Darwin
(B) De Candolle
(C) Nikolai Vavilov
(D) Carl Von Linnaeus

367. Vavilov recognized primary centres of probable origin of cultivated plants numbering:

(A) 4 (B) 5
(C) 8 (D) 10

368. Species and subdivisions in plants were designated respectively at Linneons and Jordanons by:

(A) W.F. Libby (B) L.P. Lotsy
(C) R.E. Reeves (D) M.Y. Nuttonson

369. The settled village community farming based on small grain and animal domestication occurred in Near East about ------ years ago:

(A) 5000 (B) 7500
(C) 10000 (D) 10500

370. The causal organism of black rust in wheat is:

(A) *Pucinia recondite tritici*
(B) *Puccinia streiformis*
(C) *Puccinia graminis tritici*
(D) *Ustilago nuda tritici*

371. Karnal bunt or partial bunt was first observed in experimental seed plots at:

(A) Kernal (B) Pantnagar
(C) New Delhi (D) Hisar

372. Brown stripe downy mildew of maize can be controlled by use of:

(A) Bordeaux mixture (4:450)
(B) Bavistin
(C) Karathane-45
(D) Dithane M-45

373. The causal agent of anthracnose diseases in sorghum is:

(A) *Puccinia purpurea*
(B) *Helminthosporium turcicum*
(C) *Colletotrichum graminicola*
(D) *Gleourcospora sorghi*

374. The chickpea wilt is caused by:

(A) *Fusarium oxysporum*
(B) *Ascochyta rabi*
(C) *Bortrylis cineraria*
(D) *Scelerotinia sclerotiorum*

375. Chickpea pod borer can be controlled by the spray of:

(A) Monocrotophos @ 1 ml/litre of water
(B) Aldrin @ 15 kg/ha
(C) Seed treatment by captan (0.25%) @ 4g/ha of seed
(D) BHC @ 10 kg/ha

376. The seed of safflower should be treated with Agrosan GN

(A) 30 g/kg seed
(B) 30mg/kg seed
(C) 30 decigram/kg seed
(D) 100 decigram/kg seed

377. Safflower rust can be controlled by:
(A) Spray of Zineb
(B) Seed reatment with thiram
(C) Destroying the infected material
(D) Following the 3 to 4 years crop rotation

378. Treatment of cotton seed with organo-mercurial compound (s) helps to minimize the disease:
(A) Soil borne (B) Air borne
(C) Water borne (D) Seed borne

379. The most destructive disease of sugar cane is:
(A) Red rot (B) Wilt
(C) Smut (D) Albino

380. To avoid rotting in early planting of potato, seeds should be treated with:
(A) 0.5 per cent Dithane M-45
(B) 2.0 per cent Dithane M-45
(C) 1.5 per cent Aretan
(D) 2.0 per cent Tafasan

381. Treat the sugarcane seed setts with--------- against fungaldise:
(A) Diathane M-45 (B) Zineb
(C) Thiram (D) Aretan

382. The ------- botanist Micheli (1679-1737) was the first scientist who in 1729 studied fungi and show their spores:
(A) German (B) Dutch
(C) Italian (D) Austrian

383. In --------, the French botanist Tillet published a paper on bunt or stinking smut of wheat:
(A) 1555 (B) 1655
(C) 1755 (D) 1855

384. By 1857, Berkley and Schacht discovered root knot nematodes and cyst nematode in- --------:
(A) Sweet sorghum (B) Sugarcane
(C) Potato (D) Sugarbeet

385. The discovery of -------- mosaic by Meyer in 1886 was the beginning of studies on viruses as incitants of diseases in plants:
(A) Tomato (B) Mungbean
(C) Tobacco (D) Soybean

386. --------, before 1910 initiated exhaustive study of Indian fungi and diseases caused by them:
(A) E.J. Butler (B) J.G. Moseman
(C) R.E. Nelson (D) A. Livney

387. Dr. K. C. Mehta of Agra College made an outstanding contribution to plant pathology by discovering the disease cycle of cereal in India:
(A) Smuts (B) Rusts
(C) Wilts (D) Blights

388. In bacterial brown rot and wilt of potato the amount of -------- produced by pathogens are proportional to severity of symptoms:
(A) Gibberellins (B) Lipids
(C) Proteins (D) Polysaccharides

389. In *Alternaria solani*, the conidia are formed at:
(A) Top end of the condiophores
(B) Intercalary formation
(C) Both (A) and (B)
(D) None of these

390. The infected seed of bajra smut are converted into:
(A) Sori (B) Nod
(C) Mycelium (D) None of these

391. Grain smut of jowar may cause following disease in humen beings:
(A) Hey fever (B) Cholera
(C) Cold (D) None of these

392. Loose smut is a:
(A) Internally seed borne disease
(B) Air borne disease
(C) Soil borne disease
(D) Externally seed borne disease

393. The pigeonpea wilt is caused by:
(A) *Fusarium oxysporum*
(B) *Fusarium udum*
(C) *Ascochyta rabi*
(D) *Bortrytis cineraria*

394. Mechanism of wilting can be understood by:
(A) Plugging of vessels by fungus hyphae
(B) Toxin theory-production of toxins like Fusaric acids
(C) Enzyme theory-Production of enzymes like Pectolytic enzyme
(D) All of these

395. Alternaria solani is a:
(A) Soil borne disease
(B) Seed borne disease
(C) Air borne disease
(D) None of these

396. Which of the perfect stage of red rot of sugarcane:
(A) *Glomerella tucumanensis*
(B) *Physalospora tucumansis*
(C) Both (A) and (B)
(D) None of these

397. The perfect stage of tikka disease of groundnut is:
(A) *Mycosphaerella berkeleyii*
(B) *Mycosphaerella arachidicola*
(C) Both (A) and (B)
(D) None of these

398. Rotten neck disease is known as:
(A) Blast of rice
(B) Red rot of sugarcane
(C) Leaf rust of wheat
(D) None of these

399. What is the main symptoms of red rot of sugarcane?
(A) Dull red pith
(B) Stunted growth
(C) Shorter internodes
(D) None of these

400. Which of the main symptoms of leaf blight of wheat?
(A) Small, oval, discolored, irregular lesions scattered on leaves
(B) Leaf starts drying from tip
(C) Black powdery conidia of the fungus cover the lesion under moist conditions
(D) All of these

401. Polymorphism refers to the process having:
(A) > 2 life cycle stages
(B) Only one life cycle
(C) < 2 life cycle stages
(D) All of these

402. Which of the following fruiting body is formed in red rot of sugarcane?
(A) Acervuli (B) Pycnidia
(C) Apothecia (D) All of these

403. Photosynthetic bacteria have photosynthetic pigments made of:
(A) Lamellae and chromatophores
(B) Ribosomes
(C) Endoplasmic reticulum
(D) All of these

404. Which of the following plant disease is caused by MLO's?
(A) Witch brooms of potato
(B) Mulbery dwarf
(C) Aster yellows (D) All of these

405. Which of the following feature is related to MLO.s?
(A) Can be grown on cell free media
(B) They are resistant to Penicillin and susceptible to tetracycline
(C) They can pass through anti bacterial filters
(D) All of these

406. Nimbin and Nimbidin are extracted from:
(A) Neem (B) Ashwagandha
(C) Amla (D) Corainder

407. Aflatoxin are produced by:
(A) Bacteria (B) Algae
(C) Virus (D) Fungi

408. *Bacillus thuringiensis* is commercially useful as:
(A) Bio-fungicide (B) Bio-weedicide
(C) Bio-insecticide (D) Bio-fertilizer

409. The optimum temperature for development of leaf blight of wheat is:
(A) 15^0C (B) 20^0C
(C) 25^0C (D) 30^0C

410. Literary meaning of virus is:
(A) Morbid poison (B) Acute poison
(C) Poison (D) All of these

411. Antibiotics are prepared mainly from:
(A) Actinomycetes (B) Rhizobium
(C) Pseudomonas (D) Azotobacter

412. Grain smut disease of jowar is a disease of:
(A) Internally seed borne
(B) Externally seed borne
(C) Air borne (D) Soil borne

413. Bordeaux mixture is a:
(A) Systemic fungicide
(B) Fungicide and bactericide
(C) Tropical inorganic fungicide
(D) Tropical organic fungicide

414. Which of the following does not posses mitochondria?

(A) Fungi (B) Bacteria
(C) Virus (D) MLO's

415. The virus that attacks on bacteria is called:

(A) Bacteria parasite
(B) Bacteria phages
(C) Virophages
(D) None of these

416. Viroids have only:

(A) Naked nucleic acid particles
(B) Nucleic acid protein core
(C) Only protein core
(D) None of these

417. What is optimum temperature for spread of loose smut disease?

(A) 18^{0}C (B) 23^{0}C
(C) 28^{0}C (D) 33^{0}C

418. Which of the following does not possess metabolic ability?

(A) Fungi (B) Bacteria
(C) Virus (D) All of these

419. What is causal organism of yellow rust of wheat?

(A) *Puccinia striiformis*
(B) *Puccinia recondita*
(C) *Puccinia tritici*
(D) All of these

420. Ear cockle of wheat or Sehun disease is caused by:

(A) *Anguina tritici*
(B) *Heterodera avenae*
(C) *Meloidogyne javanica*
(D) All of these

421. Molya disease of barley is caused by:

(A) Bacteria (B) Virus
(C) Nematodes (D) None of these

422. The causal organism of molya diseases is:

(A) *Heterodera avenae*
(B) *Melodogne spp.*
(C) *Anguina spp.*
(D) None of these

423. Molya disease of barley first reported at:

(A) Gujarat (B) UP
(C) Punjab (D) Rajasthan

424. Virus is found in cells in form of:

(A) Crystalline groups
(B) Liquid from
(C) Suspension form
(D) None of these

425. Which of the following crops are affected by orobanche?

(A) Tobacco (B) Brinjal
(C) Tomato (D) All of these

426. Ear cockle of wheat is often associated with:

(A) Yellow ear rot or Tundu disease
(B) Loose smut
(C) Rust
(B) None of these

427. Which of the following striga species are found in India?

(A) *Striga lutea*
(B) *Striga densiflora*
(C) *Striga euphrasiodes*
(D) All of these

428. Loranthus is parasite of:

(A) Mangoes (B) Guava
(C) Pipal (D) All of these

429. Who is the father of Indian plant pathology?

(A) E.T. Butler (B) K.C. Mehta
(C) R.S. Singh (D) R. Verma

430. Who told, 'Virus is a contagium vivum fluidum'?

(A) Beijerinck (B) Ivanowski
(C) Meyor (D) R. Kotch

431. The different spore stages of rust fungi are known as:

(A) 0, I, II, III and IV stages
(B) 1, 2, 3, 4 and 5 stages
(C) 0, I, ii, iii, iv and iv stages
(D) None of these

432. Who is associated with the study of wheat rust in India:

(A) B.B. Mundkur (B) E.J. Butler
(C) K.C. Mehta (D) J.F. Dastur

433. Which type of rust does not occur on wheat:

(A) White rust (B) Black rust
(C) Brown rust (D) Yellow rust

434. Heteroecious rust completes its life cycle on:
(A) One host (B) Two unrelated hosts
(C) Three hosts (D) None of these

435. Pustules are often arranged in rows in:
(A) Black rust (B) Brown rust
(C) Yellow rust (D) None of these

436. *Puccinia recondita* is the causal organism of:
(A) Black rust (B) Brown rust
(C) Yellow rust (D) White rust

437. Appearance of wheat rusts in India takes place through the dissemination of:
(A) Pycniospores (B) Aeciospores
(C) Uredospores (D) Teliospores

438. Fusaric acid plays an important role in pathogenesis of:
(A) Wilt (B) Blight
(C) Rust (D) Smut

439. The perfect size of *Colletotrichum falcatum* is:
(A) *Pyrenophora graminae*
(B) *Mycosphaerella arachidicola*
(C) *Glumerellatu cumanensis*
(D) *Cochliobolus miyabeanus*

440. In red rot of sugarcane, symptoms also appear on:
(A) Tip of the leaves
(B) Margin of the leaf
(C) Midrib of the leaves
(D) None of these

441. In *Colletrotrichum falcatum*, the conidia are produced in:
(A) Pycnidia (B) Acervuli
(C) Synnemata (D) Sporodochia

442. Karathane is basically used for the control of:
(A) Downey mildew
(B) Powdery mildew
(C) Wilt
(D) Blast

443. Bordeaux mixture is a:
(A) Copper fungicide
(B) Sulphur fungicide
(C) Dithiocarbamate
(D) Systemic fungicide

444. Which viruses is likely to be commercialized in India for control of *Helicoverpa (Heliothis) armigera*?
(A) Nucleopolyhedrosis viruses (NPV)
(B) Granulosis viruses (GV)
(C) Advenoviruses
(D) All of these

445. Pusa 1406 has resistance for :
(A) Bacterial leaf blight
(B) Blast
(C) Tungro (D) Grassy stunt

446. Which disease has caused the famous 'Bengal famine':
(A) Rust of wheat
(B) Red rot of sugarcane
(C) Tikka disease of groundnut
(D) Brown spot of paddy

447. The toxin 'Cochliobolin' is produced by the causative fungus of:
(A) Chickpea grey mould
(B) Sugarcane wilt
(C) Rice brown spot
(D) Rice blast

448. In blast of rice, the symptoms appear on:
(A) Leaves only (B) Leaf-sheath only
(C) Panicle only (D) On all above parts

449. A disease which occurs widely but periodically is:
(A) Epidemic disease
(B) Endemic disease
(C) Sporadic disease
(D) Pandemic disease

450. A disease which occurs at a short interval and its severity is more called:
(A) Epidemic disease
(B) Endemic disease
(C) Sporadic disease
(D) Pandemic disease

451. The pathogen of brown spot of rice is:
(A) Soil borne (B) Seed borne
(C) Soil and seed borne
(D) None of these

452. Smuts, in general are considered as:
(A) Obligate parasites
(B) Obligate saprophytes
(C) Facultative parasites
(D) Facultative saprophytes

453. The conidia of *Pyricularia oryzae* are:
(A) One celled (B) Two celled
(C) Three celled (D) Many celled

454. Mycellium of powdery mildew pathogen is:
(A) Ectophytic (B) Endophytic
(C) Intercellular (D) Intracellular

455. In red rot of sugarcane, the reddening is found in:
(A) Parenchyma (B) Pith
(C) Phloem (D) Xylem

456. In *Pyricularia oryzae*, the spots on leaves are mostly:
(A) Reniform (B) Pyriform
(C) Doliform (D) Spherical

457. In Ascochyta blight, the spots on leaves are mostly:
(A) Angular (B) Circular
(C) Oblong (D) Irregular

458. What is causal organism of yellow rust of wheat?
(A) *Puccinia striiformis*
(B) *Puccinia recondita*
(C) *Puccinia tritici*
(D) All of these

459. What is optimum temperature for germination of Urediniospores?
(A) 3-5⁰C (B) 15-16⁰C
(C) 30-35⁰C (D) 40-45⁰C

460. What is optimum temperature for spread of loose smut disease?
(A) 18⁰C (B) 23⁰C
(C) 28⁰C (D) 33⁰C

461. Grain smut disease of jowar is a disease of:
(A) Internally seed borne
(B) Externally seed borne
(C) Air borne (D) Soil borne

462. Karnal bunt is the name because:
(A) Symptoms appear in Kernal
(B) Reported from Karnal
(C) Reported by Karnal
(D) None of these

463. False smut is the name because its causal organism does not belong to:
(A) Masticomycotina
(B) Zygomycotina
(C) Ascomycotina (D) Basidiomyotina

464. Disease development is rapid in Karnal bunt due to:
(A) Frequent light rains
(B) Cloudy weather
(C) Low temperature
(D) Combination of above three factors

465. The optimum temperature for development of leaf blight of wheat is:
(A) 15⁰C (B) 20⁰C
(C) 25⁰C (D) 30⁰C

466. The causal organism of ear cockle of wheat is:
(A) *Ustilago tritici*
(B) *Puccinia triticina*
(C) *Anguina tritici*
(D) *Alternaria triticina*

467. Whip smut occurs in:
(A) Maize (B) Sorghum
(C) Bajra (D) Sugarcane

468. Silica content in leaf helps rice plant to resist against:
(A) Blast (B) False smut
(C) Bacterial blight (D) None of these

469. *Erysiphe graminis tritici* is:
(A) Obligate saprophyte
(B) Obligate parasite
(C) Facultative parasite
(D) Facultative saprophyte

470. Ear cockle of wheat is a :
(A) Fungal disease
(B) Bacterial disease
(C) Viral disease
(D) Nematode disease

471. The pathogen of loose smut of wheat survives:
(A) With the seeds (B) On the seeds
(C) Within the seeds (D) None of these

472. Which one is the abiotic disease of rice:
(A) Blast (B) False smut
(C) Khaira (D) Brown spot

473. Only the naked rachis remains in the last in case of:
(A) Covered smut of barley
(B) Loose smut of wheat
(C) Grain smut of Jowar
(D) Bajra smut

474. Solar energy treatment against loose smut of wheat was evolved by:

(A) Mehta (B) Mundkur
(C) Butler (D) Luthra and Sattar

475. Ghost light is a result of:

(A) Bacterial luminescence
(B) Phosphorescent glow
(C) Bioluminescence
(D) All of these

476. The main symptoms of black leg or soft rot of potato is:

(A) Brown – black or jet black colour of stem at soil level
(B) Black leaf coloration
(C) Black crust on tubers
(D) None of these

477. Bacterial blight of cotton is also called as:

(A) Tar leaf spot
(B) Angular leaf spot
(C) Zonate leaf spot
(D) Red leaf spot

478. Bacterial blight of cotton does not produce the symptoms on:

(A) Seedlings (B) Stems
(C) Bolls (D) Roots

479. The disease is a typical vascular wilt in rice:

(A) Blast
(B) Brown leaf spot
(C) Bacterial leaf streak
(D) Bacterial leaf blight

480. Some varieties of -------- are resistant to wilt due to the presence of hydrocyanides (HCN) in root exudates:

(A) Sunflower (B) Safflower
(C) Castor (D) Linseed (flax)

481. Oil from American elm seed contains 55 per cent capric acid toxic to *Ceratocystis ulmi,* the pathogen causing the famous disease:

(A) English elm (B) Dutch elm
(C) American elm (D) French elm

482. Root knot nematodes have higher growth at :

(A) 15-30^{0}C (B) 20-25^{0}C
(C) 25-28^{0}C (D) 30-35^{0}C

483. Which of the following disease is caused by MLO's?

(A) Little leaf of egg plant
(B) Citrus greening
(C) Purple top roll of potato
(D) All of these

484. Which of the following disease is caused by virus?

(A) Fiji disease of sugarcane
(B) Leaf curl of papaya
(C) Ratoon stunting of sugarcane
(D) None of these

485. A plant disease spreads after irregular intervals at different locations in big way, is called:

(A) Endemic disease
(B) Epidemic disease
(C) Sporadic disease
(D) All of these

486. Which of the crop is affected mostly by striga?

(A) Sugarcane (B) Maize
(C) Millets (D) All of these

487. Which of the following gas can stimulate the seeds of striga?

(A) Ethylene (B) Nitrous oxide
(C) Carbon mono oxide
(D) Oxygen

488. Strigol is used to control:

(A) Striga (B) Cuscuta
(C) Orobanche (D) Loranthus

489. Which of the following is a synthetic culture medium?

(A) Oat agar medium
(B) Plant extracts
(C) Rechard's solution
(D) None of these

490. Hill bunt of wheat is caused by:

(A) *Tiletia foetida* (B) *Neovossia indica*
(C) *Uromystis tritici*
(D) None of these

491. Which of the following is used for tree surgery?

(A) Bordeaux mixture
(B) Chaubatia paste
(C) Santar A
(D) All of these

492. A plant disease spreads on a larger area with great loss is called:
(A) Endemic disease
(B) Epidemic disease
(C) Sporadic disease
(D) All of these

493. Hill bunt can be controlled by:
(A) Seed treatment with 0.2% vitavex
(B) Use of resistant varieties
(C) Clean the field
(D) All of these

494. Litchen is a example of:
(A) Mutalistic association
(B) Symbiotic association
(C) Parasitism
(D) All of these

495. Trychoderma can be used as:
(A) Soil (B) Seed
(C) Both of these (D) None of these

496. Which one the important disease of groundnut is:
(A) Phyllody (B) Wilt
(C) Tikka (D) Cercospora leaf spot

497. Which one the important disease of sesame is:
(A) Phyllody (B) Wilt
(C) Tikka (D) Cercospora leaf spot

498. Sterility mosaic is a disease related to:
(A) Fungal (B) Bacterial
(C) Viral (D) Physiological

499. Black tip is a common disease of:
(A) Banana (B) Grape
(C) Orange (D) Mango

500. Which wheat rust normally appears earlier in plains of Uttar Pradesh:
(A) Black rust (B) Yellow rust
(C) Brown rust (D) None of these

501. Covered smut of barley is a disease of:
(A) Externally seed borne
(B) Internally seed borne
(C) Air borne
(D) None of these

502. Seed treatment of sugarcane is done by:
(A) Agalol (B) Thiram
(C) Vitavax (D) Zinc sulphate

503. The causal organism of powdery mildew of wheat is:
(A) *Erysiphe graminis var tritici*
(B) *Erysiphe polygoni*
(C) *Erysiphe aestivum*
(D) None of these

504. Bacterial disease are controlled by:
(A) Antibiotics (B) Viricide
(C) Fungicide (D) Kethane

505. Seed treatment of pulses is done with:
(A) Azotobactor (B) Rhizobium
(C) Azospirillum (D) Phosphotica

506. Dr. K.C. Mehta was associated with:
(A) Cereal rust in India
(B) Smut research
(C) Bacterial research
(D) Virus research

507. All plant pathogenic bacteria are:
(A) Gram positive
(B) Gram negative
(C) Gram positive and gram negative
(D) None of these

508. Infectious disease is caused by:
(A) Parasites (B) Non parasites
(C) Virus (D) All of these

509. Non infectious disease is caused by:
(A) Nutrient deficiency
(B) Mineral deficiency
(C) Pollutants
(D) All of these

510. Saprophytic pathogens feed on:
(A) Only dead matter
(B) Only living organisms
(C) Both (A) and (B)
(D) None of these

511. Parasites feed on:
(A) Only dead matter
(B) Only living organisms
(C) Both (A) and (B)
(D) None of these

512. Which of the following has its own metabolic activity?
(A) Fungus (B) Virus
(C) Viroid (D) Plasmids

513. In gametangial contact, ascogonia contact with antheridia with the help of a special structure known as:

(A) Trichogyne (B) Dikaryon hyphae
(C) Stroma (D) All of these

514. A plant disease spreads on a regular intervals is called:

(A) Endemic disease
(B) Epidemic disease
(C) Sporadic disease (D) All of these

515. Which of the following is a biotroph pathogen?

(A) Viruses (B) Rust fungi
(C) Downy and Powdery mildew
(D) All of these

516. The pathogen that can grow only under controlled environment is called:

(A) Ecologically obligate
(B) Bio-chemically obligate
(C) Facultative parasite
(D) All of these

517. Leaf curl is a serious disease of:

(A) Moong (B) Pea
(C) Gram (D) Lentil

518. Ear cockle disease is found in:

(A) Sugarcane (B) Wheat
(C) Cotton (D) Rice

519. Blight in pulses can be controlled by:

(A) Spray of manganese carbamate @ 2.0 kg / ha
(B) Ziram 80% @ 2.0 kg/ha before flowering
(C) Seed treatment with Benlate and Thiram
(D) All of these

520. Downey mildew in pulses can be controlled by spraying of:

(A) Mencozeb @ 0.2 %
(B) Thiofinate methyl 70% @ 0.15 %
(C) Grow resistant variety
(D) All of these

521. Powdery mildew in pea can be controlled by spraying of:

(A) Dusting of Pendimethalin @ 2.0-3.0 l/ha
(B) Karathan @ 0.06% spray
(C) Stomp @ 2.0 l/ha
(D) None of these

522. Khaira disease is caused due to deficiency of:

(A) Iron (B) Zinc
(C) Sulphur (D) Nitrogen

523. Wilt is an important disease of:

(A) Pea (B) Arhar
(C) Mustard (D) Wheat

524. The wilt resistant variety of chickpea is:

(A) K-850 (B) KWR-108
(C) Sadabahar (D) Radhey

525. England famine in 1846 was caused mainly by:

(A) Late blight of potato
(B) Helminthosporium leaf spot of rice
(C) Rust of wheat
(D) Leaf curl virus in tomato

526. One of the most important reason of Bengal famine in 1943 was:

(A) Late blight of potato
(B) Helminthosporium leaf spot of rice
(C) Rust of wheat
(D) Leaf curl virus in tomato

527. An obligate pathogen can:

(A) Grow only on living organism
(B) Grow on dead matters
(C) Grow on both living and non living matters
(D) None of these

528. Fungi is devoid of:

(A) Chlorophyll (B) Cell wall
(C) Nucleus capsid (D) None of these

529. Yellow mosaic in soybean spread by:

(A) Seeds (B) Soils
(C) Air (D) White fly

530. Loose sumt of wheat in controlled by:

(A) Seed treatment
(B) Soil solarization
(C) Crop rotation
(D) Thiram spray on crop

531. Some ------varieties of apple exude toxic wax on leaf surface to prevent germination of conidia of *Podosphaera leucotricha*:

(A) Scab resistant
(B) Leaf blight resistant
(C) Powdery mildew resistant
(D) Rust resistant

532. Homothallic fungus is:
(A) Dioecious (B) Monoecious
(C) Both of these (D) None of these

533. The fungi grown on dung of herbivorous animals are called as:
(A) Coprophilous fungi
(B) Keratinaceous fungi
(C) Amphibious fungi
(D) All of these

534. The fungi grown on dead bodies are called as:
(A) Coprophilous fungi
(B) Keratinaceous fungi
(C) Amphibious fungi
(D) All of these

535. The root like structure of fungi is called as:
(A) Rhizoid (B) Stroma
(C) Rhizomorph (D) Sclerotium

536. Karnal bunt is a disease of:
(A) Viral (B) Nematodes
(C) Fungal (D) Bacterial

537. The pigeonpea wilt is caused by:
(A) *Fusarium oxysporum*
(B) *Fusarium udum*
(C) *Ascochyta rabi*
(D) *Bortrytis cineraria*

538. Bakanae disease is caused by:
(A) Fungus (B) Alga
(C) Bacterium (D) Virus

539. *Chenopodium album* is a weed which mostly grown in the crop:
(A) Kharif (B) Rabi
(C) Zaid (D) None of these

540. The main objective of Integrated Pest management is:
(A) To reduce pest status
(B) To kill pests
(C) To prevent pest entry
(D) To reduce use of insecticides

541. Aflatoxin is produced by:
(A) *Aspergillus flavus*
(B) *Phytophthora infestans*
(C) *Albugo candida*
(D) *Puccinia graminis tritici*

542. Under integrated system of insect control we integrate:
(A) Cultural and biological methods
(B) Biological and legal methods
(C) Physical and chemical methods
(D) All methods

543. The least toxic insecticides to human being belong to the following group:
(A) Organo chlorines
(B) Organo phosphates
(C) Carbamates
(D) Synthetic pyrethroids

544. Which of the following is a facultative parasite?
(A) Fusarium (B) Rhizoctonia
(C) Alternaria (D) All of these

545. Which of the following is a facultative saprophyte?
(A) Smut
(B) Helminthosporium
(C) Leaf curl fungus
(D) All of these

546. In which of the following group of fungi has similar sporangiophore to its vegetative hyphae?
(A) Pythium (B) Albugo
(C) Permospora (D) Sclerospora

547. Aflatoxin poison is produced by:
(A) *Aspergillus flavus*
(B) *Claviceps purpurea*
(C) *Pencillium spp*
(D) All of these

548. Gibberellic acid is formed by:
(A) *Gibberella fujikuroi*
(B) *G. mosshata*
(C) *Rhizopus niger*
(D) *Aspergillus flavus*

549. The member of oomycetes are:
(A) Uniflagellate (B) Biflagellate
(C) Triflagellate (D) None of these

550. Amphigynous antheridium is found in:
(A) Pythium (B) Phytophthora
(C) Both A and B (D) None of these

551. Which of the following is a obligate parasite?
(A) Puccinia (B) Pernospora
(C) Albugo (D) All of these

552. Which of the following is not saprophyte?
(A) Erysiphe (B) Aspergillus
(C) Mucor (D) Penicillium

553. Indole acetic acid is obtained from:
(A) *Gibberella spp.*
(B) *Rhizopus niger*
(C) *Aspergillus flavus*
(D) *Claviceps purpurea*

554. Alkaloid can be obtained from:
(A) *Gibberella spp.*
(B) *Rhizopus niger*
(C) *Aspergillus flavus*
(D) *Claviceps purpurea*

555. Pin moulds are also known as:
(A) Mucorales (B) Algal fungi
(C) Yeast fungi (D) None of these

556. Seed treatment with vitavex is the mainly control :
(A) Loose smut (B) Rust
(C) Downy mildew (D) Powdery mildew

557. Loose smut of wheat can be controlled by:
(A) Seed treatment by vitavex before sowing
(B) Seed treatment with hot water
(C) Use of disease resistant varieties
(D) All of these

558. Chemically virus is a:
(A) Nucleoprotein (B) Proteir
(C) Nucleic acid (D) All of these

559. Virus has mainly two parts, one is nucleic acid core and second part is:
(A) Cell wall (B) Nucleic membrane
(C) Protein core (D) None of these

560. Which of the following disease is caused by virus?
(A) Small pox (B) Measles
(C) Common cold (D) All of these

561. Ultra microscopic particle capable of infecting live cells is called:
(A) Plasmic (B) Nucleiod
(C) Virion (D) None of these

562. Bacterial soft rot is the result of:
(A) High temperature
(B) Very low temperature
(C) Waterlogging
(D) None of these

563. Bacteria having flagella at its whole body is known as:
(A) Monotrichous
(B) Amphitrichous
(C) Petritrichous
(D) None of these

564. Bacteria having a bunch of flagella at one end is known as:
(A) Monotrichous
(B) Lophotrichous
(C) Petritrichous
(D) None of these

565. Which of the following disease is caused by bacteria?
(A) Brown rot of potato
(B) Angular leaf spot of cotton
(C) Fire blight of apple and pear
(D) All of these

566. Which of the following disease is caused by bacteria?
(A) Red stripe of sugarcane
(B) Soft rot of cotton
(C) Leaf roll of potato
(D) Stalk rot of maize

567. Which of the following is a autotrophic bacteria?
(A) *Chlorobium* (B) *Thiospirillum*
(C) *Nitrobacter* (D) All of these

568. Which of the following is a heterotrophic bacteria?
(A) *Chlorobium* (B) *Pseudomonas*
(C) *Xantthomonas* (D) All of these

569. Fungi have been originated from:
(A) Polygeneric line
(B) Phytogeneric line
(C) Polygendric line
(D) None of these

570. The carrier of leaf curl tomato virus is:
(A) White fly (B) Beetle
(C) Bacteria (D) Fungus

571. Cyst formation is very common in which of the following bacteria?
(A) Azotobacter
(B) Myxococcus
(C) Chondromyces
(D) All of these

572. Which of the following feature of gram positive bacteria?
(A) Homogenous cell wall that contains >85% mucopeptides and its polymers
(B) Teichoic acids, ribitol, glutamic acid and alanine are found in more quantity
(C) Violet or blue colour after staining with solution of crystal violet and iodine
(D) All of these

573. Which of the following feature of gram negative bacteria?
(A) Cell wall is thin and made of mainly lipoprotein and its polymers
(B) They react with counter stain safranine and become red to pink in colour
(C) Teichoic acid is not formed
(D) All of these

574. Generally plant virus posses a:
(A) Single stranded RNA
(B) Double stranded RNA
(C) Doubled stranded DNA
(D) None of these

575. Viruses are resistant to:
(A) Acids (B) Bases
(C) Salts (D) All of these

576. Virus is a:
(A) Obligate parasite
(B) Obligate saprophyte
(C) Facultative parasite
(D) Facultative saprophyte

577. Suitable bioagents for controlling of wilt disease of chickpea, lentil and pigeonpea is:
(A) *Trychoderma harianum*
(B) *Trichoderma viridi*
(C) Both of these
(D) None of these

578. Sexual reproduction in yeast occurs under:
(A) Favourable condition
(B) Unfavourable condition
(C) Both A and B
(D) None of these

579. Asperigillosis is caused by:
(A) Aspergillus fungus
(B) Bacteria
(C) Virus
(D) None of these

580. Wood is deteriorated by the species of Panicillium:
(A) *Penicillium divaricatum*
(B) *Penicillium expensum*
(C) *Penicillium glacuum*
(D) None of these

581. Rust fungi mainly a:
(A) Saprophyte
(B) Parasite
(C) Facultative parasite
(D) All of these

582. Which of the alkaloid is produced by ergot fungi?
(A) Ergotin (B) Penicillin
(C) Cutrinin (D) All of these

583. The initial stage of ergot infection is known as:
(A) Honey dew stage
(B) Scleotial stage
(C) Spore formation stage
(D) All of these

584. The sexual fruiting body of ergot fungi is:
(A) Apothecium (B) Cleistothecium
(C) Synnemata (D) Perithecium

585. The size of virus particle is:
(A) 10-300 μ (B) 10-20 μ
(C) 50-80 μ (D) 25-35 μ

586. The chemical present in Temik is:
(A) Malathion (B) Propoxur
(C) Carbofuran (D) Aldicarb

587. The chemical present in flit (finit) is:
(A) Malathion (B) DDT
(C) BHC (D) Aldicarb

588. Smut fungi are:
(A) Obligate parasite
(B) Obligate saprophyte
(C) Facultative saprophyte
(D) Facultative parasite

589. Rust fungi are:
(A) Polymorphic (B) Homothallic
(C) Monoecious (D) All of these

590. Smut means:
(A) Sooty powder (B) Wet powder
(C) Grey powder (D) All of these

591. Seed treatment with vitavex is the mainly control :
(A) Loose smut (B) Rust
(C) Downy mildew (D) Powdery mildew

592. Loose smut of wheat can be controlled by:
(A) Seed treatment by vitavex before sowing
(B) Seed treatment with hot water
(C) Use of disease resistant varieties
(D) All of these

593. Suitable bioagents for controlling of wilt disease of chickpea, lentil and pigeonpea is:
(A) *Trychoderma harianum*
(B) *Trichoderma viridi*
(C) Both of these
(D) None of these

594. Karnal bunt is a disease of:
(A) Viral (B) Nematodes
(C) Fungal (D) Bacterial

595. Eu-form rust are also known as:
(A) Long cycled rust
(B) Short cycled rust
(C) Demi cycled rust
(D) None of these

596. Which is not an example of long cycled rust?
(A) Wheat rust
(B) Sunflower rust
(C) Lucerne rust
(D) White rust of Brassica

597. The life cycle of rust fungi, in which uredinial stage is not found is known as:
(A) Macro cycle (B) Demi cycle
(C) Micro cycle (D) None of these

598. The life cycle of rust fungi, in which all five stages are not found is known as:
(A) Macro cycle or long cycle
(B) Demi cycle
(C) Micro cycle
(D) None of these

599. The life cycle of rust fungi where only dicoryotic telial stage is not found is known as:
(A) Macro cycle (B) Demi cycle
(C) Micro cycle (D) None of these

600. Autocious fungi complete their life cycle on:
(A) Single host (B) Two host
(C) Dead matter (D) None of these

601. Heterocious fungi complete their life cycle on:
(A) Single host
(B) More than one host
(C) Dead matter
(D) None of these

602. Leaf curl of tomato is caused by:
(A) Tobacco virus 16
(B) Tomato virus 16
(C) Virus No. 10
(D) Virus No.16

603. Most appropriate time of black rust attack in Northern Plains is:
(A) Late February – March
(B) December- January
(C) October-November
(D) All of these

604. Most appropriate time of black rust attack in southern Plains is:
(A) Late February – March
(B) December- February
(C) October-November
(D) All of these

605. Optimum temperature for black rust spread is:
(A) 17-20^{0}C (B) 25-30^{0}C
(C) 30-35^{0}C (D) 35-40^{0}C

606. Which is the alternate host of yellow rust?
(A) *Tholictrum* spp (B) *Berberis* spp
(C) Wild wheat (D) Not known

607. Which of the following is the collateral host of yellow rust?
(A) *Agropyron semicostatum*
(B) *Bromus catharticus*
(C) *Bromus japonicus*
(D) All of these

608. Which of the following urediniospores of rust can require the highest temperature for its germination?
(A) Black rust (B) Yellow rust
(C) Brown rust (D) All of these

609. Covered smut of barley is a disease of:
(A) Externally seed borne
(B) Internally seed borne
(C) Air borne
(D) None of these

610. Which of the following sequence of rust occurrence in UP:
(A) Leaf rust > Stripe rust > Stem rust
(B) Stripe rust > Leaf rust > Black rust
(C) Black rust > Brown rust > Yellow rust
(D) Yellow rust > Leaf rust > Black rust

611. The causal organism of powdery mildew of wheat is:
(A) *Erysiphe graminis var tritici*
(B) *Erysiphe polygoni*
(C) *Erysiphe aestivum*
(D) None of these

612. Pin moulds are also known as sugar fungi because of:
(A) They use simple carbohydrates
(B) They do not use carbohydrate
(C) They use polysaccharides
(D) All of these

613. The pathogen that can grow only under controlled environment is called:
(A) Ecologically obligate
(B) Bio-chemically obligate
(C) Facultative parasite
(D) All of these

614. Bajra smut is a:
(A) Air borne (B) Seed borne
(C) Soil borne (A) Soil and seed borne

615. The colour of bajra smut spores is:
(A) Black (B) Yellow
(C) Brownish green (D) Red

616. The sorus of bajra smut is formed usually on:
(A) Leaf (B) Stem
(C) Ovary (D) All aerial parts

617. The main symptoms of early blight of potato is:
(A) Formation of concentric ringes on infected spots
(B) Formation of cottony growth
(C) Formation of chlorosis
(D) All of these

618. Optimum temperature for development of early blight of potato is:
(A) 12-19^{0}C (B) 25-26^{0}C
(C) 30-36^{0}C (D) 38-40^{0}C

619. Which of the following spores are not formed in Fusarium wilt?
(A) Micro conidia
(B) Macro conidia
(C) Chlamydospores
(D) Onidia

620. The symptoms of false smut of rice appears on:
(A) Leaf (B) Stem
(C) Panicle (D) Root

621. The main symptoms of false smut of rice is:
(A) Apearance of round bluish spots on leaves
(B) Conversion of few seeds into scleratial bodies covering by green colour layer
(C) Formation of black powder in whole panicle
(D) All of these

622. The main symptoms of rice blast is:
(A) Round black spots on leaves
(B) Oval shaped round spots on leaves
(C) Boat shaped spots with grey red borders and grey ash colour centre
(D) None of these

623. Rice blast is also known as neck rot because of:
(A) Neck of panicle becomes shriveled due to severity of disease
(B) Neck becomes stronger as fungus produces pyricularin toxin
(C) It produces hyaline spores
(D) None of these

624. Which type of bacteria can be killed by Penicillin?
(A) Gram +ve (B) Gram – ve
(C) Both A & B (D) None of these

625. The pathogenic bacteria are generally:
(A) Gram +ve
(B) Gram –ve
(C) Both (A) & (B)
(D) None of these

626. The main function of mesosomes in bacterial cell is:
(A) To help in cell division
(B) To help in respiration
(C) To assist in photosynthesis
(D) All of these

627. Virus can be seen only with:
(A) Naked eyes
(B) Microscope
(C) Electron microscope
(D) All of these

628. Yellow Jack is a:
(A) Yellow fever (B) Small pox
(C) Swine fever (D) None of these

629. Cuscuta is a:
(A) Holoparasite on stem
(B) Semi-parasite on stem
(C) Holoparasite on root
(D) Semi-parasite on root

630. Loranthus is a:
(A) Stem Holoparasite
(B) Stem semi-parasite
(C) Root Holoparasite
(D) Root semi-parasite

631. Orobanche is a:
(A) Stem Holoparasite
(B) Stem semi-parasite
(C) Root Holoparasite
(D) Root semi-parasite

632. Striga is a:
(A) Stem Holoparasite
(B) Stem semi-parasite
(C) Root Holoparasite
(D) Root semi-parasite

633. Broom rape is also known as:
(A) Orobanche (B) Striga
(C) Loranthus (D) Cuscuta

634. Witch weed is also known as:
(A) Orobanche (B) Striga
(C) Loranthus (D) Cuscuta

635. Vrikshabhaksha is known as:
(A) Loranthus (B) Banda
(C) Giant mistletoe (D) All of these

636. Nematodes are:
(A) Saprophyte (B) Predator
(C) Pathogenic (D) All of these

637. Nematodes may cause following symptoms?
(A) Cellular hypertrophy/hyperplasia
(B) Suppression of mitosis
(C) Cell necrosis
(D) All of these

638. Plant nematodes always have:
(A) Stylet
(B) Respiratory organs
(C) Circulatory system
(D) None of these

639. Which of the following system is not present in plant nematodes?
(A) Respiratory system
(B) Circulatory system
(C) Reproductive system
(D) None of these

640. Which of the following nematicides are used to control of nematodes?
(A) Nemagon (B) Nemaphos
(C) Hexanema (D) All of these

641. Which stage of root knot nematode attacks to plant?
(A) First larvae
(B) Second larvae
(C) Third larvae
(D) Fourth Larvae

642. Male nematodes differ with female nematodes in:
(A) Male nematodes are cylindrical and filliform while female are bigger in size with round and oval shaped
(B) Male nematodes have respiratory system
(C) Females do not have stylets
(D) None of these

643. Main symptoms of root knot of tomato is:
(A) Knot formation in roots
(B) Yellowing of leaves
(C) Stunting growth
(D) All of these

644. The nematodes take their food from host with the help of:
(A) Mouth (B) Stylet
(C) Haustoria (D) None of these

645. The important feature of plant parasitic nematodes is:
(A) Body cavity absent
(B) Transparent segmentless body
(C) Respiratory and circulation systems absent
(D) All of these

646. Which of the following nematodes acts as carrier of viruses?
(A) *Xiphineiva spp*
(B) *Longidorus spp*
(C) *Trix chodorous spp*
(D) All of these

647. Cyst forming nematodes spp is:
(A) *Meloidogyne spp*
(B) *Heterodera spp*
(C) *Aphelenchoides spp*
(D) *Pratylenchus spp*

648. Leaf nematodes spp is:
(A) *Meloidogyne spp*
(B) *Heterodera spp*
(C) *Aphelenchoides spp*
(D) *Pratylenchus spp*

649. Root injury creating nematodes spp is:
(A) *Meloidogyne spp*
(B) *Heterodera spp*
(C) *Aphelenchoides spp*
(D) *Pratylenchus spp*

650. Which of the following nematode species is the causal agent for root knot disease in plants?
(A) *Meloidogyne spp*
(B) *Heterodera spp*
(C) *Aphelenchoides spp*
(D) *Pratylenchus spp*

651. Reverse transcription is found only in:
(A) Bacteria (B) Virus
(C) Protozoa (D) Fungus

652. Which of the following disease is caused by MLO's?
(A) Papaya bunchy top
(B) Sugarcane grassy shoot
(C) Witch's broom of groundnut
(D) All of these

653. Stinking smut is also known as:
(A) Hill bunt of wheat
(B) Karnal bunt
(C) Covered smut of jowar
(D) None of these

654. Which of the following is used for disinfestations of ware houses?
(A) $CuSO_4$
(B) Formaldehyde
(C) Fumigation with chloropicrin tear gas
(D) All of these

655. Which of the following gas is used for sterilization?
(A) Chlorine (B) Sulphur dioxide
(C) Formadehyde (D) All of these

656. Pink boll worm is mainly harmful to:
(A) Stem (B) Flower
(C) Bolls (D) All of these

657. Mustard aphid is a serious pest of:
(A) Rape seed and mustard
(B) Chickpea
(C) Pigeonpea
(D) Sesame

658. Scientific name of mustard aphid is:
(A) *Lipaphis erysimi*
(B) *Lipophis cruciferaum*
(C) *Athalia proxima*
(D) None of these

659. Vericides are used to control:
(A) Bacteria (B) Virus
(C) Nematodes (D) Fungus

660. Damping off of tobacco is caused by:
(A) *Pythium aphanidermatum*
(B) *Pythium infestans*
(C) *Phytophthora infestans*
(D) None of these

661. White Blister disease is caused by:
(A) *Pythiaceae* (B) *Albuginaceae*
(C) *Perouosporaceae* (D) None of these

662. Downy mildew disease is caused by:
(A) *Pythiaceae*
(B) *Albuginaceae*
(C) *Perouosporaceae*
(D) None of these

663. Late blight of potato is caused by:
(A) *Phytophthora infestans*
(B) *Albugo candida*
(C) *Peronospora pisi*
(D) None of these

664. Downy mildew of pea disease is caused by:
(A) *Phytophthora pisi*
(B) *Peronospora pisi*
(C) *Albigo candida*
(D) None of these

665. Green ear disease is caused by:
(A) *Albugo candida*
(B) *Sclerospora graminicola*
(C) *Erysiphe graminis*
(D) None of these

666. White blister of crucifers disease is caused by:
(A) *Albugo candida*
(B) *Erysiphe graminis*
(C) *Peronospora pisi*
(D) None of these

667. Downy mildew of brassica is caused by:
(A) *Albugo candida*
(B) *Peronospora brassicae*
(C) *Phytophthora infestans*
(D) None of these

668. Downey mildew of maize is caused by:
(A) *Sclerospora sacchari*
(B) *Sclerospora rayssiae var zeae*
(C) *Sclerospora philippinensis*
(D) All of these

669. Powdery mildew of cereals is caused by:
(A) *Erysiphe graminis*
(B) *Erysiphi polygoni*
(C) *Taphrina deformans*
(D) All of these

670. Stem rot of paddy is caused by:
(A) *Sclerotium oryzae*
(B) *Claviceps purpurea*
(C) *Ustilaginoidea virens*
(D) All of these

671. False smut of paddy is caused by:
(A) *Ustilaginoidea virens*
(B) *Claviceps purpurea*
(C) *Erysiphe graminis*
(D) None of these

672. Ergot of bajra is caused by:
(A) *Claviceps microcephala*
(B) *Claviceps purpurea*
(C) *Sclerotium oryzae*
(D) None of these

673. Black rust of wheat is caused by:
(A) *Puccinia graminis var tritici*
(B) *Puccinia recondita*
(C) *Puccinia striformis*
(D) *Melampsora lini*

674. Leaf rust of wheat is caused by:
(A) *Puccinia graminis*
(B) *Puccinia recondita*
(C) *Puccinia striformis*
(D) *Melampsora lini*

675. Stripe rust of wheat is caused by:
(A) *Puccinia graminis*
(B) *Puccinia recondita*
(C) *Puccinia striformis*
(D) *Melampsora lini*

676. Stem rust of wheat is also known as:
(A) Black rust (B) Brown rust
(C) Yellow rust (D) White rust

677. Leaf rust of wheat is also known as:
(A) Black rust (B) Brown rust
(C) Yellow rust (D) White rust

678. Stripe rust of wheat is also known as:
(A) Black rust (B) Brown rust
(C) Yellow rust (D) White rust

679. Causal organism of pea rust is:
(A) *Uromyces pisi*
(B) *Uromyces aestivum*
(C) *Puccinia recondita*
(D) None of these

680. Causal organism of linseed rust is:
(A) *Uromycese pisi*
(B) *Uromyces fabae*
(C) *Melampsora lini*
(D) None of these

681. Causal organism of loose smut of wheat is:
(A) *Ustilago nuda*
(B) *Ustilago nuda var tritici*
(C) *Ustilago hordei*
(D) *Tolyposporium penicillariae*

682. Causal organism of grain smut of bajra is:
(A) *Sphacelotheca sorghi*
(B) *Sphacelotheca hordi*
(C) *Tolyposporium penicillariae*
(D) None of these

683. Causal organism of smut of bajra is:
(A) *Fusarium oxysporum*
(B) *Tolyposorium penicillariae*
(C) *Ustilago nuda*
(D) None of these

684. Causal organism of early blight of potato is:
(A) *Alternaria solani*
(B) *Alternaria infestans*
(A) *Pyricularia oryzae*
(D) *Cercospora personata*

685. Causal organism of Alternaria leaf blight of wheat is:
(A) *Alternaria tritici*
(B) *Alternaria aestivum*
(C) *Pyricularia triticina*
(D) None of these

686. Causal organism of lentil wilt is:
(A) *Fusarium oxysporum*
(B) *Fusarium udum*
(C) *Fusarium solani*
(D) *Fusarium pisi*

687. Causal organism of arhar wilt is:
(A) *Fusarium oxysporum*
(B) *Fusarium udum*
(C) *Fusarium solani*
(D) *Fusarium pisi*

688. Causal organism of linseed wilt is:
(A) *Fusarium oxysporum*
(B) *Fusarium udum*
(C) *Fusarium solani*
(D) *Fusarium pisi*

689. Causal organism of tikka disease of groundnut is:
(A) *Cercospora personata*
(B) *Cercospora arachidicola*
(C) Both (A) and (B)
(D) None of these

690. Which of the following causal organism of tikka disease of groundnut is more fatal?
(A) *Cercospora personata*
(B) *Cercospora arachidicola*
(C) Both (A) and (B)
(D) None of these

691. Which of the following disease is caused by mycoplasma like organisms (MLO)?
(A) Rice yellow dwarf
(B) Sugarcane grassy shoot
(C) Potato witch's broom
(D) All of these

692. Which of the following disease is caused by Rickettsia like microorganisms (RLO's)?
(A) Ratoon stunting
(B) Aster yellow
(C) Khaira disease
(D) All of these

693. Khaira disease is a example of:
(A) Endemic disease
(B) Epidemic disease
(C) Non infectious disease
(D) Infectious disease

694. Green ear disease is a example of:
(A) Endemic disease
(B) Epidemic disease
(C) Sporodic disease
(D) Infectious disease

695. Late blight of potato disease is a example of:
(A) Endemic disease
(B) Epidemic disease
(C) Sporodic disease
(D) Infectious disease

696. The pesticide used in foundations of buildings for preventing termite attack is:
(A) DDT (B) BHC
(C) Endrin (D) Aldrin

697. Necrotrophs are pathogens that:
(A) Kill host plants
(B) Do not kill host plant
(C) Provide nutrition to host plants
(D) None of these

698. Which of the following is a example of necrotroph pathogen?
(A) Pythium (B) Rhizoctonia
(C) Soft rot bacteria
(D) All of these

699. Facultative parasites are:
(A) Primarily saprophytes but can attack on living organisms under favourable conditions
(B) Primarily saprophytes and do not attack on living organisms
(C) Primarily obligate parasites
(D) All of these

700. Main symptoms of karnal bunt is:
(A) Formation of black powder below the pericarp of seed
(B) Produces rotten fish odour
(C) Bad odour is due to Trimethylamine
(D) All of these

701. Hill bunt can be controlled by:
(A) Seed treatment with 0.2% vitavex
(B) Use of resistant varieties
(C) Clean the field
(D) All of these

702. Which of the following culture is a natural medium for isolation of pathogen?
(A) Potato dextrose agar medium
(B) Milk
(C) Egg
(D) All of these

703. Trychoderma can be used as:
(A) Soil treatment
(B) Seed treatment
(C) Both of these
(D) None of these

704. LSD is a:
(A) Stimulant (B) Pain killer
(C) Antibiotic (D) None of these

705. LSD is abbreviated as:
(A) Least Stimulant Dose
(B) d-Lysergic Acid diethylamide
(C) Least Statistical Deviation
(D) None of these

706. Irish blight is also known as:
(A) Late blight of potato
(B) White blight of mustard
(C) Yellow rust
(D) Loose smut of wheat

707. Which of the fruiting body is related to sexual reproduction?
(A) Cleistothecia (B) Perithecia
(C) Apothecia (D) All of these

708. Which of the following disease is caused by virus?
(A) Fiji disease of sugarcane
(B) Leaf curl of papaya
(C) Ratoon stunting of sugarcane
(D) None of these

709. Which of the crop is affected mostly by striga?
(A) Sugarcane (B) Maize
(C) Millets (D) All of these

710. Which of the following gas can stimulate the seeds of striga?
(A) Ethylene
(B) Nitrous oxide
(C) Carbon mono oxide
(D) Oxygen

711. Strigol is used to control:
(A) Striga (B) Cuscuta
(C) Orobanche (D) Loranthus

712. Hill bunt of wheat is caused by:
(A) *Tiletia foetida*
(B) *Neovossia indica*
(C) *Uromystis tritici*
(D) None of these

713. Which of the following is used for tree surgery?
(A) Bordeaux mixture
(B) Chaubatia paste
(C) Santar A
(D) All of these

714. Penicillin antibiotic is prepared by:
(A) *Penicillium chrysogenum*
(B) *Penicillium notatum*
(C) *Penicillium albidum*
(D) *Penicillium citrinum*

715. Citrinin is prepared by:
(A) *Penicillium citrinum*
(B) *Penicillium flavus*
(C) *Penicillium albidum*
(D) *Penicillium citrinum*

716. Penicillin B is prepared by:
(A) *Penicillium flavum*
(B) *Penicillium notatum*
(C) *Penicillium albidum*
(D) *Penicillium citrinum*

717. Which of the *penicillium spp* is harmful for food products?
(A) *Penicillium expensum*
(B) *Penicillium glaccusm*
(C) *Penicillium digitatum*
(D) All of these

718. Aspergillin antibiotic was prepared by which fungus?
(A) *Aspergillus niger*
(B) *Aspergillus fumigatus*
(C) *Aspergillus flavus* (D) All of these

719. The causal organism of powdery mildew of grape vine is:
(A) Unicellular nector
(B) Podosphaera leucotricha
(C) Sclerotium oryzae
(D) All of these

720. Most important symptoms of ergot disease is:
(A) Transformation of kernels into black structure
(B) Dispersion of spores
(C) Both (A) and (B)
(D) None of these

721. The symptoms of powdery mildew of pea appear first of all on:
(A) Leaves (B) Stem
(C) Whole plant (D) Roots

722. The main symptoms of black leg or soft rot of potato is:
(A) Brown – black or jet black colour of stem at soil level
(B) Black leaf coloration
(C) Black crust on tubers
(D) None of these

723. If the virus multiplies in host bacterium which can undergoes complete lysis to release the daughter phages, this type of virus is called:
(A) Lytic bacterio phages
(B) Lysogenic bacteriophages
(C) Transformation
(D) None of these

724. Root knot nematodes have higher growth at :
(A) 15-30^{0}C (B) 20-25^{0}C
(C) 25-28^{0}C (D) 30-35^{0}C

725. Which of the following disease is caused by MLO's?
(A) Little leaf of egg plant
(B) Citrus greening
(C) Purple top roll of potato
(D) All of these

726. Which of the following is not the collateral host of yellow rust?
(A) *Agropyron semicostatum*
(B) *Bromes cathartics*
(C) *Puccinia striiformis*
(D) All of these

727. Which of the following is the alternate host of black rust?
(A) Wheat (B) Barbary
(C) Calotropis (D) *Phalaris minor*

728. Which of the following plant is the natural alternate host of *Puccinia recondita*?
(A) *Berberis vulgaris*
(B) *Isopyron fumarioides*
(C) *Avena fatua*
(D) *Triticum aestivum*

729. Which of the following rust requires the lowest temperature for its over summe-ring?
(A) Black rust (B) Leaf rust
(C) Strip rust (D) All of these

730. Linseed rust is a:
(A) Obligate parasite
(B) Obligate saprophyte
(C) Facultative parasite
(D) Facultative saprophyte

731. Linseed rust fungus is a:
(A) Autocious
(B) Heterocious
(C) Both (A) and (B)
(D) None of these

732. Cutworms pupate:
(A) Inside the pods
(B) Inside the stem
(C) On upper surface of soil
(D) In an earthen shell inside the soil

733. *Agrotis flammatra* larvae cut the:
(A) Leaves (B) Buds
(C) Flowers (D) Stems

734. A dark smoky spot is present on the centre of fore wings is:
(A) *Helicoverpa armigera*
(B) *Agrotis ipsilon*
(C) *Etiella zinckenella*
(D) *Exelastris atomosa*

735. Which one of the following is better for the control of gram pod borer?
(A) Parathion
(B) Phosphomidon
(C) Chlorpyriphos
(D) Malathion

736. Pheromone trap is used for capturing:
(A) Larvae (B) Pupae
(C) Adult males (D) Adult females

ANSWER SHEET
14. Plant Protection Measures

1. (A)	2. (D)	3. (C)	4. (D)	5. (B)	6. (A)	7. (C)	8. (D)	9. (B)	10. (C)
11. (A)	12. (A)	13. (C)	14. (B)	15. (D)	16. (B)	17. (A)	18. (B)	19. (B)	20. (C)
21. (C)	22. (C)	23. (B)	24. (D)	25. (A)	26. (D)	27. (D)	28. (C)	29. (C)	30. (A)
31. (C)	32. (A)	33. (B)	34. (A)	35. (A)	36. (D)	37. (B)	38. (C)	39. (A)	40. (A)
41. (B)	42. (A)	43. (B)	44. (A)	45. (C)	46. (A)	47. (C)	48. (A)	49. (D)	50. (B)
51. (C)	52. (B)	53. (A)	54. (D)	55. (B)	56. (B)	57. (A)	58. (A)	59. (B)	60. (A)
61. (A)	62. (D)	63. (D)	64. (B)	65. (D)	66. (B)	67. (A)	68. (D)	69. (A)	70. (C)
71. (D)	72. (A)	73. (D)	74. (B)	75. (D)	76. (C)	77. (A)	78. (D)	79. (C)	80. (B)
81. (A)	82. (C)	83. (C)	84. (D)	85. (D)	86. (A)	87. (C)	88. (B)	89. (D)	90. (A)
91. (D)	92. (C)	93. (D)	94. (C)	95. (B)	96. (B)	97. (A)	98. (D)	99. (C)	100. (D)
101. (B)	102. (C)	103. (D)	104. (D)	105. (A)	106. (A)	107. (C)	108. (D)	109. (A)	110. (A)
111. (C)	112. (D)	113. (A)	114. (B)	115. (C)	116. (B)	117. (A)	118. (D)	119. (C)	120. (D)
121. (A)	122. (C)	123. (C)	124. (D)	125. (D)	126. (A)	127. (B)	128. (D)	129. (D)	130. (D)
131. (A)	132. (D)	133. (A)	134. (A)	135. (A)	136. (D)	137. (A)	138. (C)	139. (C)	140. (B)
141. (C)	142. (C)	143. (B)	144. (D)	145. (A)	146. (A)	147. (A)	148. (C)	149. (B)	150. (D)
151. (A)	152. (C)	153. (C)	154. (C)	155. (A)	156. (B)	157. (C)	158. (B)	159. (D)	160. (D)
161. (D)	162. (A)	163. (A)	164. (A)	165. (B)	166. (B)	167. (B)	168. (B)	169. (A)	170. (B)
171. (A)	172. (B)	173. (B)	174. (A)	175. (A)	176. (B)	177. (A)	178. (C)	179. (D)	180. (B)
181. (B)	182. (A)	183. (B)	184. (D)	185. (A)	186. (A)	187. (B)	188. (C)	189. (A)	190. (D)
191. (B)	192. (A)	193. (D)	194. (B)	195. (D)	196. (D)	197. (D)	198. (A)	199. (A)	200. (B)
201. (D)	202. (D)	203. (A)	204. (A)	205. (D)	206. (D)	207. (A)	208. (C)	209. (A)	210. (C)
211. (A)	212. (B)	213. (A)	214. (B)	215. (B)	216. (D)	217. (A)	218. (A)	219. (B)	220. (A)
221. (D)	222. (A)	223. (B)	224. (C)	225. (D)	226. (D)	227. (C)	228. (B)	229. (D)	230. (A)
231. (A)	232. (D)	233. (A)	234. (B)	235. (C)	236. (A)	237. (A)	238. (A)	239. (D)	240. (A)
241. (D)	242. (D)	243. (D)	244. (D)	245. (B)	246. (B)	247. (A)	248. (B)	249. (A)	250. (D)
251. (D)	252. (C)	253. (A)	254. (A)	255. (A)	256. (A)	257. (B)	258. (D)	259. (D)	260. (A)

261. **(A)** 262. **(C)** 263. **(D)** 264. **(B)** 265. **(B)** 266. **(A)** 267. **(A)** 268. **(D)** 269. **(A)** 270. **(A)**
271. **(B)** 272. **(A)** 273. **(D)** 274. **(A)** 275. **(A)** 276. **(D)** 277. **(C)** 278. **(A)** 279. **(C)** 280. **(D)**
281. **(C)** 282. **(D)** 283. **(B)** 284. **(B)** 285. **(D)** 286. **(B)** 287. **(D)** 288. **(D)** 289. **(D)** 290. **(D)**
291. **(D)** 292. **(C)** 293. **(A)** 294. **(B)** 295. **(D)** 296. **(D)** 297. **(C)** 298. **(D)** 299. **(D)** 300. **(D)**
301. **(D)** 302. **(A)** 303. **(A)** 304. **(B)** 305. **(C)** 306. **(B)** 307. **(A)** 308. **(B)** 309. **(A)** 310. **(A)**
311. **(A)** 312. **(D)** 313. **(A)** 314. **(C)** 315. **(A)** 316. **(D)** 317. **(D)** 318. **(A)** 319. **(C)** 320. **(D)**
321. **(A)** 322. **(D)** 323. **(D)** 324. **(A)** 325. **(B)** 326. **(B)** 327. **(B)** 328. **(B)** 329. **(B)** 330. **(C)**
331. **(A)** 332. **(D)** 333. **(D)** 334. **(A)** 335. **(A)** 336. **(C)** 337. **(D)** 338. **(A)** 339. **(C)** 340. **(D)**
341. **(A)** 342. **(B)** 343. **(B)** 344. **(D)** 345. **(C)** 346. **(A)** 347. **(B)** 348. **(D)** 349. **(B)** 350. **(C)**
351. **(D)** 352. **(A)** 353. **(A)** 354. **(B)** 355. **(D)** 356. **(B)** 357. **(C)** 358. **(A)** 359. **(A)** 360. **(C)**
361. **(D)** 362. **(A)** 363. **(B)** 364. **(A)** 365. **(B)** 366. **(C)** 367. **(C)** 368. **(B)** 369. **(D)** 370. **(A)**
371. **(A)** 372. **(A)** 373. **(C)** 374. **(A)** 375. **(A)** 376. **(C)** 377. **(B)** 378. **(D)** 379. **(A)** 380. **(A)**
381. **(D)** 382. **(C)** 383. **(C)** 384. **(D)** 385. **(C)** 386. **(A)** 387. **(C)** 388. **(D)** 389. **(A)** 390. **(A)**
391. **(A)** 392. **(A)** 393. **(B)** 394. **(D)** 395. **(A)** 396. **(C)** 397. **(C)** 398. **(A)** 399. **(A)** 400. **(D)**
401. **(A)** 402. **(A)** 403. **(A)** 404. **(D)** 405. **(D)** 406. **(A)** 407. **(A)** 408. **(C)** 409. **(C)** 410. **(A)**
411. **(A)** 412. **(B)** 413. **(C)** 414. **(A)** 415. **(B)** 416. **(A)** 417. **(B)** 418. **(C)** 419. **(A)** 420. **(A)**
421. **(C)** 422. **(A)** 423. **(D)** 424. **(A)** 425. **(D)** 426. **(A)** 427. **(D)** 428. **(D)** 429. **(A)** 430. **(A)**
431. **(A)** 432. **(C)** 433. **(A)** 434. **(B)** 435. **(C)** 436. **(B)** 437. **(C)** 438. **(A)** 439. **(C)** 440. **(C)**
441. **(B)** 442. **(B)** 443. **(A)** 444. **(A)** 445. **(A)** 446. **(D)** 447. **(C)** 448. **(D)** 449. **(A)** 450. **(B)**
451. **(C)** 452. **(D)** 453. **(C)** 454. **(A)** 455. **(B)** 456. **(B)** 457. **(B)** 458. **(A)** 459. **(B)** 460. **(B)**
461. **(B)** 462. **(B)** 463. **(D)** 464. **(D)** 465. **(C)** 466. **(C)** 467. **(D)** 468. **(A)** 469. **(B)** 470. **(D)**
471. **(C)** 472. **(C)** 473. **(B)** 474. **(D)** 475. **(D)** 476. **(A)** 477. **(B)** 478. **(D)** 479. **(D)** 480. **(D)**
481. **(B)** 482. **(C)** 483. **(D)** 484. **(D)** 485. **(C)** 486. **(D)** 487. **(A)** 488. **(A)** 489. **(C)** 490. **(A)**
491. **(D)** 492. **(B)** 493. **(D)** 494. **(A)** 495. **(C)** 496. **(C)** 497. **(A)** 498. **(C)** 499. **(D)** 500. **(B)**
501. **(A)** 502. **(A)** 503. **(A)** 504. **(A)** 505. **(B)** 506. **(A)** 507. **(C)** 508. **(A)** 509. **(D)** 510. **(A)**
511. **(B)** 512. **(A)** 513. **(A)** 514. **(A)** 515. **(D)** 516. **(A)** 517. **(A)** 518. **(B)** 519. **(D)** 520. **(D)**
521. **(B)** 522. **(B)** 523. **(B)** 524. **(B)** 525. **(A)** 526. **(B)** 527. **(A)** 528. **(A)** 529. **(D)** 530. **(A)**
531. **(C)** 532. **(B)** 533. **(C)** 534. **(B)** 535. **(A)** 536. **(C)** 537. **(B)** 538. **(A)** 539. **(B)** 540. **(D)**
541. **(A)** 542. **(D)** 543. **(D)** 544. **(D)** 545. **(D)** 546. **(A)** 547. **(A)** 548. **(A)** 549. **(B)** 550. **(B)**
551. **(A)** 552. **(A)** 553. **(B)** 554. **(D)** 555. **(A)** 556. **(A)** 557. **(D)** 558. **(A)** 559. **(C)** 560. **(D)**
561. **(C)** 562. **(C)** 563. **(C)** 564. **(B)** 565. **(D)** 566. **(C)** 567. **(D)** 568. **(D)** 569. **(B)** 570. **(A)**
571. **(D)** 572. **(D)** 573. **(D)** 574. **(A)** 575. **(D)** 576. **(A)** 577. **(C)** 578. **(B)** 579. **(A)** 580. **(A)**
581. **(B)** 582. **(A)** 583. **(A)** 584. **(D)** 585. **(A)** 586. **(D)** 587. **(A)** 588. **(C)** 589. **(A)** 590. **(A)**
591. **(A)** 592. **(D)** 593. **(C)** 594. **(C)** 595. **(A)** 596. **(D)** 597. **(B)** 598. **(A)** 599. **(C)** 600. **(A)**
601. **(B)** 602. **(A)** 603. **(A)** 604. **(C)** 605. **(A)** 606. **(D)** 607. **(D)** 608. **(D)** 609. **(A)** 610. **(A)**
611. **(A)** 612. **(A)** 613. **(A)** 614. **(A)** 615. **(C)** 616. **(C)** 617. **(A)** 618. **(B)** 619. **(D)** 620. **(C)**

621. **(B)**	622. **(C)**	623. **(D)**	624. **(A)**	625. **(B)**	626. **(A)**	627. **(C)**	628. **(A)**	629. **(A)**	630. **(B)**
631. **(C)**	632. **(D)**	633. **(A)**	634. **(B)**	635. **(D)**	636. **(D)**	637. **(D)**	638. **(A)**	639. **(A)**	640. **(D)**
641. **(B)**	642. **(A)**	643. **(D)**	644. **(B)**	645. **(D)**	646. **(D)**	647. **(B)**	648. **(C)**	649. **(D)**	650. **(A)**
651. **(C)**	652. **(D)**	653. **(A)**	654. **(D)**	655. **(D)**	656. **(C)**	657. **(A)**	658. **(A)**	659. **(B)**	660. **(A)**
661. **(B)**	662. **(C)**	663. **(A)**	664. **(B)**	665. **(B)**	666. **(A)**	667. **(B)**	668. **(B)**	669. **(A)**	670. **(A)**
671. **(A)**	672. **(A)**	673. **(A)**	674. **(B)**	675. **(C)**	676. **(A)**	677. **(B)**	678. **(C)**	679. **(A)**	680. **(C)**
681. **(B)**	682. **(A)**	683. **(B)**	684. **(A)**	685. **(A)**	686. **(A)**	687. **(B)**	688. **(A)**	689. **(C)**	690. **(A)**
691. **(D)**	692. **(A)**	693. **(C)**	694. **(C)**	695. **(B)**	696. **(D)**	697. **(A)**	698. **(D)**	699. **(A)**	700. **(D)**
701. **(D)**	702. **(D)**	703. **(C)**	704.**(A)**	705. **(B)**	706. **(A)**	707. **(D)**	708. **(D)**	709. **(D)**	710. **(A)**
711. **(A)**	712. **(A)**	713. **(D)**	714. **(A)**	715. **(A)**	716. **(B)**	717. **(D)**	718. **(A)**	719. **(A)**	720. **(A)**
721. **(A)**	722. **(A)**	723. **(A)**	724. **(C)**	725. **(D)**	726. **(C)**	727. **(B)**	728. **(B)**	729. **(C)**	730. **(A)**
731. **(A)**	732. **(D)**	733. **(A)**	734. **(A)**	735. **(D)**	736. **(C)**				

15 FARM MANAGEMENT

1. Farm management is an:
   (A) Intra farm science
   (B) Inter farm science
   (C) Inter regional farm science
   (D) International farm science
2. Farm management is:
   (A) An art (B) A science
   (C) A business (D) All of these
3. The ultimate aim of farm management is to:
   (A) Increase gross return
   (B) Reduced total cost
   (C) Optimize factor utilization
   (D) None of these
4. Which one is not a tool of farm management:
   (A) Farm planning
   (B) Farm budgeting
   (C) Farm book keeping
   (D) Production and cost function
5. Instability in agriculture is caused by:
   (A) Fluctuations in price
   (B) Fluctuations in income
   (C) Fluctuations in outcome
   (D) All of these
6. Farm management should answer the following question:
   (A) How to organize a farm
   (B) How to operate it
   (C) How to achieve the maximum netprofit
   (D) All of these
7. Farm layout means:
   (A) Physical arrangement of farm in the form of plots, building, roads, channels etc.
   (B) Planning of farm to maximize the profit
   (C) Production of crops
   (D) None of these
8. Farm layout is required because of:
   (A) Newly acquired land
   (B) Change in power and enterprises
   (C) Change in technologies
   (D) All of these
9. Farm profit is equal to:
   (A) Gross revenue – cost
   (B) Gross revenue – net returns
   (C) Gross income – gross revenue
   (D) All of these
10. Farm profit is affected by:
    (A) Total produce
    (B) Market and farm prices
    (C) Input cost (D) All of these
11. Farm budget refers to:
    (A) A tentative estimation of expenses and returns of a farm
    (B) A blueprint of production and cost
    (C) Loan and its interest
    (D) All of these
12. Complete farm budget refers to:
    (A) Farm budget for a entire farm as a single unit
    (B) Farm budget separately for each enterprise
    (C) It is a budget for a Gram Panchayat as a single unit
    (D) It is a budget for whole area of the region
13. Partial farm budget refers to:
    (A) Farm budget for a entire farm as a single unit
    (B) Farm budget separately for each enterprise
    (C) It is a budget for a Gram Panchayat as a single unit
    (D) It is a budget for whole area of the region

14. Farm inventory refers to:
(A) List of items possessed by a farm on a specific date
(B) Land under cultivation
(C) Inventory of new farm
(D) All of these

15. Farm production records indicate:
(A) Financial position of farm
(B) Production in quantity
(C) Production in quality
(D) All of these

16. Which of the following is a farm production record?
(A) Livestock record
(B) Crop cultivation record
(C) Crop production record
(D) All of these

17. The motto of land reform in India is:
(A) Land to tiller
(B) Land to landlord
(C) Land to Talukedar
(D) All of these

18. The tool for optimum utilization of the available resources in the maximization of the attainable income of the farmers is:
(A) Farm planning
(B) Resource allocation
(C) Planned change
(D) Farm layout

19. Farming system approach ensures the best use of:
(A) Land (B) Labour
(C) Capital
(D) Land+Labour+Capital + Management

20. One of the important fixed cost on a farm is:
(A) Seed cost (B) Land rent
(C) Irrigation cost (D) Cultivation

21. Agriculture differs with industry in:
(A) Nature plays an important role in agricultural production
(B) Land is the most important factor for agricultural production
(C) Elasticity of price and supply of agricultural products are lower than industrial products
(D) All of these

22. Predominant source of net domestic product in India over the years is:
(A) Industry (B) Agriculture
(C) Service sector (D) All of these

23. National income generated by agriculture is calculated by using:
(A) Income method (B) Out put method
(C) Expenditure method
(D) All of these

24. Capital formation in agriculture over the years:
(A) Decreasing (B) Increasing
(C) Costant (D) Both (A) & (B)

25. Private investment is .... to public investment in agriculture:
(A) Less than (B) More than
(C) Equal (D) Doubtfull

26. Demand for agricultural products is relatively:
(A) Less elastic (B) More elastic
(C) More inelastic (D) Less inelastic

27. When a new variety of crop is introduced:
(A) Partial budgeting is required
(B) Full budgeting is a must
(C) No budgeting is needed
(D) Only planning is needed

28. The objective of preparation of calendar of operations in:
(A) To prepare a cropping scheme
(B) To calculate the cost of production
(C) To assess the need of human and bullock power
(D) To adjust farm enterprises

29. Which one of the following refers to 'system of farming':
(A) Economic (B) Operational
(C) Organizational (D) cultural

30. The percentage share of agriculture in National economy is:
(A) 50 (B) 40
(C) 45 (D) 25

31. Economy of a developing country is marked by:
(A) Predominance of agriculture
(B) Predominance of industry
(C) Predominance of transportation
(D) None of these

32. The chairman of S.F.D.A. is:
(A) Prime Minister
(B) Chief Minister
(C) Finance Minister
(D) District Magistrate

33. The owner of loan in state farming is:
(A) Farmer (B) Capitalist
(C) Government (D) None of these

34. A farmer who has land holding between 1 and 2 hectares will be called as:
(A) Marginal farmer
(B) Small farmer
(C) Medium farmer
(D) Large farmer

35. The Regional Rural Bank was established through:
(A) RRB Act 1974
(B) RRB Act 1975
(C) RRB Act 1976
(D) None of these

36. India's working population engaged in agriculture is about:
(A) 50-52% (B) 67-69%
(C) 75-80% (D) 90-95%

37. The population of agricultural goods manufactured with agricultural contents contributes to India's export by about:
(A) 30% (B) 50%
(C) 70% (D) 90%

38. The percentage share of agricultural export in total export of India is about:
(A) 10% (B) 18%
(C) 25% (D) 30%

39. The growth of Indian economy can be sustained only if agriculture and allied activities grow at an annual growth rate of:
(A) 4% (B) 6%
(C) 8% (D) 10%

40. The scheme of agriculture in the National income has been continuously:
(A) Increasing (B) Decreasing
(C) Constant (D) None of these

41. As compared to other developing countries, the share of agriculture in National income of India is:
(A) Much lower (B) Much higher
(C) Slightly higher (D) Slightly lower

42. With the globalization of trade, the importance of the following has increased:
(A) Food processing
(B) Steel industries
(C) Machines and Tools
(D) Chemicals

43. The first five year plant in India was started on:
(A) Ist April 1958 (B) Ist April 1951
(C) Ist April 1952 (D) Ist April 1953

44. The chairman of the Planning Commission in India is:
(A) Minister of Planning
(B) Minister of Industries
(C) Prime Minister
(D) President of India

45. A marginal farmer owns cultivated land of:
(A) 1 to 2 hectares
(B) 2 to 3 hectares
(C) Below 1 hectare
(D) 3 hectares and above

46. A large farmer owns cultivated land of:
(A) 1 to 2 hectares
(B) 2 to 3 hectares
(C) Below 1 hectare
(D) Above 5 hectares

47. Which of the following organization governs international trade:
(A) U.N.O. (B) F.A.O.
(C) I.M.F. (D) W.T.O.

48. The fourth Five Year Plan gave first priority to:
(A) Industries
(B) Transport
(C) Health and Education
(D) Agriculture

49. The Agricultural Prices are recommended by:
(A) I.C.A.R. (B) C.S.I.R.
(C) C.A.C.P. (D) R.B.I.

50. The Agricultural Prices are recommended on the basis of:
(A) Average cost of production
(B) Minimum cost of production
(C) Bulk line cost
(D) Maximum cost of production

51. Which of the following organization refinance agriculture:
(A) I.D.B.I. (B) I.C.I.C.
(C) N.A.B.A.R.D. (D) S.B.I.

52. Regional Rural Banks provide credit to:
(A) Big farmers
(B) Weaker section of urban population
(C) Weaker section of rural population
(D) None of these

53. Loan for purchase of tractor is known as:
(A) Short term loan
(B) Medium term loan
(C) Long term loan
(D) Cop plan

54. First priority to industry was given during:
(A) First Five Year Plan
(B) Second Five Year Plan
(C) Third Five Year Plan
(D) Fourth Five Year Plan

55. Agricultural Produce and Marketing Act was passed in:
(A) 1937 (B) 1935
(C) 1947 (D) 1949

56. The growth rate of food production in India after 1980 has been:
(A) Increasing (B) Decreasing
(C) Constant (D) Negative

57. Increase in Indian population is due to:
(A) High birth rate
(B) Low birth rate
(C) High birth rate and low death rate
(D) Defective Family Planning

58. The export of Agricultural commodities since Nineties is:
(A) Increasing (B) Decreasing
(C) Stand still (D) Negative

59. I.R.D.P. Programme is meant for:
(A) Poverty alleviation
(B) Employment generation
(C) Disposal of agricultural produce
(D) Agro-processing

60. The agro-climatic Regional Planning approach was started in:
(A) Sixth Five Year Plan
(B) Seventh Five Year Plan
(C) Eighth Five Year Plan
(D) Ninth Five Year Plan

61. CAPART expands as:
(A) Council fro Action and people Advancement in Rural Technology
(B) Council for Advancement of People Action and Rural Technology
(C) Commission for Action and People Advancement in Rural Technology
(D) Committee for Advancement of People Action and Rural Technology

62. CAPART is an autonomous wing of:
(A) Ministry of Agriculture and Cooperation
(B) Ministry of Rural Development
(C) ICAR
(D) Ministry of HRD

63. Which decade of planning has been the best for agricultural growth:
(A) Fifties (B) Sixties
(C) Seventies (D) Eighties

64. Which one is not the function of a production manager?
(A) Planning (B) Organisation
(C) Marketing (D) Monitoring

65. Which system of farming does not involve farmer?
(A) Collective farming
(B) Dry farming
(C) State farming
(D) Diversified farming

66. Livestock insurance is included to:
(A) Fixed cost (B) Special cost
(C) Variable cost (D) Marginal cost

67. Which one of the following is a gift of nature?
(A) Land (B) Labour
(C) Capital (D) None of these

68. The efficiency of marketing channel is determined mainly on the basis of:
(A) Price spread
(B) Difference between retail and wholesale prices
(C) Regulated market
(D) Producer's share in consumer's rupees

69. One of the important fixed cost on a farm is:
(A) Seed cost (B) Land cost
(C) Irrigation cost (D) Fertilizer cost

70. If the marginal product is more than marginal cost than further investment in production is:
(A) Profitable (B) Harmful
(C) No effect (D) Equal

71. Principle of equi-marginal return is applied when:
(A) The resources are unlimited
(B) The price of resources is high
(C) The prices of resources vary
(D) The resources are limited

72. Which of the following is also referred to as "Opportunity cost"?
(A) Money cost (B) Alternate cost
(C) Social cost (D) Fixed cost

73. The law of diminishing return refers to an eventual fall in:
(A) Marginal product of the variable factor
(B) Average income of the farm
(C) Total production of all factors
(D) Total earning of the farm

74. Which one of the following is not correctly matched?

Type of market	Basis of classification
(A) Unorganised market	Organization
(B) Regulated market	Control
(C) Perfect market	Competition
(D) Terminal market	Area

75. Marketing surplus is:

(A) Total production	Total family consumption
(B) Total consumption	Marginal consumption
(C) Total cost	Marginal cost
(D) Total production	Total cost

76. Read the following statements and mark your answer according to code given below:
I. For variable input, APP = MPP in the I stage
II. For variable input, APP = MPP at the end of I stage
III. For variable input, APP = MPP at the start of II stage
IV. For variable input, APP = MPP in the II stage

Answer code
(A) I and III are correct
(B) II and III are correct
(C) I and IV are correct
(D) II and IV are correct

77. The concept of farming is best explained by:
(A) Secretarial planning approach
(B) Mixed farming approach
(C) Holistic farming approach
(D) Diversified farming approach

78. When MPP is zero, TPP is at its maximum, because:
(A) Further addition to MPP is negative
(B) Further addition to MPP is positive
(C) MC = MR
(D) Elasticity of productive is more than one

79. When marginal product increases at increasing rate, then:
(A) Total product increases at a decreasing rate
(B) Total product increases at a increasing rate
(C) Average product decreases
(D) None of these

80. According to A. Marshall the "law of diminishing return" is applicable to:
(A) Industry (B) Agriculture
(C) Service sector (D) All of these

81. The working population engaged in agriculture is very high in:
(A) European countries
(B) Developing countries
(C) Developed countries
(D) United States of America

82. The sugar industries in India depend for their raw materials on:
(A) Agriculture (B) Craft
(C) Industrial product
(D) Mines and minerals

83. The differences between value in use and value in exchange is:
(A) Consumer surplus
(B) Marketing surplus
(C) Farm production
(D) None of these

84. When marginal product increases at a increasing rate, then:
(A) Total product increases at a decreasing rate
(B) Total product increases at a increasing rate
(C) Average product decreases
(D) None of these

85. Principle of equimarginal return is applied when:
(A) The resources are unlimited
(B) The price of resources is high
(C) The prices of resources vary
(D) The resources are limited

86. The law of dimishing return refers to an eventual fall in:
(A) Marginal product of the variable factor
(B) Average income of the farm
(C) Total production of all factor
(D) Total earning of the farm

87. If the demand is elastic, for monopoly:
(A) Higher prices, more profit
(B) Lower price, more profit
(C) Higher price, lower profit
(D) None of these

88. Type of farming is based on:
(A) The law of substitution
(B) The law of diminishing returns
(C) The law of comparative advantage
(D) The law of equimarginal return

89. The cost of production of the crop can be minimized by uusing economic principle:
(A) Cost principle
(B) Law of opportunity cost
(C) Law of diminishing
(D) Principle of least cost combination

90. The resource use efficiency in Agriculture is measured by:
(A) Summation of production elasticities
(B) Ratio of opportunity cost to marginal value product
(C) Regression coefficient
(D) Least cost combination

91. Application of the law of diminishing return may be delayed in case of:
(A) Improvement of technology
(B) Improvement in the efficiency of the farmer
(C) Residual effect of early doses of input
(D) All of these

92. The concept of farming is best explained by:
(A) Sectoral planning approach
(B) Mixed farming approach
(C) Holistic farming approach
(D) Diversifie farming approach

93. List of all thee physical property of a business along with their values at a specified date is:
(A) Farm inventory records
(B) Farm income records
(C) farm business analysis
(D) Farm financial account

94. Till which plan employment strategy has been growth linked:
(A) Third plan (B) Fourth plan
(C) Fifth plan (D) Sixth plan

95. Command Area Development Programme was launched during:
(A) Fifth five year plan
(B) Sixth five year plan
(C) Seventh five year plan
(D) Eight five year plan

96. The opening of Indian economy and signing of WTO happened during:
(A) Seventh five year plan
(B) Eight five year plan
(C) Ninth five year plan
(D) Tenth five year plan

97. When marginal product in increasing at increasing rate then:
(A) Total product increases at a decreasing rate
(B) Total product increases at a increasing rate
(C) Average product decreases
(D) None of these

98. Marketing surplus is:
(A) Total production – Total family requirement
(B) Total consumption – Marginal consumption
(C) Total cost – marginal cost
(D) Total produce – Total cost

99. The term "Hindu rate of growth" was coined by:
(A) J N Bhagawathi
(B) K N Raj
(C) Raj Krishna
(D) Mahalanobis

100. Attainment of economic self reliance and removal of poverty were the main objectives of:
(A) First five year plan
(B) Fourth five year plan
(C) Fifth five year plan
(D) Sixth five year plan

101. Which of the following is the most effective way to overcome the defects of agricultural marketing?
(A) Cooperative marketing
(B) Forward marketing
(C) Regulated marketing
(D) Monopoly

102. India achieved self reliance in food production during:
(A) Fourth five year plan
(B) Fifth five year plan
(C) Sixth five year plan
(D) Seventh five year plan

103. Micro level planning is an example for:
(A) Imperative planning
(B) Five year planning
(C) Regional planning
(D) Target group planning

104. Planning in India as it is at present can be termed as:
(A) State planning
(B) Centralized planning
(C) Decentralized planning
(D) Indicative planning

105. Indicative planning started from:
(A) Eight five year plan
(B) Seventh five year plan
(C) Sixth five year plan
(D) Fifth five year plan

106. A farm plan can be prepared by using:
(A) Partial budgeting
(B) Complete budgeting
(C) Input-output model
(D) Break even point

107. Why was plan holiday necessary in the national planning:
(A) First Indo-Pakistan war
(B) Second Indo-Pakistan war
(C) Third Indo-Pakistan war
(D) Indo-China war

108. The highest priority of the first five year plan was given for:
(A) Agriculture and irrigation
(B) Medium and large industry
(C) Unemployment and poverty
(D) Communication and education

109. The goal of socialistic pattern of society was presented before the nation in the:
(A) First plan (B) Second plan
(C) Third plan (D) Fourth plan

110. The concept of mixed economy in planning was introduced during:
(A) Second plan (B) Fifth plan
(C) Seventh plan (D) Ninth plan

111. A rolling plan is a plan for:
(A) One year (B) Two year
(C) Three year (D) Year to year basis

112. The overhead cost is also known as:
(A) Variable cost (B) Fixed cost
(C) Total cost (D) Marginal cost

113. The subsidiary service of a marketing system is:
(A) Welfare (B) Assembling
(C) Dispersion (D) Processing

114. According to law of diminishing return, the optimum, profit will be at the point where:
(A) MP = AP (B) MP > AP
(C) AP > MP (D) MC = MR

115. The vertical demand curve for a commodity shows that the demand is:
(A) Perfectly inelastic
(B) Highly elastic
(C) Unit elastic
(D) Perfectly elastic

116. The demand of agricultural product is mainly dependent on:
(A) Increase in agricultural holding
(B) Increase in income
(C) Increase in population
(D) Increase in cost

117. Type of farming is based on:
(A) Law of comparative advantage
(B) Law of equi-marginal return
(C) Law of diminishing return
(D) Law of substitution

118. Land Development Banks in India are owned by whom:
(A) Govt. of India
(B) State Governments
(C) Commercial banks
(D) Cooperative banks

119. The Panchayati Raj institutions are intended to strengthen by:
(A) Structural planning
(B) Indicative planning
(C) Multilevel planning
(D) Functional planning

120. The record of farm produce is entered in:
(A) Log book
(B) Cash book
(C) Farm stock register
(D) Inventory book

121. The marginal cost –curve intersects the average cost-curve when average cost is:
(A) Maximum (B) Minimum
(C) Rising (D) Falling

122. The term Panchayati Raj was christened by:
(A) M.K. Gandhi (B) J. L. Nehru
(C) B. R. Mehta (D) M. S. Mehta

123. Panchayati Raj was introduced in India in:
(A) 1950 (B) 1952
(C) 1958 (D) 1967

124. Panchayati Raj institution is formed based on:
(A) Ashok Mehta Committee
(B) Hanumantha Rai Committee
(C) Dantiwala Committee
(D) Balwant Rai Mehta Committee

125. Panchayat Raj is a system of:
(A) Local government
(B) Local administration
(C) Local self government
(D) None of these

126. Annual growth rate in agriculture sector expected during the eleventh plan:
(A) 8 per cent (B) 6 per cent
(C) 4 per cent (D) 3 per cent

127. The increase in the production of rice expected at the end of eleventh plan:
(A) 8 million tones
(B) 10 million tones
(C) 12 million tones
(D) 15 million tones

128. The rural development programme launched on the basis of area development approach:
(A) DPAP (B) MWP
(C) RWP (D) IRDP

129. Small farmers Agri-business Consortium was set up in:
(A) 1992 (B) 1994
(C) 1971 (D) 1986

130. A farmer cooperative society among the following is:
(A) IFFCO (B) FCI
(C) ICAR (D) NIRD

131. A welfare programme for the rural poor:
(A) KVK
(B) Lab to Land
(C) National Demonstration
(D) Minikit Trial

132. The Ganga Kalyan Yojna is for:
(A) Consumption purpose of farmers
(B) Irrigation for small and marginal farmers
(C) Credit support scheme for rural women
(D) Credit support scheme for small farmers

133. The Food Corporation of India was set up in:
(A) 1960 (B) 1962
(C) 1965 (D) 1967

134. An Agricultural labourer earns more than of his wage from agriculture:
(A) 33.33% (B) 40.00%
(C) 50.00% (D) 66.66%

135. "India's food crises and steps to meet it" is a:
(A) Book on food situation in India
(B) Report of ford foundation team
(C) Formula to solve food problem
(D) A slogan for campaign

136. The farm plan would be based on the availability of efficiency of various factors of production:
(A) True (B) False
(C) Partially true
(D) Incomplete statement

137. Bhoodan movement was initiated on:
(A) 1947 (B) 1948
(C) 1951 (D) 1963

138. Bhoodan movement was started by:
(A) Acharya Vinoba Bhave
(B) Mahatma Gandhi
(C) Jai Prakash Narayan
(D) Jawaharlal Nehru

139. Bhodan movement was started at:
(A) Santhinietan (B) Wardha
(C) Panchampalli (D) Pavnar

140. The land collected in bhoodan is distributed among:
(A) Marginal farmers
(B) Landless peasants
(C) Medium farmers
(D) Krishi pandits

141. Principle of equi-marginal returns is applied hen:
(A) The resources are unlimited
(B) The resource prices are high
(C) The prices of resources are varying
(D) The resources are limited

142. Which of the following network connect agricultural produce, wholesale markets and rural markets:
(A) AFPINET (B) AGMARKNET
(C) HORTNET (D) VISTARNET

143. Which one is the land Information System Network linking all institutions involved in land and water management for agricultural productivity and production system:
(A) LISNET (B) ARINET
(C) AGRISNET (D) VISTARNET

144. Which one is the most effective method is identifying real rural needs:
(A) Election method
(B) Selection method
(C) Workshop method
(D) Socio-metric method

145. Which one is the purpose of social control:
(A) Social order
(B) Stability
(C) Custom
(D) None of these

146. Which one of the following is not an objective of India's economic planning:
(A) Population growth
(B) Industrial growth
(C) Self reliance
(D) Poverty alleviation

147. What is five year plan:
(A) Perspective plan
(B) Medium term plan
(C) Adhoc plan
(D) Contingent plan

148. Which of the following is co-operation among family members:
(A) Non formal
(B) Voluntary
(C) Formal
(D) Informal

149. The cost of production of the crop can be minimized by using economic principle:
(A) Law of diminishing return
(B) Law of opportunity cost
(C) Cost principle
(D) Principle of least cost combination

150. Allocation of funds for different sectors of programme:
(A) Controlling (B) Budgeting
(C) Financing (D) Recruiting

151. A farm plan can be prepared by using:
(A) Partial budgeting
(B) Complete budgeting
(C) Input-output model
(D) Break-even point

152. By whom the Rural infrastructural fund set up:
(A) RBI (B) NABARD
(C) IDRB (D) SBI

153. Farm budget is a tool of farm management analysis:
(A) True
(B) False
(C) Never include cost
(D) It is a futile exercise

154. Mixed farming is when:
(A) Crop production is combined with livestock enterprise and 50% : 50% income is from both
(B) Entire farm land is sown under mixed crops
(C) Only milch cattle are raised on the farm
(D) Income from milch animal is 95%

155. When marginal product is increasing at increasing rate than total product increases at a decreasing rate:
(A) Total product increases at a decreasing rate
(B) Total product increases at a increasing rate
(C) Average product decreases
(D) None of these

156. Price elasticity of supply of food products is generally:
(A) Infinite (B) Less than one
(C) More than one (D) Zero

157. The record of farm produce is entered in:
(A) Log book
(B) Inventory book
(C) Farm stock register
(D) Cash book

158. When marginal product is declining but remains more than zero, the total product will/will be:
(A) Zero
(B) Increase with increasing rate
(C) Increase with decreasing rate
(D) Initially decreasing than increasing

159. Which one of the following is not included in the factors of production:
(A) Enterprise (B) Organisation
(C) Land (D) Time

160. When $\frac{\Delta x1}{\Delta x2} = \frac{\Delta x2}{\Delta x1}$ then we can :
(A) Employ any combination of x1 and x2
(B) Do nothing
(C) Go on substituting x1 ad x2
(D) Employ x2 only

161. Which one of the following kind of budgeting would you advise to a new farm planner
(A) Enterprise budgeting
(B) Complete budgeting
(C) Partial budgeting
(D) None of these

162. Land holding of either 10 or more hectares has been classified as which of following in India:
(A) Semi-medium (B) Medium
(C) Semi large (D) Large

163. Land reforms have been equated with total:
(A) Urban development
(B) Agrarian development
(C) Industrial development
(D) Educational development

164. National demonstration is an example for:
(A) Firstline demonstration
(B) Frontline demonstration
(C) Futureline demonstration
(D) All of these

165. The Panchayati Raj institutions are intended to strengthen by which:
(A) Structural planning
(B) Indicative planning
(C) Multilevel planning
(D) Functional planning

166. The scheme of Regional Rural Banks was launched in India on:
(A) 1st Oct. 1975
(B) 15th Aug. 1978
(C) 2nd Oct. 1975
(D) 1st Jan. 1970

167. The --------- year plan sought to maintain a clear cut distinction between the minimum support price and the procurement price:
(A) Second (B) Third
(C) Fifth (D) Sixth

168. The Food Corporation of India was set up in:
(A) 1960 (B) 1962
(C) 1965 (D) 1967

169. Formula which indicates the least-cost combination point is:
(A) MR = MC
(B) $\Delta Y/\Delta x1$
(C) MRSY1, Y2
(D) MRSx1 x2 = PX2/PX1

170. The chief function of money is:
(A) Reserve base of credit expansion
(B) Providing liquidity
(C) Medium of exchange
(D) None of these

171. Support prices of agricultural commodities are worked out by:
(A) Finance ministry
(B) Agricultural Price Commission
(C) Food and Agricultural ministry
(D) None of these

172. According to the law of Diminishing Return the optimum profit will be at the point where:
(A) MP = AP
(B) MP > AP
(C) AP > MP
(D) MC = MP

173. The best level of output is obtained at point where:
(A) MR < MC
(B) MR = MC
(C) MR > MC
(D) MR = AC

174. Most prevalent "system of farming" in India is:
(A) Cooperative farming
(B) Collective farming
(C) Peasant farming
(D) State farming

175. When the depreciation of an asset is measured by dividing the first cost by its estimated life, it is called:
(A) Diminishing value method
(B) Sum of years digit method
(C) Straight line method
(D) Revaluation method

176. The commission for agricultural cost and prices is located in:
(A) Lucknow (B) Hyderabad
(C) New Delhi (D) Faridabad

177. NABARD Act was passed in---------, 1979:
(A) January (B) February
(C) March (D) April

178. NABARD came into existence on July 1, ----------:
(A) 1980 (B) 1981
(C) 1982 (D) 1983

179. If a manufacture has sold you a pressure cooker with improper safety value, which of the following right can help you to pull up the manufacturer?
(A) Rights to safety
(B) Right to Choice
(C) Right to Consumer Education
(D) Right to Representation

180. Under which right the producer is suppose to provide details about the products, its ingredient, price hatch number, date of manufacturing, etc?
(A) Right to Informed
(B) Right to Information
(C) Right to Choose
(D) Right to Represent

181. What is RTI?
(A) Right to Informed
(B) Right to Information
(C) Right to Choose
(D) Right to Represent

182. Which of the following right you will use in case you want to have information about the functioning of a government department?
(A) Right to Informed
(B) Right to Information
(C) Right to Choose
(D) Right to Seek Redressal

183. You have joined an education institution and had paid the fee for the whole year but you found that the institution is not providing what it promised. Now you want to change the institution but the management refused to refund your fee. Which of your right is being violated?
(A) Right to Informed
(B) Right to Information
(C) Right to Choose
(D) Right to Represent

184. Suppose you want to buy toothpaste, and the shop owner says that he/she can sell the toothpaste only if you buy a toothbrush. Which of your right is being violated by the shopkeeper?
(A) Right to Informed
(B) Right to Information
(C) Right to Choose
(D) Right to Seek redressal

185. Under which of the following right a consumer can represent his/her own case by himself?

(A) Right to Informed
(B) Right to Information
(C) Right to Choose
(D) Right to Represent

186. Under which of the following rights the government gives advertisements in the newspapers about the consumer's rights and print slogans like "Jago Grahak Jago":

(A) Right to Informed
(B) Right to Information
(C) Right to Education
(D) Right to Represent

187. The district level Consumer courts deals with the cases involving claims up to Rs.:

(A) 20 lakh (B) 40 lakh
(C) 60 lakh (D) 80 lakh

188. The state level Consumer courts deals with the cases involving claims up to Rs.:

(A) 20 lakh - 1 crore
(B) 40 lakh- 2 crore
(C) 60 lakh - 3 crore
(D) 80 lakh – 4 core

189. The national level Consumer courts deals with the cases involving claims up to Rs.:

(A) 20 lakh (B) 40 lakh
(C) 60 lakh (D) 1 crore

190. While buying a product from a market which of the following logo assures you for quality?

(A) ISI or Agmark (B) LTTE
(C) KCF (D) BJP

191. Which day is celebrated as the National Consumer Day in India?

(A) 24th January (B) 24th December
(C) 24th October (D) 26th January

192. Why is 24th December celebrated as the National Consumer Day?

(A) It was the day when the Indian Parliament enacted the Consumer Protection Act 1986
(B) It was the day when the Indian Parliament celebrates Gandhi Jayanti
(C) On this day the first Consumer Court was set up in Delhi
(D) On this day the Supreme Court was setup

193. A chemist sold you a medicine of expiry date. Under which consumer right you can approach the Consumer court?

(A) Right to Informed
(B) Right to Information
(C) Right to safety
(D) Right to Equality

194. To which court you will approach if your claim is more than Rs 40 lakh and less than Rs one crore?

(A) State Consumer Court
(B) National Consumer Court
(C) District Consumer Court
(D) The Supreme Court

195. Which of the following is not a Consumer Right?

(A) Right to Informed
(B) Right to Information
(C) Right to safety
(D) Right to Equality

196. Find the odd one out:

(A) Right to Informed
(B) Right to Information
(C) Right to Education
(D) Right to Equality

197. In India --------- of the population lives in villages:

(A) Half (B) One-third
(C) Three-fourth (D) Two-third

198. Elasticity of production is equal to:

(A) $\frac{\Delta Y}{\Delta X} \times \frac{X}{Y}$
(B) $\frac{MP}{AP}$
(C) $\frac{\text{Present change in output}}{\text{Present change in input}}$
(D) All of these

199. If MP, AP and EP represent marginal product, average product and elasticity of production, which of the following inference is correct?

(A) When MP = AP, Ep=1
(B) When MP>AP, Ep>1
(C) When MP<AP, Ep<1
(D) All of these

200. The least cost combination of inputs will be at:

(A) $\frac{\Delta X1}{\Delta X2} = \frac{P\times 2}{P\times 1}$

(B) $\frac{\Delta X1}{\Delta X2} > \frac{P\times 2}{P\times 1}$

(C) $\frac{\Delta P\times 1}{\Delta P\times 2} < \frac{Py2}{Py1}$

(D) None of these

201. The contribution of livestock to gross farm income in mixed farming is:

(A) At least 10% (B) At least 30%
(C) At least 40 % (D) More than 50%

202. Under diversified farming, the income from the single enterprise should not be:

(A) < 50% of total farm income
(B) > 50% of total farm income
(C) > 33% of total farm income
(D) None of these

203. Under specialized farming, the income from the single enterprise should not be:

(A) < 50% of total farm income
(B) > 50% of total farm income
(C) > 75% of total farm income
(D) None of these

204. Ranching refers to:

(A) Growing of crops only
(B) Growing of livestock only
(C) Growing of crops and livestock
(D) All of these

205. Price elasticity of demand is equal to:

(A) $\frac{\text{Percentage change in quantity demanded}}{\text{Percentage change in price}}$

(B) $\frac{\text{Percentage change in supply of quantity}}{\text{Percentage change in price}}$

(C) $\frac{\text{Percentage change in quantity demanded}}{\text{Percentage change in quantity supplied}}$

(D) All of these

206. The income elasticity is generally:

(A) 1% increase income increase only 0.2% demand
(B) 1% increase income decreases 0.2% demand
(C) Both (A) and (B)
(D) None of these

207. Labour efficiency can be increased by:

(A) Improving farm layout
(B) Farm work simplification
(C) Training of labourers
(D) All of these

208. Farm management refers to:

(A) Managing the farm
(B) The art of managing the farm
(C) The art of managing the farm successfully as measured by the test of profitability
(D) All of these

209. Farm building is important because of:

(A) It saves feed, both in quantity and quality
(B) It increases safety in farm work
(C) It helps to increase production
(D) All of these

210. Which of the following is a partial standard of measuring farm production efficiency?

(A) Crop yield index
(B) Production per unit livestock
(C) Yield per hectare of the most important crop
(D) All of these

211. Farm survey refers to:

(A) An enquiry in the composition activities and living conditions of farmers and farm
(B) Collection of information related expenses and profit of the farms
(C) Visit of farms
(D) All of these

212. Which of the following step(s) are involved in farm survey?

(A) Planning of survey and collection of data
(B) Analysis of data
(C) Presentation and interpretation of the findings
(D) All of these

213. Which of the following is not accepted as mortgage against long term loan by land development bank?

(A) Land (B) House
(C) Waterlogged soil
(D) All of these

214. Regional Rural banks were established on:
(A) 2 October, 1975
(B) 19 July, 1969
(C) 1st July, 1965
(D) 2 October, 1969

215. Market refers to:
(A) A place where producers and sellers are assembled for selling and purchasing of goods
(B) Sellers and producers have free but competitive relationship
(C) Prices of commodities remain at normal prices
(D) All of these

216. Market process refers to:
(A) Activities through which a commodity is reached from producer to final consumer
(B) Grading and storage of a commodity
(C) Middlemen through which a commodity is either sold or purchase
(D) None of these

217. Which of the following approach useful and simple to study the agricultural marketing?
(A) Functional approach
(B) Institutional approach
(C) Commodity approach
(D) Behavioral approach

218. What is difference between Dalal and Commission agent?
(A) Commission agents have more rights than Dalal
(B) Dalal does not have physical control of produce
(C) Commission agents can sell the produce and charge the commission
(D) All of these

219. Which of the following statement is correct?
(A) All regulated markets should be organized markets
(B) All organized markets should be regulated markets
(C) All regulated markets may not be organized markets
(D) All unorganized markets may not be regulated markets

220. Which of the following middlemen accept the ownership of produce during the marketing process?
(A) Itinerant traders
(B) Village traders
(C) FCI
(D) All of these

221. Which of the following is a trade practices of marketing?
(A) Personal selling
(B) Advertisement based sale
(C) Sales by inspection
(D) All of these

222. Percentage marketing margin is equal to:
(A) Sale price – purchase price
(B) Absolute margin/sale price x 100
(C) Sale price – total marketing cost
(D) None of these

223. Agricultural Production Mandi Act was passed in UP in year:
(A) 1962 (B) 1964
(C) 1966 (D) 1968

224. Which of the following is a feature of a regulated market?
(A) Proper marketing system
(B) Proper communication of marketing information
(C) Availability of optimum grading, storage and packing facilities
(D) All of these

225. What is the objective of keeping farm records?
(A) To know annual performance of farm
(B) To know the present worth
(C) To compare the performance over a period of years
(D) All of these

226. Programme related to elimination of rural poverty is:
(A) JRY (B) RLEGP
(C) IRDP (D) All of these

227. Type of farming is based on:
(A) The law of substitution
(B) The law of diminishing returns
(C) The law of competitive advantage
(D) The law of equimarginal return

228. Short term credit needs of a farmer is assessed by taking into account:
(A) Area under different crops
(B) Existing level of inputs used in crop production
(C) Recommended doses of inputs needed in crop production
(D) Recommended doses of inputs needed – (minus) existing level of inputs used in crop production

229. India's largest nationalised commercial bank is:
(A) Reserved Bank of India
(B) Central Bank of India
(C) State Bank of India
(D) Bank of India

230. Which of the following is the most effective way to overcome the defects of agricultural marketing?
(A) Cooperative marketing
(B) Forward marketing
(C) Regulated marketing
(D) Monopoly

231. Which of the following function is related to production function analysis?
(A) Total physical product
(B) Average physical product
(C) Marginal physical product
(D) All of these

232. Principle of least-cost combination explains:
(A) Output is the function of two or more than two variable inputs
(B) Use of two variable inputs in such a way to get maximum profit
(C) Use of two variable inputs in such a way to get same level of output at comparatively lowest cost
(D) All of these

233. Law of equi-marginal return states that:
(A) Each unit of input is invested in different enterprises in such a way so that a farmer can get maximum marginal return
(B) Each unit of input is invested to get maximum return
(C) Each unit of input invested to get maximum average return
(D) None of these

234. Principal of equi-marginal return states that:
(A) Maximum satisfaction is obtained from a given level of expenditure on two goods when the last penny spent on either good gives the same amount of satisfaction
(B) Equal marginal return is obtained by substitution of one good by other
(C) Both (A) and (B)
(D) None of these

235. Factor-product relationship is also known as:
(A) Input-output relationship
(B) Output-output relationship
(C) Demand-supply relationship
(D) Output-Input relationship

236. Law of equi-marginal return also known as:
(A) Law of opportunity cost
(B) Alternative cost principle
(C) Both (A) and (B)
(D) Law of equi-marginal return

237. Law of variable proportions is also known as:
(A) Law of returns
(B) Law of factor-product relationship
(C) Law of input-output relationship
(D) All of these

238. Law of variable proportions deals with:
(A) Input-output relationship where the output is the function of input
(B) Cost of output and input
(C) Supply of output
(D) All of these

239. Maximum net return from one variable input will occur when:
(A) Additional return from an additional unit of output from employing an additional unit of input is equal to the additional cost of that unit of input
(B) Value of marginal return is equal to cost of the input
(C) Marginal return x selling price of product = marginal input cost
(D) All of these

240. The production possibilities curve shows:
(A) Alternative combination of two outputs that may be produced with a given amount of resource during a period of time
(B) Combination of inputs that can produce equal level of production
(C) Combination of outputs that can be produced with same amount of input
(D) All of these

241. Indian agriculture is characterized by:
(A) Low productivity
(B) Subsistence farming
(C) Labour dominance
(D) All of these

242. The cost of production can be minimized by using the principle of:
(A) Law of return
(B) Law of opportunity cost
(C) Law of least-cost combination
(D) Cost principle

243. Who defined Economics is a wealth of science:
(A) Smith (B) Marshall
(C) Robins (D) Pigue

244. Who defined Economics as the study of human behaviour in the daily business of life?
(A) Smith (B) Marshall
(C) Robins (D) Pigue

245. Who defined economics as the study of human welfare?
(A) Smith (B) Marshall
(C) Robins (D) Pigue

246. Who defined economics as the study of choice making behaviour of human being?
(A) Smith (B) Marshall
(C) Robins (D) Pigue

247. Which of the following is a subject matter of economics?
(A) Consumption
(B) Production and distribution
(C) Public finance
(D) All of these

248. Which is not a factor of production?
(A) Land (B) Labour
(C) Capital (D) Crops

249. Principle of equi-marginal return is also known as:
(A) Product-product relationship
(B) Output-output relationship
(C) Substitution between output/products
(D) All of these

250. Law of diminishing marginal utility states that:
(A) Utility varies with time and space
(B) A commodity generates different utilities for different people and for the same person at different times
(C) Utility of additional unit of a commodity to any consumer commodity to already consumption increases
(D) All of these

251. Principle of least –cost combination is also known as:
(A) Substitution between inputs
(B) Factor-factor relationship
(C) Rate of technical substitution
(D) All of these

252. Which of the following is not a type of production function?
(A) Linear (B) Quadratic
(C) Square root (D) Rectangular

253. Maximum profit can be achieved in agri-business when marginal return (MR) is equal to:
(A) Marginal cost
(B) Average output
(C) Total output
(D) Total cost

254. Production function illustrates the relationship between:
(A) Input levels and corresponding output
(B) Input levels and corresponding marginal input
(C) Demand and supply pattern
(D) Demand and competition

255. At optimum level of production or output, the marginal return should be:
(A) Positive
(B) Negative
(C) Zero
(D) Equal to average return

256. Which of the following function is related to an economic system?
(A) What goods and services are to be produced?
(B) How to best organize and combine recourse of production?
(C) Allocation or distribution of income to owners of resources
(D) All of these

257. At point of inflection in production function curve, MR is:
(A) >AR (B) <AR
(C) MR=AR (D) All of these

258. Marginal cost is equal to:
(A) $\frac{\text{Change in total cost}}{\text{Additional output}}$
(B) $\frac{\text{Change in total revenue}}{\text{Additional output}}$
(C) $\frac{\text{Change in total value product}}{\text{Additional input}}$
(D) All of these

259. Marginal rate of substitution is equal to:
(A) $\frac{\text{Change in number of y units}}{\text{Change in number of x units}}$
(B) $\frac{\text{Change in price of x commodity}}{\text{Change in price of y commodity}}$
(C) $\frac{\text{Price of X commodity}}{\text{Price of y commodity}}$
(D) None of these

260. The main purpose of farmers is:
(A) High production
(B) Minimum cost of cultivation
(C) High living standard
(D) High net profit

261. If the increase or decrease one product increases or decreases the other product, the relationship between both products is known as:
(A) Complementary
(B) Supplementary
(C) Competitive
(D) Symbiotic

262. If the increase or decrease in the production level of one product does not affect the production level of the other product, the relationship is called as:
(A) Complementary
(B) Supplementary
(C) Competitive
(D) Symbiotic

263. If the increase in the production level of one product necessarily decreases the production level of the other product, the relationship is called as:
(A) Complementary
(B) Supplementary
(C) Competitive
(D) Symbiotic

264. Production possibility curve is also called as:
(A) Iso-revenue curve
(B) Opportunity curve
(C) Iso-return curve
(D) All of these

265. Types of farming refers to:
(A) Types of enterprises are grown or taken on a farm
(B) Ownership of farm
(C) Both A and B
(D) None of these

266. Types of farming refers to:
(A) Types of enterprises are grown or taken
(B) Ownership of farm and method or practices of production
(C) Both A and B
(D) None of these

267. Law of demand indicates:
(A) Inverse relationship between demand and price
(B) Linear relationship between demand and price
(C) Inverse relationship between market demand and marketable surplus
(D) Linear relationship between supply and demand

268. Price line or budget line indicates:
(A) A capacity of a consumer to purchase within a given budget
(B) A capacity of a consumer to provide marketable surplus
(C) The graph between demand and satisfaction
(D) None of these

269. Price elasticity of demand is always:
(A) Positive (B) Zero
(C) Negative (D) None of these

270. The demand of any commodity is affected by:
(A) Price of the commodity
(B) Taste and liking of the consumers
(C) Income of the consumers
(D) All of these

271. Which of the following item has the highest income elasticity?
(A) Bread (B) Butter
(C) Milk (D) Wine

272. Income elasticity is generally:
(A) Zero (B) Positive
(C) Negative (D) Infinite

273. The supply elasticity of perishable commodities is:
(A) Elastic (B) Inelastic
(C) Zero (D) Equal to one

274. The supply elasticity of bulky products is normally:
(A) Less elastic
(B) High elastic
(C) Complete elastic
(D) All of these

275. Supply of a commodity is governed by:
(A) Price of the commodity
(B) Price of production factors
(C) State of technology
(D) All of these

276. The price elasticity of demand for agricultural commodities is always:
(A) <1 (B) >1
(C) Zero (D) Infinite

277. The price elasticity of luxury items is always:
(A) <1 (B) >1
(C) Zero (D) Infinite

278. If the price elasticity is >1, the total revenue will be increased at:
(A) Decrease in price
(B) Increase in price
(C) No change in price
(D) Unknown

279. If the price elasticity is <1, the total revenue will be increased at:
(A) Decrease in price
(B) Increase in price
(C) No change in price
(D) Unknown

280. The law of diminishing returns states that:
(A) The output increases with increasing input levels continuously
(B) The output increases with input levels irregularily
(C) The output increases with increasing input levels upto certain level and than declines
(D) All of these

281. Specialised market refers to:
(A) Sell-buy of specialized kind of a commodity
(B) Sell-buy of different kind of a commodity
(C) Communication, transport and storage play an important role
(D) All of these

282. Terminal market are found at:
(A) Large cities and sea ports
(B) Village level
(C) Town levels
(D) District level

283. Local market generally found at:
(A) Large cities and sea ports
(B) Village level
(C) Town levels
(D) District level

284. Which of the marketing system was prevalent in the ancient time?
(A) Barter system
(B) Modern system
(C) European system
(D) None of these

285. Computer based market refers:
(A) Prices are decided on interest or phone
(B) Exchange of commodities are based on computer information
(C) Payment can also be made through computer
(D) All of these

286. Complete competitive market should have:
(A) Large number of sellers and buyers
(B) No product differentiation
(C) Dynamic factors of production
(D) All of these

287. The market that is organized under some rules and regulations made by trade body federation etc. is known as:
(A) Regulated market
(B) Organized market
(C) Economic market
(D) Unorganized market

288. The market that is under control of government rules and regulations is called as:
(A) Regulated market
(B) Organized market
(C) Economic market
(D) Unorganized market

289. Agricultural products are characterized as:
(A) Bulky raw material
(B) Perishable
(C) Essential for life
(D) All of these

290. Agricultural production is characterized as:
(A) Annual variability
(B) Seasonal variation
(C) Varying costs of production
(D) All of these

291. Which of the following statement is correct?
(A) Marketable surplus > marketed surplus
(B) Marketable surplus < Marketed surplus
(C) Marketable surplus = marketed surplus
(D) All of these

292. Fixed cost is also known as:
(A) Over head cost (B) Prime cost
(C) Alternate cost (D) All of these

293. Variable cost is also known as:
(A) Overhead cost (B) Prime cost
(C) Alternate cost (D) All of these

294. Opportunity cost is also known as:
(A) Overhead cost (B) Prime cost
(C) Alternate cost (D) All of these

295. If price elasticity is -1, it indicates that:
(A) An equal change in demand and price
(B) Percentage change on quantity in smaller than the percentage price change
(C) Percentage quantity change is greater than the percentage price change
(D) None of these

296. If the price elasticity of demand is more than one, the elasticity is known as:
(A) Elastic (B) Inelastic
(C) Infinity (D) None of these

297. If the price elasticity of demand is less than one, the elasticity is known as:
(A) Elastic (B) Inelastic
(C) Infinity (D) None of these

298. Changes in supply elasticity occurs mainly during:
(A) Short period
(B) Long period
(C) Winter season
(D) Rainy or summer season

299. The elasticity of the agricultural commodity in very long run will be always:
(A) Elastic (B) Inelastic
(C) Both (A) & (B) (D) None of these

300. Agricultural commodities have greater price fluctuations mainly because of:
(A) Inelastic demand and supply
(B) Elastic demand and supply
(C) More control of government
(D) All of these

301. Which of the following is a government measure to deal with uncertainty in agriculture?
(A) Guaranteed agricultural prices
(B) Buffer stock schemes
(C) Crop insurance
(D) All of these

302. Which of the following government agency is responsible for marketing and distribution of agricultural produce in the country?
(A) FCI
(B) State Trading Corporation Ltd
(C) Agricultural Price Commission
(D) All of these

303. Which of the following government agency is responsible for marketing and distribution of tobacco produce in the country?
(A) FCI
(B) State Trading Corporation Ltd
(C) Tobacco Board
(D) Agricultural Price Commission

304. Which of the following government agency is responsible for international trading of agricultural produce in the country?
(A) FCI
(B) State Trading Corporation Ltd
(C) Agricultural Price Commission
(D) All of these

305. Crop area production and productivity data are available with:
(A) Directorate of Economics and Statistics
(B) Forward Market Commission
(C) Central Storage Corporation
(D) All of these

306. Marketing margin can be defined as:
(A) Part of total cost that goes to marketing firms
(B) It is a difference between producer's selling cost and consumer's purchasing cost
(C) It is the produce of total services of marketing firms
(D) All of these

307. The new name of Agricultural Price Commission is:
(A) Commission on Agricultural Cost and Prices (CACP)
(B) Commission on agricultural Prices
(C) Agricultural Prices and Cost Commission
(D) None of these

308. The CACP came into existence in year:
(A) 1982 (B) 1985
(C) 1987 (D) 1990

309. Uncertainity in agriculture refers to:
(A) Yield uncertainty
(B) Price uncertainty
(C) Tenurial uncertainty
(D) All of these

310. Uncertainty can be reduced mainly at farmers level by:
(A) Diversification (B) Flexibility
(C) Capital rationing
(D) All of these

311. What is the important measure to stabilize the agricultural prices and incomes?
(A) Buffer stock scheme
(B) Crop insurance scheme
(C) Control over agricultural production
(D) All of these

312. Which of the following basis is adopted in determination of prices of agricultural commodities in our country?
(A) Production cost
(B) Parity cost
(C) Prevalent market price
(D) All of these

313. The basic objective of farm planning is:
(A) To guide a farmer to move in the right direction
(B) To indicate the destination
(C) Proper use and allocation of input
(D) All of these

314. Farm price means to:
(A) The price that farmer receives for his produce
(B) Price prevailing in the market
(C) Price declared by Govt.
(D) All of these

315. Farm plan refers to:
(A) Systematic pre-thinking of future activities related to farm business
(B) It is a tentative future plan for farm business
(C) It is related to maximize the farm profit
(D) All of these

316. Farm production efficiency can be measured with following overall standard?
(A) Input per unit of output
(B) Total production index
(C) Output per hectare
(D) All of these

317. The main objective of land reform in India is:
(A) Abolition of intermediates
(B) Confirment of ownership right to tenants
(C) Imposition of ceiling on agricultural land holdings
(D) All of these

318. The maximum limit is land holding under best soil is:
(A) 10-18 acres
(B) > 27 acres
(C) > 34 acres
(D) None of these

319. The maximum limit of land holding if the soil is poor is:
(A) 20-22 acres
(B) > 37 acres
(C) > 54 acres
(D) None of these

320. Farmers can get long term loan from:
(A) Money lenders
(B) Land Development Banks
(C) Government
(D) All of these

321. Long term loan is given mainly for:
(A) Purchase of seeds, fertilizers and weedicide
(B) Purchase of agricultural implements
(C) Purchase of land and its development
(D) All of these

322. Farmers can get short term loan from:
(A) Money lenders
(B) Land Development Banks
(C) Government
(D) All of these

323. Short term loan is given mainly for:
(A) Purchase of seeds, fertilizers and weedicide
(B) Purchase of agricultural implements
(C) Purchase of land and its development
(D) All of these

324. What is the main principle of good system of credit to agriculture?
(A) Repayment capacity
(B) Risk bearing capacity
(C) Estimation of returns to credit
(D) All of these

325. The lead Bank Scheme was started in 1969 on the recommendation of:
(A) Nariman Committee
(B) All India Rural Credit Survey Committee
(C) Committee on Cooperative Credit
(D) Expert group of RBI on credit scheme

326. Agricultural marketing involves the operations:
(A) Movement of food and raw material from field to final consumer
(B) Handling of produce at farm
(C) Processing, grading and packing
(D) All of these

327. The main objective of agricultural marketing is:
(A) To make available agricultural produce to consumer at optimum prices
(B) To help farmers to get maximum profit from their produce
(C) To maintain optimum market prices of agricultural commodities
(D) All of these

328. Functional approach of agricultural marketing deals with:
(A) Different functions
(B) Different institutions
(C) Different commodities
(D) None of these

329. Which of the following is a facilitating function of agricultural marketing:
(A) Standardization
(B) Financing
(C) Risk bearing (D) All of these

330. Agricultural holding refers to:
(A) Total area of land which is held for cultivation as a single unit
(B) Total area of land used as a single unit
(C) Total area owned by a cultivator
(D) All of these

331. Operational holding refers to:
(A) Total area of land which is held for cultivation as a single unit
(B) Total area including fallow and barren used as a single unit
(C) Total area owned by a owner
(D) All of these

332. Fragmentation of holding refers to:
(A) It is a breaking up of a single holdings into a number of small pieces situated at a distance from each other
(B) Consolidating the small holdings
(C) Breaking of holdings into economic holdings
(D) None of these

333. An optimum holding refers to the size that:
(A) Should not be more than 3 times of economic holding
(B) Should be > 3 times of economic holding
(C) Should be equal to operational holding
(D) Should be <2 ha land

334. Which of the following is physical function of agricultural marketing?
(A) Buyers and sellers
(B) Transportation
(C) Processing and storage
(D) All of these

335. Which of the following is known as a middlemen?
(A) Whole sellers
(B) Retailers
(C) Commission agents
(D) All of these

336. Income elasticity of inferior goods is:
(A) Positive
(B) Negative
(C) Between zero to one
(D) Infinite

337. An income elasticity of necessary is:
(A) 1 (B) < 1
(C) >1 (D) Between zero to one

338. Which of the following is a feature of monopolistic competition?
(A) Many firms
(B) Firm is price maker
(C) Products/services differentiated through promotion, quality and design
(D) All of these

339. Cyclic fluctuations in prices of agricultural commodities means:
(A) Ups and downs in the prices after certain intervals
(B) Increase in prices after certain intervals
(C) Decrease in prices after certain intervals
(D) None of these

340. The fluctuations in agricultural prices due to operation of Cobweb theorem shows:
(A) Cyclic fluctuations
(B) Long term fluctuations
(C) Seasonal fluctuations
(D) All of these

341. Long term fluctuations in agricultural prices and incomes are caused mainly due to:
(A) Change in population
(B) Opening of new markets
(C) Development of new technology
(D) All of these

342. Mechanisation of agriculture refers to:
(A) Use of power driven machines for various farm jobs
(B) Ploughing with tractor operated plough only
(C) Crushing of sugarcane with crusher only
(D) All of these

343. Mechanisation of agriculture is required for:
(A) Better agricultural yield
(B) Multiple cropping
(C) Generation of employment
(D) All of these

344. How much farm power is available per hectare?
(A) 0.5 KW (B) 1.0 KW
(C) 1.5 KW (D) 2.0 KW

345. The wage is not paid to:
(A) Permanent labour
(B) Contract labour
(C) Family labour
(D) Casual labour

346. Level of farm mechanization is affected by:
(A) Size of holding
(B) Labour requirement
(C) Intensification of crop production
(D) All of these

347. Farm planning refers to:
(A) Reorganization and/or alteration of present farm organization and operation in order to get maximum return from limited resources
(B) Activities for crop production on farm
(C) Physical arrangement of farm resources
(D) All of these

348. Regional rural banks provide loan to:
(A) Small and marginal farmers
(B) Landless labourers
(C) Rural artisans
(D) All of these

349. Which type of loan is supplied agricultural service societies?
(A) Short term (B) Medium term
(C) Long term (D) All of these

350. Marketing involves the activities from:
(A) Production to end consumer
(B) Production only
(C) Production to middlemen
(D) Production to storage

351. Monopoly market refers to:
(A) Single producer or seller in market
(B) Single buyer in market
(C) More number of producers and single buyers
(D) None of these

352. Monopsony market refers to:
(A) Market having only one buyer or institution
(B) Buyer is free to decide the quantity to be purchased
(C) Buyer is free to change the price policy
(D) All of these

353. Duopsony refers to:
(A) Market having only two buyers
(B) Market having only two sellers
(C) Market having only two buyers and sellers each
(D) None of these

354. Oligopsony refers to:
(A) Few buyers in a market
(B) Only one buyer in a market
(C) Only one seller in a market
(D) None of these

355. Oligopoly refers to:
(A) More than sellers in market
(B) Both product homogeneity and differentiation may be found
(C) Product substitution is necessary
(D) All of these

356. Marketable surplus refers to:
(A) Actual amount that is marketed
(B) Excess amount than requirement for consumption
(C) Both A and B
(D) None of these

357. Marketed surplus refers to:
(A) Actual amount that is marketed
(B) Excess amount than requirement for consumption
(C) Both A and B
(D) None of these

358. Normally marketed surplus is higher than marketable surplus because of:
(A) To repay the loan
(B) Poor facilities of storage at farms
(C) Inability to estimate the total family consumption
(D) All of these

359. Which of the following is a market for agricultural commodities?
(A) Whole sale market
(B) Retail market
(C) Fairs
(D) All of these

360. What will happen if the market price of a commodity is higher than normal price?
(A) Demand will be reduced and supply will be increased in the market
(B) Demand will be increased
(C) Supply will be reduced
(D) None of these

361. Marketing middlemen refers to:
(A) Institutions or persons that work between the transfer process of a commodity from producers to consumers
(B) Institutions or persons that purchase the produce from producers
(C) Institutions or persons that sell the produce to consumers
(D) None of these

362. Kachcha commission agent is related to:
(A) Collection of agricultural produce
(B) Distribution of agricultural produce
(C) Both (A) and (B)
(D) None of these

363. Packka commission agent is related to:
(A) Collection of agricultural produce
(B) Distribution of agricultural produce
(C) Both (A) and (B)
(D) None of these

364. Market structure refers to:
(A) Organizational structure that affects the competition and price in market
(B) Behavioral pattern of buyers and sellers through which prices are affected
(C) It indicates about the role of commission agents in market only
(D) All of these

365. Market conduct refers to:
(A) Organizational structure that affects the competition & price of a commodity
(B) Behavioral pattern of market that affect the commodity price
(C) Involvement of brokers in marketing
(D) All of these

366. Which of the following is a important element of market structure?
(A) Degree of seller's concentration
(B) Degree of buyer's concentration
(C) Degree of product differentiation
(D) All of these

367. Which of the following is a important element of market conduct?
(A) Rules and regulations of production firms
(B) Product policy
(C) Product promotion policy
(D) All of these

368. The production or supply of a product will be closed by a firm if:
(A) Production cost < cost of marginal product
(B) Production cost > cost of marginal product
(C) Variable cost > production cost
(D) All of these

369. Which of the following condition is required for perfect competition in the market?
(A) More number of firms in the market
(B) Homogeneous product
(C) Free entry or exit in the market
(D) All of these

370. Which of the following is a component of marketing cost?
(A) Labour
(B) Transportation
(C) Packing and storage
(D) All of these

371. Price spread is also known as:
(A) Marketing cost
(B) Marketing margin
(C) Farm price
(D) None of these

372. Which of the statement is correct for monopoly market?
(A) MC = MR (B) MC > AC
(C) MC < AC (D) None of these

373. New name of Indian Standard Institution is:
(A) Bureau of Indian Standards
(B) Institute of Indian Standards
(C) Standard Institute of India
(D) None of these

374. The Bureau of Indian Standard is situated at:
(A) Lucknow (B) New Delhi
(C) Hyderabad (D) Mumbai

375. The price which necessarily covers the variable cost and also the fixed price is:
(A) The long run price
(B) The short run price
(C) The market price
(D) The equilibrium price

376. Marginal Physical Product of a variable input is negative:
(A) Only in I stage
(B) Only in II stage
(C) Only in III stage
(D) In I and II stage

377. Average cost is equal to:
(A) Total cost + opportunity cost
(B) $\frac{\text{Total cost}}{\text{Output}}$
(C) $\frac{\text{Average variable cost}}{\text{Output}}$
(D) Cost B + Cost C

378. Fixed cost is equal to:
(A) $\frac{\text{Total fixed cost}}{\text{Output}}$
(B) $\frac{\text{Total cost}}{\text{Output}}$
(C) $\frac{\text{Total variable cost}}{\text{Output}}$
(D) None of these

379. Variable cost is equal to:

(A) $\frac{\text{Total cost}}{\text{Output}}$

(B) $\frac{\text{Total variable cost}}{\text{Unit of output}}$

(C) Cost A + Cost B + Cost C

(D) All of these

380. Marginal cost is equal to:

(A) $\frac{\text{Total cost}}{\text{Output}}$

(B) $\frac{\text{Increase in variable cost}}{\text{Increase in output}}$

(C) Cost C + Rent paid to leased rent

(D) $\frac{\text{Total variable cost}}{\text{Increase in output}}$

381. What will happen if the market price of a commodity is lower from its normal price?

(A) Demand of a commodity will be increased

(B) Supply will be decreased

(C) Market price will increase towards the normal price line

(D) All of these

382. If the price elasticity of barley is -0.5, it means:

(A) Barley demand will be decreased by 0.5%, if the price increased by 1%

(B) Barley demand will be increased by 0.5% if the price is increased by 1%

(C) If the price of barley is increased by 0.5%, the demand will also be increased by 0.5%

(D) Price and demand both either increased or decreased by 0.5%

383. If the petrol price fall from Rs. 10 to 8 for 1 litre and as a result the demand of petrol is increased from 5 units to 9 units. What will be the price elasticity of demand?

(A) -1 (B) -2

(C) -3 (D) -4

384. Cross elasticity of demand refers to:

(A) $\frac{\text{\% change in quantity demanded of X}}{\text{\% change in income}}$

(B) $\frac{\text{\% change in quantity demanded of X}}{\text{\% change in price}}$

(C) $\frac{\text{\% change in income}}{\text{\% change in quantity demanded}}$

(D) None of these

385. Income elasticity of demand refers to:

(A) $\frac{\text{\% change in quantity demanded}}{\text{\% change in income}}$

(B) $\frac{\text{\% change in quantity demanded}}{\text{\% change in price}}$

(C) $\frac{\text{\% change in income}}{\text{\% change in quantity demanded}}$

(D) None of these

386. Price elasticity of supply is equal to:

(A) $\frac{\text{\% change in quality supplied of a commodity A}}{\text{\% change in price of A}}$

(B) $\frac{\text{\% change in demand of commodity A}}{\text{\% change in price of A}}$

(C) $\frac{\text{\% change in income}}{\text{\% change in price}}$

(D) All of these

387. National income generated by agriculture is calculated by using:

(A) Income method

(B) Out put method

(C) Expenditure method

(D) All of these

388. Capital formation in agriculture over the years:

(A) Decreasing

(B) Increasing

(C) Constant

(D) Both (A) and (B)

ANSWER SHEET
15. Farm Management

1. (A) 2. (D) 3. (C) 4. (C) 5. (D) 6. (D) 7. (A) 8. (D) 9. (A) 10. (D)
11. (A) 12. (A) 13. (B) 14. (A) 15. (B) 16. (D) 17. (A) 18. (A) 19. (D) 20. (B)
21. (D) 22. (B) 23. (B) 24. (A) 25. (B) 26. (C) 27. (A) 28. (A) 29. (B) 30. (D)
31. (A) 32. (D) 33. (C) 34. (B) 35. (B) 36. (B) 37. (C) 38. (B) 39. (A) 40. (B)
41. (B) 42. (A) 43. (C) 44. (A) 45. (C) 46. (D) 47. (D) 48. (D) 49. (C) 50. (A)
51. (C) 52. (C) 53. (C) 54. (B) 55. (A) 56. (B) 57. (C) 58. (A) 59. (A) 60. (D)
61. (B) 62. (B) 63. (B) 64. (D) 65. (C) 66. (A) 67. (A) 68. (C) 69. (B) 70. (A)
71. (D) 72. (B) 73. (A) 74. (D) 75. (A) 76. (B) 77. (D) 78. (D) 79. (A) 80. (B)
81. (B) 82. (A) 83. (A) 84. (B) 85. (D) 86. (A) 87. (B) 88. (D) 89. (D) 90. (A)
91. (D) 92. (C) 93. (A) 94. (C) 95. (A) 96. (B) 97. (B) 98. (A) 99. (C) 100. (C)
101. (C) 102. (A) 103. (C) 104. (D) 105. (A) 106. (C) 107. (A) 108. (A) 109. (B) 110. (A)
111. (D) 112. (B) 113. (D) 114. (A) 115. (A) 116. (C) 117. (A) 118. (D) 119. (C) 120. (D)
121. (B) 122. (B) 123. (C) 124. (D) 125. (C) 126. (C) 127. (B) 128. (A) 129. (B) 130. (A)
131. (B) 132. (B) 133. (C) 134. (C) 135. (B) 136. (A) 137. (C) 138. (A) 139. (C) 140. (B)
141. (D) 142. (B) 143. (A) 144. (D) 145. (B) 146. (A) 147. (B) 148. (D) 149. (D) 150. (B)
151. (C) 152. (B) 153. (A) 154. (A) 155. (B) 156. (B) 157. (B) 158. (B) 159. (D) 160. (B)
161. (B) 162. (D) 163. (B) 164. (A) 165. (C) 166. (B) 167. (C) 168. (C) 169. (A) 170. (B)
171. (B) 172. (C) 173. (C) 174. (C) 175. (C) 176. (C) 177. (C) 178. (A) 179. (A) 180. (A)
181. (B) 182. (B) 183. (C) 184. (D) 185. (D) 186. (C) 187. (A) 188. (A) 189. (D) 190. (A)
191. (B) 192. (A) 193. (C) 194. (A) 195. (D) 196. (C) 197. (B) 198. (D) 199. (D) 200. (A)
201. (A) 202. (A) 203. (B) 204. (B) 205. (A) 206. (A) 207. (D) 208. (D) 209. (D) 210. (D)
211. (A) 212. (D) 213. (C) 214. (A) 215. (D) 216. (A) 217. (A) 218. (D) 219. (A) 220. (D)
221. (D) 222. (B) 223. (B) 224. (D) 225. (D) 226. (D) 227. (D) 228. (A) 229. (C) 230. (C)
231. (D) 232. (C) 233. (A) 234. (C) 235. (A) 236. (C) 237. (D) 238. (A) 239. (D) 240. (A)
241. (D) 242. (C) 243. (A) 244. (B) 245. (D) 246. (C) 247. (D) 248. (D) 249. (D) 250. (C)
251. (D) 252. (D) 253. (A) 254. (A) 255. (C) 256. (D) 257. (C) 258. (A) 259. (A) 260. (D)
261. (A) 262. (B) 263. (C) 264. (D) 265. (A) 266. (B) 267. (A) 268. (A) 269. (C) 270. (D)
271. (D) 272. (B) 273. (B) 274. (A) 275. (D) 276. (A) 277. (B) 278. (A) 279. (B) 280. (C)
281. (A) 282. (A) 283. (B) 284. (A) 285. (D) 286. (D) 287. (B) 288. (A) 289. (D) 290. (D)
291. (B) 292. (A) 293. (B) 294. (C) 295. (A) 296. (A) 297. (B) 298. (A) 299. (A) 300. (A)
301. (D) 302. (A) 303. (C) 304. (B) 305. (A) 306. (D) 307. (A) 308. (C) 309. (D) 310. (A)
311. (D) 312. (D) 313. (D) 314. (D) 315. (D) 316. (D) 317. (D) 318. (A) 319. (C) 320. (D)
321. (C) 322. (D) 323. (A) 324. (D) 325. (A) 326. (D) 327. (D) 328. (A) 329. (D) 330. (A)
331. (B) 332. (A) 333. (A) 334. (D) 335. (D) 336. (B) 337. (D) 338. (D) 339. (A) 340. (A)

341. (D) 342. (A) 343. (D) 344. (D) 345. (C) 346. (D) 347. (A) 348. (D) 349. (D) 350. (A)
351. (A) 352. (D) 353. (A) 354. (A) 355. (D) 356. (B) 357. (A) 358. (D) 359. (D) 360. (A)
361. (A) 362. (A) 363. (B) 364. (A) 365. (B) 366. (D) 367. (D) 368. (B) 369. (D) 370. (D)
371. (A) 372. (A) 373. (A) 374. (B) 375. (B) 376. (C) 377. (B) 378. (A) 379. (B) 380. (B)
381. (D) 382. (B) 383. (D) 384. (A) 385. (A) 386. (A) 387. (B) 388. (A)

16 APPENDIX

ICAR RESEARCH INSTITUTES

AGRICULTURAL ENGINEERING

1. CIAE	Central Institute of Agricultural Engineering, Nabi Bagh, Berasia road, Bhopal- 462038 (MP)
2. CIPHET	Central Institute of Post Harvest Engineering & Technology, Ludhiana- 141004 (Punjab)
3. CICT	Central Institute of Cotton Technology, P.B. No. 16640, Mumbai – 400019 (Maharashtra)
4. ILRI	Indian Lac Research Institute, Ranchi- 834010 (Jharkhand)
5. NIRJAFT	National Institute of Research on Jute & Allied Fibre Technology, Ranikuthi, Kolkata -700120 (W.B.)

AGRICULTURE EXTENSION

1. CIWA	Central Institute for Women in Agriculture, 1199 Baramunda, Bhubneshwar-751030 (Odisha)
2. DIPA	Directorate of Information and Publication of Agriculture, KAB-I, Pusa, New Delhi-110012
3. DKMA	Directorate of Knowledge Management in Agriculture, New Delhi

ANIMAL SCIENCES

1. CARI	Central Avian Research Institute, Izzatnagar, Bareili-243122(U.P)
2. CIRB	Central Institute for Research on Buffaloes, Sirsa Road, Hisar- 125001 (Haryana)
3. CIRG	Central Institute for Research on Goats, Makhdoom, Mathura-281122 (U.P)
4. IVRI	Indian Veterinary Research Institute, Izzatnagar, Bareili-243122 (U.P)
5. NAINP	National Institute of Animal Nutrition & Physiology, Adugodi, Bangalore-560030 (Karnataka)
6. CSWRI	Central Sheep and Wool Research Institute, Avikanagar- 304501 (Rajasthan)
7. NBAGR	National Bureau of Animal Genetic Resources, P.B. No. 129, Kernal-132001 (Haryana)
8. NDRI	National Dairy Research Institute, Karnal-132001(Haryana)
9. NRCC	National Research Centre on Camel, Jobber, P.B. No. 07, B ikhaner-334001 (Rajasthan)
10. NRCM	National Research Centre for Mithun, Jharnapani, Medziphema, Nagaland-500059 (A.P.)
11. NRCM	National Research Centre on Meat, Saidabad, CRIDA campus, Hyderabad- 500059 (A.P.)

12. NRCY National Research Centre on Yak, Dirang, West Kameng-790101 (Arunachal Pradesh)
13. NRCP National Research centre on Pigs, Panjabari Road, 6th mile, Guwahati-785037 (Assam)
14. NIVEDI National Institute of Veterinary Epidemiology & Disease Information, Hebbal, Bangalore-560024 (Karnataka)
15. CIRC Central institute for Research on Cattle, Grass farm Road, P.B. No. 17, Meerut Cant, Meerut- 250001 (U.P.)
16. PDFMD Project Directorate on Foot and Mouth Disease, IVRI, Mukteshwar, Kumaon- 263138 (Uttaranchal)
17. PDP Project Directorate on Poultry, Hyderabad-500030(A.P.)
18. NRCE National Research Centre for Equines, Hisar – 125001
19. NIHSAD National Institute of High Security Animal Diseases, Bhopal

CROP SCIENCES

1. CICR Central Institute for Cotton Research, P.B. No. 2, Nagpur-440010 (Maharashtra)
2. CRIJAF Central Research Institute for Jute & Allied Fiber, Nilganj, Barrackpore-700120 (W.B.)
3. CRRI Central Rice Research Institute, Cuttak, 753006 (Orrisa)
4. CTRI Central Tobacco Research Institute, Rajamundry-533105(A.P.)
5. IIMR Indian Institute of Maize Research, IARI Pusa Campus, New Delhi -110012
6. IIOR Indian Institute of Oilseed Research, Rajendranagar, Hyderabad-500030 (A.P.)
7. IIRR Indian Institute of Rice Research, Hyderabad-500030 (A.P.)
8. IIWBR Indian Institute of Wheat and Barley Research, PB No.158, Karnal -132001 (Haryana)
9. IARI Indian Agricultural Research Institute, New Delhi-110012
10. IGFRI Indian Grassland and Fodder Research Institute, Jhansi-284003 (U.P.)
11. IIPR Indian Institute of Pulses Research, Kanpur-208024
12. IISR Indian Institute of Sugarcane Research, Lucknow-226602
13. NBAIMO National Bureau of Agriculturally Important Microorganisms, Kusmaur, Mau Nath Bhanjan-275101 (U.P.)
14. NBPGR National Bureau of Plant Genetic Resources, New Delhi-110012
15. NCIPM National Centre for Integrated Pest Management, LBS Building, Pusa campus, New Delhi -110012
16. NIPB National Institute for Plant Biotechnology, Pusa campus, New Delhi-110012
17. DGR Directorate of Groundnut Research, PB No. 5, Ivnagar Road, Junagarh-362001 (Gujarat)
18. DRMR Directorate of Rapeseed and Mustard Research, Bharatpur - 321303 (Rajasthan)
19. IIMR Indian Institute of Millets Research, Rajendranagar, Hyderabad-500030 (A.P.)
20. DSR Directorate of Soybean Research, Indore-452017 (M.P.)
21. NABII National Bureau of Agriculturally Important Insect, PB.No. 2491, Hebbal, Banglore-560024 (Karnataka)
22. VPKAS Vivekanand Parvatiya Krishi Anusandhan Sansthan, Almora-263601 (Uttaranchal)
23. IISR Indian Institute of Seed Research, P.B. No. 11, Village Kushmour, P.O. Kaithuli, Mau – 275901

24. SBI — Sugarcane Breeding Institute, Coimbatore – 641007
25. IIAB — Indian Institute of Agricultural Biotechnology, Ranchi
26. NBAIR — National Bureau of Agricultural Insects Research, Bangalore
27. NIBSM — National Institute of Biotic Stress Management, Raipur

FISHERIES

1. CIBA	Central Institute for Brackish Water Aquaculture, 75 Santhome High Road, RA Puram, Chennai-600028 (T.N.)
2. CIFA	Central Institute for Fresh Water Aquaculture, Kausalyaganga, Bhubneshwar-751002 (Orissa)
3. CIFA	Central Institute for Fisheries Education, Mumbai-400061 (Maharashtra)
4. CIFRI	Central Inland Fisheries Research Institute, Barrackpore- 700120 (W.B)
5. CIFT	Central Institute of Fisheries Technology, Cochin-682029 (Kerala)
6. CMFRI	Central Marine Fisheries Research Institute, PB.No. 1603, Tatpuram P.O. Kochi-682018 (Kerala)
7. NBFGR	National Bureau of Fish Genetic Resources, P.O. Dilkusha, Lucknow-226002 (U.P)
8. DCFR	Directorate of Coldwater Fisheries Research, Saurabh Cottage, Bhimtal, Nainital-263136 (Uttaranchal)

HORTICULTURE

1. CIARI	Central Island Agricultural Research Institute, PB No. 181, Port Blair- 744101(A&N Island)
2. CIAH	Central Institute of Arid Horticulture, Sri Ganganagar Highway, Bikaner-334006(Rajasthan)
3. CISH	Central Institute for Subtropical Horticulture, Rehmankheda, PO Kakori, Lucknow-227107 (U.P)
4. CITH	Central Institute of Temperate Horticulture, Old Air Field, Rangreth-190007 (J&K)
5. CPCRI	Central Plantation Crop Research Institute, Kasargod-671124 (Kerela)
6. CPRI	Central Potato Research Institute, Shimla-171001 (H.P.)
7. CTCRI	Central Tuber Crops Research Institute, Sreekariam, Thiruvanthapuram-695017 (Kerela)
8. IIHR	Indian Institute of Horticultural Research, Hassarghatta Lake post, Banglore-560089 (KN)
9. IISR	Indian Institute of Spices Research, PB.No. 1701, Marikunnu PO, Kozhikode-673012 (Kerela)
10. IIVR	Indian Institute of Vegetable Research, PB No. 01, PO.Jakini, Sahansapuram, Varanasi- 221305 (U.P.)
11. NRCB	National Research Centre for Banana, Thogamalai Main Road, Thayar PO Thiruchirapalli -620102 (TN)
12. DCR	Directorate of Cashew Research, Puttur-574202 Dakshina Kannada (KN)
13. NRCC	National Research Centre for Citrus, PB No. 464 Shankarnagar PO, Nagpur-440010
14. NRCSS	National Research Centre for Seed Spices, Tabiji, Ajmer-305206 (Rajasthan)
15. NRCG	National Research Centre for Grapes, P B No. 3, Manjri Farm Post, Solapur Road, Pune-412307 (MH)
16. NRCL	National Research Centre for Litchi, Manchi House, Club Road, Muzaffarpur-842002 (Bihar)

17. NRCM National Research Centre for Mushroom, Chambaghat, Solan-173213 (H.P)
18. IIOR Indian Instite of Oil Palm Research, Near Jawahar Navodaya Vidyalaya Pedavegi-534450, West Godavari dist. (A.P)
19. NRCO National Research Centre for Orchids, Pakyang, Sikkim- 737106
20. NRCOG National Research Centre for Onion & Garlic, Ragurunagar-410505, Pune (MH)
21. IINRG Indian Institute of Natural Resins & Gums, Ranchi
22. DFR Directorate of Floricultural Research, Pune, Maharashtra
23. DMR Directorate of Mushroom Research, Solan, H.P.
24. NRCP National Resreach centre on Pomegranate, Solapur, Maharashtra

MANAGEMENT AND STATISTICS

1. IASRI Indian Agricultural Statistics Research Institute, Library Avenue, Pusa, New Delhi -110012
2. NAARM National Academy of Agricultural Research and Management, Hyderabad-500030 (A.P.)
3. NIAEPR National Institute of Agricultural Economics & Policy Research, Library Avenue, Pusa, New Delhi-110012

NATURAL RESOURCES MANAGEMENT

1. CAZRI Central Arid Zone Research Institute, Jodhpur-342003 (Rajasthan)
2. CRIDA Central Research Institute for Dryland Agriculture, Santhoshnagar, Hyderabad-500059 (A.P.)
3. CSSRI Central Soil Salinity Research Institute, Kernal- 132001 (Haryana)
4. IISWC Indian Institute of Soil and Water Conservation, Kaulagarh Road, Dehradun 248 145 (Uttarakhand)
5. ICARRCE ICAR Research Complex of Eastern Region, Walmi Complex, PO Phulwari Sharif, Patna-801505 (Bihar)
6. CCARI Central Coastal Agricultural Research Institute, Ela , Old Goa-403402
7. ICARRCNEH ICAR Research Complex for NEH Region, Barapani-793103 (Meghalaya)
8. IISS Indian Institute of Soil Science, Nabi bagh, Berasia Road, Bhopal-462038 (M.P.)
9. NBSSLUP National Bureau of Soil Survey and Land Use Planning, Nagpur-440010 (MH)
10. CARI National Agroforestry Research Institute, Near Pahujdam, Jhansi-284003 (UP)
11. DWRS Directorate of Weed Science Research, Maharajpur, Adhartal, Jabalpur-482004 (M.P.)
12. IIFRS Indian Institute of Farming Systems Research, Meerut-250110
13. IIWM Indian Institute of Water Management, Chandra Sekharpur, PO SE Railway, Project Complex, Bhubneswar-751023 (Orissa)
14. NIASM National Institute of Abiotic Stress Management, Malegaon, Baramati – 413115 (Pune- Maharashtra)
15. NIBSM National Institute of Biotic Stress Management, Raipur
16. CCARI Central Coastal Agricultural Research Institute, Ela, Old Goa
17. IINRG Indian Institute of Natural Resins and Gums, Ranchi

OTHER IMPORTANT NATIONAL ORGANISATIONS

1. APEDA Agricultural and Processed Food Export Development Authority, NCU 1 Building, 3 Siri Industrial Area , August Kranti marg, Hauzkhas, New Delhi-110016

2. BARC	Bhabha Atomic Research Centre, Trombay, Mumbai
3. CFTRI	Central Food Technological Research Institute, Mysore
4. CIMAP	Central Medicinal and Aromatic Plants, Lucknow
5. CPPTI	Central Plant Protection Training Institute, Hyderabad
6. CSIR	Council of Scientific and Industrial Research, New Delhi
7. CSMCRI	Central Salt and Marine Chemical Research Institute, Bhavnagar
8. FRI	Forest Research Institute, Dehradun
9. IAS	Indian Academy of Science, Pune
10. ICFRI	Indian Council of Forestry Research and Education, Dehradun
11. ICM	Institute of Cooperative Management, Ahmedabad
12. IFGTB	Institute of Forest Genetics and Tree Breeding, Coimbatore
13. IGNOU	Indira Gandhi National Open University, New Delhi
14. IIM	Indian Institute of Management, Ahmedabad
15. INSA	Indian National Science Academy, Bahadur Shah Zafar Marg, NewDelhi
16. IWST	Institute of Wood Science & Technology, Banglore
17. JARI	Jute Agricultural Research Institute, Barrackpore
18. JTRI	Jute Technology Research Institute, Kolkata
19. MANAGE	National Institute of Agricultural Extension Management, Hyderabad
20.	MS Swaminathan Research Foundation, 3rd Cross Street, Taramani Institutional Area, Chennai-600133 (TN)
21. NABARD	National Bank of Agriculture and Rural Development
22. NAFED	National Agricultural Cooperative Marketing Federation of India Ltd., NAFED House, Siddartha Enclave, Asharm Chowk, New Delhi
23. NBRI	National Botanical Research Institute, Lucknow
24. NCDC	National Cooperative Development Corporation, New Delhi
25. NCERT	National Council of Education Research & Training, New Delhi
26. NCL	National Chemical Laboratory, Pune
27. NDDB	National Dairy Development Board, Safadarjung Enclave, New Delhi
28. NIAG	National Institute of Animal Genetics, Kernal (Harayana)
29. NIN	National Institute on Nutrition, Hyderabad
30. NIRD	National Institute for Rural Development, Hyderabad
31. NOVDB	National Oil Seed & Vegetable Development Board, New Delhi
32. NPL	National Physical Laboratory, New Delhi
33. NRL	Nuclear Research Laboratory, IARI, New Delhi
34. NSC	National Seed Corporation, Bee Bhawan, Pusa, New Delhi
35. NSI	National Sugar Institute, Kalyanpur, Kanpur
36. PPII	Patash Research Institute of India, Gurgaon (Haryana)
37. RSAC	Remote Sensing Application Centre, Lucknow
38. UGC	University Grant Commission, Bahadur Shah Zafar Marg, New Delhi
39. WLMTRI	Water and Land Management Training Research Institute, Hyderabad
40. WTC	Water Technology Centre, IARI, New Delhi

INTERNATIONAL ORGANISATIONS

1. AIT	Asian Institute of Technology, P.O. Box 2734, Bankok, Thailand-10501
2. AVRDC	Asian Vegetable Research and Development Centre, P.O.Box 42, Shanhua, Taiwan 74199, Taiwan
3. BARC	Bangladesh Agricultural Research Council, Dhaka
4. CGIAR	Consultative Group on International Agriculture Research, Washington (USA)

5. CIAT Centro International de Agricultural Tropical (International Centre for Tropical Agriculture, Coli, Columbia), South Africa
6. CIDA Canadian International Development Agency, Cuba
7. CIMMYT Centro International deMajormiento de Maizy Trigo (International Centre for Improvement of Maize, Wheat and Barley),
8. CIP International Potato Centre (Centro Internationally de la Papa), Peru, Mexico
9. CSIRO Commonwealth Scientific and Industrial Research Organization, Near Perth, Australia
10. FAO Food and Agriculture Organization, Rome, Italy
11. IBPGR International Plant and Genetic Resources, Italy
12. IBSNAT International Benchmark Soil Network for Agrotechnology Transfer, Hawaii
13. ICARDA International Centre for Agricultural Research for Dryland Areas, Syrea
14. ICRAF International Centre for Research in Agroforestry, Nairobi, Kenya
15. ICRISAT International Crop Research Institute for the semi Arid Tropics, Patancheru, Hyderabad
16. IDRC International Development Research Centre, Canada
17. IFAD International Fund for Agriculture Development, Rome
18. IFDC International Fertilizer Development Centre, Muscale Shoala, Alobama (USA)
19. IFPRI International Food Policy Research Institute, Washington DC (USA)
20. IIASA International Institute for Applied Systems Analysis, Luxemberg, Vienna, Austria
21. IITA International Institute for Tropical Agriculture, Ibadan, Nigeria
22. INSFFER International Network on Soil Fertility and Fertilizer Evaluation of Rice, Manila, Philippines
23. INSNAR International Service for National Agricultural Research, The Hague, Netherland
24. IPGRI International Plant Genetic Research Institute
25. IRRI International Rice Research Institute, Manila Philippines
26. ISRIC International Soil Reference and Information Centre, Netherland
27. IWMI International Water Management Institute
28. LURI Land Use Research Institute, Aberdeen, Scotland
29. NIAS National Institute on Agricultural Science, Japan
30. PPIC Potash and Phosphate Institute, Canada
31. TPRI Tropical Pesticides Research Institute, Arushe, Tanzania
32. WARDA West Africa Rice Development Association, Manrovia, Liberia
33. WMO World Meteorological Organization, Geneva, Switzerland

STATE AGRICULTURAL UNIVERSITIES

1. AAU Anand Agricultural University, Anand (Gujarat)
2. AAU Assam Agricultural University, Jorhat (Assam)
3. ANGRAU Acharya NG Ranga Agricultural University, Rajendranagar, Hyderabad (A.P.)
4. BAU Birsa Agricultural University, Kanke, Ranchi (Jharkhand)
5. BCKV Bidhan Chandra Krishi Vishwavidhalaya, Mohanpur, Nadia (WB)
6. BSKKV Dr. Babasaheb Sawant Kokan Krishi Vidyapeeth, Dapoli, Ratnagiri, (Maharashtra)
7. CAU Central Agricultural University, P O Box No. 23, Imphal, (Manipur)
8. CCSHAU Chaudhary Charan Singh Haryana Agricultural University, Hisar (Haryana)
9. CSAUA&T Chandra Shekhar Azad University of Agriculture & Technology, Kanpur, (U.P.)

10. CSKKV	Ch. Sarwan Kumar Himachal Pradesh Krishi Vishwavidyalaya, Palampur (HP)
11. CVV&GAS	U.P. Pandit Deen Dayal Upadhyaya Pashu Chiktsha Vigyan Vishwavidhayala evam Go Anushandhan Sansthan, Mathura, (U.P.)
12. GBPUA&T	Gobind Ballabh Pant University of Agriculture & Technology, Pantnagar- 263145, Udam Singhnagar (Uttaranchal)
13. IGKV	Indira Gandhi Krishi Vishwavidyalaya, Krishak negar, Raipur-492012 (Chhattisgarh)
14.JAU	Junagarh Agricultural University, Junagarh-362001 (Gujarat)
15. JNKVV	Jawaharlal Nehru Krishi Vishwa Vidyalaya, Jabalpur – 482004 (MP)
16.KAU	Kerala Agricultural University, Vellanikara, Trichur-680656 (Kerala)
17. MAS&FU	Maharastra Animal Science & Fishery University, Seminary Hillls, Nagpur-440006 (Maharashtra)
18.VNMAU	Vasantrao Naik Marathwada Agricultural University, Parbhani-431 402 (Maharastra)
19.MPKV	Mahatma Phule Krishi Vidyapeeth, Rahuri-413 402 (Maharastra)
20. MPUA&T	Maharana Pratap University of Agriculture & Technology, Udaipur-3133001 (Rajasthan)
21. NAU	Navsari Agricultural University, Navsari (Gujarat)
22.NDAU&T	Narendra Deo University of Agriculture & Technology, Kumar Ganj, Faizabad-224 229 (UP)
23.OUA&T	Orissa University of Agriculture & Technology, Bhubneshwar-751003 (Orissa)
24.PAU	Punjab Agriculture University, Ludhiana -141004 (Punjab)
25.PDKV	Dr. Panjabrao Deshmukh Krishi Vidyapeeth, Krishinagar, Akola-444 104 (Maharashtra)
26.SKRAU	Swami Keshvanand Rajasthan Agriculture University, Bikaner-33400 (Rajasthan)
27. RAU	Rajendra Agriculture University, Pusa, Samastipur-848 125 (Bihar)
28. SBBPUA&T	Sardar Ballabh Bhai Patel University of Agriculture & Technology, Modipuram, Meerut-250 110 (UP)
29. SKDAU	Sardar Krushinagar Dantiwada Agricultural University, Sardar Krushinagar, Banskantha-385506 (Gujarat)
30. SKUAC&T	Sher-E-Kashmir University of Agricultural Sciences & Technology, Jammu -188812 (J&K)
31. SKUAC&T	Sher-E-Kashmir University of Agricultural Sciences & Technology, Kashmir -191121 (J&K)
32.TAU	Tamilnadu Agricultural University, Coimbattor-641003 (Tamilnadu)
33. TV&ASU	Tamilnadu Veterinary & Animal Sciences University, Chennai-600051 (Tamilnadu)
34. UAS	University of Agricultural Sciences GKVK Bangalore-560065 (Karnataka)
35. UAS	University of Agricultural Sciences, Dharwad-580005 (Karnataka)
36. UBKV	Uttar Banga Krishi Vishwavidyalaya, Pundibari, Cooch behar – 736 165 (WB)
37. WUA&FS	West Bengal University of Animal & Fishery Sciences, 68 K Sahani, Kolkata-700037 (WB)
38. YSPUH&F	Dr. Yashwant Singh Parmar University of Horticulture & Forestry, Solan, Nauni-173 230 (HP)
39. UUHF	UttarakhandUniversity of Horticulture and Forestry, Pauri, Uttarakhand
40. MKJUAT	Manyawar Kansiram Ji University of Agriculture and Technology, Banda (UP)
41. SVVU	Sri Venkateswara Veterinary University, Trupati – 517502 (AP)

42. YSRHU	Dr YSR Horticultural University, Venkataramanagudem, West Godawari – 534001 (AP)
43. RVSKV	Rajmata Vijayaraje Scindia Krishi Vishwavidyalaya, Gwalior – 474002 (MP)
44. MARFSU	Maharashtra Animal & Fisheries Sciences University, Nagpur – 4400006
45.GADVASU	Guru Angad Dev Veterinary And Animal Science University, Ludhiana – 141004
46.SHIATS	Sam Higginbottom Institute of Agriculture Technology and Sciences, Naini, Allahabad – 211007
47. KUFOS	Kerala University of Fisheries and Ocean Studies, Pananagad, Cochin – 682506
48. KV&ASU	Kerala Veterinary and Animal Sciences University, Pookot, Wayanand – 673576
49.LLRUVAS	Lala Lajpat Rai University of Veterinary & Animal Sciences, Hisar (Haryana)
50.RUV&AS	Rajasthan University of Veterinary and Animal Sciences, Bijay Bhawan Palace Complex, Near Pt. Deendayal Circle, Bikaner – 334006 (Rajasthan)
51. BAU	Bihar Agriculture University, Sabour, Bhagalpur – 813210 (Bihar)
52. NDVSU	Nanaji Deshmukh Veterinary Science University, Jabalpur – 482001 (MP)
53.UHS	University of Horticultural Sciences, Sector-60, Navanagar, Bagalkot – 587102
54. UAS	University of Agricultural Sciences, Raichur – 584102 (Karnataka)
55. AUJ	Agriculture University Jodhpur, Mandor, Jodhpur-342004
56. AUK	Agriculture University Kota, Borkhere, Kota-324001
57. KVAFSU	Karnataka Veterinary, Animal and Fisheries Sciences University, Nandnagar, PB No. 6, Bidar-585401 (Karnataka)
58. JTSAU	Prof. Jayashankar Talengana State Agricultural University, Rajendra nagar, Hyderabad
59. RLBCAU	Rani Laxmi Bai Central Agricultural University, Jhansi
60. TFU	Tamilnadu Fishries University, Nagapattanam-611003 (TN)
61. STSHU	Srikonda Telangana State Horticultual University, Rajendranagar, Hyderabad
62. UAS	University of Agricultural Sciences, Shimoha, Karnataka

ICAR DEEMED TO BE UNIVERSITY

IARI	Indian Agricultural Research Institute, New Delhi – 110012
IVRI	Indian Veterinary Research Institute, Izatnagar, Berrreili- 243122 (UP)
NDRI	National Dairy Research Institute, Kernal – 132001 (Haryana)
CIFE	Centrral Institute of Fisheries Education, Mumbai – 400061

Central Agricultural Universities

1.	Central Agricultural University, P.O. Box 23, Imphal - 795 004, Manipur
2.	Rani Laxmi Bai Central Agricultural University, Jhansi (U.P.)
3.	Dr. Rajendra Prasad Central Agricultural University, Pusa, Samastipur

Central Universities related to Agriculture

AMU	Aligarh Muslim University, Aligarh
BHU	Banaras Hindu University, Varanasi
VISWA	Viswa Bharti (Palli Siksha Bhawan), P.O. Sriniketan – 731236
NU	Nagaland University (School of Agricultural Sciences and Rural Development), Modziphema - 797106

ABBREVIATIONS/ACRONYMS

A3P	Accelerated Pulse Production Programme
ACIAR	Australian Council on International Agriculture Research
AGRIS	International Information System for Agricultural Science and Technology
AICRP	All India Coordinated Research Project
AOAC	Association of Official Analytical Chemists
AP Cess Fund	Agricultural Product Cess Fund
APAS	Andhra Pradesh Academy of Science
APEDA	Agriculture & Processed Products Export Authority
APWSS	Asian Pacific Weed Science Society
ARIC	Agricultural Research Information Centre
ARS	Agricultural Research Services
ASA	American Society of Agronomy
ASRB	Agricultural Scientist Recruitment Board
ASSOD	Assessment of the Status of Human –included Soil Degradation
ATIC	Agricultural Technology Information Centre
ATMA	Agriculture Technology Management Agency
BARC	Bhabha Atomic Research Centre
BHC	Benzene Hexa chloride
CAB	Commonwealth Agricultural Bureau
CCIS	Comprehensive Crop Insurance Scheme
CDD	Community Development Directorate
CEP	Central Exchange Programme
CFB	Corrugated Fibre Board
CIDA	Canadian International Development Agency
CSC	Central Seed Committee
CSSA	Crop Science Society of America
CVRC	Central Variety Release Committee
DAP	Di Ammonium Phosphate
DARE	Department of Agricultural Research and Education
DBT	Department of Bio Technology
DDP	Desert Development Programme
DDT	Dichloro Diphenyle Trichloro Ethane
DPAP	Drought Prone Area Programme
DPD	Directorate of Pulses Development
DST	Department of Science and Technology
EAS	Employment Assurance Scheme
EDB	Ethylene Di Bromide
EDCIL	Educational Development Consultant India Ltd
ESCAP	Economic and Social Commission for Asia and the Pacific
FAI	Fertilizer Association of India
FAO	Food and Agriculture Organisation
FCI	Food Corporation of India
FDA	Food and Drug Adulteration Act
FPC	Fertilizer Promotion Campaign
FPTC	Food Processing and Training Centre
FRC	Food Research Council
FSC	Farm Science Centre

FYM	Farm Yard Manure
GLASOD	Global Assessment of the Status of Human-included Soil Degradation
GOI	Government of India
GSFC	Gujarat State Fertilizer Corporation Ltd
HCN	Hydro Cyanic Acid
HETA	Horticulture Experiment & Training Centre
HIL	Hindustan Insecticide Limited
HSDC	Haryana Seed Development Corporation
HYVP	High Yielding Variety Programme
IAAP	Intensive Agriculture Area Programme
IADP	Intensive Agriculture District programme
IARC	International Agricultural Research Centre
IBRD	International Bank for Reconstruction and Development
ICDS	Integrated Child Development Scheme
ICID	International Commission on Irrigation and Drainage
IEL	Indian Explosive Limited
IFFCO	Indian Farmers Fertilizers Corporation Ltd
IFPC	Intensive Fertilizer Promotion Campaign
IFWA	International Federation for Woman Development
IJO	International Jute Organisation
ILCA	International Livestock Centre for Africa
IRDC	International Research and Development Council
IRDP	Integrated Rural Development Programme
IRTP	International Rice Testing Programme
ISA	Indian Society of Agronomy
ISOPOM	Integrated Scheme for Oilseeds, Pulses, Oilpalm and Maize
ISRIC	International Soil Reference and Information Centre
ISSS	Indian Society of Soil Science
ISSS	International Soil Science Society
ISTS	Indian Society of Tobacco Science
ITEC	Indian Technical and Economic Corporation
IUFRO	International Union of Forest Research Organisation
IWSS	Indian Weed Science Society
KGK	Krishi Gyan Kendra
KRIBHCO	Krishak Bharti Corporation Limited
KVK	Krishi Vigyan Kendra
LCU	Lac Coated Urea
LEISA	Low External Input and Sustainable Agriculture
LLP	Lab to Land Programme
LTPC	Long Term Programme of Cooperation
MCFC	Maharastra Cooperative Fertilizer and Chemical Limited
MEPDA	Marine Product Export Development Authority
MFAL	Marginal Farmers and Agricultural Labour Programme
MGREGS	Mahatma Gandhi Rural Employment Guarantee Scheme
MIS	Management Information System
NAAS	National Academy of Agriculture Science
NACA	Network of Aquaculture Centre in Asia –Pacific
NAFED	National Agriculture Cooperative Marketing Federation of India Limited

NAIP	National Agriculture Innovative Project
NARP	National Agricultural Research Project
NARS	National Agricultural Research System
NATP	National Agricultural Technology Project
NCA	National Commission on Agriculture
NCAER	National Council of Applied Economic Research
NCDC	National Cooperative Development Corporation
NCIPM	National Centre for Integrated Pest Management
NCL	National Chemical Laboratory
NCU	Neem Coated Urea
NDDB	National Dairy Development Board
NDP	National Development Programme
NET	National Eligible Test
NFL	National Fertilizer Limited
NFSM	National Food Security Mission
NGO	Non Government Organisation
NIPMA	North Indian Pesticide Manufacture Association
NICRA	National Initiative on Climate Resilient Agriculture
NOVDB	National Oilseed and Vegetable Development Board
NPC	National Productivity Council
NPL	National Physical Laboratory
NRDC	National Research Development Corporation
NREGA	National Rural Employment Guarantee Agency
NREP	National Rural Employment Programme
NSKS	Neem Kernel Seed Extract
NWDPRA	National Watershed Development Project for Rainfed Area
OFC	Orissa Fertilizer and Chemical Limited
ORP	Oilseeds Research Project
PARC	Pakistan Agriculture Research Council
PCARR	Philippines Council of Agriculture and Resources Research
PCARRD	Philippines Council for Agriculture Forestry and Natural Resources Research and Development
PCNB	Penta Chloro Nitro Benzene
PIRCOM	Project for Intensification of Regional Research on Cotton, Oilseeds and Millets
PPCL	Pyrites Phosphate and Chemical Limited
PRA	Participatory Rural Appraisal
PRAI	Planning Research and Action Institute
PSSC	Punjab State Seed Corporation
PTD	Participatory Technology Development
RAPI	Rural Agro Processing Industries
RKVY	Rastriya Krishi Vikas Yojna
RLEGP	Rural Landless Employment Guarantee Programme
RNE	Renewable Natural Energy
RRA	Rapid Rural Appraisal
SAARC	South Asian Association for Regional Cooperation
SADCC	Southern African Development Coordination Conference
SAMETI	State Agricultural Management and Extension Training Institute
SAU	State Agricultural Universities

SFCI	State Farms Corporation of India
SFDA	Small Farmers Development Agency
SIDA	Swedish International Development Agency
SISSTA	South Indian Sugarcane and Sugar Technology Association
SITE	Satellite Instructional Television Experiment
SOTER	Soils and Terrain Digital Database
SRI	System of Rice Intensification
SSDC	State Seed Development Corporation
SSP	Single Super Phosphate
SURE	Society to Uplift Rural Economy
SVRC	State Variety Release Committee
TAS	Time Allocation Study
TDC	Terai Development Corporation
TMO	Technology Mission on Oilseeds
TRIMS	Trade Related Investment Measures
TRIPS	Trade Related Intellectual Property Rights
TRYSEM	Training of Rural Youth Self Employment Programme
UGC	University Grand Commission
UNDP	United Nations Development Programme
UNEP	United Nations Environment Programme
UNESCO	United Nations Educational Scientific and Cultural Organisation
UNICEF	United Nations International Children Emergency Fund
UPCAR	Uttar Pradesh Council of Agricultural Research
USAID	United States Agency for International Development
USDA	United State Department of Agriculture
USIF	United State Indian Fund
WAPCCS	Water and Power Consultative Service India Limited
WMO	World Meteorological Organisation

Conversion factors between important primary and secondary agricultural commodities

Commodity	Conversion factor
Rice (cleaned) production	2/3 of paddy production

Cotton

Cotton Lint Production	1/3 of kapas production
Cotton seed production	2/3 of kapas production
	2 time of cotton lint production

Jute

100 yards of Hessian	54 lbs of raw jute
4148 yards of Hesian	1 ton of raw jute
	5.55 bales of raw jute (of 180 kg each)
1 Ton of sacking	1.11 tons of raw jute
	6.17 bales of raw jute (of 180 kg each)
1 Ton of Hesian	1.05 tons of raw jute
Sacking etc	5.85 bales of raw jute (of 180 kg each)

Groundnut

Kernel to nuts in shell	70 per cent
Oil to nuts in shell	28 per cent
Oil to kernels crushed	40 per cent
Cake to kernels crushed	60 per cent

Sesamum

Oil to seeds crushed	40 per cent
Cake to seeds crushed	60 per cent

Rapeseed and mustard

Oil to seeds crushed	33 per cent
Cake to seeds crushed	67 per cent

Linseed

Oil to seeds crushed	33 per cent
Cake to seeds crushed	67 per cent

Castor seed

Oil to seeds crushed	37 per cent
Cake to seeds crushed	63 per cent

Cotton seed

Oil to seeds crushed	14 to 18 per cent
Cake to seeds crushed	82 to 86 per cent

Coconut

Copra to nuts	One ton of Copra = 6773 nuts
Oil to copra crushed	62 per cent
Cake to copra crushed	38 per cent

Niger seed

Oil to seeds crushed	28 per cent
Cake to seeds crushed	72 per cent

Kardi seed

Oil to seeds crushed	40 per cent
Cake to seeds crushed	60 per cent

Mahua seed

Oil to seeds crushed	36 per cent
Cake to seeds crushed	64 per cent

Neem seed

Oil to kernels crushed	45 to 50 per cent
Cake to kernels crushed	50 to 55 per cent

Soyabean seed

Oil to soybean seed crushed	18 per cent
Meal to soyabean seed crushed	73 per cent
Hull from soybean seed crushed	8 per cent
Wastage from soyabean seed crushed	1 per cent

Sugar

Gur from Cane crushed	11.20 to 11.50 per cent
Crystal sugar from gur refined (Gur refineries)	62.5 per cent
Crystal suga from cane crushed (Cane factories)	10-20 per cent
Khandsari sugar (Sulphur & non sulphur) from stabdard gur refined	46 per cent
Molasses from cane crushed	4.0 to 4.5 per cent
Cane – Trash from cane harvested	8.0 to 12.0 per cent

Lac

Seed lac	66.0 per cent of stick lac
Shell lac	57.4 per cent of stick lac

Cashew nut

Cashew Kernel	25 per cent of cashew nuts